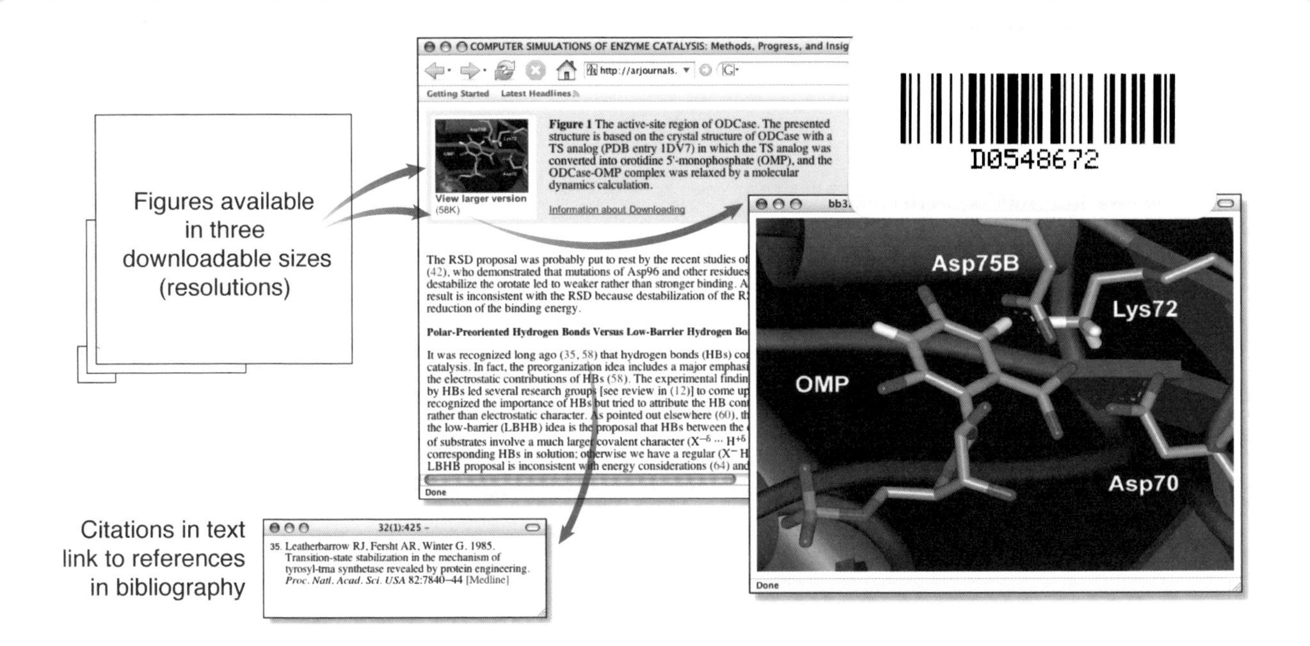

Figures available in three downloadable sizes (resolutions)
Citations in text link to references in bibliography
D0548672
COMPUTER SIMULATIONS OF ENZYME CATALYSIS: Methods, Progress, and Insig
Figure 1 The active-site region of ODCase. The presented structure is based on the crystal structure of ODCase with a TS analog (PDB entry 1DV7) in which the TS analog was converted into orotidine 5'-monophosphate (OMP), and the ODCase-OMP complex was relaxed by a molecular dynamics calculation.
View larger version (58K)
Information about Downloading
Polar-Preoriented Hydrogen Bonds Versus Low-Barrier Hydrogen Bo
35. Leatherbarrow RJ, Fersht AR, Winter G. 1985. Transition-state stabilization in the mechanism of tyrosyl-trna synthetase revealed by protein engineering. Proc. Natl. Acad. Sci. USA 82:7840–44 [Medline]
Asp75B
Lys72
OMP
Asp70

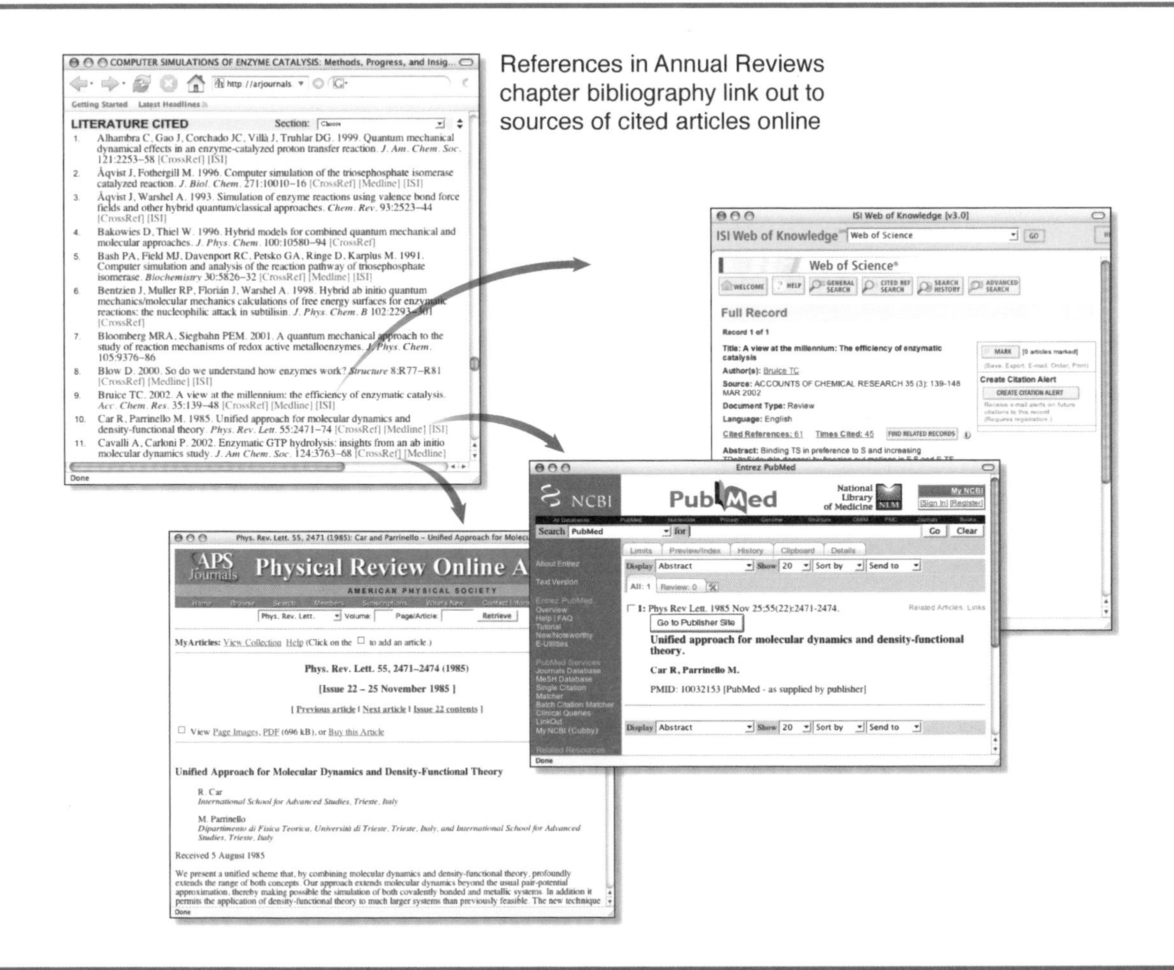

References in Annual Reviews chapter bibliography link out to sources of cited articles online
LITERATURE CITED
ISI Web of Knowledge
Web of Science
Full Record
Title: A view at the millennium: the efficiency of enzymatic catalysis
Author(s): Bruice TC
Source: ACCOUNTS OF CHEMICAL RESEARCH 35 (3): 139-148 MAR 2002
Entrez PubMed
NCBI
PubMed
National Library of Medicine
1: Phys Rev Lett. 1985 Nov 25;55(22):2471-2474.
Go to Publisher Site
Unified approach for molecular dynamics and density-functional theory.
Car R, Parrinello M.
PMID: 10032153 [PubMed - as supplied by publisher]
APS Journals
Physical Review Online A
AMERICAN PHYSICAL SOCIETY
Phys. Rev. Lett. 55, 2471–2474 (1985)
[Issue 22 – 25 November 1985]
Unified Approach for Molecular Dynamics and Density-Functional Theory
R. Car
International School for Advanced Studies, Trieste, Italy
M. Parrinello
Received 5 August 1985

Annual Review of Biophysics and Biomolecular Structure

Production Editor: Cleo X. Ray
Bibliographic Quality Control: Mary A. Glass
Electronic Content Coordinator: Suzanne K. Moses
Subject Indexer: Kyra Kitts
Illustration Editor: Doug Beckner

Annual Review of Biophysics and Biomolecular Structure

Volume 35, 2006

Douglas C. Rees, *Editor*
California Institute of Technology

Michael P. Sheetz, *Associate Editor*
Columbia University

James R. Williamson, *Associate Editor*
The Scripps Research Institute

www.annualreviews.org • science@annualreviews.org • 650-493-4400

Annual Reviews
4139 El Camino Way • P.O. Box 10139 • Palo Alto, California 94303-0139

Annual Reviews
Palo Alto, California, USA

International Standard Serial Number: 1056-8700
International Standard Book Number: 0-8243-1835-8
Library of Congress Catalog Card Number: 79-188446

♾ The paper used in this publication meets the minimum requirements of American National Standards for Information Sciences—Permanence of Paper for Printed Library Materials, ANSI Z39.48-1992.

TYPESET BY TECHBOOKS, FALLS CHURCH, VA
PRINTED AND BOUND BY FRIESENS CORPORATION, ALTONA, MANITOBA, CANADA

Annual Review of Biophysics and Biomolecular Structure

Volume 35, 2006

Contents

INDEX

ERRATA

An online log of corrections to *Annual Review of Biophysics and Biomolecular Structure* chapters (if any, 1997 to the present) may be found at http://biophys.annualreviews.org/errata.shtml

Related Articles

From the ***Annual Review of Biochemistry***, Volume 74 (2005)

Antibiotics Targeting Ribosomes: Resistance, Selectivity, Synergism, and Cellular Regulation
Ada Yonath

Cellular DNA Replicases: Components and Dynamics at the Replication Fork
Aaron Johnson and Mike O'Donnell

Copper-Zinc Superoxide Dismutase and Amyotrophic Lateral Sclerosis
Joan Selverstone Valentine, Peter A. Doucette, and Soshanna Zittin Potter

DNA Mismatch Repair
Thomas A. Kunkel and Dorothy A. Erie

Protein Families and Their Evolution—A Structural Perspective
Christine A. Orengo and Janet M. Thornton

Regulation of Protein Function by Glycosaminoglycans—As Exemplified by Chemokines
T.M. Handel, Z. Johnson, S.E. Crown, E.K. Lau, M. Sweeney, and A.E. Proudfoot

Structural Insights into Translational Fidelity
James M. Ogle and V. Ramakrishnan

Structural Studies by Electron Tomography: From Cells to Molecules
Vladan Lučić, Friedrich Förster, and Wolfgang Baumeister

Structure and Function of Fatty Acid Amide Hydrolase
Michele K. McKinney and Benjamin F. Cravatt

Structure and Physiologic Function of the Low-Density Lipoprotein Receptor
Hyesung Jeon and Stephen C. Blacklow

Structure, Function, and Formation of Biological Iron-Sulfur Clusters
Deborah C. Johnson, Dennis R. Dean, Archer D. Smith, and Michael K. Johnson

The Mammalian Unfolded Protein Response
Martin Schröder and Randal J. Kaufman

The Structural Biology of Type II Fatty Acid Biosynthesis
Stephen W. White, Jie Zheng, Yong-Mei Zhang, and Charles O. Rock

The Structure and Function of SMC and Kleisin Complexes
Kim Nasmyth and Christian H. Haering

Applications of DNA Microarrays in Biology
Roland B. Stoughton

From the ***Annual Review of Cell and Developmental Biology***, Volume 21 (2005)

Mechanisms of Apoptosis Through Structural Biology
Nieng Yan and Yigong Shi

Regulation of Protein Activities by Phosphoinositide Phosphates
Verena Niggli

Anisotropic Expansion of the Plant Cell Wall
Tobias I. Baskin

Rho GTPases: Biochemistry and Biology
Aron B. Jaffe and Alan Hall

Spatial Control of Cell Expansion by the Plant Cytoskeleton
Laurie G. Smith and David G. Oppenheimer

RNA Silencing Systems and Their Relevance to Plant Development
Frederick Meins, Jr., Azeddine Si-Ammour, and Todd Blevins

Quorum Sensing: Cell-to-Cell Communication in Bacteria
Christopher M. Waters and Bonnie L. Bassler

Pushing the Envelope: Structure, Function, and Dynamics of the Nuclear Periphery
Martin W. Hetzer, Tobias C. Walther, and Iain W. Mattaj

Integrin Structure, Allostery, and Bidirectional Signaling
M. A. Arnaout, B. Mahalingam, and J.-P. Xiong

Protein Translocation by the Sec61/SecY Channel
Andrew R. Osborne, Tom A. Rapoport, and Bert van den Berg

In Vivo Imaging of Lymphocyte Trafficking
Cornelia Halin, J. Rodrigo Mora, Cenk Sumen, and Ulrich H. von Andrian

From the ***Annual Review of Genetics***, Volume 39 (2005)

Orthologs, Paralogs, and Evolutionary Genomics
Eugene V. Koonin

Mitochondrial Morphology and Dynamics in Yeast and Multicellular Eukaryotes
Koji Okamoto and Janet M. Shaw

From the ***Annual Review of Pharmacology and Toxicology***, Volume 46 (2006)

Accessory Proteins for G Proteins: Partners in Signaling
Motohiko Sato, Joe B. Blumer, Violaine Simon, and Stephen M. Lanier

Nuclear and Mitochondrial Compartmentation of Oxidative Stress and Redox Signaling
Jason M. Hansen, Young-Mi Go, and Dean P. Jones

Molecular Mechanism of 7TM Receptor Activation—A Global Toggle Switch Model
Thue W. Schwartz, Thomas M. Frimurer, Birgitte Holst, Mette M. Rosenkilde, and Christian E. Elling

From the ***Annual Review of Physical Chemistry***, Volume 57 (2006)

Reflections on Physical Chemistry: Science and Scientists
Joshua Jortner

On a Research Rollercoaster with Friends
Robin M. Hochstrasser

4D Ultrafast Electron Diffraction, Crystallography, and Microscopy
Ahmed H. Zewail

Raman Crystallography and Other Biochemical Applications of Raman Microscopy
Paul R. Carey

From the ***Annual Review of Physiology***, Volume 68 (2006)

Commentary on the Pleasures of Solving Impossible Problems of Experimental Physiology
Watt W. Webb

Design and Function of Superfast Muscles: New Insights into the Physiology of Skeletal Muscle
Lawrence C. Rome

CNG and HCN Channels: Two Peas, One Pod
Kimberley B. Craven and William N. Zagotta

Regulation of Normal and Cystic Fibrosis Airway Surface Liquid Volume by Phasic Shear Stress
Robert Tarran, Brian Button, and Richard C. Boucher

Chronic Effects of Mechanical Force on Airways
Daniel J. Tschumperlin and Jeffrey M. Drazen

An Introduction to TRP Channels
I. Scott Ramsey, Markus Delling, David E. Clapham

Insights on TRP Channels from In Vivo Studies in *Drosophila*
Baruch Minke and Moshe Parnas

Permeation and Selectivity of TRP Channels
Grzegorz Owsianik, Karel Talavera, Thomas Voets, and Bernd Nilius

TRP Channels in *C. elegans*
Amanda H. Kahn-Kirby and Cornelia I. Bargmann

Annual Reviews is a nonprofit scientific publisher established to promote the advancement of the sciences. Beginning in 1932 with the *Annual Review of Biochemistry*, the Company has pursued as its principal function the publication of high-quality, reasonably priced *Annual Review* volumes. The volumes are organized by Editors and Editorial Committees who invite qualified authors to contribute critical articles reviewing significant developments within each major discipline. The Editor-in-Chief invites those interested in serving as future Editorial Committee members to communicate directly with him. Annual Reviews is administered by a Board of Directors, whose members serve without compensation.

Annual Reviews of

Anthropology
Astronomy and Astrophysics
Biochemistry
Biomedical Engineering
Biophysics and Biomolecular Structure
Cell and Developmental Biology
Clinical Psychology
Earth and Planetary Sciences
Ecology, Evolution, and Systematics
Entomology
Environment and Resources
Fluid Mechanics
Genetics
Genomics and Human Genetics
Immunology
Law and Social Science
Materials Research
Medicine
Microbiology
Neuroscience
Nuclear and Particle Science
Nutrition
Pathology: Mechanisms of Disease
Pharmacology and Toxicology
Physical Chemistry
Physiology
Phytopathology
Plant Biology
Political Science
Psychology
Public Health
Sociology

SPECIAL PUBLICATIONS
Excitement and Fascination of Science, Vols. 1, 2, 3, and 4

Martin Karplus

Spinach on the Ceiling: A Theoretical Chemist's Return to Biology

Martin Karplus

Department of Chemistry and Chemical Biology, Harvard University, Cambridge, Massachusetts 02138; and Laboratoire de Chimie Biophysique, ISIS, Université Louis Pasteur, F67083 Strasbourg, France; email: marci@tammy.harvard.edu

Annu. Rev. Biophys. Biomol. Struct. 2006. 35:1–47

The *Annual Review of Biophysics and Biomolecular Structure* is online at biophys.annualreviews.org

doi: 10.1146/annurev.biophys.33.110502.133350

1056-8700/06/0609-0001$20.00

Key Words

molecular dynamics, NMR coupling constants, trajectory calculations, reaction kinetics

Abstract

I was born in Vienna and came to the United States as a refugee in October 1938. This experience played an important role in my view of the world and my approach to science: It contributed to my realization that it was safe to stop working in fields that I felt I understood and to focus on different areas of research by asking questions that would teach me and others something new. I describe my experiences that led me from chemistry and physics back to my first love, biology, and outline some of the contributions I have made as part of my ongoing learning experience.

Contents

EARLY YEARS IN EUROPE

I was born in Vienna, Austria, in 1930 and seemed destined to become a physician. For several generations, there had been one or more physicians in the family, partly because medicine was a profession in which Jews in Austria could work with relatively little hindrance from discrimination. Neither my brother nor any of my numerous cousins displayed any interest in becoming a doctor, unlike me who at the age of five went around bandaging chairlegs and other substitutes for broken bones. I was fascinated by the stories of various relatives, including my Uncle Paul, who was a superb clinician. So, my family concluded that I was to become "the" doctor.

My paternal grandfather, Johann Paul Karplus, was a professor at the medical school of the University of Vienna, where his research led to the discovery of the function of the hypothalamus. My maternal grandfather, Samuel Goldstern, ran a private clinic that specialized in the treatment of rheumatological diseases by use of a mildly radioactive mud. The clinic, which I visited often because my grandparents had their apartment on an upper floor of the building, was called the *Fango Heilanstalt*. I had always assumed that *Fango* was a made-up name and learned decades later that *fango* is nothing more than the Italian word for *mud*. Many of the patients at the Fango were wealthy Arabs from the Middle East, an example of how the history of Arabs and Jews has long been intertwined and not always in the present destructive manner.

My childhood home was located in the Viennese suburb named Grinzing, well known as a wine growing area. There were small informal inns (Heurige) where we sometimes went in family groups for relaxed evenings eating and drinking (mainly the adults) the fruity young white wine of the region. The Heurige were also great places for children because between eating sausages, cheese, and bread, they could play in the garden.

In the early 1930s, when owning an automobile was still relatively rare, we already had a small car, a "Steyer Baby." One day when it was parked in front of our house, I scooted into the driver's seat and pretended to drive. I inadvertently released the brake, and the car started to roll downhill. I was terribly frightened as the car approached a pit at the end of the street. Miraculously, I steered the car so it turned just before reaching the pit and stopped. I recall the slope of the hill as being steep and the pit as very deep. Forty-five years later on a visit to Vienna with my family, we went to see my childhood home, which had been appropriated by the Nazis during the Second World War. I discovered that the "steep" hill was a very gentle slope and the "pit" a shallow ditch, so that the danger was more in my young mind than in reality.

Stories from my early childhood, most of which I know from their retelling by my parents, aunts, and uncles, indicate that I was a strong-willed independent child (in a positive sense) and a brat (in a negative sense). One such story concerns my "escape" at the age of three from a summer daycare center, where I apparently did not like the way I was being treated. One morning I simply walked out and somehow, despite the center being several miles from our house, I made my way home. Both the daycare people and my parents, who had been notified, were extremely worried about my disappearance and were searching for me. Although I was scolded, my parents were so happy to see me that my punishment consisted of a mild reprimand. The venture was justified in my mind because after that I was allowed to stay home. Then there was the infamous "spinach incident." My beloved nanny, Mitzi, told me that I must eat my spinach. (Popeye did not exist in Austria, but unfortunately spinach did.) With all the vehemence I could muster, I took a spoonful of the spinach and threw it at the ceiling. The spinach stain remained visible on the ceiling for a long time and was pointed out at appropriate moments when my parents wanted to indicate what a naughty child I was.

Traditionally we summered at an Austrian lake or on the Adriatic coast of Italy with several families that were either relatives or friends with children of similar ages. Such extended family activities were an integral part of my growing up and gave me confidence and a sense of belonging. One day at the beach, a friend of my parents picked me up and cuddled me, much to my dismay. I yelled out, "*Ich bin ein Nazi*" ("I am a Nazi"), which so shocked her that she dropped me. Clearly, I had somehow realized, presumably from listening to my parents and others, that being a Nazi was the worst possible thing to be.

Already before the Nazis entered Austria in 1938, our life had changed significantly, even from the viewpoint of an eight year old. Among our neighbors were two boys of ages comparable to my brother, Robert, and me. They were our "best friends," and we played regularly with them. In the spring of 1937, they suddenly refused to have anything to do with us and began taunting us by calling us "dirty Jew boys" when we foolishly continued to try to interact with them. Similar problems occurred at school with our non-Jewish classmates. Prior to this, my school experience in first grade and the beginning of the second had been wonderful, in part because I had a great teacher, Herr Schraik, not the least of whose outstanding attributes was that his wife ran a candy store. When my class was ready to advance to second grade, the parents petitioned that Herr Schraik be "promoted" with us and, because of his outstanding record as a teacher, this request was granted. Nevertheless, in the middle of that school year (1937–1938) he was no longer allowed to teach. He was Jewish and the authorities had decided that any contact with him would contaminate the minds of the children. The new teacher was incompetent and blatantly anti-Semitic—he constantly criticized the Jewish students like myself, independent of how well or poorly we were doing. The situation became so bad that my parents took me out of school.

One small, but not insignificant, part of my schooling in Austria had to do with my being left-handed. When I started first grade, I was obliged to learn to write with my right hand. Whatever the supposed psychological consequences of that may be, I have always been grateful that right-handedness in writing was imposed on me. This was particularly true when I first went to the United States and saw the contortions children went through to write with their left hand. Clearly everything, at least in Western European languages (not Hebrew or Arabic), is set up for right-handed people.

On March 13, 1938, the German Nazi troops crossed the border into Austria and completed the *Anschluss*, the "joining" of Austria with Nazi Germany, which had been specifically forbidden by the Versailles Treaty after the First World War. The Germans had been "invited" into Austria by the puppet Seyss-Inquart, who took over the Austrian government after Chancellor Schuschnigg was forced to resign. The night before, my family and some friends were listening to the radio in the living room, which was darkened to conform to the curfew and black-out requirements in case of air attacks. I was furious, not because of what was happening politically, but rather because my parents had planned an early birthday party (my actual birthday being March 15), and I was far from being the center of attention. Hitler entered Austria in triumph and his troops were welcomed by enthusiastic crowds. (To this day, more than 65 years later, I have mixed feelings about visiting Austria, which I rarely do, because anti-Semitism seems nearly as prevalent now as it was then. However, I recently learned that I am still officially an Austrian citizen, so I have dual nationality.) A few days after the *Anschluss*, my mother, brother, and I left Austria by train for Switzerland. My parents had been concerned about Hitler's takeover of Austria for some time. For the previous three years, my Aunt Claire, who had studied in England, had been teaching English to me and my brother, Bob. Well before March 13, train tickets had been purchased and a bed-and-breakfast "pension" had been reserved in Zurich.

The most traumatic aspect of our departure was that my father was not allowed to come with us and had to give himself up to be incarcerated in the Viennese city jail. In part, he was kept as a hostage so that any money we had would not be spirited out of the country. My mother reassured my brother and me, saying that nothing would happen to him, though of course she herself had no assurance that this was true. At that time, the Nazi government still allowed Jews to leave Austria, as long as they left their money behind. One way of spiriting money out of the country was to buy diamonds and hide them in clothing and food. During our train trip to Zurich, the guards at the German/Austrian border meticulously reduced a beautiful large sausage to thin slices, presumably searching for hidden diamonds. Fortunately, the sausage could still be eaten, and Bob and I were not too bothered by this event.

My parents hoped to go to America, but visas to the United States were granted only to applicants who had an "affidavit," a document from an American citizen guaranteeing their financial support. Many would-be immigrants would have been allowed to leave Austria (or Germany) but were not permitted to enter the United States. Some of them ended up in other countries (South America, Australia, and New Zealand), but, as history has recorded, many were not able to leave and died in concentration camps. My father's older brother Edu had immigrated to the United States some years earlier and had become chief engineer at the General Radio Corporation in Boston, where he invented the Variac, then a widely used device for continuously varying the electric voltage. The president of General Radio, Mr. Eastman, provided the required affidavit, enabling us to obtain visas for the United States.

Arranging for the journey to the United States took several months. After leaving Austria, my mother, brother and I lived in Zurich. Bob and I enrolled in a neighborhood public school where we rapidly learned *Schwyzertütsch* (the Swiss German dialect). Speaking *Schwyzertütsch* was not only a way of belonging, but it also provided us with a secret language which my mother did not understand. When summer came, we left Zurich and went to La Baule, a beach resort in Brittany, France, on the Atlantic coast, where our Uncle Ernst Papanek had established a summer colony for refugee children. The children were mainly from Jewish families, though there also were some whose parents were political refugees. During the late 1930s and early 1940s, Ernst organized a number of these children's homes, which saved the lives of many children. He described these efforts in his autobiography, *Out of the Fire*. Ernst's philosophy was that one could rely on the common sense and intelligence of children; one of the hallmarks of the homes was that they were run as cooperatives, with the children's input playing a significant role. Bob and I, and our cousins Gus and George (Ernst's children), as well as newfound friends, spent a blissful summer in La Baule swimming and building sand castles. Although food was in short supply, we were well nourished. I did not realize it then, but my mother and the other grownups were extremely worried about the future. Somehow they kept this from us and gave us a happy summer. Both Bob and I looked back on this period as a wonderful experience. Roberto Benigni's film *Life Is Beautiful* brought back memories of those days.

The visas finally arrived, passage was booked, and the three of us were ready to leave for the United States. Although there had been no news from my father, he miraculously turned up at Le Havre a few days before our ship, the *Ile de France*, was scheduled to depart for New York. From my point of view, it was exactly what my mother had told me would happen: We would all go to America together. When my father joined us in Le Havre, Bob and I asked him what jail had been like. He told us that he had been treated well in jail and cheerfully described how he had passed the time teaching the guards to play chess. One aspect of my father's personality, which strongly influenced both my brother and me, was to make something positive out of any experience.

We never learned the full details of my father's release from jail. My Uncle Edu had posted a $5000 bond for his release and after the war, several Viennese (e.g., a jailer, someone concerned with running the prisons) wrote us claiming credit for his release. There was no evidence as to who had done what, if anything. Nevertheless, my parents sent CARE packages to all the claimants and wished them well. (CARE, Cooperative for Assistance and Relief Everywhere, was a humanitarian organization that distributed nonperishable food packages after the Second World War to help overcome food shortages in the war-torn countries.) My parents directed food packages to people who might have helped free my father, as well as to many other people we knew (e.g., my nanny, Mitzi) who had survived the war.

A NEW LIFE IN AMERICA

We arrived in New York Harbor early in the morning on October 8, 1938, and I stood on the deck watching the Statue of Liberty appear out of the mist. The symbolism associated with the Statue of Liberty may seem trite today (and somewhat deceptive given our present immigration policies), but in 1938 it was special for me. Most of the immigration formalities had been taken care of by Uncle Edu, so that a few hours after our arrival we boarded a train to Boston. (My uncle had a house in an exclusive part of Belmont, a suburb of Boston, which at that time did not allow Jews to live there. We rarely were invited to visit him when we first came, and when we did we had to hide our "Jewishness"; in particular, we were not to say anything before entering the house to avoid the neighbors' being aware of our foreign accent.) During our initial weeks in the United States, we were lodged in a welcoming center in Brighton, where a large mansion had been transformed into an interim home for refugee families. We were taught about America (what it was like for foreigners to live in Boston), given lessons to improve our English, and aided in the steps required to be allowed to remain in the United States as refugees.

Soon we were ready to start a new life. My parents rented a small apartment in Brighton (part of Greater Boston), and Bob and I immediately entered the local public schools, as we had in Zurich. I was in the third grade and had the good fortune to have a teacher (I had a crush on her) who gave me special English lessons after school. At the age of eight, my English advanced so rapidly by being in school and by playing with the neighborhood children that these special lessons, alas, lasted only a few months.

I very much wanted to be accepted by my new country, and while living in Allston I was a street kid in every sense, hanging around with my friends, playing stick-ball and other such games, occasionally stealing candy just for the fun of it. For a while, I refused to speak German at home, despite my parents' limited English. One afternoon when I was playing in the street, I tripped in my haste to get out of a car's way and ended up with my foot under the car's rear tire. The driver had stopped and gotten out to see what was wrong, but with me screaming and crying it took him a while to understand that he should move the car off my foot. Once he did, thankfully my foot was only mildly sore. He wanted to drive me home and report to my parents what had happened, but I insisted that I was fine and would get home by myself. It was true that I was not hurt, but my primary concern was to keep my parents from knowing what had happened, how I spent my time, and specifically that I played in the streets.

After we came to the United States, our comfortable life in Vienna was a thing of the past. We were now relatively poor, although my father had brought some money in the form of valuable stamps, which he had purchased while we were still in Austria. Despite our economic straits, my parents did everything to ensure that our lives were as unchanged as possible. The first summer we were in the States, both of my parents worked as domestics—my father as a handyman and my mother as cook and cleaning woman—for a wealthy family who had a summer house in New Hampshire. We lived in a small house on the grounds, where Bob and I had an idyllic summer. The next summer my parents were similarly employed at a boys' camp, again to enable Bob and me to spend the summer as campers. My parents both took courses to help them find better employment. My father studied machining at the Wentworth Institute and my mother studied home economics at Simmons College. With the United States entering the War, employment was high, and my father joined an airplane pump factory after he graduated from a one-year course. He rapidly advanced in the company to become a quality inspector and worked at this company until he retired 20 years later. My father frequently told stories to Bob and me about problems he had solved or how he had suggested improvements in the pumps. The way he described his ideas (he had been educated in physics at the University of Vienna) helped arouse our scientific curiosity. After finishing an undergraduate program, my mother found a position as a hospital dietician, similar to the type of work she had done in Vienna at the Fango, her father's hospital. She continued her education and received a Master's degree at the age of 65.

Motivated by their concern for our education, my parents moved to Newton (a suburb of Boston), where the schools were recognized as superior to the Boston public schools. My parents bought a small house in a pleasant neighborhood in West Newton, and I attended the Levi F. Warren Junior High School. To my knowledge, we were the first Jewish family to live there. One Saturday, after we had lived in West Newton for about six months, an FBI agent knocked on the door and politely requested to see my father. The agent explained that he was investigating a complaint from our next-door neighbor, who had telephoned the FBI to report that every morning as he was leaving for work, my father would step out on the front porch, turn around, and make the Nazi salute while shouting "Heil Hitler." The FBI agent appeared rather embarrassed and said that he realized that such an accusation against a Jewish refugee from Nazi Austria was ridiculous. After some questioning (about our family history and present status), he got up to leave and told us that nothing further would happen.

My junior high teachers soon realized that I was bored with the regular curriculum, so they let me sit in the back of the classroom and study on my own. What made this experience particularly nice was that another student, a very pretty girl, was given the same privilege, and we worked together. The arrangement was that we could learn at our own pace without being responsible for the day-to-day material but had to take the important exams. Several dedicated teachers at Warren Junior High helped us when questions arose, particularly with science and mathematics. With this freedom, we explored whatever interested us and, of course, did much more work than we would have done if we were only concerned with passing the required subjects.

In nonacademic activities, I participated like everyone else (I was not particularly good in sports, though I enjoyed soccer a lot) and made a number of friends that formed a close-knit group throughout high school. As part of our education, we had to choose several technical courses. The two I chose were printing and home economics, the latter not because it was essentially all girls, but because the students did real cooking. I had become interested in cooking early on and used to spend time in the kitchen with my mother and grandmother, who both cooked simply but well with the freshest ingredients. The final exam had us prepare a dinner for the class, with each group responsible for one course.

BEGINNING OF SCIENTIFIC INTERESTS

Soon after we moved to Newton, Bob was given a chemistry set, which he augmented with materials from the school laboratory and drug stores. He spent many hours in the basement generating the usual bad smells and making explosives. I was fascinated by his experiments and wanted to participate, but he informed me that I was too young for such dangerous scientific research. My plea for a chemistry set of my own was vetoed by my parents because they felt that this might not be a good combination—two teenage boys generating explosives could be explosive! Instead, my father had the idea of giving me a Bausch and Lomb microscope. Initially I was disappointed—no noise, no bad smells, although I soon produced the latter with the infusions I cultured from marshes, sidewalk drains, and other sources of microscopic life. I came to treasure this microscope, and more than 60 years later it is still in my possession. One especially rewarding aspect of my working with the microscope was that my father, who was a thoughtful observer of nature (he liked to fish with a simple line, not to catch fish, but because it gave him an excuse to sit by a stream and watch the fish and observe their behavior), spent a lot of time with me and was always ready to come and look when I had discovered something new. I had found an exciting new world and looked through my microscope whenever I was free. The first time I saw a group of rotifers I was so excited by the discovery that I refused to leave them, not even taking time out for meals. They were the most amazing creatures as they swam across the microscope field with their miniature rotary motors. (The rotifers come to mind today in relation to my research on the smallest biological rotatory motor, F_1-ATPase.) My enthusiasm was sufficiently contagious that I even interested some of my friends. It was a special occasion when they came to my house and looked at the rotifers through the microscope.

This was the beginning of my interest in nature study, which was nurtured by my father and encouraged by my mother, even though it was still assumed that I would go to medical school and become a doctor. One day my closest friend, Alan MacAdam, saw an announcement of the Lowell Lecture Series (a Boston institution, originally supported by a Brahmin family—the Lowells), which organized evening courses on a wide range of subjects at the Boston Public Library that were free and open to the public. The organizers invited excellent lecturers from the many universities in the Boston area, as well as from nonacademic disciplines. The series that had caught Alan's eye was entitled "Birds and Their Identification in the Field," to be given by Ludlow Griscom, the curator of ornithology at the Museum of Comparative Zoology at Harvard University. Alan and I occasionally walked in the green areas in Newton, particularly the Newton Cemetery, and looked for birds with my father's old pair of binoculars. Together we attended the first lecture, which had a good-sized audience, although it was not clear whether most of the people came simply to have a nice warm place in winter rather than because of their interest in birds. I was enthralled by the lecture, which provided insights into bird behavior and described the large number of different species one could observe within a 50-mile radius of Boston. I was amazed that it was possible to identify a given species from "field marks" evident even from a glimpse of a bird, if one knew how and where to look. Alan did not attend the subsequent lectures, but I continued through the entire course. At the end of the fourth or fifth lecture, Griscom came up to me and asked me about myself. He then invited me to join his field trips, and a new passion was born. From that time on, my treasured microscope was relegated to a closet, and I devoted my free time to observing birds on my own, as well as with Griscom and his colleagues, with the Audubon Society, and other groups that organized field trips.

The culmination of these trips was the annual "census," usually held at the height of the bird migration in May. This was an activity sponsored by the Audubon Society, and the objective was to observe (see or hear) the largest number of different species within a given 24 hours. Each year, Griscom organized one such field trip, inviting only a select group of "birders" to participate. The census lasted a full 24 hours, starting just after midnight to find owls in the woods and rails and other aquatic species in the swamps. There was a carefully planned route, based on known habitats and recent sightings of rare species. The specific itinerary was worked out in a meeting, at which everyone contributed what interesting birds they had sighted recently, but Griscom made the final decision on how we would proceed. As the youngest (by far) in the group, I was assigned special tasks. One of these tasks—perhaps not the most pleasant—was to wade into the swamp at night (fortunately there was a moon) and scare up birds so that they would fly off and could be identified by their calls. On this census trip we found 160 or so different species, a record at the time for the area.

I became intrigued by alcids, of which the now extinct, flightless Great Auk was the most spectacular member of the family. I persuaded my family to go for summer vacation to the Gaspé Peninsula in Canada because there was a famous rocky island just offshore that had nesting colonies of two kinds of alcids, puffins and guillemots, as well as gannets, spectacular large black and white sea birds. Although one could not visit the island sanctuary (it was forbidden in order to protect the nesting birds), I borrowed a telescope from the Audubon Society to view the birds. We drove through the Gaspé Peninsula by car and spent the nights in bed-and-breakfast places. One strong impression of the trip through New Brunswick and the Gaspé area was that the houses in most of the villages were in much poorer condition than those on the coast, which were supported by the tourist trade. Moreover, each such small village was dominated by an outsized church, indicating the power of religion in these communities.

Many of the alcids go south to New England for the winter. Usually, they are far out to sea, but when there is a storm they are likely to be blown close to shore, so that such storms (particularly Nor'Easters) present the best opportunity to see rare species, such as the tiny Little Auk, which is only eight inches in length but can survive in the roughest seas. In the winter of 1944, school was closed because of a heavy snowstorm, and I took the early morning train up to Gloucester, a town on Cape Ann north of Boston, and hiked out to the shore. I sat on the steps of one of the large mansions, shuttered for the winter, and looked out to sea through my binoculars. The day started well as I quickly spotted several alcids in the cove below. As I was getting chilled after a couple of hours and ready to walk back to the train station, a car pulled up. A couple of men got out and walked toward me. At first, I naively presumed they were other birders, interested in what I had seen. I soon realized that they did not look like birders—no binoculars for one thing and not really dressed for a day in the snow. Moreover, they approached me in a rather aggressive manner, asking me what I was doing and why. They did not believe that I was sitting in such a storm looking for birds. Shortly after, they showed me their police badges and bundled me into the car, not for a ride to the train station, but instead to the Gloucester police station. I was only 14 years old and very frightened and became even more so when I realized from their questions that they thought I might be a German spy. That I was an immigrant from Austria, spoke German, and had German-made Zeiss binoculars did not help. They suspected me of signaling to submarines off the American coast, preparing to land saboteurs or whatever. It took hours of interrogation and several phone calls to the Audubon Society before the officers finally decided that I was not doing anything wrong and drove me to the train station. That was the last time I ventured on such a trip by myself.

On one of the field trips to Newburyport with Griscom, I spotted an unusual gull. When I pointed it out to him, he concluded, after looking through his telescope, that he had never seen a bird like that and that we should try to "collect" it, a euphemism for shooting the bird. He had a license to carry a "collecting gun" with a pistol grip and a long barrel that made it easier to aim. Because the bird was far away and separated from us by mud flats, only partly exposed at low tide, I was given the task of collecting the bird. I waded out fairly close to the bird and successfully shot it, even though I had never fired a gun other than at fairs. After a careful comparison with birds in the Museum of Comparative Zoology collection, Griscom was convinced that we had found something new, a hybrid between a Bonaparte's gull (common in America) and a European black-headed gull (common in Europe but rare in North America), which had somehow crossed the ocean. The bird was in the museum collection with its wing feathers, which were essential for its identification, beautifully spread out (**Figure 1**). (I say "was" because when I went recently to find it, the gull had disappeared. My reason for going to look for it was to discuss with Scott V. Edwards, the newly appointed Professor of Ornithology at Harvard, the possibility of sequencing the DNA to find out whether the result would confirm Griscom's identification.)

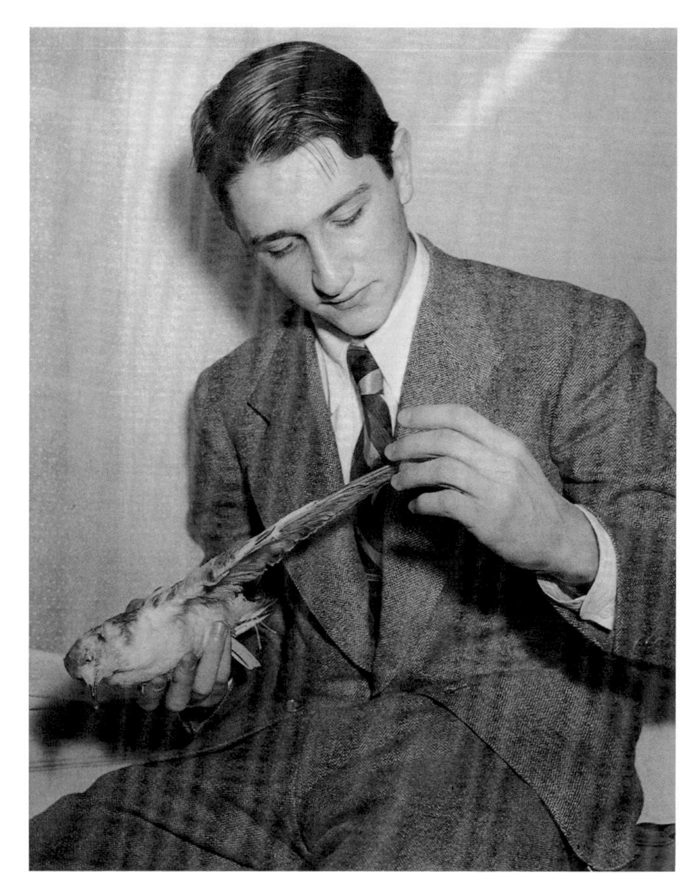

Figure 1

A photograph of me with the hybrid gull showing its wing feathers, which played the essential role in its identification.

I entered Newton High School in the fall of 1944 but soon found that I did not have the same supportive environment as in elementary and junior high school. My brother, Bob, had graduated from Newton High School two years before and had done exceedingly well. My teachers presumed that I could not measure up to the standards set by my brother. Since I had always been striving to keep up with Bob and his friends, this just reinforced my feelings of inferiority. Particularly unpleasant were my interactions with the chemistry teacher. When my brother suggested I compete in the Westinghouse Science Talent Search, the chemistry teacher, who was in charge of organizing such applications, told me that it was a waste of time for me to enter and that it was really too bad that Bob had not tried instead. However, I talked to the high school principal and he gave me permission to go ahead with the application. I managed to obtain all the necessary papers without encouragement from anyone in the school. A test was given as part of the selection process, and I found a teacher who was willing to act as proctor. I did well enough to be invited as one of the 40 finalists to Washington, DC. Each finalist had a science project for exhibition in the Statler Hotel, where we were staying. My project was on the lives of alcids, based in part on the trip to the Gaspé Peninsula and some of the field studies I had made during New England winters. The various judges spent considerable time talking with us, and the astronomer Harlow Shapley, who was the chief judge, charmed me with his apparent interest in my project. I was chosen as one of two co-winners. (At that time, there was one male and one female winner; Rada Demereck and I were co-winners.) The visit to Washington, DC, was a great experience, especially because we met President Truman, who welcomed us as the future leaders of America. Moreover, winning the Westinghouse Talent Search made up for the discouraging interactions with some of my high school teachers. Their attitude contrasted with that of my fellow classmates, who voted me "most likely to succeed."

My final forays into ornithology took place during several summers at the end of high school and after I entered college. In 1947, I had a summer internship at the Maryland Patuxent Research Refuge of the Fish and Wild Life Service, the only National Wildlife Refuge established to conduct research. Publicity about the harmful effect of DDT on bird life had prompted studies at the refuge. We collected eggshells as part of a field survey of two several-acre plots, one sprayed with DDT at the normal level and the other without DDT. My task was to analyze eggshells for their DDT contents and to determine the differences between shells from the two plots (their thickness and other features). I also conducted a census twice a day of the birds in the two areas, determined the number of nesting pairs, and collected any dead birds I found. The summer was an exciting one for me and a fine introduction to field research and laboratory work as part of a team. It was a very hot summer, and all my (older) colleagues drank beer to relax in the late afternoon, so I joined them. I did not like the bitter taste initially but rapidly learned to enjoy beer.

Thanks to a meeting with Professor Robert Galambos, who did research on the echolocation of bats in the basement of Memorial Hall at Harvard, I was invited by his collaborator, Professor Donald Griffin, to join his group in a study of bird orientation that was to take place in Alaska during the summer of 1948. (Griffin and Galombos had demonstrated in 1940 that bats used echolocation to orient themselves and find their prey, though Lazzaro Spallanzani had already suggested this in 1794, but apparently no one believed him.) Our team was based at the Arctic Research Laboratory in the town of Point Barrow, Alaska, which is located at latitude 71° north, the northern-most point of land on the North American continent. The laboratory was run by the Office of Naval Research (ONR), primarily to study how humans (presumably soldiers and sailors) can adapt to life in the arctic winter. Its director, Lawrence Irving from Swarthmore College, had a broad view of the laboratory mission and had invited our group to use the facilities. (It is worth mentioning that before there were organizations to support civilian research, like the National Science Foundation, ONR was the leading government agency in this area.) Our primary interest was in golden plovers, which nested on the tundra in areas not far from the laboratory. Atlantic and Pacific golden plovers nested together, and the two species separated in the fall to migrate over a thousand miles or more southward to their respective winter homes. We trapped some plovers of both species and attached radio transmitters to all of them and magnets to half of them. We then released the plovers 20 to 50 miles from their nests, both in the Atlantic and Pacific direction, and followed them at a distance with a small airplane. The idea was to ascertain if the birds with magnets would have a more difficult time returning to their nesting area. There were suggestive results (the birds with magnets seemed to get more disoriented than those without, though they all found their way home), but Griffin felt that what we had found was not conclusive proof that the plovers had a magnetic sense, and the work was never published.

In Umiat, an observation camp at a distance from the main laboratory, I organized my first experiment with the aid of the other scientists, who must have been amused by my youthful enthusiasm. (I was 17 years old and by far the youngest member of the research team.) The experiment involved several nesting pairs, including robins. Three people participated in the observation of the nests; each person stood a 4-h shift twice a day because there were 24 hours of daylight. We found that the parents fed the robins over the entire 24-h period and, interestingly, that the young robins left the nest earlier than did their cousins, who nested in Massachusetts. I wrote a paper (50) describing the results, with the conclusion that the survival value of the shorter time in the nests, which were highly exposed to local predators, made up for the dangers of the longer flights required to reach the summer nesting area in the Arctic. It is not clear whether my youthful conclusion was correct, although it did stimulate a number of papers, both pro and con, and the paper continues to be cited (106). Also, I noticed that at Umiat, where we were on our own, the normal 24-h day stretched to about 30 h; we stayed awake for 22 or so hours and then slept for about 8 hours.

The following summer Griffin invited me to Cornell University, where he was on the faculty before joining the Biology Department at Harvard. In addition to conducting experiments in the Griffin lab, I enjoyed "hanging out" with college students and school teachers who were taking summer courses at Cornell. I initially worked on bat echolocation and was very impressed by the way the bats were kept in the refrigerator between experiments; they went to sleep, hanging from a rod all in a row. At Griffin's suggestion, I then focused on trying to condition pigeons to respond to a magnetic field to test the results of an article that had concluded that pigeons use the earth's magnetic field to navigate. I was doubtful about the paper because I thought the analysis was flawed. My attempts at conditioning the pigeons proved elusive, but we never published our negative (to me, positive) results. Subsequently, other experiments have shown that pigeons, as well as wild birds, do use the earth's magnetic field as an aid in navigation. This experience taught me that being skeptical is essential in science, but that it is also important to be receptive to new ideas—even if you do not like them.

COLLEGE YEARS

I entered Harvard in the fall of 1947. There was never any question about my wanting to attend Harvard and I did not apply to any other school. In addition to the Westinghouse scholarship, I received a National Scholarship from Harvard to cover the cost of living on campus. Otherwise I would have had to live at home to save money. I would not have minded this, since I was not a rebellious teenager eager for independence and distance from my parents. However, as I soon discovered, much of the Harvard experience took place outside of classes at dinner and in evening discussions with friends.

At first I still intended to go to medical school but changed my mind during my freshman year. My teenage ornithological studies, fostered by Griscom and Griffin, had already introduced me to the fascinating world of research, where one is trying to discover something new (something that no one has ever known); I began to think about doing research in biology. I had concluded that to approach biology at a fundamental level (to understand life), a solid background in chemistry, physics, and mathematics was imperative, and so I enrolled in the Program in Chemistry and Physics. This program, unique to Harvard at the time, exposed undergraduates to courses in both areas at a depth that they would not have had from either one alone. It had the additional advantage, from my point of view, that it was less structured than chemistry. For example, it did not require Analytical Chemistry, certainly a good course as taught by Professor James J. Lingane, but one that did not appeal to me. Although I shopped around for advanced science courses to meet the rather loose requirements, I also enrolled in Freshman Chemistry because it was taught by Leonard Nash. A relatively new member of the Harvard faculty, Nash had the deserved reputation of being a superb teacher. Elementary chemistry in Nash's lectures was an exciting subject. A group of us (including DeWitt Goodman, Gary Felsenfeld, and John Kaplan—my "crazy" roommate, who became a law professor at Stanford) had the special privilege that Nash spent extra time discussing with us a wide range of chemical questions, far beyond those addressed in the course. The interactions in our group, though we were highly competitive at exam times, were also supportive. This freshman experience confirmed my interest in research and the decision not to go to medical school.

Harvard provided me with a highly stimulating environment as an undergraduate. One aspect was the laissez-faire policy, which allowed one to take any courses with the instructor's permission, even without having the formal prerequisites. The undergraduate dean said it was up to me to decide and, if a course turned out to be too much for me, that would be "my" problem. I enrolled in a wide range of courses, chosen partly because of the subject matter and partly because of the outstanding reputation of the lecturers; these courses included one in *Democracy and Government* and another in *Abnormal Psychology*. More related to my long-term interests were some advanced biology courses, which I registered in without having to suffer through elementary biology and biochemistry. Two memorable courses were George Wald's *Molecular Basis of Life* and Kenneth Thimann's class on plant physiology with its emphasis on the chemistry and physiology of growth hormones (auxins) in plants. Both professors were inspiring lecturers and imbued me with the excitement of the subject. These courses emphasized that biological phenomena (life itself) could be understood at a molecular level, which has been a leitmotif of my subsequent research career. Wald's course also introduced me to the mechanism of vision, which led to my first paper on a theoretical approach to a biological problem (42).

Although I remember my undergraduate career at Harvard as a formative experience that furthered my interest in the world of science, it was reminiscent of my high school days in that my brother had preceded me, had been a stellar student in the Program in Chemistry and Physics, and was in the process of completing a PhD at Harvard with E. Bright Wilson, Jr., and Julian Schwinger. I spent considerable time with Bob and his fellow graduate students in Wilson's group. I was tolerated, I suppose, as Bob's little brother, though one day I made my mark when I solved a problem (they were always "challenging" each other) before any of them did. Given the importance of this "success" in my life, I am fond of the problem and restate it here: "It is agreed that to divide a pie between two people so that both are satisfied, one is allowed to cut the pie in two and the other chooses. The problem is how to extend this concept (dividing or choosing) to three or more people, so that everyone is satisfied." There is a special solution for three people and a general solution for any number. After solving this problem, I was accepted as part of the group by my brother and his friends, and during the afternoons when I could escape from the many labs I had to do, I would often join them. Their discussions of science exposed me to new ideas that I would not have come across otherwise.

The legendary *Elementary Organic* course taught by Louis Fieser was a standard part of the Program in Chemistry and Physics, but I thought it would be a waste of time: The course had the reputation of requiring a very tedious laboratory and endless memorization. An early version of the well-known textbook by Louis Fieser and Mary Fieser was available in lecture-note form and, rather than enrolling in the course, I tried to learn organic chemistry by reading it on my own, not with complete success. After studying the lecture notes, I enrolled in Paul Bartlett's Advanced Organic because that course taught the physical basis of organic reactions. It was an excellent course, though difficult for me because one was supposed to know many organic reactions, which I had to learn as we went along. At one point, Bartlett suggested that we read Linus Pauling's *Nature of the Chemical Bond*, which had been published in 1939 based on his Baker Lectures at Cornell. The *Nature of the Chemical Bond* presented chemistry for the first time as an integrated subject that could be understood, albeit not quite derived, from its quantum chemical basis. The many insights in this book were a critical element in orienting my subsequent research in chemistry.

At the end of three years at Harvard I needed only one more course to complete the requirements for a bachelor degree. During the previous year I had done research with Ruth Hubbard and her husband, George Wald. (Although Hubbard was scientifically on par with Wald, she remained a Senior Research Associate, a nonprofessorial appointment, until very late in her career when she was finally "promoted" to Professor. This was not an uncommon fate for women in science.) I mostly worked with Hubbard on the chemistry of retinal, the visual chromophore, because she had a deeper knowledge of chemistry than Wald. When I brought up my need to find a course for graduation, Wald suggested that I enroll in the physiology course at the Marine Biological Laboratory in Woods Hole, Massachusetts. This course was one of the few non-Harvard courses that were accepted for an undergraduate degree by the Faculty of Arts and Sciences. The physiology course was widely known as a stimulating course designed for postdoctoral fellows and junior faculty. The lectures in the physiology course by scientists who were summering at Woods Hole, while doing some research and enjoying boating and swimming, offered students a state-of-the-art view of biology and biological chemistry. (The course still exists, although its subject matter has shifted toward cell biology, of greater current interest.) For me, the only undergraduate in the course, it was a wonderful experience. I not only learned a great deal of biology but I also met several people, including Jack Strominger and Alex Rich, who became lifelong friends.

Woods Hole was an exciting place. Among the famous scientists that I met there were Otto Loewi, who had received a Nobel Prize in Physiology and Medicine (1936) for the discovery of the chemical basis of the transmission of nerve impulses, and Albert Szent Gyorgi, who had won a Nobel Prize, also in Physiology and Medicine (1937) for discovering vitamin C and showing that it existed in high concentration in paprika, a staple of the Hungarian diet. His little book, *Nature of Life, A Study of Muscle* (116), helped inspire my interest in doing research in biology. Like the *Nature of the Chemical Bond*, its stress on the logic of the subject helped to arouse my interest. These scientists held court in the afternoons at the Woods Hole beach and fascinated us young people with their discussions of new experiments and scientific gossip. In addition, there was an active student life, since many of the senior researchers brought along students from their labs. One of the justifications for having the laboratory in Woods Hole was that there were a wide range of marine animals which were caught for use in experiments. A prime example is the squid, whose giant axon was the ideal system for studies of the mechanism of nerve conduction. Because most of the experiments used only a small fraction of the animal, each week or so I collected some of the leftover laboratory squids and lobsters and prepared a feast for myself and friends, who provided bread, wine, and salad.

In considering graduate school during my last year at Harvard, I had decided to go to the West Coast and had applied to chemistry at the University of California at Berkeley and to biology at the California Institute of Technology (Caltech). Accepted at both, I found it difficult to choose between them. Providentially, I visited my brother, Bob, who was working with J. R. Oppenheimer at the Institute of Advanced Studies in Princeton, New Jersey. Bob introduced me to Oppenheimer, and briefly to Einstein. When Oppenheimer asked me what I was doing, I told him of my dilemma in choosing between U.C. Berkeley and Caltech for graduate school in chemistry or biology. He had held simultaneous appointments at both institutions and strongly recommended Caltech, describing it as "a shining light in a sea of darkness." His comment influenced me to choose Caltech, and I discovered that Oppenheimer's characterization of the local environment was all too true. Pasadena itself held little attraction for a student at that time. However, camping trips in the nearby desert and mountains and the vicinity of Hollywood made up for what Pasadena lacked.

I had become very interested in films and organized a classic film series at Caltech during my time there. This enabled me to preview many films that I had always wanted to see. The series showed mainly silent films accompanied by live piano music played by fellow students. (One of them was Walter Hamilton, a crystallographer, who died very young.) In searching for films, I gained access to several production studios, where I would ask the librarians to lend me films for our nonprofit Caltech series. The high point was my visit to the Chaplin studio. The receptionist was not particularly forthcoming, but then Charlie Chaplin himself walked in and asked what I wanted. He seemed intrigued by the idea that a science student was interested in films, and I asked him about the possibility of showing Monsieur Verdoux. He said it had been withdrawn (for political reasons), but told the librarian to let me have some of his early short films which at the time were not available to the public. I had always been a Chaplin fan and this meeting was one of the very special events of my graduate career.

At Caltech, I first joined the group of Max Delbrück in biology. He had started out as a physicist but, following the advice of Niels Bohr, had switched to biology. With Salvador Luria and others, he had been instrumental in transforming phage genetics into a quantitative discipline. His research fascinated me, and I thought that working with such a person would be a perfect entrée for me to do graduate work in biology. There were many bright and lively people working with Delbrück. I particularly remember Seymour Benzer, like Delbrück, a former physicist. We had discussions of phage genetics, biology, and a variety of subjects of mutual interest. Among other things, it was Seymour who introduced me to horsemeat, which became a staple of our household in Altadena. The law in Los Angeles was such that at the supermarket horsemeat was sold only as meat for dogs. Not deterred by that, horse filet, which cost a fraction of the corresponding cut of beef, turned up in my cooking as horse stroganoff, horse Cordon Bleu, and so on. Each of the members of our household, which included Sidney Bernhard, Gary Felsenfeld, and Walter Hamilton, had specific tasks to do. Mine, with Sidney, was cooking, and the horsemeat (as well as the local seafood) was regularly served.

After I had been in the Delbrück group for a couple of months, Delbrück proposed that I present a seminar on a possible area of research. I intended to discuss my ideas for a theory of vision (how the excitation of retinal by light could lead to a nerve impulse), which I had started to develop while doing undergraduate research with Hubbard and Wald. Among those who came to my talk was Richard Feynman; I had invited him to the seminar because I was taking his quantum mechanics course and knew he was interested in biology, as well as everything else. I began the seminar confidently by describing what was known about vision but was interrupted after a few minutes by Delbrück's comment from the back of the room, "I do not understand this." The implication of his remark, of course, was that I was not being clear, and this left me with no choice but to go over the material again. As this pattern repeated itself (Delbrück saying "I do not understand" and my trying to explain), after 30 minutes I had not even finished the 10-minute introduction and was getting nervous. When he intervened yet again, Feynman turned to him and whispered loud enough so that everyone could hear, "I can understand, Max; it is perfectly clear to me." With that, Delbrück got red in the face and rushed out of the room, bringing the seminar to an abrupt end. Later that afternoon, Delbrück called me into his office to tell me that I had given the worst seminar he had ever heard. I was devastated by this and agreed that I could not continue to work with him. It was only years later that I learned from reading a book dedicated to him that what I had gone through was a standard rite of passage for his students—everyone gave the "worst seminar he had ever heard."

After the devastating exchange with Delbrück, I spoke with George Beadle, the chairman of the Biology Department. He suggested that I find someone else in the department with whom to do graduate research. However, I felt that I wanted to go "home" to chemistry and asked him to help me make the transfer. Once in the Chemistry Department, I joined the group of John Kirkwood, who was doing research on charge fluctuations in proteins, as well as on his primary concern with the fundamental aspects of statistical mechanics and its applications. I undertook work on proteins and research started out well. It was complemented by a project involving Irwin Oppenheim and Alex Rich. Kirkwood's course in *Advanced Thermodynamics* was famous for its rigor, and the three of us, with Kirkwood's encouragement, worked together to prepare a set of lecture notes for the course. Each of us was responsible for writing up some of the lectures and the other two read them over. This was very useful for our learning thermodynamics and the set of notes was circulated widely. Some years later, Irwin Oppenheim prepared an improved version of the notes and it was published as a text entitled *Chemical Thermodynamics* (75).

In the spring of 1951, as I was getting immersed in my research project, Kirkwood received an offer from Yale. Linus Pauling, who was no longer taking graduate students, asked each student who was working with Kirkwood whether he would like to stay at Caltech and work with him. I was the only one to accept and, in retrospect, I think it was a very good choice. Initially, I was rather overwhelmed by Pauling. Each day upon arriving at the lab, I found a handwritten note on a yellow piece of paper in my mailbox which always began with something like "It would be interesting to look at...." As a new student I took this as an order and tried to read all about the problem and work on it, only to receive another note the next day beginning in the same way. When I raised this concern with Alex Rich and other postdocs, they laughed, pointing out that everyone received such notes and that the best thing to do was to file them or throw them away. Pauling had so many ideas that he could not work on all of them. He would communicate them to one or another of his students, but he did not expect a response. After I got over that, my relation with Pauling developed into a constructive collaboration.

Given Pauling's interest in hydrogen bonding in peptides and proteins, he proposed that I study the different contributions to hydrogen bonding interactions for a biologically relevant system, but I felt this would be too difficult to do in a rigorous way. Because quantum mechanical calculations still had to be done with calculating machines and tables of integrals (something difficult to imagine when even log tables have followed dinosaurs into oblivion), we had to find a simple enough system to be treated by quantum mechanical theory. I chose the bifluoride ion (FHF^-) because the hydrogen bond is the strongest known, the system is symmetric, and only two heavy atoms are involved. (Today, such "strong" hydrogen bonds have become popular in analyses of enzyme catalysis, although there is no convincing evidence for their role.)

I sometimes felt intimidated when I went in to talk with Pauling, but it was wonderful to work with him and be exposed to (although not necessarily understand) his intuitive approach to chemical problems. One day I asked Pauling about the structure of a certain hydrogen bonded system (i.e., whether the hydrogen bond would be symmetric, as in FHF^-). He paused, thought for a while, and gave a prediction for the structure. When I asked him why, he thought again and offered an explanation. I left his office and soon realized that his explanation made no sense. So I went in to ask Pauling again, thinking that perhaps he would come up with a different conclusion. Instead, Pauling said that he believed that the predicted structure was correct, but he proposed an entirely different explanation. After going over his analysis, I was again dissatisfied with his rationale and caught up with him as he was leaving his office. He said that he still believed the conclusion was valid and produced yet another rationale. This one made sense to me, so I did some crude calculations which indicated that Pauling was indeed correct. What amazed me then, and still does today, is that Pauling came to the correct conclusion, apparently based on intuition, without having worked through the analysis. He "knew" the right answer, even if it took more thought to figure out why.

The research was very rewarding, all the more so because of the intellectual and social atmosphere of the Chemistry Department at Caltech. The professors—like Pauling, Verner Schomaker, and Norman Davidson—treated the graduate students and postdoctoral fellows as equals. We participated in many joint activities that included trips into the desert, as well as frequent parties held at our Altadena house, where Feynman would occasionally come and play the drums. At one such party, Pauling disappeared for a while and I discovered him out in the backyard on his knees collecting snails, which had infested our yard, for his wife Ava Helen to cook for dinner. (It was only later in France, when I collected my own snails, that I learned how complicated it was to prepare them—the snails had to fast for a week—so that now, looking back, I am not sure what the Paulings did with their snails.)

Pauling's presence attracted many postdoctoral fellows to Caltech. When I was there the group included Alex Rich, Jack Dunitz, Massimo Simonetta, Leslie Orgel, Edgar Heilbrunner, and Paul Schatz. Interacting with them (as a graduate student, I was the "baby" of the group) was a wonderful part of my Caltech education and many of them became my friends. Sadly, Massimo Simonetta, who remained a dear friend and colleague after his return to Italy and whom I visited regularly in Milan, as well as on ski trips to Courmayeur in the winter and Portofino in the summer, died suddenly from virulent leukemia in 1986.

My parents had given me their old car as a graduation present, and several times during my Caltech career I drove across the country to our home in Newton, Massachusetts, for part of the summer. Each time I took a different route, once through Canada with visits to the Banff and Jasper National Parks, and another time through the Deep South. On one such trip while driving through Texas on a very hot summer day, my friends and I decided to take a swim and cool off. We passed one swimming area, but it was full of people. Because we were unshaven and dirty from the long drive, we looked along the river for a quieter place to swim. About a mile downstream we came upon another swimming area with broken-down steps leading to the water. As it was deserted, we decided it was the perfect place. After we had been in the water for about 10 minutes, a couple of pickup trucks drove up and several men jumped out with guns at the ready. What turned out to be the local law enforcement officers ordered us out of the water and demanded to know what we thought we were doing... "white folks swimming in an area reserved for niggers." They had noticed the Massachusetts license plate on the car and had concluded we were Northern "troublemakers." After some effort we succeeded in explaining that, given our scruffy state, we had just not wanted to bother other (white) people and the officers let us go, with the admonition that we had better drive straight through Texas without stopping anywhere, which we did.

My initial attempt at a purely ab initio approach to the bifluoride ion failed and I soon realized that it was necessary to introduce experimental information concerning the atomic states to obtain a meaningful estimate of the relative contribution of covalent and ionic structures. I developed a method for doing this and completed my work only to discover that William Moffitt (who was to be my predecessor at Harvard) had just published a similar approach called the method of *Atoms in Molecules*. He had presented the method in a more general and elegant formulation (91) compared with my treatment, which focused specifically on the bifluoride ion. Although there were significant differences in the details of the methodology, I felt so discouraged by the similarities that I never published "My Great Idea," as Verner Schomaker, one of the members of my thesis committee, called it. Not too surprisingly, I was having a difficult time writing my thesis. (I retained my interest in this type of approach, and some years later Gabriel Balint-Kurti joined my group as a graduate student, and we proposed an improved version of the theory, which we called the Orthogonalized Moffitt method, and applied it to the potential energy surfaces for simple reactions (4).) Such calculations are mostly of historical interest today, when fast computers and ab initio programs are widely used without empirical corrections, except for complex systems like biomolecules.

POSTDOCTORAL SOJOURN IN OXFORD AND EUROPE

One day in October 1953, Pauling came into the office I shared with several postdocs and announced that he was leaving in three weeks for a six-month trip and that "it would be nice" if I finished my thesis and had my exam before he left. This was eminently reasonable, since I had finished the calculations some months before and I had received a National Science Foundation (NSF) postdoctoral fellowship to go to England that fall. Pauling's "request" provided just the push I needed, even though the introduction was all I had written thus far. With so much to get done, I literally wrote night and day, with my friends typing and correcting what I wrote. In this way, the thesis was finished within three weeks, and I had my final PhD exam and celebratory party. After a brief visit with my parents in Newton, I left for England and arrived shortly before Christmas 1953.

In my NSF postdoctoral application I had proposed to work with John Lennard-Jones in Cambridge, England. He, however, had left Cambridge to become Principal of Keele University in 1953, and so I had to alter my plans. Instead I joined Charles Coulson at Oxford University, where he had an active group in theoretical chemistry at the Mathematical Institute. One member was Simon

Altmann, who greatly improved my limited knowledge of group theory and who, with his wife, Bochia, "adopted" me while I was in Oxford. In addition, there were visitors such as Don Hornig and Bill Lipscomb, who were on sabbaticals.

I was 23 years old when I arrived in England. Having worked continuously all the way through graduate school, I was eager to have the sojourn in Europe provide experiences beyond science. The NSF postdoctoral fellowship provided a generous (at the time) salary of $3000 per year, which was sufficient to do considerable traveling. I took the NSF guidelines about following the customs of the institution quite literally, perhaps more so than was intended, and traveled throughout Europe outside of the three six-week terms when I was in residence in Oxford. In fact, upon my arrival in Oxford shortly before Christmas, I went in to see Coulson, got him to sign my NSF form, and immediately left on the first of many trips to Paris. Although there were few scientific interactions during these extended trips, I learned much about the peoples and their cultures, art, architecture, and cuisine, all of which continue to play an important role in my life. One trip involved driving a Volkswagen across Europe through Yugoslavia to Athens; another an extended visit to France, Switzerland, and Italy. On the latter, I first saw the Lake Annecy and the Haute Savoie region. I concluded that I wanted to own a chalet in one of the upper valleys for summer and winter vacations, but I could not afford it then and for many years to come.

I became interested in photography at that time. Upon completing my PhD, my family had given me a Leica IIIC, a superb camera, which my Uncle Alex had brought to the United States from Vienna. Throughout my travels, I took photographs, particularly of people, using a trick I had developed. The Leica had a long focus lens with a reflex viewer, which enabled me to face away from my subject; I could take photographs of crowds and individuals without their being aware of it. Although the resulting Kodachrome slides are now more than 50 years old, they remain in excellent condition. My wife, Marci, had the idea that the slides, which had hardly been looked at over the years, should be printed and enlarged. This endeavor has, in a sense, come full circle. During the academic year 1999–2000 I was Eastman Professor at Oxford. While there, we were introduced to a marvelous photographic craftsman, Paul Sims (Colourbox Techunique). He made beautiful prints from the collection of slides, and a selection was exhibited at my 75th birthday celebration at NIH in Bethesda, Maryland (101).

During the two years in Oxford as a postdoctoral fellow, I spent more time thinking about chemical problems than actually solving them. My aim was to find areas where theory could make a contribution of general utility in chemistry. I did not want to do research whose results were of interest just to theoretical chemists. Reading the literature, listening to lectures, and talking to scientists like Don Hornig and the Oxford physicist H.M.C. Pryce, I realized that magnetic resonance was a vital new area. Chemical applications of magnetic resonance were in their infancy and it seemed to me that nuclear magnetic resonance (NMR), in particular, was a field where theory could make a contribution. Chemical shifts, for example, could provide a means of testing theoretical calculations, but, of even greater import, quantum mechanical theory could aid in interpreting the available experimental results and propose new measurements.

My first paper in chemistry on the quadrupole moment of the hydrogen molecule, obtained from different approximate wavefunctions, was in this vein (51). Although this was a short paper, I spent an enormous amount of time rewriting and polishing it before I finally was ready to submit it. When students seem to face similar problems in finishing their first paper, I often tell them about what I went through and that publishing (or being ready to publish) one's work becomes easier with each succeeding paper.

FIVE YEARS AT THE UNIVERSITY OF ILLINOIS: NMR AND COUPLING CONSTANTS

As my postdoctoral fellowship in Oxford (1953–1955) neared its end, I was looking for a position to begin my academic career in the United States. With my growing interest in magnetic resonance, I focused on finding an institution that had active experimental programs in the area. One of the best schools from this point of view was the University of Illinois, where Charles Slichter in Physics and Herbert Gutowsky in Chemistry were doing pioneering work in applying NMR to chemical problems. The University of Illinois had a number of openings in Chemistry at that time because the department was undergoing a radical renovation; several professors, including the chairman Roger Adams, had retired. Pauling recommended me to the University of Illinois and the department offered me a job without an interview and without waiting for a recommendation from Coulson. The latter was fortunate, because Coulson had written that, although he had no doubt about my intellectual abilities, I had done very little work on problems he had suggested. I accepted the offer from Illinois without visiting the department, something unimaginable today with the extended courtships that have become an inherent part of the academic hiring process. The University of Illinois offered me an Instructorship at a salary of $5000 per year; the department offered nothing like the present-day start-up funds, and I did not think of asking for research support.

Although the University of Illinois was a very good institution with excellent chemistry and physics departments, it was located in a small town in the flat rural Midwest, where I could not imagine living for more than five years. Having had such a good time as a postdoctoral fellow traveling in Europe, I was ready to get to work, and Urbana-Champaign seemed like a place where I could concentrate on science with few distractions. The presence of four new instructors—Rolf Herber, Aron Kupperman, Robert Ruben, and me—plus other young scientists on the faculty, such as Doug Applequist, Lynn Belford, and E.J. Corey, led to a very interactive and congenial atmosphere.

I focused a major part of my research on theoretical methods for relating nuclear and electron spin magnetic resonance parameters to the electronic structure of molecules. The first major problem I examined was concerned with proton-proton coupling constants, which were known to be dominated by the Fermi contact interaction. What made coupling constants of particular interest was that for protons, which were not bonded, the existence of a nonzero value indicated that there was an interaction beyond that expected from localized bonds. In the valence bond framework, which I used in part because of my training with Pauling, nonzero coupling constants provide a direct measure of the deviation from the perfect-pairing approximation. To translate this qualitative idea into a quantitative model, I chose to treat the vicinal coupling constant in a molecule like ethyl alcohol, one of the first molecules to have its NMR spectrum analyzed experimentally. Specifically, I chose to study the HCC'H' fragment as a function of the HCC'H' dihedral angle, a relatively simple system consisting of six electrons (with neglect of the inner shells). I believed that it could be described with sufficient accuracy for the problem at hand by including only five covalent valence-bond structures. To calculate the contributions of the various structures, I introduced semiempirical values of the required molecular integrals. Although the HCC'H' fragment is relatively simple, the calculations for a series of dihedral angles were time consuming and it seemed worthwhile to develop a computer program. This was not as obvious in 1958 as it is now. Fortunately, the ILLIAC, a "large" digital computer at that time, had recently been built at the University of Illinois.

If I remember correctly, it had 1000 words of memory, which was enough to store my program. The actual program was written by punching holes in a paper tape. If you made a mistake, you filled in the incorrect holes with nail polish so that you could continue the program, the output appearing on spools of paper. Probably the most valuable aspect of having a program for this type of simple calculation, which could have been done on a desk calculator, was that once the program was known to be correct, a large number of calculations could be performed without having to worry about arithmetic mistakes.

Just as I finished the analysis of the vicinal coupling constants (52), I heard a lecture by R.V. Lemieux on the conformations of acetylated sugars. I do not remember why I went to the talk, because it was an organic chemistry lecture, and the chemistry department at Illinois was rigidly separated into divisions, which had a semiautonomous existence. Lemieux reported measurements of vicinal coupling constants and noted that there appeared to be a dihedral angle dependence, although the details of the behavior were not clear. The results were exciting to me because the experiments confirmed the theory, at least qualitatively, before it was even published.

E.J. Corey, who was an assistant professor at Illinois and later became a colleague at Harvard, was one of the people with whom I had dinner on a fairly regular basis at the Tea Garden, a passable Chinese restaurant in Urbana-Champaign. We often discussed recent work of mutual interest and one day I described the studies that I had made of vicinal coupling constants. Corey immediately recognized the possibility of using the results for structure determination and published what is probably the first application of my results in organic chemistry (7). Not long after, the theory appeared in a comprehensive review of the use of NMR in organic structure determinations (19), and someone introduced the name *Karplus equation* for the relationship I had developed. This proved a mixed blessing. Many people attempted to apply the equation to determine dihedral angles of organic compounds. They found some deviations of the measured coupling constants from the predicted values for known structures and published their results, commenting on the inaccuracy of the theory.

As happens too often with the application of theoretical results in chemistry, most people who used the so-called *Karplus equation* did not read the original paper (52) and thus do not know the limitations of the theory. They assumed that because the equation had been used to estimate vicinal dihedral angles, the theory said that the coupling constant depends *only* on the dihedral angle. By 1963, having realized organic chemists tend to write and read Communications to the *Journal of the American Chemical Society*, I published such a Communication (56). In it, I described various factors, other than the dihedral angle, that are expected to affect the value of the vicinal coupling constant; they include the electronegativity of substituents, the valence angles of the protons (HCC' and CC'H), and bond lengths. The main point of the paper was not to provide a more accurate equation but rather to make clear that caution had to be used in applying the equation to structural problems. My closing sentence, which has often been quoted, was the following: "Certainly with our present knowledge, the person who attempts to estimate dihedral angles to an accuracy of one or two degrees does so at his own peril."

In spite of my concerns about the limitations of the model, the use of the equation has continued, and the original paper (52) is one of the *Current Contents* "most-cited papers in chemistry"; correspondingly, the 1963 paper was recently listed as one of the most-cited papers in the *Journal of the American Chemical Society* (20a). In addition, there have been many empirical "extensions" of the equation; perhaps the most complex published form (46) uses a 12-term expression. Equations have been developed for vicinal coupling constants involving a variety of nuclei (73) (e.g., ^{13}C-CC'-H, ^{15}N-CC'-H), and they have been applied in areas ranging from inorganic to organic to biochemistry. An important more recent application is the use of these relationships as part of the data employed in structure determination of proteins by NMR (76, 89). The vicinal coupling constant model, which was developed primarily to understand deviations from perfect pairing, has been much more useful than I would have guessed. "In many ways my feeling about the uses and refinements of the *Karplus equation* is that of a proud father. I am very pleased to see all the nice things that the equation can do, but it is clear that it has grown up and now is living its own life" (59).

I continued to work on problems in NMR and ESR (electron spin resonance) because new areas of chemistry were being studied by these spectroscopic methods and it seemed worthwhile to try to provide insights from theoretical analyses of some of these applications. Examples are a study of the hyperfine interactions in the ESR spectrum of the methyl radical (53) and the contributions of π-electron delocalization to the NMR coupling constants in conjugated molecules (54). My general approach to the magnetic properties of molecules was summarized in an article entitled "Weak Interactions in Molecular Quantum Mechanics" (55). The choice of title was apt because the energies involved in coupling constants and hyperfine interactions are indeed weak relative to the electron volts of bond energies, excitation energies, and ionization potentials that are the bread and butter of quantum chemistry. However, the title also had a facetious aspect in that my brother had been working on what the physicists call "weak interactions" (72).

At Illinois, my officemate was Aron Kuppermann. Our instructorship at Illinois was the first academic position for both of us, and we discussed science, as well as politics and culture, for hours on end. We soon became fast friends. He and his wife, Roza, lived in an apartment next to mine and often invited me for dinner. Our friendship has continued for more than 50 years, even though I left Illinois to go to Columbia University and Aron moved to Caltech. We see each other only once in several years, but having Aron and Roza as friends provides a special continuity in my life.

Aron and I decided that, although we were on the faculty, we wanted to continue to learn and would teach each other. I taught Aron about molecular electronic structure theory [we published two joint papers on molecular integrals (64, 77)] and Aron taught me about chemical kinetics, his primary area of research. Aron is officially an experimentalist, but he is also an excellent theoretician, as was demonstrated by his landmark quantum mechanical study of the $H + H_2$ exchange reaction with George Schatz. This work was some years in the future (it was published in 1975) (78), but in the late 1950s we both felt that it was time to go beyond descriptions of reactions in terms of the Arrhenius formulation based on the activation energy and preexponential factor. My research in this area had to wait until I moved to Columbia University, where I would have access to the required computer facilities.

MOVE TO COLUMBIA AND FOCUS ON REACTION KINETICS

During the summer of 1960 I participated in an NSF program at Tufts University with the purpose of exposing high school and small college science teachers to faculty actively engaged in research. Our task was to present some modern chemical concepts in a way that would help the teachers in the classroom. Ben Dailey, one of the organizers of the program, asked me one day as we were standing next to each other in the washroom whether I would consider joining the chemistry faculty at Columbia University, where he was a professor. Because I had already been at Illinois for four of the five years I had planned to stay there, I responded positively. I heard from Columbia shortly thereafter and received an offer to join the IBM Watson Scientific

Laboratory with an adjunct associate professorship at Columbia.

The Watson Scientific Laboratory was an unusual institution to be financed by a company like IBM. Although the laboratory played a role in the development of IBM computers, many of the scientists there were doing fundamental research. The Lab had been founded in 1945 near the end of World War II to provide computing facilities needed by the Allies. Its director, Wallace Eckart, is perhaps best known for his highly accurate perturbation calculation of the three-body problem posed by the motion of the earth around the sun in the presence of the moon; the $H + H_2$ reaction, which I studied while at the Lab, is also a three-body problem in the Born-Oppenheimer approximation. When Eckart described the position at the Watson Lab to me, he made clear that staff members were judged by their peers for what they did in their research and not for their contribution to IBM. The presence of outstanding scientists on the staff, such as Erwin Hahn, Seymour Koenig, Alfred Redfield, and L.H. Thomas (of Thomas-Fermi fame), supported this description and made the place very attractive. Moreover, the Watson Lab had a special advantage for me in that it had an IBM 650, an early digital computer, which was much more useful than the ILLIAC because of its greater speed, larger memory, and simpler (card) input. (No more nail polish!) I was to have access to considerable amounts of time on the IBM 650 and to receive support for postdocs, as well as other advantages over a regular Columbia faculty appointment. This was a seductive offer, but I hesitated about accepting a position that, in any way, depended on a company, even a large and stable one like IBM. This was based, in part, on my political outlook, but even more so on the fact that industry has as its primary objective making a profit, and all the rest is secondary. By contrast, my primary focus was on research and teaching, which are the essential aspects of a university, but not of industry. Consequently, I replied to Columbia and the Watson Lab that the offer was very appealing, but that I would consider it only if it included a tenured position in the chemistry department, even though I agreed initially to be at the Watson Lab as well. Columbia acceded to my request, and after some further negotiation I accepted the position for the fall of 1960.

The environment at the Watson Lab was indeed fruitful, both in terms of discussions with other staff members and of the available facilities. I was able to do research there that would have been much more difficult at Columbia. However, not unexpectedly, the atmosphere gradually changed over the years, with increasing pressure from IBM to do something useful (i.e., profitable) for the company, such as visiting people at the much larger and more applied IBM laboratory in Yorktown Heights, essentially doing internal consulting. I decided in 1963 that the time had come to leave the Watson Lab, and I moved to the full-time professorial position that was waiting for me in Chemistry at Columbia. (IBM closed the Watson Lab in 1970.) Given that experience, I always warn my students and postdocs about accepting jobs in industry. They may well have an exciting environment when they first join the staff, receive a significantly higher salary than they could at a university, and not have to worry about obtaining grants to support their research. What I urge them to remember is that a new management team can take over at any time, particularly if the company is not doing well, and decide to cut down on the research budget. (Research in the short run only spends money, even if it can finally produce a profit.) This attitude has led to layoffs of individual scientists or the closing of beautifully equipped research laboratories that were built only a few years before. It is of primary importance that your objectives (in my case, teaching and research) be the same as those of the institution where you work. This requirement is ideally satisfied in a good university but cannot be guaranteed in industry.

I continued research in the area of magnetic resonance after moving to New York.

One reward of being at Columbia was the stimulation provided by interactions with new colleagues, such as George Fraenkel, Ben Dailey, Rich Bersohn, and Ron Breslow. Frequent discussions with them helped to broaden my view of chemistry. In particular, my interest in ESR was rekindled by George Fraenkel and we published several papers together (62, 66, 80), including a pioneering calculation of ^{13}C hyperfine splittings (62). Although the techniques we used were rather crude, the results provide insights concerning the electronic structure of the molecules considered and aided in understanding the measurements. Many of the weak interactions, which could be used to provide information about the electronic structure of molecules, were a real challenge to estimate in the mid-1960s. Now they can be calculated essentially by pushing a button with programs like the widely used Gaussian package. The high-level ab initio treatments that are used routinely today have the drawback that, even though the results can be accurate, the insights obtained by the earlier, simplified approaches are often lost in the complexity of the calculation.

My interest in chemical reaction dynamics had deepened at Illinois through many discussions with Aron Kuppermann, as already mentioned, but I began to do research in the area only after moving to Columbia. There were several reasons for this. There is no point in undertaking a problem if the methodology and means for solving it are not available: It is important to feel that a problem is ripe for solution. (This has been a guiding rule for much of my research—there are many exciting and important problems, but only when one feels that they are ready to be solved should one invest the time to work on them. This rule has turned out to be even more important in the application of theory to biology, as we shall see later.) Given the availability of the IBM 650 at the Watson Lab, the very simple reaction, $H + H_2 \rightarrow H_2 + H$, which involves an exchange of hydrogen atom with a hydrogen molecule, could now be studied by theory at a relatively fundamental level. Moreover, early measurements made by Farkas & Farkas in 1935 (26) of the rate of reaction over a wide temperature range provided important data for comparison with calculations. A second reason for focusing on chemical kinetics was that crossed molecular beam studies were beginning to provide much more detailed information about these reactions than had been available from gas phase or solution measurements. The pioneering experiments of Taylor & Datz opened up this new field in 1955 (117), although it was not until 2000 that Datz received the prestigious Enrico Fermi Award in recognition of this work. Many groups extended their original crossed molecular beam experiments and showed that it was possible to study individual collisions and determine whether they were reactive. Thus, calculated reaction cross sections, rather than overall rate constants, could be compared directly with experimental data.

To do a theoretical treatment of this, or any other reaction (including the protein folding reaction), a knowledge of the potential energy of the system as a function of the atomic coordinates is required; it is necessary to know the potential energy surface or energy landscape, as it is now called. Isaiah Shavitt, who was working with me as a postdoctoral fellow at the Watson Lab on quantum mechanical calculations, had developed new methods for evaluating multicenter two-electron integrals (109), and he used the $H + H_2$ potential surface as his first application (110). Even though this reaction involved only three electrons and three nuclei, the theoretical surface was expected to be useful only for determining the general features, and a more accurate surface was needed for calculating reaction attributes for comparison with experiment. (Five years later Liu (86) was able to calculate an accurate surface for the $H + H_2$ exchange reaction.) Thus, it was necessary to resort to so-called semiempirical surfaces, whose form was given by a quantum mechanical model with parameters determined from experiment.

Already in 1936, J. Hirshfelder and B. Topley, two students of H. Eyring, had attempted a trajectory calculation of the $H + H_2$ reaction with the three atoms restricted to move on a line, for simplicity (40). In a trajectory calculation one determines the forces on the atoms, here the three hydrogen atoms, from the potential energy surface and integrates Newton's equation, $F = ma$, to obtain the positions of the atoms as a function of time. Hirschfelder and Topley used a three-body potential for the reaction based on the Heitler-London method. They calculated a few steps along the trajectory but were not able to finish the calculation, so we do not know (and never will since we do not have the initial velocities) whether the trajectory was reactive. The potential had a well in the region where all three atoms were close to each other ("Lake Eyring" as it was called), which was expected to give a three-body complex under the collision conditions appropriate for the reaction. Ab initio quantum mechanical calculations, such as that of Shavitt, indicated that this was incorrect, i.e., there was a simple activation barrier. Thus, to obtain a meaningful description of the $H + H_2$ reaction, it was necessary to introduce a more realistic potential function. Moreover, what was needed was the reaction surface in full three-dimensional space, rather than restricting the hydrogen atoms to positions on a line.

Richard Porter, a graduate student with F.T. Wall, had used a surface without Lake Eyring to improve the collinear collision calculations (119). Much impressed by Porter, I invited him to join my group at Columbia as a postdoctoral fellow. At Columbia, we rapidly developed a semiempirical extension of the original Heitler-London surface for the $H + H_2$ reaction, based on the method of diatomics in molecules, and calibrated the surface with ab initio quantum calculations and experimental data for the reaction (100). This surface, which is known as the Porter-Karplus surface, has an accuracy and simplicity that led to its continued use in many reaction rate calculations by a variety of methods over the years.

Once the surface was developed, we were ready to undertake the first full three-dimensional trajectory calculation for the exchange reaction. Dick Porter was a fundamental contributor to the $H + H_2$ research, which also involved R.D. Sharma, another postdoctoral fellow. With the availability of large amounts (by the standards of the day) of computer time on the IBM 650 at the Watson Lab, we were able to calculate enough trajectories to obtain statistically meaningful results (69). Determination of the reaction cross section required the calculation of a series of trajectories with an appropriately chosen range of initial conditions, such as relative velocities and impact parameters. The trajectories start with the reactants far apart (so far that the interaction between the hydrogen atom and

Figure 2

The exchange reaction between a hydrogen atom and a hydrogen molecule. (*a*) Schematic representation of the reaction with definitions for the distances R_{AB}, R_{AC}, and R_{BC}. (*b*) Contour plot of the potential energy surface for a linear collision as a function of the distances R_{AB} and R_{BC} with $R_{AC} = R_{AB} + R_{BC}$; the minimum-energy path is shown in red. (*c*) Same as panel (*b*), but in a three-dimensional representation. (*d*) Energy along the reaction coordinate corresponding to the minimum-energy path in panels (*b*) and (*c*); the transition state is indicated in yellow. (*e*) A typical trajectory for the reactive, three-dimensional collision; the three distances R_{AB}, R_{AC}, and R_{BC} are represented as a function of time. (*f*) A typical trajectory for the nonreactive collision; as shown in panel (*e*). In both panels (*e*) and (*f*) the interactions between the three atoms are limited to a very short time period (yellow background); this mirrors the narrow potential energy barrier in panel (*d*). The figure has been reproduced, with permission, from Reference 24.

hydrogen molecule are negligible), let them collide in the presence of the interaction potential, and then follow the atoms until the products are again far apart. By looking at which atoms are close together, one can determine whether a reaction has taken place (**Figure 2**). As can be seen from the figure, the reaction (i.e., the time during which the three atoms are interacting) takes only a few femtoseconds. This illustrates a fundamental

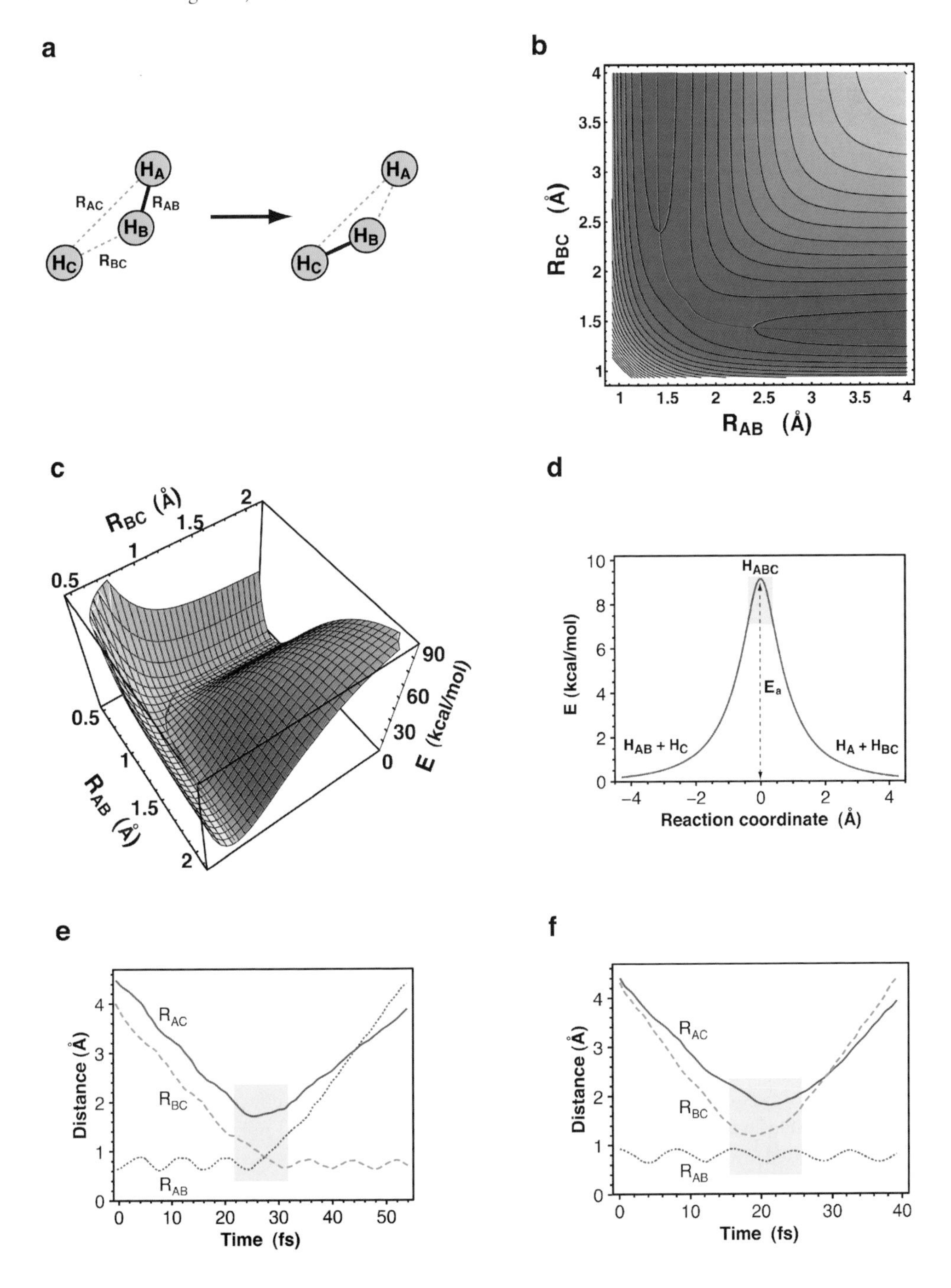

point, namely that many simple reactions have a small rate constant, not because of the elementary reaction rate, which is fast when it occurs, but because of the large activation energy, which makes most thermal collisions nonreactive. Within the approximation that classical mechanics is accurate for describing the atomic motions involved in the $H + H_2$ reaction and that the semiempirical Porter-Karplus surface is valid, a set of trajectories makes it possible to determine any and all reaction attributes, e.g., the reaction cross section as a function of the collision energy. The ultimate level of detail that can be achieved is an inherent attribute of this type of approach, which I was to exploit 15 years later in studies of the dynamics of macromolecules. Although there are significant quantum corrections for $H + H_2$, the results for the reaction cross section as a function of internal energy and the rate constant as a function of temperature provided insights concerning the fundamental nature of chemical reactions that are as valid today as they were 40 years ago when the calculations were performed.

As in many of my papers in which a new method was developed (69), I tried to present the detail necessary for the reader to reproduce and use what had been done. (We had to work through it all, so why not save others the effort?) The requirement that the work be reproducible is often cited as a standard for publishing papers, but, in practice, few papers are written in this way. Recently, I was pleased to learn that our paper was cited by George Schatz (105) as one of the key twentieth-century papers in theoretical chemistry. Schatz pointed out that what we had called the quasiclassical trajectory method (classical trajectories with quantized initial conditions, which are very important for the H_2 molecule because of its large zero-point vibrational energy, some of which is available in the transition state) was still widely used, even though now full quantum dynamics calculations for the $H + H_2$ reactions and a small number of other reactions are available. Moreover, Schatz states, "The KPS paper stimulated research in several new directions and ultimately spawned new fields."

After the $H + H_2$ reaction calculation, Dick Porter and I collaborated on the textbook *Atoms and Molecules* (68), which developed quantum chemistry at the introductory level for students of physical chemistry. It was based on a lecture course I had given at Columbia and then at Harvard University. (Teaching is a very good discipline, which forces the instructor to understand a subject well enough so it can presented in a fashion clear enough for students to understand.) We thought that writing a book based on the lecture notes would be simple to do, but in fact it was an enormous task. The book was finished only because Dick and I were together for a summer on Martha's Vineyard at a "writing camp" for scientists, sponsored by an imaginative young publisher, Bill Benjamin. Later, while spending a semester at the Weizmann Institute in Rehovot, Israel, I used part of the time to correct the proofs, which in my case inevitably led to significant rewriting. The book has been a success, particularly as a source of material for teachers of physical chemistry.

Unlike quantum dynamics calculations, the quasiclassical trajectory method was easily extended to more complex systems. One study that I remember as particularly interesting was done by Martin Godfrey, that of the $K + CH_3I \rightarrow KI + CH_3$ reaction, in which the collisions involved an orientated CH_3I molecule (63). We were stimulated to do this calculation by an ingenious experiment performed by Richard Bernstein, who was one of the outstanding contributors to the field of crossed molecular chemistry but died young. He, indeed, oriented the CH_3I reactant relative to the incoming K^+ ion so that one could study the effect of the orientation on the reaction and obtain additional information for comparison with the calculation; the two papers were published together in the *Journal of the American Chemical Society* (6).

RETURN TO HARVARD UNIVERSITY AND BIOLOGY

In 1965, it was time to move again. Columbia and New York City were stimulating places to live and work, but I felt that new colleagues in a different environment would help to keep my research productive. I had incorporated this idea into a "plan": I would change schools every five years and when I changed schools I would also change my primary area of research. It was more exciting for me to work on something new, where I had much to learn so as to stay mentally young and have new ideas.

When it became known that I was planning to leave Columbia University, numerous schools invited me to join their faculty. With considerable difficulty, I narrowed the choice down to U.C. Berkeley and Harvard and decided to visit each place for a semester during a sabbatical year (1965–1966). I enjoyed my stay in both places; discussions with colleagues were stimulating and it was very hard to decide between the two schools. I particularly enjoyed my interactions at U.C. Berkeley with Bob Harris, a gifted theoretician who 10 years before had been my very first research student while he was an undergraduate at the University of Illinois. We spent many hours together talking science and politics. This was the era of the Vietnam antiwar movement, and I was introduced to police brutality during some of the marches in Berkeley, a center of the movement, and particularly in neighboring Oakland, which supported the war, in contrast to the Berkeley citizens.

This was not my first experience with political protest, however. I had participated in the 1950s as a Caltech graduate student in meetings organized against the death penalty sentences of Julian and Ethel Rosenberg. In the mid-1970s, my brother, Bob, was initially refused a security clearance to do research at the Livermore Laboratory. Because his participation was deemed important to the laboratory, the administrators appealed and found out that the reason for the refusal was my presence at a Unitarian Church in Los Angeles at a Rosenberg protest meeting two decades previous. Once this item in his FBI file was exposed, Bob rapidly received the necessary clearance.

By contrast, the one time I needed a clearance, it was not granted. I was invited to participate in a disarmament study sponsored in Woods Hole by the National Academy of Sciences in the late 1970s. Every participant had to have a security clearance so that when the report was issued, no one could say that the people involved did not have access to the required information. I had not requested a clearance previously, and it was arranged that I apply for one. I arrived in Woods Hole and attended the opening meeting, which was public, but was not allowed into the sessions after that. While awaiting my clearance to participate in the month-long meeting, I relaxed with my family and worked on my own research in a pleasant nearby hotel with a swimming pool, courtesy of the sponsors. This also permitted me to renew my ties with the Marine Biological Laboratory. The disturbing element in my FBI record was the one that had raised problems for my brother, namely that I had attended the Rosenberg protest meeting. It is still amazing to me that the U.S. government wastes so much time and money recording the attendance at meetings that one should have the right to attend without fear of future harassment or discrimination.

Another such experience occurred while I was on the faculty of the University of Illinois in 1955. Shortly after I moved there, I began to receive a number of visits from FBI agents. There were always two of them; one was from the local (Chicago) office and the other was a different person each time. They questioned me about my political views and focused on the time I had spent in Yugoslavia while I was a postdoctoral fellow at Oxford. I had been there as a tourist and photographer, as already mentioned, but the FBI apparently had other ideas. On one such visit, the "changing" agent suddenly began speaking in a language I did not understand. It turned out to be Serbian and was apparently meant to trick me. What exactly the FBI was looking for I never discovered. The visits stopped after about six months.

A variety of factors led me to choose Harvard, one of which, in retrospect, was the fact that I had been a Harvard undergraduate. Also, I felt that the Berkeley environment and its weather were just too nice and that the distractions from work were too great, as evident from the activities of a considerable portion of the Berkeley chemistry faculty at that time. I came to Harvard as Professor in the fall of 1966 and received the title of Theodore William Richards Professor in Chemistry in 1979. Although such chairs mean little at Harvard, other than that the funds for one's salary come out of a special endowment, I was pleased to receive this title for two reasons. First, the previous holder of the chair had been E. Bright Wilson, a highly respected member of the department for his science, his humanity in dealing with students, and his high standards of intellectual honesty. Second, this chair was the only one in chemistry named after a scientist (the first American to win the Nobel Prize in Chemistry) instead of the donor of the funds.

At Harvard I continued to do research in the some of the areas that I had developed at Illinois and Columbia, including the study of hyperfine interactions in ESR (102) and the use of quasiclassical trajectory methods for the study of reactions (37). With the results from the trajectory calculations, Keiji Morokuma, B.C. Eu, and I undertook a study of the relation between reaction cross sections and transition state theory as a test of this widely used model in chemistry (93, 94). The application of many-body theory to the electronic structure of atoms and molecules (17, 31), as an extension of methods developed in physics, was also of interest to me, in part to understand better what was involved in this development. I have always found it very helpful when I see a new idea or method to apply it to some problem, even if the specific research is not that significant.

After I had been at Harvard for only a short time, I realized that if I was ever to return to my long-standing interest in biology I had to make a break with what had been thus far a successful and very busy research program in theoretical chemistry. I felt that I had grasped what was going on in elementary chemical reactions and the excitement of learning something new was no longer there. The initial qualitative insights obtained from relatively simple approaches to a new problem are often the most rewarding. This is not to imply that the field of gas-phase chemical reactions has not continued to flourish. It is still active, with ever-finer details concerning reactive collisions being elucidated. For example, I very much enjoyed attending a meeting on reaction dynamics at the Fritz Haber Institute in Berlin in 1982, where I learned about the exciting research going on; I, however, gave a lecture on protein dynamics (58). The meeting also brought home to me that other people with skills different from mine are better able to contribute to the advanced technologies now required in this area.

I planned to take a six-month leave in the fall of 1969 and chose the Weizmann Institute, in part because it had an excellent library. I was aware of Shneior Lifson's work on polymer theory and his reputation of being an open-minded scientist, as well as a marvelous storyteller. I wrote Shneior asking whether I could come for a semester, and he kindly invited me to join his group. The sabbatical gave me the leisure to read and explore a number of areas in which I hoped to do constructive research by applying my expertise in theoretical chemistry to biology. Discussions with Shneior and visitors to his group helped me in these explorations.

A key, although accidental, element in my choice of a problem for study in biology was the publication of *Structural Chemistry and Molecular Biology*, a compendium of papers in a volume dedicated to Linus Pauling for his 65th birthday. I had contributed an article entitled, "Structural Implications of Reaction Kinetics," (57) which reviewed some of the work I have already described in the context of Pauling's view that a knowledge of structure was the basis for understanding reactions. However, it is not my article that leads me to mention this volume, but rather an article by Ruth Hubbard and George Wald entitled, "Pauling and Carotenoid Stereochemistry." They reviewed Pauling's contribution to the understanding of polyenes with emphasis on the visual chromophore, retinal. The article contained a paragraph, which I reproduce here because it describes an element of Pauling's approach to science that greatly influenced my research:

"One of the admirable things about Linus Pauling's thinking is that he pursues it always to the level of numbers. As a result, there is usually no doubt of exactly what he means. Sometimes his initial thought is tentative because the data are not yet adequate, and then it may require some later elaboration or revision. But it is frequently he who refines the first formulation."

On looking through the article, it was clear to me that the theory of the electronic absorption of retinal and its geometric changes on excitation, which play an essential role in vision, had not advanced significantly since my discussions with Hubbard and Wald during my undergraduate days at Harvard. I realized, in part from my time in Oxford with Coulson, that polyenes, such as retinal, were ideal systems for study by the available semiempirical approaches; that is, if any biologically interesting system in which quantum effects are important could be treated adequately at that time, retinal was it. Barry Honig, who had received his PhD in theoretical chemistry working with Joshua Jortner, joined my research group at that time. He was the perfect candidate to work on the retinal problem. It was known that the retinal (see **Figure 3**) is not planar. It is twisted about the C_{12}–C_{13} single bond, and this was thought to play a role in the photoisomerization reaction (the C_{11}–C_{12} double bond charges, from *cis* to *trans*) that gives rise to the visual signal. No structure of retinal was available, so it was not known whether the twist led to a 12-s-*cis* or 12-s-*trans* configuration (**Figure 3**). Honig did a calculation with a Hückel one-electron Hamiltonian for the π-electron system and a pairwise nonbonded energy function for the sigma bond framework of the molecule (42). The theory predicted that the structure was 12-s-*cis*.

(a) all-trans

(b) 11-cis, 12-s-cis

(c) 11-cis, 12-s-trans

(d) 11-trans, 12-s-cis

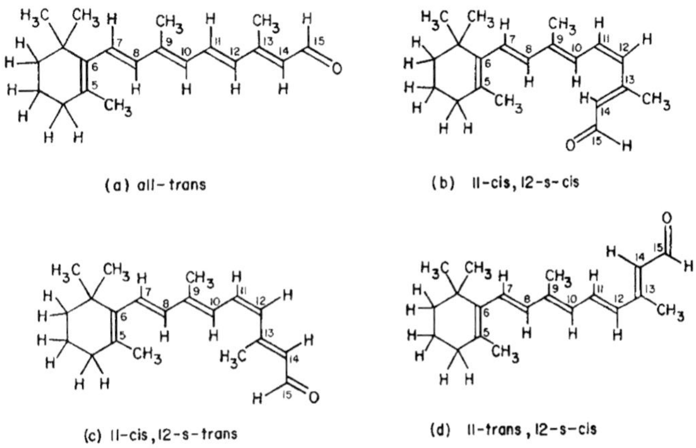

Figure 3

The different form of retinal considered in the calculations. The figure has been reproduced, with permission, from Reference 42.

We felt that this result with its implication for visual excitation was appropriate for publication in *Nature*. We submitted the paper, which received excellent reviews, but came back with a rejection letter stating that because there was no experimental evidence to support our results, it was not certain that the conclusions were correct. This was my first experience with *Nature* and with the difficulty of publishing theoretical results related to biology, particularly in high impact journals. The problem is almost as prevalent today as it was then, i.e., if theory agrees with experiment it is not interesting because the result is already known, whereas if one is making a prediction, then it is not publishable because there is no evidence that the prediction is correct. I was sufficiently upset by this reaction to our work that I called John Maddox, the insightful

Editor of *Nature*, and explained the situation to him. Apparently, I was successful, as the paper was accepted. A subsequent crystal structure verified our prediction (36). In a review of studies of the visual chromophore (43), I noted that "Theoretical chemists tend to use the word 'prediction' rather loosely to refer to any calculation that agrees with experiment, even when the latter was done before the former."

The study of the retinal chromophore gave rise to a sustained effort in my group concerned with the properties of retinal and other polyenes. The continuing effort was fostered, in part, by Bryan Kohler and Bryan Sykes, two assistant professors who had joined the Chemistry Department and were doing experiments that provided challenges for theory. They were part of a group of young faculty (it also included William Reinhardt and Roy Gordon, plus William Miller, a Junior Fellow), which made the department particularly stimulating at that time. Their offices were located along a narrow corridor on the ground floor of Converse Laboratory, with my office at one end. My daily strolls down this corridor provided many occasions for scientific discussion. A collaboration with Sykes led to one of the earliest uses of vicinal spin-spin coupling constants and the nuclear Overhauser effect in NMR to determine the conformations of a biomolecule (retinal in this case) (41). The technique, in a much more elaborate implementation, is now the basis of most protein NMR structure determinations. Kohler and his student, Bruce Hudson, were doing high resolution spectral studies of simple polyenes (e.g., hexatriene) and had observed a very weak absorption below the strongly absorbing transition, which is the analog of the one involved in retinal isomerization. They suggested that there exists a forbidden transition, which was not predicted by simple (single-excitation) models of polyene spectra, such as the one used by Honig in his study of retinal. Klaus Schulten, then a graduate student, working jointly with Roy Gordon and me, introduced double excitations into the Parisar-Parr-Pople (PPP) approximation for π-electron systems and found the low-lying (forbidden) state in hexatriene and octatetraene (108). A number of related studies followed. Arieh Warshel had joined my group at Harvard after we met at the Weizmann Institute, where he had been a graduate student with Lifson. He extended the polyene model by introducing a quantum mechanical Hamiltonian that refined the PPP method for the π-electrons and by treating the sigma-bonded framework by a molecular mechanics approach fitted to a large set of experimental data (120); the method, like the simpler model used by Honig, was an early version of the quantum mechanical/molecular mechanical (QM/MM) approach that is now widely employed for studying enzymatic reactions (28). We used the method to calculate the vibronic spectra of retinal and related molecules (120). Subsequently, a collaboration was initiated with Veronica Vaida, a member of the chemistry faculty, and her graduate student Russ Hemley. He extended the approach we had developed for excited states to molecules such as styrene (39), which Vaida and her students were studying experimentally.

In the 1970s I moved to Mallincrodt from the Converse offices, where the large amphitheatre lecture hall had been renovated into a three-story integrated space to house the physical chemistry faculty and the theoretical students. The renovated area, known as the "New Prince House" (Prince House was an old Cambridge house near the Chemistry Department where the theoretical students had offices for a number of years) promoted interaction among all occupants—senior and junior faculty and the theoretical postdocs and graduate students who had offices in the lower depths of the tri-level complex. Its lounge area equipped with an espresso coffee machine was ideal for generating discussions. Among my many interactions over the 20-year period this complex existed, none proved more fruitful than those with Chris Dobson, who was a junior faculty member in the department from 1978 to 1980 before returning to

Oxford. Our collaborations continue to this day, as described below.

HEMOGLOBIN: A REAL BIOLOGICAL PROBLEM

Another scientific question that appeared ready for a more fundamental investigation was the origin of hemoglobin cooperativity, the model system for allosteric control in biology. Although the phenomenological model of Monod, Wyman, and Changeux (92) had provided many insights, it did not attempt to make contact with the detailed structure of the molecule. I had already begun working on hemoglobin with Robert Shulman, then at Bell Labs, who had measured the paramagnetic NMR shifts of the heme protons, and we had developed an interpretation of the results on the basis of the electronic structure of the heme (111). In 1971 Max Perutz had just determined the X-ray structure of deoxy hemoglobin, which complemented his earlier results for oxy hemoglobin (98). By comparing the two structures, he was able to propose a qualitative molecular mechanism for the cooperativity. Alex Rich, now a professor at the Massachusetts Institute of Technology, had invited Perutz to present two lectures describing the X-ray data and his mechanism. After the second lecture, Alex suggested that I come to his office to have a discussion with Perutz. Perutz was sitting on a couch in Alex's office and eating his customary banana. I asked him whether he had tried to formulate a quantitative thermodynamic mechanism based on his structural analysis. He said no and seemed very enthusiastic, although I was not sure whether he had understood what I meant. Having been taught by Pauling that until one expressed an idea in quantitative terms, it was not possible to test one's results, I went away from our meeting thinking about the best way to proceed. Attila Szabo had recently joined my group as a graduate student, and the hemoglobin mechanism seemed like an ideal problem for his theoretical skills. The basic idea proposed by Perutz was that the hemoglobin molecule has two quaternary structures, R and T, in agreement with the ideas of Monod, Wyman, and Changeux; that there are two tertiary structures, liganded and unliganded for each of the subunits; and that the coupling between the two is introduced by certain salt bridges whose existence depended on both the tertiary and quaternary structures of the molecule. Moreover, some of the salt bridges depended on pH, which introduced the Bohr effect on the oxygen affinity of the subunits. These ideas were incorporated into the statistical mechanical model Szabo and I developed (115). It was a direct consequence of the formulation that the cooperativity parameter n (i.e., the Hill coefficient) varied with pH. This was in disagreement with the hemoglobin dogma at the time and led a number of the experimentalists in the field to initially disregard our model, which was subsequently confirmed by experiments.

When we began working on the model, I discussed our approach with John Edsall and Guido Guidotti, both biology professors at Harvard. Edsall was well known for his deep understanding of protein thermodynamics and Guidotti was an expert on hemoglobin. There were a number of parameters in the model and we had chosen their values by use of physical arguments. Because the values of the parameters were estimated, the results from the model gave only approximate agreement with experiment. Guidotti warned me that such results would not be accepted by the hemoglobin community, in particular, and biologists, in general. Consequently, we inverted the description of the model. We used experimental data to determine the parameters so that the agreement with experiment was excellent and then justified the values of the parameters with the physical arguments we had developed. During the formulation of our ideas, we often asked Guidotti which of certain experiments were to be trusted, since the nearly overwhelming hemoglobin literature contained sets of data that disagreed with each other, without any comment from the authors indicating which measurements were correct and why.

The paper describing the hemoglobin work was written in Paris, much of it at *Aux Deux Magots*, a left-bank café famous as a meeting place for writers and philosophers from the time of Jean-Paul Sartre. I was on sabbatical leave during 1972–1973 and officially at the Université de Paris XI in Orsay, a suburb of Paris, with the group of Jeannine Yon-Kahn, a pioneer in experimental studies of protein dynamics. However, I spent much of my time in Paris at the Institut de Biology Physico-Chimique on rue Paul et Marie Curie in the 5th Arrondissement. Having often visited Paris since my postdoctoral days and lived there on a sabbatical, I had begun to consider the possibility of moving to Paris on a permanent basis in 1970. I had been at Harvard for the canonical five years, and the idea of returning to Europe was tempting. Given the anti-Semitism and Nazi-leaning parties that still were prevalent in Austria, I had no desire to return to the country of my birth. France offered many attractive aspects of European life and culture, and I believed that I could do high-level research there in theoretical chemistry and its biological applications. After the 1968 revolution, the immense Université of Paris, with more than 300,000 students, had been divided into a dozen campuses. With true French rigor, they were named Paris I, Paris II, etc., although they now have names, instead of merely numbers. Orsay (Paris XI) was one of three science campuses, and it was certainly the best. However, it had the drawback that it was about 40 minutes by the RER (commuter rail) from Paris. If I was going to move to a Paris university, I wanted to live in Paris itself. Consequently, I focused on the two other scientific universities (Paris VI and Paris VII) that were intertwined on the Jussieu campus, a block of ugly modern buildings. Their saving grace was a central location in the area where the Halles aux Vins had been located before World War II. The neighboring streets were still dotted with good inexpensive restaurants dating back to the area's previous existence and now thriving on the faculty and student clientele.

In discussions with colleagues who had urged me to live in France, a serious obstacle became clear. I was a tenured professor at Harvard and not surprisingly was willing to move only if I was offered a permanent position in Paris. However, French university professors were civil servants and only French citizens could be civil servants. Because obtaining French citizenship without losing my American one was out of the question at the time (it is now possible and our son, Mischa, has dual citizenship), I was ready to give up the idea of moving to Paris. Many things in France were achieved then (and still are) by political influence. Jacques DuBois, a chemistry professor at Paris VII with connections to the Pompidou government, said he would try to "arrange the situation." I did not know exactly what he meant but hoped that the tenure problem could be solved. On that basis I took a leave of absence from Harvard.

With only a verbal commitment of a permanent position, I moved the major part of my research group (including David Case, Bruce Gelin, and Iwao Ohmine, among others) from Harvard in the fall of 1974. At Paris VII empty laboratory spaces awaited us. We bought office furniture and computing equipment and went to work. One thing that made the transition much easier was that Marci Hazard, who had joined the lab as secretary in May, came along. Many of the logistical problems (e.g., finding where to purchase what we needed) were solved by her, and she played a key role in the cohesion of the group. As the year went on, DuBois reported on his progress in regularizing my status. I had come as a Professeur Associé, which is an annual appointment open to non-French citizens. Finally, in January a decree was published in the official government register (much of the French government functions by decrees that do not require votes of the National Assembly). This decree exempted university professors from the citizenship requirement, which made possible my appointment as a tenured professor.

Not everything was resolved, however, and the complexity of dealing with the French administration led me to renounce my dream and return to Harvard. The decree remained valid, however, and subsequently I received a number of thank-you letters from non-French scientists who had for many years been appointed annually as Professeur Associé and suddenly received a permanent position. (Jean-Pierre Hanson, who was born in Luxembourg, told me recently that he believes he is one of the first people to have profited from "my" decree.)

During this period I started spending summer vacations with my family in the foothills of the Alps above Annecy and its stunning lake, an area which I had first seen on my postdoctoral trips in the early 1950s. My colleagues at Harvard viewed such absences from Cambridge as improper. However, I found that being away gave me a chance to think, undistracted by everyday pressures. Contemplative hikes in the mountains provided the backdrop for my reading and thinking and played an essential role in developing new areas of research. In fact, I had asked my NSF program director whether these trips were justified under the conditions of my grant. His conclusion was that, given their importance to my research, they constituted an appropriate, even if somewhat unorthodox, summer program. In 1974 I finally found a plot of land with a magnificent view in Chalmont, a small hamlet in the Manigod valley above Lac d'Annecy. A chalet was built, which has been our summer home for 30 years.

PROTEIN FOLDING

In 1969 I was challenged by the mechanism of protein folding during a visit by Chris Anfinsen to the Lifson group at the Weizmann Institute. We had many discussions of his experiments on protein folding, which had led to the realization that proteins can refold in solution, independent of the ribosome and other aspects of the cellular environment (2). [Of course, it is now known that some proteins have more complex folding mechanisms and require chaperones, such as the supramolecular complex GroEL, to fold. This molecular machine is one for which we have used molecular dynamics simulations to elucidate the mechanism (87, 118).] What most impressed me was Anfinsen's film showing the folding of a protein with "flickering helices forming and dissolving and coming together to form stable substructures." The film was a cartoon, but it led to my asking him, in the same vein as I had asked Perutz earlier about hemoglobin, whether he had thought of taking the ideas in the film and translating them into a quantitative model. Anfinsen said that he did not really know how he would do this, but to me it suggested an approach to the mechanism of protein folding. When David Weaver joined my group at Harvard, while on a sabbatical leave from Tufts, we developed what is now known as the diffusion-collision model for protein folding (70, 71). Although it is a simplified coarse-grained description of the folding process, it showed how the search problem for the native state could be solved by a divide-and-conquer approach. Formulated by Cy Levinthal, the so-called *Levinthal Paradox* points out that to find the native state by a random search of the astronomically large configuration space of a polypeptide chain would take longer than the age of the earth, while proteins fold experimentally on a timescale of microseconds to seconds. In addition to providing a conceptional answer to the question posed by Levinthal, the diffusion-collision model made possible the estimation of folding rates. The model was ahead of its time because data to test it were not available. Only relatively recently have experimental studies demonstrated that the diffusion-collision model describes the folding mechanism of many helical proteins (48), as well as some others (49).

Protein folding is an area that has continued to interest me and has led to numerous collaborations in addition to that with David Weaver. When David and I developed the diffusion-collision model in 1975, protein folding was a rather esoteric subject of interest to a very small community of scientists. The field has been completely transformed in recent years because of its assumed importance for understanding the large number of protein sequences available from genome projects and because of the realization that misfolding can lead to a wide range of human diseases (22); these diseases are found primarily in the older populations that form an ever-increasing portion of humanity. Scientists, both experimentalists and theoreticians—physicists, as well chemists and biologists—now study protein folding. Over the past decade or so the mechanism of protein folding has been resolved, in principle. It is now understood that there are multiple pathways to the native state and that the bias on the free-energy surface, due to the greater stability of native-like versus nonnative contacts, is such that only a very small fraction of the total number of conformations is sampled in each folding trajectory (24). This understanding was achieved by the work of many scientists, but a crucial element was the study of lattice models of protein folding. Such toy models, as I like to call them, are simple enough to permit many folding trajectories to be calculated to make possible an analysis of the folding process and free-energy surface sampled by the trajectories (104). However, they are sufficiently complex so that they embody the Levinthal problem, i.e., there are many more configurations than could be visited during the calculated folding trajectory. The importance of such studies was in part psychological, in that even though the lattice model uses a simplified representation, "real" folding was demonstrated on a computer for the first time. An article based on a lecture at a meeting in Copenhagen (60) describes this change in attitude as a paradigm of scientific progress.

The mechanism of protein folding and the development of methods to predict the structure of a protein from its amino acid sequence continue to be subjects under intense investigation. It is likely that calculations with programs such as CHARMM will permit folding simulations to be done at an atomic level of detail owing to the ever-increasing speed of computers (either localized multiprocessor supercomputers or delocalized grid-based access to many individual processors) before too long. However, this type of brute force approach is of less interest to me than solving the conceptual problem of protein folding, which has been accomplished by more approximate techniques.

ORIGINS OF THE CHARMM PROGRAM

When I visited Lifson's group in 1969 there was considerable interest in developing empirical potential energy functions for small molecules. The novel idea was to use a functional form that could serve not only for calculating vibrational frequencies, as did the expansions of the potential about a known or assumed minimum-energy structure, but also for determining that structure. The so-called consistent force field (CCF) of Lifson and his coworkers, particularly Arieh Warshel, included nonbonded interaction terms so that the minimum-energy structure could be found after the energy terms had been appropriately calibrated (84). The possibility of using such energy functions for larger systems struck me as potentially very important for understanding biological macromolecules like proteins, though I did not begin working on this immediately.

Once Attila Szabo had finished the statistical mechanical model of hemoglobin cooperativity, I realized that his work raised a number of questions that could be explored only with a method for calculating the energy of hemoglobin as a function of the atomic positions. No way of doing such a calculation existed. Bruce Gelin, a new graduate student, had begun theoretical research in my group in 1967. He started out by studying the application of the random-phase approximation to two-electron systems, such as the helium atom. This was still the Vietnam War era and after two years at Harvard, Gelin was drafted.

He was assigned to the military police in a laboratory concerned with drug usage (e.g., LSD). Paradoxically, this work aroused his interest in biology, and when he returned to finish his degree Gelin wanted to change his area of research to a biologically related problem. We decided the time was ripe to try to develop a program that would make it possible to take a given amino acid sequence (e.g., that of the hemoglobin alpha chain) and a set of coordinates (e.g., those obtained from the X-ray structure of deoxy hemoglobin) and to use this information to calculate the energy of the system and its derivatives as a function of the atomic positions. This could be used for perturbing the structure (e.g., by binding oxygen to the heme group) and finding a new structure by minimizing the energy. Developing the program was a major task, but Gelin had the right combination of abilities to carry it out (33).

The result was pre-CHARMM, although it did not have a name at that time. While not trivial to use, the program was applied to a variety of problems, including Gelin's pioneering study of aromatic ring flips in the bovine pancreatic trypsin inhibitor (BPTI) (34), as well as his primary project on hemoglobin. The idea was to introduce the effect of ligand binding on the heme group as a perturbation (undoming of the heme) and to use energy minimization to determine the response of the protein to the perturbation. To attempt to do such a calculation on the available computers (an IBM 7090 at Columbia University was our work-horse at the time, because computing at the Harvard Computer Center was too expensive) required considerable courage, but Gelin's efforts were successful. His work introduced a new dimension to theoretical approaches to understanding protein structure and function. Gelin showed how the effect of undoming of the heme induced by the binding of oxygen was transmitted to the interface between the hemoglobin subunits. The analysis provided an essential element in the cooperative mechanism in its demonstration at an atomic level of detail how communication between the subunits occurred (35). Another application of pre-CHARMM was Dave Case's simulation of ligand escape after photodissociation from myoglobin (16); a study that was followed by the work of Ron Elber (25), which gave rise to the locally enhanced sampling (LES) and multiple copy simultaneous search (MCSS) methods. The latter was developed by Andrew Miranker as a fragment-based approach to drug design (90).

Gelin would have faced an almost insurmountable task in developing pre-CHARMM if there had not been prior work by others on protein energy calculations. Although many persons have contributed to the development of empirical potentials, the two major inputs to our work came from Schneior Lifson's group at the Weizmann Institute and Harold Scheraga's group at Cornell University (107). As I already mentioned, Warshel had come to Harvard and had brought his CFF program with him. His presence and the availability of the CFF program were important resources for Gelin, who was also aware of Michael Levitt's pioneering energy calculations for proteins (82).

Gelin's program has been considerably restructured and has continued to evolve over the intervening years. In preparing to publish a paper on the program in the early 1980s, mainly to give credit to the dominant contributors at the time, we felt we needed a name. Bob Bruccoleri came up with HARMM (HARvard Macromolecular Mechanics), which seemed to me not to be the ideal choice. However, Bob's suggestion inspired the addition of a "C" for chemistry, resulting in the name CHARMM. I sometimes wonder if Bruccoleri's original suggestion would have served as a useful warning to inexperienced scientists working with the program. The CHARMM program is now being developed by a wide group of contributors, most of whom were students or postdoctoral fellows in my group; the program is distributed worldwide in both academic and commercial settings.

THE FIRST MOLECULAR DYNAMICS SIMULATION OF A BIOMOLECULE

Given that pre-CHARMM could calculate the forces on the atoms of a protein, the next step was to use these forces in Newton's equation to calculate the dynamics. This fundamental development was introduced in the mid-1970s when Andy McCammon joined my group. An essential element that encouraged us in this attempt was the existence of molecular dynamics simulation methods for simpler systems. Molecular dynamics had followed two pathways, which come together in the study of biomolecule dynamics. One pathway concerns trajectory calculations for simple chemical reactions. My own research in this area had served as preparation for the many-article problem posed by biomolecules. The other pathway in molecular dynamics concerns physical rather than chemical interactions and the thermodynamic and dynamic properties of large numbers of particles rather than detailed trajectories of a few particles. Although the basic ideas go back to van der Waals and Boltzmann, the modern era began with the work of Alder and Wainright (1) on hard-sphere liquids in the late 1950s. The paper by Rahman (103) in 1964 on a molecular dynamics simulation of liquid argon with a soft-sphere (Lennard-Jones) potential represented an essential next step. Simulations of more complex fluids followed; the now classic study of liquid water by Stillinger and Rahman was published in 1974 (114), shortly before our protein simulations.

The background I have outlined set the stage for the development of molecular dynamics of biomolecules. The size of an individual molecule, composed of 500 or more atoms for even a small protein, is such that its simulation in isolation can serve to obtain approximate equilibrium properties, as in the molecular dynamics of fluids. Concomitantly, detailed aspects of the atomic motions are of considerable interest, as in trajectory calculations. A basic assumption in initiating such studies was that potential functions could be constructed which were sufficiently accurate to give meaningful results for systems as complex as proteins or nucleic acids. In addition, it was necessary to assume that for these inhomogeneous systems, in contrast to the homogeneous character of even complex liquids like water, simulations of an attainable timescale (10 to 100 ps) could provide a useful sample of the phase space in the neighborhood of the native structure. There was no compelling evidence for either assumption in the early 1970s. When I discussed my plans with chemistry colleagues, they thought such calculations were impossible, given the difficulty of treating few atom systems accurately; biology colleagues felt that even if we could do such calculations, they would be a waste of time. By contrast, the importance of molecular dynamics simulations in biology was supported by Richard Feynman's prescient statement in the well-known volumes based on his physics lectures at Caltech:

"Certainly no subject or field is making more progress on so many fronts at the present moment, than biology, and if we were to name the most powerful assumption of all, which leads one on and on in an attempt to understand life, *it is that all things are made of atoms* (italics in the original), and that *everything that living things do can be understood in terms of the jigglings and wigglings of atoms.* (27; italics added)

More than 25 years have passed since the first molecular dynamics simulation of a macromolecule of biological interest was published (**Figure 4**) (88). This study has stood the test of time, and, perhaps more significantly, it has served to open a new field that is now the focus of the research of an ever-growing number of scientists (10). The original simulation, published in 1977 (88), concerned the bovine pancreatic trypsin inhibitor (BPTI), which has served as the "hydrogen molecule" of protein dynamics because of its small size, high stability, and a relatively accurate X-ray structure (21); interestingly, the physiological function of BPTI

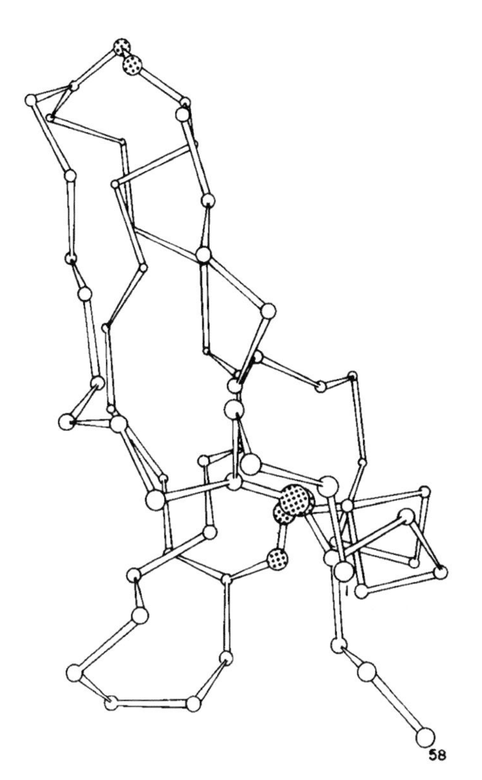

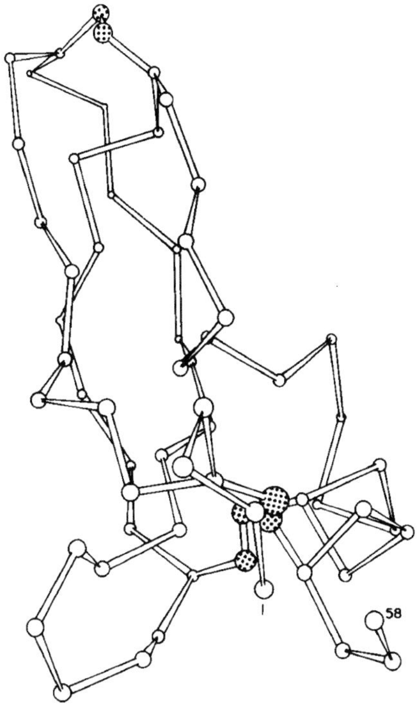

Figure 4

The peptide backbone (α carbons) and disulfide bonds of the bovine pancreatic trypsin inhibitor drawn by Bruce Gelin. (*a*) X-ray structure. (*b*) Time-evolved structure after 3.2 ps of dynamical simulation. The figure has been reproduced, with permission, from Reference 88.

remains unknown. In the mid-1970s it was difficult to obtain the computer time required to do such simulations in the United States. However, CECAM (Center Européen Calcul Atomique et Moléculaire), whose founding director and guiding spirit was Carl Moser, had access to a large computer that was available for scientific research. (Equivalent computers in the United States were found only in the defense agencies and were not generally accessible.) A CECAM Workshop (a two-month *workshop* worthy of its name) was organized by Herman Berendsen in 1976 with the title "Models for Protein Dynamics." As his introduction to the workshop states: "Thus, the simulation of water was a first topic to be studied. The application to proteins was then not foreseen in five or ten years to come." Realizing that the workshop was a great opportunity to do the required calculations, Andy McCammon and Bruce Gelin worked extremely hard to prepare and test a program for the molecular dynamics simulation for the workshop. The initial simulation (88), which was performed at the workshop, introduced many others now active in the field (including H. Berendsen, W. van Gunsteren, M. Levitt, and J. Hermans) to the possibility of doing such calculations (5).

Although the original simulation was done in a vacuum with a crude molecular mechanics potential and lasted for only 9.2 ps, the results were instrumental in replacing the view of proteins as relatively rigid structures [in 1981, Sir D. L. Phillips commented, "Brass models of DNA and a variety of proteins dominated the scene and much of the thinking" (99)] with the realization that they were dynamic systems whose internal motions play a functional role. Of course, there were already experimental data, such as the hydrogen exchange experiments of Linderstrom-Lang and his coworkers (44, 85), pointing in this direction. It is now recognized that the X-ray structure of a protein provides the average atomic positions but that the atoms exhibit fluid-like motions of sizable amplitudes about these averages. Protein dynamics subsumes the static picture. The average positions are essential for the discussion of many aspects of biomolecular function in the language of structural chemistry, but the recognition of the importance of fluctuations opened the way for more sophisticated and accurate interpretations of functional properties.

The conceptual changes resulting from the early studies make one marvel at how much of great interest could be learned with so little—such poor potentials, such small systems, so little computer time. This is, of course, one of the great benefits of taking the initial, somewhat faltering steps in a new field in which the questions are qualitative rather than quantitative and any insights, even if crude, are better than none at all.

APPLICATIONS OF MOLECULAR DYNAMICS

Molecular dynamics simulations of proteins and nucleic acids, as of other systems composed of particles (e.g., liquids, galaxies), can in principle provide the ultimate details of motional phenomena. The primary limitation of simulation methods is that they are approximate. Here experiment plays an essential role in validating the simulation methods; that is, comparisons with experimental data serve to test the accuracy of the calculated results and provide criteria for improving the methodology. Although the statistical errors can be calculated (122), estimates of the systematic errors inherent in the simulations have not been possible, e.g., the errors introduced by the use of empirical potentials are difficult to quantify. When experimental comparisons indicate that the simulations are meaningful, their capacity for providing detailed results often makes it possible to examine specific aspects of the atomic motions far more easily than by using laboratory measurements.

Two years after the BPTI simulation, it was recognized (3, 30) that thermal (B) factors determined in X-ray crystallographic refinement could provide information about the internal motions of proteins. Plots of estimated mean-square fluctuations versus residue number [introduced in the original BPTI paper (88)] have become a standard part of papers on high-resolution structures, even though the contribution to the B factors of overall translation and rotation and crystal disorder persist as a concern in their interpretation (79). During the subsequent decade, a range of phenomena were investigated by molecular dynamics simulations of proteins and nucleic acids. A plethora of experimental data were just waiting for molecular dynamics simulations to elucidate them. Most of these early studies were made by my students at Harvard and focused on the physical aspects of the internal motions and the interpretation of experiments. They include the analysis of fluorescence depolarization of tryptophan residues (45), the role of dynamics in measured NMR parameters (23, 83, 97) and inelastic neutron scattering (20, 113), and the effect of solvent and temperature on protein structure and dynamics (11, 29, 95). The now widely used simulated annealing methods for X-ray structure refinement (14, 15) and NMR structure determination (12, 96) also originated in this period. Simultaneously, a number of applications demonstrated the importance of internal motions in biomolecular function, including the hinge bending modes for opening and closing active sites (9, 18), the flexibility of tRNA (38), the induced conformation change in the activation of trypsin (13), the fluctuations required for ligand entrance and exit in heme proteins (16, 25), and the role of configurational entropy in the stability of proteins and nucleic acids (8, 47). Many of these studies, which were done about two decades ago, seem to have been forgotten; at least, they are rarely cited in the current literature. Of course, when the studies are redone, the more accurate potential functions and much longer simulations (nanoseconds instead of picoseconds) now possible yield improved results, but generally they confirm the earlier work.

Two attributes of molecular dynamics simulations have played an essential part in the explosive growth in the number of studies based on such simulations. As already mentioned, simulations provide the ultimate detail concerning individual particle motions as a function of time. For many aspects of biomolecule function, it is these details that are of interest (e.g., by what pathways does oxygen enter into and exit from the heme pocket in myoglobin). The other important aspect of simulations is that, although the potentials employed in simulations are approximate, they are completely under the user's control. By removing or altering specific contributions, their role in determining a given property can be examined. This is most graphically demonstrated by the use of computer alchemy—transmuting the potential from that representing one system to another during a simulation—in calculated free-energy differences (32, 112, 121).

There are three types of applications of simulation methods in the macromolecular area, as well as in other areas involving mesoscopic systems. The first uses the simulation simply as a means of sampling configuration space. This is involved in the utilization of molecular dynamics, often with simulated annealing protocols, to determine or refine structures with data obtained from experiments. The second uses simulations to determine equilibrium averages, including structural and motional properties (e.g., atomic mean-square fluctuation amplitudes) and the thermodynamics of the system. For such applications, it is necessary that the simulations adequately sample configuration space, as in the first application, with the additional condition that each point be weighted by the appropriate Boltzmann factor. The third application employs simulations to examine the actual dynamics. Here not only is adequate sampling of configuration space with appropriate Boltzmann weighting required, but it must be done so as to properly represent the time development of the system. Monte Carlo simulations, as well as molecular dynamics, can be utilized for the first two applications. By contrast, in the third application, where the motions and their time developments are of interest, only molecular dynamics can provide the necessary information.

FUTURE OF MOLECULAR DYNAMICS

Most of the motional phenomena examined during the first 10 years after the BPTI simulation paper was published continue to be studied both experimentally and theoretically. The increasing scope of molecular dynamics due to improvements in methodology and the tremendous increase of the available computer power is making possible the study of systems of greater complexity on ever-increasing timescales. There are many recent examples of the use of molecular dynamics to obtain information concerning the function of biological macromolecules. At present I am most interested in molecular machines, such as GroEL (87, 118), which require simulations for their understanding. Nature has designed these machines to function through ligand-induced conformational changes encoded in the structure. One of the most fascinating machines is F_0F_1-ATPase, which synthesizes ATP and H_2O, the energy currency of most living cells, from ADP and $H_2PO_4^-$ (**Figure 5**) (32a). Numerous applications by my group and others have been reviewed recently, so that I will not detail them here (61, 65, 67).

It is my hope that molecular dynamics simulations will become a tool, like any other, to be used by experimentalists as part of their arsenal for solving problems. The simulated annealing method for determining X-ray structures originally proposed in 1987 by Axel Brünger, John Kuriyan, and me (14, 15), and so ably developed by Brünger, is now an essential part of structural biology. Without this method, the high-throughput structure determination initiatives would be in considerable difficulty. However, a universal acceptance of molecular dynamics simulations methods for extending experimental data to

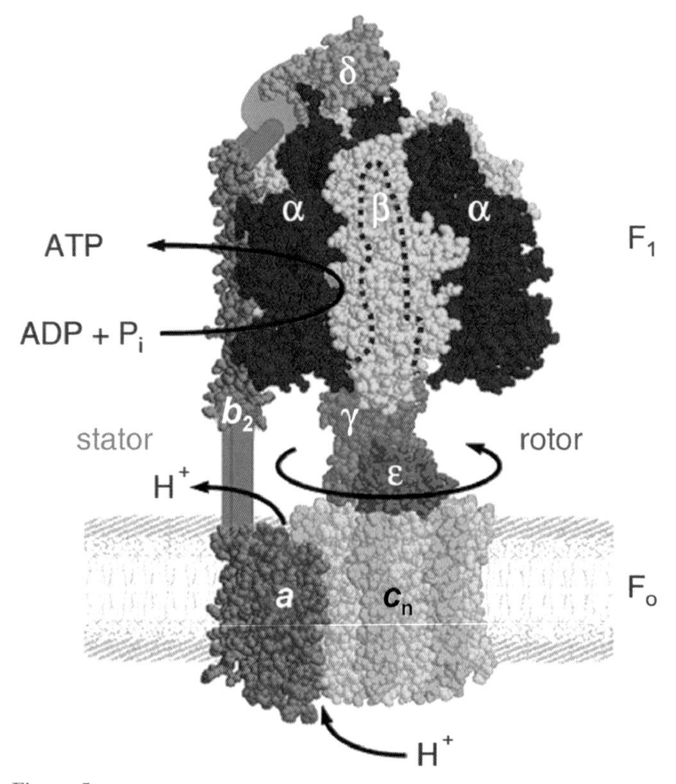

Figure 5

Structural model of F_0F_1-ATP synthase. The figure has been reproduced from Senior AE, Nadanaciva S, Weber J. 2002. *Biochim. Biophys. Acta Bioenerg.* 1553:188–211

learn how biomolecules function is still in the future. The determination of the reaction path involved in conformational change by simulations, is a clear case where, given the structural end point from experiment, simulations are essential, as some crystallographers have already recognized (123).

I still marvel at the insights that simulation methods can provide concerning the functions of biomolecules. Claude Poyart, a dear friend with whom I worked on hemoglobin (81), characterized molecular dynamics simulations of proteins with a beautiful image. He likened the X-ray structures of proteins to a tree in winter, beautiful in its stark outline but lifeless in appearance. Molecular dynamics gives life to this structure by clothing the branches with leaves that flutter because of the thermal winds.

EPILOGUE

As I read through what I have written, I see what a fragmentary picture it provides of my life, even my scientific life. Missing are innumerable interactions, most of which were constructive but some not so, that have played significant roles in my career. The more than 200 graduate students and postdoctoral fellows who at one time or another have been members of the group are listed in **Table 1**. Many have gone on to faculty positions and become leaders in their fields of research. They in turn are training students, so I now have scientific children, grandchildren, and great-grandchildren all over the world. I treasure my contribution to their professional and personal careers, as much as the scientific advances we have made together.

Contributing to the education of so many people in their formative years is a cardinal aspect of university life. My philosophy in graduate and postgraduate education has been to provide an environment where young scientists, once they have proved their ability, can develop their own ideas, as refined in discussions with me and aided by other members of the group. This fostered independence has been, I believe, an important element in the fact that so many of my students are now themselves outstanding researchers and faculty members. My role has been to guide them when problems arose and to instill in them the necessity of doing things in the best possible way, not to say that I succeeded with all of them.

Discussing my scientific family makes me realize that another missing element is my personal family, an irreplaceable part of my life. Reba and Tammy, my two daughters whose mother, Susan, died in 1982, both became physicians (thereby fulfilling my destined role); Reba lives in Jerusalem and Tammy lives on the West Coast. My wife, Marci, and our son, Mischa, who is presently in law school at Boston University, complete my immediate family. As many people know, Marci also plays the pivotal role as the

Table 1 Karplusians: 1955–2005

R. J. Aerni	Kevin Gaffney	Jianpeng Ma	Andrej Sali
David H. Anderson	Jiali Gao	Alexander D. MacKerell, Jr.	Michael Schaefer
Ioan Andricioaei	Yi Qin Gao	Christoph Maerker	Michael Schlenkrich
Yasuhide Arata	Bruce Gelin	Paul Maragkakis	David M. Schrader
Georgios Archontis	R. Benny Gerber	Marc Martí-Renom	John C. Schug
Gabriel G. Balint-Kurti	Paula M. Getzin	Jean-Louis Martin	Klaus Schulten
Christian Bartels	Debra Giammona	Carla Mattos	Eugene Shakhnovich
Paul Bash	Martin Godfrey	J. Andrew McCammon	Moshe Shapiro
Donald Bashford	Andrei Golosov	H. Keith McDowell	Ramesh D. Sharma
Oren M. Becker	David M. Grant	Jorge A. Medrano	Isaiah Shavitt
Robert Best	Daniel Grell	Morten Meeg	Henry H.-L. Shih
Anton Beyer	Peter Grootenhuis	Marcus Meuwly	Bernard Shizgal
Robert Birge	Hong Guo	Olivier Michielin	David M. Silver
Ryan Bitetti-Putzer	Robert Harris	Stephen Michnick	Jeremy Smith
Arnaud Blondel	Karen Haydock	Fredrick L. Minn	Sung-Sau So
Stefan Boresch	Russell J. Hemley	Andrew Miranker	Michael Sommer
John Brady	Jeffrey C. Hoch	Keiji Morokuma	Ojars J. Sovers
Bernard Brooks	Gary G. Hoffman	A. Mukherji	Martin Spichty
Charles L. Brooks III	L. Howard Holley	Adrian Mulholland	David J. States
Thomas H. Brown	Barry Honig	David Munch	Richard M. Stevens
Robert E. Bruccoleri	Victor Hruby	Petra Munih	Roland Stote
Paul W. Brumer	Rod E. Hubbard	Robert Nagle	John Straub
Axel T. Brünger	Robert P. Hurst	Setsuko Nakagawa	Collin Stultz
Rafael P. Brüschweiler	Vincent B.-H. Huynh	Eyal Neria	Neena Summers
Matthias Buck	Toshiko Ichiye	John-Thomas C. Ngo	Henry Suzukawa
Amedeo Caflisch	K.K. Irikura	Lennart Nilsson	S. Swaminathan
William J. Campion	Alfonso Jaramillo	Dzung Nguyen	Attila L. Szabo
William Carlson	Diane Joseph-McCarthy	Iwao Ohmine	Kwong-Tin Tang
David A. Case	Sunhee Jung	Barry Olafson	Bruce Tidor
Leo Caves	C. William Kern	E.T. Olejniczak	Hideaki Umeyama
Thomas C. Caves	Burton S. Kleinman	Kenneth W. Olsen	Arjan van der Vaart
John-Marc Chandonia	G.W. Koeppl	Neil Ostlund	Wilfred van Gunsteren
Ta-Yuan Chang	H. Jerrold Kolker	Emanuele Paci	Herman van Vlijmen
Rob D. Coalson	Yifei Kong	Yuh-Kang Pan	Michele Vendruscuolo
François Colonna-Cesari	Lewis M. Koppel	C.S. Pangali	Dennis Vitkup
Michael R. Cook	J. Kottalam	Richard W. Pastor	Shunzhou Wan
Qiang Cui	Felix Koziol	Lee Pedersen	Iris Shih-Yung Wang
Annick Dejaegere	Christoph Kratky	David Perahia	Ariel Warshel
Philippe Derreumaux	Serguei Krivov	Robert Petrella	Masakatsu Watanabe
Aaron Dinner	Krzysztof Kuczera	B. Montgomery Pettitt	David Weaver
Uri Dinur	John Kuriyan	Ulrich Pezzeca	Paul Weiner
Roland L. Dunbrack, Jr.	Joseph N. Kushick	Richard N. Porter	Michael A. Weiss
Chizuko Dutta	Peter W. Langhoff	Jay M. Portnow	Joanna Wiórkiewicz-K.
Nader Dutta	Antonio C. Lasaga	Carol B. Post	George Wolken

(Continued)

Table 1 (*Continued*)

Claus Ehrhardt	Frankie T.K. Lau	Lawrence R. Pratt	Youngdo Won
Ron Elber	Themis Lazaridis	Martine Prévost	Yudong Wu
Byung Chan Eu	Fabrice LeClerc	Blaise Prod'hom	Robert E. Wyatt
Jeffrey Evanseck	Angel Wai-mun Lee	Dagnija Lazdins Purins	Wei Yang
Erik Evensen	Irwin Lee	Lionel M. Raff	Robert Yelle
Jeffrey Evenson	Sangyoub Lee	Mario Raimondi	Swarna Yeturu Reddy
Thomas C. Farrar	Ronald M. Levy	Walter E. Reiher III	Darrin York
Martin Field	Xiaoling Liang	Nathalie Reuter	Hsiang-ai Yu
Stefan Fischer	Carmay Lim	Bruno Robert	Vincent Zoete
David L. Freeman	Xabier Lopez	Peter J. Rossky	Yaoqi Zhou
Thomas Frimurer	Paul Lyne	Benoît Roux	

Laboratory Administrator, adding a spirit of continuity for the group and making possible our commuting between the Harvard and Strasbourg labs. Without my family, my life would have been an empty one, even with scientific success.

ACKNOWLEDGMENTS

Much of the research described here was supported in part by grants from the National Science Foundation, the National Institutes of Health, the Atomic Energy Commission, and the Department of Energy, as well as the royalty income from the CHARMM Development Project. As is evident from the references, the research would not have been possible without my many coworkers, only a few of whom are mentioned in the text. Frontispiece was kindly provided by Nathan Pitt, Department of Chemistry, University of Cambridge, Cambridge CB2 1EW, United Kingdom.

LITERATURE CITED

1. Alder BJ, Wainwright TE. 1957. Phase transition for a hard sphere system. *J. Chem. Phys.* 27:1208–9
2. Anfinsen CB. 1973. Principles that govern folding of protein chains. *Science* 181:223–30
3. Artymiuk PJ, Blake CCF, Grace DEP, Oatley SJ, Phillips DC, Sternberg MJE. 1979. Crystallographic studies of the dynamic properties of lysozyme. *Nature* 280:563–68
4. Balint-Kurti GG, Karplus M. 1969. Multistructure valence-bond and atoms-in-molecules calculations for LiF, F_2, and F_{2-}. *J. Chem. Phys.* 50:478–88
5. Berendsen H. 1976. Report of CECAM Workshop: models for protein dynamics. Orsay, May 24–July 17
6. Beuhler RJ, Bernstein RB, Kramer KH. 1966. Observation of reactive asymmetry of methyl iodide crossed beam study of reaction of rubidium with oriented methyl iodide molecules. *J. Am. Chem. Soc.* 88:5331–32
7. Bradshaw WH, Conrad HE, Corey EJ, Gunsalus IC, Lednicer D. 1959. Microbiological degradation of (+)-camphor. *J. Am. Chem. Soc.* 81:5507
8. Brooks BR, Karplus M. 1983. Harmonic dynamics of proteins: normal modes and fluctuations in bovine pancreatic trypsin inhibitor. *Proc. Natl. Acad. Sci. USA* 80:6571–75
9. Brooks BR, Karplus M. 1985. Normal modes for specific motions of macromolecules: application to the hinge-bending mode of lysozyme. *Proc. Natl. Acad. Sci. USA* 82:4995–99

10. Brooks CL III, Karplus M, Pettitt BM. 1988. *Proteins: A Theoretical Perspective of Dynamics, Structure, and Thermodynamics*. New York: Wiley
11. Brünger AT, Brooks CL III, Karplus M. 1985. Active site dynamics of ribonuclease. *Proc. Natl. Acad. Sci. USA* 82:8458–62
12. Brünger AT, Clore GM, Gronenborn AM, Karplus M. 1986. Three-dimensional structure of proteins determined by molecular dynamics with interproton distance restraints: application to crambin. *Proc. Natl. Acad. Sci. USA* 83:3801–5
13. Brünger AT, Huber R, Karplus M. 1987. Trypsinogen-trypsin transition: a molecular dynamics study of induced conformational change in the activation domain. *Biochemistry* 26:5153–62
14. Brünger AT, Karplus M. 1991. Molecular dynamics simulations with experimental restraints. *Acc. Chem. Res.* 24:54–61
15. Brünger AT, Kuriyan J, Karplus M. 1987. Crystallographic *R* factor refinement by molecular dynamics. *Science* 235:458–60
16. Case DA, Karplus M. 1979. Dynamics of ligand binding to heme proteins. *J. Mol. Biol.* 132:343–68
17. Caves TC, Karplus M. 1969. Perturbed Hartree-Fock theory. I. Diagrammatic double-perturbation analysis. *J. Chem. Phys.* 50:3649–61
18. Colonna-Cesari F, Perahia D, Karplus M, Ecklund H, Brändén CI, Tapia O. 1986. Interdomain motion in liver alcohol dehydrogenase: structural and energetic analysis of the hinge bending mode. *J. Biol. Chem.* 261:15273–80
19. Conroy H. 1960. Nuclear magnetic resonance in organic structural elucidation. *Adv. Org. Chem.* Vol. II, p. 265
20. Cusack S, Smith J, Finney J, Karplus M, Trewhella J. 1986. Low frequency dynamics of proteins studied by neutron time-of-flight spectroscopy. *Physica* 136B:256–59
20a. Dalton L. 2003. Karplus Equation. *Chem. Eng. News* 81:37–39
21. Deisenhofer J, Steigemann W. 1975. Crystallographic refinement and the structure of the bovine pancreatic trypsin inhibitor at 1.5 Å resolution. *Acta Crystallogr. B* 31:238–50
22. Dobson CM. 2003. Protein folding and misfolding. *Nature* 426:884–90
23. Dobson CM, Karplus M. 1986. Internal motion of proteins: nuclear magnetic resonance measurements and dynamic simulations. *Methods Enzymol.* 131:362–89
24. Dobson CM, Sali A, Karplus M. 1998. Protein folding: a perspective from theory and experiment. *Angew. Chem. Int. Ed.* 37:868–93
25. Elber R, Karplus M. 1990. Enhanced sampling in molecular dynamics: use of the time-dependent Hartree approximation for a simulation of carbon monoxide diffusion through myoglobin. *J. Am. Chem. Soc.* 112:9161–75
26. Farkas A, Farkas L. 1935. Experiments on heavy hydrogen. V. The elementary reactions of light and heavy hydrogen. The thermal conversion of ortho-deuterium and the interaction of hydrogen and deuterium. *Proc. R. Soc. London A* 152:124–51
27. Feynman RP, Leighton RB, Sands M. 1963. The Feynman Lectures in Physics. Addison-Wesley
28. Field MJ, Bash PA, Karplus M. 1990. A combined quantum mechanical and molecular mechanical potential for molecular dynamics simulations. *J. Comp. Chem.* 11:700–33
29. Frauenfelder H, Hartmann H, Karplus M, Kuntz ID Jr, Kuriyan J, et al. 1987. Thermal expansion of a protein. *Biochemistry* 26:254–61
30. Frauenfelder H, Petsko GA, Tsernoglou D. 1979. Temperature-dependent x-ray diffraction as a probe of protein structural dynamics. *Nature* 280:558–63

31. Freeman DL, Karplus M. 1976. Many-body perturbation theory applied to molecules: analysis and calculation correlation energy calculation for Li_2, N_2, and H_3. *J. Chem. Phys.* 64:2461–59
32. Gao J, Kuczera K, Tidor B, Karplus M. 1989. Hidden thermodynamics of mutant proteins: a molecular dynamics analysis. *Science* 244:1069–72
32a. Gao YQ, Yang W, Karplus M. 2005. A structure-based model for synthesis and hydrolysis of ATP by F_1ATPase. *Cell* 123:195–205
33. Gelin BR. 1976. *Application of empirical energy functions to conformational problems in biochemical systems*. PhD thesis. Harvard Univ.
34. Gelin BR, Karplus M. 1975. Sidechain torsional potentials and motion of amino acids in proteins: bovine pancreatic trypsin inhibitor. *Proc. Natl. Acad. Sci. USA* 72:2002–6
35. Gelin BR, Karplus M. 1977. Mechanism of tertiary structural change in hemoglobin. *Proc. Natl. Acad. Sci. USA* 74:801–5
36. Gilardi R, Karle IL, Karle J, Sperling W. 1971. Crystal structure of visual chromophores, 11-cis and all-trans retinal. *Nature* 232:187
37. Godfrey M, Karplus M. 1968. Theoretical investigation of reactive collisions in molecular beams: K+Br_2. *J. Chem. Phys.* 49:3602–9
38. Harvey SC, Prabhakaran M, Mao B, McCammon JA. 1984. Phenylalanine transfer RNA: molecular dynamics simulation. *Science* 223:1189–91
39. Hemley RJ, Dinur U, Vaida V, Karplus M. 1985. Theoretical study of the ground and excited singlet states of styrene. *J. Am. Chem. Soc.* 107:836–44
40. Hirschfelder JA, Eyring H, Topley B. 1936. Reactions involving hydrogen molecules and atoms. *J. Chem. Phys.* 4:170–77
41. Honig B, Hudson B, Sykes BD, Karplus M. 1971. Ring orientation in β-ionone and retinals. *Proc. Natl. Acad. Sci. USA* 68:1289–93
42. Honig B, Karplus M. 1971. Implications of torsional potential of retinal isomers for visual excitation. *Nature* 229:558–60
43. Honig B, Warshel A, Karplus M. 1975. Theoretical studies of the visual chromophore. *Acc. Chem. Res.* 8:92–100
44. Hvidt A, Nielsen SO. 1966. Hydrogen exchange in proteins. *Adv. Protein Chem.* 21:287–86
45. Ichiye T, Karplus M. 1983. Fluorescence depolarization of tryptophan residues in proteins: a molecular dynamics study. *Biochemistry* 22:2884–93
46. Imai K, Osawa E. 1989. An extension of multiparameteric Karplus equation. *Tetrahedron Lett.* 30:4251–54
47. Irikura KK, Tidor B, Brooks BR, Karplus M. 1985. Transition from B to Z DNA: contribution of internal fluctuations to the configurational entropy difference. *Science* 229:571–72
48. Islam SA, Karplus M, Weaver DL. 2002. Application of the diffusion-collision model to the folding of three-helix bundle proteins. *J. Mol. Biol.* 318:199–215
49. Islam SA, Karplus M, Weaver DL. 2004. The role of sequence and structure in protein folding kinetics: the diffusion-collision model applied to proteins L and G. *Structure* 12:1833–45
50. Karplus M. 1952. Bird activity in the continuous daylight of arctic summer. *Ecology* 33:129
51. Karplus M. 1956. Charge distribution in the hydrogen molecule. *J. Chem. Phys.* 25:605–6
52. Karplus M. 1959. Contact electron-spin interactions of nuclear magnetic moments. *J. Chem. Phys.* 30:11–15
53. Karplus M. 1959. Interpretation of the electron-spin resonance spectrum of the methyl radical. *J. Chem. Phys.* 30:15–18

54. Karplus M. 1960. Theory of proton coupling constants in unsaturated molecules. *J. Am. Chem. Soc.* 82:4431
55. Karplus M. 1960. Weak interactions in molecular quantum mechanics. *Rev. Mod. Phys.* 32:455–60
56. Karplus M. 1963. Vicinal proton coupling in nuclear magnetic resonance. *J. Am. Chem. Soc.* 85:2870
57. Karplus M. 1968. Structural implications of reaction kinetics. In *Structural Chemistry and Molecular Biology: A Volume Dedicated to Linus Pauling by His Students, Colleagues, and Friends*, ed. A Rich, N Davidson, pp. 837–47. San Francisco: Freeman
58. Karplus M. 1982. Dynamics of proteins. *Ber. Bunsen-Ges. Phys. Chem.* 86:386–95
59. Karplus M. 1996. Theory of vicinal coupling constants. In *Encyclopedia of Nuclear Magnetic Resonance*. Vol. 1: *Historical Perspectives*, ed. DM Grant, RK Harris, pp. 420–22. New York: Wiley
60. Karplus M. 1997. The Levinthal Paradox: yesterday and today. *Fold. Des.* 2:569–76
61. Karplus M. 2002. Molecular dynamics simulations of biomolecules. *Acc. Chem. Res.* 35:321–23
62. Karplus M, Fraenkel GK. 1961. Theoretical interpretation of carbon-13 hyperfine interactions in electron spin resonance spectra. *J. Chem. Phys.* 35:1312–23
63. Karplus M, Godfrey M. 1966. Quasiclassical trajectory analysis for the reaction of potassium atoms with oriented methyl iodide molecules. *J. Am. Chem. Soc.* 88:5332
64. Karplus M, Kuppermann A, Isaacson LM. 1958. Quantum-mechanical calculation of one-electron properties. I. General formulation. *J. Chem. Phys.* 29:1240–46
65. Karplus M, Kuriyan J. 2005. Molecular dynamics and protein function. *Proc. Natl. Acad. Sci. USA* 102:6679–85
66. Karplus M, Lawler RG, Fraenkel GK. 1965. Electron spin resonance studies of deuterium isotope effects. A novel resonance-integral perturbation. *J. Am. Chem. Soc.* 87:5260
67. Karplus M, McCammon JA. 2002. Molecular dynamics simulations of biomolecules. *Nat. Struct. Biol.* 9:646–52
68. Karplus M, Porter RN. 1970. *Atoms and Molecules: An Introduction for Students of Physical Chemistry*. Menlo Park, CA: Benjamin Cummins
69. Karplus M, Porter RN, Sharma RD. 1965. Exchange reactions with activation energy. I. Simple barrier potential for (H,H_2). *J. Chem. Phys.* 43:3259–87
70. Karplus M, Weaver DL. 1976. Protein-folding dynamics. *Nature* 260:404–6
71. Karplus M, Weaver DL. 1994. Folding dynamics: the diffusion-collision model and experimental data. *Protein Sci.* 3:650–68
72. Karplus R, Kroll NM. 1950. Fourth-order corrections in quantum electrodynamics and the magnetic moment of the electron. *Phys. Rev.* 77:536–49
73. Karplus S, Karplus M. 1972. Nuclear magnetic resonance determination of the angle ψ in peptides. *Proc. Natl. Acad. Sci. USA* 69:3204–6
74. Deleted in proof
75. Kirkwood JG, Oppenheim I. 1961. *Chemical Thermodynamics*. New York: McGraw Hill
76. Kline AD, Braun W, Wüthrich K. 1988. Determination of the complete 3-dimensional structure of the alpha-amylase inhibitor tendamistat in aqueous-solution by nuclear magnetic resonance and distance geometry. *J. Mol. Biol.* 204:675–724
77. Kuppermann A, Karplus M, Isaacson LM. 1959. The quantum-mechanical calculation of one-electron properties. II. One-and two-center moment integrals. *Zeit. Nat.* 14a:311–18
78. Kuppermann A, Schatz GC. 1975. Quantum-mechanical reactive scattering: accurate 3-dimensional calculation. *J. Chem. Phys.* 62:2502–4

79. Kuriyan J, Weis WI. 1991. Rigid protein motion as a model for crystallographic temperature factors. *Proc. Natl. Acad. Sci. USA* 88:2773–77
80. Lawler RG, Bolton JR, Karplus M, Fraenkel GK. 1967. Deuterium isotope effects in the electron spin resonance spectra of naphthalene negative ions. *J. Chem. Phys.* 47:2149–65
81. Lee A-W, Karplus M, Poyart C, Bursaux E. 1988. Analysis of proton release in oxygen binding by hemoglobin: implications for the cooperative mechanism. *Biochemistry* 27:1285–301
82. Levitt M, Lifson S. 1969. Refinement of protein conformations using a macromolecular energy minimization procedure. *J. Mol. Biol.* 46:269–79
83. Levy RM, Karplus M, Wolynes PG. 1981. NMR relaxation parameters in molecules with internal motion: exact Langevin trajectory results compared with simplified relaxation models. *J. Am. Chem. Soc.* 103:5998–6011
84. Lifson S, Warshel A. 1969. Consistent force field for calculations of conformations vibrational spectra and enthalpies of cycloalkanes and n-alkane molecules. *J. Chem. Phys.* 49:5116–29
85. Linderstrom-Lang K. 1955. Deuterium exchange between peptides and water. *Chem. Soc.* Spec. Publ. 2, p. 1.
86. Liu B. 1973. Ab-initio potential-energy surface for linear H-3. *J. Chem. Phys.* 58:1925–37
87. Ma J, Sigler PB, Xu Z, Karplus M. 2000. A dynamic model for the allosteric mechanism of GroEL. *J. Mol. Biol.* 302:303–13
88. McCammon JA, Gelin BR, Karplus M. 1977. Dynamics of folded proteins. *Nature* 267:585–90
89. Mierke DF, Huber T, Kessler H. 1994. Coupling-constants again: experimental restraints in structure refinement. *J. Comp. Aided Mol. Des.* 8:29–40
90. Miranker A, Karplus M. 1991. Functionality maps of binding sites: a multiple copy simultaneous search method. *Proteins Struct. Funct. Genet.* 11:29–34
91. Moffitt W. 1954. Atomic valence states and chemical binding. *Rep. Prog. Phys.* 17:173–200
92. Monod J, Wyman J, Changeux JP. 1965. On nature of allosteric transitions: a plausible model. *J. Mol. Biol.* 12:88–118
93. Morokuma K, Eu BC, Karplus M. 1969. Collision dynamics and the statistical theories of chemical reactions. I. Average cross section from transition-state theory. *J. Chem. Phys.* 51:5193–203
94. Morokuma K, Karplus M. 1971. Collision dynamics and the statistical theories of chemical reactions. II. Comparison of reaction probabilities. *J. Chem. Phys.* 55:63–75
95. Nadler W, Brünger AT, Schulten K, Karplus M. 1987. Molecular and stochastic dynamics of proteins. *Proc. Natl. Acad. Sci. USA* 84:7933–37
96. Nilsson L, Clore GM, Gronenborn AM, Brünger AT, Karplus M. 1986. Structure refinement of oligonucleotides by molecular dynamics with nuclear Overhauser effect interproton distance restraints: application to 5′ $d(C\text{-}G\text{-}T\text{-}A\text{-}C\text{-}G)_2$. *J. Mol. Biol.* 188:455–75
97. Olejniczak ET, Dobson CM, Levy RM, Karplus M. 1984. Motional averaging of proton nuclear Overhauser effects in proteins. Predictions from a molecular dynamics simulation of lysozyme. *J. Am. Chem. Soc.* 106:1923–30
98. Perutz M. 1971. Stereochemistry of cooperative effects in haemoglobin. *Nature* 232:408–13
99. Phillips DC. 1981. Closing remarks. In *Biomolecular Stereodynamics*, ed. RH Sarma, 2:497–48. Guilderland, NY: Adenine
100. Porter RN, Karplus M. 1964. Potential energy surface for H_3. *J. Chem. Phys.* 40:1105–15
101. Post CB, Dobson CM. 2005. Meeting review frontiers in computational biophysics: a symposium in honor of Martin Karplus. *Structure* 13:949–52

102. Purins D, Karplus M. 1969. Spin delocalization and vibrational-electronic interaction in the toluene ion-radicals. *J. Chem. Phys.* 50:214–33
103. Rahman A. 1964. Correlations in motion of atoms in liquid argon. *Phys. Rev.* 136:A405–11
104. Sali A, Shakhnovich E, Karplus M. 1994. How does a protein fold? *Nature* 369:248–51
105. Schatz GC. 2000. Perspective on "Exchange reactions with activation energy. I. Simple barrier potential for (H, H_2)": Karplus M, Porter RN, Sharma RD (1965) *J. Chem. Phys.* 43:3259–3287. *Theor. Chem. Acc.* 103:270–72
106. Schekkerman H, Tulp I, Piersma T, Visser GH. 2003. Mechanisms promoting higher growth rate in arctic than in temperate shorebirds. *Ecophysiology* 134:332–42
107. Scheraga HA. 1968. Calculations of the conformations of small molecules. *Adv. Phys. Org. Chem.* 6:103–84
108. Schulten K, Karplus M. 1972. On the origin of a low-lying forbidden transition in polyenes and related molecules. *Chem. Phys. Lett.* 14:305–9
109. Shavitt I, Karplus M. 1962. Multicenter integrals in molecular quantum mechanics. *J. Chem. Phys.* 36:550–51
110. Shavitt I, Stevens RM, Minn FL, Karplus M. 1968. Potential-energy surface for H_3. *J. Chem. Phys.* 48:2700–13
111. Shulman RG, Glarum SH, Karplus M. 1971. Electronic structure of cyanide complexes of hemes and heme protein. *J. Mol. Biol.* 57:93–115
112. Simonson T, Archontis G, Karplus M. 2002. Free energy simulations come of age: protein-ligand recognition. *Acc. Chem. Res.* 35:430–37
113. Smith J, Cusack S, Pezzeca U, Brooks BR, Karplus M. 1986. Inelastic neutron scattering analysis of low frequency motion in proteins: a normal mode study of the bovine pancreatic trypsin inhibitor. *J. Chem. Phys.* 85:3636–54
114. Stillinger FH, Rahman A. 1974. Improved simulation of liquid water by molecular-dynamics. *J. Chem. Phys.* 60:1545–57
115. Szabo A, Karplus M. 1972. A mathematical model for structure-function relations in hemoglobin. *J. Mol. Biol.* 72:163–97
116. Szent-Györgyi A. 1948. *Nature of Life, A Study of Muscle*. New York: Academic. 102 pp.
117. Taylor EH, Datz S. 1955. Study of chemical reaction mechanisms with molecular beams: the reaction of K with HBr. *J. Chem. Phys.* 23:1711–18
118. van der Vaart A, Ma J, Karplus M. 2004. The unfolding action of GroEL on a protein substrate. *Biophys. J.* 87:562–73
119. Wall FT, Porter RN. 1963. Sensitivity of exchange-reaction probabilities to potential-energy surface. *J. Chem. Phys.* 39:311
120. Warshel A, Karplus M. 1974. Calculation of $\pi\pi^*$ excited state conformations and vibronic structure of retinal and related molecules. *J. Am. Chem. Soc.* 96:5677–89
121. Wong CF, McCammon JA. 1986. Dynamics and design of enzymes and inhibitors. *J. Am. Chem. Soc.* 108:3830–32
122. Yang W, Bitetti-Putzer R, Karplus M. 2004. Free energy simulations: use of reverse cumulative averaging to determine the equilibrated region and the time required for convergence. *J. Chem. Phys.* 120:2618–28
123. Young MA, Gonfloni S, Superti-Furga G, Roux B, Kuriyan J. 2001. Dynamic coupling between the SH2 and SH3 domains of c-Src and hck underlies their inactivation by C-terminal tyrosin phosphorylation. *Cell* 105:115–26

Computer-Based Design of Novel Protein Structures

Glenn L. Butterfoss and Brian Kuhlman

Department of Biochemistry and Biophysics, University of North Carolina at Chapel Hill, Chapel Hill, North Carolina 27599-7260; email: bkuhlman@email.unc.edu, butter@email.unc.edu

Annu. Rev. Biophys. Biomol. Struct. 2006. 35:49–65

First published online as a Review in Advance on December 7, 2005

The *Annual Review of Biophysics and Biomolecular Structure* is online at biophys.annualreviews.org

doi: 10.1146/annurev.biophys.35.040405.102046

1056-8700/06/0609-0049$20.00

Key Words

computational protein design, negative design, flexible backbone design, protein stability, molecular modeling

Abstract

Over the past 10 years there has been tremendous success in the area of computational protein design. Protein design software has been used to stabilize proteins, solubilize membrane proteins, design intermolecular interactions, and design new protein structures. A key motivation for these studies is that they test our understanding of protein energetics and structure. De novo design of novel structures is a particularly rigorous test because the protein backbone must be designed in addition to the amino acid side chains. A priori it is not guaranteed that the target backbone is even designable. To address this issue, researchers have developed a variety of methods for generating protein-like scaffolds and for optimizing the protein backbone in conjunction with the amino acid sequence. These protocols have been used to design proteins from scratch and to explore sequence space for naturally occurring protein folds.

Contents

INTRODUCTION

De novo protein design: protein design from scratch

Ultimately, one would like to use protein design methodology to create never-before-seen proteins that have valuable applications in medicine, research, and industrial processes. Because a protein's function is determined by its structure, learning how to create proteins of predefined structure is a key step in this process. The first successes in protein design were based on manual inspection and heuristics gleaned from examining naturally occurring proteins (10, 63). It was noticed early on that many of these designs differed from naturally occurring proteins in that they did not contain well-packed side chains in their interior. To address this problem, computational procedures were developed for searching for amino acid sequences that could pack well on a target protein backbone. These algorithms have been successful and have been used to stabilize proteins, solubilize membrane proteins, redesign protein-protein interactions, create new enzymes, and design novel protein structures (6, 15, 35, 50, 72).

In most cases, computational design has been used to redesign already-existing proteins. This is an important problem and novel functional proteins have been created with this approach (15, 47), but in the long run it will be advantageous to create proteins of arbitrary shape. Designing novel protein structures is intrinsically more difficult than protein redesign because a priori it is not known if the target structure is designable. In the recent design of a protein with a novel α/β-fold, it was found that most putative scaffolds could not be designed with packing energies comparable to those observed for naturally occurring proteins (35). Traditionally, protein design has been considered the reverse of protein structure prediction. It now appears that success in protein design is closely tied to methods in protein structure prediction. Protocols for low-resolution structure prediction can be used to generate good starting structures for design, and methods for high-resolution structure refinement can be combined with sequence optimization protocols to search for low-energy sequence/structure combinations. Here, we divide de novo protein design into three steps: (*a*) generating the design scaffold, (*b*) finding low-energy sequences for that scaffold, and (*c*) coupling sequence design with backbone optimization.

STEP 1: BUILDING THE SCAFFOLD

To design a protein from scratch, it is necessary to first define the target structure or complex. This might be a naturally occurring protein fold, a novel fold, or a new protein-protein interaction. Because most randomly generated protein structures are not designable, it is crucial that the target structure mimics many of the defining characteristics of naturally occurring proteins. Backbone polar groups should be primarily hydrogen bonding with other backbone groups, the backbone torsion angles should occupy the allowed regions of the Ramachandran plot, and the spacing between units of secondary structure should be set to allow for tight packing between amino acid side chains. Here, we summarize the various approaches that have been used to solve this problem.

Perhaps the most straightforward target structures are small units of protein secondary structure. Idealized α-helices and β-strands can be assembled by picking phi and psi angles from the appropriate region of the Ramachandran plot. Many studies have looked at β-hairpin design (37). The turn residues are often modeled in one of the four canonical turn types (I, I′, II, or II′) that are specified by established patterns of phi and psi angles (69). Once the initial model of the hairpin is built, the backbone torsion angles in the two strands and the turn are often varied to optimize the hydrogen bonding between the two strands. Generally, this requires only small perturbations in phi and psi angles, and optimization can be performed with gradient-based minimization techniques.

For target structures that contain several segments of secondary structure, it is necessary to specify the spacing between the segments and the loops that connect them. In cases in which the target fold resembles a naturally occurring protein, one approach is to superimpose idealized segments of secondary structure onto examples of the naturally occurring fold. DeGrado and coworkers (46) have shown that the backbone structures of many metalloproteins can be reconstructed with idealized helices and hairpins and that the best structural alignments deviate from the naturally occurring structure by less than 1 Å rmsd. In many cases the target folds are symmetric and under such circumstances the design template can be defined with few parameters. Many four-helix bundle proteins can be described by a 222-symmetrical arrangement of equivalent helices that allows the entire protein backbone (minus any connecting loops) to be defined by six adjustable parameters: three values to specify the displacement of the helical monomer and three to specify its orientation (79). A variety of four-helix diiron proteins can be rebuilt with idealized helices to less than 1.5 Å rmsd by optimizing only these six parameters (76). The values taken from all the superpositions can be used to construct a consensus four-helix bundle protein, or they could be used to create a large set of four-helix bundle templates. Large sets of design templates become especially useful when the design target includes a bound ligand and it is expected that many of the templates will not support amino acid side chains that can make low-energy contacts with the ligand. DeGrado and coworkers (4) recently published a review of the diiron proteins they have designed from idealized models of four-helix bundles.

rmsd: root mean square deviation

Coiled-coils are special because an analytic expression can be used to describe their allowed geometries. This property of coiled-coils was first explored by Francis Crick (7) and more recently has been used to design coiled-coils (19, 20) and predict the effects of mutations on coiled-coil stability (26). In these cases, backbone parameterization was used to build the starting scaffold as well as vary the backbone position during the simulation (see below).

The geometric properties of β-sheets and β-barrels are also well established (40, 41, 53, 54) and were recently used in the design of an idealized α/β-barrel protein (55). First, the overall geometry of the β-barrel was specified by setting the radius of the barrel and the tilt of each strand relative to an axis running through the center of the barrel. Second, the β-strands were optimized with a conjugate gradient minimization protocol to remove clashes and improve hydrogen bonding between the strands. Helices were placed outside of the barrel by using five geometric parameters that were determined by examining naturally occurring TIM barrels as

PDB: protein data bank

well as requiring that the helices connect to the strands with turn motifs commonly found in naturally occurring TIM barrels. The single backbone generated with this protocol was then used as the template for the sequence optimization program, ORBIT.

Building Starting Structures with Rosetta

Most of the template-building procedures described so far are hierarchical in nature. Starting from idealized segments of secondary structure, the template is built by orienting these segments and then connecting them with the appropriate sized loops. This approach works especially well for symmetric structures in which the segments can be connected with well-characterized loop motifs. However, when the target structure becomes more irregular, it is no longer straightforward to connect the various secondary structure segments. One possibility is to design each loop separately by searching for loops from the Protein Data Bank (PDB) that could connect the segments (36). However, in many cases it may not be possible to find such a loop, or often a loop can be found but does not interact well with the rest of the protein. Because loops are often an integral part of the protein structure and contribute residues to the protein core, it may not be best to build them after the fact. Ideally, loop structures should be optimized when the rest of the protein structure is built. One set of tools that build structures in their entirety are the various programs and algorithms that have been developed for protein structure prediction. The Baker laboratory recently used the structure prediction program Rosetta to build the starting template for the design of a novel α/β protein called Top7 (35).

The Rosetta program was developed initially for de novo structure prediction (64, 70). It builds structures from nine- and three-residue fragments taken from the PDB. Using a Monte Carlo optimization procedure, Rosetta assembles fragments into structures that maximize hydrophobic burial and satisfy the hydrogen-bonding potential of β-strands. For de novo structure prediction the fragments are picked using the query sequence and its predicted secondary structure. For protein design there is no query sequence, but the fragments can be picked using the desired secondary structure at each residue position. In some cases it may not be clear where one element of secondary structure should begin and another should end; in these cases fragments can be picked for each type of possible secondary structure. Approximately 25 fragments are considered for each nine- or three-residue segment. Distance constraints can be used to direct the fragment assembly toward a target fold. To create starting structures for Top7, short-range-distance constraints ($<$4 Å) between backbone nitrogens and carbonyl oxygens were used to specify strand pairing as well as strand register. Interestingly, although the target fold contained two $\beta/\alpha/\beta$-motifs, no constraints were needed to force the helices to the intended side of the interface. Because almost all $\beta/\alpha/\beta$-motifs in nature are right-handed, the fragments used to connect the strands to the helices were already biased to form the desired right-handed connections. In general, by building backbones from small pieces of naturally occurring proteins, it is ensured that most of the local structural motifs in the target structure will be designable.

In addition to terms for hydrogen bonding and hydrophobic burial, the Rosetta scoring function contains knowledge-based potentials that dictate strand-strand and strand-helix interactions (71). These specify the optimal distances of interaction as well as preferred orientations. This term ensures that β-sheets have the naturally occurring right-handed twist and makes sure that there are appropriate sized spaces between the secondary structure elements to allow for hydrophobic packing. Rosetta can generate structures fairly rapidly for 100 residue proteins, and therefore this approach can be used to generate thousands of structures that adopt the target fold but do not share the same exact local or

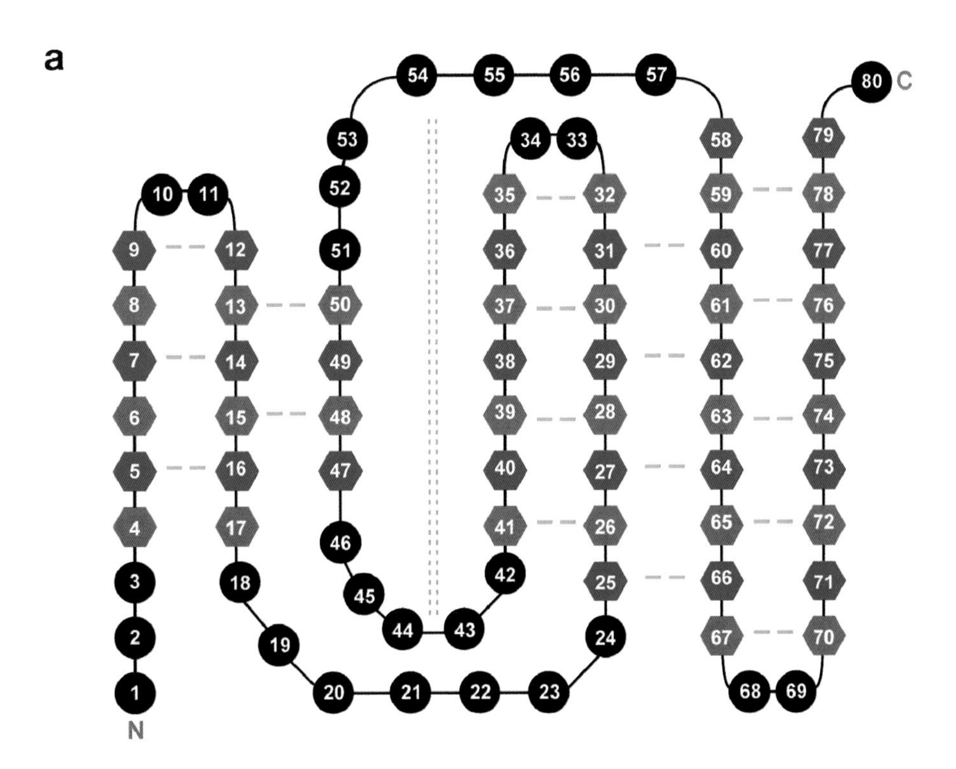

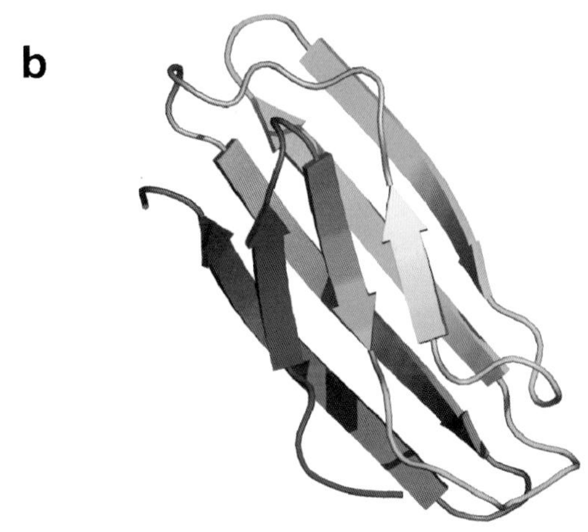

Figure 1

(*a*) Scheme for using Rosetta to build design scaffolds that adopt a fibronectin fold. Residues built from β-strand fragments are shown as hexagons, and residues built from loop fragments are shown as circles. Hydrogen-bonding constraints (*green dashes*) define the topology of the structure. (*b*) An example of one of the fibronectin scaffolds built with Rosetta using the scheme from panel *a*.

tertiary interactions. **Figure 1** shows a recent example of a β-sheet scaffold that we have built with Rosetta.

Starting Structures for the De Novo Design of Protein-Protein Interactions

Aside from the design of symmetrical helical bundle proteins, the de novo design of protein-protein interactions is largely an unsolved problem. Building starting complexes that are set up for making tight interactions between monomers is an important step toward solving this problem. Toward this end, the Mayo laboratory has adopted a fast Fourier transform-based docking algorithm for protein design (23). Side chains on the surfaces of the proteins are represented in a sequence-independent fashion as spheres that approximate the average size of an average

LJ: Lennard-Jones potential

FDPB: finite difference Poisson-Boltzmann

amino acid. The docking algorithm then uses the fast Fourier transform along with a simple scoring scheme for detecting favorable contacts and bumps to search for surfaces on the two proteins that are complementary in shape. To test the protocol, 121 naturally occurring homodimers were split apart, reoriented, and docked with simplified side chains. Forty-five of the models had less than 1 Å rmsd relative to the native structure. It is anticipated that other docking algorithms will also be useful for protein interface design (73), including protocols that use Monte Carlo optimization.

STEP 2: DESIGNING A SEQUENCE FOR A FIXED PROTEIN BACKBONE

Identifying low-energy sequences for a target protein backbone is the central problem of protein design, and many laboratories have developed computational procedures for solving this problem. All of these programs share two common components: (*a*) an energy function for evaluating the favorability of a particular sequence for a particular structure and (*b*) a procedure for searching for low-energy sequences. The common energy functions and search protocols for protein design have been reviewed previously (17, 24, 51, 56, 58, 77a, 78).

Energy Functions

In general, protein design energy functions are constructed to favor close packing between amino acids, satisfy hydrogen-bonding potential, partition hydrophobic amino acids to the core of a protein and polar amino acids to the surface, and favor low-energy torsion angles. Packing is often evaluated with a Lennard-Jones (LJ) potential. The attractive portion of the LJ potential models van der Waals forces and draws atoms near each other. The repulsive portion of the potential ensures that the atoms do not become too close. Because protein design simulations are often performed with rigid protein backbones and limited side chain flexibility, it is common to dampen the repulsion term. This can be achieved by reducing the radius of the atoms or by explicitly weakening the repulsive portion of the LJ potential (9, 47, 60). The second approach is probably preferable because the location of the most favorable distance between the two atoms is not perturbed. Instead of using an LJ potential to model packing interactions, Liang & Grishin (43, 44) used a grid-based approach to explicitly determine contacting surface areas (favorable) and the volume of atomic overlaps (unfavorable). This approach performed better than a LJ potential in a side chain prediction test.

Side chain torsion energies are typically evaluated using molecular mechanics potentials or are derived from the probability of observing a particular side chain conformation in the PDB (14). Hydrogen bonds are generally scored with an explicit hydrogen-bonding term or are accounted for by an electrostatics potential (17, 31). One limitation of the electrostatics-based approach is that it does not correctly predict the orientational dependence of hydrogen bonds (52). A variety of approaches have been used to model electrostatics. These include Coulomb potentials with a distance-dependent dielectric or an environment-dependent dielectric, the electrostatic term from the generalized Born model, a modified Tanford-Kirkwood model, and amino acid pair potentials derived from the PDB (21, 59, 71, 81).

A solvation term is required to disfavor placing polar amino acids in the core of a protein. However, because polar amino acids are not uniformly restricted from protein interiors, this term must not overwhelm other terms in the energy function. In particular, the relative strengths of desolvation energies and hydrogen-bonding energies determine how many polar amino acids are placed in the core during a design simulation. Two general types of solvation models are currently in use: empirically derived potentials and continuum potentials based on the finite difference Poisson-Boltzmann equation (FDPB)

(18, 21, 42, 59, 74, 81). The empirical potentials are parameterized to fit experimental solvation energies, and the continuum potentials are parameterized to best fit the FDPB equation. Fast pair-wise additive approximations have been developed for both approaches and have been used to successfully design proteins.

Reference Energies

To optimize the stability of a protein, it is necessary to have a model for the energy of the unfolded state. One simple approach is to assign reference energies to each amino acid type that represents its typical energy in the unfolded state. These reference energies can be determined empirically or set equal to the average energy of each amino type calculated from model peptides (33, 44, 60, 80). One approach for empirically setting the reference values is to adjust them to reproduce the naturally occurring amino acid frequencies in protein design simulations. Pokala & Handel (60) recently found that native amino acid sequences can be recovered with high identity without explicitly optimizing the reference values. They calculated the average energy of each amino acid type when placed in the center of peptides with random conformations taken from the PDB. This approach worked well for reproducing sequences in protein cores but left large hydrophobic patches on the surfaces of proteins. To improve surface design, they used a new set of reference values for surface positions that were derived from the average energy of amino acids at protein surfaces. They found that in naturally occurring protein surfaces hydrophobics were well packed, and therefore these new reference values required that a hydrophobic residue be well packed to be favorable on the protein surface.

Search Protocols

The number of possible sequences for only a 50-residue protein is enormous, and therefore a rapid optimization protocol must be used for efficiently scanning through this space. One important simplification that most protein designers use is to consider only amino acids in a limited set of most-preferred side chain conformations called rotamers (14, 49, 61). A wide variety of methods have been used to pack side chain rotamers on a protein backbone. They can be broadly divided into stochastic and deterministic models. Deterministic searches including dead-end elimination and self-consistent mean-field models, in some manner, account for the entire search space. Dead-end elimination, if converged, guarantees the global optimal solution for the model (13, 16, 48). Self-consistent mean-field calculations can return the relative preference of each amino acid at each sequence position (27, 29). Stochastic searches, which include genetic algorithms and Monte Carlo searches, "walk" through search space to find optimal solutions (11). Although stochastic searches do not necessarily return a global optimum, they can be considerably faster than their deterministic counterparts for large search problems. Recent results from Pokala & Handel suggest that in many cases Monte Carlo searches will converge results similar to those obtained with dead-end elimination (60). The speed of the rotamer search becomes most critical if one wants to couple sequence optimization with backbone optimization.

Rotamer: a conformational isomer of an amino acid

Results in Fixed Backbone Design

Studies from a variety of laboratories suggest that sequence optimization protocols are good if presented with a designable protein backbone. The first landmark success in this regard was the computational design of a new sequence that adopts the zinc-finger fold (8). Since then computational design has been used to design new sequences for α-, β-, and α/β-proteins (34). In many cases the designed sequence resembles the wild-type sequence, in effect suggesting that the protein backbone has "remembered" the wild-type sequence. Because of this memory, the redesign of naturally occurring protein backbones is not a strong test of protein design methodology or protein design energy functions, but rather it is a step toward the more ambitious goal of designing novel structures and complexes.

Flexible backbone design: design methods which allow for the exploration of backbone conformational space

STEP 3: FLEXIBLE BACKBONE DESIGN

Although fixed backbone protein design has proven to be successful in several studies, in many cases a true flexible backbone model may be desired or necessary. Many experimental results have demonstrated that the backbone makes adjustments in response to mutations (1, 45), and therefore the energetic and structural consequences of a mutation cannot be predicted accurately without allowing for backbone flexibility. A danger of adding backbone flexibility to the model is that one becomes more dependent on the energy function, but several results suggest that backbone relaxation can be predicted with some accuracy. Desjarlais & Handel (12) used a flexible backbone model to more correctly repack mutations in the core of T4 lysozyme. Keating and coworkers predicted with high accuracy the effects of mutations on the stability and structure of a coiled-coil by considering alternative backbone conformations (26).

The two reports of the rational design of novel proteins folds (tri- and tetrameric right-handed coiled-coils and the Top7 α/β-protein) incorporated an exploration of backbone conformational space, as did the first experimentally verified redesign of a β-sheet protein (19, 32, 35, 57). In the construction of Top7, Kuhlman and coworkers found that the LJ energies resulting from fixed backbone designs on 173 initial backbone templates were significantly higher than the average LJ energies of the corresponding amino acids in similar environments in the PDB. After several rounds of flexible backbone design, the energies became significantly better, with many designs having average residue energies lower than the PDB average (35). To design novel biosensors and enzymes, Hellinga's group has found that it is advantageous to consider multiple backbone templates and alternative placements of the target ligand (15, 47).

An innovative use of flexible backbone design is to utilize it to explore the sequence space compatible with a particular protein fold. Larson and coworkers have built on work by Levitt in creating a "reverse BLAST" methodology to complement homology modeling of protein structures (28, 33, 38, 39). In contrast to traditional design methodologies, which seek a small set of sequences that optimally stabilize a target structure, reverse BLASTing identifies a large number of possible sequences for templates. These sequences can then be scanned against the sequence database to search for remote homologs of the design template. The researchers found that designing over ensembles of backbone templates, centered on the native structure, significantly expands the variety of designed sequences and thus increased the breath of subsequent homology searches. Saunders & Baker (66) have used a flexible backbone design protocol in a similar manner to reproduce evolutionary relationships among protein families, again finding that flexible backbone models produce sequences corresponding to natural families better than fixed backbone designs do. Wollacott & Desjarlais (82) used design to explore potential peptide sequences (virtual interaction profiles) that likely bind a given domain. Beginning with experimental protein-peptide complexes (of PZD, SH3, and other domains), small perturbations were applied to the backbones of bound peptides and sequences were designed. The profiles compared favorably with experimental results.

Methods for Flexible Backbone Design

The published methods for flexible backbone protein design can essentially be split into two major classes: protocols explicitly separating sequence selection and backbone movement and protocols that integrate searches of both spaces into the same trajectory. The former has the advantage of being relatively fast and is easily parallelized to multiple processors, and the latter allows for energy transfer between the backbone structure and packed side chains during the design process.

One method to separate design from backbone conformational searches is to generate large ensembles of closely related structures. The reverse BLAST simulations conducted by the Pande group began with a Monte Carlo exploration of local backbone conformational space to generate families of 100 template backbones, restrained such that the final backbones had 1.0 Å rmsd from the original structure (38, 39). The template structures were then individually submitted to the Genome@home distributed grid system for fixed backbone design. Given the computational limitations of distributed computing, ensemble design is an excellent tool for allowing backbone variety.

Kraemer-Pecore et al. (32) also used a structural ensemble method to redesign a WW domain, a common β-sheet fold. Again, a Monte Carlo simulation generated 30 structures within 0.3 Å rmsd of the starting backbone configuration and fixed backbone design was run on each. However, in a second step, each residue position in each template structure was then mutated through the defined residue/rotamer space while the rest of each structure was held fixed in the originally designed configuration. The calculated energies were collected into a partition function expressing the probability of each residue at each position across the set of designed backbones and sequences. One of the two sequences determined by the highest probability at each position (slightly different algorithms were used to collect the partition function) was experimentally found to fold into a WW domain. Although this method separates backbone movements from design in execution, it collects energetic information from the global system. As the authors suggest, it is also a potential tool for intelligently seeding combinatorial libraries.

A second group of studies falling under the same umbrella (of splitting backbone motions from design steps) applied algebraic parameterization to define geometric movements of the backbone, wherein the backbone geometry (or a region thereof) is expressed as a small set of parameters, which may be varied to rapidly and systematically explore well-defined regions of conformational space. Parametric algebra has been used most thoroughly in the design of coiled-coils by applying equations described originally by Crick (7). Harbury and coworkers (19, 20, 57) designed (and experimentally verified) novel tri- and tetrameric right-handed coiled-coils, as well as variations of "traditional" left-handed coiled-coils. To design the right-handed coiled-coils, the helices were constructed with all possible combinations of the defined side chain/rotamer library, and algebraic parameters were varied to find the optimal backbone trace for each configuration. Experimental structures of designed sequences selected for optimal stability and specificity agreed with the predicted models in atomic detail (19, 57).

The Mayo group applied a similar method to produce various orientations of a helix in Gβ1 and then redesigned the core of the protein on the new templates holding the backbone fixed (65, 75). NMR studies indicated that six of seven sequences tested formed well-folded structures in solution. However, a detailed analysis of one protein designed with a highly displaced helix suggested a backbone conformation more like the native than the designed template.

The second general class of flexible backbone design methods integrates backbone optimization with sequence optimization. This may allow for more accurate representation of local relaxations but requires force field components describing backbone deformation.

Desjarlais & Handel (12) have integrated explicit backbone flexibility into a genetic/Monte Carlo search algorithm (SoftROC). The method involves generating an

Rotamer library: a database of probable rotamer conformations used to efficiently explore various packing arrangements

Negative design: these protocols explicitly account for one or more undesired states that may compete with the target in a thermal ensemble

initial population of backbones by applying small phi, psi, and omega variations to the starting structure and decorating each with randomly selected residue identities and rotamers. The population evolves through several generations of recombination and mutation, with "breeding rights" of each replicant weighted by its total energy. Recombination involves exchange of sequence, rotamer, and backbone torsional information between two replicants, and mutation consists of changing residue identities, adjusting rotamers, and small alterations of backbone torsions. The genetic stage is followed by a Monte Carlo refinement including small backbone adjustments. The protocol was tested by comparing the experimental melting temperatures of 434 cro and T4 lysozyme mutants to energies of repacked structures with the same sequences. In general, it was found that the predictive capacity of rigid backbone design only approached (or slightly exceeded) that of the flexible model when the former used a much larger search of possible rotamers. However, flexible backbone design was much better at predicting rotamer conformations in mutants having backbone rmsd >0.3 Å (from wild type) relative to fixed backbone design using the wild type as a template. The original Amber/OPLS force field seemed to "underconstrain" the backbone, allowing larger than expected changes in conformation, and considerably better results were achieved with a restraint potential based on root mean squared deviation.

A second example of iterated flexible backbone design was used in the design of the novel globular protein, Top7. Here, the design proceeded through an iterated Monte Carlo search of both sequence and conformational space using the Rosetta program. Flexible backbone design in Rosetta works in a manner similar to that described above for building novel backbone traces in Rosetta (outlined in **Figure 2**). The objective was to allow for realistic shifts in backbone conformation that are iterated with packing/design steps. The backbone motions consist of either small random perturbations of phi and psi angles of up to five contiguous residues or substitutions of small backbone regions with conformations collected from experimental structures in the PDB. The substitutions are followed by a minimization step to reduce the downstream propagation of the change. Any side chains in the protein with a higher energy (when compared with side chains before the initial perturbation) are then repacked or redesigned with optimal rotamers selected from the library. Finally, a 10-residue window surrounding the site of the initial backbone conformational change is then subject to energy minimization. The new sequence/structure is accepted or rejected according to the Metropolis criterion.

In their study of protein family homology relationships, Saunders & Baker (66) implemented a more aggressive search of conformational space to this procedure, including a melting step at the beginning, an energy term to allow for omega angle minimization, and more perturbing substitutions of backbone angles. These adjustments were advantageous for sampling the local minima in sequence/structure space.

NEGATIVE DESIGN

Negative design methods explicitly consider the energies of possible competing conformational states and tailor the optimization to identify sequences exhibiting both low energy in the target conformation and high energy in the competing conformation. These protocols have particular use in systems in which similar conformations or configurations are likely to compete in the thermal ensemble, such as interface design, where both high affinity and specificity for a particular target are desired. Several studies have demonstrated the feasibility and utility of using negative design.

Kim and coworkers (19) used a negative filter design in the creation of the right-handed coiled-coils after systematically exploring each sequence in di-, tri-, and tetrameric coiled-coil states. Sequences were selected for verification, in part, based on which showed high specificity for a particular oligomeric state. The mean energy of all sequences for each oligomeric state was calculated and specificity was determined on the basis of the relative difference between the target oligomer and the two competing conformations.

DeGrado and coworkers (77) integrated negative design more directly into the search to generate diiron binding A_2B_2 four-helix bundle proteins. They optimized the difference in energy between a desired and undesired conformation of the A_2B_2 complex. The energy function was a simple formula that considered the number of favorable/unfavorable electrostatic contacts from a subset of residues (either Lys or Glu). A top-scoring design was experimentally verified.

In a more complex implementation of negative design, Havranek & Harbury (22) used a multistate approach to design several sets of two-helical sequences that formed either homo- or hetero-specific coiled-coil dimers in solution. The search was built on a genetic algorithm that simultaneously accounted for several states: the folded dimer target structure (hetero or homo), the competing dimer conformation, aggregated homodimers, and the unfolded state. The search favored sequences with optimal fitness, i.e., low energy in the first configuration and high energy in the last three states, as defined by

$$\text{Fitness} = -RT \ln \sum_{competitors} \exp^{-Ac/RT} - A_{target},$$

where A_{target} is the free energy of the target configuration and Ac is the free energy of the competitor states. The aggregated state was used to optimize solubility and was represented by evaluating the free energy of the folded structure in a reduced dielectric. Test sets of designs that omitted one or more of the undesired states yielded sequences that were poorer, upon visual inspection, than the full multistate calculation. The designs with the best fitness scores were synthesized and tested experimentally. Binding assays demonstrated the predicted specificity, as well as free energies of transfer between dimers that correlated well with the calculated free energies.

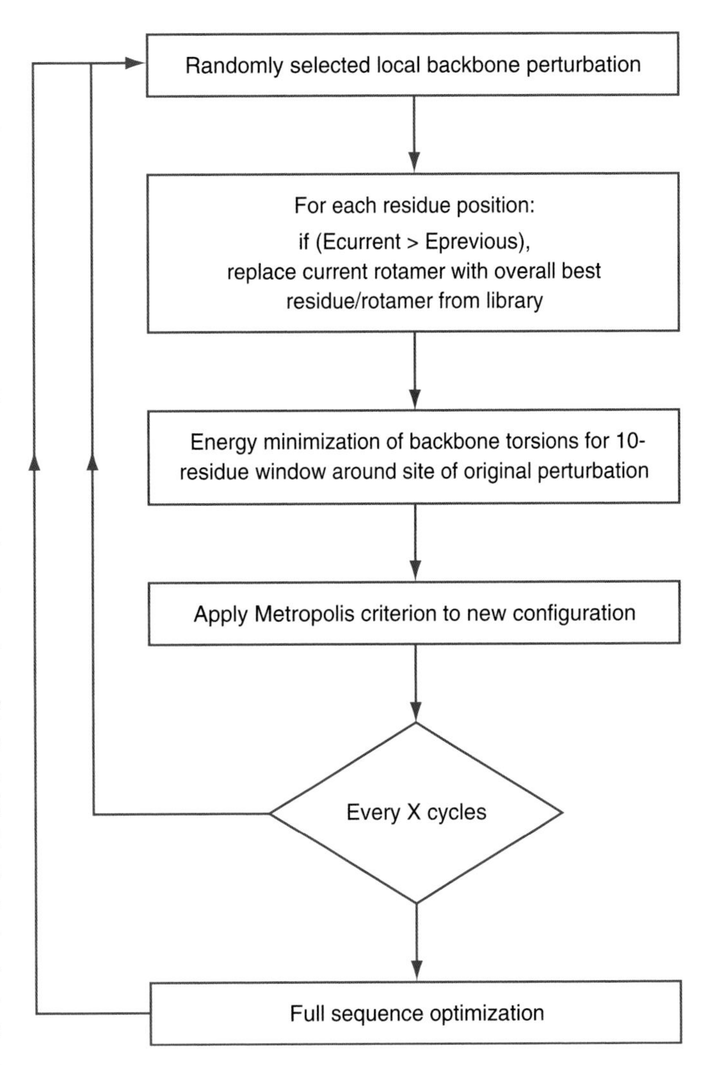

Figure 2
Flow chart of the flexible backbone methodology used in the design of Top7 (35).

Jin et al. (25) used multistate design to explicitly optimize the energy gap between a target three-helix bundle structure and an ensemble of alternative structures. To make this problem computationally tractable, a simplified model of amino acid side chains was used. NMR and circular dichroism studies suggest that the designed protein was well folded.

Despite the published successes of negative design, the question of the importance of using negative design to generate specificity has remained. Several studies have produced specificity using only positive design. For example, Shifman & Mayo (67) found that simply using positive design enhanced specificity of a particular calmodulin variant for myosin heavy chain. Shimaoka et al. (68) stabilized different conformations of the Mac-1 I domain using only positive design. Reina et al. (62) designed PDZ domains with significant specificity for novel target peptides.

Recently, Bolon et al. (2) have directly addressed this by comparing the stability and specificity of asymmetric dimer interfaces predicted by either negative or positive design simulations. Interface residues of the SspB dimer were redesigned asymmetrically using the ORBIT program (2). One simulation was a simple optimization of the total energy (designing for stability). The second was an optimization of the same residues while increasing the energy of the two competing homodimer configurations.

They compared the chemical unfolding curves of the AA, BB, and AB dimers, which were designed asymmetrically for either stability or specificity. The authors did indeed find that the AB complex designed for specificity was significantly more stable than the respective AA or BB complexes. However, the absolute stability was lower than the AB complex designed solely for stability.

Negative design may likely turn out to be critical in efficiently engineering specificity into systems with obvious and similar competing states. However, when the alternative states are not clear or closely related, such as the design of globular proteins, positive design may often be sufficient (35). In addition, as Bolon et al. (2) note, "[p]erhaps the greatest challenge in negative design is to model accurately the energetic effects of destabilizing mutations in competing states that likely involve conformational relaxation."

UNSOLVED PROBLEMS IN COMPUTER-BASED DE NOVO PROTEIN DESIGN

Although the progress of rational computational protein design has been encouraging, several major challenges remain. A significant advance would be the design of novel protein-protein or protein-DNA interfaces. The former case should, intuitively, be within the grasp of current models, given the successes of protein design. Indeed, steps have been taken in the design of interfaces (30). However, the holy grail remains the design of a protein that will bind to a naturally occurring target protein. The particular problems encountered in the design of interfaces are the frequent burial of hydrophilic surfaces, the need to couple docking (and possibly internal backbone) movements with design, and the requirement that the partners be soluble on their own. Because the strength of hydrogen bonds depends strongly on distance and orientation, we believe that extensive sampling of backbone conformation and orientation is needed to design interfaces in which hydrogen-bonding potential is satisfied. The design of DNA-protein interfaces creates additional challenges, in that the entire interface is, by necessity, polar. Successful design likely depends on the careful development and adjustment of electrostatic and hydrogen-bonding models.

A second major step is the design of dynamic properties into proteins. Many biologic processes are mediated by subtle shifts in protein structure and by the set of low-energy structures that proteins may adopt. Design of dynamic behavior into proteins will probably be rather system dependent, given the variety of such systems, for example, allosteric shifts, large conformational "switching" changes, and particular "breathing motions." The multistate and negative design protocols reviewed above provide a foundation to build upon. The unique challenges faced in the advanced design of dynamic behavior remain to be described.

The very concept of protein design is somewhat self-limiting. Ultimately, one would desire the ability to engineer a molecule to provide a desired function within a set of constraints, regardless of whether the molecule is a proper protein. As in vitro production of proteins has grown more sophisticated, a reasonable starting point for exploring additional chemical space is the addition of nonnatural amino acids (NAAs). NAAs promise a great deal of additional flexibility to the computational protein design process, such as spectroscopic probes, receptor ligands, new hydrogen bonds, and more exact packing interactions. NAAs need not be limited to altered side chains. β-amino acids (having an additional carbon in the backbone) form regular secondary structures and are highly protease resistant (5). The major barrier to including NAAs in design algorithms is accurate representation in the energy function. However, terms derived from molecular mechanics and quantum mechanics calculations may allow internally consistent models for new residues and for the derivation of rotamer libraries.

NAA: nonnatural amino acid

CONCLUSION

Although computer-based protein design is, arguably, still in its infancy, progress in the field has been significant. Several results suggest that our ability to design proteins from scratch is coupled with our ability to accurately sample conformational space. Progress in high-resolution structure prediction and docking is critical to designing novel protein structures and complexes. Encouragingly, recent studies by Bradley et al. (3) indicate that for many small proteins the accuracy of structure prediction is determined more by the amount of conformational sampling that is performed and less by the details of the energy function. We know that hydrogen bonds should be satisfied and that there should be tight packing between amino acids. The challenge is to find structures and sequences that satisfy these constraints.

ACKNOWLEDGMENT

We thank Xiaozhen Hu of the Kuhlman lab for providing unpublished work for Figure 1.

LITERATURE CITED

1. Baldwin EP, Hajiseyedjavadi O, Baase WA, Matthews BW. 1993. The role of backbone flexibility in the accommodation of variants that repack the core of T4 lysozyme. *Science* 262:1715–18
2. **Bolon DN, Grant RA, Baker TA, Sauer RT. 2005. Specificity versus stability in computational protein design. *Proc. Natl. Acad. Sci. USA* 102:12724–29**

Directly compares positive and negative design for generating specificity and stability.

3. Bradley P, Misura KM, Baker D. 2005. Toward high-resolution de novo structure prediction for small proteins. *Science* 309:1868–71
4. Calhoun JR, Nastri F, Maglio O, Pavone V, Lombardi A, DeGrado WF. 2005. Artificial diiron proteins: from structure to function. *Biopolymers* 80:264–78
5. Cheng RP, Gellman SH, DeGrado WF. 2001. β-Peptides: from structure to function. *Chem. Rev.* 101:3219–32
6. Chevalier BS, Kortemme T, Chadsey MS, Baker D, Monnat RJ, Stoddard BL. 2002. Design, activity, and structure of a highly specific artificial endonuclease. *Mol. Cell* 10:895–905
7. Crick FHC. 1953. The Fourier transform of a coiled-coil. *Acta Crystallogr.* 6:685–89
8. **Dahiyat BI, Mayo SL. 1997. De novo protein design: fully automated sequence selection. *Science* 278:82–87**

The first report of a verified full computational sequence design.

9. Dahiyat BI, Mayo SL. 1997. Probing the role of packing specificity in protein design. *Proc. Natl. Acad. Sci. USA* 94:10172–77
10. DeGrado WF, Wasserman ZR, Lear JD. 1989. Protein design, a minimalist approach. *Science* 243:622–28
11. Desjarlais JR, Handel TM. 1995. De novo design of the hydrophobic cores of proteins. *Protein Sci.* 4:2006–18
12. Desjarlais JR, Handel TM. 1999. Side-chain and backbone flexibility in protein core design. *J. Mol. Biol.* 290:305–18
13. Desmet J, Maeyer MD, Hazes B, Lasters I. 1992. The dead-end elimination theorem and its use in protein side-chain positioning. *Nature* 356:539–41
14. Dunbrack RL Jr, Cohen FE. 1997. Bayesian statistical analysis of protein side-chain rotamer preferences. *Protein Sci.* 6:1661–81
15. **Dwyer MA, Looger LL, Hellinga HW. 2004. Computational design of a biologically active enzyme. *Science* 304:1967–71**

The authors successfully designed a triose-phosphate isomerase enzyme.

16. Gordon DB, Hom GK, Mayo SL, Pierce NA. 2003. Exact rotamer optimization for protein design. *J. Comput. Chem.* 24:232–43
17. Gordon DB, Marshall SA, Mayo SL. 1999. Energy functions for protein design. *Curr. Opin. Struct. Biol.* 9:509–13
18. Guerois R, Nielsen JE, Serrano L. 2002. Predicting changes in the stability of proteins and protein complexes: a study of more than 1000 mutations. *J. Mol. Biol.* 320:369–87
19. **Harbury PB, Plecs JJ, Tidor B, Alber T, Kim PS. 1998. High-resolution protein design with backbone freedom. *Science* 282:1462–67**

Account of the first experimentally verified design of a novel structure, a coiled-coil with a right-handed superhelical twist.

20. Harbury PB, Tidor B, Kim PS. 1995. Repacking protein cores with backbone freedom: structure prediction for coiled coils. *Proc. Natl. Acad. Sci. USA* 92:8408–12
21. Havranek JJ, Harbury PB. 1999. Tanford-Kirkwood electrostatics for protein modeling. *Proc. Natl. Acad. Sci. USA* 96:11145–50
22. **Havranek JJ, Harbury PB. 2003. Automated design of specificity in molecular recognition. *Nat. Struct. Biol.* 10:45–52**

This study used "multi-state" design, integrating negative design into the sequence search, to design coiled-coils.

23. Huang PS, Love JJ, Mayo SL. 2005. Adaptation of a fast Fourier transform-based docking algorithm for protein design. *J. Comput. Chem.* 26:1222–32
24. Jaramillo A, Wernisch L, Hery S, Wodak SJ. 2001. Automatic procedures for protein design. *Comb. Chem. High Throughput Screen.* 4:643–59
25. Jin W, Kambara O, Sasakawa H, Tamura A, Takada S. 2003. De novo design of foldable proteins with smooth folding funnel: automated negative design and experimental verification. *Structure* 11:581–90
26. Keating AE, Malashkevich VN, Tidor B, Kim PS. 2001. Side-chain repacking calculations for predicting structures and stabilities of heterodimeric coiled coils. *Proc. Natl. Acad. Sci. USA* 98:14825–30
27. Koehl P, Delarue M. 1994. Application of a self-consistent mean field theory to predict protein side-chains conformation and estimate their conformational entropy. *J. Mol. Biol.* 239:249–75
28. Koehl P, Levitt M. 1999. De novo protein design. II. Plasticity in sequence space. *J. Mol. Biol.* 293:1183–93
29. Kono H, Saven JG. 2001. Statistical theory for protein combinatorial libraries. Packing interactions, backbone flexibility, and the sequence variability of a main-chain structure. *J. Mol. Biol.* 306:607–28

30. Kortemme T, Baker D. 2004. Computational design of protein-protein interactions. *Curr. Opin. Chem. Biol.* 8:91–97
31. Kortemme T, Morozov AV, Baker D. 2003. An orientation-dependent hydrogen bonding potential improves prediction of specificity and structure for proteins and protein-protein complexes. *J. Mol. Biol.* 326:1239–59
32. **Kraemer-Pecore CM, Lecomte JT, Desjarlais JR. 2003. A de novo redesign of the WW domain. *Protein Sci.* 12:2194–205**
33. Kuhlman B, Baker D. 2000. Native protein sequences are close to optimal for their structures. *Proc. Natl. Acad. Sci. USA* 97:10383–88
34. Kuhlman B, Baker D. 2004. Exploring folding free energy landscapes using computational protein design. *Curr. Opin. Struct. Biol.* 14:89–95
35. **Kuhlman B, Dantas G, Ireton GC, Varani G, Stoddard BL, Baker D. 2003. Design of a novel globular protein fold with atomic-level accuracy. *Science* 302:1364–68**
36. Kuhlman B, O'Neill JW, Kim DE, Zhang KY, Baker D. 2002. Accurate computer-based design of a new backbone conformation in the second turn of protein L. *J. Mol. Biol.* 315:471–77
37. Lacroix E, Kortemme T, Lopez de la Paz M, Serrano L. 1999. The design of linear peptides that fold as monomeric beta-sheet structures. *Curr. Opin. Struct. Biol.* 9:487–93
38. Larson SM, England JL, Desjarlais JR, Pande VS. 2002. Thoroughly sampling sequence space: large-scale protein design of structural ensembles. *Protein Sci.* 11:2804–13
39. Larson SM, Garg A, Desjarlais JR, Pande VS. 2003. Increased detection of structural templates using alignments of designed sequences. *Proteins* 51:390–96
40. Lasters I, Wodak SJ, Alard P, van Cutsem E. 1988. Structural principles of parallel beta-barrels in proteins. *Proc. Natl. Acad. Sci. USA* 85:3338–42
41. Lasters I, Wodak SJ, Pio F. 1990. The design of idealized alpha/beta-barrels: analysis of beta-sheet closure requirements. *Proteins* 7:249–56
42. Lazaridis T, Karplus M. 2000. Effective energy functions for protein structure prediction. *Curr. Opin. Struct. Biol.* 10:139–45
43. Liang S, Grishin NV. 2002. Side-chain modeling with an optimized scoring function. *Protein Sci.* 11:322–31
44. Liang S, Grishin NV. 2004. Effective scoring function for protein sequence design. *Proteins* 54:271–81
45. Lim WA, Hodel A, Sauer RT, Richards FM. 1994. The crystal structure of a mutant protein with altered but improved hydrophobic core packing. *Proc. Natl. Acad. Sci. USA* 91:423–27
46. Lombardi A, Summa CM, Geremia S, Randaccio L, Pavone V, DeGrado WF. 2000. Inaugural article: retrostructural analysis of metalloproteins: application to the design of a minimal model for diiron proteins. *Proc. Natl. Acad. Sci. USA* 97:6298–305
47. Looger LL, Dwyer MA, Smith JJ, Hellinga HW. 2003. Computational design of receptor and sensor proteins with novel functions. *Nature* 423:185–90
48. Looger LL, Hellinga HW. 2001. Generalized dead-end elimination algorithms make large-scale protein side-chain structure prediction tractable: implications for protein design and structural genomics. *J. Mol. Biol.* 307:429–45
49. Lovell SC, Word JM, Richardson JS, Richardson DC. 2000. The penultimate rotamer library. *Proteins* 40:389–408
50. Malakauskas SM, Mayo SL. 1998. Design, structure and stability of a hyperthermophilic protein variant. *Nat. Struct. Biol.* 5:470–75

The authors redesigned a small β-sheet motif.

Reports the first successful design of a globular protein with a fold not previously observed in nature.

51. Mendes J, Guerois R, Serrano L. 2002. Energy estimation in protein design. *Curr. Opin. Struct. Biol.* 12:441–46
52. Morozov AV, Kortemme T, Tsemekhman K, Baker D. 2004. Close agreement between the orientation dependence of hydrogen bonds observed in protein structures and quantum mechanical calculations. *Proc. Natl. Acad. Sci. USA* 101:6946–51
53. Murzin AG, Lesk AM, Chothia C. 1994. Principles determining the structure of beta-sheet barrels in proteins. I. A theoretical analysis. *J. Mol. Biol.* 236:1369–81
54. Murzin AG, Lesk AM, Chothia C. 1994. Principles determining the structure of beta-sheet barrels in proteins. II. The observed structures. *J. Mol. Biol.* 236:1382–400
55. Offredi F, Dubail F, Kischel P, Sarinski K, Stern AS, et al. 2003. De novo backbone and sequence design of an idealized alpha/beta-barrel protein: evidence of stable tertiary structure. *J. Mol. Biol.* 325:163–74
56. Park S, Yang X, Saven JG. 2004. Advances in computational protein design. *Curr. Opin. Struct. Biol.* 14:487–94
57. Plecs JJ, Harbury PB, Kim PS, Alber T. 2004. Structural test of the parameterized-backbone method for protein design. *J. Mol. Biol.* 342:289–97
58. Pokala N, Handel TM. 2001. Review: protein design—where we were, where we are, where we're going. *J. Struct. Biol.* 134:269–81
59. Pokala N, Handel TM. 2004. Energy functions for protein design. I. Efficient and accurate continuum electrostatics and solvation. *Protein Sci.* 13:925–36
60. Pokala N, Handel TM. 2005. Energy functions for protein design: adjustment with protein-protein complex affinities, models for the unfolded state, and negative design of solubility and specificity. *J. Mol. Biol.* 347:203–27
61. Ponder JW, Richards FM. 1987. Tertiary templates for proteins. Use of packing criteria in the enumeration of allowed sequences for different structural classes. *J. Mol. Biol.* 193:775–91
62. Reina J, Lacroix E, Hobson SD, Fernandez-Ballester G, Rybin V, et al. 2002. Computer-aided design of a PDZ domain to recognize new target sequences. *Nat. Struct. Biol.* 9:621–27
63. Richardson JS, Richardson DC. 1989. The de novo design of protein structures. *Trends Biochem. Sci.* 14:304–9
64. Rohl CA, Strauss CE, Misura KM, Baker D. 2004. Protein structure prediction using Rosetta. *Methods Enzymol.* 383:66–93
65. Ross SA, Sarisky CA, Su A, Mayo SL. 2001. Designed protein G core variants fold to native-like structures: Sequence selection by ORBIT tolerates variation in backbone specification. *Protein Sci.* 10:450–54
66. Saunders CT, Baker D. 2005. Recapitulation of protein family divergence using flexible backbone protein design. *J. Mol. Biol.* 346:631–44
67. Shifman JM, Mayo SL. 2003. Exploring the origins of binding specificity through the computational redesign of calmodulin. *Proc. Natl. Acad. Sci. USA* 100:13274–79
68. Shimaoka M, Shifman JM, Jing H, Takagi J, Mayo SL, Springer TA. 2000. Computational design of an integrin I domain stabilized in the open high affinity conformation. *Nat. Struct. Biol.* 7:674–78
69. Sibanda BL, Thornton JM. 1985. Beta-hairpin families in globular proteins. *Nature* 316:170–74
70. Simons KT, Kooperberg C, Huang E, Baker D. 1997. Assembly of protein tertiary structures from fragments with similar local sequences using simulated annealing and Bayesian scoring functions. *J. Mol. Biol.* 268:209–25

71. Simons KT, Ruczinski I, Kooperberg C, Fox BA, Bystroff C, Baker D. 1999. Improved recognition of native-like protein structures using a combination of sequence-dependent and sequence-independent features of proteins. *Proteins* 34:82–95
72. Slovic AM, Kono H, Lear JD, Saven JG, DeGrado WF. 2004. Computational design of water-soluble analogues of the potassium channel KcsA. *Proc. Natl. Acad. Sci. USA* 101:1828–33
73. Smith GR, Sternberg MJ. 2002. Prediction of protein-protein interactions by docking methods. *Curr. Opin. Struct. Biol.* 12:28–35
74. Street AG, Mayo SL. 1998. Pairwise calculation of protein solvent-accessible surface areas. *Fold. Des.* 3:253–58
75. Su A, Mayo SL. 1997. Coupling backbone flexibility and amino acid sequence selection in protein design. *Protein Sci.* 6:1701–7
76. Summa CM, Lombardi A, Lewis M, DeGrado WF. 1999. Tertiary templates for the design of diiron proteins. *Curr. Opin. Struct. Biol.* 9:500–8
77. **Summa CM, Rosenblatt MM, Hong JK, Lear JD, DeGrado WF. 2002. Computational de novo design, and characterization of an A(2)B(2) diiron protein. *J. Mol. Biol.* 321:923–38**

Reports the successful design of an A_2B_2 four-helix-bundle diiron protein.

77a. Ventura S, Serrano L. 2004. Designing proteins from the inside out. *Proteins* 56:1–10
78. Voigt CA, Gordon DB, Mayo SL. 2000. Trading accuracy for speed: a quantitative comparison of search algorithms in protein sequence design. *J. Mol. Biol.* 299:789–803
79. Weber PC, Salemme FR. 1980. Structural and functional diversity in 4-alpha-helical proteins. *Nature* 287:82–84
80. Wernisch L, Hery S, Wodak SJ. 2000. Automatic protein design with all atom force-fields by exact and heuristic optimization. *J. Mol. Biol.* 301:713–36
81. Wisz MS, Hellinga HW. 2003. An empirical model for electrostatic interactions in proteins incorporating multiple geometry-dependent dielectric constants. *Proteins* 51:360–77
82. Wollacott AM, Desjarlais JR. 2001. Virtual interaction profiles of proteins. *J. Mol. Biol.* 313:317–42

Lessons from Lactose Permease

Lan Guan[1] and H. Ronald Kaback[1,2,3]

[1]Departments of Physiology, and [2]Microbiology, Immunology & Molecular Genetics, [3]Molecular Biology Institute, University of California, Los Angeles, California 90095-1662; email: LanGuan@mednet.ucla.edu; RKaback@mednet.ucla.edu

Annu. Rev. Biophys. Biomol. Struct. 2006. 35:67–91

First published online as a Review in Advance on December 12, 2005

The *Annual Review of Biophysics and Biomolecular Structure* is online at biophys.annualreviews.org

doi: 10.1146/annurev.biophys.35.040405.102005

1056-8700/06/0609-0067$20.00

Key Words

membrane transport, membrane protein structure, transport mechanism

Abstract

An X-ray structure of the lactose permease of *Escherichia coli* (LacY) in an inward-facing conformation has been solved. LacY contains N- and C-terminal domains, each with six transmembrane helices, positioned pseudosymmetrically. Ligand is bound at the apex of a hydrophilic cavity in the approximate middle of the molecule. Residues involved in substrate binding and H^+ translocation are aligned parallel to the membrane at the same level and may be exposed to a water-filled cavity in both the inward- and outward-facing conformations, thereby allowing both sugar and H^+ release directly into either cavity. These structural features may explain why LacY catalyzes galactoside/H^+ symport in both directions utilizing the same residues. A working model for the mechanism is presented that involves alternating access of both the sugar- and H^+-binding sites to either side of the membrane.

Contents

INTRODUCTION

The mechanism of energy transduction in biological membranes is an important, fascinating problem. It has been recognized for some time that the driving force for a variety of seemingly unrelated phenomena (e.g., secondary active transport, oxidative phosphorylation, and rotation of the bacterial flagellar motor) is a bulk-phase, transmembrane electrochemical ion gradient. However, insight into the molecular mechanisms by which free energy stored in such gradients is transduced into work or into chemical energy has just begun. On the other hand, gene sequencing and analyses of deduced amino acid sequences indicate that many biological machines involved in energy transduction—membrane transport in particular—fall into families encompassing proteins from archaea to the mammalian central nervous system (**http://www.tcdb.org/**), thereby raising the possibility that the members may have common basic structural features and mechanisms. In addition, many of these proteins play important roles in human disease (e.g., cystic fibrosis, resistance to antibiotics and chemotherapeutic drugs, gastric ulcer, glucose/galactose malabsorption and some forms of drug abuse), as well as the mechanism of action of a number of drugs.

Transport involves substrate-specific membrane proteins that catalyze equilibration and/or uphill translocation of solute across a membrane. These proteins are called a variety of interchangeable names—symporters, cotransporters, transporters, carriers, or permeases. Three main categories of systems are involved in active transport, and each utilizes a distinct energy source: (*a*) The phospho*enol*pyruvate:sugar phosphotransferase system (PTS) is a multicomponent system that catalyzes vectorial phosphorylation of various sugars and sugar alcohols in certain bacteria exclusively (40, 76). Thus, in *Escherichia coli*, for example, glucose and certain other sugars and sugar alcohols are translocated across the membrane by PTS-catalyzed phosphorylation (i.e., glucose appears on the cytoplasmic side of the membrane as glucose-6-phosphate). The glucose PTS also plays a key role in the phenomenon of catabolite repression. (*b*) Transporters of the ATP-binding cassette (ABC) system (19) are found in both prokaryotes and eukaryotes and have a common global organization with two integral membrane components, each of which has multiple-transmembrane helices, and two cytoplasmic components, each of which has one ATP-binding cassette (19, 61). ABC transporters may be in the form of a monomer or various combinations of fused components. These systems utilize the energy released from the hydrolysis of ATP to drive accumulation or efflux unidirectionally against a concentration gradient. Many bacterial ABC systems also require a binding protein on the outside surface of the membrane. (*c*) Transporters that utilize an electrochemical ion gradient (46, 120) are also known as secondary transporters. Primary systems consist of pumps that generate an electrochemical ion gradient. Most ion-coupled transport proteins are composed of 12 to 14 membrane-spanning helices (**http://www.tcdb.org/**), and for those studied intensively, a single polypeptide utilizes the free energy released from the energetically downhill movement of a cation (mainly H^+ or Na^+) in response to an electrochemical ion gradient to catalyze transport of substrate against concentration gradient. However, neurotransmitter reuptake transporters often utilize anions, as well as cations (51, 70). Unlike the PTS, with ion-gradient-driven transport, substrate enters the cell in an unchanged form. Mechanistically and energetically, only the ion-gradient-driven permeases catalyze solute/ion symport in both directions across the membrane (influx and efflux), unlike the other systems described.

With regard to ion-gradient-driven permeases, the chemiosmotic hypothesis of Peter Mitchell (66–68) has been supported strongly by the experiments of West (112) and West & Mitchell (113, 114) and demonstrated quantitatively in bacterial membrane vesicles (41–43). Thus, accumulation of a wide variety of solutes against a concentration gradient is driven by an electrochemical H^+ gradient ($\Delta\bar{\mu}_{H^+}$) composed of an electrical component ($\Delta\Psi$; interior negative) and/or a pH gradient (ΔpH; interior alkaline). This review focuses on the lactose permease of *E. coli* (LacY), a galactoside/H^+ symporter, as this membrane protein is arguably the most intensively studied secondary transporter.

BACKGROUND

The existence of a transport system for galactosidic sugars was first inferred in 1955 by Cohen & Rickenberg (10) and subsequently found to be part of by the famous *lac* operon (69). In addition to regulatory loci, the *lac* operon contains three structural genes: (*a*) the *Z* gene encoding β-galactosidase, a cytosolic enzyme that catalyzes cleavage of lactose into galactose and glucose once it enters the cell; (*b*) the *Y* gene encoding LacY, the galactoside/H^+ symporter or lactose permease; and (*c*) the *A* gene encoding galactoside transacetylase, a cytosolic enzyme that catalyzes acetylation of mainly thio-β-D-galactopyranosides with acetyl-CoA as the acetyl donor and has an unknown physiological function.

The *lacY* gene was the first gene encoding a membrane transport protein to be cloned into a recombinant plasmid (100) and sequenced (6). This success in the early days of molecular biology opened the study of secondary active transport at the molecular level. LacY has been used as a paradigm for secondary transport proteins to explore the mechanism of energy transduction. Overexpression of *lacY* was combined with use of a highly specific photoaffinity probe (49) and functional reconstitution into proteoliposomes (21, 71, 72, 107). It was then shown (62, 107) that LacY catalyzes all the translocation reactions typical of the transport system in vivo with comparable turnover numbers. Thus, the product of the *lacY* gene is responsible solely for all the translocation reactions catalyzed by LacY.

LacY belongs to the major facilitator superfamily (MFS) (**http://www.tcdb.org/**), a large group of transport proteins thought to be evolutionarily related. LacY is selective for disaccharides containing a D-galactopyranosyl ring, as well as D-galactose, but has no affinity for D-glucopyranosides or D-glucose (92). LacY carries out the coupled stoichiometric translocation of a galactoside with an H^+ (i.e., galactoside/H^+ symport), utilizing the free energy released from downhill translocation of H^+ to drive accumulation of galactosides against a concentration gradient (**Figure 1*a***). Notably, in the absence of $\Delta\bar{\mu}_{H^+}$, LacY also catalyzes the converse reaction, utilizing free energy released from downhill translocation of sugar to drive uphill translocation of H^+ with generation of

Electrochemical H^+ gradient ($\Delta\bar{\mu}_{H^+}$): when two aqueous phases are separated by a membrane, the electrochemical potential difference of H^+ between the two phases is expressed as $\Delta\bar{\mu}_{H^+}/F = \Delta\Psi - 2.3RT/F\Delta pH$

$\Delta\Psi$: the electrical potential across the membrane

ΔpH: the pH gradient across the membrane

LacY: lactose permease

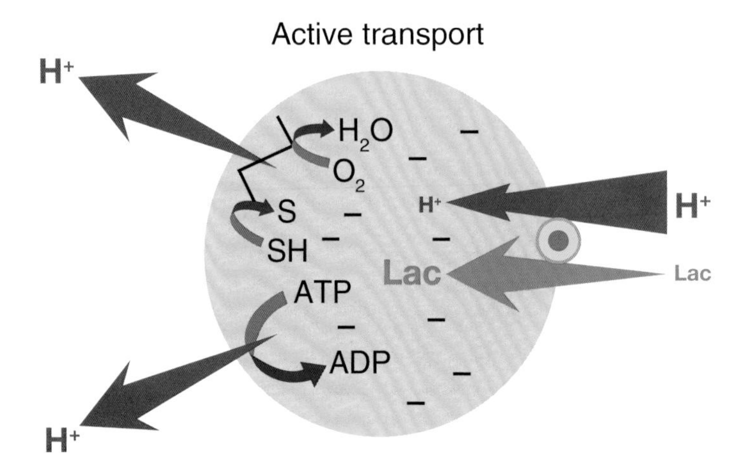

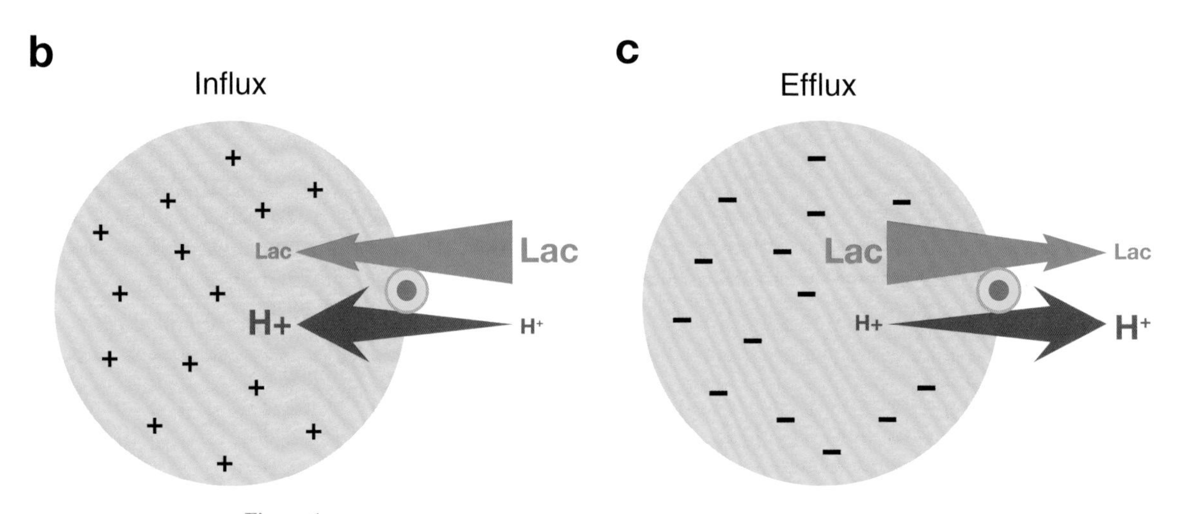

Figure 1

Lactose/H^+ symport. In the absence of substrate, LacY does not translocate H^+ in the presence of $\Delta\bar{\mu}_{H^+}$. (*a*) Free energy released from the downhill movement of H^+ is coupled to the uphill accumulation of lactose. (*b*, *c*) Substrate gradients generate electrochemical H^+ gradients, the polarity of which depends upon the direction of the substrate concentration gradient.

$\Delta\bar{\mu}_{H^+}$, the polarity of which depends upon the direction of the substrate concentration gradient (**Figure 1*b,c***). In the absence of substrate, LacY does not translocate H^+; however, LacY catalyzes exchange or counterflow of sugar without translocation of H^+. Because substrate gradients by themselves generate a $\Delta\bar{\mu}_{H^+}$ of either polarity, it seems intuitively likely that the primary driving force for turnover is binding and dissociation of sugar on either side of the membrane.

To obtain insight into the mechanism of transport, it is essential to identify side chains that are crucial, to delineate their function and relationship to one another, and to obtain structural and dynamic information. The use of molecular biological techniques to engineer LacY for site-directed biochemical and biophysical approaches has provided important information about mechanism, as well as structure, and these methods have been applied to the study of many other membrane proteins. Moreover, an X-ray structure of LacY has been solved recently to a resolution of 3.5 Å (2). The structure provides critical information regarding the overall fold, the sugar-binding site, and the position of residues involved in H^+ translocation. Importantly, the structure confirms many conclusions derived from biochemical and biophysical studies and also reveals a number of unexpected, novel findings. Remarkably, at the same time, the structure of another member of the MFS, the phosphate/glycerol-3-P antiporter (GlpT), was solved at 3.3 Å (59). The folds of LacY and GlpT are similar (1, 59), supporting the notion that all members of the MFS may have similar structures.

LacY IS FUNCTIONAL AS A MONOMER

One particularly difficult problem to resolve with hydrophobic membrane proteins is their functional oligomeric state. Although indirect early evidence (43) suggested that oligomerization might be important for activity, LacY was shown to be functional as a monomer by rotational diffusion measurements with eosinylmaleimide-labeled LacY or by freeze-fracture electron microscopy. Using the former approach, Dornmair et al. (15) utilized fluorescence anisotropy to demonstrate that purified, reconstituted *LacY* exhibits the rotational diffusion constant of a 46.5-kDa particle and that the diffusion constant is not altered in the presence of $\Delta\bar{\mu}_{H^+}$. Regarding the microscopic approach, purified LacY was reconstituted into proteoliposomes under conditions where the protein is fully functional and shown to be a monomer in the absence or presence of $\Delta\bar{\mu}_{H^+}$ (13). Moreover, the initial rate of $\Delta\bar{\mu}_{H^+}$-driven lactose transport in proteoliposomes varies linearly with the ratio of LacY to phospholipid. If more than a single molecule of LacY is required for active lactose transport, a sigmoidal relationship should be observed, particularly at low LacY/phospholipid ratios.

Notwithstanding evidence that LacY is functional as a monomer, certain paired in-frame deletion mutants complement functionally (4). Although cells expressing the deletions individually do not catalyze active transport, cells simultaneously expressing specific pairs of deletions catalyze transport up to 60% as well as cells expressing wild-type LacY, and it is clear that the phenomenon occurs at the protein and *not* at the DNA level. Remarkably, complementation is observed only with pairs of LacY molecules containing large deletions separated by at least two hypothetical transmembrane helices and not with missense mutations or point deletions. Although the mechanism of complementation is unclear, it is likely related to the phenomenon whereby independently expressed N- and C-terminal fragments of LacY interact to form a functional complex (3, 121, 122). In any case, the observation that certain pairs of deletion mutants complement functionally rekindled concern regarding the functional oligomerization state of wild-type LacY.

To resolve the problem, Sahin-Tóth et al. (91) engineered a fusion protein that contains two LacY molecules covalently linked in tandem (LacY dimer). The covalently linked LacY dimer is inserted into the membrane in a functional state, and negative dominance is not observed by either mutation or chemical modification of either half of the dimer. In order to test the caveat that oligomerization between dimers might account for the findings, a LacY dimer containing different deletion mutants, which complement each other when expressed as untethered molecules, did not catalyze lactose accumulation. Therefore, it is unlikely that LacY dimers oligomerize.

Finally, single-Cys replacement mutants in the transmembrane helices or loops of LacY dimerize in a stochastic manner (18, 31, 99). Moreover, the asymmetric unit of the LacY crystal is composed of an artificial dimer with two molecules oriented in opposite directions (2). Taken as a whole, the results argue strongly that LacY functions as a monomer.

Second-site suppressor analysis: involves selection of a second mutation that corrects (suppresses) the phenotype of an original inactivation mutation, which can be done by random or site-directed mutagenesis

PRIMARY AND SECONDARY STRUCTURE

Primary

LacY is composed of 417 amino acid residues and has a molecular mass of 46,517 Da. However, like most hydrophobic membrane proteins, LacY electrophoreses with a relative molecular mass of ~33,000 Da. Electrospray ionization–mass spectrometry (ESI-MS) has been applied successfully to LacY (54, 57, 111, 115, 116), as well as other hydrophobic membrane proteins. The molecular weight reconstruction from ESI-MS of LacY with a 6-His affinity tag at the C terminus reveals that the purified protein is homogeneous and that the computed mass is within 0.01% of that calculated from the DNA sequence with a formyl group on the initiating methionine. Although the formyl group is normally removed from native LacY (17), overexpression may saturate the deformylase.

Secondary

The α-helical content of LacY was examined by circular dichroism (85%–90%) (20), laser Raman (~70%) (108), and attenuated total reflectance–Fourier transform spectroscopy (ATR-FTIR) (~70%) (56, 75). Interestingly, relative to the X-ray structure, which has a helical content of 86% (80% within the membrane), circular dichroism is the most accurate of the spectroscopic methods used.

Hydropathy analysis was first applied, predicting 12 hydrophobic transmembrane domains (20). The number and orientation of each individual transmembrane domain was studied experimentally by *phoA* fusion analysis (7), which is consistent with 12 transmembrane domains traversing the membrane in zigzag fashion connected by relatively hydrophilic loops with both N and C termini on the cytoplasmic face. The overall topology model has been confirmed by the three-dimensional structure (**Figure 2**).

Generally, the boundaries of the transmembrane domains are difficult to predict. With LacY specifically, the secondary structure model was modified because of the identification of charge pairs between certain helices. Initially, hydropathy analysis placed Asp-237 in loop VII/VIII. However, second-site suppression analyses (52) indicated that Asp-237 interacts with Lys-358 (helix XI), a conclusion supported strongly by site-directed mutagenesis, chemical modification (16, 86), and site-directed spin labeling (101). In addition, Asp-240, which was also originally placed in loop VII/VIII, is charge-paired with Lys-319 in helix X (16, 58, 86). Therefore, the whole region from Cys-234 to Phe-247 was moved into transmembrane VII, which approximates the X-ray structure very well (2). Glu-126 and Arg-144 were also initially placed in loop IV/V. However, studies utilizing single amino acid deletions (117, 119), as well as site-directed spin labeling (124), indicate that both residues are in the helices IV and V.

Helix Packing and Functionally Irreplaceable Side Chains

Use of molecular biology approaches to engineer LacY for site-directed biochemical and biophysical studies has provided important information about its structure and mechanism (46). A possible helix packing model was proposed from about 100 distance constraints obtained from thiol cross-linking experiments and engineered Mn(II)-binding sites (97). Additional methods that approximate distance were also utilized qualitatively, among which are site-directed mutagenesis with the explicit purpose of engineering LacY for

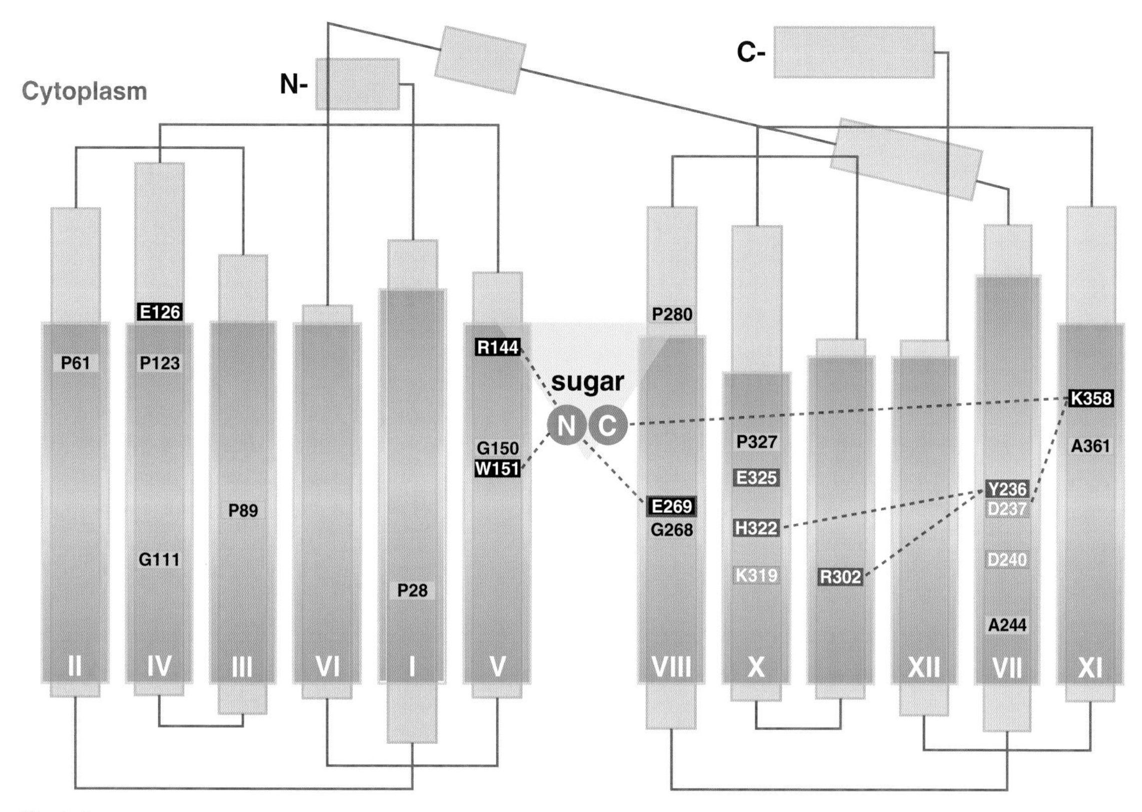

Figure 2

Secondary structure model of LacY derived from the X-ray structure. The helices (*rectangles*) traverse the membrane in a zigzag fashion connected by hydrophilic domains external to the membrane. The helices are drawn to scale according to the X-ray structure, and dark green areas represent the portions of the helices that are within the membrane. The loops depict connectivity only. Residues at the kinks in the transmembrane helical domains are shown as light blue rectangles; residues in black rectangles are involved in substrate binding; residues in red are involved in H^+ translocation. Glu-269 (*black rectangle bordered in red*) is involved in both substrate binding and H^+ translocation. Salt-bridged residues are shown as orange rectangles. The large hydrophilic cavity is designated by an inverted yellow triangle, and sugar is depicted by two green circles, with N and C representing those moieties that interact with the N- and C-terminal halves of LacY, respectively.

site-directed biochemical and biophysical techniques. In most instances, construction of mutants begins with a cassette *lacY* gene (EMBL X-56,095) containing a unique restriction site about every 100 bp that encodes LacY with a single Cys residue at virtually every position (cysteine-scanning mutagenesis) (27). With a library of mutants containing a single Cys residue at almost every position, it is a simple operation to construct paired Cys replacement mutants by restriction fragment replacement. Furthermore, by inserting the biotin acceptor domain from a *Klebsiella pneumoniae* oxalacetate decarboxylase or 6 to 10 His residues in either the middle cytoplasmic loop or at the C terminus, the mutant proteins are readily purified in a single step by monovalent avidin affinity or metal affinity chromatography (12, 96), respectively. Most of the methodology involves

Cysteine-scanning mutagenesis: the use of site-directed mutagenesis to replace individual residues in a given protein, or a region of that protein, with cysteine

Excimer fluorescence: an excited-state dimer observed in the fluorescence-emission spectrum, which emits at a higher wavelength than the monomer fluorescence

Divalent metal-binding site: the simplest consists of two imidazole side chains (His residues) within close proximity; Mn(II) binding is then measured directly by electron paramagnetic resonance

Hydrogen/deuterium exchange: the rate at which backbone amide protons exchange with deuterium

site-directed measurements—pyrene excimer fluorescence (38), engineered divalent metal-binding sites (paired His residues) (34, 35, 39), electron paramagnetic resonance (11), and thiol cross-linking (47). Although important global aspects of the structure are not revealed (e.g., the large inward-facing hydrophilic cavity or helix packing at the cytoplasmic face; see below), many of the local interactions (29, 32) are detected by using the techniques described. In view of the X-ray structure of LacY, it is not surprising that global aspects of the structure are not revealed by these approaches. However, many of the techniques are useful for dynamic information, which is difficult to obtain from a crystal structure.

Functional analysis of mutants at virtually every position has led to the following three observations (27, 46, 84). (*a*) Only six side chains are irreplaceable with respect to active transport: Glu-126 (helix IV) and Arg-144 (helix V), which are crucial for substrate binding; Glu-269 (helix VIII), which is likely involved in both substrate binding and H^+ translocation; and Arg-302 (helix IX), His-322 (helix X), and Glu-325 (helix X), which play irreplaceable roles in H^+ translocation. (*b*) Residues such as Trp-151 and Tyr-236 are important, but not irreplaceable. (*c*) Substrate-induced changes in the reactivity of side chains with various chemical modification reagents, site-directed fluorescence, and spin labeling suggest widespread conformational changes during turnover.

TERTIARY STRUCTURE

The first X-ray structure of LacY was obtained with a thermostable mutant (C154G) that binds sugar as well as the wild-type, but exhibits little or no transport activity (64, 96). Therefore it was surmised that the C154G mutant might be suitable for crystallization because it strongly favors a single conformation. This notion came to fruition, and a crystal structure was obtained in collaboration with Jeff Abramson and So Iwata at Imperial College London (2).

In a side view (**Figure 3*a***), the monomer is heart-shaped with an internal cavity open on the cytoplasmic side, and the largest dimensions of the molecule are 60 × 60 Å. In a cytoplasmic view (**Figure 3*b***), the molecule has a distorted oval shape, with dimensions of 30 × 60 Å. Remarkably, the structure features a large interior hydrophilic cavity, which is open only to the cytoplasmic side, with dimensions of 25 × 15 Å, indicating that the structure represents the inward-facing conformational state of LacY (**Figure 3*a,b***). Within the cavity, a single sugar-binding site is observed at the apex of the cavity in the approximate middle of the molecule.

The monomer has 12 transmembrane helices as described above. The molecule is organized as two six-helix bundles connected by a long loop between helices VI and VII (loop VI-VII, **Figure 2**). The N- and C-terminal six-helix domains have the same topology and exhibit twofold pseudosymmetry. Similar to GlpT (59) and OxlT (17), within each domain, twofold pseudosymmetry is observed, indicating that the N- and C-terminal domains have the same genetic origin, as postulated by Pao et al. (74). The hydrophilic cavity is lined with helices I, II, IV, and V of the N-terminal domain and helices VII, VIII, X, and XI of the C-terminal domain. Helices III, VI, IX, and XII are embedded largely in the bilayer and not exposed to solvent, as suggested by Guan et al. (31).

Many of the helices are distorted (**Figure 3**). The distorted helices and the large hydrophilic cavity provide a ready explanation for the high rate of backbone hydrogen/deuterium exchange observed with LacY, as determined by ATR-FTIR (90% within 10 to 20 min relative to ~50% in 2 to 3 h with the potassium channel SliK) (55). In addition to large-scale conformation changes, the H-bonds should be unstable in distorted helices and the hydrophilic cavity should allow access of deuterium to much of the backbone amide groups. Therefore, it is unlikely that the two six-helix domains move as rigid bodies.

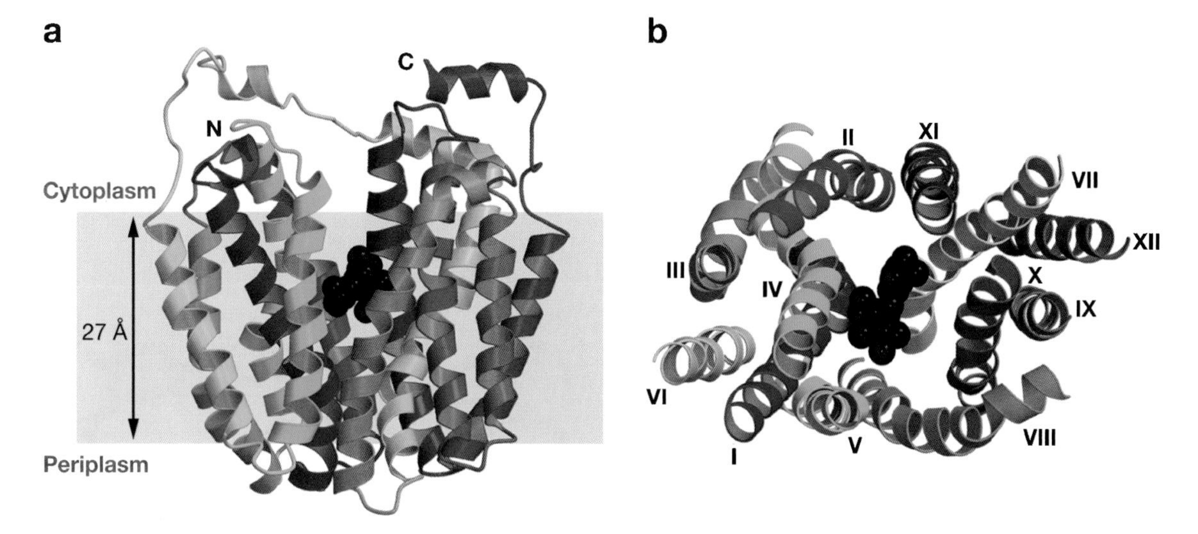

Figure 3

The figure is based on the C154G mutant structure with bound TDG. (*a*) Ribbon representation of LacY viewed parallel to the membrane. The 12 transmembrane helices from the N and C termini are colored from dark blue (N-terminal bundles) to red (C-terminal bundles), and TDG is represented by black spheres. (*b*) Ribbon representation of LacY viewed along the membrane normal from the cytoplasmic side. For clarity, the loop regions have been omitted. The color scheme is the same as in panel (*a*), and the 12 transmembrane helices are labeled with Roman numerals. Reproduced, with permission, from Reference 2.

Most recently, wild-type LacY has yielded crystals after more than a decade of effort (L. Guan & H.R. Kaback, unpublished data). Although the crystals of wild-type LacY are difficult to obtain because of rapid aggregation, a structure of unmodified, native LacY was obtained at a resolution of 4 Å. Importantly, the overall fold of wild-type LacY exhibits an inward-facing conformation and is similar to that of the C154G mutant. Because wild-type GlpT and LacY, as well as C154G LacY, have similar overall folds and are in the inward-facing conformation, this conformation likely reflects the lowest free-energy form. However, it is not clear whether this is the preferred form in the membrane or in a crystal lattice. In any event, the hydrophilic cavity is sufficiently large that it is visualized by freeze-fracture electron microscopy of proteoliposomes reconstituted with purified LacY (13, 14) or in filamentous arrays of LacY (60).

THE SUGAR-BINDING SITE

Indirect Studies

NEM: *N*-ethylmaleimide

Galactoside transport is inactivated by *N*-ethylmaleimide (NEM), and LacY can be selectively labeled with radioactive NEM by substrate protection against alkylation (22). The substrate-protected residue was shown to be Cys-148 (5), thus providing initial evidence that the substrate-binding site may involve helix V. However, when Cys-148 is replaced with various side chains, neither binding nor active transport is abolished (37). From the X-ray structure (see below), it is apparent that Cys-148 is near the galactosyl end of the ligand but does not make direct contact with sugar (**Figure 4**). Thus the inactivating effect of alkylation is due to a steric effect, and conversely, binding of a galactoside protects against alkylation. A similar steric effect is observed when Ala-122 (helix IV) is replaced with Cys. However, it is remarkable that

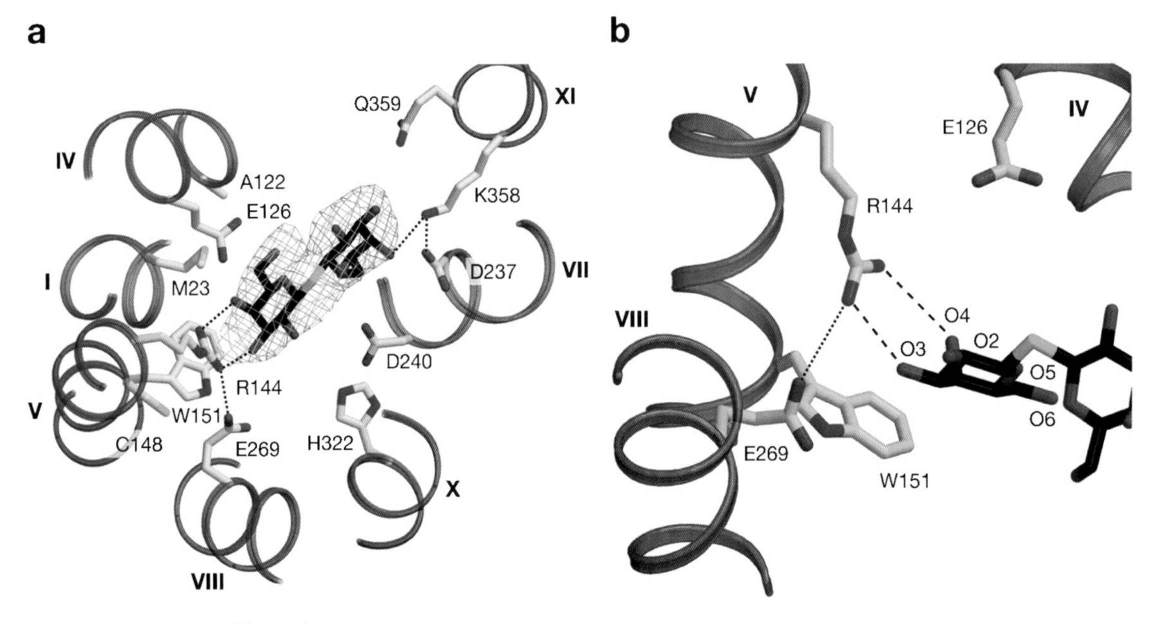

Figure 4

Substrate-binding site of LacY. Possible H-bonds and salt bridges are represented by broken blue lines. (*a*) Residues involved in TDG binding viewed along the membrane normal from the cytoplasmic side. TDG is depicted as a stick model. (*b*) Close up of the N-terminal domain of the TDG-binding site. Reproduced, with permission, from Reference 2.

NPG: *p*-nitrophenyl α-D-galactopyranoside

TDG: β-D-galactopyranosyl 1-thio-β-D-galactopyranoside

alkylation of mutant A122C causes LacY to become specific for binding and transport of the monosaccharide galactose (32, 53).

Cysteine-scanning mutagenesis reveals that replacement of Glu-126 (helix IV) or Arg-144 (helix V) with neutral amino acyl side chains abolishes transport, and activity is not observed with double-neutral substitutions or when the residues are interchanged (24). Only mutants E126D and R144K exhibit any activity whatsoever. Mutant E126D accumulates lactose at a lower rate to a normal steady state, whereas mutant R144K transports lactose at a negligible rate to a miniscule steady state (24). In addition, lactose-induced H^+ translocation is observed at a slow rate with E126D permease, but not with any of the other Glu-126 or Arg-144 mutants.

Glu-126 and Arg-144 are postulated to form a charge pair. Replacement of either residue with Ala in a LacY mutant containing a single-Cys residue at position 148 markedly decreases NEM labeling of Cys-148, but the double-Ala mutant labels normally. Thus, an unpaired charge causes a conformational perturbation that decreases the reactivity of Cys-148, whereas double-neutral replacement with Ala, replacement of Arg-144 with Lys, or interchanging Glu-126 and Arg-144 has no such effect (105). Further evidence for charge pairing was obtained by spontaneous disulfide cross-linking (118) and site-directed spin labeling (125). Direct binding assays show that none of the Arg-144 mutants, including R144K, binds lactose or the high-affinity substrate analogue *p*-nitrophenyl α-D-galactopyranoside (NPG), and that neutral replacements for Glu-126 do not bind NPG or β-D-galactopyranosyl 1-thio-β-D-galactopyranoside (TDG), but mutant E126D exhibits significant binding of TDG (93). In addition, there is no substrate protection against NEM labeling when Glu-126 and Arg-144 are interchanged. As a whole, the results demonstrate that a carboxyl group at position 126 and a guanidino group at position 144 are absolute requirements for substrate binding and suggest that the two residues may be charge-paired. Recent studies using ESI-MS also support the interaction of Arg-144 and Glu-126 by covalent modification of the guanidino group with the Arg-specific reagent butane 2,3-dione (BD) (109). The reactivity of Arg-144 with BD is low and reduced further in the presence of ligand. Interestingly, replacement of Glu-126 with Ala results in an increase in the reactivity of Arg-144, consistent with a charge pair between Arg-144 and Glu-126 in the absence of sugar that is disrupted upon ligand binding.

Trp-151, two turns of helix V from Arg-144, also plays an important role in substrate binding (29), although mutants W151Y and W151F catalyze active lactose transport with time courses similar to those of the wild-type. Mutant W151F or W151Y binds NPG and TDG relatively poorly, but surprisingly, there is relatively little change in the kinetics of lactose transport. In addition, amino acid replacements with an alkyl side chain exhibit little or no transport and no significant binding affinity.

The fluorescent properties of Trp-151 from a fully functional mutant devoid of all other Trp residues were investigated (103). The steady-state fluorescence spectrum of Trp-151 and fluorescence quenching experiments with water-soluble quenchers demonstrate that Trp-151 is in a hydrophilic environment. Furthermore, substrate binding leads to a blue shift in the fluorescence spectrum and reduction in accessibility to polar quenchers, indicating that Trp-151 becomes less exposed to aqueous solvent. In addition, the phosphorescence spectrum of Trp-151 is red-shifted in the presence of substrate, indicating a direct stacking interaction between the galactopyranosyl and indole rings. Finally, studies with N-bromosuccinimide (NBS) show that in the presence of ligand reaction with Trp-151 is protected (104).

Glu-269 (helix VIII), another irreplaceable residue, is also critical for substrate recognition and may be involved in H^+ translocation as well (see below) (23, 33, 90, 102, 105, 110). Neutral replacements abolish lactose transport, and replacement with Asp leads to decreased affinity for TDG or NPG. Significant transport of TDG is observed, and there is about a threefold increase in H^+/TDG stoichiometry (23, 102). Furthermore, Glycine-scanning mutagenesis of mutant E269D shows that it can be rescued with respect to binding and all translocation reactions, suggesting that positioning of the carboxyl group at position 269 in the binding site is critical (110). Analysis of the CNBr fragment containing Glu-269 by ES-MSI demonstrates that this carboxyl group is protected by substrate against reaction with hydrophobic carbodiimides. Together with other evidence, the findings suggest that Glu-269 may H-bond with the O_3 of the D-galactopyranosyl ring (111).

Direct Evidence from the Crystal Structure

Visualization of the substrate-binding site in the X-ray structure demonstrates a primary interaction between the irreplaceable residue Arg-144 (helix V) and the O_3 and O_4 atoms of the galactopyranosyl ring via a bidentate H-bond (2), as suggested by the biochemical findings discussed above (**Figure 4**). The irreplaceable residue Glu-126 (helix IV) is in proximity to Arg-144 and may interact with the O_4, O_5, or O_6 atoms of the galactopyranosyl ring via water molecules. A direct interaction between Arg-144 and Glu-126 is not observed in the ligand-bound structure, consistent with the idea that the salt bridge is absent in the presence of ligand (105, 109). Satisfyingly, a hydrophobic interaction between the bottom of the galactopyranosyl ring and the indole ring of Trp-151 (helix V) is confirmed. The C-6 atom of the galactopyranosyl ring also appears to interact hydrophobically with Met-23 (helix I), but mutagenesis of Met-23 has no effect on ligand binding (I. Smirnova & H. R. Kaback, unpublished data).

The binding site in the N-terminal domain is similar to those of many other galactoside- and sugar-binding proteins (69, 75, 95). Glu-269 in helix VIII in the C-terminal domain appears to form a salt bridge with Arg-144 and is in close proximity to Trp-151. Fluorescence studies with NBS suggest an H-bond between the carboxyl group at position 269 and the indole N of Trp-151 in the absence of ligand (104). Thus, contacts between Glu-269 in the C-terminal domain and Arg-144 and Trp-151 in the N-terminal domain may be key to providing the important energetic link between the two helical bundles.

Fewer interactions are observed with the sugar and the C-terminal domain. Helices VII (Asp-237) and XI (Lys-358) (**Figure 4**), which are symmetrically related to helices I and V, respectively, are also involved in TDG binding. However, these residues likely play only a supporting role relative to the N-terminal primary binding site by providing additional affinity for disaccharide substrates. This explains why the monosaccharide galactose has poor affinity for LacY but behaves like any other substrate with respect to transport and protection of Cys-148 against alkylation (85). It is critical to understand that galactose is the most specific substrate for LacY but has low affinity, which is increased markedly by various adducts, particularly if they are hydrophobic, at the anomeric carbon (85). In contrast to Arg-144 and Glu-126, which are absolutely required, the charge pair between Asp-237 and Lys-358 is interchangeable, and both can be replaced simultaneously by neutral side chains with little effect on activity (16, 86, 88). Therefore, the interaction of Lys-358 with ligand cannot be absolutely required for galactoside binding, although the charge pair between Lys-358 and Asp-237 is important for efficient insertion of LacY into the membrane (16, 26). The essential portion of the substrate-binding site with respect to specificity is in the N-terminal domain, and the residues in the C-terminal domain that interact with the adduct on the anomeric carbon of galactopyranoside increase affinity but have little to do with specificity. Furthermore, although the C_2, C_3, and C_6 OH groups on the galactopyranosyl ring play roles in H-bonding, the C_4 OH is unequivocally the most important determinant for specificity (92). The hydrophobic interaction between the galactopyranosyl ring and Trp-151 is likely to orient the ring so that important H-bonds can be realized (29).

As discussed above, highly specific photolabeling of LacY with NPG (49) was useful for following the protein during purification (71). Subsequent proteolysis experiments (28) demonstrate that the photolabel is in the C-terminal half of LacY. However, because it is clear that all the determinants for sugar specificity are in the N-terminal half of the molecule, there is an apparent conundrum. The nongalactosyl end of NPG (which is photo-activated) is proximal to helices VII and XI (i.e., the C-terminal half of the molecule). Thus, although NPG labels LacY in a highly specific manner, misleading information is obtained regarding the important part of the binding site and may represent a general caveat regarding the use of photoaffinity probes.

RESIDUES INVOLVED IN H^+ TRANSLOCATION AND COUPLING

Biochemistry

One fundamentally important problem is the identification of the residues involved in H^+ binding and the mechanism of coupling between sugar and H^+ translocation. As opposed to an initial proposal that sugar binds to unprotonated LacY (44), current evidence indicates that LacY is protonated prior to ligand binding (90). Most recently, it has been shown directly (J. Vazquez-Ibar, S. Schuldiner & H.R. Kaback, unpublished data) that addition of TDG to a concentrated solution of purified, detergent-solubilized LacY induces no change in pH, whereas a positive control with the antiporter EmrE under identical conditions releases 1 H^+/mol EmrE upon addition of tetraphenylphosphonium (98).

It is difficult to study the mechanism of H^+ translocation. However, in addition to $\Delta\bar{\mu}_{H^+}$-driven active transport, LacY catalyzes other modes of translocation that are important for coupled H^+ translocation. Because individual steps in the overall translocation cycle cannot be delineated by studying $\Delta\bar{\mu}_{H^+}$-driven active transport, LacY-mediated efflux down a chemical gradient, equilibrium exchange, and entrance counterflow are used to probe the mechanism (48, 50). Efflux, exchange, and counterflow with wild-type LacY are explained by a simple kinetic scheme (**Figure 5**). H^+-coupled efflux consists of five steps: [1] binding of H^+ and [2] binding of lactose to LacY at the inner surface of the membrane; [3] a conformational change in LacY that results in translocation of lactose and H^+ to the outer surface of the membrane; [4] release of substrate; [5] release of H^+; [6] a conformational change corresponding to return of unloaded LacY to the inner surface of the membrane. Alternatively, during exchange and counterflow LacY does not deprotonate and only steps 1, 2, and 3 are involved.

Many enzyme reactions involve H^+ transfer in the rate-limiting step, and as a result, these reactions may exhibit a solvent isotope effect when studied in deuterium oxide (D_2O). In brief, such reactions proceed slower in D_2O because of differences in the zero-point stretch vibrations of bonds to protium relative to deuterium (36, 94). With right-side-out (RSO) vesicles or proteoliposomes reconstituted with purified LacY, over threefold slowing of the rate of H^+-coupled downhill lactose influx or efflux is observed in D_2O from pH 5.5 to 7.5, with no effect on $\Delta\bar{\mu}_{H^+}$-driven active transport, exchange, counterflow, or affinity for sugar (9, 50, 106). These and other observations indicate that reactions involved in protonation or deprotonation are not rate determining for $\Delta\bar{\mu}_{H^+}$-driven active transport, exchange, or counterflow, whereas protonation or deprotonation

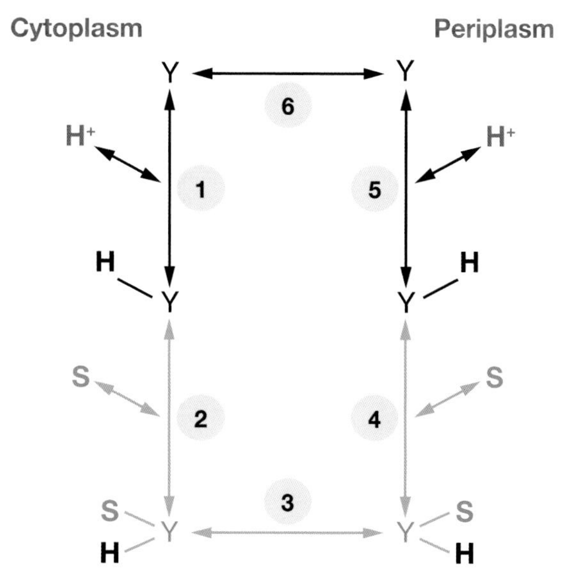

Figure 5

Kinetic scheme for galactoside/H^+ symport, exchange, and counterflow. Y represents LacY, and S is substrate (lactose). Steps involved in exchange and counterflow (steps 2, 3, and 4) are in orange. During downhill symport in the absence of $\Delta\bar{\mu}_{H^+}$, the rate-limiting step is deprotonation (step 1 or 5); in the presence of $\Delta\bar{\mu}_{H^+}$, dissociation of sugar is rate limiting (step 2 or 4).

are rate limiting when a lactose gradient drives H^+ translocation (**Figure 1*b,c***).

Efflux, exchange, and counterflow are blocked in His-322 mutants, and replacement with Asn or Gln results in a 50-fold decrease in affinity for TDG (90). However, the mutants catalyze lactose influx down a concentration gradient at a slow rate (73, 77, 78) without H^+ translocation. In contrast, Glu-325 mutants are specifically defective in all steps involving net H^+ translocation but catalyze lactose exchange and counterflow as well as or better than wild-type LacY (43). Thus, Glu-325 is required for deprotonation (step 4). Glu-325 mutations mimic D_2O or mAb 4B1 (8, 9), but affinity is unaffected by neutral replacements, D_2O, or mAb 4B1. Similarly, replacement of Arg-302 with Ala or Ser maintains exchange and counterflow, although active transport is completely inhibited (63, 89). Extensive mutagenesis and functional characterization reveal that neutral replacements for Glu-269 cause LacY to become defective

RSO: right-side-out

in all translocation reactions. Even replacement of Glu-269 with Asp yields LacY that hardly catalyzes active lactose transport, efflux down a concentration gradient, or equilibrium exchange. However, as discussed above, mutant E269D accumulates TDG with an increase in H^+/TDG stoichiometry (23, 102) and markedly decreased affinity for TDG (33, 90, 105, 110). Therefore, Glu-269 is probably involved in H^+ translocation and sugar binding, as well as coupling between sugar and H^+ translocation.

Interestingly, when Tyr-236 (helix VII) is replaced with Phe, active transport and efflux are abolished, and the mutant catalyzes significant equilibrium exchange (84). In contrast, mutants with Cys (27) or Ala (S. Frillingos & H.R. Kaback, unpublished data) in place of Tyr-236 catalyze significant active transport. A possible interpretation of this behavior is that in the Y236C or Y236A mutants, water replaces the OH group of Tyr. It is intriguing that the X-ray structure (see below) shows Tyr-236 to be within H-bond distance of Arg-302 and His-322.

X-Ray Structure

In the structure, a complex salt bridge/H-bond network (**Figure 6**) composed of residues from helix VII (Tyr-236 and Asp-240), helix X (Lys-319, His-322, and Glu-325), and helix IX (Arg-302) is observed. As discussed above, biochemical analyses indicate that His-322, Glu-325, and Arg-302 are probably directly involved in H^+ translocation. Note that Glu-325 is embedded in a hydrophobic milieu formed by Met-299 and Ala-295 (helix IX), Leu-329 (helix X), and Tyr-236 (helix VII), which is consistent with the notion that Glu-325 is protonated in this conformation (46). Therefore, the structure represents the protonated inward-facing conformation with bound substrate. It has been suggested (89) that Arg-302 could interact with Glu-325 to drive deprotonation. On the one hand, in the structure shown, the side chain of Arg-302 is ~7 Å away from Glu-325, suggesting that a large conformational rearrangement may occur. On the other hand, LacY with two Cys residues at positions 302 and 325 exhibits excimer fluorescence (38),

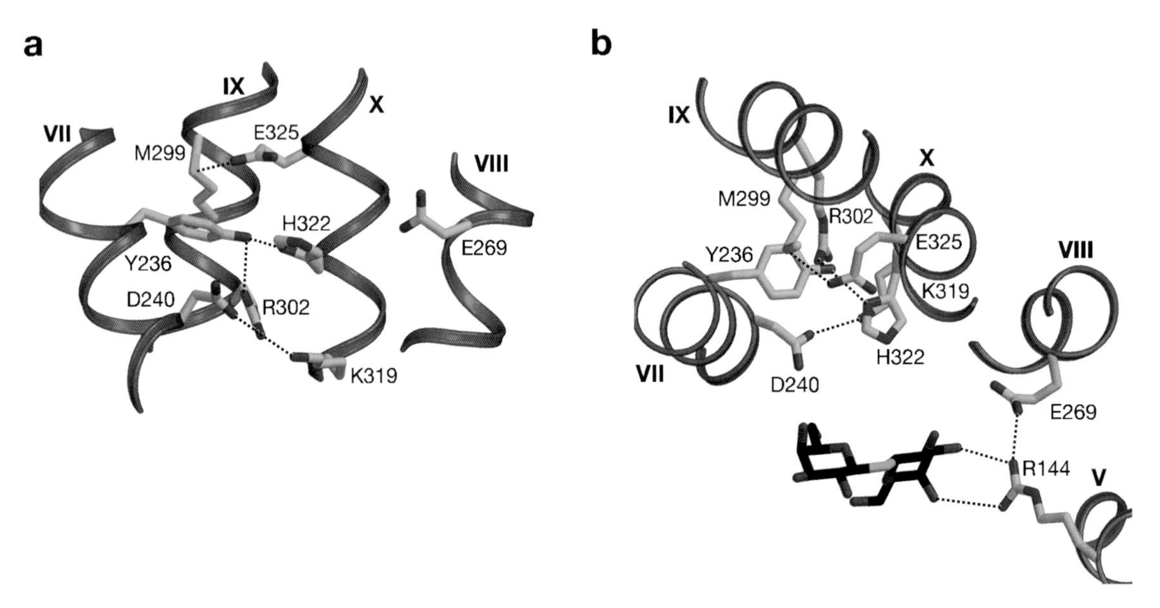

Figure 6

Residues involved in H^+ translocation and coupling. H-bonds are represented by black broken lines. (*a*) View parallel to the membrane. (*b*) View along the membrane normal from the cytoplasmic side. Reproduced, with permission, from Reference 2.

and with two His residues, an Mn(II)-binding site is observed (34). The structural data combined with biochemical/molecular biological studies discussed above provide support for the suggestion that His-322 may be the immediate H^+ donor to Glu-325. Because mutants with simultaneous neutral replacements for Asp-240 and Lys-319 maintain significant transport activity (86, 88), it is unlikely that this salt bridge is directly involved in H^+ translocation; however, the two residues could be involved in stabilization of the salt bridge/H-bond network. Interestingly, transport is abolished when Asp-240 and Lys-319 are reversed (86) or replaced with Cys residues and then cross-linked (123).

The closest distance between this network and the sugar-binding site is more than 6 Å, indicating that the network does not interact directly with the sugar-binding site in the inward-facing conformation. Glu-269 is the only irreplaceable residue in the C-terminal half of LacY that interacts with the N-terminal half and has been postulated to be involved in both substrate and H^+ translocation. Glu-269 is in the vicinity of His-322 (closest distance 5.8 Å). On the basis of biophysical studies, it has also been proposed that Glu-269 makes contact with His-322 in another conformation or via a water molecules(s) (33, 38, 39). The results are consistent with the notion that Glu-269 is critical for coupling between sugar and H^+ translocation.

Although it is generally thought that H^+ translocation across membranes involves movement of H^+ through a pathway perpendicular to the plane of the membrane and parallel to the transmembrane helices, surprisingly, this does not appear to be the case for LacY (**Figure 7**). As shown, the residues involved in H^+ translocation in LacY are at about the same level as the residues that form the sugar-binding site (**Figure 7*a***), and they are essentially parallel to the plane of the membrane and perpendicular to the transmembrane helices (**Figure 7*b***). Thus, it seems likely that H^+ translocation in LacY may involve a delocalized H^+-binding site. By this means, H^+ translocation may occur much like sugar translocation, involving alternating access to both binding sites to either side of the membrane. Such a scenario would explain the long-standing problem of how sugar-driven H^+ translocation can occur in either direction across the membrane utilizing the same residues.

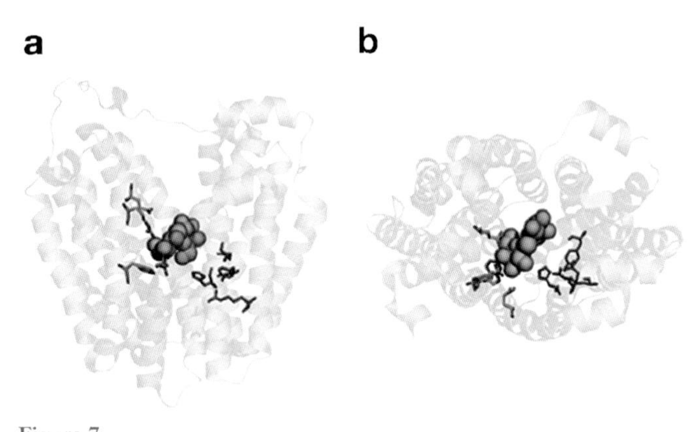

Figure 7

Configuration of residues involved in sugar binding (green) and H^+ translocation (orange). (*a*) Viewed parallel to the membrane. (*b*) Viewed along the membrane normal from the cytoplasmic side.

PROPOSED MECHANISM OF LACTOSE/H^+ SYMPORT

The mechanism of galactoside/H^+ symport can be explained by a simple scheme (**Figure 8**) in which the structure corresponds to the protonated, inward-facing conformation with bound substrate (**Figure 8*d***). LacY in the outward-facing conformation may be unstable (**Figure 8*a***) and protonated immediately as postulated (**Figure 8*b***) (46). In this state, the H^+ is shared between Glu-269 and His-322. A galactoside is then recognized initially by Trp-151, Arg-144, and Glu-126, which disrupts the salt bridge between Arg-144 and Glu-126 and brings His-322 in contact with Glu-325. These changes may induce H^+ transfer from His-322 to Glu-325 as Glu-269 is recruited to complete the binding site by forming a salt bridge with Arg-144 (**Figure 8*c***). This process may also cause

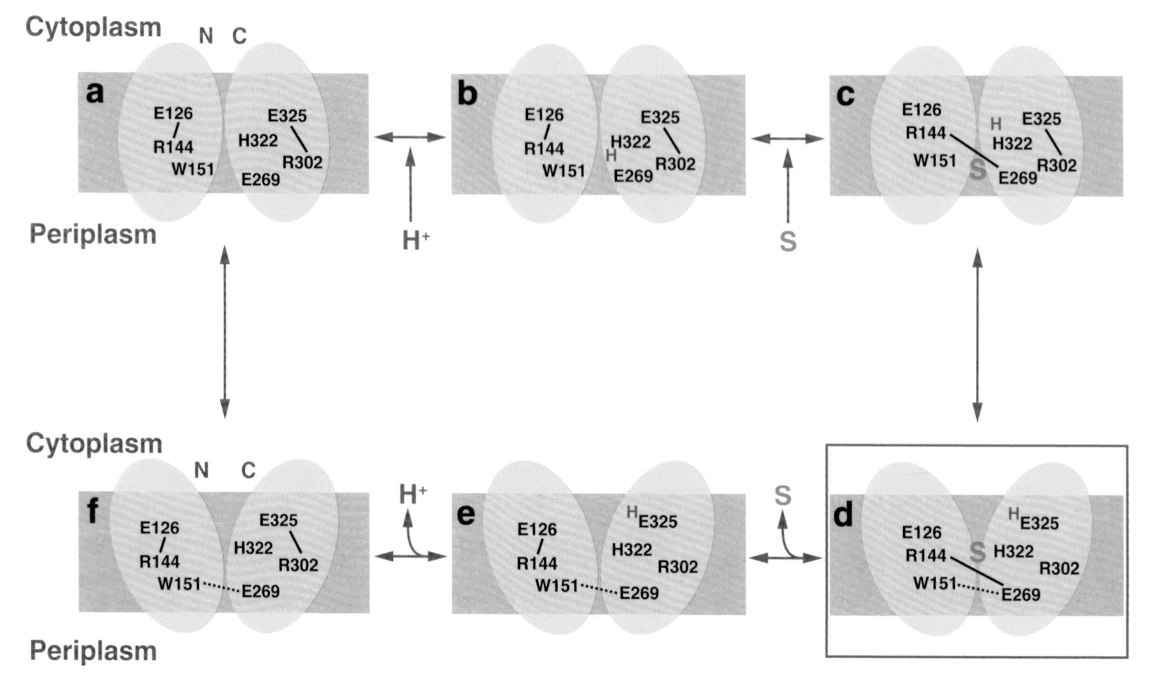

Figure 8

A postulated mechanism for lactose/H^+ symport. Key residues are labeled and charge pairs are represented as solid black lines; H-bonds are depicted as broken black lines. The H^+ and the substrate are shown in red and green, respectively.

ISO: inside-out

rapid transition to the inward-facing conformation (**Figure 8*d***). Substrate is then released into the cytoplasm (**Figure 8*e***), and the salt bridge between Arg-144 and Glu-126 is reestablished. The H^+ is then released from Glu-325 (**Figure 8*f***) probably because of a decrease in pK_a caused either by approximation to Arg-302 (46, 89), exposure to solvent in the aqueous cavity (cytoplasmic pH is constant at 7.6), or both. After releasing the H^+ inside, transition back to the outward-facing conformation is induced.

The structure of LacY exhibits a single sugar-binding site at the apex of a hydrophilic cavity open to the cytoplasm, and it has been postulated from the structure (2) that the binding site has alternating access to either side of the membrane during turnover, as shown in **Figure 9**. However, it is not clear whether $\Delta\bar{\mu}_{H^+}$ changes binding affinity, particularly with LacY, as $\Delta\Psi$ and ΔpH have quantitatively the same kinetic (83) and thermodynamic (80, 81) effects on transport.

Although substrate protection against alkylation of Cys-148 by NEM is particularly useful for obtaining apparent binding constant values (K_D) of LacY for various substrates over a wide range of concentrations (25, 85, 87, 89, 90, 93, 105), it is difficult to obtain true K_D values on each side of the membrane for a transport protein in the presence of $\Delta\bar{\mu}_{H^+}$ because the ligands used are translocated across the membrane and may accumulate in RSO vesicles, thereby leading to underestimation of K_D. However, in a recent series of experiments (30), lactose or TDG protection of Cys-148 against alkylation by NEM were carried out on ice, which decreases substrate accumulation drastically (45). Under these experimental conditions, in the absence of $\Delta\bar{\mu}_{H^+}$, both ice-cold RSO and ISO (inside-out) vesicles likely equilibrate with the

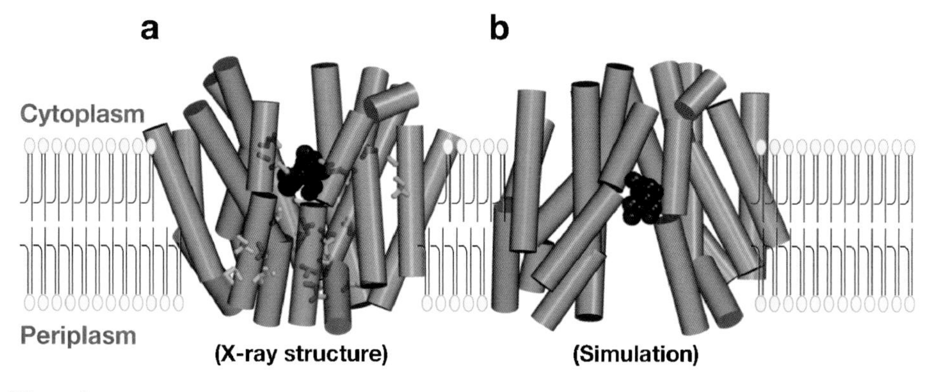

Figure 9

Possible structural changes between inward- and outward-facing conformations. Transmembrane helices in the N- and C-terminal halves are shown as blue and red cyclinders, respectively. (*a*) Inward-facing conformation (i.e., the crystal structure) viewed parallel to the membrane. Cys-replacement mutants of residues colored in yellow exhibit increased reactivity with NEM upon substrate binding. (*b*) Suggested model for outward-facing conformation based on chemical modification and thiol cross-linking. The model was obtained by applying a relatively rigid body rotation of ~60° (around the axis passing near TDG parallel to the membrane) to the N- and C-terminal domains. Reproduced, with permission, from Reference 2.

external medium. In the presence of $\Delta\bar{\mu}_{H^+}$, RSO vesicles may still be able to accumulate lactose or TDG two- to threefold, even though the reactions are carried out on ice for only 5 min. Therefore, the measured K_D values for RSO vesicles in the presence of $\Delta\bar{\mu}_{H^+}$ may be underestimated by two- to threefold. However, this is unlikely, as ISO vesicles in the presence of ATP generate a $\Delta\bar{\mu}_{H^+}$ of opposite polarity (interior positive and/or acid) (**Figure 10**) (82) that causes a decrease in the intravesicular concentration of ligand relative to the concentration in the medium (50). Remarkably, results with both lactose and TDG demonstrate that the K_D manifested by ISO vesicles exhibits less than a twofold change in the absence or presence of $\Delta\bar{\mu}_{H^+}$. Moreover, the K_D values observed with RSO or ISO vesicles in the absence or presence of $\Delta\bar{\mu}_{H^+}$ are similar within experimental error. The results provide a strong indication that $\Delta\bar{\mu}_{H^+}$ has little or no effect on binding affinity, a conclusion that raises a number of interesting considerations regarding the mechanism by which $\Delta\bar{\mu}_{H^+}$ drives accumulation.

In the presence of $\Delta\bar{\mu}_{H^+}$ (interior negative and/or alkaline), wild-type LacY can accumulate lactose against a ~100-fold concentration gradient. Without a significant decrease in binding affinity on the inside of the membrane, how does $\Delta\bar{\mu}_{H^+}$ drive lactose accumulation against a concentration gradient? Because of the effect of D_2O on various translocation reactions, Viitanen et al. (106) have postulated that the rate-limiting step for downhill transport in the absence of $\Delta\bar{\mu}_{H^+}$ is deprotonation, which precedes return of unloaded LacY to the outer surface of the membrane; in contrast, in the presence of $\Delta\bar{\mu}_{H^+}$, dissociation of sugar is rate limiting. Note that the primary kinetic effect of $\Delta\Psi$ or ΔpH on transport is a dramatic decrease in K_m (83). Therefore, it seems reasonable to suggest that $\Delta\bar{\mu}_{H^+}$ enhances the rate of deprotonation on the inner surface of the membrane and thereby allows unloaded LacY to return more rapidly to the outward-facing conformation. Thus, the major effect of $\Delta\bar{\mu}_{H^+}$ on active transport by LacY appears to be kinetic with little or no change in affinity for sugar.

Although biochemical and biophysical studies, as well as a single structure at a resolution of 3.5 Å, can give clues to how the overall conformational change in LacY may

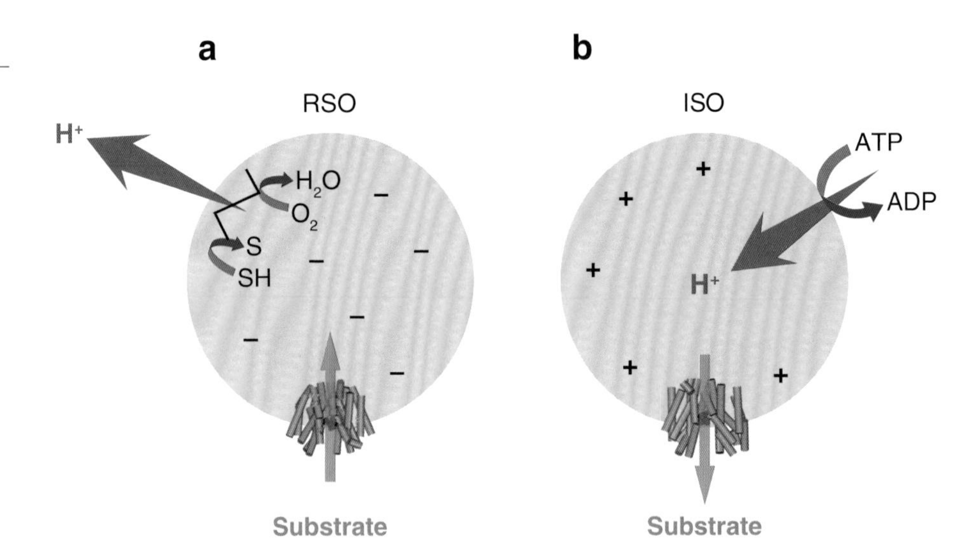

Figure 10

Effect of $\Delta\bar{\mu}_{H^+}$ of opposite polarities on substrate translocation in RSO or ISO vesicles. (*a*) $\Delta\bar{\mu}_{H^+}$ with RSO vesicles ($\Delta\Psi$, interior negative); substrate is accumulated. (*b*) $\Delta\bar{\mu}_{H^+}$ in ISO vesicles (interior positive and acid); substrate effluxes from the vesicles. Reproduced, with permission, from Reference 30.

be coupled to sugar binding and H^+ translocation, many fundamental questions remain. Why is protonation of LacY important for sugar binding? What is the detailed mechanism of coupling between binding and H^+ translocation? What is the time of occupancy of LacY in the outward-facing and inward-facing conformations? Therefore, it is essential to obtain higher-resolution structures in different conformations, as well as dynamics, in order to understand fully the mechanism of substrate/ H^+ symport by LacY.

LESSONS

It is clear from the studies summarized in this review that *local* interactions in LacY can be obtained by indirect means. However, it is also apparent that important global aspects of structure are difficult to delineate in this manner. For example, the residues in LacY involved in specificity and sugar binding were identified initially by site-directed mutagenesis and functional studies, and the conclusions are supported largely by the X-ray structure. In contrast, it is highly unlikely that the large, inward-facing hydrophilic cavity, as well as the highly distorted helices, could have been deduced from the indirect techniques utilized.

Although unexpected, the residues involved in H^+ translocation are aligned parallel to the plane of the membrane at the same level as the sugar-binding site and may be exposed to a water-filled cavity in both the inward- and outward-facing conformations. Therefore, like the galactosidic sugar, the H^+ may be released directly into either cavity during turnover. In any event, it is apparent that H^+ translocation through LacY cannot involve a water-filled channel through the molecule. These structural features may also explain why LacY catalyzes lactose/H^+ symport in both directions across the membrane utilizing the same residues.

Structural biology of membrane proteins is finally becoming a reality, as evidenced by the increasing number of structures that are now appearing (**http://www.mpibp-frankfurt.mpg.de/michel/public/memprotstruct.html**). However, it is important to realize that an X-ray structure, even at high resolution, does not provide dynamic information, and with transport proteins like LacY, the time of occupancy in different conformations may be an essential part of the transport mechanism. Therefore, use of molecular biology to engineer membrane proteins for dynamic studies with biochemical and biophysical techniques is critical.

SUMMARY

The lactose permease of *E. coli* (LacY) couples the free energy released from downhill translocation of H^+ in response to an H^+ electrochemical gradient to drive the stoichiometric accumulation of D-galactopyranosides against a concentration gradient. An X-ray structure in an inward-facing conformation has been solved, which confirms many conclusions from biochemical and biophysical studies. LacY contains N- and C-terminal domains, each with six transmembrane helices, positioned pseudosymmetrically. A large hydrophilic cavity is exposed to the cytoplasm, and ligand is bound at the twofold axis of symmetry at the apex of the hydrophilic cavity in the approximate middle of the molecule. By combining a large body of experimental data derived from systematic studies of site-directed mutants, residues involved in substrate binding and H^+ translocation have been identified, and on the basis of the functional properties of the mutants and the X-ray structure, a working model for the mechanism that involves alternating access of the binding site has been postulated.

LITERATURE CITED

1. Abramson J, Kaback HR, Iwata S. 2004. Structural comparison of lactose permease and the glycerol-3-phosphate antiporter: members of the major facilitator superfamily. *Curr. Opin. Struct. Biol.* 14:413–19
2. Abramson J, Smirnova I, Kasho V, Verner G, Kaback HR, Iwata S. 2003. Structure and mechanism of the lactose permease of *Escherichia coli*. *Science* 301:610–15
3. Bibi E, Kaback HR. 1990. In vivo expression of the *lacY* gene in two segments leads to functional *lac* permease. *Proc. Natl. Acad. Sci. USA* 87:4325–29
4. Bibi E, Kaback HR. 1992. Functional complementation of internal deletion mutants in the lactose permease of *Escherichia coli*. *Proc. Natl. Acad. Sci. USA* 89:1524–28
5. Bieseler B, Prinz H, Beyreuther K. 1985. Topological studies of lactose permease of *Escherichia coli* by protein sequence analysis. *Ann. N. Y. Acad. Sci.* 456:309–25
6. Büchel DE, Gronenborn B, Müller-Hill B. 1980. Sequence of the lactose permease gene. *Nature* 283:541–45
7. Calamia J, Manoil C. 1990. *lac* permease of *Escherichia coli*: topology and sequence elements promoting membrane insertion. *Proc. Natl. Acad. Sci. USA* 87:4937–41
8. Carrasco N, Tahara SM, Patel L, Goldkorn T, Kaback HR. 1982. Preparation, characterization, and properties of monoclonal antibodies against the *lac* carrier protein from *Escherichia coli*. *Proc. Natl. Acad. Sci. USA* 79:6894–98
9. Carrasco N, Viitanen P, Herzlinger D, Kaback HR. 1984. Monoclonal antibodies against the *lac* carrier protein from *Escherichia coli*. 1. Functional studies. *Biochemistry* 23:3681–87
10. Cohen GN, Rickenberg HV. 1955. Etude directe de la fixation d'un inducteur de la β-galactosidase par les cellules d'*Escherichia coli*. *Comptes Rendu.* 240:466–68
11. Columbus L, Hubbell WL. 2002. A new spin on protein dynamics. *Trends Biochem. Sci.* 27:288–95
12. Consler TG, Persson BL, Jung H, Zen KH, Jung K, et al. 1993. Properties and purification of an active biotinylated lactose permease from *Escherichia coli*. *Proc. Natl. Acad. Sci. USA* 90:6934–38
13. Costello MJ, Escaig J, Matsushita K, Viitanen PV, Menick DR, Kaback HR. 1987. Purified *lac* permease and cytochrome *o* oxidase are functional as monomers. *J. Biol. Chem.* 262:17072–82

14. Costello MJ, Viitanen P, Carrasco N, Foster DL, Kaback HR. 1984. Morphology of proteoliposomes reconstituted with purified *lac* carrier protein from *Escherichia coli*. *J. Biol. Chem.* 259:15579–86
15. Dornmair K, Corni AF, Wright JK, Jähnig F. 1985. The size of the lactose permease derived from rotational diffusion measurements. *EMBO J.* 4:3633–38
16. Dunten RL, Sahin-Tóth M, Kaback HR. 1993. Role of the charge pair formed by aspartic acid 237 and lysine 358 in the lactose permease of *Escherichia coli*. *Biochemistry* 32:3139–45
17. Ehring R, Beyreuther K, Wright JK, Overath P. 1980. In vitro and in vivo products of *E. coli* lactose permease gene are identical. *Nature* 283:537–40
18. Ermolova N, Guan L, Kaback HR. 2003. Intermolecular thiol cross-linking via loops in the lactose permease of *Escherichia coli*. *Proc. Natl. Acad. Sci. USA* 100:10187–92
19. Fetsch EE, Davidson AL. 2003. Maltose transport through the inner membrane of *Escherichia coli*. *Front. Biosci.* 8:d652–60
20. Foster DL, Boublik M, Kaback HR. 1983. Structure of the *lac* carrier protein of *Escherichia coli*. *J. Biol. Chem.* 258:31–34
21. Foster DL, Garcia ML, Newman MJ, Patel L, Kaback HR. 1982. Lactose-proton symport by purified *lac* carrier protein. *Biochemistry* 21:5634–38
22. Fox CF, Kennedy EP. 1965. Specific labeling and partial purification of the M protein, a component of the β-galactoside transport system of *Escherichia coli*. *Proc. Natl. Acad. Sci. USA* 54:891–99
23. Franco PJ, Brooker RJ. 1994. Functional roles of Glu-269 and Glu-325 within the lactose permease of *Escherichia coli*. *J. Biol. Chem.* 269:7379–86
24. Frillingos S, Gonzalez A, Kaback HR. 1997. Cysteine-scanning mutagenesis of helix IV and the adjoining loops in the lactose permease of *Escherichia coli*: Glu126 and Arg144 are essential. *Biochemistry* 36:14284–90
25. Frillingos S, Kaback HR. 1996. Probing the conformation of the lactose permease of *Escherichia coli* by in situ site-directed sulfhydryl modification. *Biochemistry* 35:3950–56
26. Frillingos S, Sahin-Tóth M, Lengeler JW, Kaback HR. 1995. Helix packing in the sucrose permease of *Escherichia coli*: properties of engineered charge pairs between helices VII and XI. *Biochemistry* 34:9368–73
27. Frillingos S, Sahin-Tóth M, Wu J, Kaback HR. 1998. Cys-scanning mutagenesis: a novel approach to structure function relationships in polytopic membrane proteins. *FASEB J.* 12:1281–99
28. Goldkorn T, Rimon G, Kaback HR. 1983. Topology of the *lac* carrier protein in the membrane of *Escherichia coli*. *Proc. Natl. Acad. Sci. USA* 80:3322–26
29. Guan L, Hu Y, Kaback HR. 2003. Aromatic stacking in the sugar binding site of the lactose permease. *Biochemistry* 42:1377–82
30. Guan L, Kaback HR. 2004. Binding affinity of lactose permease is not altered by the H^+ electrochemical gradient. *Proc. Natl. Acad. Sci. USA* 101:12148–52
31. Guan L, Murphy FD, Kaback HR. 2002. Surface-exposed positions in the transmembrane helices of the lactose permease of *Escherichia coli* determined by intermolecular thiol cross-linking. *Proc. Natl. Acad. Sci. USA* 99:3475–80
32. Guan L, Sahin-Tóth M, Kaback HR. 2002. Changing the lactose permease of *Escherichia coli* into a galactose-specific symporter. *Proc. Natl. Acad. Sci. USA* 99:6613–18
33. He MM, Kaback HR. 1997. Interaction between residues Glu269 (helix VIII) and His322 (helix X) of the lactose permease of *Escherichia coli* is essential for substrate binding. *Biochemistry* 36:13688–92

34. He MM, Voss J, Hubbell WL, Kaback HR. 1995. Use of designed metal binding sites to study helix proximity in the lactose permease of *Escherichia coli*. 2. Proximity of helix IX (Arg302) with helix X (His322 and Glu325). *Biochemistry* 34:15667–70
35. He MM, Voss J, Hubbell WL, Kaback HR. 1995. Use of designed metal-binding sites to study helix proximity in the lactose permease of *Escherichia coli*. 1. Proximity of helix VII (Asp237 and Asp240) with helices X (Lys319) and XI (Lys358). *Biochemistry* 34:15661–66
36. Jenks WP. 1969. *Catalysis in Chemistry and Enzymology*. New York: McGraw-Hill
37. Jung H, Jung K, Kaback HR. 1994. Cysteine 148 in the lactose permease of *Escherichia coli* is a component of a substrate binding site. I. Site-directed mutagenesis studies. *Biochemistry* 33:12160–65
38. Jung K, Jung H, Wu J, Privé GG, Kaback HR. 1993. Use of site-directed fluorescence labeling to study proximity relationships in the lactose permease of *Escherichia coli*. *Biochemistry* 32:12273–78
39. Jung K, Voss J, He M, Hubbell WL, Kaback HR. 1995. Engineering a metal binding site within a polytopic membrane protein, the lactose permease of *Escherichia coli*. *Biochemistry* 34:6272–77
40. Kaback HR. 1968. The role of the phosphoenolpyruvate-phosphotransferase system in the transport of sugars by isolated membrane preparations of *Escherichia coli*. *J. Biol. Chem.* 243:3711–24
41. Kaback HR. 1976. Molecular biology and energetics of membrane transport. *J. Cell Physiol.* 89:575–93
42. Kaback HR. 1983. The *lac* carrier protein in *Escherichia coli*: from membrane to molecule. *J. Membr. Biol.* 76:95–112
43. Kaback HR. 1989. Molecular biology of active transport: from membranes to molecules to mechanism. *Harvey Lect.* 83:77–103
44. Kaback HR. 1997. A molecular mechanism for energy coupling in a membrane transport protein, the lactose permease of *Escherichia coli*. *Proc. Natl. Acad. Sci. USA* 94:5539–43
45. Kaback HR, Barnes EM Jr. 1971. Mechanisms of active transport in isolated membrane vesicles. II. The mechanism of energy coupling between D-lactic dehydrogenase and β-galactoside transport in membrane preparations from *Escherichia coli*. *J. Biol. Chem.* 246:5523–31
46. Kaback HR, Sahin-Tóth M, Weinglass AB. 2001. The kamikaze approach to membrane transport. *Nat. Rev. Mol. Cell Biol.* 2:610–20
47. Kaback HR, Wu J. 1997. From membrane to molecule to the third amino acid from the left with the lactose permease of *Escherichia coli*. *Q. Rev. Biophys.* 30:333–64
48. Kaczorowski GJ, Kaback HR. 1979. Mechanism of lactose translocation in membrane vesicles from *Escherichia coli*. 1. Effect of pH on efflux, exchange, and counterflow. *Biochemistry* 18:3691–97
49. Kaczorowski GJ, Leblanc G, Kaback HR. 1980. Specific labeling of the *lac* carrier protein in membrane vesicles of *Escherichia coli* by a photoaffinity reagent. *Proc. Natl. Acad. Sci. USA* 77:6319–23
50. Kaczorowski GJ, Robertson DE, Kaback HR. 1979. Mechanism of lactose translocation in membrane vesicles from *Escherichia coli*. 2. Effect of imposed delta psi, delta pH, and delta mu H^+. *Biochemistry* 18:3697–704
51. Kanner BI, Kavanaugh MP, Bendahan A. 2001. Molecular characterization of substrate-binding sites in the glutamate transporter family. *Biochem. Soc. Trans.* 29:707–10
52. King SC, Hansen CL, Wilson TH. 1991. The interaction between aspartic acid 237 and lysine 358 in the lactose carrier of *Escherichia coli*. *Biochem. Biophys. Acta* 1062:177–86

53. Kwaw I, Zen KC, Hu Y, Kaback HR. 2001. Site-directed sulfhydryl labeling of the lactose permease of *Escherichia coli*: helices IV and V that contain the major determinants for substrate binding. *Biochemistry* 40:10491–99
54. le Coutre J, Kaback HR. 2000. Structure-function relationships of integral membrane proteins: membrane transporters vs channels. *Biopolymers* 55:297–307
55. le Coutre J, Kaback HR, Patel CK, Heginbotham L, Miller C. 1998. Fourier transform infrared spectroscopy reveals a rigid alpha-helical assembly for the tetrameric *Streptomyces lividans* K^+ channel. *Proc. Natl. Acad. Sci. USA* 95:6114–17
56. le Coutre J, Narasimhan LR, Patel CK, Kaback HR. 1997. The lipid bilayer determines helical tilt angle and function in lactose permease of *Escherichia coli*. *Proc. Natl. Acad. Sci. USA* 94:10167–71
57. le Coutre J, Whitelegge JP, Gross A, Turk E, Wright EM, et al. 2000. Proteomics on full-length membrane proteins using mass spectrometry. *Biochemistry* 39:4237–42
58. Lee JL, Hwang PP, Hansen C, Wilson TH. 1992. Possible salt bridges between transmembrane α-helices of the lactose carrier of *Escherichia coli*. *J. Biol. Chem.* 267:20758–64
59. Lemieux MJ, Song J, Kim MJ, Huang Y, Villa A, et al. 2003. Three-dimensional crystallization of the *Escherichia coli* glycerol-3-phosphate transporter: a member of the major facilitator superfamily. *Protein Sci.* 12:2748–56
60. Li J, Tooth P. 1987. Size and shape of the *Escherichia coli* lactose permease measured in filamentous arrays. *Biochemistry* 26:4816–23
61. Locher KP, Lee AT, Rees DC. 2002. The *E. coli* BtuCD structure: a framework for ABC transporter architecture and mechanism. *Science* 296:1091–98
62. Matsushita K, Patel L, Gennis RB, Kaback HR. 1983. Reconstitution of active transport in proteoliposomes containing cytochrome *o* oxidase and *lac* carrier protein purified from *Escherichia coli*. *Proc. Natl. Acad. Sci. USA* 80:4889–93
63. Menick DR, Carrasco N, Antes L, Patel L, Kaback HR. 1987. *lac* permease of *Escherichia coli*: arginine-302 as a component of the postulated proton relay. *Biochemistry* 26:6638–44
64. Menick DR, Sarkar HK, Poonian MS, Kaback HR. 1985. Cys154 is important for *lac* permease activity in *Escherichia coli*. *Biochem. Biophys. Res. Commun.* 132:162–70
65. Merritt EA, Sarfaty S, Feil IK, Hol WG. 1997. Structural foundation for the design of receptor antagonists targeting *Escherichia coli* heat-labile enterotoxin. *Structure* 5:1485–99
66. Mitchell P. 1963. Molecule, group and electron transport through natural membranes. *Biochem. Soc. Symp.* 22:142–68
67. Mitchell P. 1967. Translocations through natural membranes. *Adv. Enzymol.* 29:33–87
68. Mitchell P. 1968. *Chemiosmotic Coupling and Energy Transduction*. Bodmin, UK: Glynn Res. Ltd.
69. Müller-Hill B. 1996. *The* lac *Operon: A Short History of A Genetic Paradigm*. Berlin/New York: de Gruyter
70. Nelson N. 1998. The family of Na^+/Cl^- neurotransmitter transporters. *J. Neurochem.* 71:1785–803
71. Newman MJ, Foster DL, Wilson TH, Kaback HR. 1981. Purification and reconstitution of functional lactose carrier from *Escherichia coli*. *J. Biol. Chem.* 256:11804–8
72. Newman MJ, Wilson TH. 1980. Solubilization and reconstitution of the lactose transport system from *Escherichia coli*. *J. Biol. Chem.* 255:10583–86
73. Padan E, Sarkar HK, Viitanen PV, Poonian MS, Kaback HR. 1985. Site-specific mutagenesis of histidine residues in the *lac* permease of *Escherichia coli*. *Proc. Natl. Acad. Sci. USA* 82:6765–68

74. Pao SS, Paulsen IT, Saier MH Jr. 1998. Major facilitator superfamily. *Microbiol. Mol. Biol. Rev.* 62:1–32
75. Patzlaff JS, Moeller JA, Barry BA, Brooker RJ. 1998. Fourier transform infrared analysis of purified lactose permease: A monodisperse lactose permease preparation is stably folded, alpha-helical, and highly accessible to deuterium exchange. *Biochemistry* 37:15363–75
76. Postma PW, Lengeler JW, Jacobson GR, eds. 1996. Echerichia coli *and* Salmonella typhimurium: *Cellular and Molecular Biology*. Washington, DC: ASM. 1149 pp.
77. Püttner IB, Kaback HR. 1988. *lac* permease of *Escherichia coli* containing a single histidine residue is fully functional. *Proc. Natl. Acad. Sci. USA* 85:1467–71
78. Püttner IB, Sarkar HK, Poonian MS, Kaback HR. 1986. *lac* permease of *Escherichia coli:* His-205 and His-322 play different roles in lactose/H^+ symport. *Biochemistry* 25:4483–85
79. Quiocho FA, Vyas NK, eds. 1999. *Bioorganic Chemistry: Carbohydrates*. Oxford, UK: Oxford Univ. Press. 441 pp.
80. Ramos S, Kaback HR. 1977. The relationship between the electrochemical proton gradient and active transport in *Escherichia coli* membrane vesicles. *Biochemistry* 16:854–59
81. Ramos S, Schuldiner S, Kaback HR. 1976. The electrochemical gradient of protons and its relationship to active transport in *Escherichia coli* membrane vesicles. *Proc. Natl. Acad. Sci. USA* 73:1892–96
82. Reenstra WW, Patel L, Rottenberg H, Kaback HR. 1980. Electrochemical proton gradient in inverted membrane vesicles from *Escherichia coli*. *Biochemistry* 19:1–9
83. Robertson DE, Kaczorowski GJ, Garcia ML, Kaback HR. 1980. Active transport in membrane vesicles from *Escherichia coli*: The electrochemical proton gradient alters the distribution of the *lac* carrier between two different kinetic states. *Biochemistry* 19:5692–702
84. Roepe PD, Kaback HR. 1989. Site-directed mutagenesis of tyrosine residues in the *lac* permease of *Escherichia coli*. *Biochemistry* 28:6127–32
85. Sahin-Tóth M, Akhoon KM, Runner J, Kaback HR. 2000. Ligand recognition by the lactose permease of *Escherichia coli*: Specificity and affinity are defined by distinct structural elements of galactopyranosides. *Biochemistry* 39:5097–103
86. Sahin-Tóth M, Dunten RL, Gonzalez A, Kaback HR. 1992. Functional interactions between putative intramembrane charged residues in the lactose permease of *Escherichia coli*. *Proc. Natl. Acad. Sci. USA* 89:10547–51
87. Sahin-Tóth M, Gunawan P, Lawrence MC, Toyokuni T, Kaback HR. 2002. Binding of hydrophobic D-galactopyranosides to the lactose permease of *Escherichia coli*. *Biochemistry* 41:13039–45
88. Sahin-Tóth M, Kaback HR. 1993. Properties of interacting aspartic acid and lysine residues in the lactose permease of *Escherichia coli*. *Biochemistry* 32:10027–35
89. Sahin-Tóth M, Kaback HR. 2001. Arg-302 facilitates deprotonation of Glu-325 in the transport mechanism of the lactose permease from *Escherichia coli*. *Proc. Natl. Acad. Sci. USA* 98:6068–73
90. Sahin-Tóth M, Karlin A, Kaback HR. 2000. Unraveling the mechanism of lactose permease of *Escherichia coli*. *Proc. Natl. Acad. Sci. USA* 97:10729–32
91. Sahin-Tóth M, Lawrence MC, Kaback HR. 1994. Properties of permease dimer, a fusion protein containing two lactose permease molecules from *Escherichia coli*. *Proc. Natl. Acad. Sci. USA* 91:5421–25
92. Sahin-Tóth M, Lawrence MC, Nishio T, Kaback HR. 2001. The C-4 hydroxyl group of galactopyranosides is the major determinant for ligand recognition by the lactose permease of *Escherichia coli*. *Biochemistry* 43:13015–19

93. Sahin-Tóth M, le Coutre J, Kharabi D, le Maire G, Lee JC, Kaback HR. 1999. Characterization of Glu126 and Arg144, two residues that are indispensable for substrate binding in the lactose permease of *Escherichia coli*. *Biochemistry* 38:813–19
94. Schowen RL. 1977. *Isotope Effects on Enzyme-Catalyzed Reactions*. Baltimore, MD: Univ. Park Press. 64 pp.
95. Sixma TK, Pronk SE, Kalk KH, van Zanten BA, Berghuis AM, Hol WG. 1992. Lactose binding to heat-labile enterotoxin revealed by X-ray crystallography. *Nature* 355:561–64
96. Smirnova IN, Kaback HR. 2003. A mutation in the lactose permease of *Escherichia coli* that decreases conformational flexibility and increases protein stability. *Biochemistry* 42:3025–31
97. Sorgen PL, Hu Y, Guan L, Kaback HR, Girvin ME. 2002. An approach to membrane protein structure without crystals. *Proc. Natl. Acad. Sci. USA* 99:14037–40
98. Soskine M, Adam Y, Schuldiner S. 2004. Direct evidence for substrate-induced proton release in detergent-solubilized EmrE, a multidrug transporter. *J. Biol. Chem.* 279:9951–55
99. Sun J, Kaback HR. 1997. Proximity of periplasmic loops in the lactose permease of *Escherichia coli* determined by site-directed cross-linking. *Biochemistry* 36:11959–65
100. Teather RM, Müller-Hill B, Abrutsch U, Aichele G, Overath P. 1978. Amplification of the lactose carrier protein in *Escherichia coli* using a plasmid vector. *Mol. Gen. Genet.* 159:239–48
101. Ujwal ML, Jung H, Bibi E, Manoil C, Altenbach C, et al. 1995. Membrane topology of helices VII and XI in the lactose permease of *Escherichia coli* studied by *lacY-phoA* fusion analysis and site-directed spectroscopy. *Biochemistry* 34:14909–17
102. Ujwal ML, Sahin-Tóth M, Persson B, Kaback HR. 1994. Role of glutamate-269 in the lactose permease of *Escherichia coli*. *Mol. Membr. Biol.* 11:9–16
103. Vazquez-Ibar JL, Guan L, Svrakic M, Kaback HR. 2003. Exploiting luminescence spectroscopy to elucidate the interaction between sugar and a tryptophan residue in the lactose permease of *Escherichia coli*. *Proc. Natl. Acad. Sci. USA* 100:12706–11
104. Vazquez-Ibar JL, Guan L, Weinglass AB, Verner G, Gordillo R, Kaback HR. 2004. Sugar recognition by the lactose permease of *Escherichia coli*. *J. Biol. Chem.* 279:49214–21
105. Venkatesan P, Kaback HR. 1998. The substrate-binding site in the lactose permease of *Escherichia coli*. *Proc. Natl. Acad. Sci. USA* 95:9802–7
106. Viitanen P, Garcia ML, Foster DL, Kaczorowski GJ, Kaback HR. 1983. Mechanism of lactose translocation in proteoliposomes reconstituted with *lac* carrier protein purified from *Escherichia coli*. 2. Deuterium solvent isotope effects. *Biochemistry* 22:2531–36
107. Viitanen P, Garcia ML, Kaback HR. 1984. Purified reconstituted *lac* carrier protein from *Escherichia coli* is fully functional. *Proc. Natl. Acad. Sci. USA* 81:1629–33
108. Vogel H, Wright JK, Jähnig F. 1985. The structure of the lactose permease derived from Raman spectroscopy and prediction methods. *EMBO J.* 4:3625–31
109. Weinglass A, Whitelegge JP, Faull KF, Kaback HR. 2004. Monitoring conformational rearrangements in the substrate-binding site of a membrane transport protein by mass spectrometry. *J. Biol. Chem.* 279:41858–65
110. Weinglass AB, Sondej M, Kaback HR. 2002. Manipulating conformational equilibria in the lactose permease of *Escherichia coli*. *J. Mol. Biol.* 315:561–71
111. Weinglass AB, Whitelegge JP, Hu Y, Verner GE, Faull KF, Kaback HR. 2003. Elucidation of substrate binding interactions in a membrane transport protein by mass spectrometry. *EMBO J.* 22:1467–77

112. West IC. 1970. Lactose transport coupled to proton movements in *Escherichia coli*. *Biochem. Biophys. Res. Commun.* 41:655–61
113. West IC, Mitchell P. 1972. Proton-coupled β-galactoside translocation in non-metabolizing *Escherichia coli*. *J. Bioenerg.* 3:445
114. West IC, Mitchell P. 1973. Stoichiometry of lactose-H^+ symport across the plasma membrane of *Escherichia coli*. *Biochem. J.* 132:587–92
115. Whitelegge JP, Gundersen CB, Faull KF. 1998. Electrospray-ionization mass spectrometry of intact intrinsic membrane proteins. *Protein Sci.* 7:1423–30
116. Whitelegge JP, le Coutre J, Lee JC, Engel CK, Privé GG, et al. 1999. Toward the bilayer proteome, electrospray ionization-mass spectrometry of large, intact transmembrane proteins. *Proc. Natl. Acad. Sci. USA* 96:10695–98
117. Wolin C, Kaback HR. 1999. Estimating loop-helix interfaces in a polytopic membrane protein by deletion analysis. *Biochemistry* 38:8590–97
118. Wolin CD, Kaback HR. 2000. Thiol cross-linking of transmembrane domains IV and V in the lactose permease of *Escherichia coli*. *Biochemistry* 39:6130–35
119. Wolin CD, Kaback HR. 2001. Functional estimation of loop-helix boundaries in the lactose permease of *Escherichia coli* by single amino acid deletion analysis. *Biochemistry* 40:1996–2003
120. Wright EM, Turk E. 2004. The sodium/glucose cotransport family SLC5. *Pflugers Arch.* 447:510–18
121. Wrubel W, Stochaj U, Sonnewald U, Theres C, Ehring R. 1990. Reconstitution of an active lactose carrier in vivo by simultaneous synthesis of two complementary protein fragments. *J. Bacteriol.* 172:5374–81
122. Zen KH, McKenna E, Bibi E, Hardy D, Kaback HR. 1994. Expression of lactose permease in contiguous fragments as a probe for membrane-spanning domains. *Biochemistry* 33:8198–206
123. Zhang W, Guan L, Kaback HR. 2002. Helices VII and X in the lactose permease of *Escherichia coli*: proximity and ligand-induced distance changes. *J. Mol. Biol.* 315:53–62
124. Zhao M, Zen K-C, Hernandez-Borrell J, Altenbach C, Hubbell WL, Kaback HR. 1999. Nitroxide scanning electron paramagnetic resonance of helices IV, V and the intervening loop in the lactose permease of *Escherichia coli*. *Biochemistry* 38:15970–77
125. Zhao M, Zen K-C, Hubbell WL, Kaback HR. 1999. Proximity between Glu126 and Arg144 in the lactose permease of *Escherichia coli*. *Biochemistry* 38:7407–12

Evolutionary Relationships and Structural Mechanisms of AAA+ Proteins

Jan P. Erzberger and James M. Berger

Department of Molecular and Cell Biology, University of California, Berkeley, California 94720; email: jmberger@berkeley.edu

Annu. Rev. Biophys. Biomol. Struct. 2006. 35:93–114

The *Annual Review of Biophysics and Biomolecular Structure* is online at biophys.annualreviews.org

doi: 10.1146/annurev.biophys.35.040405.101933

1056-8700/06/0609-0093$20.00

Key Words

ATPase, remodeling, molecular machines, motors

Abstract

Complex cellular events commonly depend on the activity of molecular "machines" that efficiently couple enzymatic and regulatory functions within a multiprotein assembly. An essential and expanding subset of these assemblies comprises proteins of the ATPases associated with diverse cellular activities (AAA+) family. The defining feature of AAA+ proteins is a structurally conserved ATP-binding module that oligomerizes into active arrays. ATP binding and hydrolysis events at the interface of neighboring subunits drive conformational changes within the AAA+ assembly that direct translocation or remodeling of target substrates. In this review, we describe the critical features of the AAA+ domain, summarize our current knowledge of how this versatile element is incorporated into larger assemblies, and discuss specific adaptations of the AAA+ fold that allow complex molecular manipulations to be carried out for a highly diverse set of macromolecular targets.

Contents

INTRODUCTION

The AAA+ Superfamily of Proteins

From its initial characterization and subsequent rapid expansion, the AAA+ family of proteins has always been characterized by the remarkable functional heterogeneity of its members (5, 54, 72). AAA+ factors themselves represent one of the most ubiquitous and functionally diverse groups within the vast family of "P-loop"-type nucleoside triphosphate (NTP)-binding proteins (45). The functional variety within the AAA+ family is reflected in their extensive number of accessory domains and factors (19), the organizational diversity of their oligomeric assemblies, and the heterogeneity of amino acid residues that define key nucleotide-interacting motifs. These adaptations influence the rate of nucleotide binding and hydrolysis, regulate the stability of AAA+ oligomers, and determine the specificity of AAA+ assemblies for distinct substrates and the mechanisms by which conformational changes are coupled within the AAA+ assembly. This review focuses on recent structural, computational, and biochemical studies that have begun to identify key characteristics and properties that underlie the versatility of AAA+ proteins.

P-loop or Walker A motif (W-A): consensus sequence GxxGxGKT/S between β1 and α1 of the ASCE core

Walker B motif (W-B): consensus sequence D(D/E) at the apex of β3

ASCE: additional strand, conserved E

AAA+: ATPases associated with diverse cellular activities

The AAA+ Module in the Context of Other P-Loop NTPases

Early phylogenetic analyses of NTPases by Walker and colleagues (98) identified two distinct signature sequences, now commonly called the Walker A (W-A) and Walker B (W-B) motifs, that define a broad superfamily of nucleotide-binding proteins including the F_1-ATPase, myosin, and various kinases. More recent studies have comprehensively expanded this analysis and have outlined further the evolutionary relationships between the protein families that share these motifs (59, 61). A subset of these NTPases have been distinguished as the additional strand conserved E family (ASCE) by Aravind and colleagues (46, 59) (**Figure 1*a***), based in part on the insertion of a β-strand that lies between the W-A and W-B motifs and the presence of a second conserved acidic residue in the W-B motif (46, 59).

Within ASCE proteins, the nucleotide-binding pocket lies at the apex of three adjacent, parallel β-strands in a compact αβα-fold (**Figure 1*b*,*c***). This arrangement defines the basic topology present in all ASCE ATPases. Because the first ASCE protein to be characterized structurally was the RecA protein (91), many proteins that possess this core topology historically have been described as RecA-like. However, RecA and its close structural relatives form a distinct subtype in the ASCE family and contain specific structural elements not observed in more distantly

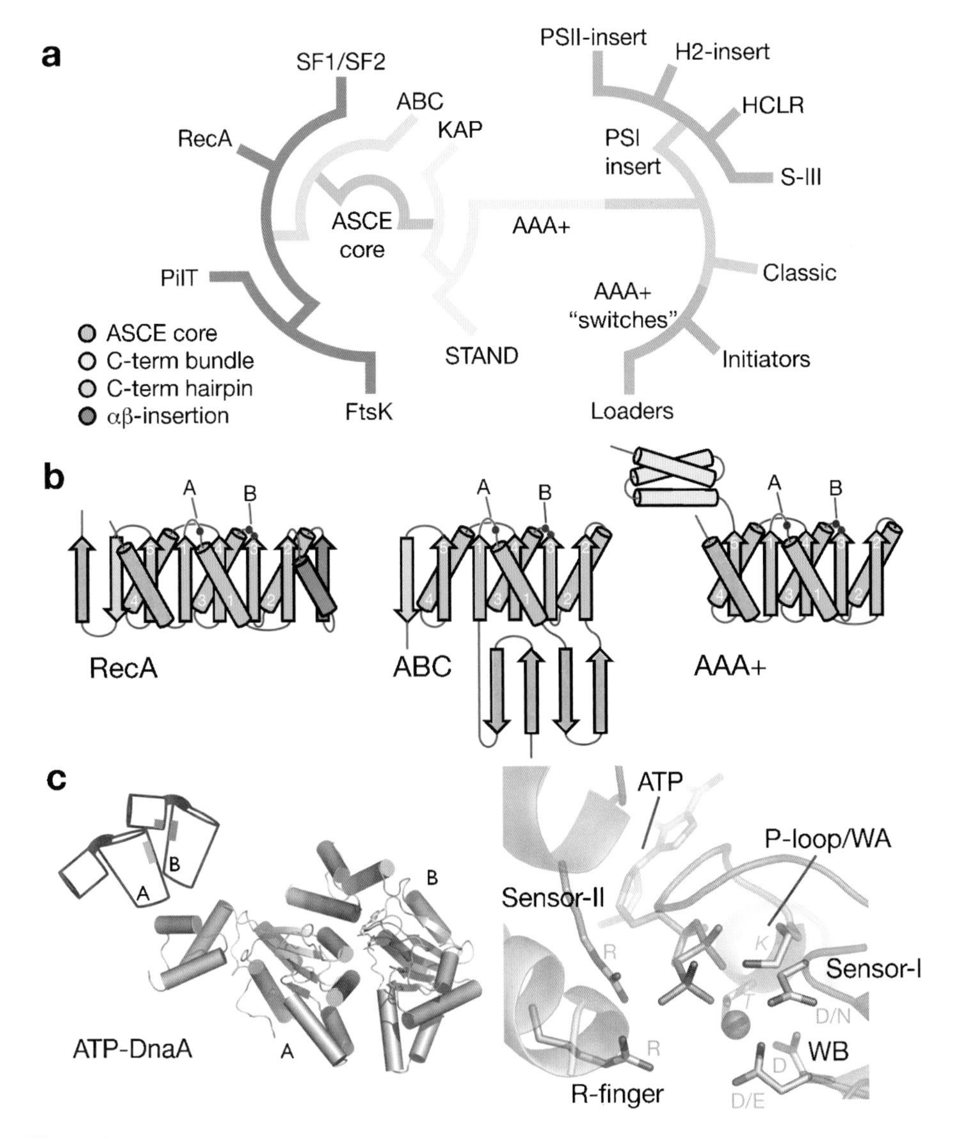

Figure 1

Classification and features of ASCE and AAA+ ATPases. (*a*) Schematic representation of main ASCE subgroups (adapted from Reference 46). The main branches are defined and colored according to their defining structural elements. Within the AAA+ family, the clades defined by Aravind and colleagues and used throughout this review are diagrammed, highlighting the pre-sensor-I insert superclade in red. (*b*) Topology diagrams of ASCE members. The core ASCE fold is colored green, and unique structural features that define each subgroup are colored as in panel *a*. The nucleotide-interacting motifs characteristic of the ASCE family are marked as red dots and labeled (A, Walker A; B, Walker B). (*c*) Example of an active interaction between neighboring AAA+ modules. (*left*) Schematic showing the contributions of each module to the bipartite ATP-binding sites, highlighted in green and red. (*right*) Detail of the active site of ATP-DnaA showing the position of nucleotide-interacting motifs and ATP (J. P. Erzberger, M. Mott & J.M. Berger, manuscript in preparation). The coloring reflects subunit contributions.

SRH: second region of homology

Sensor-I motif (S-I): polar residue (predominantly N; D, T, and H also are observed) at the apex of β4

Sensor-II motif (S-II): arginine residue at the base of α7 in the helical lid

ATP: adenosine triphosphate

Arginine (ARG)-finger: arginine (occasionally lysine), generally at end of α5

related ASCE proteins. In light of these observations, we have adopted the nomenclature and classification criteria proposed by Aravind and coworkers (45, 46) to describe the relationship of AAA+ proteins to other ASCE members.

The spatial positioning of nucleotide-interacting motifs is highly conserved among all ASCE proteins, with rmsd's between Cα atoms in the ASCE core typically ranging from 1.5 to 2.5 Å (99). Aside from superfamily I and II helicases (which contain an internal duplication of the ASCE core), ASCE ATPases are active only as oligomeric assemblies in which pairs of subunits combine to form a fully functional nucleotide-binding site. The differences between ASCE protein families arise from small insertions and rearrangements of secondary structural elements that are specific to distinct groups, and that help determine the relative organization of subunits within higher-order assemblies. For example, an auxiliary helical domain found in AAA+ proteins, as well as the specific β-strand inserts in RecA molecules, mediate interprotomer contacts (62, 88).

Although many ASCE proteins assemble into hexamers, there are fundamental differences in the relative orientations of the core αβα-fold with respect to the central axis of each hexameric assembly. Indeed, rotational variations of up to 90° are observed in the subunit/subunit packing of different ASCE assemblies, indicating that some of these protein families may operate by significantly different mechanisms (99). As a direct result of these differences, the elements on one subunit that interact with the nucleotide-binding pocket of a neighboring protomer are also family specific. These adjoining elements frequently contain a conserved activating arginine residue that is often called an arginine (Arg)-finger based on an analogous interaction used by the activating proteins of small GTPases (103). One of the results of the divergent oligomerization modes observed among ASCE proteins is that the location of the Arg-finger can vary markedly between ASCE subfamilies (30, 62, 88, 108).

The primary distinguishing feature of AAA+ proteins is an absence of β-strand additions to the ASCE core, together with a small helical bundle fused to the C terminus of the central αβα-fold. However, several additional features are characteristic of this family. For example, the Arg-finger of AAA+ proteins typically occupies a defined position within the ASCE core that has been variously termed the Box VII, second region of homology (SRH) or SRC motif (5, 31, 72). In addition, Kuriyan and colleagues (31) have defined two other nucleotide-interaction motifs termed the sensor-I (S-I) and sensor-II (S-II) elements. The S-I motif, which resides at the top of strand β4 (**Figure 1*b,c***), is typically an Asn, although other polar residues such as Ser, Thr, or Asp are also found. This residue is believed to act in concert with the second acidic residue of the W-B motif to properly orient a water molecule for nucleophilic attack on the γ-phosphate of ATP in a manner analogous to a conserved Glu residue in RecA family ASCE proteins (31, 88, 91). S-II usually contains an arginine at the base of helix α7 that interacts with the γ-phosphate of ATP. The importance of these residues for ATP turnover and function has been validated extensively in a number of AAA+ systems (7, 14, 23, 33, 66, 74, 77, 92).

The AAA+ Module as a Molecular Switch

AAA+ proteins function by linking nucleotide-mediated conformational changes within an oligomeric assembly to specific chemo-mechanical motions that are transduced to a target macromolecule (33, 52, 55, 100). A cycle of ATP-binding and hydrolysis therefore defines a switch between at least two distinct conformational states of the protein (16, 80). Because of the intersubunit coupling that exists within AAA+ assemblies, conformational events can be propagated from one active site interface to the next (27, 40, 100). A remarkable feature of the AAA+ module is that this conformational coupling mechanism is highly tunable. For example, certain assemblies, for which ATP binding and hydrolysis are slow, highly regulated events, appear to function as bimodal switches, becoming activated upon nucleotide binding and then resetting after hydrolysis and ADP release (13, 38, 56, 84). At the other extreme, the conformational coupling between AAA+ modules is coordinated to perpetually propagate cycles, or waves, of nucleotide binding and release among individual AAA+ modules, resulting in an efficient and processive molecular motor. In between the two extremes lie a number of remodeling factors that possess intermediate levels of activity that may be particularly tuned to a given biological function (39). Finally, there exists a rapidly expanding category of AAA+ modules that have degenerated into proteins that lack ATP-binding and hydrolysis activity altogether. Examples of this class include the δ and δ' subunits of the bacterial clamp loader (31, 48), the Orc4 and Orc5 subunits of the eukaryotic origin recognition complex (ORC), and the hexamerization domains of p97 and *N*-ethymaleimide-sensitive fusion protein (NSF) (62, 109). These modules typically serve as amplifiers or modulators between active AAA+ ATPase subunits within an oligomeric complex (10, 49).

Classification of AAA+ Modules

Iyer et al. (45) have recently described a rigorous classification of AAA+ proteins into defined groups based on specific structural elements (45). This analysis, which combined sequence and structural information, resulted in the definition of a number of AAA+ "clades." The differences between clades arise from the insertion of secondary structural elements at defined places within and around the core AAA+ fold (**Figures 2** and **3**). This framework provides a useful guide for discriminating between various AAA+ assemblies and understanding some of their unique functional differences.

Conformational coupling: within an AAA+ array, changes induced by nucleotide binding or hydrolysis that are coordinated between subunits through allostery

ADP: adenosine diphosphate

AAA+ core: the $\alpha\beta\alpha$ fold that defines the conserved ASCE core of all AAA+ domains and that contains the W-A, W-B and S-I ATP-binding motifs

Clamp Loader Proteins form a Spiral Assembly that Guides Clamp Deposition to Primer-Template Junctions

The clamp loader clade of AAA+ proteins (clade 1) is one of the most specialized and best characterized AAA+ assemblies (8). This pentameric complex is a critical component of DNA replication, recognizing primer-template junctions generated during lagging-strand synthesis and promoting the opening and loading of the ring-shaped polymerase processivity clamp onto DNA at these sites. The basic architecture of loader assemblies is remarkably conserved in all three domains of life (9, 15, 48). The core assembly is pentameric and formed as a hetero-oligomer from different subunits. An oligomerization domain C-terminal to the AAA+ module, often called the collar, forms a ring that stabilizes the complex and enables highly dynamic interactions among the various AAA+ modules to take place. Within the AAA+ domains themselves, only a subset of AAA+ active sites are generally functional for ATP binding and hydrolysis.

The recent structure of yeast replication factor C (RFC) bound to the proliferating cell nuclear antigen (PCNA) clamp revealed that the assembly adopts a spiral structure that is thought to interact with the 3′ end of a primer-template junction much like a screw-on bottle cap (**Figure 4*a***) (8, 9). This finding, together with biochemical experiments, has suggested that the unique architecture of the clamp loader family probably evolved as a substrate-specificity determinant, ensuring that clamps are delivered precisely to primer-template junctions (29). The structural elements postulated to interact with the DNA grooves are the N-terminal portions of helices $\alpha2$ and $\alpha3$ of the AAA+ core module

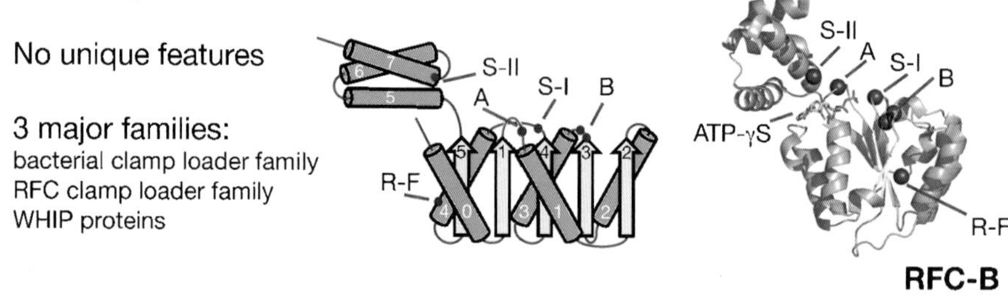

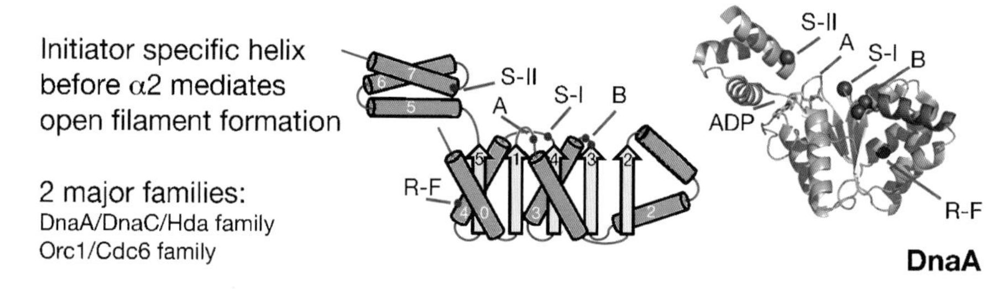

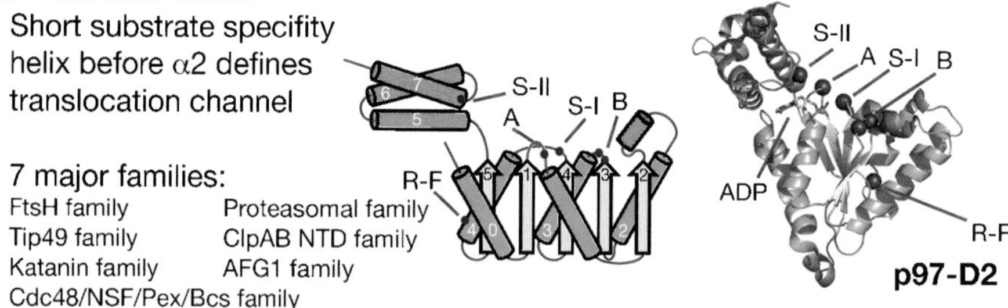

Figure 2

Basic AAA+ clades. The first three AAA+ clades show few structural changes relative to the basic ASCE fold. For each clade, its defining structural feature (if any) is highlighted, along with the major family members, topology diagram, and a representative structure (RFC-B – 1SXJ; DnaA – 1L8Q; P97-D2–1R7R) (9, 22, 43) showing conserved catalytic motifs (*red dots*; labeled as in **Figure 1**). Basic AAA+ secondary structure elements (*blue and yellow*) as well as clade-specific structural features (*red*) are depicted.

(29). By contrast, the C-terminal portion of helix α2 marks the beginning of the PCNA clamp interaction motif (8). It is therefore perhaps not unsurprising that the clamp loader clade represents a particularly undecorated form of AAA+ protein (**Figure 2**), because any of the structural additions found in other AAA+ clades would disrupt essential interactions either with substrate or with adjoining subunits within the clamp loader assembly.

Initiators Form Helical AAA+ Filaments that Impose Specific Topological Constraints on Origins of Replication

DNA replication is initiated at distinct chromosomal sites known as replication origins. Specific *trans*-acting factors, termed initiators, are responsible for properly recognizing replication origins and for enabling the subsequent assembly of the replication machinery (18, 50, 69). Clade 2, the initiator clade, includes all cellular origin-processing proteins, as well as the bacterial, archaeal, and eukaryal helicase-loading proteins. This group is defined by the presence of an extra α-helix inserted between the second and third strands of the central β-sheet (45) (**Figure 2**). Unlike the clamp loader and most other AAA+ proteins, bacterial and archaeal AAA+ initiators are monomeric in solution and assemble into oligomeric structures when bound to specific target DNA sequences at origins (12). By contrast, eukaryotes use a preformed ORC that contains at least three and possibly as many as

4 - Superfamily III helicase clade

Lacks AAA+ lid domain, unique helical bundle composed of N- and C-terminal elements.

2 major families:
DNA virus family
RNA virus family

5 - HCLR clade

No features beyond PS-I β-hairpin

4 major families:
HslU/ClpX family
ClpAB-CTD family
Lon family
RuvB family

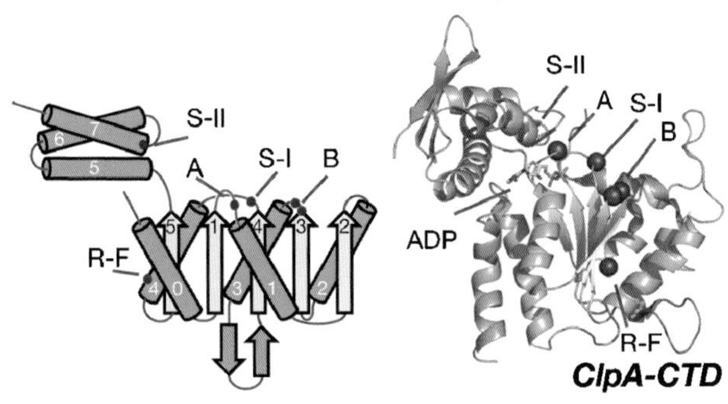

6 - H2-insert clade

β-hairpin insertion in helix 2

2 major families:
NtrC family
McrB family

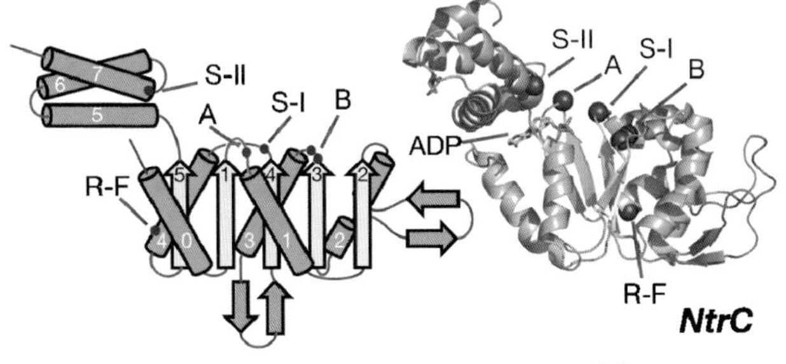

7 - PS-II insert clade

Helical insertion after α5 repositions the Sensor-II motif

3 major families:
MCM family
MoxR family
Chelatase/YifB family
Dynein/Midasin family

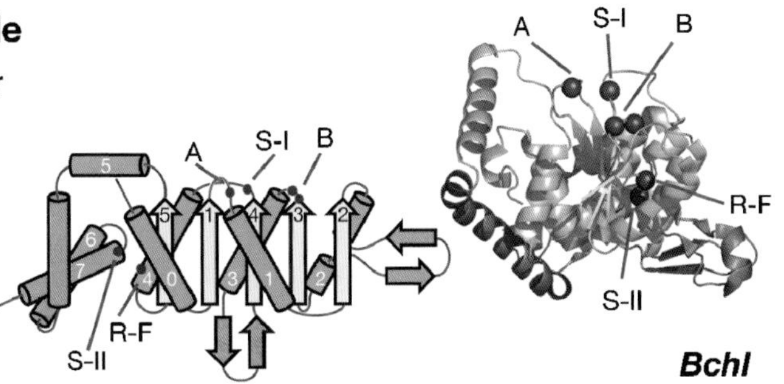

Figure 3

Pre-sensor I insert superclade. PS-I superclade members share a common β-hairpin insertion but are also distinguished by additional clade-specific features. Figure organization and labels are as in **Figure 2**. The representative structures are SV40 (1SVM), ClpACTD (1KSF), NtrC (1NY5) and BchI (1G8P) (27, 57, 24).

five AAA+ proteins as part of its heterohexameric assembly (95; J.P. Erzberger & J.M. Berger, unpublished data).

The organization of the monomeric ground state of these proteins has been observed in several structures of archaeal and bacterial initiators bound to ADP (22, 66, 86). The recent structure of an activated, ATP-bound structure of DnaA has revealed how ATP induces conformational changes that drive oligomeric assembly. Remarkably, this oligomerized initiator forms a helical

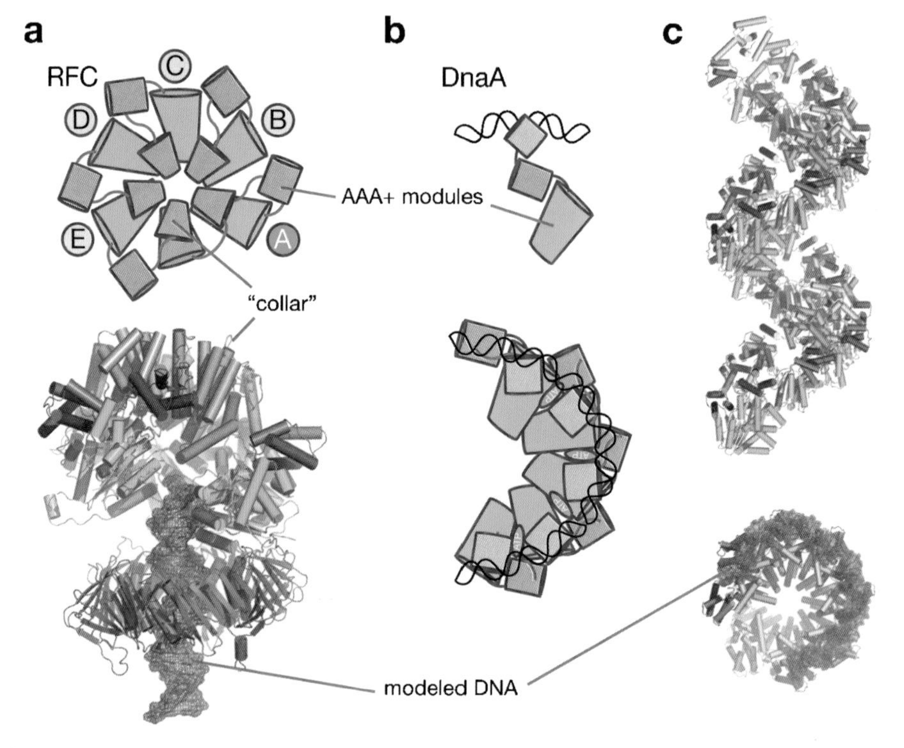

Figure 4

Clamp loader and initiator assemblies. (*a*) (*Top*) Clamp loaders are composed of a pentameric assembly of AAA+ proteins held together by a C-terminal collar domain (shown in *red*). (*Bottom*) DNA modeled into the structure of the RFC/PCNA complex (1SXJ) (9). The spiral arrangement of the RFC assembly acts as a specificity determinant, ensuring that the processivity clamp is delivered to primer-template junctions. Residues involved in clamp and DNA interactions are on opposite ends of helix $\alpha2$ and may allow for the coupling of clamp release upon binding to the correct DNA substrate. (*b*) Helical filament formation by the bacterial replication initiator DnaA. ATP-bound DnaA molecules bind specific sequences in the origin and cooperatively assemble into a right-handed helix. (*c*) (*Top*) Crystal structure of the ATP-bound bacterial DnaA filament, which displays a 14 Å rise per monomer (J.P. Erzberger, M. Mott & J.M. Berger, manuscript in preparation). (*Bottom*) The architecture of the DnaA filament allows it to organize DNA into a positive solenoidal wrap. The helical insert that characterizes the initiator clade is a critical mediator of the intermodular interactions that enable this AAA+ array to be a filamentous helix.

filament, the only such architecture observed among AAA+ proteins (J.P. Erzberger, M. Mott & J.M. Berger, manuscript in preparation) (**Figure 4*b***). The helical pitch of the DnaA filament is set up by a lateral displacement at subunit interfaces that prevent the formation of a closed-ring structure. The initiator-specific helix drives this repositioning, and its presence in various initiators suggests that an open-ring spiral is the likely assembly mode adopted by all initiators (J.P. Erzberger, M. Mott & J.M. Berger, manuscript in preparation). In ATP-bound DnaA, the filamentous arrangement places DNA-binding domains appended to the C terminus of the AAA+ module on the outside of the filament. This orientation suggests that the interaction with origin DNA sequences may lead to the introduction of right-handed helical wrap to this region, destabilizing adjacent DNA elements as a prelude to origin melting and replisome assembly (J.P. Erzberger, M. Mott & J.M. Berger, manuscript in preparation) (**Figure 4*c***).

Unlike RFC or other AAA+ proteins, the helical arrangement of DnaA leaves a pair of AAA+ interaction surfaces accessible at the boundaries of the filament, exposing an Arg-finger motif at one end and a nucleotide-binding pocket at the other. In *Escherichia coli* and also in ORC, there is evidence that suggests that these surfaces may serve as docking sites for auxiliary replication factors that also contain AAA+ domains. For example, in *E. coli*, initiation by DnaA is negatively regulated through an interaction between the ATP-binding pocket of DnaA and the Arg-finger of Hda, a regulatory AAA+ factor (51, 93). In archaea and eukaryotes, critical interactions between initiators and AAA+-type helicase loader proteins likewise occur, although direct evidence of associations between the ATPase modules of these proteins has not yet been described (64, 70, 102).

Interestingly, the formation of helical arrays may also be an important event during the establishment of silenced chromatin regions in *Saccharomyces cerevisiae*. Sir3, a silencing factor related to eukaryotic Orc1, is part of a complex that spreads along chromatin to silence transcription in the presence of *O*-acetyl-ADP-ribose (81, 65). A filament formed by Sir3 protomers could provide a scaffold that helps direct the Sir assembly to cover extended regions and regulate histone modifications underlying chromatin silencing.

The Classic Remodelers: Versatility and Adaptability of the Basic Hexameric AAA+ Assembly

Clade 3, which includes the founding members of the AAA+ family (11, 54), represents a closely related group of proteins that form closed hexameric assemblies (25, 26, 67). United by a common protein remodeling function, the classic AAA+ proteins are nevertheless involved in processes as diverse as protein degradation, microtubule severing (3, 36), membrane fusion (10), and peroxisome biogenesis (94). Classic AAA+ modules share a small helical insertion before helix $\alpha 2$ of the ASCE fold (**Figure 2**). On the basis of conservation patterns and mutagenesis studies, this structural element, which lies close to the central axis of the hexamer, has been suggested to mediate the recognition of target substrates (41, 106). However, this motif is also critical in mediating substrate interactions within the pre-sensor I (PS-I) superclade (see below) (86, 100) and therefore may not constitute a clear-cut clade-defining feature. Perhaps more important is that there is a remarkable level of conservation among the five key nucleotide interaction motifs of classic AAA+ proteins and that these sequence patterns are notably different from those seen in other AAA+ clades (**Figure 5*b***) (5). For example, structure-based alignments reveal that classic AAA+ proteins do not possess a conserved S-II arginine at the base of helix $\alpha 7$ (**Figure 5*b***) (74), a feature that characterizes all other active AAA+ proteins (with the exception of the viral helicases in clade 4). The classic clade is also unique in that members of the family possess two conserved arginine residues in their box VII region instead of the single residue typically seen in other AAA+ proteins. The processive conformational coupling that occurs within the closed hexameric assembly of clade 3 AAA+ proteins is guided by interactions of these sensor arginines with the nucleotide in the neighboring subunit (74). The unique, conserved interaction ensemble present at the interface of classic AAA+ arrays suggests that a distinct conformational coupling mechanism, rather than a specific structural feature, defines the members of this AAA+ clade.

If a common, conserved "engine" defines the classic clade of AAA+ proteins, components outside the ATPase core are most likely determinants of the functional diversity within this clade. Over the past few years, a number of factors have been identified that interact with classic AAA+ proteins to influence the activity and specificity of their remodeling function. These cofactors can be subdivided into functional partners, which

PS-I: pre-sensor I

Processive conformational coupling: through successive cycles of ATP turnover, closed, hexameric AAA+ rings can perpetually cycle through various conformational states to processively act on target substrates

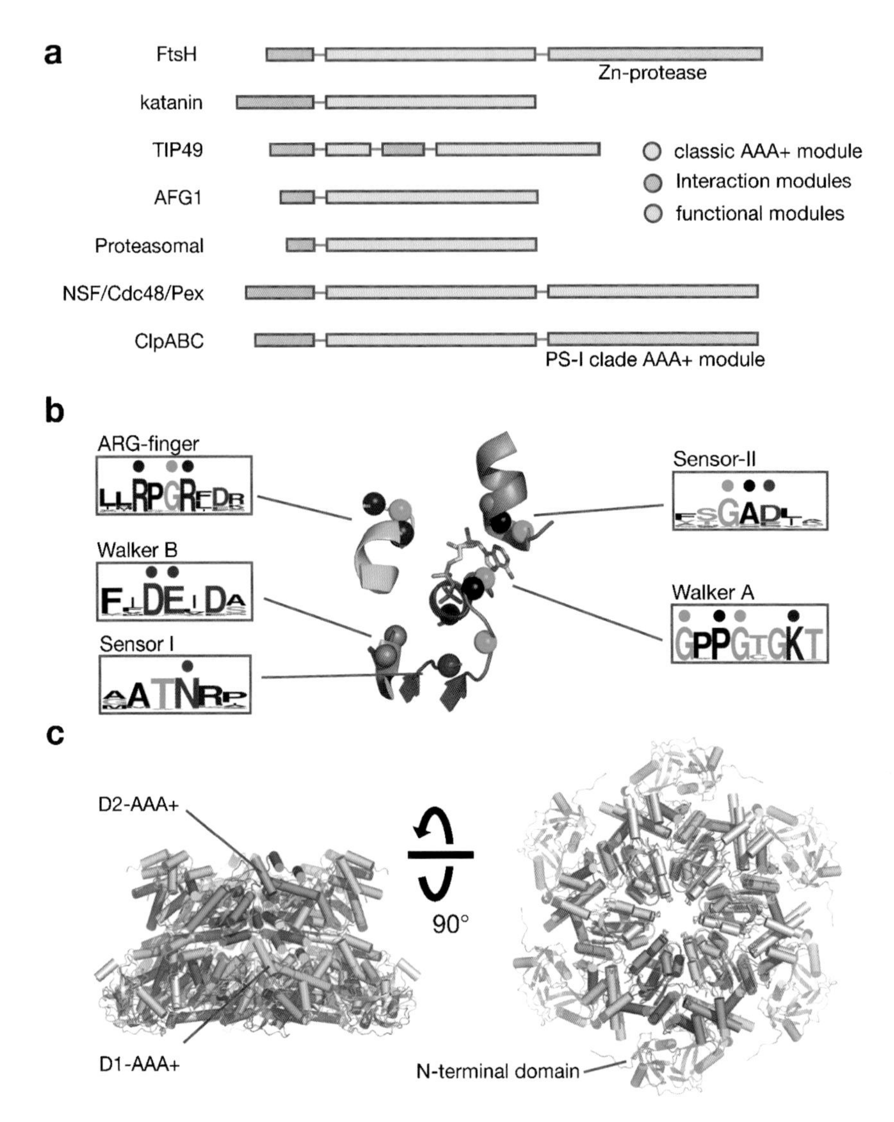

Figure 5

Conservation of catalytic residues in the classic clade. (*a*) Domain organization of the protein families within the classic clade. The proteins in this clade have highly specialized functions in spite of the similarities within their AAA+ modules. (*b*) Classic AAA+ proteins possess unique, conserved sequence features. Letter size is proportional to level of conservation in the five nucleotide interaction motifs among ~300 classic AAA+ proteins. Key residues are marked as colored spheres and their Cα positions are shown in the cartoon depiction of the nucleotide binding pocket of p97 (1E32) (43). (*c*) Side and top views of the full-length structure of hexameric p97, which contains tandemly arrayed AAA+ domains (*light* and *dark blue*) (1R7R) (43).

contribute an additional catalytic activity to the assembly, and adaptor proteins, which mediate substrate recognition (19, 20, 71, 83).The FtsH family within clade 3 offers an excellent example of how a classic AAA+ module can be integrated into a more complex assembly that encompasses several discrete functions (**Figure 5*a***). The primary biological function of FtsH is to maintain the integrity of the cellular envelope through the removal of misfolded membrane proteins (44). The organization of the FtsH polypeptide reflects this specialized function. The N terminus of the protein contains a membrane-targeting domain that promotes the engagement of membrane-embedded proteins. Its AAA+ module, whose monomeric structure has been solved (53, 73), resides in the middle of the protein and is followed by a protease domain at the C terminus. The assembled FtsH hexamer therefore contains three interconnected functionalities. The N-terminal domain serves as an anchoring element and promotes the assembly of the hexamer (2), and the AAA+ module translocates misfolded polypeptides through the central cavity of the complex for degradation by the adjacent protease domain.

In addition to FtsH, four other subfamilies in the classic clade contain single AAA+ modules (21, 35, 58, 104). Unlike FtsH, however, these factors do not possess large domains with clear functional roles adjoining their AAA+ modules (**Figure 5*a***). Instead, these proteins possess small N-terminal domains or insertions that help mediate interactions with accessory factors that control the substrate specificity of the assembly. For example, the microtubule severing factor katanin requires a second exogenous factor, p80, for appropriate targeting to its substrate (35). Interestingly, katanin is a dimer in solution and assembles into a functional enzyme only on microtubules (36). In this sense, the assembly of the active katanin protein resembles that of the bacterial initiator DnaA, which must bind to replication origins to assemble into its activated form (69).

The remaining subfamilies of the classic clade are characterized by tandem arrays of AAA+ modules (42, 94) (**Figure 5*a,c***). In the NSF/Cdc48/Pex family, the two AAA+ domains within each subunit arose from an internal duplication event (45). For Cdc48 and NSF, one of these modules has a degenerate nucleotide-binding motif that lacks significant ATPase activity and appears to function primarily as a hexamerization domain. Surprisingly, the nonfunctional AAA+ modules are swapped in Cdc48 and NSF, one of many subtle differences between these related proteins (10).

The other family to contain a tandemly arrayed AAA+ fold comprises the ClpABC proteins. Unlike the NSF/Cdc48 group, however, only the N-terminal AAA+ domain (ClpABC-NTD) shares the signature motifs of the classic clade; the C-terminal module (ClpABC-CTD) instead contains a β-hairpin insertion upstream of the S-I motif that is characteristic of the AAA+ proteins in clade 5. It therefore appears that the ClpABC family has originated from the fusion of two more distantly related AAA+ modules rather than through a gene duplication event. Some evidence from mutagenesis studies of Hsp104, the yeast ClpAB homolog, suggests that the classic domain AAA+ fold forms the active subunit and that the other AAA+ module promotes complex assembly (37).

HCLR: HslU, ClpABC-CTD, LonAB, RuvB

AAA+ lid: the lid is another term for the C-terminal helical bundle that sits atop the nucleotide-binding pocket of the ASCE core

The Pre-Sensor I Insertion Superclade

The PS-I superclade is composed of four clades of AAA+ modules that share a characteristic β-hairpin insertion between helix 3 and strand 4 (**Figure 3**) (45). Clade 5, composed of the HslU, ClpAB-CTD, LonAB, and RuvB families (HCLR clade), represents the central group of the PS-I superclade, as the PS-I insertion is the only feature that distinguishes it from clade 1 (**Figures 2** and **3**). Other members of the PS-I superclade differ from HCLR proteins because they have replaced the AAA+ lid with an unusual bundle formed by elements N- and C-terminal to the ASCE core (clade 4), or because they contain an additional β-hairpin insertion that disrupts helix $\alpha2$ (clades 6 and 7).

One intriguing question raised by these insertions is whether their roles in the various PS-I clades are conserved, or if they have been adapted and modified to mediate family-specific interactions. The PS-I and

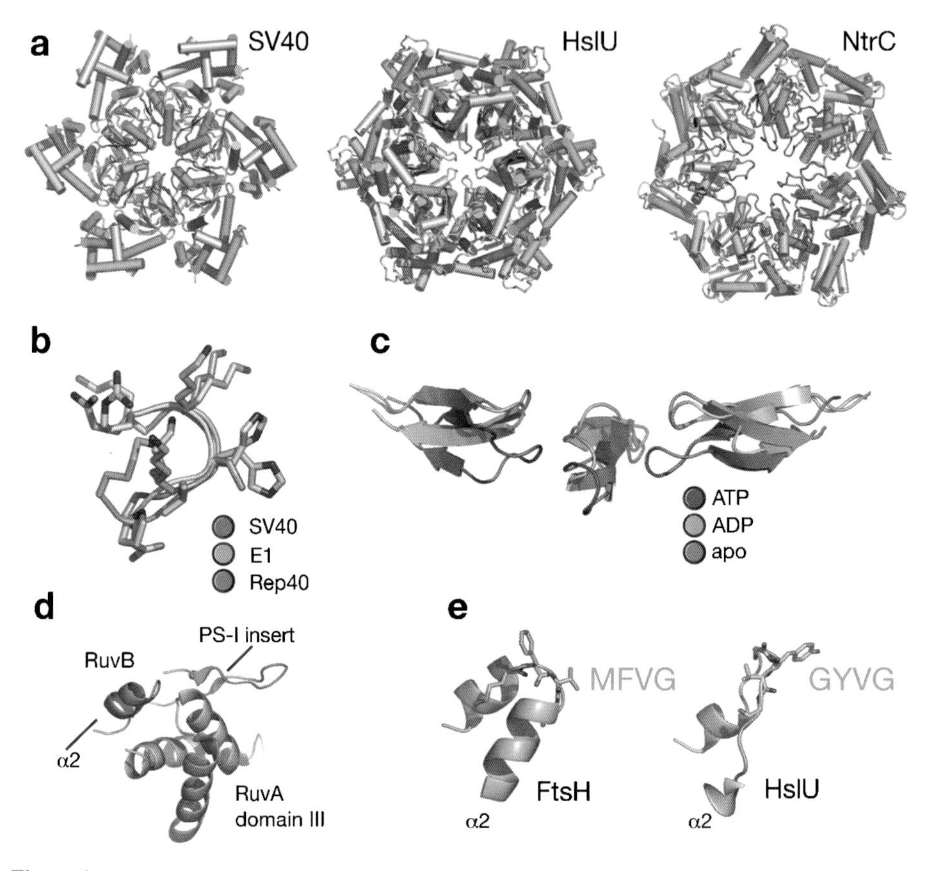

Figure 6

Role of pore loops in the PS-I superfamily clade. (*a*) Oligomeric ring structures of SV40 (clade 4–1SVM) (27), HslU (clade 5–1DO0) (6), and NtrC (Clade6–1NY5) (57), highlighting the location of the structural elements: H2-loop (*green*), PS-I insert (*red*), and H2 insert (*blue*). (*b*) Overlay of the PS-I end loop from three superfamily 3 helicases (1S9H, 1UOJ, and 1SVM) (1, 27, 47). Residues are shown as sticks and colored according to the legend. (*c*) Nucleotide-dependent motion of the PS-I insert of the SV40 helicase. The overlayed domains of three of the six subunits are shown and colored according to the legend. (*d*) The PS-I insert of RuvB mediates interactions with RuvA domain III (1ISX) (105). (*e*) The distantly related unfoldases FtsH (clade 3–1LV7) (53) and HslU (clade 5–1DO0) (6) have a similar, functionally essential sequence motif at the top of helix α2.

H2: helix 2

helix 2 (H2) β-hairpin inserts are located in the vicinity of the central pore of their respective assemblies (**Figure 6*a***) and may have evolved to be additional substrate-interaction elements. This hypothesis has been investigated in structural and biochemical studies of the SV40 large T-antigen and the RuvB Holliday junction resolvase (27, 32, 85, 105). In the SV40 helicase, the β-hairpin juts prominently into the central pore of the AAA+ ring, particularly in the ATP-bound structure (27). Residues at the end of the loop are conserved among the superfamily 3 helicases that make up clade 4 (1, 47, 63) (**Figure 6*b***). When mutated in the SV40 helicase, the DNA-binding and helicase activity of the protein is lost, whereas the ATPase activity remains largely unaffected (85). The PS-I β-hairpin thus appears to link ATP turnover to DNA translocation, a model supported by the paddle-like motions of this region seen in the structures of the SV40 helicase in the apo-, ADP-, and ATP-bound forms (**Figure 6*c***) (27).

The RuvB protein, a clade 5 member, also translocates along double-stranded DNA to promote branch migration during bacterial recombination events. Because of this functional similarity to the SV40 helicase, a similar function may have been assumed for the RuvB β-hairpin. However, mutations in the RuvB β-hairpin residues do not disrupt interactions with DNA, but rather abolish its interaction with the RuvA protein (32). This finding was corroborated in the structure of a complex between RuvA and RuvB, which showed that the PS-I β-hairpin interacts with domain III of RuvA (105) (**Figure 6*d***).

The role of the β-hairpin in the HslU protease has not been directly investigated because this element is more distal from the pore (**Figure 6*a***). Interestingly, the pore-defining elements of HslU, ClpX, Lon, and ClpA-CTD are derived from the loop linking strand β2 and helix α2 of the AAA+ core (86, 89). This loop is also the structural element responsible for substrate interaction in FtsH, a clade 3 protein (76, 106). Indeed, the residues located at the apex of this loop are remarkably similar in these distantly related proteins (**Figure 6*e***). It therefore appears that this element represents a form of convergence between these distantly related unfoldases, as they are not conserved in more closely related homologs.

Oligomeric architectural information for clade 6, characterized by a second β-hairpin insert in helix α2, is currently restricted to the NtrC transcriptional activator (57). The structure of this assembly reveals an unusual heptameric arrangement that departs from the more traditional hexameric state seen with ring AAA+ proteins. Although such higher-order assemblies have been seen in other AAA+ proteins, such as the archaeal MCM protein (110), the functional relevance of this state is uncertain. An inspection of the NtrC structure nevertheless reveals that the prominent feature within the central pore is the H2 helix insert (**Figure 6*a***). Mutations in the H2-insert sequence disrupt the interaction of NtrC and its related proteins with their remodeling target, the σ54 RNA polymerase (57).

A Unique Repositioning of the Sensor-II Domain Defines a Distinct AAA+ Clade

Iyer et al. (45) originally defined three clades within the larger PS-I superclade: the viral superfamily III helicases (clade 4), the HCLR clade (clade 5), and the H2 insert family of proteins (**Figures 2** and **3**). However, two structures of proteins from the H2 insert family, the NtrC and the BchI Mg^{2+} chelatase, reveal significant organizational differences between the two proteins (24, 57). Most notably, the crystal structure of BchI, solved as a monomer, revealed an unusual configuration between the AAA+ core of the protein and its associated C-terminal helical bundle. Rather than occupying a position typically seen for this lid domain, whereby the helical bundle clamps over the top of the αβα-fold, this region is interrupted by a long helical insertion that repositions helices α6 and α7 to the back of the protein (**Figure 7*a***). One proposed explanation for this arrangement is that BchI may undergo a conformational change that would reorient the lid to a more canonical position (45).

However, a close inspection of the BchI structure, together with phylogenetic analysis, indicates that this structural organization of the lid may be a specific feature that defines a seventh clade, referred to hereafter as the pre-sensor II (PS-II) insertion clade. An important line of evidence in this reasoning is that the helical insert and the displaced lid make extensive hydrophobic contacts with the ASCE core, arguing that this conformational state is not transient. Furthermore, when two monomers of BchI are aligned onto a functional dimer of NtrC (**Figure 7**) or HslU (24), helices α6 and α7 appear "domain swapped" onto an adjacent monomer, reconstituting a functional lid with the neighboring α5 helix (**Figure 7**) (34). This novel arrangement

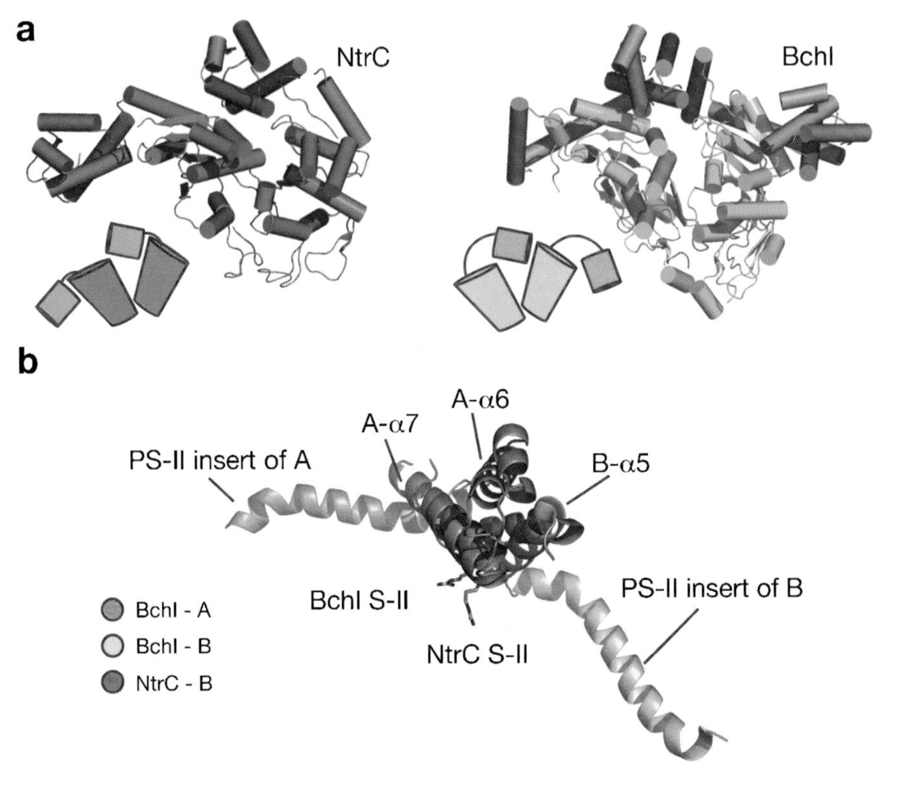

Figure 7

Clade 7 is defined by an unusual rearrangement of the AAA+ lid. (*a*) Schematic drawing of the intraprotomer domain rearrangement observed in the crystal structure of the Mg^{2+} chelatase BchI (1G8P) (24), compared with its closest structural homolog, NtrC (*left*, 1NY5) (57). BchI contains a helical insertion after helix 5 (PS-II insert, **Figure 3**), which repositions helices 6 and 7 to the "back" of the protein. The AAA+ core and AAA+ lid are colored blue and red and green and orange for BchI and NtrC, respectively. (*Right*) Model of a BchI dimer superimposed on NtrC. Two monomers of BchI were superimposed by least squares fitting of the core ASCE elements onto adjacent monomers of NtrC within the oligomeric NtrC assembly. Domains are colored according to the color scheme described for panel *a*. The model shows that the basic AAA+ arrangement is preserved, with the displaced domain acting as a lid for the neighboring monomer. (*b*) Detail of BchI lid arrangement within the oligomeric model. The BchI lid is a hybrid of helix 5 contributed by one subunit (B; *yellow*) and helices 6 and 7 from the adjacent subunit (A; *orange*). Helices 5 and 6 are separated by a long linker helix, the PS-II insert. The lid domain of a single NtrC monomer is shown superposed in gray as a reference.

PS-II: pre-sensor II

of helices appears to correctly position the S-II arginine onto the active site, albeit now as an essential *trans*-acting element (34) (**Figure 7*b***). Rather than affecting ATP turnover directly, this arrangement more likely is used to affect the stability or coordination of subunits in the AAA+ array by changing the contacts and extent of buried surface area between adjacent protomers.

We have examined H2-insert clade members for the presence of a helical insertion and used this criterion to place the MCM, MoxR, YifB, and dynein AAA+ families in clade 7, keeping the NtrC and McrB (75) families as the major groups within clade 6. In this regard, clade 7 encompasses a series of remarkably divergent members. MCM proteins are postulated to be archaeal and eukaryal helicases and one of the central players in DNA replication (96). MoxR, although not well characterized, appears to act as a chaperone in the assembly of specific enzymatic complexes

(97). Dynein and its related protein Midasin possess six tandemly arrayed AAA+ modules within a single polypeptide chain, all of which appear to have the PS-II insertion (28, 82). Dynein is a molecular motor responsible for transporting vesicles, organelles, and chromosomes along microtubules, as well as powering eukaryotic flagella. Midasin likewise has a transport function, guiding the movement of the eukaryotic 60S ribosome through the nuclear pore complex (28).

AAA+ Proteins: Common Mechanism Versus Specialized Adaptations

One of the central problems surrounding AAA+ protein function is to determine the mechanistic overlap between disparate clades. Given the multitude of catalytic rates and substrate specificities characterized for this superfamily to date, it appears likely that we have now reached the point where discussions of the commonalities among AAA+ proteins will cease to be the dominant trend. Instead, a full understanding of the functional properties and adaptations that have evolved within specific clades is needed to tease apart clade-specific biological differences. These distinctions are only now beginning to emerge from joint structural, bioinformatic, and biochemical studies of various AAA+ assemblies.

An emerging difference is that the oligomerized state of clade 1 and 2 AAA+ proteins differs significantly from that of other AAA+ proteins, most of which form closed hexameric rings. This divergence may relate to the fact that these assemblies are not processive molecular machines, but rather switches whose remodeling functions are carried out in a single, highly regulated event. The structural changes mediated by these assemblies are not dependent on ATP turnover per se (38, 69). Instead, ATP binding appears to flip these proteins into an activated state by influencing the conformation of the AAA+ modules and enabling them to remodel their respective targets. In this model, nucleotide hydrolysis and ADP release merely reset the switch for the next catalytic event (8, 13, 69).

Thus, on one level, clear mechanistic distinctions can be made between AAA+ proteins, such as RFC and DnaA, which form open-ring structures and behave in a simple, switch-like fashion, and toroidal AAA+ assemblies. For closed-ring oligomers, one central question that remains is the relationship between nucleotide hydrolysis and the conformational changes between AAA+ subunits. Some answers are now beginning to emerge from biochemical and X-ray crystallographic studies. Structural work on the SV40 helicase by Chen and colleagues (27) indicated that structural rearrangements among the six nucleotide-binding sites may occur in a concerted manner. This mechanism posits that a cycle of six simultaneous ATP binding and hydrolysis events generates paddle-like movements of the PS-I insert (**Figure 6*c***). By contrast, biochemical studies of the ClpX protease, a close homolog of HslU, revealed that distinct nucleotide-binding sites can function independently within the oligomerized assembly of the protein (40). Although such behavior could be taken as evidence of a rotary type of mechanism, in which ATP binding and hydrolysis cycles around the hexameric assembly, recent studies of an engineered ClpX protein that contains various combinations of active and inactive subunits fused together in a single polypeptide indicate that ATP hydrolysis and substrate engagement are stochastic and to some extent independent of the nucleotide status of neighboring subunits in the complex (68). Consistent with these observations, multiple nucleotide-binding states have been observed in the crystal structures of the HslU protease (6, 89, 90, 100, 101). Distinct conformational changes are also observed in electron microscopy and scattering-angle X-ray spectroscopy structural studies of the Cdc48/p97 tandem AAA+ array (4, 16, 79). These findings, along with recent X-ray structures of the full-length p97 assembly, should pave the way to distinguish the type of conformational coupling observed among tandemly arrayed AAA+ proteins (17, 43).

A common theme to emerge from these findings is that there is little correlation between AAA+ protein subtypes and a specific remodeling activity. This may suggest that the evolution of AAA+ machines involved the initial emergence of a small, defined number of AAA+ clades that subsequently expanded and adapted to allow the processing of a wide variety of targets, in some cases reconverging on common solutions for specific tasks. The information derived from the continued study of AAA+ proteins will help clarify these issues and will influence our understanding of how molecular assemblies evolve and adapt.

NOTE ADDED IN PROOF

Recently, Aravind and colleagues (60) described the phylogenetic characterization of a subfamily of ASCE NTPases. Members of this group, the STAND proteins, are related to the AAA+ family in overall topology but lack any clear Sensor-I, Sensor-II or box VII sequences. Recently, the structures of two STAND ATPases, the cell death protease activating factors Apaf-1 and Ced-4, were described (78, 107), allowing the AAA+/STAND relationship to be examined more carefully. The structures of Apaf-1 and Ced-4 are most closely related to the Cdc6/Orc1 subfamily of clade 2 and share the helical insert that characterizes these proteins. There are significant differences, however, in their oligomeric interactions; in addition to lacking classic AAA+ motifs, the dimerization interface of Ced-4 involves structural elements, such as a conserved N-terminal helix, that are different from those observed in AAA+ proteins. These findings suggest that although STAND proteins are likely evolutionarily related to clade 2 AAA+ proteins, they have evolved a distinct mode of assembly. Further structural and biochemical work will help define the precise relationship between STAND and AAA+ proteins.

SUMMARY POINTS

1. AAA+ proteins contain a conserved ATP-binding module that is activated by the formation of an oligomeric assembly.
2. Seven major clades of AAA+ proteins have been defined on the basis of sequence alignments and structural information
3. Clades 1 and 2 are characterized by open-ring assemblies that act as single-turnover allosteric switches.
4. Pore-lining motifs in the central channel of hexameric AAA+ assemblies are highly adaptable structural elements that can interact with a variety of substrates and accessory proteins.
5. Both concerted and sequential nucleotide-binding and hydrolysis mechanisms may be employed by AAA+ hexamers.
6. Clade 7 AAA+ proteins, which include the MCMs, dynein, and metal chelatases, appear to share an unusual, domain-swapped lid conformation.

ACKNOWLEDGMENTS

We would like to thank L. Aravind, Cynthia Wolberger, John Kuriyan, Jasper Rine, and Berger Lab members for helpful comments and suggestions. We gratefully acknowledge the support of the G. Harold and Leila Y. Mathers Charitable Foundation and the NIH (GM071747).

LITERATURE CITED

1. Abbate EA, Berger JM, Botchan MR. 2004. The X-ray structure of the papillomavirus helicase in complex with its molecular matchmaker E2. *Genes Dev.* 18:1981–96
2. Akiyama Y, Ito K. 2000. Roles of multimerization and membrane association in the proteolytic functions of FtsH (HflB). *EMBO J.* 19:3888–95
3. Baas PW, Karabay A, Qiang L. 2005. Microtubules cut and run. *Trends Cell Biol.* 15:518–24
4. Beuron F, Flynn TC, Ma J, Kondo H, Zhang X, Freemont PS. 2003. Motions and negative cooperativity between p97 domains revealed by cryo-electron microscopy and quantised elastic deformational model. *J. Mol. Biol.* 327:619–29
5. Beyer A. 1997. Sequence analysis of the AAA protein family. *Protein Sci* 6:2043–58
6. Bochtler M, Hartmann C, Song HK, Bourenkov GP, Bartunik HD, Huber R. 2000. The structures of HsIU and the ATP-dependent protease HsIU-HsIV. *Nature* 403:800–5
7. Bowers JL, Randell JC, Chen S, Bell SP. 2004. ATP hydrolysis by ORC catalyzes reiterative Mcm2–7 assembly at a defined origin of replication. *Mol. Cell* 16:967–78
8. Bowman GD, Goedken ER, Kazmirski SL, O'Donnell M, Kuriyan J. 2005. DNA polymerase clamp loaders and DNA recognition. *FEBS Lett.* 579:863–67
9. **Bowman GD, O'Donnell M, Kuriyan J. 2004. Structural analysis of a eukaryotic sliding DNA clamp-clamp loader complex. *Nature* 429:724–30**
10. Brunger AT, DeLaBarre B. 2003. NSF and p97/VCP: similar at first, different at last. *FEBS Lett.* 555:126–33
11. Confalonieri F, Duguet M. 1995. A 200-amino acid ATPase module in search of a basic function. *Bioessays* 17:639–50
12. Cunningham EL, Berger JM. 2005. Unraveling the early steps of prokaryotic replication. *Curr. Opin. Struct. Biol.* 15:68–76
13. Davey MJ, Fang L, McInerney P, Georgescu RE, O'Donnell M. 2002. The DnaC helicase loader is a dual ATP/ADP switch protein. *EMBO J.* 21:3148–59
14. Davey MJ, Indiani C, O'Donnell M. 2003. Reconstitution of the Mcm2–7p heterohexamer, subunit arrangement, and ATP site architecture. *J. Biol. Chem.* 278:4491–99
15. Davey MJ, Jeruzalmi D, Kuriyan J, O'Donnell M. 2002. Motors and switches: AAA+ machines within the replisome. *Nat. Rev. Mol. Cell Biol.* 3:826–35
16. Davies JM, Tsuruta H, May AP, Weis WI. 2005. Conformational changes of p97 during nucleotide hydrolysis determined by small-angle X-ray scattering. *Structure* 13:183–95
17. **DeLaBarre B, Brunger AT. 2003. Complete structure of p97/valosin-containing protein reveals communication between nucleotide domains. *Nat. Struct. Biol.* 10:856–63**
18. Donovan S, Diffley JF. 1996. Replication origins in eukaryotes. *Curr. Opin. Genet. Dev.* 6:203–7
19. Dougan DA, Mogk A, Zeth K, Turgay K, Bukau B. 2002. AAA+ proteins and substrate recognition, it all depends on their partner in crime. *FEBS Lett.* 529:6–10
20. Dreveny I, Pye VE, Beuron F, Briggs LC, Isaacson RL, et al. 2004. p97 and close encounters of every kind: a brief review. *Biochem. Soc. Trans.* 32:715–20

9. The crystal structure of RFC bound to PCNA suggested a mechanism for DNA engagement by the AAA+ clamp loader assembly.

17. The first structure of a tandemly arrayed, classic AAA+ protein.

21. Dubiel W, Ferrell K, Rechsteiner M. 1995. Subunits of the regulatory complex of the 26S protease. *Mol. Biol. Rep.* 21:27–34
22. Erzberger JP, Pirruccello MM, Berger JM. 2002. The structure of bacterial DnaA: implications for general mechanisms underlying DNA replication initiation. *EMBO J.* 21:4763–73
23. Felczak MM, Kaguni JM. 2004. The box VII motif of *Escherichia coli* DnaA protein is required for DnaA oligomerization at the *E. coli* replication origin. *J. Biol. Chem.* 279:51156–62
24. Fodje MN, Hansson A, Hansson M, Olsen JG, Gough S, et al. 2001. Interplay between an AAA module and an integrin I domain may regulate the function of magnesium chelatase. *J. Mol. Biol.* 311:111–22
25. Frickey T, Lupas AN. 2004. Phylogenetic analysis of AAA proteins. *J. Struct. Biol.* 146:2–10
26. Frohlich KU. 2001. An AAA family tree. *J. Cell Sci.* 114:1601–2
27. **Gai D, Zhao R, Li D, Finkielstein CV, Chen XS. 2004. Mechanisms of conformational change for a replicative hexameric helicase of SV40 large tumor antigen. *Cell* 119:47–60**

27. A series of structures of the SV40 helicase suggests a concerted mechanism for nucleotide binding and hydrolysis.

28. Garbarino JE, Gibbons IR. 2002. Expression and genomic analysis of midasin, a novel and highly conserved AAA protein distantly related to dynein. *BMC Genomics* 3:18
29. Goedken ER, Kazmirski SL, Bowman GD, O'Donnell M, Kuriyan J. 2005. Mapping the interaction of DNA with the *Escherichia coli* DNA polymerase clamp loader complex. *Nat. Struct. Mol. Biol.* 12:183–90
30. Gomis-Ruth FX, Moncalian G, Perez-Luque R, Gonzalez A, Cabezon E, et al. 2001. The bacterial conjugation protein TrwB resembles ring helicases and F1-ATPase. *Nature* 409:637–41
31. Guenther B, Onrust R, Sali A, O'Donnell M, Kuriyan J. 1997. Crystal structure of the delta' subunit of the clamp-loader complex of *E. coli* DNA polymerase III. *Cell* 91:335–45
32. Han YW, Iwasaki H, Miyata T, Mayanagi K, Yamada K, et al. 2001. A unique beta-hairpin protruding from AAA+ ATPase domain of RuvB motor protein is involved in the interaction with RuvA DNA recognition protein for branch migration of Holliday junctions. *J. Biol. Chem.* 276:35024–28
33. Hanson PI, Whiteheart SW. 2005. AAA+ proteins: have engine, will work. *Nat. Rev. Mol. Cell. Biol.* 6:519–29
34. Hansson A, Willows RD, Roberts TH, Hansson M. 2002. Three semidominant barley mutants with single amino acid substitutions in the smallest magnesium chelatase subunit form defective AAA+ hexamers. *Proc. Natl. Acad. Sci. USA* 99:13944–49
35. Hartman JJ, Mahr J, McNally K, Okawa K, Iwamatsu A, et al. 1998. Katanin, a microtubule-severing protein, is a novel AAA ATPase that targets to the centrosome using a WD40-containing subunit. *Cell* 93:277–87
36. Hartman JJ, Vale RD. 1999. Microtubule disassembly by ATP-dependent oligomerization of the AAA enzyme katanin. *Science* 286:782–85
37. Hattendorf DA, Lindquist SL. 2002. Cooperative kinetics of both Hsp104 ATPase domains and interdomain communication revealed by AAA sensor-1 mutants. *EMBO J.* 21:12–21
38. Henneke G, Gueguen Y, Flament D, Azam P, Querellou J, et al. 2002. Replication factor C from the hyperthermophilic archaeon *Pyrococcus abyssi* does not need ATP hydrolysis for clamp-loading and contains a functionally conserved RFC PCNA-binding domain. *J. Mol. Biol.* 323:795–810

39. Herman C, Prakash S, Lu CZ, Matouschek A, Gross CA. 2003. Lack of a robust unfoldase activity confers a unique level of substrate specificity to the universal AAA protease FtsH. *Mol. Cell* 11:659–69
40. Hersch GL, Burton RE, Bolon DN, Baker TA, Sauer RT. 2005. Asymmetric interactions of ATP with the AAA+ ClpX6 unfoldase: allosteric control of a protein machine. *Cell* 121:1017–27
41. Hinnerwisch J, Fenton WA, Furtak KJ, Farr GW, Horwich AL. 2005. Loops in the central channel of ClpA chaperone mediate protein binding, unfolding, and translocation. *Cell* 121:1029–41
42. Hoskins JR, Sharma S, Sathyanarayana BK, Wickner S. 2001. Clp ATPases and their role in protein unfolding and degradation. *Adv. Protein Chem.* 59:413–29
43. **Huyton T, Pye VE, Briggs LC, Flynn TC, Beuron F, et al. 2003. The crystal structure of murine p97/VCP at 3.6 A. *J. Struct. Biol.* 144:337–48**

43. Another glimpse of the full-length p97 structure.

44. Ito K, Akiyama Y. 2005. Cellular functions, mechanism of action, and regulation of FtsH protease. *Annu. Rev. Microbiol.* 59:211–31
45. **Iyer LM, Leipe DD, Koonin EV, Aravind L. 2004. Evolutionary history and higher order classification of AAA+ ATPases. *J Struct Biol* 146:11–31**

45. A highly informative classification scheme for AAA+ proteins based on sequence and structural information in which several AAA+ clades are defined.

46. Iyer LM, Makarova KS, Koonin EV, Aravind L. 2004. Comparative genomics of the FtsK-HerA superfamily of pumping ATPases: implications for the origins of chromosome segregation, cell division and viral capsid packaging. *Nucleic Acids Res.* 32:5260–79
47. James JA, Escalante CR, Yoon-Robarts M, Edwards TA, Linden RM, Aggarwal AK. 2003. Crystal structure of the SF3 helicase from adeno-associated virus type 2. *Structure* 11:1025–35
48. Jeruzalmi D, O'Donnell M, Kuriyan J. 2001. Crystal structure of the processivity clamp loader gamma (gamma) complex of *E. coli* DNA polymerase III. *Cell* 106:429–41
49. Jeruzalmi D, Yurieva O, Zhao Y, Young M, Stewart J, et al. 2001. Mechanism of processivity clamp opening by the delta subunit wrench of the clamp loader complex of *E. coli* DNA polymerase III. *Cell* 106:417–28
50. Kaguni JM. 1997. Escherichia coli DnaA protein: the replication initiator. *Mol. Cell* 7:145–57
51. Kato J, Katayama T. 2001. Hda, a novel DnaA-related protein, regulates the replication cycle in *Escherichia coli*. *EMBO J* 20:4253–62
52. Kenniston JA, Baker TA, Fernandez JM, Sauer RT. 2003. Linkage between ATP consumption and mechanical unfolding during the protein processing reactions of an AAA+ degradation machine. *Cell* 114:511–20
53. Krzywda S, Brzozowski AM, Verma C, Karata K, Ogura T, Wilkinson AJ. 2002. The crystal structure of the AAA domain of the ATP-dependent protease FtsH of *Escherichia coli* at 1.5 A resolution. *Structure* 10:1073–83
54. Kunau WH, Beyer A, Franken T, Gotte K, Marzioch M, et al. 1993. Two complementary approaches to study peroxisome biogenesis in *Saccharomyces cerevisiae*: forward and reversed genetics. *Biochimie* 75:209–24
55. Langer T. 2000. AAA proteases: cellular machines for degrading membrane proteins. *Trends Biochem. Sci.* 25:247–51
56. Lee DG, Bell SP. 2000. ATPase switches controlling DNA replication initiation. *Curr. Opin. Cell Biol.* 12:280–85
57. Lee SY, De La Torre A, Yan D, Kustu S, Nixon BT, Wemmer DE. 2003. Regulation of the transcriptional activator NtrC1: structural studies of the regulatory and AAA+ ATPase domains. *Genes Dev.* 17:2552–63

58. Lee YJ, Wickner RB. 1992. AFG1, a new member of the SEC18-NSF, PAS1, CDC48-VCP, TBP family of ATPases. *Yeast* 8:787–90
59. Leipe DD, Koonin EV, Aravind L. 2003. Evolution and classification of P-loop kinases and related proteins. *J. Mol. Biol.* 333:781–815
60. Leipe DD, Koonin EV, Aravind L. 2004. STAND, a class of P-loop NTPases including animal and plant regulators of programmed cell death: multiple, complex domain architectures, unusual phyletic patterns, and evolution by horizontal gene transfer. *J. Mol. Biol.* 343:1–28
61. Leipe DD, Wolf YI, Koonin EV, Aravind L. 2002. Classification and evolution of P-loop GTPases and related ATPases. *J. Mol. Biol.* 317:41–72
62. Lenzen CU, Steinmann D, Whiteheart SW, Weis WI. 1998. Crystal structure of the hexamerization domain of N-ethylmaleimide-sensitive fusion protein. *Cell* 94:525–36
63. Li D, Zhao R, Lilyestrom W, Gai D, Zhang R, et al. 2003. Structure of the replicative helicase of the oncoprotein SV40 large tumour antigen. *Nature* 423:512–18
64. Liang C, Weinreich M, Stillman B. 1995. ORC and Cdc6p interact and determine the frequency of initiation of DNA replication in the genome. *Cell* 81:667–76
65. Liou GG, Tanny JC, Kruger RG, Walz T, Moazed D. 2005. Assembly of the SIR complex and its regulation by O-acetyl-ADP-ribose, a product of NAD-dependent histone deacetylation. *Cell* 121:515–27
66. Liu J, Smith CL, DeRyckere D, DeAngelis K, Martin GS, Berger JM. 2000. Structure and function of Cdc6/Cdc18: implications for origin recognition and checkpoint control. *Mol. Cell* 6:637–48
67. Lupas AN, Martin J. 2002. AAA proteins. *Curr Opin Struct Biol* 12:746–53
68. **Martin A, Baker TA, Sauer RT. 2005. Rebuilt AAA+ motors reveal operating principles for ATP-fuelled machines. *Nature* 437:1115–20**

68. First biochemical evidence of a stochastic ATP hydrolyzing mechanism within a clade 5 AAA+ assembly.

69. Messer W. 2002. The bacterial replication initiator DnaA. DnaA and oriC, the bacterial mode to initiate DNA replication. *FEMS Microbiol. Rev.* 26:355–74
70. Mizushima T, Takahashi N, Stillman B. 2000. Cdc6p modulates the structure and DNA binding activity of the origin recognition complex in vitro. *Genes Dev.* 14:1631–41
71. Mogk A, Dougan D, Weibezahn J, Schlieker C, Turgay K, Bukau B. 2004. Broad yet high substrate specificity: the challenge of AAA+ proteins. *J. Struct. Biol.* 146:90–98
72. Neuwald AF, Aravind L, Spouge JL, Koonin EV. 1999. AAA+: a class of chaperone-like ATPases associated with the assembly, operation, and disassembly of protein complexes. *Genome Res.* 9:27–43
73. Niwa H, Tsuchiya D, Makyio H, Yoshida M, Morikawa K. 2002. Hexameric ring structure of the ATPase domain of the membrane-integrated metalloprotease FtsH from *Thermus thermophilus* HB8. *Structure* 10:1415–23
74. Ogura T, Whiteheart SW, Wilkinson AJ. 2004. Conserved arginine residues implicated in ATP hydrolysis, nucleotide-sensing, and inter-subunit interactions in AAA and AAA+ ATPases. *J. Struct. Biol.* 146:106–12
75. Panne D, Muller SA, Wirtz S, Engel A, Bickle TA. 2001. The McrBC restriction endonuclease assembles into a ring structure in the presence of G nucleotides. *EMBO J.* 20:3210–17
76. Park E, Rho YM, Koh OJ, Ahn SW, Seong IS, et al. 2005. Role of the GYVG pore motif of HslU ATPase in protein unfolding and translocation for degradation by HslV peptidase. *J. Biol. Chem.* 280:22892–98

77. Putnam CD, Clancy SB, Tsuruta H, Gonzalez S, Wetmur JG, Tainer JA. 2001. Structure and mechanism of the RuvB Holliday junction branch migration motor. *J. Mol. Biol.* 311:297–310
78. Riedl SJ, Li W, Chao Y, Schwarzenbacher R, Shi Y. 2005. Structure of the apoptotic protease-activating factor 1 bound to ADP. *Nature* 434:926–33
79. Rouiller I, Butel VM, Latterich M, Milligan RA, Wilson-Kubalek EM. 2000. A major conformational change in p97 AAA ATPase upon ATP binding. *Mol. Cell* 6:1485–90
80. Rouiller I, DeLaBarre B, May AP, Weis WI, Brunger AT, et al. 2002. Conformational changes of the multifunction p97 AAA ATPase during its ATPase cycle. *Nat. Struct. Biol.* 9:950–57
81. Rusche LN, Kirchmaier AL, Rine J. 2003. The establishment, inheritance, and function of silenced chromatin in *Saccharomyces cerevisiae*. *Annu. Rev. Biochem.* 72:481–516
82. Sakato M, King SM. 2004. Design and regulation of the AAA+ microtubule motor dynein. *J. Struct. Biol.* 146:58–71
83. Sauer RT, Bolon DN, Burton BM, Burton RE, Flynn JM, et al. 2004. Sculpting the proteome with AAA(+) proteases and disassembly machines. *Cell* 119:9–18
84. Sekimizu K, Bramhill D, Kornberg A. 1987. ATP activates dnaA protein in initiating replication of plasmids bearing the origin of the *E. coli* chromosome. *Cell* 50:259–65
85. Shen J, Gai D, Patrick A, Greenleaf WB, Chen XS. 2005. The roles of the residues on the channel beta-hairpin and loop structures of simian virus 40 hexameric helicase. *Proc. Natl. Acad. Sci. USA* 102:11248–53
86. Siddiqui SM, Sauer RT, Baker TA. 2004. Role of the processing pore of the ClpX AAA+ ATPase in the recognition and engagement of specific protein substrates. *Genes Dev.* 18:369–74
87. Singleton MR, Morales R, Grainge I, Cook N, Isupov MN, Wigley DB. 2004. Conformational changes induced by nucleotide binding in Cdc6/ORC from *Aeropyrum pernix*. *J. Mol. Biol.* 343:547–57
88. Singleton MR, Sawaya MR, Ellenberger T, Wigley DB. 2000. Crystal structure of T7 gene 4 ring helicase indicates a mechanism for sequential hydrolysis of nucleotides. *Cell* 101:589–600
89. Song HK, Hartmann C, Ramachandran R, Bochtler M, Behrendt R, et al. 2000. Mutational studies on HslU and its docking mode with HslV. *Proc. Natl. Acad. Sci. USA* 97:14103–8
90. Sousa MC, Trame CB, Tsuruta H, Wilbanks SM, Reddy VS, McKay DB. 2000. Crystal and solution structures of an HslUV protease-chaperone complex. *Cell* 103:633–43
91. Story RM, Steitz TA. 1992. Structure of the recA protein-ADP complex. *Nature* 355:374–76
92. Su'etsugu M, Kawakami H, Kurokawa K, Kubota T, Takata M, Katayama T. 2001. DNA replication-coupled inactivation of DnaA protein in vitro: a role for DnaA arginine-334 of the AAA+ Box VIII motif in ATP hydrolysis. *Mol. Microbiol.* 40:376–86
93. Su'etsugu M, Shimuta TR, Ishida T, Kawakami H, Katayama T. 2005. Protein associations in DnaA-ATP hydrolysis mediated by the Hda-replicase clamp complex. *J. Biol. Chem.* 280:6528–36
94. Tamura S, Shimozawa N, Suzuki Y, Tsukamoto T, Osumi T, Fujiki Y. 1998. A cytoplasmic AAA family peroxin, Pex1p, interacts with Pex6p. *Biochem. Biophys. Res. Commun.* 245:883–86

95. Tugal T, Zou-Yang XH, Gavin K, Pappin D, Canas B, et al. 1998. The Orc4p and Orc5p subunits of the *Xenopus* and human origin recognition complex are related to Orc1p and Cdc6p. *J. Biol. Chem.* 273:32421–29
96. Tye BK. 1999. MCM proteins in DNA replication. *Annu. Rev. Biochem.* 68:649–86
97. Van Spanning RJ, Wansell CW, De Boer T, Hazelaar MJ, Anazawa H, et al. 1991. Isolation and characterization of the moxJ, moxG, moxI, and moxR genes of *Paracoccus denitrificans*: inactivation of moxJ, moxG, and moxR and the resultant effect on methylotrophic growth. *J. Bacteriol.* 173:6948–61
98. Walker JE, Saraste M, Runswick MJ, Gay NJ. 1982. Distantly related sequences in the alpha- and beta-subunits of ATP synthase, myosin, kinases and other ATP-requiring enzymes and a common nucleotide binding fold. *EMBO J.* 1:945–51
99. Wang J. 2004. Nucleotide-dependent domain motions within rings of the RecA/AAA(+) superfamily. *J. Struct. Biol.* 148:259–67
100. Wang J, Song JJ, Franklin MC, Kamtekar S, Im YJ, et al. 2001. Crystal structures of the HslVU peptidase-ATPase complex reveal an ATP-dependent proteolysis mechanism. *Structure* 9:177–84
101. Wang J, Song JJ, Seong IS, Franklin MC, Kamtekar S, et al. 2001. Nucleotide-dependent conformational changes in a protease-associated ATPase HslU. *Structure* 9:1107–16
102. Weinreich M, Liang C, Stillman B. 1999. The Cdc6p nucleotide-binding motif is required for loading Mcm proteins onto chromatin. *Proc. Natl. Acad. Sci. USA* 96:441–46
103. Wittinghofer A, Scheffzek K, Ahmadian MR. 1997. The interaction of Ras with GTPase-activating proteins. *FEBS Lett.* 410:63–67
104. Wood MA, McMahon SB, Cole MD. 2000. An ATPase/helicase complex is an essential cofactor for oncogenic transformation by c-Myc. *Mol. Cell* 5:321–30
105. Yamada K, Miyata T, Tsuchiya D, Oyama T, Fujiwara Y, et al. 2002. Crystal structure of the RuvA-RuvB complex: a structural basis for the Holliday junction migrating motor machinery. *Mol. Cell* 10:671–81
106. Yamada-Inagawa T, Okuno T, Karata K, Yamanaka K, Ogura T. 2003. Conserved pore residues in the AAA protease FtsH are important for proteolysis and its coupling to ATP hydrolysis. *J. Biol. Chem.* 278:50182–87
107. Yan N, Chai J, Lee ES, Gu L, Liu Q, et al. 2005. Structure of the CED-4-CED-9 complex provides insights into programmed cell death in *Caenorhabditis elegans*. *Nature* 437:831–87
108. Yeo HJ, Savvides SN, Herr AB, Lanka E, Waksman G. 2000. Crystal structure of the hexameric traffic ATPase of the *Helicobacter pylori* type IV secretion system. *Mol. Cell* 6:1461–72
109. Yu RC, Hanson PI, Jahn R, Brunger AT. 1998. Structure of the ATP-dependent oligomerization domain of *N*-ethylmaleimide sensitive factor complexed with ATP. *Nat. Struct. Biol.* 5:803–11
110. Yu X, VanLoock MS, Poplawski A, Kelman Z, Xiang T, et al. 2002. The *Methanobacterium thermoautotrophicum* MCM protein can form heptameric rings. *EMBO Rep.* 3:792–97

Symmetry, Form, and Shape: Guiding Principles for Robustness in Macromolecular Machines

Florence Tama and Charles L. Brooks, III

Department of Molecular Biology, The Scripps Research Institute, La Jolla, California 92037; email: brooks@scripps.edu

Annu. Rev. Biophys. Biomol. Struct. 2006. 35:115–33

First published online as a Review in Advance on January 13, 2006

The *Annual Review of Biophysics and Biomolecular Structure* is online at biophys.annualreviews.org

doi: 10.1146/annurev.biophys.35.040405.102010

1056-8700/06/0609-0115$20.00

Key Words

normal mode, conformational change, hybrid methods, elastic network, molecular machines, ribosome

Abstract

Computational studies of large macromolecular assemblages have come a long way during the past 10 years. With the explosion of computer power and parallel computing, timescales of molecular dynamics simulations have been extended far beyond the hundreds of picoseconds timescale. However, limitations remain for studies of large-scale conformational changes occurring on timescales beyond nanoseconds, especially for large macromolecules. In this review, we describe recent methods based on normal mode analysis that have enabled us to study dynamics on the microsecond timescale for large macromolecules using different levels of coarse graining, from atomically detailed models to those employing only low-resolution structural information. Emerging from such studies is a control principle for robustness in Nature's machines. We discuss this idea in the context of large-scale functional reorganization of the ribosome, virus particles, and the muscle protein myosin.

Contents

INTRODUCTION

It is now well established that large-scale structural rearrangements in proteins and nucleic acids are important for a variety of functions including catalysis and regulation of activity. The recent developments in experimental methods such as cryo-electron microscopy (cryo-EM), small angle X-ray scattering (SAXS), fluorescence resonance energy transfer (FRET), and electron tomography have revealed structures and structural transitions of large molecular assemblies at low resolution. In particular, through the use of these techniques, large functionally important conformational changes have been characterized in assemblies such as the ribosome (23, 96, 99), the RNA polymerase (14), molecular motor proteins (98), GroEL (68), and viruses (24, 42). While most of the information on these dynamical transitions is based on experiment, computational methods must be employed to complement experimental observations. Indeed, by using theory to explore functionally important rearrangements observed in such low-resolution experiments, it is possible to gain insights into the mechanism of these transformations that are presently inaccessible to the experiments.

Cryo-EM: cryo-electron microscopy

SAXS: small-angle X-ray scattering

Ribosome: protein-RNA molecular motor responsible for protein synthesis

NMA: normal mode analysis

B-factor: empirical determined amplitude of thermal motion

To understand these large conformational changes at the atomic level, one must employ theoretical and computational techniques that can allow long timescales to be reached. Here, the use of standard molecular dynamics simulation is limited, as the timescales of these conformational changes are on the order of microseconds. In addition, it is often the case that a high-resolution structure is not available and only low-resolution structural information is available; therefore molecular dynamics simulations cannot be performed. To study motions of macromolecules on longer timescales, normal mode analysis (NMA) provides an alternative to molecular dynamics. In NMA, molecular motions are decomposed into vibrational modes. This approach extends the timescale accessible to theoretical work and is useful for studying collective motions and mechanical properties of biological systems (7, 26).

Recent studies are indicative of the utility of normal modes for studying the dynamics of molecules of biological interest and include B-factor refinement in crystallography (36–38, 94), enzyme specificity (33, 55), the study of large-scale conformational changes of proteins (8, 25, 28, 52, 64, 70, 71) and viruses (67, 74, 81, 101), NMR order parameters (9, 78, 79), electron transfer (3, 4, 59), and vibrational energy transfer (62). NMA has also been used to investigate DNA and RNA dynamics (19, 20, 53, 54). Of particular interest is that the exploration of the normal modes of a molecular system can yield insights, at an atomic level, into the mechanism and pathway of large-scale rearrangements of protein/protein complexes that occur upon ligand/protein binding (8, 16, 25, 28, 50, 52, 64, 66, 69, 71, 88, 89). Studies employing NMA generally focus on a few large-amplitude/low-frequency normal modes, which represent the most facile motions occurring in those systems and are expected to be relevant to function (88).

Although NMA has a long history in its application to investigate dynamics of biological molecules, we have limited the scope of this review to a description of recently developed techniques and their applications to the study of large macromolecular assemblies. This limited focus ignores the contributions of many others to this field. We note that more comprehensive overviews and discussions of the normal mode technique can be found in Reference 6; Ma (49) recently commented on the usefulness and limitations of NMA to model dynamics of biomolecular assemblies. We refer readers to these works for a more comprehensive overview.

We concentrate particularly on the nature of conformational change occurring in large biological machines and on the biological insights that have been gained through the use of NMA. Techniques to calculate normal modes from different levels of structure description, ranging from all-atom calculations to pseudoatomic representations, are presented and illustrated by successful studies that have confirmed or predicted functional motions. In addition, we also discuss the applicability of each of these techniques to specific biological problems and show that by combining NMA methods with experimental data one can deduce atomic-level structural information for large biological complexes. Finally, we summarize the general principle that has emerged from these studies as a determinant for predicting functional motion, i.e., the shape of a biological object determines its mechanical properties.

NORMAL MODE ANALYSIS APPLIED TO LARGE BIOLOGICAL SYSTEMS

Normal Mode Analysis Versus Molecular Dynamics Simulations

Several computational techniques are available to study the dynamics of biological molecules. The standard approach would be molecular dynamics simulations. However, even though computational techniques have evolved and computer power has increased, which enable simulations of more than 1 million atoms (N. Go, personal communication) and over longer timescales (34), to date the timescales of functional motions in large macromolecular assemblies are still computationally intensive and practically inaccessible. Indeed, Schulten and coworkers (80) showed that for a system with ~100,000 atoms, 1 week on 32 processors (1.3 GHz) was necessary to reach 1 ns, which is a timescale that is not yet relevant to large conformational changes that occur during function. Coarser-grained models have been introduced to simulate the dynamics of biological molecules that enable microsecond timescales to be reached for small proteins (95). In these simulations, bead models are used, typically the Cα and P atoms are considered, which reduce considerably the number of atoms to simulate. However, such simulation techniques are still computationally expensive for large macromolecular assemblies. In the case of the ribosome, it took 20 days on 16 processors to reach 500 ns (97), which is likely insufficient to observe the full conformational rearrangements of the ribosome. Finally, long timescale and large-amplitude conformational changes (e.g., protein folding) can be explored using conformational biasing methods such as umbrella sampling (72) or assuming barrier crossing is a Poisson process and coupling this idea with distributed computing (109). However, such calculations are far from routine.

An alternative technique to access longer timescale dynamics is NMA. This is a

GHz: gigahertz (10^9 hertz)

Hessian: the matrix of second-derivatives of the energy function

DIMB: diagonalization in a mixed basis

RTB: rotation-translation block

relatively mature technique (7, 26, 43), which has in recent years piqued the interest of researchers owing to new algorithmic developments that enabled larger systems to be accessed. NMA is based on the harmonic approximation of the potential energy, and the dynamics of the molecule is then described as a collection of independent harmonic oscillators on the basis of normal mode eigenvectors (an orthonormal set of directional vectors that represent the uncoupled motions of the system). Low-frequency oscillators represent collective, large-amplitude motions and describe the most facile deformations of the structure. The low-frequency modes often correlate well with experimentally observed conformational changes associated with function. The standard approach for NMA requires the minimization of the potential energy followed by the diagonalization of the Hessian, which is the 3N x 3N matrix of second derivatives of the potential energy (where N is the number of atoms in the system). The diagonalization procedure has represented a significant bottleneck for the computation of normal modes for some time and, as recently as the early 1990s, has limited studies to proteins of about 300 amino acids (10) (**Figure 1**).

When comparing NMA with molecular dynamics, one can argue that the applicability of NMA is limited because of the harmonic approximation required in the development of the Hessian. However, as we discuss further below, this approach presents some advantage when dealing with large conformational changes of biomolecular assemblies (49).

Diagonalization Techniques

In **Figure 1** we depict a history of NMA applied to biological molecules and the progress that has been made during the past few years. The first application of NMA to a protein was to study the atomic fluctuations of bovine pancreatic trypsin inhibitor (BPTI), which is a protein of 56 residues (26). Subsequent studies on larger proteins (however, still less than 300 residues) such as lysozyme (25) and deoxymyoglobin (70, 71) as well as others have been performed. As techniques in X-ray crystallography have progressed, structures of larger biological molecules have become available and new techniques for NMA were needed to address dynamics of these larger molecules. To accomplish this, researchers need to reduce the number of degrees of freedom considered in the calculation. One approach to address this problem is the use of alternative coordinate representations, such as the space of dihedral angles rather than a Cartesian coordinate space (39). This reduction in dimensionality has been a useful methodology that has extended the application of NMA to larger systems. Nevertheless, until Perahia and colleagues (63, 66) introduced the diagonalization in mixed basis (DIMB) method, which performs an iterative diagonalization utilizing partial solutions to the entire problem, to perform calculations on large systems, normal mode studies remained limited to proteins of ~300 residues (10). The first successful application of the DIMB method was to study the conformational change between the T and R state of the aspartyl transcarbamylase, which is a large macromolecular assembly comprising 2760 residues (91, 92). More recently, the rotation translation block (RTB) method (21, 45, 84) was introduced. Compared with the DIMB approach, the RTB method yields approximate low-frequency normal modes that are obtained by dividing the protein into blocks (one or a few residues per block). However, this is sufficient to describe the global dynamics of a number of biological molecules (84), which is of the highest interest for the analysis of biological function. Using the RTB methodology, one can perform NMA on large biological systems with minimal computational time, in contrast to the DIMB approach. The RTB technology has been employed in studies performed on several large biological machines such as the ribosome (89), RNA polymerase (16), and the F_1-ATPase (13), among others. By combining symmetry

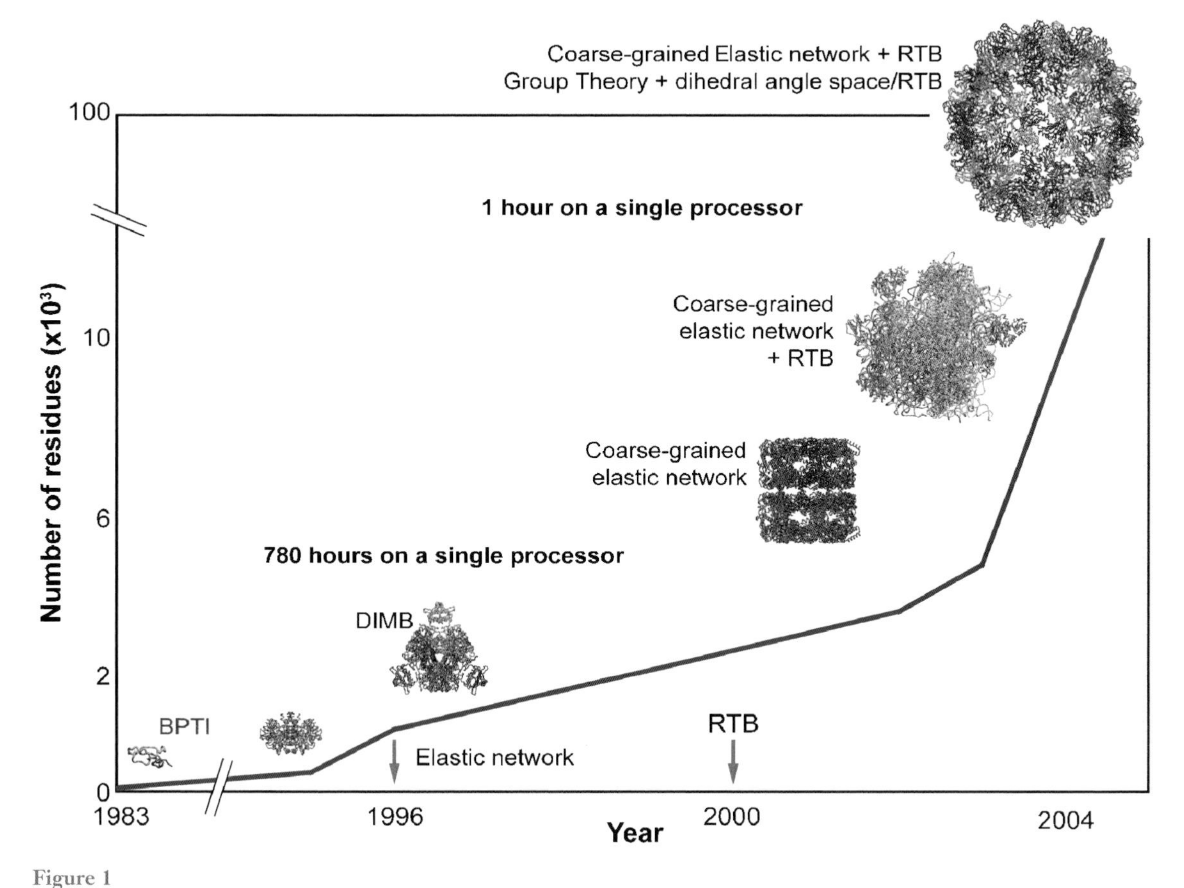

Figure 1

Evolution of NMA techniques used to study the dynamics of biological molecules. The first application of NMA was on BPTI, a small protein comprising 56 residues. Evolution in diagonalization techniques such as DIMB and RTB combined with the coarse-grained elastic network model has enabled calculations on large macromolecular assemblies to be achieved with minimal computational expense in recent years. This figure illustrates the interdependence of the system size that can be studied via NMA and the algorithmic and model advancements.

with the RTB method or with a dihedral angle space basis, all-atom calculations for large systems are now accessible, as has been demonstrated for icosahedral viruses (101).

All-Atom Approaches

In the standard approach for NMA, the protein model used in the calculation consists of classical points of mass with typically one atom per point. The energy terms for interactions between these atoms are defined by semiempirical (molecular mechanics) force fields. Because NMA is based on the harmonic approximation (27), a minimization of the potential energy function is first required to perform the diagonalization process. This minimization process can be cumbersome for large biological molecules, because the quality of the modes is dependent on reaching a true minimum on the potential energy surface. Typically, minimization is performed until the final conformation reaches a root-mean-square gradient on the order of 10^{-6} kcal/(mol Å) (51), which is sufficient to ensure real frequencies for all modes not related to translation or rotation of the entire molecule. Recent studies have used a less strict criterion

GB: gigabyte (10^9 bytes)

EN: elastic network

[10^{-2} kcal/(mol Å)] to minimize large deviations from the X-ray structure (13, 45). These procedures are not computer intensive, compared with molecular dynamics simulation, as the combined minimization and diagonalization to obtain the lowest-frequency normal modes for a system such as RNA polymerase (~40,000 atoms) does not exceed 40 h on a desktop machine running with a 1.2 GHz AMD Athlon processor with 1 GB memory (102). Such calculation is sufficient to obtain motions that correlate well with experimentally observed conformational changes. Using a detailed force field provides information on the frequency of these slow normal modes. Using diagonalization techniques such as the RTB method shifts the frequencies to values higher than those in the full model; however, this shift occurs in a consistent manner (44, 84) and can therefore be adjusted to determine the natural frequencies of these modes. From the normal mode frequencies one can determine the vibrational contributions to thermodynamic properties (6).

Coarse-Grained Models

In 1997, Tirion (93) introduced a simplified representation of the potential energy for use with NMA of biological systems. In this representation, which is closely aligned to continuum elastic theory, the biological system is described as a three-dimensional elastic network (EN) on the basis of the equilibrium distribution of atomic distances in an experimentally determined structure. The EN model is reminiscent of the ideas expressed in earlier work by Suezaki & Go (75), who explored the relationship between the low-frequency motions of continuous elastic bodies and globular proteins. These studies established a link between the mechanical properties of globular proteins, e.g., α-helices, and representative elastic objects. For example, by assuming that the mechanical strength of a molecule was similar to that of an elastic rod, for which the Young modulus and vibrational frequency could be derived, they showed that analogous motions occurred in both objects and that the elastic modulus could be deduced from the atomic representation (75, 76).

In the EN model introduced more recently, particles or atoms are used to identify the junctions of the network. These junctions are representative of the mass distribution of the system and are connected together via a simple harmonic restoring force:

$$E(\vec{r}_a, \vec{r}_b) = \begin{cases} \frac{k}{2}\left(|\vec{r}_a - \vec{r}_b| - |\vec{r}_a^{\,0} - \vec{r}_b^{\,0}|\right)^2 & for \quad |\vec{r}_a^{\,0} - \vec{r}_b^{\,0}| \le R_C \\ 0 \quad for\, |\vec{r}_a^{\,0} - \vec{r}_b^{\,0}| > R_C \end{cases}, \qquad 1.$$

where $\vec{r}_a - \vec{r}_b$ denotes the vector connecting pseudoatoms a and b, the zero superscript indicates the initial configuration of the pseudoatoms, and R_C is a spatial cutoff for interconnections between the particles. The strength of the potential k is a phenomenological constant assumed to be the same for all interacting pairs; different types of atoms or residues are not assigned different values of k.

The total potential energy of the molecule is expressed as the sum of elastic strain energies:

$$E_{System} = \sum_{a,b} E(\vec{r}_a, \vec{r}_b). \qquad 2.$$

Note that this energy function, E, is a minimum for any chosen configuration of any system, thus eliminating the need for minimization prior to NMA. Consequently, NMA can be performed directly on crystallographic or NMR structures (93).

When using the EN model for NMA, information on the frequency of the low-frequency normal modes is lost because no physical information is included in the potential. As suggested by Tirion (93), one can derive the approximate frequency by shifting the

B-factors obtained from NMA so that they agree with experimental B-factors. Because most modern X-ray structures are solved at 100 K, it is difficult to obtain a meaningful shift because experimental B-factors obtained at 100 K are not representative of fluctuations at 300 K. One alternative is to study a set of proteins, all solved at 300 K, to derive a universal force constant (41). However, note that in such cases, the energetic properties derived from the EN remain qualitative only in nature.

Despite its simplicity, several studies have shown that this Hookean potential is sufficient to reproduce the low-frequency normal modes, i.e., the nature of the motions, of proteins as produced by a more complete potential energy function (88). Although this agreement tends to break down at high frequencies, there have been many cases demonstrating that collective motions found in the low-frequency modes characterize biologically relevant conformational changes (88).

One advantage of the EN approach over an all-atom approach is that no prior minimization is necessary to perform NMA, which reduces considerably the computational expense [(~4 h for the RNA polymerase (102)] and makes NMA an ideal candidate for exploration of conformational change pathways between two conformational states, as illustrated by several studies (60, 61, 83). In this approach, the direction given by low-frequency normal modes can be followed to deform the structure from its original conformation toward other known (functionally informed) conformations. Because of the nonlinearity of the conformational change (60), one needs to deform the structure iteratively, which may require hundreds of NMA. Such deformations in an all-atom approach would cause the structure to move from its minimum, and therefore a minimization procedure would be necessary to perform NMA on the deformed structure, whereas with the EN one can directly perform NMA without additional minimization. This approach presents a clear computational advantage when dealing with large systems.

More importantly, in contrast to the all-atom approach, the EN model does not require a fully atomic-level description of the system to model the mechanical properties. Amino acids or base pairs may be represented in full atomic detail or at a more coarse-grained level. For example, one mass point per residue (30), only Cα atoms (1, 2, 88), or more coarse-grained pseudo particle-based models (18) may be used to identify the junctions of the network. This reduces the computational complexity of the system and allows calculations to be done for large systems.

The introduction of the EN model for proteins together with NMA has had a tremendous influence in the field by increasing the size of the systems that can be explored as well as the extent of the application of NMA. Large systems (21, 84) and large macromolecular assemblies such as viruses, the ribosome, or muscle proteins can now be studied efficiently by NMA with modest expenditure of computational time (16, 35, 81, 83, 89, 103). These combined techniques are also useful to the electron microscopy community, providing a means to examine dynamical properties of biological systems directly from cryo-EM data (11, 56, 57, 90), as we describe below.

Elastic Networks for Pseudo-Atomic Models

As large biomolecular assemblies are difficult to study by X-ray crystallography, structural information is often obtained by cryo-EM, which provides a low-resolution description of the biological molecules. In addition, cryo-EM has been helpful in unveiling the dynamics of large biological molecules and assemblies (68). Despite the amount of available low-resolution structural information, few theoretical methods have been developed to assist the interpretation of inferred dynamical transitions in large biological assemblies characterized by cryo-EM. The introduction of the EN model for NMA has allowed researchers to study conformational transitions using low-resolution structural information.

Indeed, until recently, reduced representations of proteins for NMA were limited to the use of Cα atoms. Greater reductions in the representation have since been considered in the study of the influenza virus hemagglutinin (18). NMA based on EN models from coarse-grained structures containing $N/2$, $N/10$, $N/20$, and $N/40$ (where N is the number of residues) Cα atoms have been performed, and such coarse-graining was sufficient to capture the slow dynamics of proteins with high accuracy. Similarly, one can represent the biological molecule by a lattice and use points of this lattice to define junctions of the EN. This approach demonstrated that the nature of the low-frequency normal modes is robust (17).

Because these highly reduced representations are sufficient to provide dynamical information on X-ray and NMR protein structures, one can assume that there is no need for an atomic description of the molecule to obtain a description of its low-frequency normal modes. Therefore one can study mechanical properties for biological systems from low-resolution structural information as obtained from cryo-EM.

In particular, information from the shape and density distribution as present in the density map can be extracted using a vector quantization approach, which represents the structure with a set of landmark points (104, 105). Using the EN model with the landmark points as a representation of the electron microscopy map, researchers can perform NMA (56, 90). The normal modes obtained by this approach are found to agree well with those obtained directly from the X-ray structure of the whole-protein system (56, 90). These studies have shown that the global dynamical properties of a biological object can be extracted not only from its detailed structure but also from a low-resolution representation of its mass (electron) distribution.

NORMAL MODE ANALYSIS APPLICATIONS TO LARGE BIOLOGICAL SYSTEMS

Functional Conformational Changes

The techniques we have just described have enabled studies of large macromolecular assemblies. One of the first large biomolecular assembly studies using an EN was the GroEL-GroES complex, which comprises about 8000 residues (35). This study revealed the rather complex nature of dynamics in the GroEL-GroES complex. More specifically, rotations of the two GroEL rings with different behavior and longitudinal elongation of the complex, which might be related to the dissociation of the GroES upon ATP binding, were observed.

To date, viruses are the largest systems that have been studied with the EN NMA approach (67, 81, 82, 101). These systems are peculiar, as they obey icosahedral symmetry. In the case of recent all-atom calculations, group theory (100) needs to be considered in conjunction with coarse-graining to dihedral angle space (40) or utilizing the RTB approach to be feasible. EN approaches have been also considered for viruses (67, 81, 82). In these cases the level of coarse-graining was much higher, due to the block definition used in the RTB method; nevertheless, the characteristics of the low-frequency normal modes, and in particular of an expansion mode, seem well captured (82).

Motor proteins involved in critical processes such as cell motility, muscle contraction, and vesicle trafficking have also received a great deal of attention owing to NMA. More particularly, the myosin molecule, which mechanically moves along the actin filament upon ATP conversion, is highly dynamic. Several studies have been performed on the myosin head (~1000 residues). Using an all-atom approach in conjunction with the RTB method, Li & Cui (45) found that the modes and functional motions were in agreement; in particular the rigid-body rotation of the converter domain was identified by NMA. A comparative study of three myosin X-ray structures with the coarse-grained (Cα – EN) approach has revealed that the motion of the lever arm is rigidified by the light chains (65). The interface between the motor domain and the lever arm, and more particularly its dynamics, appears to be affected by the presence and/or absence of the ligand within the nucleotide binding pocket. This later observation is reinforced by a recent study, also based on the EN model, that showed that the small perturbations in the nucleotide binding pocket are elastically coupled to more global changes in the myosin molecule (111, 113). Finally, a pathway generation method based on the EN model and NMA was used to study the conformational change between the inhibited state and active smooth muscle myosin (~3500 residues). Structural properties of the active myosin molecule were probed, in particular the region that connects the myosin head with the coil-coiled region, and were in agreement with a recent high-resolution structure (46). The length of the coil-coiled region affects the ability of the myosin molecule to accomplish its conformational change from active to inhibited myosin (83). Taken together, these studies provide a near-atomic-level understanding of the mechanical properties of myosin molecule.

In the case of the ribosome, the molecular machine that synthesizes new proteins, connecting motions obtained from NMA based on its X-ray structure (108), together with experimentally observed conformational changes (23, 99), has shed light on specific features related to ribosome dynamics during translocation (89). The ribosome is composed of two subunits and several protuberances are apparent. The role of one of these protuberances, the L1 stalk, was highlighted by NMA. Similarly, a classification of the bridges, i.e., the network of interactions between the two subunits, was made possible by NMA. Furthermore, one should note that even though these calculations were performed using a coarse-grained model (only the Cα and P atoms were considered), the predictions from these calculations provide a starting point for detailed energetic calculations performed on the EN NMA-derived structures to characterize in greater detail the nature of these motions (96). Therefore by complementing coarse-grained models with more accurate physical calculations, one can obtain an overall picture both in a mechanical sense and in an atomically detailed sense of the functional conformational changes occurring in large biological machines.

Other biomolecular assemblies that have been extensively studied using NMA are molecules in the polymerase family, which are critical for DNA replication and transcription. Studies employing both high- and low-resolution structures have been conducted (16, 102, 107). These studies have revealed in the case of the RNA polymerase a large conformational change, the opening and/or closing of the main DNA/RNA channel (**Figure 2**), which is in agreement with experimental data (12, 14). **Figure 2** represents the normal mode that is the most similar to the experimentally observed functional motion when NMA is performed with an all-atom EN, a Cα atom model, and from low-resolution structural information. One can observe that the same type of motion is obtained independent of the type of model that is used to represent the RNA polymerase, which further validates the coarse-grained approach. A more detailed picture of the RNA polymerase function has emerged from a study using a full-atom approach of several of its conformational states, including a DNA-bound conformation. Using an atomically detailed model, Van Wynsberghe et al. (102) commented on the difference in the flexibility of the RNA polymerase, and in particular of its different subunits, between several conformational states. Finally, Zheng et al. (112) have proposed a methodology based on NMA and highly conserved residues to isolate residues

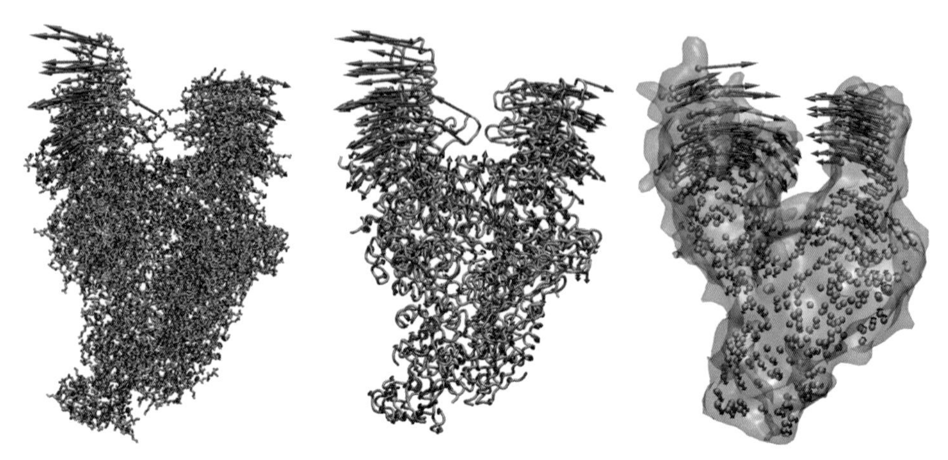

Figure 2

RNA polymerase exists in two different functionally important conformational states: a closed conformation (12) and an open structure (14). A high-resolution X-ray structure exists for the closed state, and only lower-resolution data from single particle cryo-EM is available for the open conformation. The conformational change of RNA polymerase as obtained from EN NMA using the high-resolution all-atom structure, a coarse-grained model based on the Cα atoms of the high-resolution structure, and 1000 codebook vectors used to represent the low-resolution data obtained from cryo-EM experiments are compared. Cut-off values of 5, 8, and 25 Å, respectively, were used to construct the EN models. Each arrow represents the direction and amplitude of motion of the atoms or pseudoatoms along the normal mode that best represents the conformational change between the open and closed conformations. The good visual correspondence between these data suggests that all levels of the model yield similar motions. All the graphics were produced using VMD (32).

in the RNA polymerase that are mechanically important in accomplishing the transition between its open and closed conformations, which is of primary interest because these residues could affect the function of enzyme while not necessarily in the active site.

CCT: cytoplasmic chaperonin containing T-complex polypeptide-1

Dynamics for Low-Resolution Structures

One advantage of the EN model over a full-atomic normal mode approach is that the NMA techniques can be extended to low-resolution structural data. Indeed, applications of NMA to low-resolution structural information have revealed functionally important rearrangements of several macromolecular assemblies (11). NMA based on a discrete representation of an electron microscopy map of the ribosome at 25 Å resolution can display motions similar to the ratchet-like reorganization observed in cryo-EM experiments (23) and NMA of the 70S atomic structure (89, 103). In the case of the *Escherichia coli* RNA polymerase, the lowest-frequency normal mode obtained from a discretized representation of the open low-resolution structure at 15 Å reveals a large conformational rearrangement of the clamp domains, more specifically a closure, that is consistent with several studies (12, 14). **Figure 2** shows the closing of the clamp obtained from the low-resolution structure. The motion is rather similar to the motion obtained from the closed high-resolution structure, which shows an opening of the clamp. NMA performed from a discretized representation of a map at 27 Å resolution of the chaperonin CCT (cytoplasmic chaperonin containing T-complex polypeptide-1) suggests large flexibility of the apical domain, which is involved in the substrate binding. This result is in agreement with the known structural variability of the chaperonin (47). Finally, application of these techniques to the 19 Å resolution electron density map of the human fatty acid synthase reveals specific bridges that are important for its dynamics (57).

Hybrid Methods

Structural information comes from a variety of experiments that provide either high-resolution or low-resolution structural models. Several studies have now shown that NMA can be useful when applied to hybrid methods, i.e., in which a combination of experimental data (X-ray, cryo-EM, SAXS, fluorescence data) with a known high-resolution structure are employed to obtain new structural insights.

Cryo-EM is a powerful technique used to observe large macromolecular machines in several conformational states. If a high-resolution structure of the assembly is known, it is of significant interest to combine the low-resolution structural information with the known high-resolution structure to determine, at higher resolution, the conformation of each distinct conformational state and possibly the paths between them. Because the normal modes can be used to predict the conformational changes of these macromolecular assemblies, one can use the directions of these low-frequency normal modes in an iterative manner to deform the structure so that they optimally fit (i.e., conform to some metric of agreement with the experimental observable) the low-resolution structural data. This concept was introduced in the normal mode flexible fitting (NMFF) method (85, 86) and is reminiscent of the use of normal modes in X-ray crystallography refinement (36). NMFF has since been applied successfully to investigate dynamics of GroEL upon ligand binding (22) and the protein-conducting channel bound to a translating ribosome (58). In addition, the structure of the lethal factor bound to the protective antigen (F. Tama, G. Ren, S.H. Leppla, C.L. Brooks III & A.K. Mitra, manuscript submitted) as well as the structure of the sarcoplasmic reticulum Ca^{2+}-ATPase (31) have been elucidated. Finally, using NMFF, structural transitions between native and EDTA-treated Red clover necrotic mosaic virions have been characterized, providing a possible mechanism for RNA release (M.B. Sherman, R.H. Guenther, F. Tama, C.L. Brooks II, A.M. Mikhailov, E.V. Orlova, T.S. Baker & S.A. Lommel, manuscript submitted). Other refinement protocols that represent the system and density in reciprocal space have been introduced, which are similar to crystallographic refinement, and have been presented for both cryo-EM data and SAXS experiments (15).

NMFF: normal model flexible fitting

R-factor: metric of goodness of fit between computed and measured X-ray density in reciprocal space X-ray structure determination

Other applications of NMA used to interpret experimental data are related to the molecular replacement technique used in X-ray crystallography (77). Structure of unknown macromolecules can be solved using data of known structural fragments or suitable templates. By exploring the dynamics of these templates using NMA, one can generate several conformations that can subsequently be screened as a possible solution for molecular replacement. Similarly, in the case of fiber diffraction data, by including a few low-frequency normal modes as adjustable parameters to take into account the conformational variability of the actin molecule, a better fit to the experimental data, as indicated by a lower R-factor, can be obtained (106). Finally, NMA has been recently used as a tool for predicting conformational changes from a subset of distance changes (110), which should be helpful to interpret fluorescence experiments that measure distance changes in biological molecules.

Three-Dimensional Image Reconstruction

NMA appears to be a promising tool for image reconstruction in cryo-EM experiments. Indeed, because biological molecules often undergo large conformational changes, the particle image data collected from cryo-EM experiments may capture the molecule in different conformational states. Therefore the reconstruction of a three-dimensional structure from data set may lead to a structure with a resolution lower than expected owing to the conformational variability of the molecule. Ideally, one would classify the particle image data set into several classes, and each of these classes would correspond to one of the conformational states that the molecule can adopt. As we have illustrated above, NMA can be used to illuminate the dynamics of low-resolution structures; structures derived from this analysis may then be used to classify the image data according to different models, i.e., conformational states of the molecule. Each subset of data could then be used to generate the low-resolution structure. Encouraging preliminary work has been undertaken with the 19 Å resolution electron density data for the human fatty acid synthase (5).

By progressing from an all-atom description to a coarse-grained model, the application of NMA can be extended effectively to larger biological molecules and address motions as observed in lower-resolution structural data. This advance provides a powerful complement to the emerging facility with which lower-resolution techniques (particularly cryo-EM and tomography) are becoming more prevalent to study both the structure and dynamics of large biological systems.

SUMMARY: SHAPE GOVERNS DYNAMICS

The evolution of the techniques and models for NMA, from an all-atom representation of the biological molecule to multiresolution EN models, has shed light on and new insights into the role of shape and form in controlling the motions of biological molecules and their assemblies. Indeed, the high degree of accord between the motions inferred from individual, or collections of, normal modes constructed from these methods and experimental data suggests that low-frequency normal modes, which are the most facile motions, are the dominate carriers of functional reorganization. Theory has demonstrated that such modes are predominantly a property of the shape of the molecular system (48, 89).

As we illustrate in **Figure 3**, using either an EN model constructed from the near-atomic-resolution X-ray structure of the ribosome or from the pseudoatom model of the RNA polymerase, the key property that is captured is the overall shape of the molecule as well as the relative interaction between its distant parts. The EN captures not only the two subunits of the ribosome and a network of interactions between them but also distinctive features such as the L1 stalk or the spur region. From this description of the molecule, one can obtain functional motions such as the ratchet-like reorganization of the ribosome or the opening and/or closing of the clamp in the RNA polymerase, both of which have been observed experimentally. This further reinforces the fact that the robustness of functional modes is encoded in the shape of the underlying molecular structure.

In summary, studies based on such multiresolution (from atomic levels of detail to highly coarse-grained) models show that a key to understanding the function of biological systems appears to lie in the shape-dependent dynamical properties of their complex architecture. These observations support the notion that nature builds robustness into the functioning of these machines by assembling particular shapes, and that it is this shape that dominates the character of the most facile motions used in achieving function in such assemblies. Thus, symmetry, form, and shape provide a robustness to the functional reorganization of biological assemblies that is necessary for their function.

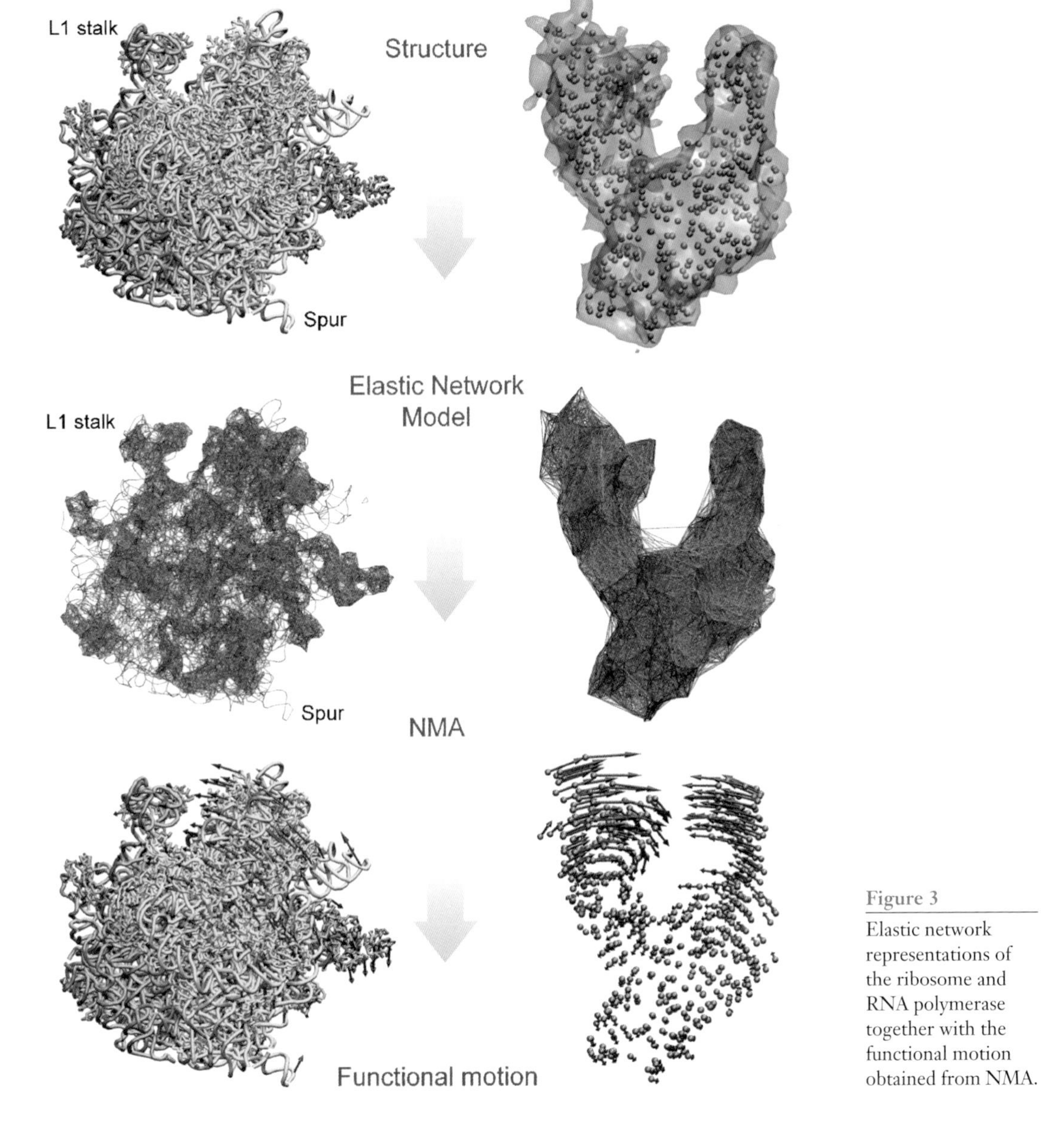

Figure 3

Elastic network representations of the ribosome and RNA polymerase together with the functional motion obtained from NMA.

ACKNOWLEDGMENTS

Financial support from the Center for the Development of Multi-Scale Modeling Tools for Structural Biology (MMTSB) through the National Institutes of Health (RR12255) is greatly appreciated. Present address for Dr. Florence Tama: Department of Biochemistry and Molecular Biophysics, University of Arizona, P.O. Box 210088, Tucson, Arizona 85721.

LITERATURE CITED

1. Atilgan AR, Durell SR, Jernigan RL, Demirel MC, Keskin O, Bahar I. 2001. Anisotropy of fluctuation dynamics of proteins with an elastic network model. *Biophys. J.* 80:505–15
2. Bahar I, Atilgan AR, Erman B. 1997. Direct evaluation of thermal fluctuations in proteins using a single-parameter harmonic potential. *Fold Des.* 2:173–81
3. Basu G, Kitao A, Kuki A, Go N. 1998. Protein electron transfer reorganization energy spectrum from normal mode analysis. 1. Theory. *J. Phys. Chem. B* 102:2076–84
4. Basu G, Kitao A, Kuki A, Go N. 1998. Protein electron transfer reorganization energy spectrum from normal mode analysis. 2. Application to Ru-modified cytochrome c. *J. Phys. Chem. B* 102:2085–94
5. Brink J, Ludtke SJ, Kong YF, Wakil SJ, Ma JP, Chiu W. 2004. Experimental verification of conformational variation of human fatty acid synthase as predicted by normal mode analysis. *Structure* 12:185–91
6. Brooks BR, Janezic D, Karplus M. 1995. Harmonic-analysis of large systems. 1. Methodology. *J. Comput. Chem.* 16:1522–42
7. Brooks BR, Karplus M. 1983. Harmonic dynamics of proteins: normal mode and fluctuations in bovine pancreatic trypsin inhibitor. *Proc. Natl. Acad. Sci. USA* 80:6571–75
8. Brooks BR, Karplus M. 1985. Normal modes for specific motions of macromolecules: application to the hinge-bending mode of lysozyme. *Proc. Natl. Acad. Sci. USA* 82:4995–99
9. Bruschweiler R, Case DA. 1994. Collective NMR relaxation model applied to protein dynamics. *Phys. Rev. Lett.* 72:940–43
10. Case DA. 1994. Normal-mode analysis of protein dynamics. *Curr. Opin. Struct. Biol.* 4:285–90
11. Chacon P, Tama F, Wriggers W. 2003. Mega-Dalton biomolecular motion captured from electron microscopy reconstructions. *J. Mol. Biol.* 326:485–92
12. Cramer P, Bushnell DA, Kornberg RD. 2001. Structural basis of transcription: RNA polymerase II at 2.8 angstrom resolution. *Science* 292:1863–76
13. Cui Q, Li GH, Ma JP, Karplus M. 2004. A normal mode analysis of structural plasticity in the biomolecular motor F-1-ATPase. *J. Mol. Biol.* 340:345–72
14. Darst SA, Opalka N, Chacon P, Polyakov A, Richter C, et al. 2002. Conformational flexibility of bacterial RNA polymerase. *Proc. Natl. Acad. Sci. USA* 99:4296–301
15. Delarue M, Dumas P. 2004. On the use of low-frequency normal modes to enforce collective movements in refining macromolecular structural models. *Proc. Natl. Acad. Sci. USA* 101:6957–62
16. Delarue M, Sanejouand YH. 2002. Simplified normal mode analysis of conformational transitions in DNA-dependent polymerases: the Elastic Network Model. *J. Mol. Biol.* 320:1011–24
17. Doruker P, Jernigan RL. 2003. Functional motions can be extracted from on-lattice construction of protein structures. *Proteins* 53:174–81
18. Doruker P, Jernigan RL, Bahar I. 2002. Dynamics of large proteins through hierarchical levels of coarse-grained structures. *J. Comput. Chem.* 23:119–27
19. Duong TH, Zakrzewska K. 1997. Calculation and analysis of low frequency normal modes for DNA. *J. Comput. Chem.* 18:796–811
20. Duong TH, Zakrzewska K. 1998. Sequence specificity of bacteriophage 434 repressor-operator complexation. *J. Mol. Biol.* 280:31–39

21. Durand P, Trinquier G, Sanejouand YH. 1994. New approach for determining low-frequency normal-modes in macromolecules. *Biopolymers* 34:759–71
22. Falke S, Tama F, Brooks CL, Gogol EP, Fisher MT. 2005. The 13 angstrom structure of a chaperonin GroEL-protein substrate complex by cryo-electron microscopy. *J. Mol. Biol.* 348:219–30
23. Frank J, Agrawal RK. 2000. A ratchet-like inter-subunit reorganization of the ribosome during translocation. *Nature* 406:318–22
24. Gan L, Conway JF, Firek BA, Cheng NQ, Hendrix RW, et al. 2004. Control of crosslinking by quaternary structure changes during bacteriophage HK97 maturation. *Mol. Cell* 14:559–69
25. Gibrat JF, Go N. 1990. Normal mode analysis of human lysozyme: study of the relative motion of the two domains and characterization of the harmonic motion. *Proteins* 8:258–79
26. Go N, Noguti T, Nishikawa T. 1983. Dynamics of a small globular proteins in terms of low-frequency vibrational modes. *Proc. Natl. Acad. Sci. USA* 80:3696
27. Goldstein H. 1950. *Classical Mechanics*. Reading, MA: Addison-Wesley
28. Harrison W. 1984. Variational calculation of the normal modes of a large macromolecule: methods and some initial results. *Biopolymers* 23:2943–49
29. Deleted in proof
30. Hinsen K. 1998. Analysis of domain motions by approximate normal mode calculations. *Proteins* 33:417–29
31. Hinsen K, Reuter N, Navaza J, Stokes DL, Lacapere JJ. 2005. Normal mode-based fitting of atomic structure into electron density maps: application to sarcoplasmic reticulum Ca-ATPase. *Biophys. J.* 88:818–27
32. Humphrey W, Dalke A, Schulten K. 1996. VMD: visual molecular dynamics. *J. Mol. Graph.* 14:33–38
33. Ishida H, Jochi Y, Kidera A. 1998. Dynamic structure of subtilisin-eglin c complex studied by normal mode analysis. *Proteins* 32:324–33
34. Karplus M, McCammon JA. 2002. Molecular dynamics simulations of biomolecules. *Nat. Struct. Biol.* 9:646–52
35. Keskin O, Bahar I, Flatow D, Covell DG, Jernigan RL. 2002. Molecular mechanisms of chaperonin GroEL-GroES function. *Biochemistry* 41:491–501
36. Kidera A, Go N. 1990. Refinement of protein dynamic structure: normal mode refinement. *Proc. Natl. Acad. Sci. USA* 87:3718–22
37. Kidera A, Go N. 1992. Normal mode refinement: crystallographic refinement of protein dynamic structure. 1. Theory and test by simulated diffraction data. *J. Mol. Biol.* 225:457–75
38. Kidera A, Inaka K, Matsushima M, Go N. 1992. Normal mode refinement: crystallographic refinement of protein dynamic structure. 2. Application to human lysozyme. *J. Mol. Biol.* 225:477–86
39. Kitao A, Go N. 1991. Conformational dynamics of polypeptides and proteins in the dihedral angle space and in the Cartesian coordinate space: normal mode analysis of deca-alanine. *J. Comput. Chem.* 12:359–68
40. Kitao A, Hayward S, Go N. 1994. Comparison of normal-mode analyses on a small globular protein in dihedral angle space and Cartesian coordinate space. *Biophys. Chem.* 52:107–14
41. Kundu S, Melton JS, Sorensen DC, Phillips GN. 2002. Dynamics of proteins in crystals: comparison of experiment with simple models. *Biophys. J.* 83:723–32

42. Lata R, Conway JF, Cheng NQ, Duda RL, Hendrix RW, et al. 2000. Maturation dynamics of a viral capsid: visualization of transitional intermediate states. *Cell* 100:253–63
43. Levitt M, Sander C, Stern PS. 1985. Protein normal-mode dynamics: trypsin inhibitor, crambin, ribonuclease and lysozyme. *J. Mol. Biol.* 181:423–47
44. Li GH, Cui Q. 2002. A coarse-grained normal mode approach for macromolecules: an efficient implementation and application to Ca2+-ATPase. *Biophys. J.* 83:2457–74
45. Li GH, Cui Q. 2004. Analysis of functional motions in Brownian molecular machines with an efficient block normal mode approach: myosin-II and Ca2+-ATPase. *Biophys. J.* 86:743–63
46. Li Y, Brown JH, Reshetnikova L, Blazsek A, Farkas L, et al. 2003. Visualization of an unstable coiled coil from the scallop myosin rod. *Nature* 424:341–45
47. Llorca O, Martin-Benito J, Ritco-Vonsovici M, Grantham J, Hynes GM, et al. 2000. Eukaryotic chaperonin CCT stabilizes actin and tubulin folding intermediates in open quasi-native conformations. *EMBO J.* 19:5971–79
48. Lu M, Ma J. 2005. The role of shape in determining molecular motions. *Biophys. J.* 89:2395–401
49. Ma JP. 2005. Usefulness and limitations of normal mode analysis in modeling dynamics of biomolecular complexes. *Structure* 13:373–80
50. Ma JP, Karplus M. 1997. Ligand-induced conformational changes in ras p21: a normal mode and energy minimization analysis. *J. Mol. Biol.* 274:114–31
51. Ma JP, Karplus M. 1998. The allosteric mechanism of the chaperonin GroEL: a dynamic analysis. *Proc. Natl. Acad. Sci. USA* 95:8502–7
52. Marques O, Sanejouand YH. 1995. Hinge-bending motion in citrate synthase arising from normal mode calculations. *Proteins* 23:557–60
53. Matsumoto A, Go N. 1999. Dynamic properties of double-stranded DNA by normal mode analysis. *J. Chem. Phys.* 110:11070–75
54. Matsumoto A, Tomimoto M, Go N. 1999. Dynamical structure of transfer RNA studied by normal mode analysis. *Eur. Biophys. J. Biophys. Lett.* 28:369–79
55. Miller DW, Agard DA. 1999. Enzyme specificity under dynamic control: a normal mode analysis of alpha-lytic protease. *J. Mol. Biol.* 286:267–78
56. Ming D, Kong YF, Lambert MA, Huang Z, Ma JP. 2002. How to describe protein motion without amino acid sequence and atomic coordinates. *Proc. Natl. Acad. Sci. USA* 99:8620–25
57. Ming DM, Kong YF, Wakil SJ, Brink J, Ma JP. 2002. Domain movements in human fatty acid synthase by quantized elastic deformational model. *Proc. Natl. Acad. Sci. USA* 99:7895–99
58. Mitra K, Schaffitzel C, Shaikh T, Tama F, Jenni S, et al. 2005. Structure of the *E. coli* protein-conducting channel bound to a translating ribosome. *Nature* 438:318–24
59. Miyashita O, Go N. 1999. Pressure dependence of protein electron transfer reactions: theory and simulation. *J. Phys. Chem. B* 103:562–71
60. Miyashita O, Onuchic JN, Wolynes PG. 2003. Nonlinear elasticity, proteinquakes, and the energy landscapes of functional transitions in proteins. *Proc. Natl. Acad. Sci. USA* 100:12570–75
61. Miyashita O, Wolynes PG, Onuchic JN. 2005. Simple energy landscape model for the kinetics of functional transitions in proteins. *J. Phys. Chem. B* 109:1959–69
62. Moritsugu K, Miyashita O, Kidera A. 2000. Vibrational energy transfer in a protein molecule. *Phys. Rev. Lett.* 85:3970–73

63. Mouawad L, Perahia D. 1993. Diagonalization in a mixed basis: a method to compute low-frequency normal-modes for large macromolecules. *Biopolymers* 33:599–611
64. Mouawad L, Perahia D. 1996. Motions in hemoglobin studied by normal mode analysis and energy minimization: evidence for the existence of tertiary T-like, quaternary R-like intermediate structures. *J. Mol. Biol.* 258:393–410
65. Navizet I, Lavery R, Jernigan RL. 2004. Myosin flexibility: structural domains and collective vibrations. *Proteins* 54:384–93
66. Perahia D, Mouawad L. 1995. Computation of low-frequency normal-modes in macromolecules: improvements to the method of diagonalization in a mixed basis and application to hemoglobin. *Comput. Chem.* 19:241–46
67. Rader AJ, Vlad DH, Bahar I. 2005. Maturation dynamics of bacteriophage HK97 capsid. *Structure* 13:413–21
68. Saibil HR. 2000. Conformational changes studied by cryo-electron microscopy. *Nat. Struct. Biol.* 7:711–14
69. Sanejouand YH. 1996. Normal-mode analysis suggests important flexibility between the two N-terminal domains of CD4 and supports the hypothesis of a conformational change in CD4 upon HIV binding. *Protein. Eng.* 9:671–77
70. Seno Y, Go N. 1990. Deoxymyoglobin studied by the conformational normal mode analysis. 1. Dynamics of globin and the heme-globin interaction. *J. Mol. Biol.* 216:95–109
71. Seno Y, Go N. 1990. Deoxymyoglobin studied by the conformational normal mode analysis. 2. The conformational change upon oxygenation. *J. Mol. Biol.* 216:111–26
72. Shea JE, Brooks CL III. 2001. From folding theories to folding proteins: a review and assessment of simulation studies of protein folding and unfolding. *Annu. Rev. Phys. Chem.* 52:499–535
73. Deleted in proof
74. Simonson T, Perahia D. 1992. Normal-modes of symmetrical protein assemblies: application to the tobacco mosaic-virus protein disk. *Biophys. J.* 61:410–27
75. Suezaki Y, Go N. 1975. Breathing mode of conformational fluctuations in globular proteins. *Int. J. Pept. Protein Res.* 7:333–34
76. Suezaki Y, Go N. 1976. Fluctuations and mechanical strength of alpha-helices of polyglycine and poly(L-alanine). *Biopolymers* 15:2137–53
77. Suhre K, Sanejouand YH. 2004. On the potential of normal-mode analysis for solving difficult molecular-replacement problems. *Acta Crystallogr.* D 60:796–99
78. Sunada S, Go N. 1996. Calculation of nuclear magnetic resonance order parameters in proteins by normal mode analysis. 2. Contribution from localized high frequency motions. *J. Chem. Phys.* 105:6560–64
79. Sunada S, Go N, Koehl P. 1996. Calculation of nuclear magnetic resonance order parameters in proteins by normal mode analysis. *J. Chem. Phys.* 104:4768–75
80. Tajkhorshid E, Aksimentiev A, Balabin I, Gao M, Isralewitz B, et al. 2003. Large scale simulation of protein mechanics and function. In *Protein Simulations*, ed. FM Richards, DS Eisenberg, J Kuriyan, pp. 195–247. New York: Elsevier Academic
81. Tama F, Brooks CL III. 2002. The mechanism and pathway of pH induced swelling in cowpea chlorotic mottle virus. *J. Mol. Biol.* 318:733–47
82. Tama F, Brooks CL III. 2005. Diversity and identity of mechanical properties of icosahedral viral capsids studied with elastic network normal mode analysis. *J. Mol. Biol.* 345:299–314
83. Tama F, Feig M, Liu J, Brooks CL III, Taylor KA. 2005. The requirement for mechanical

coupling between head and S2 domains in smooth muscle myosin ATPase regulation and its implications for dimeric motor function. *J. Mol. Biol.* 345:837–54
84. Tama F, Gadea FX, Marques O, Sanejouand YH. 2000. Building-block approach for determining low-frequency normal modes of macromolecules. *Proteins* 41:1–7
85. Tama F, Miyashita O, Brooks CL III. 2004. Flexible multi-scale fitting of atomic structures into low-resolution electron density maps with elastic network normal mode analysis. *J. Mol. Biol.* 337:985–99
86. Tama F, Miyashita O, Brooks CL III. 2004. NMFF: flexible high-resolution annotation of low-resolution experimental data from cryo-EM maps using normal mode analysis. *J. Struct. Biol.* 147:315–26
87. Deleted in proof
88. Tama F, Sanejouand YH. 2001. Conformational change of proteins arising from normal mode calculations. *Protein. Eng.* 14:1–6
89. Tama F, Valle M, Frank J, Brooks CL III. 2003. Dynamic reorganization of the functionally active ribosome explored by normal mode analysis and cryo-electron microscopy. *Proc. Natl. Acad. Sci. USA* 100:9319–23
90. Tama F, Wriggers W, Brooks CL III. 2002. Exploring global distortions of biological macromolecules and assemblies from low-resolution structural information and elastic network theory. *J. Mol. Biol.* 321:297–305
91. Thomas A, Field MJ, Mouawad L, Perahia D. 1996. Analysis of the low frequency normal modes of the T-state of aspartate transcarbamylase. *J. Mol. Biol.* 257:1070–87
92. Thomas A, Field MJ, Perahia D. 1996. Analysis of the low-frequency normal modes of the R state of aspartate transcarbamylase and a comparison with the T state modes. *J. Mol. Biol.* 261:490–506
93. Tirion MM. 1996. Large amplitude elastic motions in proteins from a single-parameter, atomic analysis. *Phys. Rev. Lett.* 77:1905–8
94. Tirion MM, Benavraham D, Lorenz M, Holmes KC. 1995. Normal-modes as refinement parameters for the F-actin model. *Biophys. J.* 68:5–12
95. Tozinni V, McCammon JA. 2005. A coarse grained model for the dynamics of.ap opening in HIV-1 protease. *Chem. Phys. Lett.* 413:123–28
96. Trylska J, Konecny R, Tama F, Brooks CL, McCammon JA. 2004. Ribosome motions modulate electrostatic properties. *Biopolymers* 74:423–31
97. Trylska J, Tozinni V, McCammon JA. 2004. A coarse-grained model for the ribosome: Molecular dynamics simulations. *Protein Sci.* 13:121
98. Vale RD, Milligan RA. 2000. The way things move: looking under the hood of molecular motor proteins. *Science* 288:88–95
99. Valle M, Zavialov A, Sengupta J, Rawat U, Ehrenberg M, Frank J. 2003. Locking and unlocking of ribosomal motions. *Cell* 114:123–34
100. van Vlijmen HWT, Karplus M. 2001. Normal mode analysis of large systems with icosahedral symmetry: application to (dialanine) (60) in full and reduced basis set implementations. *J. Chem. Phys.* 115:691–98
101. van Vlijmen HWT, Karplus M. 2005. Normal mode calculations of icosahedral viruses with full dihedral flexibility by use of molecular symmetry. *J. Mol. Biol.* 350:528–32
102. Van Wynsberghe A, Li GH, Cui Q. 2004. Normal-mode analysis suggests protein flexibility modulation throughout RNA polymerase's functional cycle. *Biochemistry* 43:13083–96
103. Wang YM, Rader AJ, Bahar I, Jernigan RL. 2004. Global ribosome motions revealed with elastic network model. *J. Struct. Biol.* 147:302–14

104. Wriggers W, Milligan RA, McCammon JA. 1999. Situs: a package for docking crystal structures into low-resolution maps from electron microscopy. *J. Struct. Biol.* 125:185–95
105. Wriggers W, Milligan RA, Schulten K, McCammon JA. 1998. Self-organizing neural networks bridge the biomolecular resolution gap. *J. Mol. Biol.* 284:1247–54
106. Wu YH, Ma JP. 2004. Refinement of F-actin model against fiber diffraction data by long-range normal modes. *Biophys. J.* 86:116–24
107. Yidirim Y, Doruker P. 2004. Collective motions of RNA polymerases. Analysis of core enzyme, elongation complex and holoenzyme. *J. Biomol. Struct. Dyn.* 22:267–80
108. Yusupov MM, Yusupova GZ, Baucom A, Lieberman K, Earnest TN, et al. 2001. Crystal structure of the ribosome at 5.5 A resolution. *Science* 292:883–96
109. Zagrovic B, Snow CD, Shirts MR, Pande VS. 2002. Simulation of folding of a small alpha-helical protein in atomistic detail using worldwide-distributed computing. *J. Mol. Biol.* 323:927–37
110. Zheng WJ, Brooks BR. 2005. Normal-modes-based prediction of protein conformational changes guided by distance constraints. *Biophys. J.* 88:3109–17
111. Zheng WJ, Brooks BR. 2005. Probing the local dynamics of nucleotide-binding pocket coupled to the global dynamics: myosin versus kinesin. *Biophys. J.* 89:167–78
112. Zheng WJ, Brooks BR, Doniach S, Thirumalai D. 2005. Network of dynamically important residues in the open/closed transition in polymerases is strongly conserved. *Structure* 13:565–77
113. Zheng WJ, Doniach S. 2003. A comparative study of motor-protein motions by using a simple elastic-network model. *Proc. Natl. Acad. Sci. USA* 100:13253–58

Fusion Pores and Fusion Machines in Ca^{2+}-Triggered Exocytosis

Meyer B. Jackson[2] and Edwin R. Chapman[1,2]

[1]Howard Hughes Medical Institute, [2]Department of Physiology, University of Wisconsin, Madison, Wisconsin 53706; email: mjackson@physiology.wisc.edu; chapman@physiology.wisc.edu

Annu. Rev. Biophys. Biomol. Struct. 2006. 35:135–60

First published online as a Review in Advance on January 13, 2006

The *Annual Review of Biophysics and Biomolecular Structure* is online at biophys.annualreviews.org

doi: 10.1146/annurev.biophys.35.040405.101958

1056-8700/06/0609-0135$20.00

Key Words

membrane, SNARE, synaptotagmin, syntaxin, SNAP-25, synaptobrevin

Abstract

Exocytosis is initiated within a highly localized region of contact between two biological membranes. Small areas of these membranes draw close, molecules on the two surfaces interact, and structural transformations take place. Membrane fusion requires the action of proteins specialized for this task, and these proteins act as a fusion machine. At a critical point in this process, a fusion pore forms within the membrane contact site and then expands as the spherical vesicle merges with the flat target membrane. Hence, the operation of a fusion machine must be realized through the formation and expansion of a fusion pore. Delineating the relation between the fusion machine and the fusion pore thus emerges as a central goal in elucidating the mechanisms of membrane fusion. We summarize present knowledge of fusion machines and fusion pores studied in vitro, in neurons, and in neuroendocrine cells, and synthesize this knowledge into some specific and detailed hypotheses for exocytosis.

Contents

Hemifusion: an intermediate in lipid bilayer fusion in which the proximal leaflets of two fusing bilayers have merged, leaving the distal leaflets in direct apposition as a new bilayer

ENERGY BARRIERS AND MEMBRANE FUSION

Two opposing membranes must overcome large energy barriers in order to fuse. Bringing two lipid bilayers together requires energetically costly removal of water and the dehydration of the polar phospholipid headgroups. In protein-free model systems the proximal leaflets of two bilayers make the initial contact and merge before the distal leaflets, thus forming a hemifusion intermediate with the distal leaflets in direct contact as a new bilayer (**Figure 1*a3***, in which proteins pull the two bilayers together, but the site of contact is purely lipidic). As membranes fuse the bilayers bend and deform; such contortions meet strong opposition, as bilayers prefer to form planar structures that optimize the attractive dispersion forces between the acyl chains. Furthermore, sharp curvature could create energetically unfavorable void spaces within the bilayer core (**Figure 1*a2, a3***). Bending the surface of a lipid bilayer exposes the lipid hydrocarbon chains to water, adding an additional energy cost (148). Surmounting these barriers can be facilitated by proteins with unique capabilities for action on lipid bilayers. The thermodynamics of bilayer fusion has been discussed in detail in a number of excellent recent reviews (32, 36, 87), and these energetic factors help define the demands placed on the proteins that catalyze membrane fusion in vivo.

LIPID-LINED AND PROTEIN-LINED FUSION PORES

Two models are widely invoked in studies of protein-mediated membrane fusion (90, 103, 111). In the first model, proteins act as a scaffold, pulling membranes into close apposition (**Figure 1*a***). Fusion then occurs spontaneously once the bilayers come within a critical distance (i.e., the proximity model). The precise mechanism by which proximity is coupled to fusion is not entirely clear. A variety of factors including the structure of lipid molecules dictate the precise energy requirements for fusion (25, 66, 86, 99, 101, 102). A recent molecular dynamics simulation study illustrates current thinking on protein-free membrane fusion (102). This coarse-grained lipid model used small liposomes 15 nm in diameter. The liposomes were first brought together and small fluctuations in one of the monolayers triggered the merger of a few lipid headgroups with the opposing bilayer. The site of contact then radially expanded to form a stalk intermediate (**Figure 1*a2***) that progressed to a hemifusion diaphragm

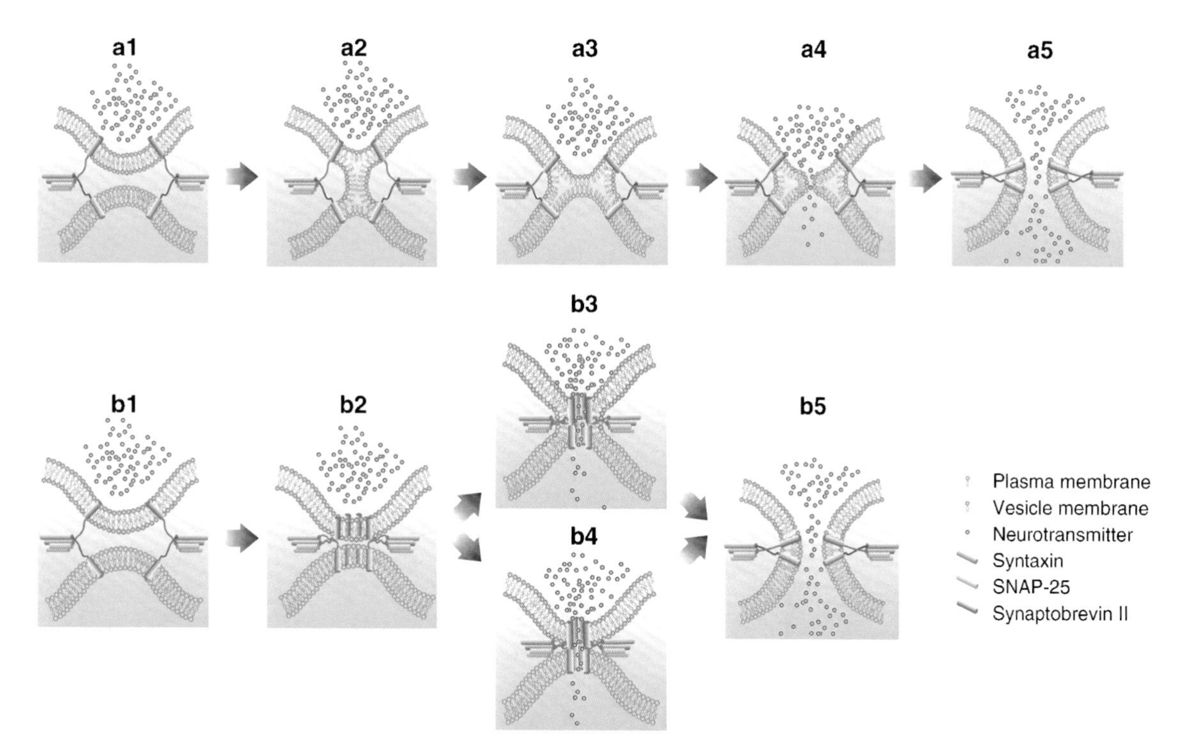

Figure 1

Models for membrane fusion. Two general models for membrane fusion are depicted, using SNAREs to illustrate the roles of proteins. (*a*) The "proximity" model shows a fusion pore lined by lipids. Proteins draw two lipid bilayers together, driving hemifusion and then full fusion. The void spaces in the bilayers are included in the figure, but as outlined in the text, these spaces are likely filled by "tilted" phospholipids. (*b*) The "protein-pore" model shows a fusion pore formed from transmembrane segments (TMS) of proteins. When the fusion pore dilates, the TMSs separate laterally, lipids are incorporated into the pore, and the bilayers merge. A protein-lined pore could be surrounded by the hemifused proximal leaflet shown in panel *b3*.

(**Figure 1*a3***). A small lipidic fusion pore is formed through the bilayer (**Figure 1*a4***), leading to complete fusion (**Figure 1*a5***). To overcome the dehydration barrier, the liposomes were forced into close proximity (1 to 1.5 nm). Unfavorable void spaces were avoided by tilting the hydrophobic tails of the phospholipids. A key element of this model is a purely lipidic fusion pore at the moment of inception.

X-ray diffraction has shown stable stalk structures between bilayers of diphytanoyl phosphatidylcholine upon dehydration (170–172). Recent studies suggest that intracellular membrane fusion might also proceed via a hemifusion intermediate, consistent with the transient existence of a stalk or hemifusion diaphragm (98, 125, 168).

The proximity model has the advantage of making relatively modest demands on the proteins. They need to have a firm membrane anchor in the form of a transmembrane segment (TMS) and they need to interact with one another to exert force and pull the membranes together. These forms of activity are well documented in the protein literature. The proximity idea can be extended to include curving or "puckering" the membrane to promote fusion. Now the proteins are more than a scaffold; they induce curvature by distributing the mass of their TMSs unequally between the two bilayer leaflets. A convex lipid bilayer,

Fusion pore: the first aqueous connection that forms between two membrane-delimited compartments

TMS: transmembrane segment

SNARE: soluble NSF attachment protein receptor

NSF: *N*-ethylmaleimide sensitive factor

with its polar headgroups spread out, exposes its hydrophobic core and thus promotes fusion. This extension of the proximity model still entails a hemifusion intermediate and a lipidic fusion pore.

A key question concerning the proximity model is which steps are reversible. If membrane tension and stress drive each step after the critical proximity and curvature are achieved, reversal should be energetically prohibitive. As discussed in detail below, fusion pores can close and the membranes can reseal. If these reversals occur spontaneously, how can we explain this in terms of the physical properties of lipid bilayers? If these reversals require the expenditure of chemical energy, what proteins transduce this energy and how do they interact with membranes?

A second model posits the initiation of fusion by a protein-lined pore (**Figure 1*b***). This protein-lined pore is analogous to a gap junction, in which members of the connexin family of proteins form two hemichannels, one in each membrane (90). Two hemichannels combine to form a pore that spans two lipid bilayers (**Figure 1*b3*, *b4***). After a protein-lined fusion pore opens, it could in principle either close again or incorporate lipid to allow the pore to expand and the membranes to merge completely (**Figure 1*b5***).

Although the fusion pore in this model is protein, it need not obstruct lipid flow between the two bilayers. For example, hydrophobic segments that span each bilayer could provide a surface that would allow the proximal leaflets to mix, as in hemifusion (**Figure 1*b3***). If each hemichannel were composed of separate subunits, their lateral separation coordinated with lipid incorporation between them would initiate the expansion of the fusion pore (**Figure 1*b5***). This model makes demands on the proteins far beyond the capacity of gap junction channels. Fusion pore proteins must interact dynamically with lipids and provide conduits to direct the movement of lipids along selected paths. Proteins with these capabilities have yet to be established.

It is interesting that both the proximity model and the protein-pore model are consistent with a hemifusion intermediate (**Figure 1*a3*, *b3***). The essential difference concerns the structure of the initial open state of the pore. In one case the pore is lined with lipid, and in the other it is lined with proteins. Importantly, the protein-lined pore is envisioned as an unstable intermediate. Driven by membrane tension and stress imposed on a highly curved lipid bilayer, a protein-pore must give way to lipid in order to expand. Once expansion has started, the two models become indistinguishable. Thus, a major challenge is to capture and study the earliest forms of the pore in order to probe its structure before it expands.

SNAREs: ESSENTIAL PROTEINS FOR MEMBRANE FUSION

Over the past two decades, a number of proteins that play critical roles in synaptic vesicle exocytosis have been identified (31, 72). However, determining whether these proteins function during the final stages of membrane fusion, by mediating the opening and/or dilation of fusion pores, poses a challenging problem.

Three completely independent approaches led to the identification of a conserved set of proteins that play key roles in membrane fusion. Rothman and coworkers (129) used a Golgi transport assay to identify a number of soluble factors essential for membrane trafficking. The first factor isolated was an ATPase that was unusually sensitive to inactivation by *N*-ethylmaleimide (NEM), and was named NEM-sensitive factor (NSF) (18, 56). Adaptor proteins required for NSF function were subsequently purified and named soluble NSF attachment proteins, or SNAPs (35). There are three forms: α-, β-, and γ-SNAP. The Rothman group then used immobilized α-SNAP and NSF to isolate the SNAP receptors (SNAREs) from brain detergent extracts. In a landmark study, they found that the SNAP receptor was a trimeric complex composed of three previously identified proteins: the synaptic vesicle protein synaptobrevin II (syb, or VAMP2), the plasma membrane proteins syntaxin (syx, isoform 1A and 1B), and synaptosome-associated protein of 25 kDa (SNAP-25) (141).

Simultaneously, other laboratories showed that the clostridial neurotoxins (i.e., tetanus neurotoxin, TeNT; and botulinum neurotoxin, BoNT, A–G) are zinc-dependent proteases that block exocytosis by cleaving one or more of the SNARE proteins. This work established that SNAREs are essential for secretion (16, 17, 136, 137) at a step following vesicle docking (70).

Moreover, the genetic dissection of membrane trafficking in yeast, pioneered by the Schekman laboratory, revealed that yeast versions of SNAP, NSF, and SNARE proteins mediate membrane transport in vivo (47, 118). Hence, the SNARE complex—syb/syx/SNAP-25—emerged as the putative core of a membrane fusion machine that is conserved throughout eukaryotic cells (19). Because syb is found on vesicles (12, 150), it was termed a v-SNARE. syx and SNAP-25 reside on the target membrane (13, 71, 119) and are called t-SNAREs (**Figure 2*a***). In yeast vacuolar fusion, v- and t-SNAREs must be present on opposing membranes for efficient fusion (113).

The crucial function of SNARE proteins in numerous membrane trafficking pathways has been established by a wide range of studies in various organisms including yeast, flies, worms, squid, and mice. Different SNARE proteins are targeted to specific and distinct intracellular compartments where they play key roles in membrane fusion (19). In all cases, at least one SNARE protein is anchored to the vesicular or donor membrane while another SNARE is anchored to the target or acceptor membrane (120). By binding one another, SNAREs can tightly link these membranes together (63), potentially driving membrane fusion in vivo.

ASSEMBLY OF SNARE COMPLEXES: PART I

The cytoplasmic domains of the SNARE proteins assemble into a SNARE complex composed of four α-helices in a parallel bundle (**Figure 2*a***) (5, 77, 146). The associating segments are called SNARE motifs and contain 60 to 70 residues. In neurons, SNAP-25 contributes two helices to the SNARE complex while syx and syb each contribute one helix. These recombinant complexes are *cis* and would form in vivo only when all three proteins are anchored to the same bilayer, during or after fusion. Little is known regarding the *trans*-SNARE complexes that form between bilayers prior to fusion. *cis*-SNARE complexes are highly stable and must be heated to 80 to 90°C to melt (43, 64). Hence, disassembly of *cis*-SNARE complexes requires chemical energy. NSF must hydrolyze ATP, and together with SNAP these proteins drive unfolding of *cis*-SNARE complexes both in vitro (65, 140) and in vivo (92, 135).

The stability of SNARE complexes prompted the idea that their formation can provide the driving force for fusion. Because the SNARE motifs form a parallel rather than antiparallel helical bundle (63, 89, 124, 146), SNARE complex formation should pull the C-terminal TMSs of syx and syb into close proximity. A widely discussed idea is that *trans*-SNARE complexes begin to assemble at the membrane distal N termini and "zipper" down toward the C-terminal TMSs of syb and syx, pulling the membranes together. This close proximity could drive membrane fusion, as outlined above. Although this idea is appealing, there is little direct evidence to support it, either in vivo or in vitro. Recent single-molecule measurements of recombinant SNAREs suggest that these complexes can assemble into both parallel and antiparallel bundles in vitro (162). Fluorescence resonance energy transfer (FRET) studies of labeled SNARE proteins in living cells hold promise for addressing the structure and relative orientation of SNARE proteins in situ (4, 165).

syb: synaptobrevin

syx: syntaxin

TeNT: tetanus neurotoxin

BoNT: botulinum neurotoxin

SNARE complex: a heterotrimeric protein complex formed by interactions between proteins on the vesicle membrane (e.g. synaptobrevin) and the target membranes (e.g. syntaxin and SNAP-25); thought to form the core of a conserved membrane fusion complex

Fusion machine: an assembly of proteins that catalyzes the fusion of biological membranes

FRET: fluorescence resonance energy transfer

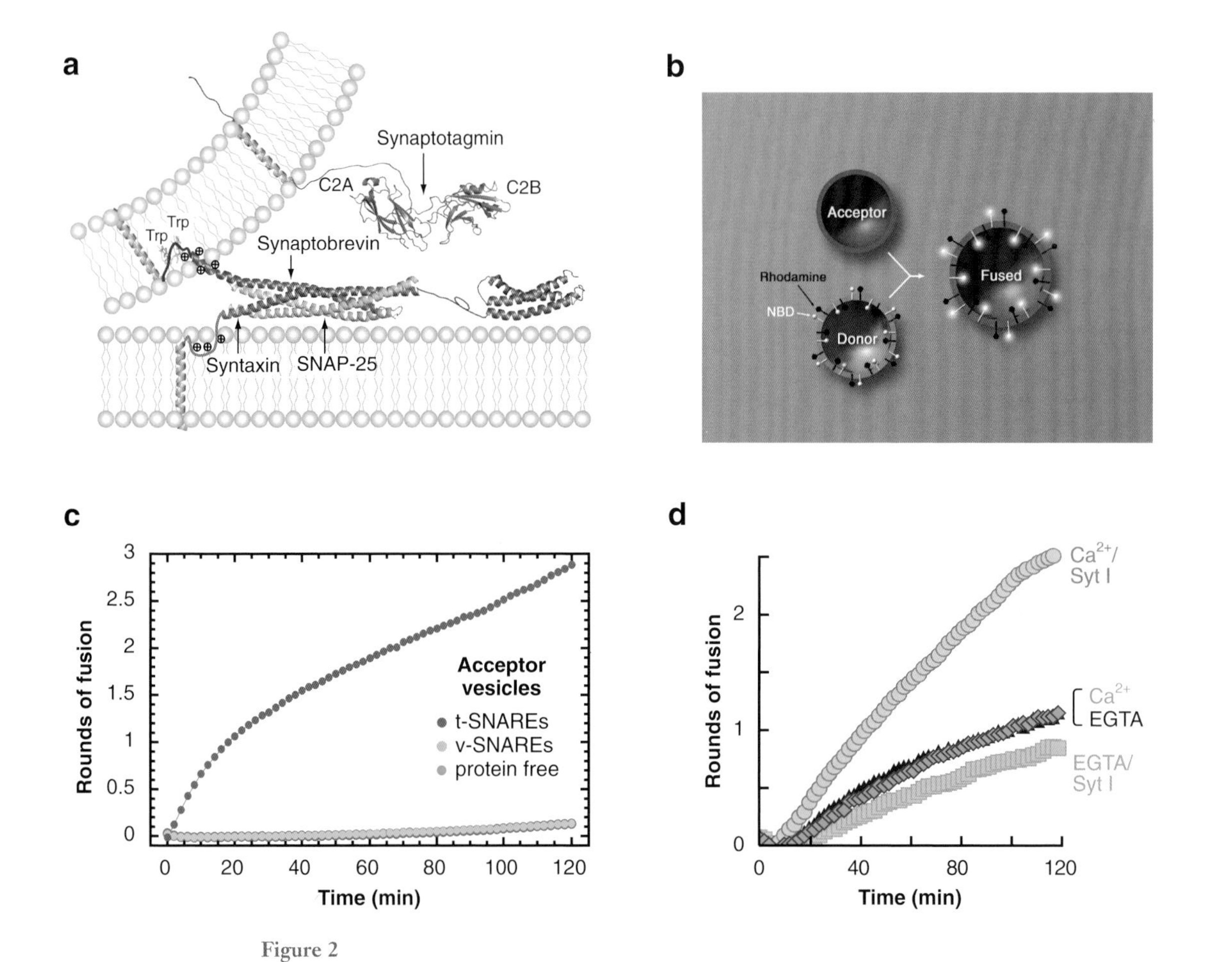

Figure 2

Reconstitution of Ca^{2+}-triggered membrane fusion. (*a*) Model of a *trans*-SNARE complex and syt (modified from Reference 91). (*b*) In an in vitro fusion assay, v-SNARE vesicles (containing syb II) labeled with a FRET pair, NBD- and rhodamine-labeled phosphatidylethanolamine, fuse with t-SNARE vesicles (containing syx and SNAP-25), diluting the v-SNARE vesicle fluorophores and increasing the emission of NBD, the FRET donor. (*c*) Example of a membrane fusion reaction catalyzed by neuronal SNARE proteins. Donor vesicles (as outlined in panel *b*) fuse only when presented with t-SNARE vesicles; no fusion occurred with protein-free or v-SNARE vesicles. Increases in NBD fluorescence were converted to rounds of fusion using a standard curve (121). (*d*) Reconstitution of Ca^{2+}-regulated membrane fusion. Membrane fusion mediated by neuronal SNARE proteins was assayed in the presence and absence of the cytoplasmic domain of syt I, in either 1 mM Ca^{2+} or 0.2 mM EGTA (15, 153).

Perhaps the best evidence for zippering stems from biochemical and structural studies indicating that the membrane proximal region of t-SNAREs becomes more structured upon binding of the v-SNARE (48, 108). However, this remains an open issue, because a more recent study revealed that neuronal t-SNARE complexes are not disordered at the N terminus (174). A handful of additional studies are consistent with the zipper model but leave room for alternative interpretations. One such study used clostridial neurotoxins, which cannot cleave SNAREs once they have assembled into SNARE complexes (64). TeNT binds at a site toward the N terminus of syb and BoNT/B binds closer to the C terminus. This study showed that BoNT/B was able to cleave syb in the crayfish neuromuscular junction in the absence of vesicle recycling (69). In contrast, syb resisted TeNT for hours; cleavage occurred only after stimulation to drive vesicle recycling, and presumably SNARE disassembly. From these data it was suggested that the N terminus of syb, but not the C terminus, was zippered into SNARE complexes under resting conditions. However, it is not known whether the resistance of SNAREs to toxins in vivo is due to SNARE complex assembly, or to interactions of individual SNAREs with other molecules. It is notable that the crayfish neuromuscular junction differs markedly from chromaffin cells, in which TeNT cleaves syb within minutes in the absence of stimulus-induced vesicle cycling (166).

In another study, an antibody that prevents SNAP-25 assembly into SNARE complexes inhibited a slow sustained component of exocytosis in chromaffin cells. This component was attributed to a slower assembly of fusion complexes and thus should depend on SNAREs that have yet to assemble. The antibody thus blocks this process by binding to free SNAP-25 and preventing the assembly of new SNARE complexes. However, the antibody also blocked the fastest component of release from these cells, consistent with the idea that it can impair SNAP-25 function during a relatively late stage in fusion in these cells (167). The authors present a model in which SNARE complexes exist in two states, loose and tight, and proposed that the antibody can interact with one of these.

Experiments in permeabilized PC12 cells suggest that the SNARE complex does not assemble until the arrival of the Ca^{2+} trigger (30). This was discovered by "snipping" out one strand of SNAP-25 from the four-helix bundle of the SNARE complex using BoNT/E. Exocytosis could be rescued by adding this strand back, but only if it were presented at the same time as the Ca^{2+} trigger. Whether this is an upstream or downstream effect is not clear, especially since the overall rates of exocytosis from PC12 cells are slow (41, 158), with latencies in the range of seconds (75, 156).

In summary, direct evidence for the zippering model is lacking. Note that coil-to-helix transitions can occur on nanosecond to microsecond timescales (82), so it is not necessary to invoke staged-zippering to mediate rapid fusion. Indeed, recent EPR studies suggest that assembly of SNARE complexes is rapid and concerted (175). Although SNAREs play essential roles in membrane fusion, it remains unanswered whether they assemble into *trans* complexes prior to or during exocytosis, or whether assembly is directional in situ.

RECONSTITUTION OF THE MEMBRANE FUSION MACHINE

It is well established that SNAREs play essential roles in membrane trafficking, but do they directly mediate membrane fusion? In 1998, Weber et al. addressed this question using a lipid mixing assay (**Figure 2*b***) (161). Purified syb II was reconstituted into one population of vesicles (v-SNARE liposomes), and purified t-SNARE heterodimers composed of syx pre-complexed with SNAP-25 were reconstituted into a separate population of vesicles (t-SNARE liposomes). The v-SNARE liposomes were labeled with NBD-phosphatidylethanolamine (PE) and rhodamine-PE at levels sufficient to give rise to FRET from the NDB (donor) to the rhodamine (acceptor). Lipid mixing of v- and t-SNARE vesicles then dilutes the donor- and acceptor-labeled lipids, thus diminishing FRET and increasing donor NBD fluorescence. When the v- and t-SNARE proteoliposomes were incubated together at 37°C, extensive lipid mixing was observed. Owing to an excess of v-SNAREs per liposome, v-SNARE vesicles fused with the equivalent of three or more t-SNARE vesicles (**Figure 2*c***) (161).

Either hemifusion of the outer leaflets or full fusion of both leaflets can produce a lipid mixing signal in this assay. To discriminate between these two possibilities, Weber et al. (161) quenched the outer leaflet of the syb liposomes using dithionite so that only the inner leaflet contained the FRET pair. The inner leaflet still exhibited the same degree of dequenching, indicating that full fusion took place. One concern about this experiment was the possibility that the proteoliposomes used were leaky (114), making it unclear whether dithionite quenched only the outer leaflet. Complete fusion, as opposed to hemifusion, was further established when it was shown that fluid-phase markers within the v-SNARE and t-SNARE proteoliposomes also mix (114). In addition, visualization of the fusion products by electron microscopy revealed an increase in the size of the proteoliposomes (139). In summary, this work established SNAREs as the minimal fusion machine sufficient for complete membrane fusion.

The reconstituted fusion assay has been used to address a wide range of structure-function relationships among SNARE proteins. For example, early on it was proposed that the specificity of membrane trafficking was encoded, at least in part, by the specificity of *trans*-SNARE pairing (130, 140). This idea was challenged by biochemical studies showing that recombinant cytoplasmic domains of v- and t-SNAREs assemble promiscuously with little specificity (42, 169). However, removal of the TMS from SNARE proteins profoundly affects their properties (17, 88, 128), and subsequent functional assays using full-length SNARE proteins reconstituted into liposomes established a high degree of specificity for SNARE pairing in membrane fusion reactions (105). Nevertheless, there seem to be some exceptions to this rule, as some SNAREs can function in more than one trafficking pathway (57, 95). Tethering proteins, small G-proteins, and other factors are also likely to contribute to compartmental specificity of membrane transport (20).

The architecture of a productive SNARE complex has also been explored in detail. A key observation is that in all SNARE complexes, at least one protein on the transport vesicle membrane, and one protein on the target membrane, must harbor a TMS (120). A number of experiments point to essential functions for the TMS in vivo (117). For example, replacement of the TMSs of either v- or t-SNAREs with lipid anchors, which span one leaflet of the bilayer, inhibits the exocytosis of docked vesicles (59) and disrupts fusion between vacuoles in yeast (128). In the reconstituted proteoliposome fusion assay, replacement of the TMS of syb II or syx with a single phospholipid anchor was sufficient for vesicle-vesicle docking but failed to support fusion. However, when a polyisoprenoid anchor was long enough to span the entire bilayer, syb and syx regained fusion activity (105). Apparently, in some cases, a lipid anchor can replace the protein TMS if the anchor is able to span both leaflets of the bilayer. These TMS replacement experiments suggest that merely tethering vesicles closely together does not result in efficient membrane fusion; proximity alone is insufficient.

ASSEMBLY OF SNARE COMPLEXES: PART II

While mounting evidence indicates that SNARE proteins directly mediate membrane fusion, it remains unclear how the assembly of *trans*-SNARE complexes bring about the opening of a fusion pore. It has been suggested that zippering of the four-helix bundle might progress all the way down through the TMSs (100). Fusion would occur as the TMSs of the v- and t-SNAREs coalesced into the same bilayer. There is evidence that the TMSs of syb II and syx 1A directly associate into hetero- and homodimers (85, 100, 131), although some of these interactions have been questioned (22). The introduction of proline residues to disrupt helices between the SNARE motif and the TMS of syb II had no effect on fusion activity in vitro (106). Thus, helical continuity between the SNARE motif and the TMS is not essential for SNAREs to drive membrane fusion. Furthermore, an 11-amino-acid linker between the TMS and SNARE motif of syb had little effect on fusion, but the same modification of syx reduced the extent of fusion by one half (106). However, it is not known what rate is limiting in this system. Given the slow kinetics of fusion in this assay, it is possible that these mutations could have more pronounced effects on fast fusion reactions that either did not occur or were not detected in these experiments.

The linker studies indicate that the SNARE motifs are somewhat autonomous from the TMSs. However, electron spin resonance experiments indicate that in neuronal SNAREs the juxtamembrane regions of both syb II and syx 1A are tightly associated with the bilayer (78, 83, 84). Whether this interaction is tight enough to transduce work from an assembling SNARE motif to the TMS is not entirely clear, and a function for the linker in force transduction is difficult to reconcile with the linker experiments described above, as the linker sequences would not be expected to associate as strongly with the membrane.

The reconstituted fusion assay has been used to address the zipper model. A clever series of experiments made use of two peptides derived from the sequence of syb II: One peptide corresponded to the N-terminal half of the SNARE motif, and the other peptide corresponded to the C-terminal half of the SNARE motif. Both were tested for inhibition in the proteoliposome fusion assay (108). If zippering occurs in an N-to-C direction, one might expect the fusion reaction to become resistant to the N-terminal peptide first, followed by resistance to the C-terminal peptide. Indeed, within a few minutes, the reaction was refractory to the N-terminal peptide. However, the C-terminal peptide failed to inhibit the reaction at any stage; instead, this peptide enhanced fusion. Hence, rather than resolve the zippering question, the experiments with these peptides raise new questions about the state of the SNARE motifs during fusion.

Is the slow kinetics of SNARE-mediated membrane fusion (161) due to inefficient docking, or is the fusion step itself rate limiting? A second-generation fusion assay based on total internal reflection microscopy (TIRF) (50) is making it possible to track separately the kinetics of proteoliposome "landing" on planar lipid bilayers from the kinetics of fusion after landing (94). Surprisingly, in TIRF studies using proteoliposomes and supported lipid bilayers, SNAP-25 was not required for fusion (23, 94), perhaps because even relatively weak helical bundles composed of syb and syx alone are sufficient to drive fusion under some conditions. In this assay system, both docking and fusion occur with rapid kinetics (94).

TIRF: total internal reflection microscopy

ACTIVATION OF SNARE-MEDIATED MEMBRANE FUSION BY Ca^{2+}

It appears that SNAREs constitute the core of a conserved membrane fusion machine; however, additional accessory proteins must serve to regulate the fusion reaction. One especially important accessory protein is needed to sense Ca^{2+}, the signal that triggers exocytosis from neurons and neuroendocrine cells (76). The delay between a rapid rise in Ca^{2+} and a postsynaptic response can be as short as 60 to 200 μs (96, 134), placing strong kinetic constraints on the transduction pathway that culminates in secretion.

There is now strong evidence that synaptotagmin (syt) functions as the Ca^{2+} sensor that regulates SNARE-mediated exocytosis (8, 26, 81, 123, 153). syt I is an abundant constituent of synaptic vesicles and large, dense core vesicles; it spans the vesicle membrane once near its N terminus and possesses a short intravesicular domain (**Figure 2*a***). Its large cytoplasmic domain contains tandem Ca^{2+} binding C2 domains called C2A and C2B (26, 123).

syt: synaptotagmin

Each C2-domain binds Ca^{2+} via a set of five conserved acidic residues that reside in two flexible loops (44, 145, 154). Upon binding Ca^{2+}, these loops partially penetrate into lipid bilayers that contain anionic phospholipids and, in particular, phosphatidylserine and phosphatidylinositol 4,5-bisphosphate (9, 11, 28, 38). Recent evidence indicates that anionic lipids are essential effectors for the action of Ca^{2+}•syt I during fusion (15). Some isoforms of syt oligomerize in the presence of Ca^{2+}, providing another avenue through which Ca^{2+} could regulate exocytosis (152, 164). In addition, the tandem C2-domains synergize to bind the t-SNAREs syx and SNAP-25 (27, 38, 153). Mutations that selectively disrupt binding of syt to t-SNAREs reduce the rate of secretion and destabilize fusion pores in PC12 cells (10). These findings suggest that syt•t-SNARE interactions regulate the opening and closing of fusion pores.

Disruption of syt I in mice, flies, and worms results in impaired synaptic transmission (40, 54, 93, 115, 116). It was argued that in mice this phenotype results from the inability of docked synaptic vesicles to fuse in response to elevated Ca^{2+} (54), but in flies and mice there were also indications that docking and vesicle recycling were impaired (74, 126). Furthermore, it has become evident that mutations in syt I result in phenotypes that are mimicked by mutations in other genes that do not encode Ca^{2+} binding proteins (81).

To determine whether Ca^{2+} and syt I are necessary and sufficient to couple Ca^{2+} to fusion, experiments were performed with the reconstituted fusion assay (161). In this defined system, the cytoplasmic domain of syt I conferred Ca^{2+} regulation to SNARE-mediated fusion (**Figure 2*d***) (15, 153). Interestingly, the cytoplasmic domain of syt I failed to stimulate fusion when liposomes contained a high number of copies of syb [e.g., 600 to 800 copies per vesicle (161)]; stimulation occurred only with a physiological density of ~100 copies of syb per vesicle. Thus, in the presence of the cytoplasmic domain of syt I, supraphysiological levels of SNARE proteins are not required to drive membrane fusion.

Armed with a brief description of the proteins that mediate membrane fusion, we now turn our attention to biophysical studies directed toward understanding how these proteins participate in fusion pore opening and expansion in living cells.

BIOPHYSICAL MEASUREMENTS OF FUSION PORES IN CELLS

The earliest attempts to characterize the fusion pore came from the ultrastructural studies of Heuser & Reese (67). They focused on the smallest omega-figures in nerve terminals where exocytosis had been captured by rapid freezing. The narrowest fluid connections between the vesicle lumen and the extracellular fluid had dimensions of 20 nm. This number is an upper bound as earlier fusion events could not be observed. Glimpses of even smaller and presumably earlier fusion pore structures were provided by two electrophysiological techniques, capacitance and amperometry, and by imaging vesicles with fluorescent labels. These techniques have revealed some of the basic properties of fusion pores, as well as some interesting clues about molecular structure and regulation.

Capacitance

Capacitance is proportional to membrane area (53); therefore, recording capacitance monitors the addition of membrane to the cell surface during fusion (55). In recordings from chromaffin cells the fusion of single vesicles produced small stepwise increases in capacitance of ~1 fF (112). Occasionally, upward steps were rapidly followed by a downward step of the same size. The beige mouse mast cells have large vesicles such that the capacitance steps are also large (~16 fF), and when the distribution of upward steps was compared with the distribution of downward steps, the partial similarity strengthened the case for reversals (45). These observations suggested that the fusion apparatus forms a transient structure that is not irrevocably committed to the full fusion of the two membranes. This is closely related to the hypothesis of "kiss-and-run," in which vesicles expel their content without full merger of the two membranes. By avoiding membrane mixing, the vesicle can be reused almost immediately. From the perspective of the fusion pore, this suggests the reversal of pore opening. By contrast, when membranes fully fuse, the fusion pore dilates and expands, making reversal more difficult. The transient capacitance steps thus hint at the existence of a distinct structural intermediate that lasts long enough for electrophysiological detection.

The basic observation of reversal of upward capacitance steps was extended from dense-core vesicles of chromaffin cells and mast cells to small clear vesicles that are morphologically identical to synaptic vesicles (80). Upward steps were occasionally followed in less then one second by downward steps of the same amplitude (**Figure 3*a1*, *a2***). Thus, the size of the vesicle remains the same while the fusion pore opens (**Figure 3*a3*, *a4***). The lack of exchange of membrane between the vesicle and plasma membrane argues for the presence of a barrier to lipid flow, as one would expect for a protein-pore. But in mast cells the downward steps were smaller than the upward steps, indicating that these larger fusion pores were more fluid and permitted membrane movement (110).

The first measurement of the conductance of a fusion pore was made in mast cells by analyzing a spike-like current discharge concomitant with single-vesicle capacitance steps (24). This current resulted from the equalization of the vesicle potential when the fusion pore opened, and indicated that the fusion pore has a conductance comparable to that of a large ion channel. This represents a basic similarity to ion channels. Perhaps a very large conductance could be judged as too large to be formed by membrane proteins in the usual size range, but ion channels of a few nanosiemens are known, such that a measurement of the conductance is not useful in addressing the fundamental question of the composition of a fusion pore.

By extending capacitance recording to measure complex admittance, one obtains the full dynamic current response to a time-varying voltage. Differences in phase lag for the charging of different elements of membrane contain more information, and by assuming a simple and plausible equivalent circuit, one can compute the conductance of the fusion pore (39, 97). Indeed, fusion pore conductances in different types of cells and vesicles vary by more than an order of magnitude, indicating that the fusion pore is a diverse structure without a truly unique molecular composition (80, 90). In addition to being diverse, fusion pore conductances can fluctuate over a considerable range during one long-lasting event and still close (142). This raises an interesting question about whether reversal is possible even after the fusion pore incorporates lipid (see Which Steps Are Reversible? below).

PSF: pre-spike foot

Amperometry

Another critical technique in the study of fusion pores is amperometry (33). This electrochemical method detects readily oxidized substances, such as the biogenic amines, and has the sensitivity to detect the exocytosis of single vesicles (163). Amperometric recording from chromaffin cells of norepinephrine release from a single vesicle revealed an early event in which the catecholamine leaks out of a vesicle slowly (34). This slow component of release is followed by a spike, as the remaining content of the vesicle is expelled almost instantaneously (**Figure 3*b1***). The initial slow phase, the pre-spike foot (PSF), reflects an early stage in the fusion pore. The amplitude of the PSF is related to the size of the open fusion pore and the duration is a basic kinetic measure of its stability. Feet can occur without resolving to spikes when a fusion pore opens and then closes. This provides another independent demonstration of kiss-and-run

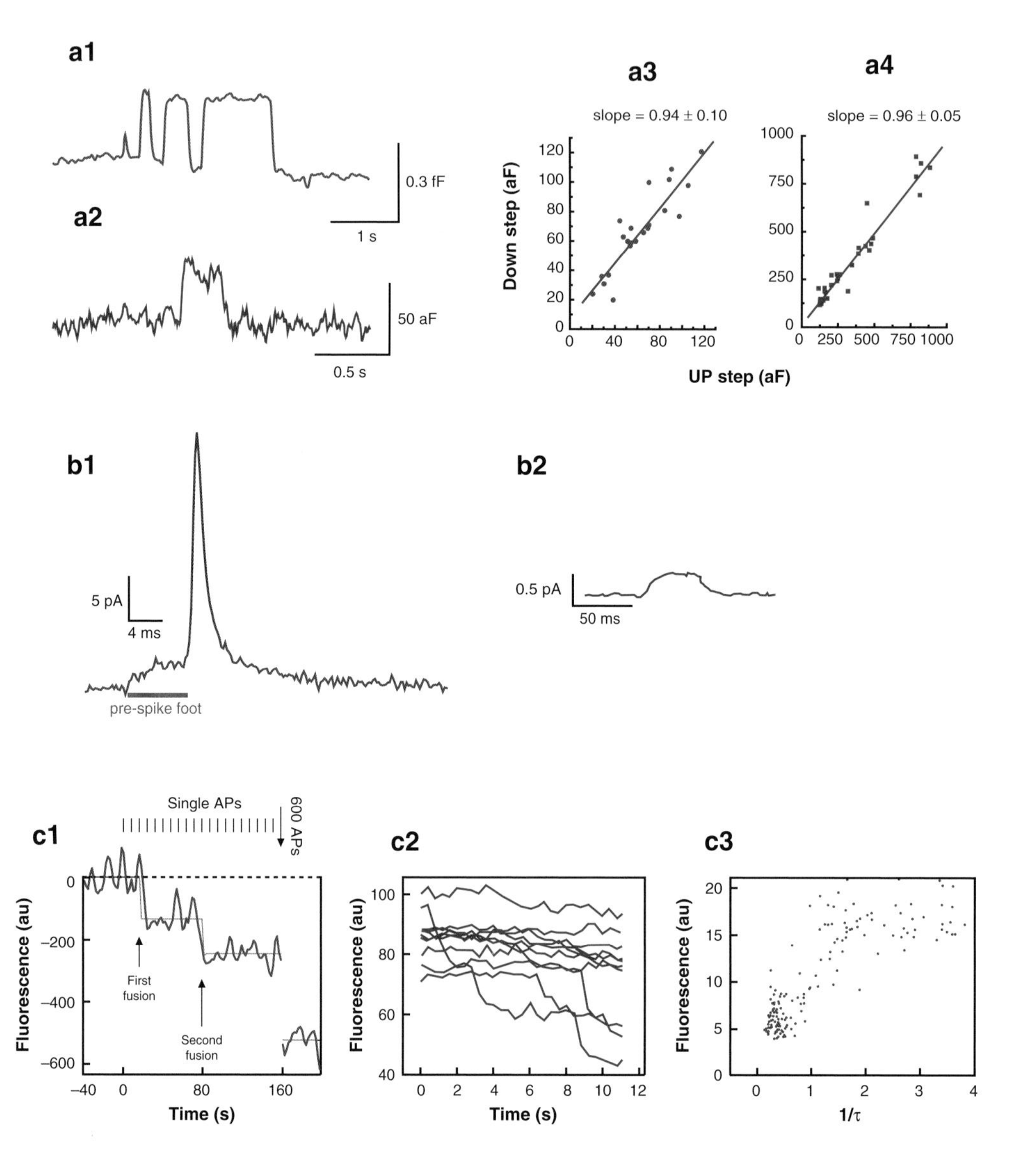

Figure 3

(*a*) Single-vesicle capacitance steps in posterior pituitary nerve terminals. (*a1*) Large capacitance steps are produced by exocytosis of large dense-core vesicles (*a2*), and small capacitance steps are produced by small clear vesicles. The subsequent step down indicates kiss-and-run and is tightly correlated with the up-step for both small clear (*a3*) vesicles and large dense-core (*a4*) vesicles. The slope of about 1 indicates no membrane transfer through the fusion pore (80). (*b*) Amperometry traces of norepinephrine release from a PC12 cell. (*b1*) A single-vesicle release event shows two phases. First the fusion pore opens to produce a pre-spike foot (*dark gray bar*), and then the dilation of the fusion pore initiates the spike. Recording from cells transfected with a SNAP-25 mutant with alanine at residues 178 and 181 (176). (*b2*) An amperometry recording of a small kiss-and-run event (157). (*c*) Single-vesicle FM1-43 destaining events in a hippocampal nerve terminal. (*c1*) A single vesicle fused twice as action potentials were applied at 10-s intervals. The second drop was smaller than the first (6). (*c2*) Full fusion reduced fluorescence by ~17 au in these records. Kiss-and-run events produce smaller steps. (*c3*) A plot of the step size versus one over the step decay time constant shows that smaller steps have slower decays owing to slower dye efflux through fusion pores (127).

(3). Feet with different amplitudes indicate fusion pores with different sizes. In PC12 cells, tiny kiss-and-run events with a flux of about one fifth the flux of the fusion pores associated with spikes indicated a distinct class of very small fusion pores that are incapable of dilating (**Figure 3*b2***) (159). Thus, amperometry recording supports the picture drawn from capacitance studies of the diversity of fusion pore structures.

The PSF duration varies much in the same way that the lifetime of an open ion channel varies, suggesting that fusion pore dynamics are governed by stochastic processes. The PSF lifetime distribution is well described by a single exponential (34, 156), indicating that fusion pore transitions to other states are governed by a Markov process in which the transition probability depends only on the momentary state of the system and not on earlier events. This is an important result because it defines the initial state of the fusion pore as a static structure. Analogous to the single-channel kinetic studies of ion channels, this dynamic behavior of fusion pores suggests that they open, possibly close, and dilate in discrete transitions. These discrete changes in state are then another feature shared with ion channels and thus favor the protein-pore model (90).

Fluorescent Imaging

A number of fluorescent labels can be used to study exocytosis. The lipophilic fluorescent dye FM1-43 and its variants can be used to load recycling synaptic vesicles. Measurements of fluorescence then track dynamic changes in the synaptic vesicle distribution (14, 133). This technique is unique in bringing a quantitative method of analysis to the study of vesicle dynamics in presynaptic nerve terminals. Like capacitance and amperometry, FM1-43 fluorescence supports the view that neurotransmitter is released in discrete packets (132).

The first studies with FM1-43 focused on recycling dynamics and interpreted dye uptake into and loss from nerve terminals as the creation and loss of synaptic vesicles. In an early effort to probe the mechanism of exocytosis, the kinetics of destaining for three related dyes with different lipid-water partition coefficients was compared. The slower destaining of more lipophilic dyes was proposed to reflect a rapid and selective reuptake of the vesicle membrane before it could mix with the plasma membrane (79). This is consistent with kiss-and-run, but in another study identical destaining kinetics was reported for different dyes (46), and the interpretation of destaining kinetics has become controversial. Some forms of stimulation evoke neurotransmitter release without loss of FM1-43, indicating that the fusion pore closes without reaching a size large enough for the dye to escape (143). However, this raises questions about how neurotransmitter can be expelled rapidly enough to produce synaptic currents with the same time course as the synaptic currents associated with full fusion (80).

In experiments in which only a small fraction of the vesicles were labeled, fluorescence measurements produced clearer evidence for kiss-and-run. With only a few labeled vesicles, the reduced background revealed single-vesicle fusion events as discrete drops in fluorescence (**Figure 3*c1*, *c2***) (6, 127). When two steps occur in rapid succession, the second was generally smaller than the first (**Figure 3*c1***). This suggests that the two events arise from the same vesicle; the second event is smaller because the vesicle has already lost a fraction of its dye. Steps fell into two size groups: large quantal steps and smaller subquantal steps. The amplitudes of the large steps were similar to estimates of the amount of label in a single vesicle, and the subquantal steps were attributed to kiss-and-run, where the transient pore opening allowed a fraction of the dye in a vesicle to escape. Furthermore, with subquantal steps the fluorescence fell more slowly than with quantal steps (**Figure 3*c2*, *c3***), indicating that dye sees a restricted escape route as expected for a narrow fusion pore that does not allow lateral diffusion of the dye from the vesicle membrane to the plasma membrane

(127). Additional single-vesicle imaging experiments with the pH-sensitive label synaptophluorin reported multiple components of endocytosis, and the fastest of these was attributed to kiss-and-run (52).

In contrast to the studies in cultured hippocampal neurons (6, 127), FM1-43 destaining in ribbon synapses in goldfish retinal neurons is rapid and generally proceeds with full quantal steps (173). High-resolution imaging of individual events revealed the dispersal of dye into the plasma membrane within about 100 ms. This could be due to rapid departitioning through a fusion pore that has expanded to a size such that permeation through the fusion "neck" was not rate limiting (127). Alternatively, this time course could represent the time it takes for the vesicle membrane to mix with the plasma membrane, a process that would require the incorporation of lipid into the fusion pore.

Labeling of the vesicle lumen, or content labeling, provides an alternative to lipid labels such as FM1-43. In pancreatic β cells, larger extracellular labels entered vesicles at later times, suggesting that fusion pore expansion proceeds from an initial diameter of 1.4 to 12 nm in a few seconds (147). In a simultaneous study of content and membrane label in PC12 cells, loss of membrane label occurred at the same rate whether the content label was lost or retained (149). This experiment suggests that the membranes can mix during kiss-and-run and thus implies the existence of a lipid pathway between the two membranes. These experiments were conducted using a vesicle-targeted GFP fusion protein, and recent studies with a number of these content labels indicate that the destaining time course can be variable and dependent on the choice of label (109).

MOLECULAR COMPOSITION AND REGULATION

The biophysical methods used to characterize fusion pores have been combined with molecular manipulations in an effort to identify the roles of specific proteins in fusion pore structure and function. In amperometric recordings of norepinephrine release from PC12 cells, overexpression of different syt isoforms altered the mean lifetime of PSF, indicating an effect on the stability of open fusion pores (156). Open fusion pores can either close or dilate, and this result indicates that syt interacts with fusion pores to alter the rates of one or both of these processes. syt I, an isoform already abundant in PC12 cells, prolonged PSF, indicating that increasing the proportion of this endogenous protein stabilized the fusion pore (156). syt IV, an isoform present in only trace amounts, shortened the duration of PSF, indicating that this isoform has the opposite effect. Studies with mutants of syt I provided further evidence for a regulatory role in fusion pore dynamics (10, 177).

syt transduces Ca^{2+} signals to trigger exocytosis, and these results suggest that the fusion pore is a target of this regulation. Thus, some of the actions of syt take place at the instant of fusion. However, because syt binds to both proteins and lipids in response to Ca^{2+}, this experiment cannot distinguish between protein and lipidic fusion pores.

syt IV is an especially interesting isoform. It is upregulated in brain by psychoactive drugs and by seizure activity, and the mammalian protein has no known Ca^{2+}-dependent effector interactions (29, 157). In PC12 cells this protein induces a unique form of exocytosis that is exclusively kiss-and-run (157). Small fusion pores remain open for approximately 100 times longer than the mean duration of PSF and always close without dilating (an example is shown in **Figure 3*b2***). This indicates the existence of a specialized fusion pore with a small size and with a limited functional capacity. There is also evidence that syt IV can influence the choice between full fusion and kiss-and-run in pancreatic β cells (151) and at the *Drosophila* neuromuscular junction (122). syt may have even more diverse roles in the regulation of fusion pores. For example, syt VII appears to regulate expansion of fusion pores formed between

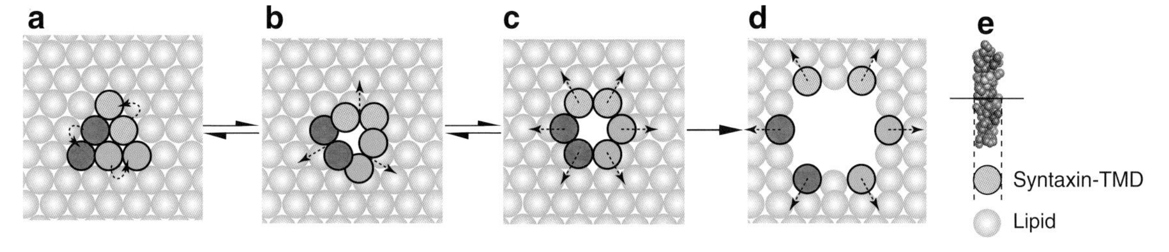

Figure 4

Model of a fusion pore lined by syx TMSs. Analysis of the fusion pore conductance by Han et al. (61) hypothesized a pore formed from five to eight TMSs. This model uses six TMSs grouped in three pairs (*gray, green, or pink circles*). (*a*) A closed fusion pore forms first. (*b* and *c*) The next step is opening, and grouping the TMSs into three pairs allows one to envision opening as a rotation of pairs around the line of TMS intersection in each pair. (*d*) The open fusion pore dilates as the syx TMSs spread and lipids come between them. This figure shows only a flat plane without the curving structural rearrangements shown in **Figure 1**. (*e*) A space-filling model of the syx 1A TMS is shown with residues that influence fusion pore flux highlighted in red ($P < 0.01$) and yellow ($P < 0.05$). Circular cross-sections of a syx TMS and a phospholipid molecule show that they occupy similar areas of membrane.

lysosomes and the plasma membrane of fibroblasts. Disruption of syt VII facilitates the exocytosis of large fluorescently labeled dextran molecules that are otherwise retained in lysosomes, suggesting that syt VII normally acts to impede expansion of the fusion pore (73).

In an effort to test the hypothesis that SNARE proteins form the fusion pore, the amino acids in the TMS of syx were mutated to tryptophan (61). The rationale for these experiments was that if a residue lines the pore, then a large side chain should occlude the pore and retard the flux of neurotransmitter. In PC12 cells, this tryptophan-scanning approach identified three sites that influence the flux through an open fusion pore. Complex admittance measurements with mutants at two of these sites confirmed that the fusion pore conductance was also reduced. In an α-helical wheel of the syx TMS, the three residues that influenced fusion pore permeation fell along the same face (residues indicated in **Figure 4*e***). These results support a structural model of fusion pores formed by a circular arrangement of several syx TMSs, with each forming an α-helix as it spans the membrane. On the basis of a simple model of the fusion pore as a conducting cylinder, the measured fusion pore conductance of 100 pS gave an estimated fusion pore diameter of 1 nm. This can be achieved by a circular arrangement of five to eight syx TMSs. A model for opening and dilation is shown for a fusion pore consisting of six syx TMSs (**Figure 4**). Pairwise rotations provide an easy way to envision conversion from a closed state to an open state. Insertion of lipids between adjacent TMSs initiates dilation.

This work was extended by changing the electrical charge of syx pore-lining residues (60). Norepinephrine is positively charged, and with pore-lining residues changed to arginine, the fusion pore flux was reduced. Conversely, changing pore-lining residues to aspartate increased the fusion pore flux. These changes could not be accounted for by changes in side chain volume and showed a direct correlation with side chain pK.

The demonstration that the syx TMS is a structural component of the fusion pore establishes a physical link with the SNARE complex. Thus, there is structural continuity between the fusion pore and the fusion machine. Now we can ask how structural changes in the SNARE complex exert forces on the TMSs to effect the opening, closing, and dilation of fusion pores (**Figure 4*a–d***).

Indeed, different v-SNAREs produce fusion pores with different lifetimes (21); mutations throughout the SNARE complex also alter fusion pore lifetimes (176). SNARE binding proteins should also exert forces on the SNARE complex, and the results will depend on how these forces are transduced to the fusion pore. For example, as noted above, when the SNARE binding capacity of syt is reduced, fusion pores can still open but have a stronger tendency to close without dilating (10).

WHICH STEPS ARE REVERSIBLE?

It is easy to envision a reversal of exocytosis immediately after the opening of a proteinaceous fusion pore; as proteinaceous ion channels open and close through conformational transitions. However, the incorporation of lipid into a fusion pore makes reversal harder. Incorporation of lipids should bring membrane tension and curvature stress into the picture, and these forces will drive the vesicle membrane to flatten out and mix with the plasma membrane. Reversal of such processes would require some form of scaffold and a means by which chemical energy can shape lipid bilayers. Although more difficult to envision, such processes almost certainly do occur, as mentioned above.

Spruce et al. (142) saw the conductance of mast cell fusion pores grow by up to 50-fold and then close, and pancreatic β cells show similar behavior (147). Monck et al. (110) and Taraska et al. (149) saw membrane flux or mixing in events that subsequently reversed. Ales et al. (1) saw a sudden burst of release just as a fusion pore closed, and suggested that this reflected a rapid transient expansion that would require extreme flexibility. Molecular manipulations of munc18 (49), cysteine string protein (58), and complexin II (7) altered the shape and size of amperometric spikes. Similar results were obtained with several syt and SNARE mutants (C.-T. Wang, X. Han, E. R. Chapman & M. B. Jackson, unpublished observations). Because the amperometric spike marks a time when the fusion pore is rapidly expanding, these changes in size and kinetics probably indicate closure of the fusion pore following the onset of dilation. How proteins control and reverse the expansion of a lipidic fusion pore remains a major conceptual challenge.

Although these results point to a proximity model, fluid fusion pores could start out as protein, with the fluctuations and membrane flow occurring after an expansion step involving the incorporation of lipid. The more important issue is that these fusion pores can expand and incorporate lipid without losing the ability to close. Cells need a mechanism of maintaining the narrow lipid stalk between the vesicle and plasma membrane and pinching it shut. Because the final step of endocytosis is similar to the reversal of the dilation step of exocytosis, we can look to the molecular machinery of endocytosis for clues about how to constrict and close off a lipid stalk. Proteins such as dynamin can harvest the energy of GTP hydrolysis to overcome the energy barriers encountered in membrane fission (37, 138), and this protein or a homolog with similar activity could drive the closure of fusion pores.

FUTURE PERSPECTIVES

A central question in membrane fusion concerns the structure of the fusion pore. At present, two studies suggest that the TMS of syx lines the fusion pore in PC12 cells, at least during the first 1 to 2 ms of exocytosis (60, 61). Is the TMS of syb the other half of a fusion pore hemichannel? How are lipids subsequently intercalated between SNARE TMSs to mediate pore expansion? Do similar mechanisms operate at chemical synapses? That is, do the slow partial FM1-43 destaining events in hippocampal neurons report a form of kiss-and-run, analogous to that of dense core vesicles, in which the fusion pore opens and closes without membrane merger (6, 127)? Does the mode of release depend on the stimulus intensity? This idea formed the basis of the original proposal for kiss-and-run (107) and is receiving experimental support from work in cultured hippocampal neurons (52) and chromaffin cells (51). What is the physiological relevance of kiss-and-run in synaptic transmission? And can it evoke a rapid postsynaptic response, or does it allow a mere trickle of neurotransmitter that serves to desensitize postsynaptic receptors (157)?

Numerous fundamental issues remain open questions. Even at the level of SNARE complex assembly and disassembly it has not yet been resolved whether SNAPs and NSF disassemble *trans*-SNARE pairs (155) or whether SNARE pairing irreversibly commits the membranes to fusion (160). How is SNARE assembly regulated, and what is the temporal relationship between SNARE assembly and the opening and dilation of a fusion pore? Numerous additional proteins are essential for regulated membrane fusion. How do these proteins, including the vacuolar proton pump (68, 103), complexin (104), and nsec1/munc18 (47), regulate membrane fusion? What is the molecular basis for the diversity of fusion pore sizes and kinetics—does it involve differences in SNARE isoforms or copy numbers?

While many proteins that play key roles in exocytosis have been identified, the actual function of these proteins is, for the most part, completely unknown. Hence, this is an exciting period in the study of the biophysics of biological membrane fusion: The players have been largely identified, and sensitive biophysical tools are now in place. The critical challenge in the coming years will be to combine these molecular and biophysical tools to gain a precise understanding of fusion pore opening, closing, and dilation. Once these questions are resolved, it should become possible to understand how changes in fusion pores affect synaptic physiology.

SUMMARY POINTS

1. Membrane fusion must overcome energy barriers.
2. SNAREs are essential proteins for membrane fusion.
3. How do SNARE proteins assemble into SNARE complexes?
4. Reconstitution of liposome fusion with SNAREs, and of $Ca2^{+}$ regulated fusion by the addition of synaptotagmin.
5. Biophysical measurements provide insight into the structure and dynamics of fusion pores in cells.
6. What is the molecular composition of fusion pores; are they made of protein or lipid or both?
7. How are the opening, closing, and dilation of fusion pores regulated?
8. Which steps of exocytosis are reversible?

ACKNOWLEDGMENTS

We thank E. Hui and W. Almers for providing artwork in **Figures 1** and **4**. **Figure 2*b*** was created by Craig Foster Medical Communications. We thank T. Weber and J. Rothman for **Figure 2*c*** and A. Aravanis and R. Tsien for **Figure 3*c1***. Additional unpublished data were provided by A. Bhalla (**Figure 2*d***) and by X. Han and C.-T. Wang (**Figure 3*b***). We also thank X. Han, J.M. Edwardson, and C. Dean for comments on this manuscript. This study was supported by grants from the NIH (GM56827 and MH61876 to E.R.C.; NS30016 and NS 44057 to M.B.J.) and the AHA (0440168N to E.R.C.). E.R.C. is an Investigator of the Howard Hughes Medical Institute.

LITERATURE CITED

1. Ales E, Tabares L, Poyato JM, Valero V, Lindau M, Alvarez de Toledo G. 1999. High calcium concentrations shift the mode of exocytosis to the kiss-and-run mechanism. *Nat. Cell Biol.* 1:40–44
2. Almers W, Tse FW. 1990. Transmitter release from synapses: Does a preassembled fusion pore initiate exocytosis? *Neuron* 4:813–18
3. Alvarez de Toledo G, Fernandez-Chacon R, Fernandez JM. 1993. Release of secretory products during transient vesicle fusion. *Nature* 363:554–58
4. An SJ, Almers W. 2004. Tracking SNARE complex formation in live endocrine cells. *Science* 306:1042–46
5. Antonin W, Fasshauer D, Becker S, Jahn R, Schneider TR. 2002. Crystal structure of the endosomal SNARE complex reveals common structural principles of all SNAREs. *Nat. Struct. Biol.* 9:107–11
6. Aravanis AM, Pyle JL, Tsien RW. 2003. Single synaptic vesicles fusing transiently and successively without loss of identity. *Nature* 423:643–47
7. Archer DA, Graham ME, Burgoyne RD. 2002. Complexin regulates the closure of the fusion pore during regulated vesicle exocytosis. *J. Biol. Chem.* 277:18249–52
8. Augustine GJ. 2001. How does calcium trigger neurotransmitter release? *Curr. Opin. Neurobiol.* 11:320–26
9. Bai J, Tucker WC, Chapman ER. 2004. PIP_2 increases the speed-of-response of synaptotagmin and steers its membrane penetration activity toward the plasma membrane. *Nat. Struct. Mol. Biol.* 11:36–44
10. Bai J, Wang CT, Richards DA, Jackson MB, Chapman ER. 2004. Fusion pore dynamics are regulated by synaptotagmin*t-SNARE interactions. *Neuron* 41:929–42
11. Bai J, Wang P, Chapman ER. 2002. C2A activates a cryptic Ca^{2+}-triggered membrane penetration activity within the C2B domain of synaptotagmin I. *Proc. Natl. Acad. Sci. USA* 99:1665–70
12. Baumert M, Maycox PR, Navone F, De Camilli P, Jahn R. 1989. Synaptobrevin: an integral membrane protein of 18,000 daltons present in small synaptic vesicles of rat brain. *EMBO J.* 8:379–84
13. Bennett MK, Calakos N, Scheller RH. 1992. Syntaxin: a synaptic protein implicated in docking of synaptic vesicles at presynaptic active zones. *Science* 257:255–59
14. Betz WJ, Bewick GS. 1992. Optical analysis of synaptic vesicle recycling at the frog neuromuscular junction. *Science* 255:200–3
15. Bhalla A, Tucker WT, Chapman ER. 2005. Synaptotagmin isoforms couple distinct ranges of Ca2+, Ba2+ and Sr2+ concentration to SNARE-mediated membrane fusion. *Mol. Biol. Cell.* 16:4755–64
16. Blasi J, Chapman ER, Link E, Binz T, Yamasaki S, et al. 1993. Botulinum neurotoxin A selectively cleaves the synaptic protein SNAP-25. *Nature* 365:160–63
17. Blasi J, Chapman ER, Yamasaki S, Binz T, Niemann H, Jahn R. 1993. Botulinum neurotoxin C1 blocks neurotransmitter release by means of cleaving HPC-1/syntaxin. *EMBO J.* 12:4821–28

18. Block MR, Glick BS, Wilcox CA, Wieland FT, Rothman JE. 1988. Purification of an N-ethylmaleimide-sensitive protein catalyzing vesicular transport. *Proc. Natl. Acad. Sci. USA* 85:7852–56
19. Bock JB, Matern HT, Peden AA, Scheller RH. 2001. A genomic perspective on membrane compartment organization. *Nature* 409:839–41
20. Bonifacino JS, Glick BS. 2004. The mechanisms of vesicle budding and fusion. *Cell* 116:153–66
21. Borisovska M, Zhao Y, Tsytsyura Y, Glyvuk N, Takamori S, et al. 2005. v-SNAREs control exocytosis of vesicles from priming to fusion. *EMBO J.* 24:2114–26
22. Bowen ME, Engelman DM, Brunger AT. 2002. Mutational analysis of synaptobrevin transmembrane domain oligomerization. *Biochemistry* 41:15861–66
23. Bowen ME, Weninger K, Brunger AT, Chu S. 2004. Single molecule observation of liposome-bilayer fusion thermally induced by soluble N-ethyl maleimide sensitive-factor attachment protein receptors (SNAREs). *Biophys. J.* 87:3569–84
24. Breckenridge LJ, Almers W. 1987. Currents through the fusion pore that forms during exocytosis of a secretory vesicle. *Nature* 328:814–17
25. Burgess SW, McIntosh TJ, Lentz BR. 1992. Modulation of poly(ethylene glycol)-induced fusion by membrane hydration: importance of interbilayer separation. *Biochemistry* 31:2653–61
26. Chapman ER. 2002. Synaptotagmin: a Ca^{2+} sensor that triggers exocytosis? *Nat. Rev. Mol. Cell Biol.* 3:498–508
27. Chapman ER, An S, Edwardson JM, Jahn R. 1996. A novel function for the second C2 domain of synaptotagmin. Ca^{2+}-triggered dimerization. *J. Biol. Chem.* 271:5844–49
28. Chapman ER, Davis AF. 1998. Direct interaction of a Ca^{2+}-binding loop of synaptotagmin with lipid bilayers. *J. Biol. Chem.* 273:13995–4001
29. Chapman ER, Desai RC, Davis AF, Tornehl CK. 1998. Delineation of the oligomerization, AP-2 binding, and synprint binding region of the C2B domain of synaptotagmin. *J. Biol. Chem.* 273:32966–72
30. Chen YA, Scales SJ, Patel SM, Doung YC, Scheller RH. 1999. SNARE complex formation is triggered by Ca^{2+} and drives membrane fusion. *Cell* 97:165–74
31. Chen YA, Scheller RH. 2001. SNARE-mediated membrane fusion. *Nat. Rev. Mol. Cell Biol.* 2:98–106
32. Chernomordik LV, Kozlov MM. 2003. Protein-lipid interplay in fusion and fission of biological membranes. *Annu. Rev. Biochem.* 72:175–207
33. Chow RH, von Rüden L. 1995. Electrochemical detection of secretion from single cells. See Ref. 134a, pp. 245–75
34. Chow RH, von Rüden L, Neher E. 1992. Delay in vesicle fusion revealed by electrochemical monitoring of single secretory events in adrenal chromaffin cells. *Nature* 356:60–63
35. Clary DO, Griff IC, Rothman JE. 1990. SNAPs, a family of NSF attachment proteins involved in intracellular membrane fusion in animals and yeast. *Cell* 61:709–21
36. Cohen FS, Melikyan GB. 2004. The energetics of membrane fusion from binding, through hemifusion, pore formation, and pore enlargement. *J. Membr. Biol.* 199:1–14
37. Danino D, Hinshaw JE. 2001. Dynamin family of mechanoenzymes. *Curr. Opin. Cell Biol.* 13:454–60
38. Davis AF, Bai J, Fasshauer D, Wolowick MJ, Lewis JL, Chapman ER. 1999. Kinetics of synaptotagmin responses to Ca^{2+} and assembly with the core SNARE complex onto membranes. *Neuron* 24:363–76

39. Debus K, Lindau M. 2000. Resolution of patch capacitance recordings and of fusion pore conductance in small vesicles. *Biophys. J.* 78:2983–97
40. DiAntonio A, Schwarz TL. 1994. The effect on synaptic physiology of synaptotagmin mutations in *Drosophila*. *Neuron* 12:909–20
41. Earles CA, Bai J, Wang P, Chapman ER. 2001. The tandem C2 domains of synaptotagmin contain redundant Ca^{2+} binding sites that cooperate to engage t-SNAREs and trigger exocytosis. *J. Cell Biol.* 154:1117–23
42. Fasshauer D, Antonin W, Margittai M, Pabst S, Jahn R. 1999. Mixed and non-cognate SNARE complexes. Characterization of assembly and biophysical properties. *J. Biol. Chem.* 274:15440–46
43. Fasshauer D, Antonin W, Subramaniam V, Jahn R. 2002. SNARE assembly and disassembly exhibit a pronounced hysteresis. *Nat. Struct. Biol.* 9:144–51
44. Fernandez I, Arac D, Ubach J, Gerber SH, Shin O, et al. 2001. Three-dimensional structure of the synaptotagmin 1 C2B-domain. Synaptotagmin 1 as a phospholipid binding machine. *Neuron* 32:1057–69
45. Fernandez JM, Neher E, Gomperts BD. 1984. Capacitance measurements reveal stepwise fusion events in degranulating mast cells. *Nature* 312:453–55
46. Fernandez-Alfonso T, Ryan TA. 2004. The kinetics of synaptic vesicle pool depletion at CNS synaptic terminals. *Neuron* 41:943–53
47. Ferro-Novick S, Jahn R. 1994. Vesicle fusion from yeast to man. *Nature* 370:191–93
48. Fiebig KM, Rice LM, Pollock E, Brunger AT. 1999. Folding intermediates of SNARE complex assembly. *Nat. Struct. Biol.* 6:117–23
49. Fisher RJ, Pevsner J, Burgoyne RD. 2001. Control of fusion pore dynamics during exocytosis by Munc 18. *Science* 291:875–78
50. Fix M, Melia TJ, Jaiswal JK, Rappoport JZ, You D, et al. 2004. Imaging single membrane fusion events mediated by SNARE proteins. *Proc. Natl. Acad. Sci. USA* 101:7311–16
51. Fulop T, Radabaugh S, Smith C. 2005. Activity-dependent differential transmitter release in mouse adrenal chromaffin cells. *J. Neurosci.* 25:7324–32
52. Gandhi SP, Stevens CF. 2003. Three modes of synaptic vesicular recycling revealed by single-vesicle imaging. *Nature* 423:607–13
53. Gentet LJ, Stuart GJ, Clements JD. 2000. Direct measurement of specific membrane capacitance in neurons. *Biophys. J.* 79:314–20
54. Geppert M, Goda Y, Hammer RE, Li C, Rosahl TW, et al. 1994. Synaptotagmin I: a major Ca^{2+} sensor for transmitter release at a central synapse. *Cell* 79:717–27
55. Gillis KD. 1995. Techniques for membrane capacitance measurement. See Ref. 134a, pp. 155–98
56. Glick BS, Rothman JE. 1987. Possible role for fatty acyl-coenzyme A in intracellular protein transport. *Nature* 326:309–12
57. Gotte M, von Mollard GF. 1998. A new beat for the SNARE drum. *Trends Cell Biol.* 8:215–18
58. Graham ME, Burgoyne RD. 2000. Comparison of cysteine string protein (Csp) and mutant alpha-SNAP overexpression reveals a role for csp in late steps of membrane fusion in dense-core granule exocytosis in adrenal chromaffin cells. *J. Neurosci.* 20:1281–89
59. Grote E, Baba M, Ohsumi Y, Novick PJ. 2000. Geranylgeranylated SNAREs are dominant inhibitors of membrane fusion. *J. Cell Biol.* 151:453–66
60. Han X, Jackson MB. 2005. Electrostatic interactions between the syntaxin membrane anchor and neurotransmitter passing through the fusion pore. *Biophys. J.* 88:L20–22

61. Han X, Wang CT, Bai J, Chapman ER, Jackson MB. 2004. Transmembrane segments of syntaxin line the fusion pore of Ca2+-triggered exocytosis. *Science* 304:289–92
62. Deleted in proof
63. Hanson PI, Roth R, Morisaki H, Jahn R, Heuser JE. 1997. Structure and conformational changes in NSF and its membrane receptor complexes visualized by quick-freeze/deep-etch electron microscopy. *Cell* 90:523–35
64. Hayashi T, McMahon H, Yamasaki S, Binz T, Hata Y, et al. 1994. Synaptic vesicle membrane fusion complex: action of clostridial neurotoxins on assembly. *EMBO J.* 13:5051–61
65. Hayashi T, Yamasaki S, Nauenburg S, Binz T, Niemann H. 1995. Disassembly of the reconstituted synaptic vesicle membrane fusion complex in vitro. *EMBO J.* 14:2317–25
66. Helm CA, Israelachvili JN, McGuiggan PM. 1992. Role of hydrophobic forces in bilayer adhesion and fusion. *Biochemistry* 31:1794–805
67. Heuser JE, Reese TS. 1981. Structural changes after transmitter release at the frog neuromuscular junction. *J. Cell Biol.* 88:564–80
68. Hiesinger PR, Fayyazuddin A, Mehta SQ, Rosenmund T, Schulze KL, et al. 2005. The v-ATPase V0 subunit a1 is required for a late step in synaptic vesicle exocytosis in Drosophila. *Cell* 121:607–20
69. Hua SY, Charlton MP. 1999. Activity-dependent changes in partial VAMP complexes during neurotransmitter release. *Nat. Neurosci.* 2:1078–83
70. Hunt JM, Bommert K, Charlton MP, Kistner A, Habermann E, et al. 1994. A post-docking role for synaptobrevin in synaptic vesicle fusion. *Neuron* 12:1269–79
71. Inoue A, Obata K, Akagawa K. 1992. Cloning and sequence analysis of cDNA for a neuronal cell membrane antigen, HPC-1. *J. Biol. Chem.* 267:10613–19
72. Jahn R, Lang T, Sudhof TC. 2003. Membrane fusion. *Cell* 112:519–33
73. Jaiswal JK, Chakrabarti S, Andrews NW, Simon SM. 2004. Synaptotagmin VII restricts fusion pore expansion during lysosomal exocytosis. *PLoS Biol.* 2:E233
74. Jorgensen EM, Hartwieg E, Schuske K, Nonet ML, Jin Y, Horvitz HR. 1995. Defective recycling of synaptic vesicles in synaptotagmin mutants of Caenorhabditis elegans. *Nature* 378:196–99
75. Kasai H. 1999. Comparative biology of Ca^{2+}-dependent exocytosis: implications of kinetic diversity for secretory function. *Trends Neurosci.* 22:88–93
76. Katz B. 1969. *The Release of Neural Transmitter Substances*. Springfield, IL: Thomas
77. Katz L, Hanson PI, Heuser JE, Brennwald P. 1998. Genetic and morphological analyses reveal a critical interaction between the C-termini of two SNARE proteins and a parallel four helical arrangement for the exocytic SNARE complex. *EMBO J.* 17:6200–9
78. Kim CS, Kweon DH, Shin YK. 2002. Membrane topologies of neuronal SNARE folding intermediates. *Biochemistry* 41:10928–33
79. Klinglauf J, Kavalali ET, Tsien RW. 1998. Kinetics and regulation of fast endocytosis at hippocampal synapses. *Nature* 394:581–85
80. Klyachko VA, Jackson MB. 2002. Capacitance steps and fusion pores of small and large-dense-core vesicles in nerve terminals. *Nature* 418:89–92
81. Koh TW, Bellen HJ. 2003. Synaptotagmin I: a Ca^{2+} sensor for neurotransmitter release. *Trends Neurosci.* 26:413–22
82. Kubelka J, Hofrichter J, Eaton WA. 2004. The protein folding 'speed limit'. *Curr. Opin. Struct. Biol.* 14:76–88
83. Kweon DH, Kim CS, Shin YK. 2002. The membrane-dipped neuronal SNARE complex: a site-directed spin labeling electron paramagnetic resonance study. *Biochemistry* 41:9264–68

84. Kweon DH, Kim CS, Shin YK. 2003. Insertion of the membrane-proximal region of the neuronal SNARE coiled coil into the membrane. *J. Biol. Chem.* 278:12367–73
85. Laage R, Rohde J, Brosig B, Langosch D. 2000. A conserved membrane-spanning amino acid motif drives homomeric and supports heteromeric assembly of presynaptic SNARE proteins. *J. Biol. Chem.* 275:17481–87
86. Lee J, Lentz BR. 1998. Secretory and viral fusion may share mechanistic events with fusion between curved lipid bilayers. *Proc. Natl. Acad. Sci. USA* 95:9274–79
87. Lentz BR, Malinin V, Haque ME, Evans K. 2000. Protein machines and lipid assemblies: current views of cell membrane fusion. *Curr. Opin. Struct. Biol.* 10:607–15
88. Lewis JL, Dong M, Earles CA, Chapman ER. 2001. The transmembrane domain of syntaxin 1A is critical for cytoplasmic domain protein-protein interactions. *J. Biol. Chem.* 276:15458–65
89. Lin RC, Scheller RH. 1997. Structural organization of the synaptic exocytosis core complex. *Neuron* 19:1087–94
90. Lindau M, Almers W. 1995. Structure and function of fusion pores in exocytosis and ectoplasmic membrane fusion. *Curr. Opin. Cell Biol.* 7:509–17
91. Littleton JT, Bai J, Vyas B, Desai R, Baltus AE, et al. 2001. Synaptotagmin mutants reveal essential functions for the C2B domain in Ca^{2+}-triggered fusion and recycling of synaptic vesicles *in vivo*. *J. Neurosci.* 21:1421–33
92. Littleton JT, Barnard RJ, Titus SA, Slind J, Chapman ER, Ganetzky B. 2001. SNARE-complex disassembly by NSF follows synaptic-vesicle fusion. *Proc. Natl. Acad. Sci. USA* 98:12233–38
93. Littleton JT, Stern M, Perin M, Bellen HJ. 1994. Calcium dependence of neurotransmitter release and rate of spontaneous vesicle fusions are altered in *Drosophila* synaptotagmin mutants. *Proc. Natl. Acad. Sci. USA* 91:10888–92
94. Liu TT, Tucker WC, Bhalla A, Chapman ER, Weisshaar JC. 2005. SNARE-driven, 25-millisecond vesicle fusion in vitro. *Biophys. J.* 89:2458–72
95. Liu Y, Barlowe C. 2002. Analysis of Sec22p in endoplasmic reticulum/Golgi transport reveals cellular redundancy in SNARE protein function. *Mol. Biol. Cell.* 13:3314–24
96. Llinas R, Steinberg IZ, Walton K. 1981. Relationship between presynaptic calcium current and postsynaptic potential in squid giant synapse. *Biophys. J.* 33:323–51
97. Lollike K, Borregaard N, Lindau M. 1995. The exocytotic fusion pore of small granules has a conductance similar to an ion channel. *J. Cell Biol.* 129:99–104
98. Lu X, Zhang F, McNew JA, Shin Y-K. 2005. Membrane fusion induced by neuronal SNAREs transits through hemifusion. *J. Biol. Chem.* 280:30538–41
99. Malinin VS, Lentz BR. 2004. On the analysis of elastic deformations in hexagonal phases. *Biophys. J.* 86:3324–28
100. Margittai M, Otto H, Jahn R. 1999. A stable interaction between syntaxin 1a and synaptobrevin 2 mediated by their transmembrane domains. *FEBS Lett.* 446:40–44
101. Markin VS, Albanesi JP. 2002. Membrane fusion: stalk model revisited. *Biophys. J.* 82:693–712
102. Marrink SJ, Mark AE. 2003. The mechanism of vesicle fusion as revealed by molecular dynamics simulations. *J. Am. Chem. Soc.* 125:11144–45
103. Mayer A. 2001. What drives membrane fusion in eukaryotes? *Trends Biochem. Sci.* 26:717–23
104. McMahon HT, Missler M, Li C, Sudhof TC. 1995. Complexins: cytosolic proteins that regulate SNAP receptor function. *Cell* 83:111–19
105. McNew JA, Parlati F, Fukuda R, Johnston RJ, Paz K, et al. 2000. Compartmental specificity of cellular membrane fusion encoded in SNARE proteins. *Nature* 407:153–59

106. McNew JA, Weber T, Engelman DM, Sollner TH, Rothman JE. 1999. The length of the flexible SNAREpin juxtamembrane region is a critical determinant of SNARE-dependent fusion. *Mol. Cell* 4:415–21
107. Meldolesi J, Ceccarelli B. 1981. Exocytosis and membrane recycling. *Philos. Trans. R. Soc. London Sci. Ser. B* 296:55–65
108. Melia TJ, Weber T, McNew JA, Fisher LE, Johnston RJ, et al. 2002. Regulation of membrane fusion by the membrane-proximal coil of the t-SNARE during zippering of SNAREpins. *J. Cell Biol.* 158:929–40
109. Michael DJ, Geng X, Cawley NX, Loh YP, Rhodes CJ, et al. 2004. Fluorescent cargo proteins in pancreatic beta-cells: Design determines secretion kinetics at exocytosis. *Biophys. J.* 87:L03–5
110. Monck JR, Alvarez de Toledo G, Fernandez JM. 1990. Tension in secretory granule membranes causes extensive membrane transfer through the exocytotic fusion pore. *Proc. Natl. Acad. Sci. USA* 87:7804–8
111. Monck JR, Fernandez JM. 1994. The exocytotic fusion pore and neurotransmitter release. *Neuron* 12:707–16
112. Neher E, Marty A. 1982. Discrete changes of cell membrane capacitance observed under conditions of enhanced secretion in bovine adrenal chromaffin cells. *Proc. Natl. Acad. Sci. USA* 79:6712–16
113. Nichols BJ, Ungermann C, Pelham HR, Wickner WT, Haas A. 1997. Homotypic vacuolar fusion mediated by t- and v-SNAREs. *Nature* 387:199–202
114. Nickel W, Weber T, McNew JA, Parlati F, Sollner TH, Rothman JE. 1999. Content mixing and membrane integrity during membrane fusion driven by pairing of isolated v-SNAREs and t-SNAREs. *Proc. Natl. Acad. Sci. USA* 96:12571–76
115. Nishiki T, Augustine GJ. 2004. Synaptotagmin I synchronizes transmitter release in mouse hippocampal neurons. *J. Neurosci.* 24:6127–32
116. Nonet ML, Grundahl K, Meyer BJ, Rand JB. 1993. Synaptic function is impaired but not eliminated in C. elegans mutants lacking synaptotagmin. *Cell* 73:1291–305
117. Nonet ML, Saifee O, Zhao H, Rand JB, Wei L. 1998. Synaptic transmission deficits in Caenorhabditis elegans synaptobrevin mutants. *J. Neurosci.* 18:70–80
118. Novick P, Field C, Schekman R. 1980. Identification of 23 complementation groups required for post-translational events in the yeast secretory pathway. *Cell* 21:205–15
119. Oyler GA, Higgins GA, Hart RA, Battenberg E, Billingsley M, et al. 1989. The identification of a novel synaptosomal-associated protein, SNAP-25, differentially expressed by neuronal subpopulations. *J. Cell Biol.* 109:3039–52
120. Parlati F, McNew JA, Fukuda R, Miller R, Sollner TH, Rothman JE. 2000. Topological restriction of SNARE-dependent membrane fusion. *Nature* 407:194–98
121. Parlati F, Weber T, McNew JA, Westermann B, Sollner TH, Rothman JE. 1999. Rapid and efficient fusion of phospholipid vesicles by the alpha-helical core of a SNARE complex in the absence of an N-terminal regulatory domain. *Proc. Natl. Acad. Sci. USA* 96:12565–70
122. Pawlu C, DeAntonio A, Heckmann M. 2004. Postfusional control of quantal current shape. *Neuron* 42:607–18
123. Perin MS, Fried VA, Mignery GA, Jahn R, Sudhof TC. 1990. Phospholipid binding by a synaptic vesicle protein homologous to the regulatory region of protein kinase C. *Nature* 345:260–63
124. Poirier MA, Xiao W, Macosko JC, Chan C, Shin YK, Bennett MK. 1998. The synaptic SNARE complex is a parallel four-stranded helical bundle. *Nat. Struct. Biol.* 5:765–69

125. Reese C, Heise F, Mayer A. 2005. Trans-SNARE pairing can precede a hemifusion intermediate in intracellular membrane fusion. *Nature* 436:410–14
126. Reist NE, Buchanan J, Li J, DiAntonio A, Buxton EM, Schwarz TL. 1998. Morphologically docked synaptic vesicles are reduced in synaptotagmin mutants of *Drosophila*. *J. Neurosci.* 18:7662–73
127. Richards DA, Bai J, Chapman ER. 2005. Two modes of exocytosis at hippocampal synapses revealed by rate of FM1-43 efflux from individual vesicles. *J. Cell Biol.* 168:929–39
128. Rohde J, Dietrich L, Langosch D, Ungermann C. 2003. The transmembrane domain of Vam3 affects the composition of cis- and trans-SNARE complexes to promote homotypic vacuole fusion. *J. Biol. Chem.* 278:1656–62
129. Rothman JE. 1994. Intracellular membrane fusion. *Adv. Second Messenger Phosphoprotein Res.* 29:81–96
130. Rothman JE, Warren G. 1994. Implications of the SNARE hypothesis for intracellular membrane topology and dynamics. *Curr. Biol.* 4:220–33
131. Roy R, Laage R, Langosch D. 2004. Synaptobrevin transmembrane domain dimerization-revisited. *Biochemistry* 43:4964–70
132. Ryan TA, Reuter H, Smith SJ. 1997. Optical detection of a quantal presynaptic membrane turnover. *Nature* 388:478–82
133. Ryan TA, Reuter H, Wendland B, Schweizer FE, Tsien WR, Smith SJ. 1993. The kinetics of synaptic vesicle recycling measured at single presynaptic boutons. *Neuron* 11:713–24
134. Sabatini BL, Regehr WG. 1996. Timing of neurotransmission at fast synapses in the mammalian brain. *Nature* 384:170–72
134a. Sakmann B, Neher E, eds. 1995. *Single-Channel Recording*. New York: Plenum
135. Sanyal S, Tolar LA, Pallanck L, Krishnan KS. 2001. Genetic interaction between shibire and comatose mutations in Drosophila suggest a role for snap-receptor complex assembly and disassembly for maintenance of synaptic vesicle cycling. *Neurosci. Lett.* 311:21–24
136. Schiavo G, Benfenati F, Poulain B, Rossetto O, Polverino de Laureto P, et al. 1992. Tetanus and botulinum-B neurotoxins block neurotransmitter release by proteolytic cleavage of synaptobrevin. *Nature* 359:832–35
137. Schiavo G, Matteoli M, Montecucco C. 2000. Neurotoxins affecting neuroexocytosis. *Physiol. Rev.* 80:717–66
138. Schmid SL, McNiven MA, De Camilli P. 1998. Dynamic and its partners: a progress report. *Curr. Opin. Cell Biol.* 10:504–12
139. Schuette CG, Hatsuzawa K, Margittai M, Stein A, Riedel D, et al. 2004. Determinants of liposome fusion mediated by synaptic SNARE proteins. *Proc. Natl. Acad. Sci. USA* 101:2858–63
140. Sollner T, Bennett MK, Whiteheart SW, Scheller RH, Rothman JE. 1993. A protein assembly-disassembly pathway *in vitro* that may correspond to sequential steps of synaptic vesicle docking, activation, and fusion. *Cell* 75:409–18
141. Sollner T, Whiteheart SW, Brunner M, Erdjument-Bromage H, Geromanos S, et al. 1993. SNAP receptors implicated in vesicle targeting and fusion. *Nature* 362:318–24
142. Spruce AE, Breckenridge LJ, Lee AK, Almers W. 1990. Properties of the fusion pore that forms during exocytosis of a mast cell secretory vesicle. *Neuron* 4:643–54
143. Stevens CF, Williams JH. 2000. "Kiss and run" exocytosis at hippocampal synapses. *Proc. Natl. Acad. Sci. USA* 97:12828–33
144. Struck DK, Hoekstra D, Pagano RE. 1981. Use of resonance energy transfer to monitor membrane fusion. *Biochemistry* 20:4093–99

145. Sutton RB, Davletov BA, Berghuis AM, Sudhof TC, Sprang SR. 1995. Structure of the first C2 domain of synaptotagmin I: a novel Ca^{2+}/phospholipid-binding fold. *Cell* 80:929–38
146. Sutton RB, Fasshauer D, Jahn R, Brunger AT. 1998. Crystal structure of a SNARE complex involved in synaptic exocytosis at 2.4 A resolution. *Nature* 395:347–53
147. Takahashi N, Kishimoto T, Nemoto T, Kadowaki T, Kasai H. 2002. Fusion pore dynamics and insulin granule exocytosis in the pancreatic islet. *Science* 297:1349–52
148. Tanford C. 1980. *The Hydrophobic Effect: Formation of Micelles and Biomembranes.* New York: Wiley
149. Taraska JW, Perrais D, Ohara-Imaizumi M, Nagamatsu S, Almers W. 2003. Secretory granules are recaptured largely intact after stimulated exocytosis in cultured endocrine cells. *Proc. Natl. Acad. Sci. USA* 100:2070–75
150. Trimble WS, Cowan DM, Scheller RH. 1988. VAMP-1: a synaptic vesicle-associated integral membrane protein. *Proc. Natl. Acad. Sci. USA* 85:4538–42
151. Tsuboi T, Rutter GA. 2003. Multiple forms of "kiss-and-run" exocytosis revealed by evanescent wave microscopy. *Curr. Biol.* 13:563–67
152. Tucker WC, Chapman ER. 2002. Role of synaptotagmin in Ca^{2+}-triggered exocytosis. *Biochem. J.* 366:1–13
153. Tucker WC, Weber T, Chapman ER. 2004. Reconstitution of Ca^{2+}-regulated membrane fusion by synaptotagmin and SNAREs. *Science* 304:435–38
154. Ubach J, Zhang X, Shao X, Sudhof TC, Rizo J. 1998. Ca^{2+} binding to synaptotagmin: How many Ca^{2+} ions bind to the tip of a C2-domain? *EMBO J.* 17:3921–30
155. Ungermann C, Sato K, Wickner W. 1998. Defining the functions of trans-SNARE pairs. *Nature* 396:543–48
156. Wang CT, Grishanin R, Earles CA, Chang PY, Martin TF, et al. 2001. Synaptotagmin modulation of fusion pore kinetics in regulated exocytosis of dense-core vesicles. *Science* 294:1111–15
157. Wang CT, Lu JC, Bai J, Chang PY, Martin TF, et al. 2003. Different domains of synaptotagmin control the choice between kiss-and-run and full fusion. *Nature* 424:943–47
158. Wang P, Chicka MC, Bhalla A, Richards D, Chapman ER. 2005. Synaptotagmin VII is targeted to secretory organelles in PC12 cells where it functions as a high affinity calcium sensor. *Mol. Cell Biol.* 25:8693–702
159. Wang P, Wang CT, Bai J, Jackson MB, Chapman ER. 2003. Mutations in the effector binding loops in the C2A and C2B domains of synaptotagmin I disrupt exocytosis in a non-additive manner. *J. Biol. Chem.* 278:47030–37
160. Weber T, Parlati F, McNew JA, Johnston RJ, Westermann B, et al. 2000. SNAREpins are functionally resistant to disruption by NSF and alphaSNAP. *J. Cell Biol.* 149:1063–72
161. Weber T, Zemelman BV, McNew JA, Westermann B, Gmachl M, et al. 1998. SNAREpins: minimal machinery for membrane fusion. *Cell* 92:759–72
162. Weninger K, Bowen ME, Chu S, Brunger AT. 2003. Single-molecule studies of SNARE complex assembly reveal parallel and antiparallel configurations. *Proc. Natl. Acad. Sci. USA* 100:14800–5
163. Wightman RM, Jankowski JA, Kennedy RT, Kawagoe KT, Schroeder TJ, et al. 1991. Temporally resolved catecholamine spikes correspond to single vesicle release from individual chromaffin cells. *Proc. Natl. Acad. Sci. USA* 88:10754–58
164. Wu Y, He Y, Bai J, Ji SR, Tucker WC, et al. 2003. Visualization of synaptotagmin I oligomers assembled onto lipid monolayers. *Proc. Natl. Acad. Sci. USA* 100:2082–87

165. Xia Z, Zhou Q, Lin J, Liu Y. 2001. Stable SNARE complex prior to evoked synaptic vesicle fusion revealed by fluorescence resonance energy transfer. *J. Biol. Chem.* 276:1766–71
166. Xu T, Binz T, Niemann H, Neher E. 1998. Multiple kinetic components of exocytosis distinguished by neurotoxin sensitivity. *Nat. Neurosci.* 1:192–200
167. Xu T, Rammner B, Margittai M, Artalejo AR, Neher E, Jahn R. 1999. Inhibition of SNARE complex assembly differentially affects kinetic components of exocytosis. *Cell* 99:713–22
168. Xu Y, Zhang F, Su Z, McNew JA, Shin YK. 2005. Hemifusion in SNARE-mediated membrane fusion. *Nat. Struct. Mol. Biol.* 12:417–22
169. Yang B, Gonzalez L Jr, Prekeris R, Steegmaier M, Advani RJ, Scheller RH. 1999. SNARE interactions are not selective. Implications for membrane fusion specificity. *J. Biol. Chem.* 274:5649–53
170. Yang L, Ding L, Huang HW. 2003. New phases of phospholipids and implications to the membrane fusion problem. *Biochemistry* 42:6631–35
171. Yang L, Huang HW. 2002. Observation of a membrane fusion intermediate structure. *Science* 297:1877–79
172. Yang L, Huang HW. 2003. A rhombohedral phase of lipid containing a membrane fusion intermediate structure. *Biophys. J.* 84:1808–17
173. Zenisek D, Steyer J, Feldman ME, Almers W. 2002. A membrane marker leaves synaptic vesicles in milliseconds after exocytosis in retinal bipolar cells. *Neuron* 35:1085–97
174. Zhang F, Chen Y, Kweon DH, Kim CS, Shin YK. 2002. The four-helix bundle of the neuronal target membrane SNARE complex is neither disordered in the middle nor uncoiled at the C-terminal region. *J. Biol. Chem.* 277:24294–98
175. Zhang F, Chen Y, Su Z, Shin YK. 2004. SNARE assembly and membrane fusion, a kinetic analysis. *J. Biol. Chem.* 279:38668–72
176. Han X, Jackson MB. 2006. Structural transitions in the synaptic SNARE complex during Ca^{2+}-triggered exocytosis. *J. Cell Biol.* In press
177. Wang CT, Bai J, Chang PY, Chapman ER, Jackson MB. 2006. Synaptotagmin · Ca^{2+} triggers two sequential steps in regulated exocytosis in rat PC12 cells: fusion pore opening and fusion pore dilation. *J. Physiol.* 570:295–307

RNA Folding During Transcription

Tao Pan[1] and Tobin Sosnick[1,2]

[1]Department of Biochemistry and Molecular Biology, [2]Institute for Biophysical Dynamics, University of Chicago, Chicago, Illinois 60637;
email: taopan@uchicago.edu, trsosnic@uchicago.edu

Annu. Rev. Biophys. Biomol. Struct.
2006. 35:161–75

First published online as a Review in Advance on January 13, 2006

The *Annual Review of Biophysics and Biomolecular Structure* is online at biophys.annualreviews.org

doi: 10.1146/annurev.biophys.35.040405.102053

1056-8700/06/0609-0161$20.00

Key Words

pausing, riboswitch, ribozyme, RNA polymerase

Abstract

The evolution of RNA sequence needs to satisfy three requirements: folding, structure, and function. Studies on folding during transcription are related directly to folding in the cell. Understanding RNA folding during transcription requires the elucidation of structure formation and structural changes of the RNA, and the consideration of intrinsic properties of the RNA polymerase and other proteins that interact with the RNA. This review summarizes the research progress in this area and outlines the enormous challenges facing this field. Significant advancement requires the development of new experimental methods and theoretical considerations in all aspects of transcription and RNA folding.

Contents

INTRODUCTION

Research on RNA folding can be broadly categorized into three main areas. Studies on folding pathways and structure aim to understand physicochemical principles of how RNA sequence dictates the formation of three-dimensional structures. This area has been investigated intensely and many recent reviews are available (12, 31, 49, 53, 61, 62). Investigations of dynamics and conformational switches aim to illuminate how structural changes are related to RNA function. A particularly powerful advance in this area involves single-molecule studies (17, 18, 36, 67, 69).

Studies on folding during transcription and assembly of ribonucleoprotein complexes aim to reveal how folding in cellular contexts leads to the formation of functional RNA structures and RNA-protein complexes. A well-studied system in the assembly of ribonucleoprotein complexes is the formation of ribosomes, which requires the action of many proteins. The ribosomal proteins remain associated with the ribosomal RNAs (rRNAs), while helicases use ATP (adenosine 5′-triphosphate) hydrolysis to catalyze RNA conformational changes during assembly. Recent reviews on protein-assisted and protein-catalyzed RNA folding are also available (8, 42, 43, 58).

Folding during transcription can be modulated by properties of the RNA polymerase that carry out the synthesis reaction. Three particular properties are relevant to RNA folding during transcription: the speed of elongation, site-specific pausing of the RNA polymerase, and cotranscriptional interaction of the nascent RNA with proteins (**Figure 1**). This review provides a comprehensive description and future perspectives on studies of RNA folding during transcription.

BASIC STEPS IN TRANSCRIPTION

Transcription describes RNA synthesis by the RNA polymerase enzyme (**Figure 2**) using a DNA template. The three basic steps are initiation, elongation, and termination. In the initiation step, RNA polymerase binds to a specific "promoter" sequence upstream of the DNA template coding for the RNA sequence. Recognition of the promoter allows the RNA polymerase to initiate transcription from a

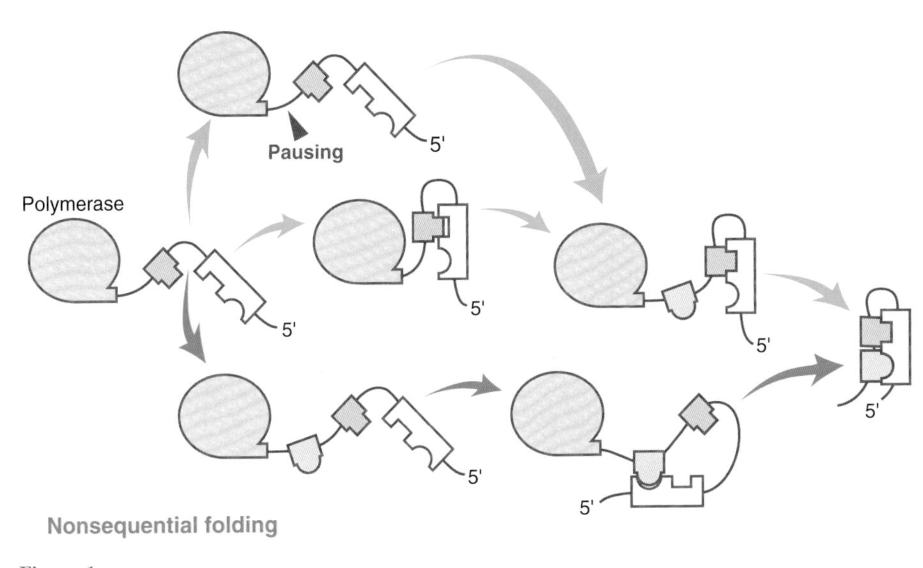

Figure 1

Properties in transcription that influence folding. Upper two pathways represent sequential folding, which is more likely with pausing at the designated site (*triangle*), or when transcription speed is slower. Nonsequential folding is more likely (*bottom pathway*) at fast transcription speed. RNA-protein interactions can occur at any point of the reaction and favor the transit along either pathway.

RNA folding: the problem of understanding the relationship between RNA sequence and its three-dimensional structure

Transcription: biosynthesis of RNA carried out by the RNA polymerase enzyme

rRNA: ribosomal RNA

ATP: adenosine 5′-triphosphate

RNA polymerase: the protein enzyme that carries out the biosynthesis of RNA

Pausing: temporal stops of the RNA polymerase during transcription

designated start-site. After the synthesis of ~10 polynucleotides, RNA polymerase leaves the promoter and enters the elongation stage. In the elongation stage, RNA polymerase synthesizes the RNA in a processive manner at a rapid speed, from 20–80 nt s^{-1} for a bacterial RNA polymerase to up to 200 nt s^{-1} for a phage RNA polymerase, until a termination signal is reached. In the termination step, transcription is halted, RNA polymerase dissociates from the DNA template, and the RNA transcript is released from the polymerase. For the purpose of RNA folding during transcription, the elongation step is the most relevant. It is primarily during this step that the nascent RNA molecule undergoes structure formation and structural changes, processes that can be influenced strongly by the specific RNA polymerase carrying out the transcription reaction.

Transcription speed or elongation rate is often considered a passive parameter to RNA folding, although varying speed can provide crucial time windows to influence folding. Because RNA bases and backbone have a high propensity to interact with each other, a major issue of RNA folding is to avoid the formation of alternative, nonnative structures that form

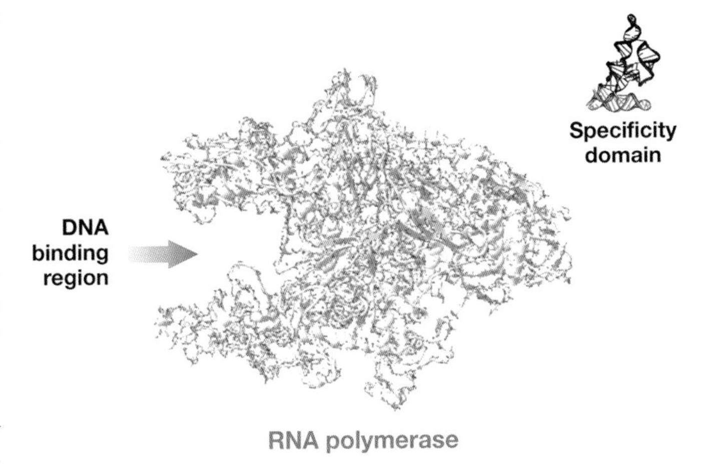

Figure 2

Crystal structure of a bacterial RNA polymerase (33, 66). The RNA transcript exits at the right site, where it can interact with regions on the RNA polymerase. The folded structure of a 154-residue transcript (the specificity domain of a bacterial RNase P RNA, ~48 kDa) is shown on the upper right for comparison in size.

RNA-protein interaction: cellular proteins that bind to RNA or alter RNA conformation using the energy from ATP hydrolysis

from nearby segments. Consider the formation of a native helix containing residues that are 100 nt apart. During transcription, the 5′ region of the helix is synthesized 0.5 to 5 seconds prior to the 3′ portion, depending upon the polymerase. During this period, the 5′ half of the helix may become a part of a nonnative structure involving either previously synthesized regions or the newly synthesized RNA between the two halves. For folding to proceed, this nonnative structure has to be disrupted, a process that can take seconds, minutes, or longer.

Transcriptional pausing refers to momentary stops at specific locations by the elongating RNA polymerase. The duration of the pause can vary from tens of milliseconds to seconds. Pausing is an intrinsic property of bacterial RNA polymerases. It is derived from several complex factors including interactions of the RNA polymerase with the nascent RNA, the DNA template, the incoming nucleotide triphosphate at the site of polymerization, and so on (24, 32). Of particular interest for RNA folding are those pause sites caused largely by the interactions between the nascent RNA and the RNA polymerase. Such interaction at appropriate sites can lead to significant differences in RNA folding.

Cotranscriptional interactions between the nascent RNA and RNA binding proteins or RNA helicases can influence RNA folding in two ways. In a passive mechanism, protein binding to a specific structure or sequence in the nascent RNA prevents the formation of nonnative structures involving this binding site. In an active mechanism, protein binding or helicase action fueled by ATP hydrolysis facilitates the interaction between two regions in the RNA to promote the formation of the native structure.

In a cellular environment, all three processes—speed, pausing, and RNA-protein interactions—play some role in dictating how RNA folds during transcription. The integration of these processes is possible when protein binding occurs during transcription. Because protein binding to a nascent RNA is a bimolecular, second-order reaction, the free protein concentration needs to be high enough so that binding can occur during the transcription process. Transcriptional pausing may also play an important role by increasing the available time window for protein binding.

Alternatively, transcriptional pausing at some sites can be caused by interactions between the nascent RNA and the elongating RNA polymerase itself, or polymerase-associated factors such as the NusA protein in bacteria (25). These interactions can be present in an elongating RNA polymerase complex and therefore no longer depend on the concentration of the free polymerase or factors. These intracomplex interactions can directly influence RNA folding during transcription. An RNA folding mechanism that depends on such interactions can be thought of as polymerase-assisted folding, with the RNA polymerase acting as a folding chaperone.

RNA FOLDING DURING TRANSCRIPTION AND TRANSCRIPTIONAL ANTITERMINATION

A classic case of RNA folding during transcription is the process of antitermination or attenuation, which controls the transcription of numerous amino acid biosynthetic operons in bacteria (for recent reviews see References 15, 21, 28). These operons contain 5′-untranslated regions of several hundred nucleotides prior to the coding regions for the amino acid synthetic enzymes. The same 5′-untranslated region can form two different structures; one has a single hairpin loop and the other has two hairpin loops. Formation of the single hairpin loop structure results in antitermination of transcription, allowing the RNA polymerase to continue transcription into the coding regions. The formation of the two hairpin loops structure results in termination of transcription, so that the coding regions are not transcribed. Whether the antitermination or the termination structure forms depends on several factors including transcription speed, specific pause sites, the rate of translation of a leader peptide whose coding region encompasses a portion of the upstream hairpin, and the binding of uncharged tRNAs.

FOLDING OF THE *TETRAHYMENA* GROUP I INTRON

The *Tetrahymena* group I intron was the first autocatalytic RNA discovered and has been studied extensively as a model system for RNA folding (6, 11, 53, 61). In the ciliate *Tetrahymena*, this intron is located in the gene for the large ribosomal RNA subunit. The autocatalyzed splicing reaction takes just a few seconds and occurs within the same time frame of synthesis by the RNA polymerase (4). This result suggests that the catalytically active structure of this ribozyme is folded as soon as the RNA is transcribed. This rapid folding in vivo is much faster compared with its folding in vitro upon Mg^{2+} addition to the fully synthesized and denatured RNA, where folding is limited by the disruption of prematurely formed nonnative structures (53, 61). Hence, either transcription by the host RNA polymerase or solvent conditions in vivo eliminated the formation of nonnative structures that dominate in vitro folding kinetics. A third possibility is that an unknown protein in *Tetrahymena* is involved in facilitating the folding of the intron.

To distinguish these possibilities, Woodson and coworkers (23, 34, 65) inserted the *Tetrahymena* intron into several locations in the 23S rRNA of *Escherichia coli* and monitored its self-splicing in *E. coli*. When the intron was inserted into location in the *E. coli* 23S rRNA corresponding to that in the *Tetrahymena* rRNA, splicing was fast and occurred as soon as its synthesis was completed. When the intron was inserted into other locations in the *E. coli* 23S rRNA, however, splicing became markedly slower. These results indicate that folding of this intron in *E. coli* as well as in *Tetrahymena* may be tightly coupled to the assembly of the large ribosomal subunit. It is clear that a *Tetrahymena*-specific protein is not needed for rapid folding of the intron. However, the linkage of folding to ribosome assembly may be influenced by conserved ribosomal proteins whose interactions with the ribosome or even transient interactions with the intron may be necessary for rapid folding in vivo.

Ribozyme: RNA molecules that catalyze chemical reactions

Folding during transcription in vitro of several circularly permuted *Tetrahymena* ribozymes was analyzed in an effort to understand how changes in the synthesis order influenced folding (20). Circular permutation describes a molecular isomerization in which the natural 5′ and 3′ termini are covalently linked followed by breaking the phosphodiester backbone elsewhere to create new termini. For example, the synthesis order changes from A—Z to G—Z-A—F, where A and F are the 5′ ends of the naturally and newly created circularly permuted RNA, respectively. For circularly permuted group I ribozymes, the folding rate changed during transcription by either the phage T7 or the *E. coli* RNA polymerase. These results suggest that the synthesis order can play a role in determining the rate of folding, or perhaps even the folding pathways of nascent RNA transcripts.

FOLDING OF A BACTERIAL RNase P RNA DURING TRANSCRIPTION

Ribonuclease P is a ribonucleoprotein enzyme that catalyzes the 5′ maturation of tRNAs through a site-specific endonucleolytic cleavage reaction (1, 14). The RNA component is the catalytic subunit of this enzyme. The bacterial RNase P is made of a large RNA subunit of 330 to 450 nt (>100 kDa) and a small protein of ~120 amino acids (~14 kDa). Under high ionic conditions, bacterial RNase P RNA catalyzes the cleavage reaction in vitro in the absence of the protein at high efficiency. Under physiological conditions, however, the protein subunit is required for optimal catalytic activity.

Unlike self-cleaving group I introns where reaction occurs in *cis*, RNase P catalyzes its reaction in *trans*. Hence, studies of its folding during transcription are hampered by the lack of self-splicing as an indicator of folding to the native RNA, as utilized in the intron studies. It is not yet clear how bacterial RNase P is assembled in vivo: For example, is the RNA or the protein subunit synthesized first? In vitro, the presence of the protein subunit has no effect on the folding of the *Bacillus subtilis* RNase P RNA during transcription by either *E. coli* or *B. subtilis* RNA polymerase (T. Pan & T. Sosnick, unpublished results). For this reason, studies on folding during transcription have so far been carried out in vitro without the protein subunit (37, 60).

A main focus of studies on folding during transcription of RNase P RNA was to understand how properties of the RNA polymerase influence folding (**Figure 3**). Folding of a circularly permuted RNase P RNA had the same rate and similar pathways during transcription by either the single-subunit, phage T7 or the multi-subunit, core *E. coli* RNA polymerase. At 37°C, the catalytic domain folded three to four times faster than the specificity domain. This folding rate and presumably the folding pathway were maintained even when the transcription speed varied from 3 to 200 nt s^{-1}.

Folding behavior of this circularly permuted RNase P RNA changed dramatically when folding was measured during transcription by the *E. coli* RNA polymerase in the presence of an elongation factor, the *E. coli* NusA protein (37). In the absence of NusA, folding of the specificity domain was several times slower than the folding of the catalytic domain. In the presence of NusA, folding of the specificity domain was at least three times faster than the folding of the catalytic domain (**Figure 3**). NusA protein exerted two effects on *E. coli* RNA polymerase transcription: It altered the duration of pausing at specific sites and it decreased the transcription speed. The altered folding during transcription by the NusA protein was not due to the change in the transcription speed, as slowing transcription speed via the reduction of the nucleotide triphosphate concentration did not generate a similar outcome.

The presence of the NusA protein significantly increased the duration of pausing at a specific site located in the catalytic domain,

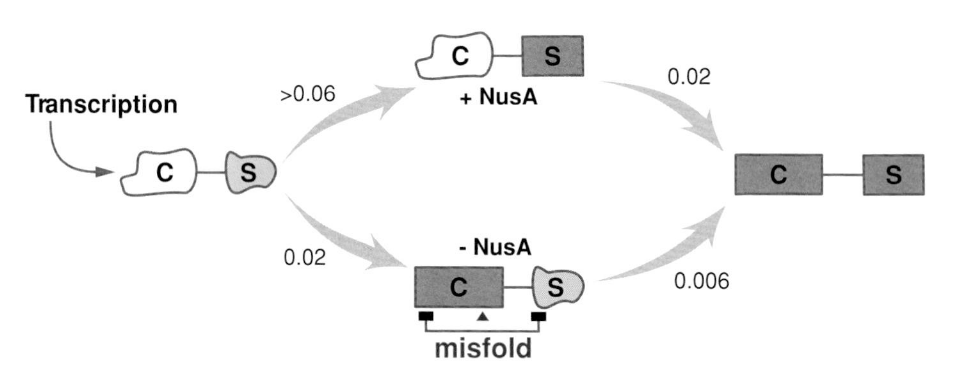

Figure 3

Effect of NusA protein induced pausing (*filled triangle*) on the folding of a circularly permuted RNase P RNA during transcription. Folding rates in $seconds^{-1}$ are shown above the arrows. Shaded boxes indicate folded catalytic and specificity domains, and pink and yellow shapes represent not yet folded domains. In the absence of the NusA protein (*lower pathway*), this RNA folds through an intermediate containing a folded catalytic domain and an interdomain misfold indicated by two connected thick bars. In the presence of the NusA protein (*upper pathway*), this RNA folds through a different intermediate containing a folded specificity domain.

U55, or the 225th nucleotide of the circular permutated transcript. The change in pausing at this site altered folding during transcription as demonstrated in two ways. Folding did not respond to the presence of the NusA protein when pausing was ablated either by mutations in RNase P RNA or the use of a pausing deficient *E. coli* RNA polymerase.

The order of synthesis of the circularly permuted RNase P RNA is the catalytic domain (255 residues) followed by the specificity domain (154 residues). Pausing at the 225th position in the catalytic domain therefore occurs before the specificity domain starts to be synthesized. How then does pausing change the folding of an RNA structure that is not present even when pausing takes place? The mechanism of how pausing at this site altered folding was revealed in a subsequent study (60). First, a strong context dependence of pause site effects was observed. When the synthesis order was changed from catalytic domain first, specificity domain second (C-to-S) to specificity domain first (S-to-C) or starting from one third of the catalytic domain (5′C-to-S-to-3′C, the natural P RNA), pausing at U55 no longer had an effect on folding. Second, certain mutations at a specific loop region located immediately 3′ to the catalytic domain greatly increased or decreased folding rates during transcription. This result, together with further probing of the folding with oligonucleotides, suggests that this loop region was involved in some kind of misfolded structure that slows the folding of the specificity domain. Third, a stem-loop region in the catalytic domain was identified as interacting with the aforementioned loop in the specificity domain. Apparently, the NusA-induced pausing at the site in the catalytic domain prevented the formation of this nonnative interaction, allowing more rapid folding of the specificity domain.

Folding during transcription does not always accelerate folding. Transcription, just like Mg^{2+}-induced folding of the fully synthesized RNA, can produce alternative nonnative structures that need to be resolved for the folding to proceed. When investigated for folding initiated by Mg^{2+} addition, the catalytic domain alone folds rapidly at 6 s^{-1} (13). Folding during transcription of the same catalytic domain by *E. coli* RNA polymerase occurred at ~0.02 s^{-1} in the absence and presence of the NusA protein (T. Pan & T. Sosnick, unpublished results). For the catalytic domain embedded in the circularly permuted RNase P RNA, folding initiated by Mg^{2+} addition can occur at >0.2 s^{-1} (38), whereas its folding during transcription occurs at least 10 times more slowly.

FOLDING OF SELF-CLEAVING SMALL RIBOZYMES DURING TRANSCRIPTION

Several self-cleaving small ribozymes derived from viroid RNAs have been studied extensively as model systems for RNA catalysis. Because of their relatively slow cleavage rates (less than 0.05 s^{-1}), it has been difficult to unambiguously separate folding from catalysis. By using single-molecule techniques, it was possible to separate the folding step (docking between two internal loops) from the catalysis step of the hairpin ribozyme (3, 50, 68). Such sophisticated measurements have yet to be applied to its folding during transcription.

Folding during transcription in vitro for the hairpin and delta ribozymes focused on the effect of upstream and downstream regions that could interfere with the folding of the catalytically active structure. In the case of the hairpin ribozyme (26), complementary insertions of an interfering stem-loop upstream of the ribozyme inhibited folding more severely than did an insertion downstream of the ribozyme. In the case of the delta ribozyme (10), folding during transcription was inhibited strongly by the presence of a downstream portion of the viroid RNA. Folding of the delta ribozyme was more efficient upon decreasing the transcription speed. These results are consistent with a kinetically driven folding scheme in which folding is determined by the

Riboswitch: RNA sequences that form different structures upon binding of small-molecule metabolites

FMN: flavin mononucleotide

sequential formation of the RNA structures that fold in the order they are transcribed.

However, transcription in yeast cells in the presence of either the upstream or the downstream insertions appreciably slowed the folding of the hairpin ribozyme (26). This unexpected result indicates that stem-loops can form nonsequentially in vivo, either during or immediately after transcription. Potentially, proteins bind nascent single-stranded RNA during synthesis, enabling them to adopt lower energy structures after transcription. Apparently, the cellular environment can enhance the importance of thermodynamic stability relative to sequential folding mechanisms observed in vitro.

A theoretical study on folding of the delta ribozyme during transcription postulated the sequential formation of the secondary structure on a realistic timescale (22). The modeling relied on computing the free energies of regular helices and pseudoknots in the framework of a kinetic Monte Carlo code. Base-pairing is assumed to be in fast equilibrium relative to synthesis rates, which enables a thermodynamic approximation at each synthesis step. It was predicted that several non-native helices would form, some of which trap the ribozyme to result in slow folding, and some others may guide sequential folding to the native structure. Although these predictions have yet to be tested experimentally to our knowledge, the ability to carry out these kinds of theoretical simulations represents a major advance in understanding folding during transcription.

FOLDING OF RIBOSWITCHES DURING TRANSCRIPTION

Riboswitches are RNA sequences capable of forming two distinct structures in the presence and absence of small-molecule metabolites or coenzymes (27, 59). These RNAs are generally of bacterial origin and are located upstream of coding regions of enzymes involved in the metabolism of the molecules they bind to. Binding of the metabolite results in the formation of elaborate tertiary structures, which in turn permits the arrangement of a downstream structure required for transcriptional termination or translational repression. Folding during transcription of riboswitch structures should be particularly relevant because its formation involves bimolecular interactions coupled to transcription.

Folding during transcription of a flavin mononucleotide (FMN)-dependent riboswitch in vitro was analyzed in an elegant study by Crothers, Breakers and coworkers (57). The importance of the transcription speed, two major pause sites, and the kinetics of FMN binding were determined. The results are summarized below:

- Folding of this riboswitch structure is kinetically driven with a strong dependence on the transcription speed and ligand binding kinetics. The ligand binding region does not attain thermodynamic equilibrium with FMN before the RNA polymerase decides to continue or terminate transcription.
- Two major pause sites, each with a duration of up to 60 s, are present between the riboswitch structure and the termination hairpin. The duration of the pauses is longer in the presence of the elongation factor NusA protein. These pause sites provide additional "breathing time" for the formation of the riboswitch structure.
- The association rate constant (k_{on}) of FMN with the riboswitch structure is $\sim 10^5\ M^{-1}s^{-1}$. This magnitude of k_{on} is important for the effectiveness of the riboswitch, which depends on intracellular FMN concentrations. The FMN dissociation constant (k_{off}) is $\sim 10^{-3}\ s^{-1}$, hence riboswitch formation during transcription is practically irreversible once FMN binds.
- This riboswitch is from the bacterium *B. subtilis*, a distant relative of *E. coli*. Nevertheless, transcription by both *B. subtilis* and *E. coli* RNA polymerase is effective in promoting its formation.

- Depending on the transcription conditions in vitro, 10% to 40% of the transcripts terminate in the absence of FMN. An additional 10% to 50% of the transcripts terminate in the presence of saturating amounts of FMN. This result suggests that in vivo, additional factors may be needed to improve the effectiveness of this "on-off" switch, or alternatively, this termination fraction is sufficient for regulating the expression of these metabolic enzymes.

FOLDING DURING TRANSCRIPTION OF STRUCTURES EMBEDDED IN EUKARYOTIC mRNAs

Eukaryotic transcription is a complex process carried out by three RNA polymerases. RNA polymerase II is composed of more than 10 subunits and transcribes mRNA genes. Transcription of mRNA is coupled with several RNA processing events, 5′ capping, intron splicing, and 3′-end formation (39, 40). The coupling of transcription and RNA processing is derived from the association of protein factors that carry out these processing reactions with the largest subunit of RNA polymerase II. Here, both the transcription speed of RNA polymerase II (10 to 20 nt s^{-1} on average) and strategically placed pause sites can influence the processing efficiency and the splicing pattern of the mRNA products. Coupled transcription and processing is currently depicted as being driven by protein factors and spliceosomes but is yet to be linked to the formation of secondary or tertiary RNA structures.

When an RNA structure is embedded within an mRNA, the location of this structure influences folding in vivo (19). When the *Tetrahymena* group I intron was inserted into five different locations in a luciferase mRNA in mammalian cells, the efficiency of self-splicing varied by 10-fold. Insertion of this exogenous intron in mammalian cells also significantly decreased the expression of the luciferase proteins, suggesting a link between mRNA maturation and folding of this group I intron during transcription.

A self-cleaving ribozyme was discovered in 2004 in the mRNA precursor of human β-globin (51). This ribozyme is located downstream of the polyadenylation site, and its cotranscriptional cleavage is required for the transcriptional termination of the RNA polymerase II. Autolytic cleavage of the ribozyme produces a free 5′ hydroxyl end in the 3′ cleavage product of the RNA transcript. The product becomes the substrate for a processive exonuclease (56). Here, the function of this ribozyme requires its rapid folding during transcription.

FUTURE CHALLENGES AND PERSPECTIVES

Challenge One: Selection of RNA Targets

Most folding studies today are carried out with ribozymes and other RNAs that function after transcription is finished. These "hard" structures, once formed, can remain functional in the cell for a long time. Because the function of such RNAs is not coupled to transcription, the efficiency of their folding during transcription may simply be a matter of abundance. More rapid folding may allow these RNAs to accumulate faster in the cell, resulting in increased cell fitness for growth and adaptation.

Many RNA structures, however, function only during transcription. A typical example is a riboswitch whose function is to halt or resume transcription by the RNA polymerase. These "soft" structures are relevant only during transcription. Hence, folding during transcription is an integral function of these RNA structures.

The recent discovery of the existence of $\sim 10^6$ human RNA transcripts longer than 150 nt (7) illustrates the need for learning about RNA folding. Given the high propensity of RNA to form structures through base-pairing and stacking, a fraction of these

FRET: fluorescence resonance energy transfer

FCS: fluorescence correlation spectroscopy

transcripts will undoubtedly contain functional structures. Cotranscriptional formation for some of these structures may be necessary for their function in such processes as chromatin remodeling or DNA methylation.

Challenge Two: Better Experimental Methods

Most current methods employed to study folding during transcription are low-tech, using gels and radioactive labels. High-tech single-molecule methods are superb tools for monitoring the movements and properties of the RNA polymerase and identifying the order of events in the reaction (9, 52, 64). But such methods generally do not provide atomic-level details of the RNA structures that form during transcription.

A significant challenge is to study RNA folding during transcription by observing RNA folding directly in real time. Because of their high signal level, fluorescence probes are most promising for monitoring folding. In addition to the application of fluorescence resonance energy transfer (FRET) in studying folding during transcription, fluorescence correlation spectroscopy (FCS) may be applied to observe oligonucleotide hybridization to an exposed RNA region in real time (16, 44, 55). FCS measures the fluctuation of the fluorescence signal in a confocal volume of less than a femtoliter. The fluctuation is derived from diffusion of a few fluorescent molecules through the confocal volume. A 1-s acquisition is sufficient to obtain the concentration and the diffusion coefficients of the fluorescent molecules. When a fluorescent oligonucleotide binds to its much larger RNA target, its diffusion becomes significantly slower.

Other desirable methods to be developed would increase our ability to map the details of folding pathways and to model the structures of key folding intermediates during transcription. These methods would detect structural changes at the nucleotide resolution on the millisecond-to-second timescale of transcription. The measurement of the protection from hydroxyl radical cleavage provides residue-level information at the appropriate timescale (5). Unfortunately, the assay is rather stringent, and many folding intermediates often exhibit little protection. Chemical probes that selectively react in less than 1 s with each nucleotide type in a conformational-dependent manner would be highly desirable.

Challenge Three: RNA Polymerase and Elongation Factors

In bacteria, RNA sequences have been coevolving with the RNA polymerase and its elongation factors for billions of years. Conceivably, properties of each "cognate" RNA polymerase, such as varying transcription speeds or pausing patterns, may play a significant role in the folding process for RNA derived from the same organism. Indeed, RNA polymerases from the bacteria *E. coli* and *B. subtilis* exhibit different pausing patterns, presumably through recognition of distinct sequences and structures in the nascent RNA transcript (2). This result suggests that folding during transcription of an *E. coli* RNA (e.g., RNase P RNA) could be different when transcription is performed with the cognate *E. coli* RNA polymerase versus noncognate *B. subtilis* RNA polymerase. Similarly, folding of the *B. subtilis* RNase P RNA may vary depending on whether the cognate (*B. subtilis*) or noncognate (*E. coli*) RNA polymerase carries out transcription.

Challenge Four: RNA Binding Proteins and Helicases

There are many RNA binding proteins in the cell; some bind specifically to an RNA structure, some bind selectively to homologous RNA sequences, and some others have little selectivity. Several RNA binding proteins influence RNA folding in vitro, through either prevention of alternative structures or promotion of strand exchange (42). Some of these proteins have been termed RNA chaperones to indicate their role in RNA folding. RNA helicases actively modulate RNA structure formation using the energy from ATP hydrolysis. Helicases participate in mRNA splicing, rRNA maturation, replication of viral RNA genomes, and so on (45, 54).

The action of a particular RNA binding protein or RNA helicase can be specific or selective to only a subset of cellular RNA. This behavior is in contrast to the action of RNA polymerase, which has to accommodate the structure/function requirements for all RNA transcripts. A major challenge of understanding the role of these proteins is to decipher the specificity and selectivity of their authentic cellular RNA substrates.

Challenge Five: Connecting Theory and Experiment

Most theoretical work in RNA folding deals with thermodynamics of fully synthesized RNA or folding kinetics of small RNA upon the addition of Mg^{2+} (30, 31, 35, 41, 46–48). The negatively charged ribosephosphate backbone substantially elevates the importance of electrostatic interactions. The large electrostatic component represents a severe challenge to theoretical treatments of the thermodynamics or the kinetics of fully synthesized RNAs (12, 62).

Kinetics of RNA structure formation is the dominating factor in folding during transcription in vitro. The polarity of RNA synthesis allows incremental consideration of structural formation of nascent RNA chains, rather than dealing with the fully synthesized RNA at all times. These issues reduce the complexity of theoretical analysis of folding during transcription. For instance, a theoretical evaluation suggests that an evolutionary driving force of RNA sequence appears to minimize the chance of forming alternate secondary structures (29). The aforementioned theoretical predication of secondary structure formation in folding during transcription of the delta ribozyme (22) demonstrates the feasibility of modeling folding pathways during transcription.

These theoretical predictions, however, still await experimental verification that is not yet feasible using the existing methods. Most experimental assays on folding during transcription take advantage of the properties of the fully folded product. Although convenient for identifying the acquisition of the native state, the events during transcription are not determined. In principle, structural details may be assessed by oligonucleotide hybridization or chemical and enzymatic footprinting upon synchronization of transcriptional initiation. Unfortunately, the current reagents typically require reaction times on a timescale longer than that of the transcription process.

As for theory, the properties of the RNA polymerase and other proteins that interact with RNA should be accounted for in modeling cotranscriptional folding. Because these interactions are likely idiosyncratic for each RNA sequence/structure, incorporating such protein-induced constraints in a unifying framework will be a huge challenge.

An additional challenge will be the inclusion of tertiary interactions, particularly those involving metal ions. Such interactions can change relative proximity of alternative base pair combinations, thereby influencing both the kinetics and thermodynamics of their formation. In addition, Mg^{2+}-induced tertiary interactions can outweigh nonnative secondary structure formation present at lower ionic conditions (63).

CONCLUDING REMARKS

The evolution of a biological RNA sequence correlates with its function, structure, and folding. Folding during transcription relates directly to RNA folding in the cell. However, studies in this area are still in an early stage. Further developments of experimental methods and theoretical considerations in all aspects of transcription and RNA folding are necessary to significantly advance this field.

SUMMARY POINTS

1. A biological RNA sequence evolves according to its function, structure, and folding. RNA has great propensity to fold while it is being transcribed. This review summarizes studies on folding during transcription of enzymatic and regulatory RNAs. Both the speed of transcription and the pausing pattern can influence folding pathways. Future challenges in this area include the development of new methods and theoretical tools to significantly advance our understanding.

ACKNOWLEDGMENTS

We thank Nathan Baird and Terrence Wong for comments and stimulating discussions. This work is funded by a grant from the NIH (GM57880).

LITERATURE CITED

1. Altman S, Kirsebom L. 1999. Ribonuclease P. In *The RNA World, Second Edition*, ed. RF Gesteland, TR Cech, JF Atkins, pp. 351–80. Cold Spring Harbor, NY: Cold Spring Harbor Lab. Press
2. Artsimovitch I, Svetlov V, Anthony L, Burgess RR, Landick R. 2000. RNA polymerases from *Bacillus subtilis* and *Escherichia coli* differ in recognition of regulatory signals in vitro. *J. Bacteriol.* 182:6027–35
3. Bokinsky G, Rueda D, Misra VK, Rhodes MM, Gordus A, et al. 2003. Single-molecule transition-state analysis of RNA folding. *Proc. Natl. Acad. Sci. USA* 100:9302–7
4. Brehm SL, Cech TR. 1983. Fate of an intervening sequence ribonucleic acid: excision and cyclization of the *Tetrahymena* ribosomal ribonucleic acid intervening sequence in vivo. *Biochemistry* 22:2390–97
5. Brenowitz M, Chance MR, Dhavan G, Takamoto K. 2002. Probing the structural dynamics of nucleic acids by quantitative time-resolved and equilibrium hydroxyl radical "footprinting". *Curr. Opin. Struct. Biol.* 12:648–53
6. Cech TR, Bass BL. 1986. Biological catalysis by RNA. *Annu. Rev. Biochem.* 55:599–629
7. **Cheng J, Kapranov P, Drenkow J, Dike S, Brubaker S, et al. 2005. Transcriptional maps of 10 human chromosomes at 5-nucleotide resolution. *Science* 308:1149–54**
8. Culver GM. 2003. Assembly of the 30S ribosomal subunit. *Biopolymers* 68:234–49
9. Davenport RJ, Wuite GJ, Landick R, Bustamante C. 2000. Single-molecule study of transcriptional pausing and arrest by *E. coli* RNA polymerase. *Science* 287:2497–500
10. Diegelman-Parente A, Bevilacqua PC. 2002. A mechanistic framework for co-transcriptional folding of the HDV genomic ribozyme in the presence of downstream sequence. *J. Mol. Biol.* 324:1–16
11. Doudna JA, Cech TR. 2002. The chemical repertoire of natural ribozymes. *Nature* 418:222–28
12. Draper DE, Grilley D, Soto AM. 2005. Ions and RNA folding. *Annu. Rev. Biophys. Biomol. Struct.* 34:221–43
13. Fang X, Pan T, Sosnick TR. 1999. Mg2+-dependent folding of a large ribozyme without kinetic traps. *Nat. Struct. Biol.* 6:1091–95
14. Frank DN, Pace NR. 1998. Ribonuclease P: unity and diversity in a tRNA processing ribozyme. *Annu. Rev. Biochem.* 67:153–80

7. Describes the discovery of several hundred thousand human RNA transcripts using high-resolution DNA microarrays that may function without being translated into proteins, i.e., noncoding RNAs.

15. Gollnick P, Babitzke P, Antson A, Yanofsky C. 2005. Complexity in regulation of tryptophan biosynthesis in *Bacillus subtilis*. *Annu. Rev. Genet.* 39:47–68
16. Gosch M, Rigler R. 2005. Fluorescence correlation spectroscopy of molecular motions and kinetics. *Adv. Drug Deliv. Rev.* 57:169–90
17. Ha T. 2004. Structural dynamics and processing of nucleic acids revealed by single-molecule spectroscopy. *Biochemistry* 43:4055–63
18. Ha T, Zhuang X, Kim HD, Orr JW, Williamson JR, Chu S. 1999. Ligand-induced conformational changes observed in single RNA molecules. *Proc. Natl. Acad. Sci. USA* 96:9077–82
19. Hagen M, Cech TR. 1999. Self-splicing of the *Tetrahymena* intron from mRNA in mammalian cells. *EMBO J.* 18:6491–500
20. Heilman-Miller SL, Woodson SA. 2003. Effect of transcription on folding of the *Tetrahymena* ribozyme. *RNA* 9:722–33
21. Henkin TM, Yanofsky C. 2002. Regulation by transcription attenuation in bacteria: how RNA provides instructions for transcription termination/antitermination decisions. *Bioessays* 24:700–7
22. **Isambert H, Siggia ED. 2000. Modeling RNA folding paths with pseudoknots: application to hepatitis delta virus ribozyme. *Proc. Natl. Acad. Sci. USA* 97:6515–20**
23. **Koduvayur SP, Woodson SA. 2004. Intracellular folding of the *Tetrahymena* group I intron depends on exon sequence and promoter choice. *RNA* 10:1526–32**
24. Landick R. 1997. RNA polymerase slides home: pause and termination site recognition. *Cell* 88:741–44
25. Liu K, Zhang Y, Severinov K, Das A, Hanna MM. 1996. Role of *Escherichia coli* RNA polymerase alpha subunit in modulation of pausing, termination and anti-termination by the transcription elongation factor NusA. *EMBO J.* 15:150–61
26. **Mahen EM, Harger JW, Calderon EM, Fedor MJ. 2005. Kinetics and thermodynamics make different contributions to RNA folding in vitro and in yeast. *Mol. Cell* 19:27–37**
27. Mandal M, Breaker RR. 2004. Gene regulation by riboswitches. *Nat. Rev. Mol. Cell Biol.* 5:451–63
28. Merino E, Yanofsky C. 2005. Transcription attenuation: a highly conserved regulatory strategy used by bacteria. *Trends Genet.* 21:260–64
29. Meyer IM, Miklos I. 2004. Co-transcriptional folding is encoded within RNA genes. *BMC Mol. Biol.* 5:10
30. Misra VK, Draper DE. 2002. The linkage between magnesium binding and RNA folding. *J. Mol. Biol.* 317:507–21
31. Misra VK, Shiman R, Draper DE. 2003. A thermodynamic framework for the magnesium-dependent folding of RNA. *Biopolymers* 69:118–36
32. Mooney RA, Artsimovitch I, Landick R. 1998. Information processing by RNA polymerase: recognition of regulatory signals during RNA chain elongation. *J. Bacteriol.* 180:3265–75
33. Murakami KS, Masuda S, Campbell EA, Muzzin O, Darst SA. 2002. Structural basis of transcription initiation: an RNA polymerase holoenzyme-DNA complex. *Science* 296:1285–90
34. Nikolcheva T, Woodson SA. 1999. Facilitation of group I splicing in vivo: misfolding of the *Tetrahymena* IVS and the role of ribosomal RNA exons. *J. Mol. Biol.* 292:557–67
35. Nivon LG, Shakhnovich EI. 2004. All-atom Monte Carlo simulation of GCAA RNA folding. *J. Mol. Biol.* 344:29–45
36. Onoa B, Dumont S, Liphardt J, Smith SB, Tinoco I Jr, Bustamante C. 2003. Identifying kinetic barriers to mechanical unfolding of the *T. thermophila* ribozyme. *Science* 299:1892–95

22. Describes theoretical modeling of potential folding intermediates of the delta ribozyme during transcription.

23. Describes intracellular folding of the *Tetrahymena* group I intron in *E. coli*. The identity of the rRNA exons contributes significantly to the folding rate and the partitioning of the pre-RNA into multiple pathways.

26. Describes the comparative study of the hairpin ribozyme folding during transcription in vitro and its folding in yeast cells. In vitro, folding is sequential and follows the order of RNA synthesis. In vivo, folding is nonsequential and the stability of competing RNA structures plays a bigger role.

37. Describes the folding analysis of a bacterial RNase P RNA and shows that pausing by the RNA polymerase at a specific site alters the folding pathway.

51. Describes the discovery of a catalytic RNA embedded in the mRNA of the human β-globin. Autocatalytic cleavage is required for the transcriptional termination of the RNA polymerase.

57. Describes the folding analysis of a flavin mononucleoside-dependent riboswitch during transcription. Structural formation of this riboswitch is kinetically driven, which may be common for the function of other riboswitches.

37. Pan T, Artsimovitch I, Fang X, Landick R, Sosnick TR. 1999. Folding of a large ribozyme during transcription and the effect of the elongation factor NusA. *Proc. Natl. Acad. Sci. USA* 96:9545–50

38. Pan T, Fang X, Sosnick TR. 1999. Pathway modulation, circular permutation and rapid RNA folding under kinetic control. *J. Mol. Biol.* 286:721–31

39. Proudfoot N. 2000. Connecting transcription to messenger RNA processing. *Trends Biochem. Sci.* 25:290–93

40. Proudfoot NJ, Furger A, Dye MJ. 2002. Integrating mRNA processing with transcription. *Cell* 108:501–12

41. Russell R, Millett IS, Tate MW, Kwok LW, Nakatani B, et al. 2002. Rapid compaction during RNA folding. *Proc. Natl. Acad. Sci. USA* 99:4266–71

42. Schroeder R, Barta A, Semrad K. 2004. Strategies for RNA folding and assembly. *Nat. Rev. Mol. Cell Biol.* 5:908–19

43. Schroeder R, Grossberger R, Pichler A, Waldsich C. 2002. RNA folding in vivo. *Curr. Opin. Struct. Biol.* 12:296–300

44. Sei-Iida Y, Koshimoto H, Kondo S, Tsuji A. 2000. Real-time monitoring of in vitro transcriptional RNA synthesis using fluorescence resonance energy transfer. *Nucleic Acids Res.* 28:E59

45. Silverman E, Edwalds-Gilbert G, Lin RJ. 2003. DExD/H-box proteins and their partners: helping RNA helicases unwind. *Gene* 312:1–16

46. Sorin EJ, Engelhardt MA, Herschlag D, Pande VS. 2002. RNA simulations: probing hairpin unfolding and the dynamics of a GNRA tetraloop. *J. Mol. Biol.* 317:493–506

47. Sorin EJ, Nakatani BJ, Rhee YM, Jayachandran G, Vishal V, Pande VS. 2004. Does native state topology determine the RNA folding mechanism? *J. Mol. Biol.* 337:789–97

48. Sorin EJ, Rhee YM, Pande VS. 2005. Does water play a structural role in the folding of small nucleic acids? *Biophys. J.* 88:2516–24

49. Sosnick TR, Pan T. 2003. RNA folding: models and perspectives. *Curr. Opin. Struct. Biol.* 13:309–16

50. Tan E, Wilson TJ, Nahas MK, Clegg RM, Lilley DM, Ha T. 2003. A four-way junction accelerates hairpin ribozyme folding via a discrete intermediate. *Proc. Natl. Acad. Sci. USA* 100:9308–13

51. Teixeira A, Tahiri-Alaoui A, West S, Thomas B, Ramadass A, et al. 2004. Autocatalytic RNA cleavage in the human beta-globin pre-mRNA promotes transcription termination. *Nature* 432:526–30

52. Tolic-Norrelykke SF, Engh AM, Landick R, Gelles J. 2004. Diversity in the rates of transcript elongation by single RNA polymerase molecules. *J. Biol. Chem.* 279:3292–99

53. Treiber DK, Williamson JR. 2001. Beyond kinetic traps in RNA folding. *Curr. Opin. Struct. Biol.* 11:309–14

54. von Hippel PH, Delagoutte E. 2003. Macromolecular complexes that unwind nucleic acids. *Bioessays* 25:1168–77

55. Vukojevic V, Pramanik A, Yakovleva T, Rigler R, Terenius L, Bakalkin G. 2005. Study of molecular events in cells by fluorescence correlation spectroscopy. *Cell Mol. Life Sci.* 62:535–50

56. West S, Gromak N, Proudfoot NJ. 2004. Human 5′ → 3′ exonuclease Xrn2 promotes transcription termination at co-transcriptional cleavage sites. *Nature* 432:522–25

57. Wickiser JK, Winkler WC, Breaker RR, Crothers DM. 2005. The speed of RNA transcription and metabolite binding kinetics operate an FMN riboswitch. *Mol. Cell* 18:49–60

58. Williamson JR. 2003. After the ribosome structures: How are the subunits assembled? *RNA* 9:165–67
59. Winkler WC, Breaker RR. 2003. Genetic control by metabolite-binding riboswitches. *Chembiochemistry* 4:1024–32
60. Wong T, Sosnick TR, Pan T. 2005. Mechanistic insights on the folding of a large ribozyme during transcription. *Biochemistry* 44:7535–42
61. Woodson SA. 2002. Folding mechanisms of group I ribozymes: role of stability and contact order. *Biochem. Soc. Trans.* 30:1166–69
62. Woodson SA. 2005. Metal ions and RNA folding: a highly charged topic with a dynamic future. *Curr. Opin. Chem. Biol.* 9:104–9
63. Wu M, Tinoco I Jr. 1998. RNA folding causes secondary structure rearrangement. *Proc. Natl. Acad. Sci. USA* 95:11555–60
64. Wuite GJ, Smith SB, Young M, Keller D, Bustamante C. 2000. Single-molecule studies of the effect of template tension on T7 DNA polymerase activity. *Nature* 404:103–6
65. Zhang F, Ramsay ES, Woodson SA. 1995. In vivo facilitation of *Tetrahymena* group I intron splicing in *Escherichia coli* pre-ribosomal RNA. *RNA* 1:284–92
66. Zhang G, Campbell EA, Minakhin L, Richter C, Severinov K, Darst SA. 1999. Crystal structure of *Thermus aquaticus* core RNA polymerase at 3.3 A resolution. *Cell* 98:811–24
67. Zhuang X. 2005. Single-molecule RNA science. *Annu. Rev. Biophys. Biomol. Struct.* 34:399–414
68. Zhuang X, Ha T, Kim HD, Centner T, Labeit S, Chu S. 2000. Fluorescence quenching: a tool for single-molecule protein-folding study. *Proc. Natl. Acad. Sci. USA* 97:14241–44
69. Zhuang X, Rief M. 2003. Single-molecule folding. *Curr. Opin. Struct. Biol.* 13:88–97

Roles of Bilayer Material Properties in Function and Distribution of Membrane Proteins

Thomas J. McIntosh[1] and Sidney A. Simon[2,3]

Departments of [1]Cell Biology and [2]Neurobiology, and [3]Center for Neuroengineering, Duke University Medical Center, Durham, North Carolina 27710;
email: T.McIntosh@cellbio.duke.edu

Annu. Rev. Biophys. Biomol. Struct. 2006. 35:177–98

The *Annual Review of Biophysics and Biomolecular Structure* is online at biophys.annualreviews.org

doi: 10.1146/annurev.biophys.35.040405.102022

1056-8700/06/0609-0177$20.00

Key Words

peptides, channels, rafts, microdomains, binding

Abstract

Structural, compositional, and material (elastic) properties of lipid bilayers exert strong influences on the interactions of water-soluble proteins and peptides with membranes, the distribution of transmembrane proteins in the plane of the membrane, and the function of specific membrane channels. Theoretical and experimental studies show that the binding of either cytoplasmic proteins or extracellular peptides to membranes is regulated by the presence of charged lipids and that the sorting of transmembrane proteins into or out of membrane microdomains (rafts) depends on several factors, including bilayer material properties governed by the presence of cholesterol. Recent studies have also shown that bilayer material properties modify the permeability of membrane pores, formed either by protein channels or by cell-lytic peptides.

Contents

INTRODUCTION

Rafts: transient microdomains in plasma or bilayer membranes thought to contain specific lipids and proteins

For many years the structure of cell membranes was often characterized by the fluid mosaic model (157), which depicted membranes in terms of proteins embedded and freely diffusing in a lipid sea composed of a fluid bilayer. In this paradigm the bilayer was considered a uniform semipermeable barrier that served as a passive matrix for membrane proteins. However, the ideas of membrane structure and biological function have evolved and now depict a more prominent role for lipids. In this regard, over the past years substantial evidence has accumulated implicating the importance of the lipid milieu for the proper function, distribution, and organization of proteins in membranes (4, 41, 62, 66, 159, 167, 182). Moreover, as emphasized in this review, recent research has provided new insights into the mechanisms by which lipids contribute to membrane functions.

The roles of lipids in protein function and organization arise from the specific chemical properties of individual lipid molecules, as well as from the structural and material properties of the bilayer. This review summarizes some recent experimental and theoretical studies showing the importance of these features of lipid molecules and bilayer assemblies in the following selected areas of protein-lipid interactions: (*a*) the binding of proteins and peptides to membranes, (*b*) the functioning of ion channels, and (*c*) the formation of functional membrane microdomains (rafts).

ROLE OF BILAYER PROPERTIES IN PEPTIDE BINDING

The interactions between water-soluble proteins or peptides and lipid bilayers are of biological significance for several reasons. First, many water-soluble proteins bind to membrane bilayers, where they perform roles in signal transduction, vesicle trafficking, and energy conversion (33, 47, 72, 121, 124, 131), and peptides from the binding sites of specific proteins have been studied as probes of binding and protein function (8, 77, 82, 126). Second, peptide toxins bind to bilayers and modulate the function of membrane channels (90, 161). Third, secreted membrane-lytic and antimicrobial peptides involved in the host-defense system bind to and permeabilize bilayers in bacteria (61, 186).

Here we consider peptides or proteins that bind to the bilayer and accumulate in the bilayer's interfacial or acyl chain regions. The magnitude of the binding can be expressed by the membrane-water (mw) partition coefficient (K_p) that is related to the free energy of transfer (ΔG_{mw}) from the aqueous phase to the membrane by $K_p = \exp(-\Delta G_{mw}/RT)$. Contributions to ΔG_{mw} include terms due to electrostatic interactions, the hydrophobic effect, conformational changes of the peptide, and effects of lipid deformations (15, 75, 163, 176). Depending on the initial aqueous concentration and the value of ΔG_{mw}, molecular concentrations as high as several millimolars can be found at the bilayer surface or in the bilayer (90). Depending on the peptide's charge and structure, this accumulation can change the bilayer's surface charge density, thickness, or material properties, which in turn can affect the functioning of ion channels.

Bilayer Surface Charge: Electrostatics

The traditional manner in which the electrostatic properties of membrane systems have been analyzed is the Gouy-Chapman (G-C) theory, which assumes that the charges are smeared uniformly over the membrane surface and that the dielectric constant is that of the bulk aqueous phase (118). The G-C theory, which relates the peptide concentration at the bilayer surface to the concentration in the aqueous phase, predicts for a peptide with a net positive charge that (*a*) increasing the bilayer negative surface charge density increases binding, (*b*) increasing the net charge on the peptide increases binding, and (*c*) increasing the ionic strength decreases binding (118). The G-C model provides quantitative descriptions of binding, but only when the shape and charge distribution of the molecules are neglected (64, 150).

Although useful, the G-C analysis is unable to accurately describe electrostatic properties that depend on localized effects produced by multivalent lipids such as phosphatidylinositol 4,5-bisphosphate (PIP2), or proteins or peptides with clusters of positive charges. For systems in which such molecular details are important, the FDPB method is now used (16, 17, 52, 120, 121, 126), which determines finite-difference solutions to the nonlinear Poisson-Boltzmann (NLPB) equation: $\nabla[\varepsilon(r)\,\nabla\Psi(r)] - \varepsilon_r\,\kappa(r)^2 \sinh[\Psi(r)] + e^2\,\rho^f(r)/(\varepsilon_o RT) = 0$, where $\varepsilon(r)$ is the dielectric constant, e is the electronic charge, ρ^f is the distribution of fixed charges or charge density, $\kappa(r)$ is the Debye-Huckel parameter, and $\Psi(r)$ is the electrostatic potential. After the peptide is sparsely mapped into a three-dimensional lattice of points and each atom of the peptide and lipid is assigned a radius and a partial charge, the numerical solution to this equation gives the electrostatic potential patterns of peptide-membrane interaction and the mean distribution of the ions. The electrostatic contribution to ΔG_{mw} is then calculated as the difference in free energy when the peptide and lipid are together and when they are far apart. This energy can be further decomposed into the purely attractive Coulombic energy and the repulsive desolvation energy by calculating the interactions at all distances after discharging the lipid and peptide. When present, hydrophobic interactions can drive the peptide from outside the bilayer headgroup region into the acyl chain region, which increases the potential (114). In such cases the binding is also affected by bilayer material properties (see below).

PIP2: phosphatidylinositol 4,5-bisphosphate

FDPB: finite-difference solutions to the non-linear Poisson-Boltzmann equation

MARCKS: myristoylated alanine-rich C kinase substrate

Recent studies using the FDPB treatment have shed light on how electrostatic interactions affect the binding of polyvalent peptides to bilayers with various mole fractions and types of acidic lipids. Peptides that have been investigated include polybasic peptides such as polylysines and polyalanines with and without fatty acids (17, 132), positively charged toxins such as charybdotoxin (16), and domains of proteins such as the myristoylated alanine-rich C kinase substrate (MARCKS) (8, 52), growth-associated protein-43 (121), and protein tyrosine kinases (120).

Some important findings have been obtained from these studies. First, the electrostatic (es) and hydrophobic (hp) contributions to the binding or partition coefficient (K_p) can be expressed as $K_p = \alpha\, K_{es}\, K_{hp}$, where α depends on the distance between the hydrophobic and electrostatic binding sites, and for short distances $\alpha \cong 1$ (125). Second, the valence of an unstructured polybasic peptide or molecule is the major factor in determining the strength of the electrostatic binding (K_{es}) to a vesicle containing negatively charged lipids (125, 172). Third, some polycationic proteins, such as MARCKS, bind to PIP2, preventing ATP-dependent phospholipases from cleaving the PIP2 into its

Area expansion/compression modulus: the fractional change in membrane area produced by an isotropic membrane stress; related to the work required to change the area per lipid molecule in a bilayer

Spontaneous radius of curvature: a measure of the tendency of lipid molecules to form curved (nonbilayer) structures

second messenger products, inositol trisphosphate and diacylglycerol (57). Fourth, polybasic peptides can be disassociated from the bilayer by the activation of lipases that remove acyl chain(s) to reduce K_{hp} (76, 119) and/or by kinases that phosphorylate the compound to reduce its positive charge and thus reduce K_{es} (76, 81). Fifth, the binding of an acidic calcium/calmodulin complex to an adsorbed peptide can produce a charge reversal that causes the peptide to dissociate from the bilayer (120). For example, in the case of a tyrosine kinase of the epidermal growth factor receptor (EGFR/ERbB) family, when the calmodulin complex ($z = -16e$) binds to the polybasic ($z = +8$) juxtamembrane (JM) domain of ErB1 it reverses its charge and dissociates it from the membrane (120). The disassociation of the JM domain also induces a tethered kinase domain to dissociate, permitting it to become active. Finally, relevant to the following section, the disassociation of the JM domain releases electrostatically sequestered PIP2, thus making it available for action by phospholipases.

Recent studies have also shown that, depending on the valence of the phospholipids, the binding of polybasic peptides to bilayers can sequester acidic lipids. Two of the most important acidic lipids that have been tested are phosphatidylserine (PS) and PIP2, which differ in their net charge (–1e and –3e, respectively, at pH 7.4) and also in their concentration in the membrane (the cytoplasmic monolayers of many plasma membranes contain about 25% PS and 1% PIP2). Importantly, PIP2 plays key roles in many cell-signaling pathways involving Ca^{2+}, inositol trisphosphate, and diacylglycerol (21, 109) and is electrostatically sequestered until it is ready to be utilized (30, 52).

The mechanism by which PIP2 is sequestered, even in the presence of larger amounts of PS, has been addressed both experimentally and theoretically by McLaughlin and colleagues (52, 173). Their studies and analysis show the following. When a highly basic peptide, such as the MARCKS domain (residues 151–175), adsorbs to a negatively charged bilayer, the local potential near the peptide is positive (about +30 mV), even though the potential far from the peptide is approximately that of a bilayer containing 25 mol% PS (about –30 mV). PIP2 is sequestered about 1000-fold more effectively than PS. Thus, unlike in the case of peptide binding, sequestration is not strongly dependent on the valence of the peptide, but it does strongly depend on the valence of the lipid.

Bilayer Material Properties

Lipid bilayers do not support shear stresses and hence are true liquids. Bilayers have been characterized by a number of parameters, including the isothermal area expansion/compression modulus (K_T), which is related to the work necessary to change the area per lipid molecule (48, 141); the bilayer bending modulus (B), which is related to the energy necessary to bend the bilayer (84, 141); the tilt modulus (K), which is related to the energy necessary to tilt the acyl chains in the same direction without changing molecular area (84); and the spontaneous radius of curvature (R_o), which is a measure of the tendency of the lipid molecules to form curved (nonbilayer) structures (60, 91) (**Figure 1**). Calculations indicate that the energy to deform a monolayer by area compression is significantly greater than that needed to deform it by bending or chain tilting (84). Many of the studies reviewed here use measured values of these macroscopic material constants for bilayers with different lipid compositions with the assumption that the microscopic and macroscopic parameters are proportional. It also must be realized that these material parameters are coupled (141), such that the addition of specific lipids can change more than one of the parameters.

Experimental studies have shown that the binding of protein or peptides to bilayers depends on these structural and material properties. To separate the contributions to binding of bilayer material properties from those of electrostatic interactions, experiments have been performed with bilayers containing electrically neutral phospholipids, such as phosphatidylcholine (PC), phosphatidylethanolamine, or sphingomyelin (SM). The binding of proteins or peptides depends on the composition of the lipid hydrocarbon chains (14, 55, 93, 131), the amount of cholesterol in the bilayer (2, 5, 18), and the type of phospholipid headgroup (23, 65, 93).

Some of these results can be rationalized in terms of the monolayer spontaneous curvature (23, 37, 93, 115). Because R_o depends on the shape of the lipid molecule, or on the excluded area of the lipid headgroup compared with the excluded area of the hydrocarbon chain region (**Figure 1**), both the headgroup size and the hydrocarbon chain unsaturation modify R_o. For the channel-forming peptide alamethicin, Lewis & Cafiso (93) found that the free energy of peptide binding is linearly dependent on $1/R_o$ (**Figure 2**), indicating that at least some of the effects of hydrocarbon chain unsaturation and headgroup size can be explained in terms of changes in the spontaneous curvature.

Several studies have shown that the presence of cholesterol in the membrane has a marked effect on peptide or protein binding to bilayers (2, 19, 34, 44, 151). One explanation for these observations is that the addition of cholesterol to a phospholipid bilayer increases the area expansion modulus (K_T). To incorporate a peptide into a lipid bilayer, one must increase the bilayer surface area to provide room for the peptide, and the energy to create a vacancy in a bilayer is directly proportional to K_T (190). Thus, it should require about seven times more work to create a vacancy in a bilayer composed of equimolar SM:cholesterol [$K_T \sim 1700$ dyn/cm (127)] than in a dioleoylPC bilayer [$K_T \sim 230$ dyn/cm (141)]. To determine quantitatively how peptide binding depends on bilayer cohesive properties, Allende et al. (5) measured the free energy of binding of two amphipathic peptides to bilayers as a function of K_T. As shown in **Figure 3**, for both peptides the magnitude of the binding decreased with increasing values of K_T, demonstrating that K_T plays a critical role in the binding of these peptides to bilayers.

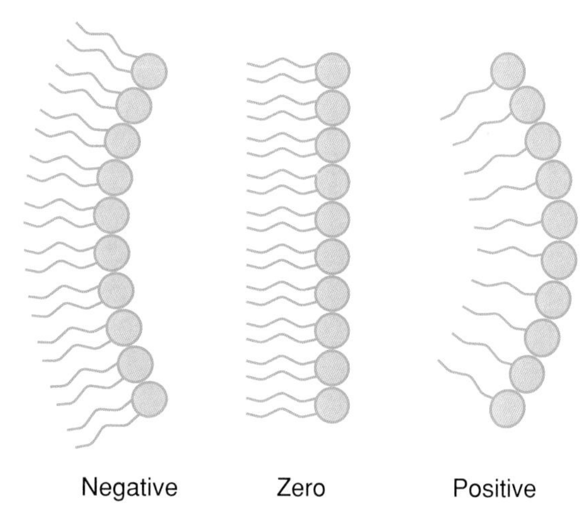

Figure 1

Schematic drawings from Reference 3 illustrating monolayer spontaneous curvature. Lipids that have approximately the same excluded area in their headgroup and acyl chain regions form structures with near zero membrane curvature ($1/R_o \sim 0$). Lipids whose excluded headgroup area is smaller than the area of its acyl chains form structures with negative curvature ($1/R_o < 0$), whereas lipids whose headgroup area is larger than its acyl chain area have positive curvature ($1/R_o > 0$).

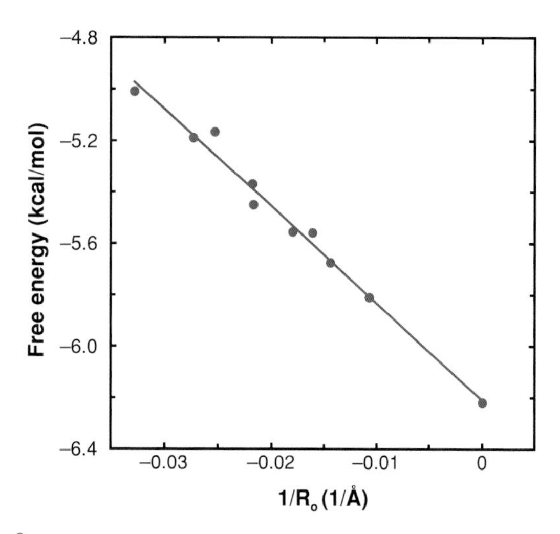

Figure 2

Plot of the free energy of binding of alamethicin to electrically neutral bilayers as a function of bilayer $1/R_o$. Data are taken from Reference 93.

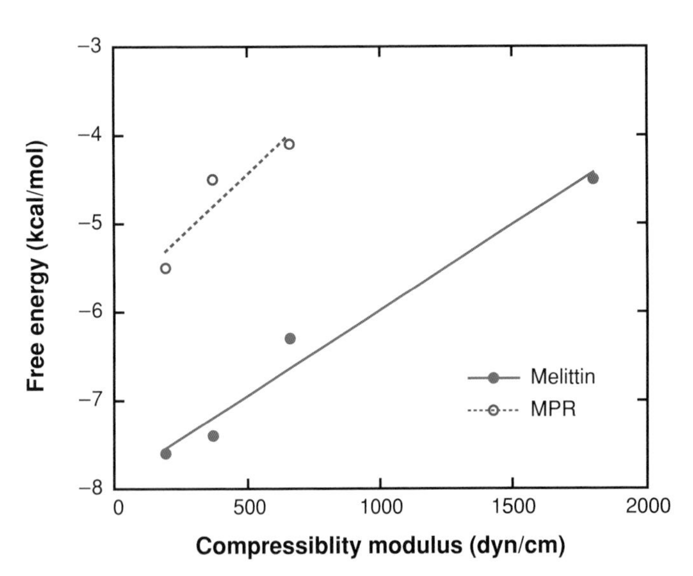

Figure 3

Plot of the free energy of binding versus the bilayer area compressibility modulus for two amphipathic peptides, the presequence of the mitochondrial protein rhodanese (MPR) and the cell lytic peptide melittin. Data are taken from Reference 5.

ROLES OF BILAYERS IN THE FUNCTIONING OF ION CHANNELS

Several studies have shown how bilayer properties can influence the functions of ion channels, including channels composed of transmembrane proteins and channels formed by small peptides.

PC: phosphatidylcholine

SM: sphingomyelin

Transient receptor potential (TRP) family: a family of calcium-permeable channels activated by several types of physical and chemical stimuli

Membrane Channels

As outlined below, bilayer effects on transmembrane channels can arise from (*a*) PIP2 interacting directly with polybasic sites on channel subunits to stabilize a conformational state (67, 70, 179, 189) or (*b*) bilayer thickness or material properties modulating conformational changes between the channel's conducting and nonconducting states (113, 135, 161).

The effects of PIP2 have been demonstrated on both voltage-gated potassium channels and on channels of the transient receptor potential (TRP) family. A recent study on potassium channels involved a lipid-induced conversion between noninactivating delayed rectifier (IK) channels and rapidly inactivating (IA) channels (129). PIP2 electrostatically immobilized the positively charged inactivation domains of IA currents to produce nonactivating or slowly inactivating IK currents, whereas the addition of negatively charged arachidonic acid caused IK currents to exhibit rapid voltage-dependent inactivation. It was suggested that this control of the gating machinery of potassium channels by lipids provides a mechanism for the dynamic regulation of electrical signaling in the nervous system (129).

Several subtypes of TRP receptors are modulated by their direct interaction with PIP2. For the TRPV1 receptor, which is activated by capsaicin and high temperatures, Prescott & Julius (137) identified a positively charged region in the C-terminal domain that is required for PIP2-mediated inhibition of channel gating. A subsequent study localized the PIP2 binding site to the terminal 62 amino acids (96). For the TRPM8 channel, which is activated by menthol and cooling, PIP2 both activated the channel directly and restored rundown activity (95).

The role of the lipid bilayer in ion channel function has been demonstrated by experiments that use exogenously added molecules to modify bilayer structural or material properties. Activators or inhibitors of channels include peptides, toxins, or amphipathic ringed structures such as nicotine, capsaicin, and genistein. In some cases there appears to be specific binding between the molecule and the channel, as the enantiomer or analogues of the molecule have reduced activity (27, 68). However, in other cases, described below, molecular analogues have similar activities, indicating that their effect on ion channels likely arises from bilayer deformation rather than from direct binding to the channel. The compound's effect on bilayer deformation and its modification of ion channel properties depend on its concentration and location (depth) in the bilayer (101).

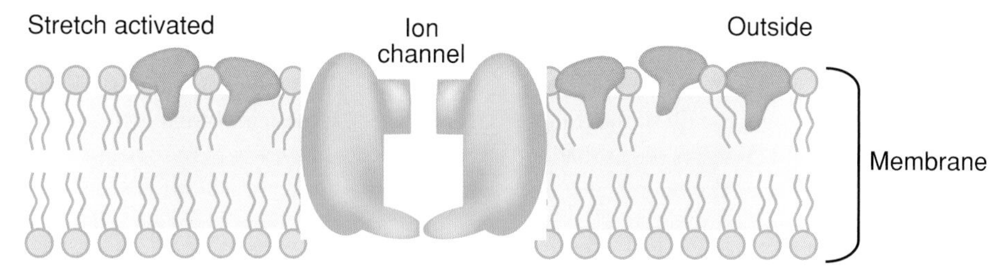

Figure 4

Schematic drawing taken from Reference 54 illustrating how exogenously added peptides can inhibit stretch-activated cation channels. The peptide GsMTx4 (*purple*) acts by altering the bilayer lipid packing near the channel (161).

The GsMTx4 peptide isolated from the venom of the tarantula was believed to inhibit stretch-activated channels (SACs) by a selective binding (lock-and-key) mechanism (59, 130). However, GsMTx4 and its enantiomer, enGsMTx4, interact with SACs in an almost identical manner (161), which suggests that these two molecules produce their effect by perturbing the lipid packing adjacent to a channel and thereby altering the line tension of the SAC/bilayer interface (**Figure 4**). When the bilayer's hydrophobic thickness differs from the channel's hydrophobic length, the bilayer must adjust locally to match the channel's length (**Figure 5*a***). To show that GsMTx4 and enGsMTx4 affect well-characterized channels in bilayers in a similar manner, Suchyna et al. (161) also investigated planar bilayers containing gramicidin A (gA) channels, which have proven to be excellent probes of bilayer structural and material properties (6, 103, 106). Suchyna et al. (161) found that the addition of either GsMTx4 or enGsMTx4 produced an increase in the gramicidin channel appearance rate and open channel lifetime. These data from both membrane SAC and gA channels are difficult to reconcile with lock-and-key mechanisms and suggest that GsMTx4 produces its effects by altering the lipids adjacent to the channel (161).

The gating characteristics of some bacterial SACs can also be modulated by the addition of amphiphiles including anesthetics and lysophospholipids, as well as by changes in bilayer thickness (112, 113, 133, 135). Some SACs can be opened by a change in membrane tension in bilayers lacking a cytoskeleton, implying that the tension can be transduced directly from the lipid molecules to the protein (62, 135, 162). When reconstituted into a bilayer system, the mechanosensitive channel of large conductance (MscL) can be opened from its closed state(s) by the addition of

SAC: stretch-activated channel

gA: gramicidin A

MscL: mechanosensitive channel of large conductance

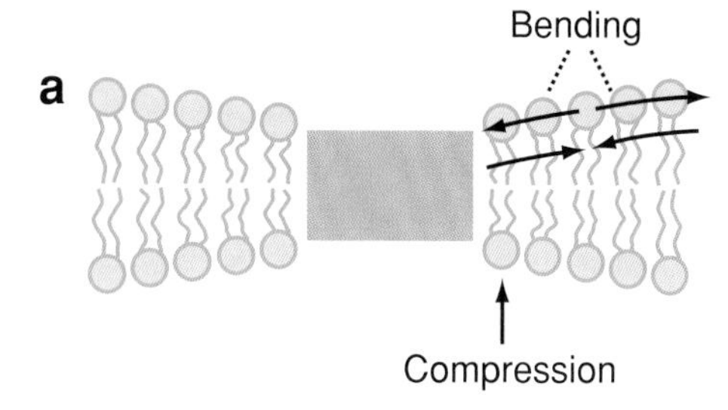

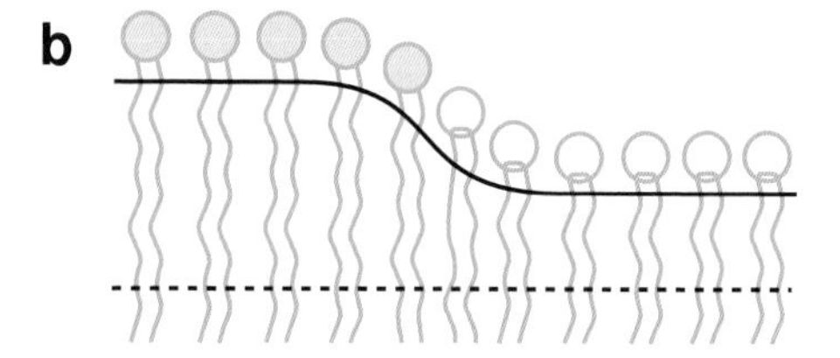

Figure 5

Schematic drawings illustrating bilayer deformation (*a*) adjacent to a transmembrane protein (*gray rectangle*) and (*b*) at the boundary between a thick raft bilayer (*blue headgroups*) and a thinner nonraft bilayer (*white headgroups*). Taken from References 104 and 84, respectively.

nAChR: nicotinic acetylcholine channel receptor

PTK: protein tyrosine kinase

lysophosphatidylcholine to one monolayer of the bilayer, which causes an asymmetry in the tension across the bilayer (133). This increase in tension changes the tilt of the transmembrane helices and increases the probability that the channel is in the open state. An implication of these experiments is that the channel's hydrophobic thickness may differ in its open and closed states, such that changing the acyl chain thickness should stabilize one of these states. To this point, Perozo et al. (133, 134) have measured the free energy of the channel opening of 4, 9, and 20 RT for PC bilayers with acyl chain lengths of 16, 18, and 20 carbons, respectively. Moreover, in a recent elegant study Wiggins & Phillips (177) analyzed the bilayer deformation energies for the MscL channel under similar conditions and found that the free energies of deformation induced by lipid-protein interactions are of the same order as the measured free-energy differences between conductance states. That is, they calculated the energies for areal deformation, spontaneous curvature, and thickness deformation of approximately 10 RT. On the basis of these energy calculations, Wiggins and Phillips concluded that the structural and elastic properties of the bilayer play essential roles in determining the conformation and function of mechanosensitive ion channels.

Other studies have shown that molecules such as nicotine, capsaicin, and genistein can affect ion channels in ways that indicate the bilayer is being deformed. Many of these molecules accumulate at the bilayer-water interface (88), where they could produce bending moments and/or torques on the channels (177). Some of these molecules activate specific receptors; nicotine activates the nicotinic acetylcholine channel receptor (nAChR). However, nicotine can also sensitize TRPV1 receptors and inhibit voltage-dependent sodium channels in cells that do not contain nAChRs (100). The TRPV1 channel agonist, capsaicin, inhibits nonspecifically voltage-gated sodium, potassium, and calcium channels (39, 98, 106). For sodium channels, many compounds produce a hyperpolarizing shift in the inactivation-voltage relationship, with the magnitude of the shift related to the effect the compound has on bilayer stiffness. Capsaicin, which increases bilayer fluidity, causes a hyperpolarizing shift, whereas cholesterol, which stiffens bilayers, produces a depolarizing shift (39, 98, 105, 106). Antagonists are frequently structurally similar to agonists and may also produce nonspecific effects on ion channels. Such is the case for the TRPV1 receptor antagonist, capsazepine, which inhibits voltage-dependent sodium and calcium channels as well as nAChRs (40, 97, 106).

Genistein is commonly used as an inhibitor of protein tyrosine kinases (PTKs). Genistein also modulates a variety of voltage-gated sodium, potassium, and calcium channels (see table 4 in Reference 73), with the direct involvement of PTK inhibition shown in a few cases, but not in others. Genistein has two structural analogues, daidzein, which is much less active, and genistin, which is inactive, in inhibiting PTK. These structure-activity relations could indicate that the interactions with the PTKs are specific. However, Hwang et al. (73) found that the ability of genistein and its analogues to alter gA channel function in planar bilayers was in the order genistein > daidzein > genistin ~0, the same order that Liu et al. (99) found for inhibition of voltage-dependent ion channels in neurons. Because the planar bilayer systems do not have the cellular machinery to phosphorylate or dephosphorylate ion channels, these results suggest that genistein alters the material properties of the lipids surrounding the channel (73). Owing to their differences in structure, genistein and its analogues may partition into the bilayer at different locations. Thus, they could produce different conformational changes in the transbilayer helices that could differentially modulate the voltage-gated parameters. The salient point of these studies with nicotine, capsaicin, and genistein is that when hydrophobic or amphipathic compounds are added to membranes one must consider whether

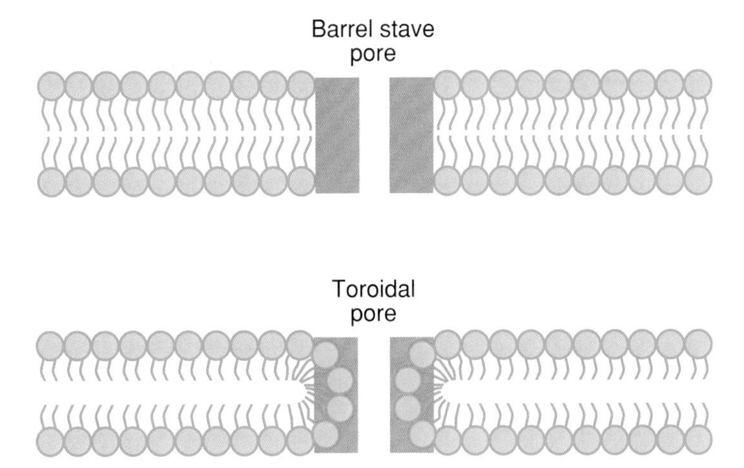

Figure 6

Schematic drawings of "barrel stave" and "toroidal pore" models for peptide-induced pores (3). Peptides are depicted as rectangular gray boxes, and lipids are shown with circular headgroups and wavy hydrocarbon chains.

the observed effects on channels are due to activating a specific protein binding site or changing the material properties of the bilayer.

Another way that the bilayer can play a role is to concentrate exogenous molecules at a binding site on the channel protein. Lee & MacKinnon (90) have found that the peptide toxin VSTX1 reaches the voltage-sensitive potassium channel by partitioning into the lipid bilayer. Thus, the high affinity of binding of VSTX1 to the bilayer concentrates the toxin in the membrane and promotes its interaction with the channel's transmembrane voltage sensor.

Peptide Channel and Pores

For a wide variety of small membrane-lytic peptides, including gramicidin (107, 113, 123), alamethicin (29, 78), magainin (36, 115), cathelicidin (13), sticholysin (164), and melittin (3, 19, 140), the properties of the peptide-induced permeability depend on the lipids composing the bilayer. In the case of alamethicin (160) and gA (58, 63, 73, 113, 123) the channel properties depend on lipid bilayer thickness. In contrast, melittin-induced permeability is not modified by bilayer thickness (3). The permeability increases induced by several peptides, including alamethicin (78), melittin (3, 19, 140), sticholysin (164), magainin (36, 115, 116), and cathelicidin (13), depend on modifications to bilayer organization caused by the addition of cholesterol or lipids with different values of spontaneous curvature.

The gramicidin and alamethicin aqueous pores are thought to be lined by peptides oriented approximately perpendicular to the bilayer surface. In the case of gramicidin there is a transbilayer dimerization of peptides from apposing monolayers with a pore through the middle of the dimer (22, 58), whereas alamethicin is thought to form a "barrel-stave" structure (**Figure 6**) composed of several transbilayer alamethicin molecules that enclose an aqueous pore across the bilayer (12, 89). Both of these structural models are consistent with the observations that the peptide-induced permeability depends on bilayer thickness. For gramicidin there must be an approximate match of the length of the gramicidin to the thickness of each lipid monolayer to provide for optimal dimerization in the center of the bilayer (58). In the case of alamethicin, hydrophobic matching of the peptide length to the bilayer hydrocarbon thickness (**Figure 5*a***) provides the most favorable energy state (181).

Different molecular models have been proposed for the pores formed by other lytic

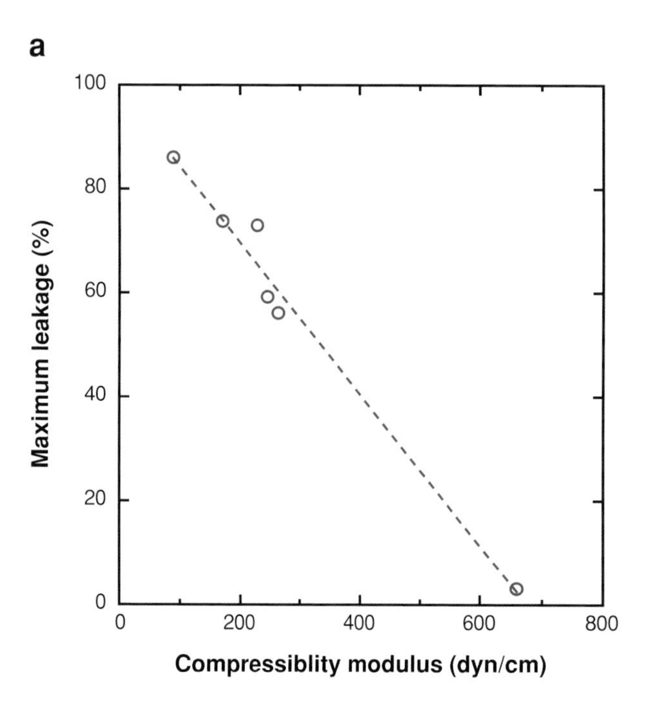

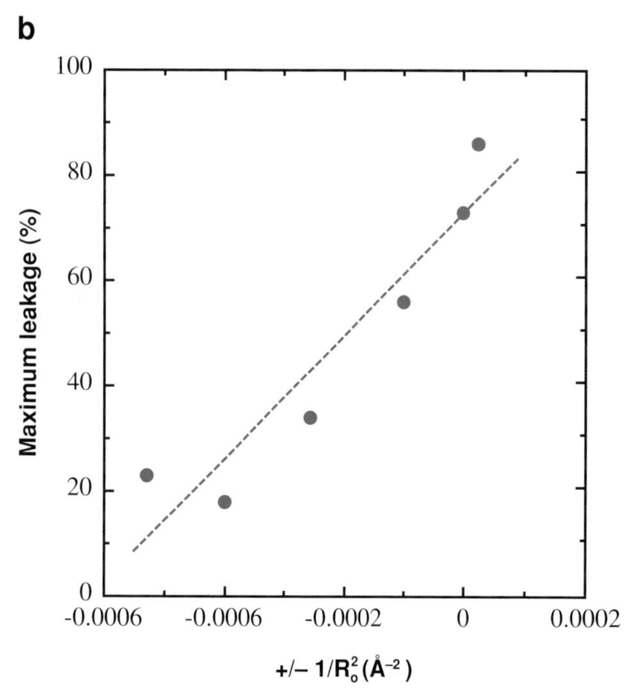

Figure 7

Plots of melittin-induced leakage versus (*a*) bilayer area compressibility modulus (K_T) and (*b*) $1/R_o^2$, which is proportional to the curvature elastic energy. Data are taken from Reference 3.

peptides such as magainin, cathelicidin, and melittin. In these models, exemplified by the "toroidal pore" model (102, 181) shown in **Figure 6**, the lipid bilayer sharply bends so that the lipid headgroups form part of the lining of the pore. Because in the toroidal pore model the lipids must sharply bend and deform to line the pore, the peptide-induced leakage should critically depend on bilayer material properties. Indeed, Matsuzaki et al. (115) and Basanez et al. (13) showed that the bilayer curvature properties do modify the dye leakage induced by magainin 2 and hagfish cathelicidin, respectively, providing strong evidence that these peptides induce formation of lipid-containing pores.

Moreover, for melittin quantitative relationships have recently been obtained that correlate dye leakage to bilayer material properties that are related to the energy needed to deform bilayers. Theoretical treatments predict that the energy of bilayer deformation contains terms involving changes in the bilayer's area and curvature that are proportional to K_T and $1/R_o^2$, respectively (71, 104). As noted above, K_T can be increased by incorporating cholesterol (127), and R_o can be systematically altered by the addition of lipids with negative or positive curvature (**Figure 1**). **Figure 7*a,b*** shows that the melittin-induced leakage is proportional to both K_T and $1/R_o^2$, indicating that bilayer deformation is critical to pore leakage. All of these data, in addition to other studies with cholesterol-containing bilayers (19, 34, 139), are consistent with the toroidal pore model for melittin (3). Moreover, theoretical (187) and molecular dynamics (94) studies are in agreement with the formation of toroidal pores.

Thus, there is strong experimental evidence that toroidal pores are formed by several small membrane-lytic peptides, including magainin, cathelicidin, and melittin. In addition, because the pore-forming activity of colicin E1 depends on bilayer curvature, it has been argued that this protein also forms toroidal pores (158).

ROLE OF LIPID PROPERTIES IN MEMBRANE LATERAL ORGANIZATION

Present evidence indicates that plasma and Golgi membranes are not uniform in structure, but rather contain dynamic lipid/protein microdomains, or rafts (24, 56, 153), with lateral dimensions ranging from 4 to 700 nm (7, 43). These rafts are enriched in cholesterol and sphingolipids, which have a more ordered (liquid-ordered) hydrocarbon chain region than the surrounding liquid-disordered nonraft bilayers do (25, 26, 153). As detailed below, membrane rafts contain several types of proteins (25, 51, 108, 136, 167). Owing to their ability to sequester specific lipids and proteins and exclude others, rafts have been postulated to perform roles in a number of normal cellular processes, such as signal transduction (50, 155, 184), membrane fusion (31, 86), organization of the cytoskeleton (30, 87), lipid sorting (154, 156), and protein trafficking/recycling (74, 152, 153), as well as pathological events such as the cellular invasion of influenza, Ebola, and human immunodeficiency viruses (HIV-1) (28, 42, 85, 110) and formation of the plaques associated with Alzheimer's disease (45, 169).

Lipid Raft Formation

A focus of current research is to determine the factors involved in the formation and organization of rafts. Many experiments have shown that lipid microdomains form in bilayers containing mixtures of SM, PC, and cholesterol that approximate the lipid composition of plasma membranes (38, 142, 145, 168). Thus, lipid-lipid interactions are undoubtedly involved in the formation of membrane rafts. A key factor appears to be the strong preference of cholesterol for SM rather than for lipids with more highly unsaturated chains (1, 24, 26, 92, 128).

X-ray diffraction experiments have shown that raft bilayers enriched in SM and cholesterol are about 25% thicker than nonraft bilayers enriched in dioleoylPC (53), and micropipette experiments have shown that K_T is about seven times higher for SM/cholesterol bilayers than for liquid-disordered PC bilayers (127, 141). Such differences in bilayer thickness and material properties give rise to a line tension at the boundary of raft and nonraft bilayers (to alleviate exposure of hydrocarbon chain to water) that is critical in determining the lateral dimensions and lifetimes of microdomains (84) (**Figure 5*b***).

Toroidal pore: a model for the pore formed in membranes by several cell-lytic peptides in which the bilayer bends sharply so that the lipid headgroups comprise part of the pore wall

Mechanisms of Sequestration of Proteins to Rafts

Although some membrane proteins are excluded from rafts (184, 185), several classes of proteins are associated with rafts, including acylated proteins (10, 69, 122, 124), glycosylphosphatidylinositol (GPI)-anchored proteins (20, 148, 153), certain transmembrane enzymes and receptors (9, 11, 24, 25, 32, 50, 180, 188), as well as several types of membrane channels, including connexons (149), aquaporin-0 (144), and specific classes of potassium channels (111, 143, 178). Recent studies have provided information on the mechanisms by which some of these proteins are sorted to rafts. Several proteins, such as G-proteins, that have covalently attached saturated hydrocarbon chains are preferentially found in rafts because of the strong van der Waals interaction between the ordered saturated chains and the liquid-ordered hydrocarbon chains of rafts (79, 124, 147). Experiments with model systems have shown that lipidated peptides containing ordered saturated fatty acids preferentially bind to liquid-ordered phases, whereas peptides containing the more disordered prenyl groups or unsaturated acyl chains preferentially interact with liquid-disordered bilayers (122, 124, 174). Likewise, in the case of GPI-anchored proteins, the ordered hydrocarbon chains of PI preferentially interact with lipid rafts (20, 148).

The sorting of transmembrane proteins to rafts is currently less well understood

TMD: transmembrane domain of protein

Hydrophobic mismatch: the mismatch between the thickness of the hydrophobic core of a bilayer and the length of the hydrophobic domain of a transmembrane protein

and is a topic of current research (32). Several possible contributing factors include protein structure, protein-protein interactions, and lipid-protein interactions that depend on bilayer structural and elastic properties. In terms of protein structure, specific amino acid sequences in the protein's N-terminal region (171), C-terminal region (32, 35), or transmembrane region (146) can target proteins to rafts. In terms of protein-protein interactions, interactions between proteins and the cholesterol-binding protein caveolin have been demonstrated in caveolae, a specialized membrane raft (32, 46, 149), and interactions between the acylated postsynaptic density protein 95 (PSD-95) can recruit voltage-gated potassium channels into rafts (178). In addition, cross-linking proteins with ligands can affect the size of membrane rafts and the distribution of transmembrane proteins (7, 11, 50, 83, 138, 166).

With regard to lipid-protein interactions, possible contributing factors include differences in both bilayer thickness (53) and material properties between raft and nonraft membranes. This thickness difference may be relevant due to effects of hydrophobic matching (80, 175) between protein transmembrane domains (TMDs) and bilayer hydrocarbon regions, whereas the difference in material properties may be important because the energy to incorporate a protein depends on the bilayer compressibility modulus (see above). Theoretical treatments predict that, although hydrophobic mismatch is a factor, the difference in material properties between liquid-disordered and liquid-ordered phases makes it energetically unfavorable for most trans-bilayer helices to partition from nonraft to raft bilayers (104). These predictions have been verified by experimental studies (49, 117, 165, 170) showing that peptides with single TMDs are largely excluded from rafts. Lundbaek et al. (104) argue that the cholesterol-induced changes in bilayer properties allow for effective sorting of membrane proteins and could be a proofreading mechanism for the exclusion from rafts of proteins with short TMDs.

CONCLUSIONS

From the extensive work reviewed here, it is now clear that the bilayer should not simply be considered as an inert thin layer of oil whose primary purpose is to provide a barrier to ions. There is now overwhelming evidence that the lipid composition of bilayers, which governs bilayer structural and material properties, can have a profound effect on several processes important to the normal functioning of cells. These include the sequestering and releasing of intracellular regulators, the proper working of ion channels, and the in-plane sorting of membrane-associated proteins between raft and nonraft domains.

Bilayer deformation is a common factor involved in all of these processes. There must be local bilayer deformation in order to (*a*) respond to the partitioning of peptides or exogenously added molecules, (*b*) accommodate transmembrane proteins whose hydrophobic thicknesses do not match that of the bilayer, and (*c*) form bilayer rafts. Theoretical descriptions of deformation energies contain terms that depend quadratically (like a spring) on the difference in hydrophobic thicknesses between the undeformed bilayer and lipids near a transmembrane ion channel (**Figure 5*a***) or between membrane microdomains (**Figure 5*b***).

SUMMARY POINTS

1. Lipid bilayer charge and material properties critically affect the binding of proteins and peptides to membranes.

2. Bilayer structure and elastic properties modulate membrane channel functions.
3. Material properties of different classes of lipids mediate the formation of the membrane microdomains (rafts) that perform several key physiological functions.

ACKNOWLEDGMENTS

This work was supported by grant GM27278 from the National Institutes of Health and by grants from Philip Morris USA Inc. and Philip Morris International.

LITERATURE CITED

1. Ahmed SN, Brown DA, London E. 1997. On the origin of sphingolipid/cholesterol-rich detergent-insoluble cell membranes: Physiological concentrations of cholesterol and sphingolipid induce formation of a detergent-insoluble, liquid-ordered lipid phase in model membranes. *Biochemistry* 36:10944–53
2. Allende D, McIntosh TJ. 2003. Lipopolysaccharides in bacterial membranes act like cholesterol in eukaryotic plasma membranes in providing protection against melittin-induced bilayer lysis. *Biochemistry* 42:1101–8
3. **Allende D, Simon SA, McIntosh TJ. 2005. Melittin-induced bilayer leakage depends on lipid material properties: evidence for toroidal pores. *Biophys. J.* 88:1828–37**
4. Allende D, Vidal A, McIntosh TJ. 2004. Jumping to rafts: gatekeeper role of bilayer elasticity. *Trends Biochem. Sci.* 29:325–30
5. Allende D, Vidal A, Simon SA, McIntosh TJ. 2003. Bilayer interfacial properties modulate the binding of amphipathic peptides. *Chem. Phys. Lipids* 122:65–76
6. Andersen OS, Nielsen C, Maer AM, Lundbaek JA, Goulian M, Koeppe RE 2nd. 1999. Ion channels as tools to monitor lipid bilayer-membrane protein interactions: gramicidin channels as molecular force transducers. *Methods Enzymol.* 294:208–24
7. Anderson RG, Jacobson K. 2002. A role for lipid shells in targeting proteins to caveolae, rafts, and other lipid domains. *Science* 296:1821–25
8. Arbuzova A, Wang L, Wang J, Hangyas-Mihalyne G, Murray D, et al. 2000. Membrane binding of peptides containing both basic and aromatic residues. Experimental studies with peptides corresponding to the scaffolding region of the caveolin and the effector region of MARCKS. *Biochemistry* 39:10330–39
9. Arnaoutova I, Smith AM, Coates LC, Sharpe JC, Dhanvantari S, et al. 2003. The prohormone processing enzyme PC3 is a lipid raft-associated transmembrane protein. *Biochemistry* 42:10445–55
10. Arni S, Keilbaugh SA, Ostermeyer AG, Brown DA. 1998. Association of GAP-43 with detergent-resistant membranes requires two palmitoylated cysteine residues. *J. Biol. Chem.* 273:28748–85
11. Baird B, Sheets ED, Holowka D. 1999. How does the plasma membrane participate in cellular signaling by receptors for immunoglobulin E? *Biophys. Chem.* 82:109–19
12. Barranger-Mathys M, Cafiso DS. 1996. Membrane structure of voltage-gated channel forming peptides revealed by site-directed spin labeling. *Biochemistry* 35:498–505
13. Basanez G, Shinnar AE, Zimmerberg J. 2002. Interaction of hagfish cathelicidin antimicrobial peptides with model lipid membranes. *FEBS Lett.* 532:115–20

3. This paper provides quantitative data demonstrating that bilayer elastic properties modify the binding and permeability properties of pore-forming peptides.

14. Basaran N, Doebler RW, Goldston H, Holloway PW. 1999. Effect of lipid unsaturation on the binding of native and a mutant form of cytochrome b5 to membranes. *Biochemistry* 38:15245–52
15. Ben-Tal N, Ben-Shaul A, Nicolls A, Honig B. 1996. Free-energy determinants of alpha-helix insertion into lipid bilayers. *Biophys. J.* 70:1803–12
16. Ben-Tal N, Honig B, Miller C, McLaughlin S. 1997. Electrostatic binding of proteins to membranes. Theoretical predictions and experimental results with charybdotoxin and phospholipid vesicles. *Biophys. J.* 73:1717–27
17. Ben-Tal N, Honig B, Peitzsch RM, Denisov G, McLaughlin S. 1996. Binding of small basic peptides to membranes containing acidic lipids: theoretical models and experimental results. *Biophys. J.* 71:561–75
18. Benachir T, Lafleur M. 1995. Study of vesicle leakage induced by melittin. *Biochim. Biophys. Acta* 1235:452–60
19. Benachir T, Monette M, Grenier J, Lafleur M. 1997. Melittin-induced leakage from phosphatidylcholine vesicles is modulated by cholesterol: a property used for membrane targetting. *Eur. Biophys. J.* 25:201–10
20. Benting J, Rietveld A, Ansorge I, Simons K. 1999. Acyl and alkyl chain length of GPI-anchors is critical for raft association in vitro. *FEBS Lett.* 462:47–50
21. Berridge MJ, Irvine RF. 1989. Inositol phosphates and cell signalling. *Nature* 341:197–205
22. Bradley RJ, Urry DW, Okamoto K, Rapaka R. 1978. Channel structures of gramicidin: characterization of succinyl derivatives. *Science* 200:435–37
23. Brink-van der Laan Evd, Dalbey RE, Demel RA, Killian JA, deKruijff D. 2001. Effect of non-bilayer lipids on membrane binding and insertion of the catalytic domain of leader peptidase. *Biochemistry* 40:9677–84
24. Brown DA, London E. 1998. Functions of lipid rafts in biological membranes. *Annu. Rev. Cell Dev. Biol.* 14:111–36
25. Brown DA, London E. 2000. Structure and function of sphingolipid- and cholesterol-rich membrane rafts. *J. Biol. Chem.* 275:17221–24
26. Brown RE. 1998. Sphingolipid organization in biomembranes: what physical studies of model membranes reveal. *J. Cell Sci.* 111:1–9
27. **Campagna JA, Miller KW, Forman SA. 2003. Mechanisms of actions of inhaled anesthetics. *N. Engl. J. Med.* 348:2110–24**

27. An outstanding review of how small hydrophobic molecules interact with lipids and channels.

28. Campbell SM, Crowe SM, Mak J. 2001. Lipid rafts and HIV-1: from viral entry to assembly of progeny virions. *J. Clin. Virol.* 22:217–27
29. Cantor RS. 2002. Size distribution of barrel-stave aggregates of membrane peptides: influence of the bilayer lateral pressure profile. *Biophys. J.* 82:2520–25
30. Caroni P. 2001. New EMBO members' review: actin cytoskeleton regulation through modulation of PI(4,5)P(2) rafts. *EMBO J.* 20:4332–36
31. Chamberlain LH, Burgoyne RD, Gould GW. 2001. Snare proteins are highly enriched in lipid rafts in PC12 cells: implications for the spatial control of exocytosis. *Proc. Natl. Acad. Sci. USA* 98:5619–24
32. Chini B, Parenti M. 2004. G-protein coupled receptors in lipid rafts and caveolae: How, when and why do they go there? *J. Mol. Endocrinol.* 32:325–38
33. Cho W, Stahelin RV. 2005. Membrane-protein interactions in cell signaling and membrane trafficking. *Annu. Rev. Biophys. Biomol. Struct.* 34:119–51
34. Constantinescu I, Lafleur M. 2004. Influence of the lipid composition on the kinetics of concerted insertion and folding of melittin in bilayers. *Biochim. Biophys. Acta* 1667:26–37

35. Crossthwaite AJ, Seebacher T, Masada N, Ciruela A, Dufraux K, et al. 2005. The cytosolic domains of Ca2+-sensitive adenylyl cyclases dictate their targeting to plasma membrane lipid rafts. *J. Biol. Chem.* 280:6380–91
36. Dathe M, Nikolenko H, Meyer J, Beyermann M, Bienert M. 2001. Optimization of the antimicrobial activity of magainin peptides by modification of charge. *FEBS Lett.* 501:146–50
37. Davies SM, Epand RM, Kraayenhof R, Cornell RB. 2001. Regulation of CTP:phosphocholine cytidylyltransferase activity by the physical properties of lipid membranes: an important role for the stored curvature strain energy. *Biochemistry* 40:10522–31
38. Dietrich C, Bagatolli LA, Volovyk ZN, Thompson NL, Levi M, et al. 2001. Lipid rafts reconstituted in model membranes. *Biophys. J.* 80:1417–28
39. Docherty RJ, Robertson B, Bevan S. 1991. Capsaicin causes prolonged inhibition of voltage-activated calcium currents in adult rat dorsal root ganglion neurons in culture. *Neuroscience* 40:513–21
40. Docherty RJ, Yeats JC, Piper AS. 1997. Capsazepine block of voltage-activated calcium channels in adult rat dorsal root ganglion neurones in culture. *Br. J. Pharmacol.* 121:1461–67
41. Dowhan W. 1997. Molecular basis for membrane phospholipid diversity: Why are there so many lipids? *Annu. Rev. Biochem.* 66:199–232
42. Duncan MJ, Shin JS, Abraham SN. 2002. Microbial entry through caveolae: variations on a theme. *Cell. Microbiol.* 4:783–91
43. Edidin M. 2003. The state of lipid rafts: from model membranes to cells. *Annu. Rev. Biophys. Biomol. Struct.* 32:257–83
44. Egashira M, Gorbenko G, Tanaka M, Saito H, Molotkovsky J, et al. 2002. Cholesterol modulates interaction between an amphipathic class A peptide, AC-18A-NH2, and phosphatidylcholine bilayers. *Biochemistry* 41:4165–72
45. Ehehalt R, Keller P, Haass C, Thiele C, Simons K. 2003. Amyloidogenic processing of the Alzheimer beta-amyloid precursor protein depends on lipid rafts. *J. Cell Biol.* 160:113–23
46. Elliott MH, Fliesler SJ, Ghalayini AJ. 2003. Cholesterol-dependent association of caveolin-1 with the transducin alpha subunit in bovine photoreceptor rod outer segments: disruption by cyclodextrin and guanosine 5′-O-(3-thiotriphosphate). *Biochemistry* 42:7892–903
47. Ellson CD, Andrews S, Stephens LR, Hawkins PT. 2002. The PX domain: a new phosphoinositide-binding module. *J. Cell Sci.* 115:1099–105
48. Evans E, Needham D. 1987. Physical properties of surfactant bilayer membranes: thermal transitions, elasticity, rigidity, cohesion, and colloidal interactions. *J. Phys. Chem.* 91:4219–28
49. Fastenberg ME, Shogomori H, Xu X, Brown DA, London E. 2003. Exclusion of a transmembrane-type peptide from ordered-lipid domains (rafts) detected by fluorescence quenching: extension of quenching analysis to account for the effects of domain size and domain boundaries. *Biochemistry* 42:12376–90
50. Field KA, Holowka D, Baird B. 1997. Compartmentalized activation of the high affinity immunoglobulin E receptor within membrane domains. *J. Biol. Chem.* 272:4276–80
51. Galbiati F, Razani B, Lisanti MP. 2001. Emerging themes in lipid rafts and caveolae. *Cell* 106:403–11
52. Gambhir A, Hangyas-Mihalyne G, Zaitseva I, Cafiso DS, Wang J, et al. 2004. Electrostatic sequestration of PIP(2) on phospholipid membranes by basic/aromatic regions of proteins. *Biophys. J.* 86:2188–207

53. Gandhavadi M, Allende D, Vidal A, Simon SA, McIntosh TJ. 2002. Structure, composition, and peptide binding properties of detergent soluble bilayers and detergent resistant rafts. *Biophys. J.* 82:1469–82
54. Garcia ML. 2004. Ion channels: gate expectations. *Nature* 430:153–55
55. Giorgione JR, Kraayenhof R, Epand RM. 1998. Interfacial membrane properties modulate protein kinase c activation: role of the position of acyl chain unsaturation. *Biochemistry* 37:10956–60
56. Gkantiragas I, Brugger B, Stuven E, Kaloyanova D, Li X-Y, et al. 2001. Sphingomyelin-enriched microdomains at the Golgi complex. *Mol. Biol. Cell* 12:1819–33
57. Glaser M, Wanaski S, Buser CA, Boguslavsky V, Rashidzada W, et al. 1996. Myristoylated alanine-rich C kinase substrate (MARCKS) produces reversible inhibition of phospholipase C by sequestering phosphatidylinositol 4,5-bisphosphate in lateral domains. *J. Biol. Chem.* 271:26187–93
58. Goforth RL, Chi AK, Greathouse DV, Providence LL, Koeppe RE 2nd, Andersen OS. 2003. Hydrophobic coupling of lipid bilayer energetics to channel function. *J. Gen. Physiol.* 121:477–93
59. Gottlieb PA, Suchyna TM, Ostrow LW, Sachs F. 2004. Mechanosensitive ion channels as drug targets. *Curr. Drug Targets CNS Neurol. Disord.* 3:287–95
60. Gruner S, Lenk RP, Janoff AS, Ostro MJ. 1985. Novel multilayered lipid vesicles: comparison of physical characteristics of multilamellar liposomes and stable plurilamellar vesicles. *Biochemistry* 24:2833–42
61. Gura T. 2001. Innate immunity: ancient system gets new respect. *Science* 291:2068–71
62. Hamill OP, Martinac B. 2001. Molecular basis of mechanotransduction in living cells. *Physiol. Rev.* 81:685–740
63. Harroun TA, Heller WT, Weiss TM, Yang L, Huang HW. 1999. Experimental evidence for hydrophobic matching and membrane-mediated interactions in lipid bilayers containing gramicidin. *Biophys. J.* 76:937–45
64. Heimberg T, Marsh D. 1996. Thermodynamics of the interaction of proteins with lipid membranes. In *Biological Membranes: A Molecular Perspective from Computation to Experiment*, ed. KM Merz, B Roux, pp. 405–62. Boston: Birkhauser
65. Heller WT, He K, Ludtke SJ, Harround TA, Huang HW. 1997. Effect of changing the size of lipid headgroups on peptide insertion into membranes. *Biophys. J.* 73:239–44
66. Helms JB, Zurzolo C. 2004. Lipids as targeting signals: lipid rafts and intracellular trafficking. *Traffic* 5:247–54
67. Hilgemann DW, Feng S, Nasuhoglu C. 2001. The complex and intriguing lives of PIP2 with ion channels and transporters. *Sci. STKE* 2001:RE19
68. Hille B. 1993. *Ionic Channels of Excitable Membranes*. Sunderland, MA: Sinauer
69. Hiol A, Davey PC, Osterhout JL, Waheed AA, Fischer ER, et al. 2003. Palmitoylation regulates regulators of G-protein signaling (RGS) 16 function. I. Mutation of amino-terminal cysteine residues on RGS16 prevents its targeting to lipid rafts and palmitoylation of an internal cysteine residue. *J. Biol. Chem.* 278:19301–8
70. Huang CL, Feng S, Hilgemann DW. 1998. Direct activation of inward rectifier potassium channels by PIP2 and its stabilization by Gbetagamma. *Nature* 391:803–6
71. Huang HW. 1986. Deformation free energy of bilayer membrane and its effect on gramicidin channel lifetime. *Biophys. J.* 50:1061–70
72. Hurley JH, Meyer T. 2001. Subcellular targeting by membrane lipids. *Curr. Opin. Cell Biol.* 13:146–52

73. **Hwang TC, Koeppe RE 2nd, Andersen OS. 2003. Genistein can modulate channel function by a phosphorylation-independent mechanism: importance of hydrophobic mismatch and bilayer mechanics. *Biochemistry* 42:13646–58**
74. Ikonen E. 2001. Roles of lipid rafts in membrane transport. *Curr. Opin. Cell Biol.* 13:470–77
75. Jahnig F. 1983. Thermodynamics and kinetics of protein incorporation into membranes. *Proc. Natl. Acad. Sci. USA* 80:3691–95
76. Johnson JE, Cornell RB. 1999. Amphitropic proteins: regulation by reversible membrane interactions. *Mol. Membr. Biol.* 16:217–35
77. Johnson JE, Rao NM, Hui S-W, Cornell RB. 1998. Conformation and lipid binding properties of four peptides derived from membrane-binding domain of CTP:phosphocholine cytidyltransferase. *Biochemistry* 37:9509–20
78. Keller SL, Bezrukov SM, Gruner SM, Tate MW, Vodyanoy I, Parsegian VA. 1993. Probability of alamethicin conductance states varies with nonlamellar tendency of bilayer phospholipids. *Biophys. J.* 65:23–27
79. Khan TK, Yang B, Thompson NL, Maekawa S, Epand RM, Jacobson K. 2003. Binding of NAP-22, a calmodulin-binding neuronal protein, to raft-like domains in model membranes. *Biochemistry* 42:4780–86
80. Killian JA. 1998. Hydrophobic mismatch between proteins and lipids in membranes. *Biochim. Biophys. Acta* 1376:401–16
81. Kim J, Blackshear PJ, Johnson JD, McLaughlin S. 1994. Phosphorylation reverses the membrane association of peptides that correspond to the basic domains of MARCKS and neuromodulin. *Biophys. J.* 67:227–37
82. Kuner T, Tokumaru H, Augustine GJ. 2002. Peptides as probes of protein-protein interactions involved in neurotransmitter release. In *Peptide-Lipid Interactions*, ed. SA Simon, TJ McIntosh, pp. 551–70. San Diego, CA: Academic
83. Kusumi A, Koyama-Honda I, Suzuki K. 2004. Molecular dynamics and interactions for creation of stimulation-induced stabilized rafts from small unstable steady-state rafts. *Traffic* 5:213–30
84. **Kuzmin PI, Akimov SA, Chizmadzhev YA, Zimmerberg J, Cohen FS. 2005. Line tension and interaction energies of membrane rafts calculated from lipid splay and tilt. *Biophys. J.* 88:1120–33**
85. Lafont F, Tran Van Nhieu G, Hanada K, Sansonetti P, van der Goot FG. 2002. Initial steps of *Shigella* infection depend on the cholesterol/sphingolipid raft-mediated CD44-Ipab interaction. *EMBO J.* 21:4449–57
86. Lang T, Bruns D, Wenzel D, Riedel D, Holroyd P, et al. 2001. SNARES are concentrated in cholesterol-dependent clusters that define docking and fusion sites for exocytosis. *EMBO J.* 20:2202–13
87. Laux T, Fukami K, Thelen M, Golub T, Frey D, Caroni P. 2000. GAP43, MARCKS, and CAP23 modulate PI(4,5)P(2) at plasmalemmal rafts, and regulate cell cortex actin dynamics through a common mechanism. *J. Cell Biol.* 149:1455–72
88. Lee AG. 2003. Lipid-protein interactions in biological membranes: a structural perspective. *Biochim. Biophys. Acta* 1612:1–40
89. Lee MT, Chen FY, Huang HW. 2004. Energetics of pore formation induced by membrane active peptides. *Biochemistry* 43:3590–99
90. Lee SY, MacKinnon R. 2004. A membrane-access mechanism of ion channel inhibition by voltage sensor toxins from spider venom. *Nature* 430:232–35

73. A study showing that a molecule thought to be a tyrosine kinase inhibitor actually produces its effects by altering the structural and elastic properties of bilayers.

84. A detailed theoretical analysis of raft size and stability.

91. Leikin S, Kozlov MM, Fuller NL, Rand RP. 1996. Measured effects of diacylglycerol on structural and elastic properties of phospholipid membranes. *Biophys. J.* 71:2623–32
92. Leventis R, Silvius JR. 2001. Use of cyclodextrins to monitor transbilayer movement and differential lipid affinities of cholesterol. *Biophys. J.* 81:2257–67
93. Lewis JR, Cafiso DS. 1999. Correlation of the free energy of a channel-forming voltage-gated peptide and the spontaneous curvature of bilayer lipids. *Biochemistry* 38:5932–38
94. Lin JH, Baumgaertner A. 2000. Stability of a melittin pore in a lipid bilayer: a molecular dynamics study. *Biophys. J.* 78:1714–24
95. Liu B, Qin F. 2005. Functional control of cold- and menthol-sensitive TRPM8 ion channels by phosphatidylinositol 4,5-bisphosphate. *J. Neurosci.* 25:1674–81
96. Liu B, Zhang C, Qin F. 2005. Functional recovery from desensitization of vanilloid receptor TRPV1 requires resynthesis of phosphatidylinositol 4,5-bisphosphate. *J. Neurosci.* 25:4835–43
97. Liu L, Simon SA. 1997. Capsazepine, a vanilloid receptor antagonist, inhibits nicotinic acetylcholine receptors in rat trigeminal ganglia. *Neurosci. Lett.* 228:29–32
98. Liu L, Simon SA. 2003. Modulation of IA currents by capsaicin in rat trigeminal ganglion neurons. *J. Neurophysiol.* 89:1387–401
99. Liu L, Yang T, Simon SA. 2004. The protein tyrosine kinase inhibitor, genistein, decreases excitability of nociceptive neurons. *Pain* 112:131–41
100. Liu L, Zhu W, Zhang ZS, Yang T, Grant A, et al. 2004. Nicotine inhibits voltage-dependent sodium channels and sensitizes vanilloid receptors. *J. Neurophysiol.* 91:1482–91

100. This paper demonstrates that nicotine can affect sodium and TRPV1 channels by altering membrane bilayer properties.

101. Liu Z, Xu Y, Tang P. 2005. Molecular dynamics simulations of C2F6 effects on gramicidin A: implications of the mechanisms of general anesthesia. *Biophys. J.* 88:3784–91
102. Ludtke SJ, He K, Heller WT, Harroun TA, Yang L, Huang HW. 1996. Membrane pores induced by magainin. *Biochemistry* 35:13723–28
103. Lundbaek JA, Andersen OS. 1999. Spring constants for channel-induced lipid bilayer deformations. Estimates using gramicidin channels. *Biophys. J.* 76:889–95
104. Lundbaek JA, Andersen OS, Werge T, Nielsen C. 2003. Cholesterol-induced protein sorting: an analysis of energetic feasibility. *Biophys. J.* 84:2080–89

104. This theoretical analysis demonstrates that the elastic properties of cholesterol-containing bilayers are critical in the sorting of transmembrane proteins in the plane of the membrane.

105. Lundbaek JA, Birn P, Hansen AJ, Sogaard R, Nielsen C, et al. 2004. Regulation of sodium channel function by bilayer elasticity: the importance of hydrophobic coupling. Effects of micelle-forming amphiphiles and cholesterol. *J. Gen. Physiol.* 123:599–621
106. Lundbaek JA, Birn P, Tape SE, Toombes GE, Sogaard R, et al. 2005. Capsaicin regulates voltage-dependent sodium channels by altering lipid bilayer elasticity. *Mol. Pharmacol.* 68:680–89
107. Lundbaek JA, Maer AM, Andersen OS. 1997. Lipid bilayer electrostatic energy, curvature stress, and assembly of gramicidin channels. *Biochemistry* 36:5695–701
108. Maekawa S, Iino S, Miyata S. 2003. Molecular characterization of the detergent-insoluble cholesterol-rich membrane microdomain (raft) of the central nervous system. *Biochim. Biophys. Acta* 1610:261–70
109. Majerus PW, Connolly TM, Deckmyn H, Ross TS, Bross TE, et al. 1986. The metabolism of phosphoinositide-derived messenger molecules. *Science* 234:1519–26
110. Manes S, del Real G, Martinez AC. 2003. Pathogens: raft hijackers. *Nat. Rev. Immunol.* 3:557–68
111. Martens JR, O'Connell K, Tamkun M. 2004. Targeting of ion channels to membrane microdomains: localization of Kv channels to lipid rafts. *Trends Pharmacol. Sci.* 25:16–21

112. Martinac B. 2004. Mechanosensitive ion channels: molecules of mechanotransduction. *J. Cell Sci.* 117:2449–60

113. Martinac B, Hamill OP. 2002. Gramicidin A channels switch between stretch activation and stretch inactivation depending on bilayer thickness. *Proc. Natl. Acad. Sci. USA* 99:4308–12

114. Mathias RT, Baldo GJ, Manivannan K, McLaughlin S. 1992. Discrete charges on biological membranes. In *Electrified Interfaces in Physics, Chemistry, and Biology*, ed. R Guidelli, pp. 473–90. Dordrecht, The Netherlands: Kluwer

115. Matsuzaki K, Sugishita K, Ishibe N, Ueha M, Nakata S, et al. 1998. Relationship of membrane curvature to the formation of pores by magainin 2. *Biochemistry* 37:11856–63

116. Matsuzaki K, Sugishita K-I, Fujii N, Miyajima K. 1995. Molecular basis for membrane selectivity of an antimicrobial peptide, magainin 2. *Biochemistry* 34:3423–29

117. McIntosh TJ, Vidal A, Simon SA. 2003. Sorting of lipids and transmembrane peptides between detergent-soluble bilayers and detergent-resistant rafts. *Biophys. J.* 85:1656–66

118. McLaughlin S. 1989. The electrostatic properties of membranes. *Annu. Rev. Biophys. Biophys. Chem.* 18:113–36

119. McLaughlin S, Aderem A. 1995. The myristoyl-electrostatic switch: a modulator of reversible protein-membrane interactions. *Trends Biochem. Sci.* 20:272–76

120. McLaughlin S, Smith SO, Hayman MJ, Murray D. 2005. An electrostatic engine model for autoinhibition and activation of the epidermal growth factor receptor (egfr/erbb) family. *J. Gen. Physiol.* 126:41–53

120. An experimental and theoretical analysis investigating how charges on cytoplasmic proteins that bind to plasma membranes can affect the function of kinases.

121. McLaughlin S, Wang J, Gambhir A, Murray D. 2002. Pip(2) and proteins: interactions, organization, and information flow. *Annu. Rev. Biophys. Biomol. Struct.* 31:151–75

122. Melkonian KA, Ostermeyer AG, Chen JZ, Roth MG, Brown DA. 1999. Role of lipid modifications in targeting proteins to detergent-resistant membrane rafts. Many raft proteins are acylated, while few are prenylated. *J. Biol. Chem.* 274:3910–17

123. Mobashery N, Nielsen C, Andersen OS. 1997. The conformational preference of gramicidin channels is a function of lipid bilayer thickness. *FEBS Lett.* 412:15–20

124. Moffett S, Brown DA, Linder ME. 2000. Lipid-dependent targeting of G proteins into rafts. *J. Biol. Chem.* 275:2191–98

125. Murray D, Arbuzova A, Honig B, McLaughlin S. 2002. The role of electrostatic and nonpolar interactions in the association of peripheral proteins with membranes. In *Peptide-Lipid Interactions*, ed. SA Simon, TJ McIntosh, pp. 277–307. San Diego: Academic

126. Murray D, Arbuzova A, Mihaly G, Ghambir A, Ben-Tal N, et al. 1999. Electrostatic properties of membranes containing acidic lipids and adsorbed basic peptides: theory and experiment. *Biophys. J.* 77:3176–88

127. Needham D, Nunn RS. 1990. Elastic deformation and failure of lipid bilayer membranes containing cholesterol. *Biophys. J.* 58:997–1009

128. Niu SL, Litman BJ. 2002. Determination of membrane cholesterol partition coefficient using a lipid vesicle-cyclodextrin binary system: effect of phospholipid acyl chain unsaturation and headgroup composition. *Biophys. J.* 83:3408–15

129. Oliver D, Lien CC, Soom M, Baukrowitz T, Jonas P, Fakler B. 2004. Functional conversion between A-type and delayed rectifier K+ channels by membrane lipids. *Science* 304:265–70

130. Ostrow KL, Mammoser A, Suchyna T, Sachs F, Oswald R, et al. 2003. CDNA sequence and in vitro folding of GsMTx4, a specific peptide inhibitor of mechanosensitive channels. *Toxicon* 42:263–74

131. Pande AH, Qin S, Tatulian SA. 2005. Membrane fluidity is a key modulator of membrane binding, insertion, and activity of 5-lipoxygenase. *Biophys. J.* 88:4084–94
132. Peitzsch RM, McLaughlin S. 1993. Binding of acylated peptides and fatty acids to phospholipid vesicles: pertinence to myristoylated proteins. *Biochemistry* 32:10436–43
133. Perozo E, Cortes DM, Sompornpisut P, Kloda A, Martinac B. 2002. Open channel structure of MscL and the gating mechanism of mechanosensitive channels. *Nature* 418:942–48
134. Perozo E, Kloda A, Cortes DM, Martinac B. 2002. Physical principles underlying the transduction of bilayer deformation forces during mechanosensitive channel gating. *Nat. Struct. Biol.* 9:696–703
135. Perozo E, Rees DC. 2003. Structure and mechanism in prokaryotic mechanosensitive channels. *Curr. Opin. Struct. Biol.* 13:432–42
136. Pike LJ. 2004. Lipid rafts: heterogeneity on the high seas. *Biochem. J.* 378:281–92
137. Prescott ED, Julius D. 2003. A modular PIP2 binding site as a determinant of capsaicin receptor sensitivity. *Science* 300:1284–88
138. Pyenta PS, Holowka D, Baird B. 2001. Cross-correlation analysis of inner-leaflet-anchored green fluorescent protein co-redistributed with IgE receptors and outer leaflet lipid raft components. *Biophys. J.* 80:2120–32
139. Raghuraman H, Chattopadhyay A. 2004. Interaction of melittin with membrane cholesterol: a fluorescence approach. *Biophys. J.* 87:2419–32
140. Raghuraman H, Chattopadhyay A. 2005. Cholesterol inhibits the lytic activity of melittin in erythrocytes. *Chem. Phys. Lipids* 134:183–89
141. Rawicz W, Olbrich KC, McIntosh T, Needham D, Evans E. 2000. Effect of chain length and unsaturation on elasticity of lipid bilayers. *Biophys. J.* 79:328–39
142. Rinia HA, Snel MME, van der Eerden JPJM, deKruijff B. 2001. Visualizing detergent resistant domains in model membranes with atomic force microscopy. *FEBS Lett.* 501:92–96
143. Romanenko VG, Fang Y, Byfield F, Travis AJ, Vandenberg CA, et al. 2004. Cholesterol sensitivity and lipid raft targeting of Kir2.1 channels. *Biophys. J.* 87:3850–61
144. Rujoi M, Jin J, Borchman D, Tang D, Yappert MC. 2003. Isolation and lipid characterization of cholesterol-enriched fractions in cortical and nuclear human lens fibers. *Invest. Ophthalmol. Vis. Sci.* 44:1634–42
145. Samsonov AV, Mihalyov I, Cohen FS. 2001. Characterization of cholesterol-sphingomyelin domains and their dynamics in bilayer membranes. *Biophys. J.* 81:1486–500
146. Scheiffele P, Roth MG, Simons K. 1997. Interaction of influenza virus haemagglutinin with sphingolipid-cholesterol membrane domains via its transmembrane domain. *EMBO J.* 16:5501–8
147. Schroeder H, Leventis R, Rex S, Schelhaas M, Nagele E, et al. 1997. S-acylation and plasma membrane targeting of the farnesylated carboxyl-terminal peptide of N-Ras in mammalian fibroblasts. *Biochemistry* 36:13102–9
148. Schroeder R, London E, Brown D. 1994. Interactions between saturated acyl chains confer detergent resistance on lipids and glycosylphosphatidylinositol (GPI)-anchored proteins; GPI-anchored proteins in liposomes and cells show similar behavior. *Proc. Natl. Acad. Sci. USA* 91:12130–34
149. Schubert AL, Schubert W, Spray DC, Lisanti MP. 2002. Connexin family members target to lipid raft domains and interact with caveolin-1. *Biochemistry* 41:5754–64
150. Seelig J. 1997. Titration calorimetry of lipid-peptide interactions. *Biochim. Biophys. Acta* 1331:103–16

151. Semple SC, Chonn A, Cullis PR. 1996. Influence of cholesterol on the association of plasma proteins with liposomes. *Biochemistry* 35:2521–25
152. Sharma DK, Choudhury A, Singh RD, Wheatley CL, Marks DL, Pagano RE. 2003. Glycosphingolipids internalized via caveolar-related endocytosis rapidly merge with the clathrin pathway in early endosomes and form microdomains for recycling. *J. Biol. Chem.* 278:7564–72
153. Simons K, Ikonen E. 1997. Functional rafts in cell membranes. *Nature* 387:569–72
154. Simons K, Ikonen E. 2000. How cells handle cholesterol. *Science* 290:1721–26
155. Simons K, Toomre D. 2000. Lipid rafts and signal transduction. *Nat. Rev. Mol. Cell Biol.* 1:31–39
156. Simons K, van Meer G. 1988. Lipid sorting in epithelial cells. *Biochemistry* 27:6197–202
157. Singer SJ, Nicolson GL. 1972. The fluid mosaic model of the structure of cell membranes. *Science* 175:720–31
158. Sobko AA, Kotova EA, Antonenko YN, Zakharov SD, Cramer WA. 2004. Effect of lipids with different spontaneous curvature on the channel activity of colicin E1: evidence in favor of a toroidal pore. *FEBS Lett.* 576:205–10
159. Spector AA, Yorek MA. 1985. Membrane lipid composition and cellular function. *J. Lipid Res.* 26:1015–35
160. Stankowski S, Schwarz G. 1989. Lipid dependence of peptide-membrane interactions. Bilayer affinity and aggregation of the peptide alamethicin. *FEBS Lett.* 250:556–60
161. Suchyna TM, Tape SE, Koeppe RE 2nd, Andersen OS, Sachs F, Gottlieb PA. 2004. Bilayer-dependent inhibition of mechanosensitive channels by neuroactive peptide enantiomers. *Nature* 430:235–40

161. Experimental demonstration that specific venoms modify membrane channel properties by perturbing lipid packing in the bilayer adjacent to the channel.

162. Sukharev SI, Blount P, Martinac B, Blattner FR, Kung C. 1994. A large-conductance mechanosensitive channel in *E. coli* encoded by MscL alone. *Nature* 368:265–68
163. Tamm L. 1994. Physical studies of peptide-bilayer interactions. In *Membrane Protein Structure: Experimental Approaches*, ed. SH White, pp. 283–313. New York: Oxford Univ. Press
164. Valcarcel CA, Dalla Serra M, Potrich C, Bernhart I, Tejuca M, et al. 2001. Effects of lipid composition on membrane permeabilization by sticholysin I and II, two cytolysins of the sea anemone *Stichodactyla helianthus*. *Biophys. J.* 80:2761–74
165. van Duyl BY, Rijkers DT, de Kruijff B, Killian JA. 2002. Influence of hydrophobic mismatch and palmitoylation on the association of transmembrane alpha-helical peptides with detergent-resistant membranes. *FEBS Lett.* 523:79–84
166. van Meer G. 2004. Invisible rafts at work. *Traffic* 5:211–12
167. van Meer G, Sprong H. 2004. Membrane lipids and vesicular traffic. *Curr. Opin. Cell Biol.* 16:373–78
168. Veatch SL, Polozov IV, Gawrisch K, Keller SL. 2004. Liquid domains in vesicles investigated by NMR and fluorescence microscopy. *Biophys. J.* 86:2910–22
169. Vetrivel KS, Cheng H, Lin W, Sakurai T, Li T, et al. 2004. Association of gamma-secretase with lipid rafts in post-Golgi and endosome membranes. *J. Biol. Chem.* 279:44945–54
170. Vidal A, McIntosh TJ. 2005. Transbilayer peptide sorting between raft and nonraft bilayers: comparisons of detergent extraction and confocal microscopy. *Biophys. J.* 89:1102–8
171. Walmsley AR, Zeng F, Hooper NM. 2003. The N-terminal region of the prion protein ectodomain contains a lipid raft targeting determinant. *J. Biol. Chem.* 278:37241–48
172. Wang J, Gambhir A, Hangyas-Mihalyne G, Murray D, Golebiewska U, McLaughlin S. 2002. Lateral sequestration of phosphatidylinositol 4,5-bisphosphate by the basic effector

domain of myristoylated alanine-rich C kinase substrate is due to nonspecific electrostatic interactions. *J. Biol. Chem.* 277:34401–12

173. Wang J, Gambhir A, McLaughlin S, Murray D. 2004. A computational model for the electrostatic sequestration of PI(4,5)P(2) by membrane-adsorbed basic peptides. *Biophys. J.* 86:1969–86

174. Wang T-Y, Leventis R, Silvius JR. 2001. Partitioning of lipidated peptide sequences into liquid-ordered lipid domains in model and biological membranes. *Biochemistry* 40:13031–40

175. Webb RJ, East JM, Sarma RP, Lee AG. 1998. Hydrophobic mismatch and the incorporation of peptides into lipid bilayers: a possible mechanism for retention in the Golgi. *Biochemistry* 37:673–79

176. White SH, Wimley WC. 1999. Membrane protein folding and stability: physical principles. *Annu. Rev. Biophys. Biomol. Struct.* 28:319–65

177. Wiggins P, Phillips R. 2005. Membrane-protein interactions in mechanosensitive channels. *Biophys. J.* 88:880–902

178. Wong W, Schlichter LC. 2004. Differential recruitment of Kv1.4 and Kv4.2 to lipid rafts by PSD-95. *J. Biol. Chem.* 279:444–52

179. Wu L, Bauer CS, Zhen XG, Xie C, Yang J. 2002. Dual regulation of voltage-gated calcium channels by PtdIns(4,5)P2. *Nature* 419:947–52

180. Xue M, Vines CM, Buranda T, Cimino DF, Bennett TA, Prossnitz ER. 2004. N-formyl peptide receptors cluster in an active raft-associated state prior to phosphorylation. *J. Biol. Chem.* 279:45175–84

181. Yang L, Harroun TA, Weiss TM, Ding L, Huang HW. 2001. Barrel-stave model or toroidal model? A case study on melittin pores. *Biophys. J.* 81:1475–85

182. Yeagle PL. 1989. Lipid regulation of cell membrane structure and function. *FASEB J.* 3:1833–42

183. Yokota T, Satoh T. 2001. Three-dimensional estimation of the distribution and size of putative functional units in rat gustary cortex. *Brain Res. Bull.* 54:575–84

184. Young RM, Holowka D, Baird B. 2003. A lipid raft environment enhances Lyn kinase activity by protecting the active site tyrosine from dephosphorylation. *J. Biol. Chem.* 278:20746–52

185. Young RM, Zheng X, Holowka D, Baird B. 2005. Reconstitution of regulated phosphorylation of FcepsilonRI by a lipid raft-excluded protein-tyrosine phosphatase. *J. Biol. Chem.* 280:1230–35

186. Zasloff M. 2002. Antimicrobial peptides of multicellular organisms. *Nature* 415:389–96

187. Zemel A, Fattal DR, Ben-Shaul A. 2003. Energetics and self-assembly of amphipathic peptide pores in lipid membranes. *Biophys. J.* 84:2242–55

188. Zhang CF, Dhanvantari S, Lou H, Loh YP. 2003. Sorting of carboxypeptidase E to the regulated secretory pathway requires interaction of its transmembrane domain with lipid rafts. *Biochem. J.* 369:453–60

189. Zhang L, Lee JK, John SA, Uozumi N, Kodama I. 2004. Mechanosensitivity of GIRK channels is mediated by protein kinase C-dependent channel-phosphatidylinositol 4,5-bisphosphate interaction. *J. Biol. Chem.* 279:7037–47

190. Zhelev DV. 1998. Material property characteristics for lipid bilayers containing lysolipid. *Biophys. J.* 75:321–30

173. Demonstration that positively charged peptides or proteins can sequester the critical lipid molecule PIP2 in the plane of the membrane.

177. A theoretical treatment showing that the free energies of lipid properties and lipid-protein interactions are comparable to the measured free energies of channel conductance states. Provides quantitative basis for the effect of lipid structure on protein activity.

Electron Tomography of Membrane-Bound Cellular Organelles

Terrence G. Frey,[1] Guy A. Perkins,[2] and Mark H. Ellisman[2]

[1]Department of Biology, San Diego State University, San Diego, California 92182-4614; email: tfrey@sunstroke.sdsu.edu

[2]National Center for Microscopy and Imaging Research, University of California, San Diego, La Jolla, California 92093-0608; email: perkins@ncmir.ucsd.edu

Annu. Rev. Biophys. Biomol. Struct. 2006. 35:199–224

First published online as a Review in Advance on January 13, 2006

The *Annual Review of Biophysics and Biomolecular Structure* is online at biophys.annualreviews.org

doi: 10.1146/annurev.biophys.35.040405.102039

1056-8700/06/0609-0199$20.00

Key Words

electron microscopy, mitochondria, Golgi, organelle structure, image analysis

Abstract

Electron microscope tomography produces three-dimensional reconstructions and has been used to image organelles both isolated and in situ, providing new insight into their structure and function. It is analogous to the various tomographies used in medical imaging. Compared with light microscopy, electron tomography offers an improvement in resolution of 30- to 80-fold and currently ranges from 3 to 8 nm, thus filling the gap between high-resolution structure determinations of isolated macromolecules and larger-scale studies on cells and tissues by light microscopy. Here, we provide an introduction to electron tomography and applications of the method in characterizing organelle architecture that also show its power for suggesting functional significance. Further improvements in labeling modalities, imaging tools, specimen preparation, and reconstruction algorithms promise to increase the quality and breadth of reconstructions by electron tomography and eventually to allow the mapping of the cellular proteomes onto detailed three-dimensional models of cellular structure.

Contents

TEM: transmission electron microscopy

Electron tomography (ET): the process of calculating the three-dimensional structure of a specimen from a tilt series of electron micrographs representing two-dimensional projections of the three-dimensional structure

INTRODUCTION

Transmission electron microscopy (TEM) has been largely responsible for shaping our views of organelle architecture, as it provides the highest resolution within a spectrum of complementary tools used in the structural study of whole organelles (**Figure 1**). When combined with techniques to detect specific molecules, TEM is the only technique with sufficient resolution to localize proteins in subregions of organelles, and the ease with which newer TEM instruments can be used has brought TEM back to mainstream cell biology. TEM imaging of cellular organelles involves several steps, some of which are automated. A sample, either embedded in plastic or frozen hydrated, is supported by a metal grid that is clamped into a removable stage and inserted in the column of the microscope. A high-energy electron beam passes down this column with a velocity and wavelength that depend on the accelerating voltage of the microscope. Electrons that pass through the sample are focused to an image, magnified, and recorded using a charge-coupled device camera or onto photographic film.

Owing to the large depth of focus in TEM, images are two-dimensional projections of the specimen, and features from different levels are superimposed and as a result electron micrographs are sometimes hard to interpret. In conventional thin-section TEM, the sections are generally much thinner than the specimen, and overlap of features is less of a problem. Establishing the three-dimensional structure of complex cellular organelles from individual sections that are much thinner than the specimen is, however, problematic. Three-dimensional reconstructions from serial thin sections of cells help in the interpretation, but the practical limitations in obtaining serial sections sufficiently thin (5–10 nm) needed for accurate visualization of complex membrane topology limit resolution. As higher-voltage electron microscopes became available, thicker sections could be examined; however, the overlapping details in two-dimensional images recorded from thicker sections were more difficult to interpret accurately. Electron tomography (ET) overcomes many of the limitations associated with serial thin-section reconstruction by using sections thick enough (200–3000 nm) to contain a significant fraction of the organelle within the section volume. The superposition of features is then resolved by computing a three-dimensional reconstruction where digital slices are only a few nanometers thick (4, 19, 67, 68).

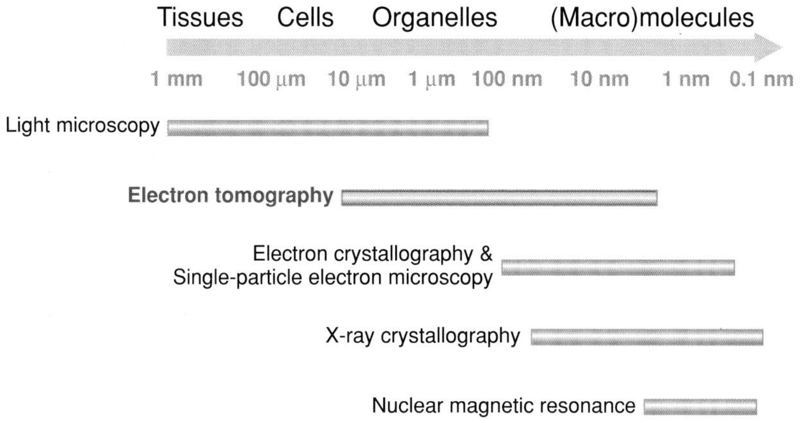

Figure 1

The structural realm of electron microscope tomography in relation to other three-dimensional structural modalities. At the top from left to right is a decreasing progression of sizes on a logarithmic scale of biological complexity ranging from tissues to molecules. On the left are commonly used techniques for structure determination. The bar corresponding to each technique represents the useful range in linear dimension for examining structural domains. Notice that electron tomography bridges the resolution gap between light microscopy and higher-resolution techniques, e.g., crystallography and nuclear magnetic resonance, and is well situated to study the structure of organelles.

ELECTRON TOMOGRAPHY

Microscope Considerations

Tomography first gained prominence in the 1970s through the tomographic imaging of human patients for X-ray diagnosis, an advance recognized by the 1979 Nobel Prize in Physiology or Medicine to Hounsfield and Cormack (101). More recently, its cousin, ET, was named a runner-up breakthrough of the year by *Science* magazine in 2002 (100) for its emergence as the leading method for the elucidation of three-dimensional ultrastructure in the 3- to 20-nm resolution range. This highest-resolution form of tomography is uniquely suited to generate three-dimensional reconstructions of pleiomorphic objects such as cells, organelles, or supramolecular assemblies (45, 65). The principal limitation with TEM employing accelerating voltages of approximately 100 kV for ET is the necessity of using relatively thin samples to avoid chromatic aberration. Higher-voltage TEM confers the ability to image specimens thick enough to contain organelles or even small cells without limiting chromatic aberration. In general intermediate high-voltage (200–600 kV) microscopes provide adequate performance and are more common than ultra-high voltage (≥800 V) microscopes that are significantly more expensive to purchase and maintain (68).

Tilt series: a sequence of electron micrographs recorded from the same sample over an angular range at defined angles of tilt perpendicular to the optic axis

In ET a tilt series is obtained by rotating the specimen holder incrementally about a fixed axis perpendicular to the electron beam and collecting projection images of the object of interest at one- to two-degree angular increments over a range of ±60 to 80° (**Figure 2**, left). Object reconstruction is based on a principle first described mathematically almost a century ago by Radon that uses back-projection of tilt images to fill in a three-dimensional density map. Before the three-dimensional density map can be calculated, usually by a weighted backprojection algorithm, the projection images must be mutually aligned (4). Defining resolution as the smallest separation between two structural components that can be observed, the resolution of a reconstruction depends directly on

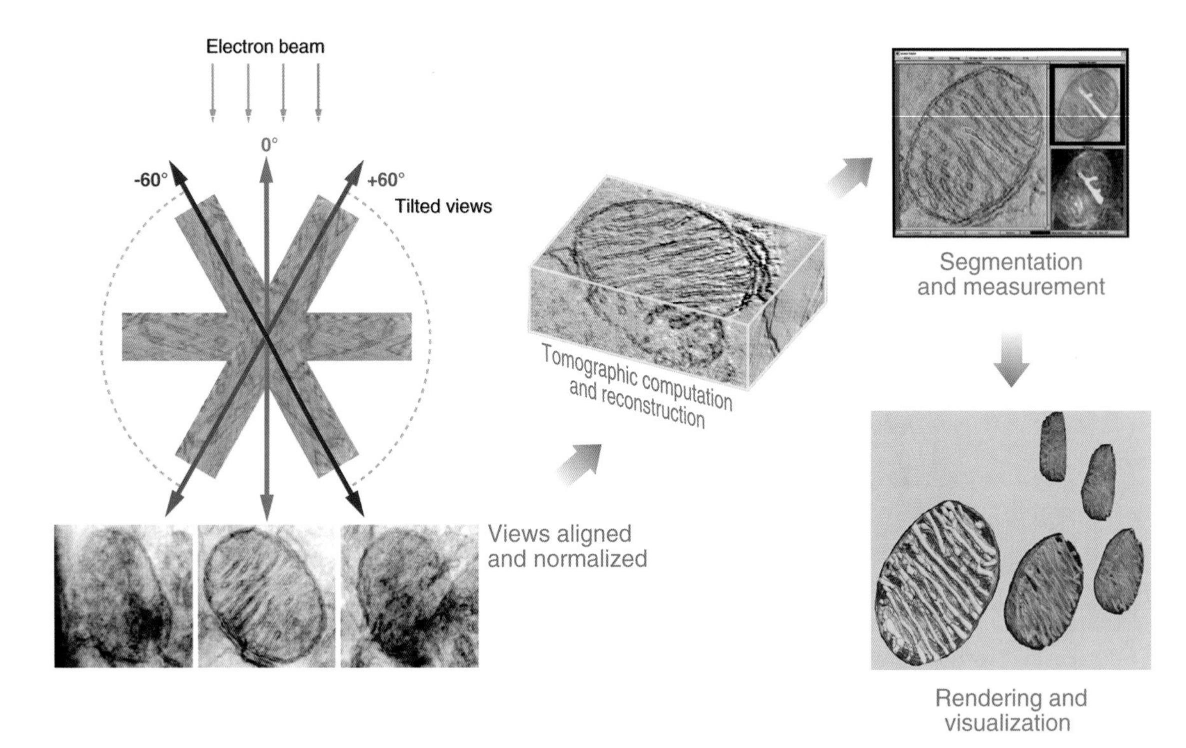

Figure 2

Schematic showing the stages in electron microscope tomography from image collection to visualization, using an example of how a mitochondrion in a 0.5-micron-thick section of rat cerebellum is ushered through the process. The section, situated on a grid in the electron microscope, is rotated typically ± 60°, with images collected every 1 to 2° producing a tilt series (*left*). The images of this tilt series are aligned to a common reference, usually the 0-degree image, and the optical density distribution in each image is normalized. Afterward, the volume is reconstructed using computational algorithms such as backprojection or iterative techniques (*center*). Although a direct volume visualization may be informative for some simple structures and analytical purposes, it is more often the case, especially with organelle structures, that segmentation methods (*top right*) for defining and dissecting components of the structure are required to create a three-dimensional model (*bottom right*), facilitating interpretation, the discovery of interrelationships, and measurement. Also see the movies *Mito Tilt Series*, *Mito Tomogram*, *Segmented Model*, *and Crista+ER Model* (follow the Supplemental Material link from the Annual Reviews home page at **http://www.annualreviews.org**).

the thickness of the specimen and inversely on the number of projections and tilt-angle range, so thicker objects require more projections to achieve the same resolution (10, 101). The resolution that one can achieve is not as straightforward as it seems, however, because one must reconcile the need to record many tilted images over a wide range of angles with the need to minimize exposure of the specimen to the electron beam in order to minimize radiation damage. The development of automated tilt-series acquisition, in which object centering and focusing are done by computer control and large-area charge-coupled device cameras, has enabled a dramatic reduction in beam exposure with improved tomographic throughput. With this improvement electron dose "overhead," i.e., the dose used on searching, re-centering, and focusing, can be kept as low as 3% of the total (4), and almost all electrons are used during the actual collection of the tilt series.

Double-Tilt Tomography

Because the path of the electron beam through the specimen increases as the specimen is tilted, it is generally not feasible to obtain good-quality images from sections at tilt angles above 60 to 70°, at which point there may also be interference by the grid holder. The limited tilt range means that traditional "single-tilt" series have a wedge of missing data corresponding to the angular range between the maximal tilt angle used in the reconstruction and 90°. This missing wedge leads to a reduction in axial resolution within the three-dimensional reconstruction (39). The missing information can be reduced by rotating the specimen by 90° around the optical axis and collecting a second tilt series at this orientation, thus reducing the missing data from a wedge to a pyramid with smaller volume. Double-tilt ET, however, is not problem-free. When a plastic section is irradiated by electrons, it thins at a variable rate along the beam axis (and to a lesser extent perpendicular to the beam axis) that decreases almost exponentially with additional exposure. "Cooking" the specimen before a tilt series is collected is usually practiced so that the greatest shrinkage occurs before data collection. Even so, the additional radiation used to collect the second tilt series induces further change (shrinkage and warping) in the specimen, so that the successive reconstructions are not from an identical specimen. Post-reconstruction software was developed to "dewarp" one reconstruction with respect to its 90° rotated version and merge the two "halves" of the double-tilt data (63).

Thicker Specimens: Energy-Filtering and Serial Tomography

Energy filtering applied to ET is a way to image thicker specimens (>0.5 microns) by reducing or eliminating altogether the blurring from chromatic aberration that results from inelastic scattering in which energy is transferred from the incident electrons to the specimen. Energy filters operating in either "zero-loss" mode or "most-probable-loss" mode select electrons within a narrow energy range providing significantly higher signal-to-noise ratio (8). Energy filtering has been valuable for ET of whole ice-embedded cells and isolated organelles 0.5 to 1 μm thick (45). Energy filtering can also be used to map the three-dimensional distributions of specific elements in cells by analysis of reconstructions from tilt series recorded above and below characteristic edges in energy-loss spectra of specific atomic species; an example is phosphorus ET of freeze-substituted tissue to reveal the subcellular locations of DNA, RNA, and ribosomes (43). Another method to obtain tomograms of thicker specimens is by using an ultrahigh-voltage TEM (UHVEM) (60). The use of higher-voltage microscopes increases the mean free electron path, allowing for thicker specimens to be imaged before degradation of the image from inelastically scattered electrons becomes noticeable. One drawback of UHVEM is that currently only a few microscopes with accelerating voltage greater than 1.0 MeV are available for study of biological samples with thicknesses ranging from 1 to 5 microns (**Table 1**). Greater access to UHVEM is being addressed through the development of "telemicroscopy" for remote control of some of these microscopes (103).

The technique of serial sectioning can be combined with ET to derive a three-dimensional reconstruction with a large depth dimension. Serial semithick or thick sections are cut through the structure of interest, with the thickness determined by the desired resolution. Tilt series are collected from the same object in each section and tomographic volume reconstructions are computed for each tilt series. Afterward, the resulting series of volumes are aligned and combined to form a single large volume (41, 99). In this way, serial-section ET complements traditional serial thin-section reconstruction by providing much higher resolution in the depth dimension (65).

Serial tomography: calculating the electron tomograms from a sequence of thick or semithick sections taken from a specimen followed by alignment of the sequence of tomograms to produce a three-dimensional reconstruction of the volume of sample contained within the sequence of sections

Energy filtering: as applied to electron microscopy this is formation of an image using only electrons with a narrow range of energy (wavelength)

Table 1 Facilities with TEMs that can be used for ET of cellular organelles[a]

Facility	TEM voltage capability
Australian National University	300 kV
Baylor University	300 kV
Florida State University	300 kV
Korean Basic Sciences Institute	1250 kV
Max-Planck Institute of Biochemistry	300 kV
Osaka University	3000 kV
University of California, Berkeley	300 kV, 400 kV
University of California, San Diego	300 kV EF, 400 kV
University of Colorado	300 kV
University of Utrecht	200 kV EF
Wadsworth Center	400 kV, 1200 kV
Weizmann Institute of Science	300 kV EF

[a]Only energy filter (EF) TEMs with 200 kV or higher voltages or conventional TEMs with 300 kV or higher voltages are included in this table.

SPECIMEN PREPARATION

Conventional Chemical Fixation and Plastic Embedding

Chemical fixation: stabilizing the structure of a sample by treatment with chemicals such as formaldehyde, glutaraldehyde, and osmium tetroxide

High-pressure freezing (HPF): freezing a sample under high pressure allowing bulk samples to be cryofixed in vitreous ice

Cryofixation: stabilizing the structure of a sample by rapid freezing that converts liquid water to vitreous ice

The most common methods for preserving cell structure for TEM and ET are based on chemical fixation using aldehydes, formaldehyde and/or glutaraldehyde, for primary fixation and osmium tetroxide for secondary fixation. Following fixation, water is replaced by an organic solvent that is then replaced with a resin monomer subsequently polymerized by heating or UV irradiation to form a plastic-embedded specimen that can be sectioned. Typically the sections are stained with heavy metals, such as uranyl acetate and lead salts, in order to increase the contrast. For more than 50 years, considerable effort has been spent on optimizing chemical fixation for the preservation of isolated organelles, cells, and tissues. However references for what constitutes good sample preservation are not always available. Currently, cryo-preparation is used to verify structural preservation (45), and leakage or dissolution of some cytoplasmic components can occur with chemical fixation and dehydration. During the seconds to minutes it takes a specimen to become well fixed, alterations to the native state may take place (73). Structural rearrangements, most noticeably kinked membranes, occur during dehydration (88). With a desire to preserve cells and tissues in a near-native state, researchers have developed alternative approaches to specimen preparation that rely for the most part on rapid freezing of samples.

Rapid Freezing and Freeze Substitution

The cryo method of choice for examining cells or tissues by ET is currently rapid high-pressure freezing (HPF) or cryofixation, which allows subcellular structures to be immobilized within milliseconds, followed by freeze substitution, in which water is substituted with an organic solvent containing chemical fixatives at temperatures around −80°C (rapid freezing and freeze substitution, RF/FS). The temperature is then slowly raised to permit the chemical fixatives to act while the structure remains physically fixed by low temperature, after which the specimen is embedded in plastic (69). Sectioning and post-staining are done as with conventionally prepared samples. The principal improvement of RF/FS over conventional sample preparation is that dehydration at room temperature is avoided, resulting in structures that are generally better preserved, with the most noticeable improvement being straight, smooth membranes. The speed of cryo-fixation also allows the investigation of processes occurring at a timescale of seconds, the rate-limiting step being the loading of the specimen into the HPF holder. Compared with the use of frozen hydrated specimens (see below), freeze-substituted specimens exhibit greater contrast and are easier to image because they don't require a cryo-specimen holder. On the other hand, the contrast is generated by added heavy atom stains that may not accurately reflect the high-resolution macromolecular structure. The frozen-hydrated, freeze substitution, and conventional fixation methods can be complementary as long as molecular resolution is not required. Once it is shown that a structure has essentially the same features by using a combination of these preparation schemes, the less-demanding RF/FS or conventional methods can be used for the analysis (65).

Frozen-Hydrated Specimens

Ideally one would like to image cells and organelles in their native hydrated state; however, the ultrahigh vacuum required for electron microscopy is not compatible with hydrated specimens, except with very special setups not suitable for high-resolution work. If a hydrated specimen is frozen rapidly enough, then the liquid water is transformed into a vitreous solid state with a structure and density similar to those of the liquid; these are called frozen-hydrated specimens, and this technique is commonly used to study the structures of macromolecules and viruses by TEM (14). One way to obtain frozen-hydrated samples of organelles in thin suspension or in small thin cells is to plunge them into liquid ethane or liquid propane, reducing the temperature at a speed of tens of thousands of degrees per second, which transforms liquid water into vitreous ice, preserving the structure in a near-native state. The key to good specimen preservation is to freeze rapidly enough to reach the temperature below which crystalline ice forms ($\leq 140^\circ$C), forming amorphous ice that is similar to liquid but with high viscosity. With the best preservation, rapid freezing immobilizes all constituents of a specimen before significant rearrangement takes place. Rapidly frozen samples can be examined directly in a microscope equipped with a cold stage that maintains the temperature below -160°C, and ET has been used to reconstruct frozen-hydrated samples (45). Cryo-ET of frozen-hydrated samples encounters two difficulties. First, as no stains are used, the image contrast is low because it is based on small differences in mass density between macromolecules and the surrounding cytosol. Although membranes are often visible, it is hard to visualize the finer details of cellular substructure (68). To partially ameliorate this problem the objective lens of the microscope is operated at extreme levels of underfocus to maximize phase contrast. Second, frozen-hydrated samples are sensitive to the electron beam and cannot tolerate the higher electron doses required for high resolution. Without automation, ET of these types of samples is simply not feasible.

Cryo-Sectioning: Frozen-Hydrated Sections

Many prokaryotic cells are small enough and a few eukaryotic cells can be grown flat enough to be studied by cryo-ET without the need of sectioning. However, as cryo-ET has yet to be combined with UHVEM, samples thicker than 1 micron cannot be used for ET with current cryo-microscopes and must be cryo-sectioned before tomographic analysis. Cryo-sectioning combines the advantages of rapid freezing with the use of bulk material; it is applicable to whole cells and tissues and with higher resolution could be used for generating proteome maps described below (69). Sections of frozen-hydrated material are cut at low temperature by using specially designed cryo-ultramicrotomes. Cryo-sectioning is still a challenging endeavor despite recent progress (22). It invariably produces surface defects, such as knife marks, cracks, or crevasses, that tend to be smoothed out upon irradiation with the electron beam (36). If these cutting defects do not penetrate far into the section, they are less of a concern for ET. Of more concern is the observation of significant compression of the section in the direction of cutting that would affect measurements of compartment sizes and spacing and interrelationships between components. Because irradiation leads to sublimation of the vitreous ice, resulting in shrinkage of the section and movement relative to the support film that holds the gold particles, alignment based on gold fiducial particles loses its accuracy, and this loss of

Freeze substitution: replacing at low temperature the ice in cryofixed sample with an organic solvent containing chemical fixatives followed by slow warming of the sample

RF/FS: rapid freezing/freeze substitution

Frozen-hydrated sample: a sample frozen in vitreous ice that is examined in the electron microscope using a special holder that keeps the temperature below -160°C

QUANTUM DOTS IN ELECTRON TOMOGRAPHY

Quantum dot (QD) technology is now being used for correlative light microscopy and ET (25). When conjugated with streptavidin, QDs are used as a tag to recognize a biotinylated antibody (43). QDs have novel properties that make them attractive for correlative studies: (*a*) They cannot be photobleached, thus allowing for extensive three-dimensional (z-section) imaging with confocal or multiphoton microscopies. (*b*) Because of their electron-dense core, their contrast is only slightly less than the colloidal gold particles commonly used in immunoTEM and for fiducials in ET and so they are easily visualized by TEM. (*c*) Their emission properties are narrow and tied to their size and therefore multilabeling for both light microscopy and TEM is practical. The size range extends from less than 5 nm to greater than 100 nm; they are ellipsoid and so can usually be distinguished from gold particles. (*d*) The cadmium core can be mapped with energy-filtering TEM and so even the smallest QDs (even those within thick sections) can be visualized for ET.

alignment accuracy adversely affects the resolution of the reconstruction. Further advances are needed to overcome these challenges. A promising new innovation is the oscillating diamond knife, which has been reported to diminish compression when cryo-sectioning (113). Currently, cryo-sectioning is an art form practiced in only a handful of labs (1). However, with the push in the ET community to develop the capability to map the proteome, we expect this technique to receive more attention in the future, because it would expand the mapping to a much wider range of specimens.

Labeling for Correlated Light and Electron Microscopy

An advantage of light microscopy over TEM is that it can monitor processes in living cells and is amenable to monitoring movement of specific macromolecules and/or changes in physiological states of organelles through the use of fluorescently tagged antibodies and fluorophores that migrate to specific environments. If a process can be arrested in a desired physiological state by rapid fixation or freezing, it can be investigated by TEM and ET at a much higher resolution than is possible by light microscopy. A more powerful approach would be specific labeling that can be followed from the light microscope level to TEM and ET; however, getting tagged antibodies into thick or semithick sections for examination by ET while preserving ultrastructure has proven difficult. Innovative advances based on noninvasive genetic manipulation are now in use to correlate dynamic physiological events monitored by light microscopy with higher-resolution structural studies conducted with ET. Griffin et al. (29) reported the use of a biarsenical fluorophore that binds tightly to four cysteines appropriately placed within a short peptide hairpin motif, causing the dye to fluoresce strongly. The utility of the tetracysteine system in TEM and ET derives from the ability of some biarsenical fluorophores to drive the oxidation of diaminobenzidine when illuminated into an osmiophilic polymer ("photooxidation"), producing an electron-dense stain that is easily discernible by TEM (24). Because the tetracysteine motif can be engineered into most proteins, this new technology for correlated light and electron microscopy is proving useful for the three-dimensional localization by TEM and ET of specific proteins (also see sidebar).

COMPUTATIONAL TECHNIQUES

Following image acquisition, image processing is divided into three phases. The first two phases are generally termed alignment and the last phase is the three-dimensional reconstruction proper. The first phase is to precisely locate a set of image features (often colloidal gold particles added to the specimen) consistent across the image series. The second phase is to find geometric correspondences between the configurations of features in the images and, if necessary, to transform the images geometrically to bring the features into better alignment. This process can be done automatically once feature tracking is accomplished. The third process is to reconstruct the object itself from the image data and the projection transforms followed by segmenting the features of interest. For an overview of the sequence of computational steps in calculating and analyzing an electron tomogram see **Figure 2** and the supplemental movies.

Alignment

The alignment of the individual images of a tilt series is a critical step in obtaining high-quality reconstructions. The two most common alignment techniques are (*a*) cross-correlation alignment and (*b*) fiducial mark tracking, usually using colloidal gold particles applied to the specimen. The advantages of cross-correlation alignment are that it requires no application of gold particles and it is a fast procedure that requires little input from the user. The disadvantage of this method is that it currently is not capable of adequate correction for image rotation or magnification changes between images of the tilt series. The advantage of fiducial mark tracking is that corrections for all of the above can be made, in addition to dewarping (correction of image and/or specimen distortions) the images on the basis of tracking a large number of fiducial marks through the tilt series.

Three-Dimensional Reconstruction

Most reconstruction methods currently in use are an extension of the standard orthogonal backprojection process using projections filtered by a modified r-weighting (39). Weighted backprojection has been one of the most popular methods in the field of ET owing to its computational simplicity and predictable outcome (16). Most alternative reconstruction techniques fall under the umbrella of "iterative techniques," and the two most commonly used are the algebraic reconstruction technique (ART) and the simultaneous iterative reconstruction technique (SIRT), both of which rely on optimizing the reconstruction by iterative comparison between projections of the reconstruction and images of the "raw" tilt series (20, 86).

Segmentation

Segmentation: the process of representing the three-dimensional structure within a tomogram as a series of geometrical objects such as spheres, lines, and contours in order to analyze and display it as a three-dimensional model

Although examination of a three-dimensional reconstruction alone may be informative for simple structures, in the case of large complex structures volume segmentation, i.e., decomposition into structural components, is required to facilitate interpretation, communication of results to others, and measurements. After segmentation, the features of interest are viewed using surface- or volume-rendering tools. Exploration of the shapes and spatial distribution of a feature or class of objects and their relation to other objects in the volume is aided by visualization from many angles of view. Comprehension of complex structures is enhanced greatly by the capability of the software to assign individual characteristics such as color and the level of transparency to different structural components and to display selectively a subset of the structural components (86). Manual segmentation generally requires the user to trace the features of interest with either a mouse or a pen directly on the screen to isolate them from the rest of the reconstruction volume (65). This step is usually the rate-limiting step in ET, because often several hundred slices are present in the tomographic volume. However, for purposes of visualization and analysis and because the oversampling during digitization exceeds the resolution present in the reconstruction, not every slice needs to be traced, and tracing every other or even every fourth slice often suffices.

Because of the tracing bottleneck, more sophisticated methods for automated or semiautomated segmentation would greatly facilitate what is currently a tedious process (53). Computer-automated or semiautomated segmentation is, in principle, more objective than manual tracing and has been used mostly to extract the surfaces of membranes and trajectories of filamentous structures. However, manual segmentation of the features of interest is more commonly used because in most cases human intelligence has proven superior to the currently available segmentation algorithms. Four new tools for autosegmentation have been made available recently. One uses thresholding and connectivity, which have been mainstays of traditional autosegmentation approaches, and was applied to the segmentation of molecular complexes present in the synaptic cleft (45). Another uses a specialized sequence of segmentation operations with the goal of improving the reliability and spatial resolution in the rendered volumes; this approach was tested with ET reconstructions of the neuromuscular junction (93). Automatic boundary and skeletonization algorithms accurately segmented fibrillar components (2). Perhaps the most used autosegmentation tool relies on a watershed algorithm (107); however, it does an incomplete segmentation with organelles, and so manual segmentation is more accurate.

Software

There has been a recent upsurge in the number of software suites designed for ET. The most complete packages are IMOD (39), SPIDER (21), EM3D (33, 93), XMIPP (98), and TOM (75). A partial list of specialty software for ET includes reconstruction by maximum entropy (97); filtering and denoising using wavelets or nonlinear anisotropic diffusion (16, 18); Tinkerbell (44), watershed (107), and IVOI (93) for segmentation; visualization (2); focus-focus-gradient correction (110); automated tilt series collection (111, 112); automated correlation and averaging of reconstructions (76); and energy-filtered ET (7).

Template Matching and Averaging

Cryo-ET is a promising approach to visualize macromolecules in their natural working environment and in combination with advanced pattern-recognition techniques to identify molecules and map their distributions inside organelles and cells. The field is now poised to bridge the resolution gap between cellular and molecular structure determinations with a resolution of 4 to 8 nm, but with the prospect to reach 2 to 3 nm (22, 65, 102). With molecular resolution, cryo-reconstructions of organelles or cells are essentially three-dimensional maps of the respective "proteomes" (4), raising the possibility of identifying specific macromolecular complexes within their cellular or organelle context by virtue of their structural signature alone by using templates from high-resolution structures to search the reconstructed volume (6, 65). This "template-matching" will make it possible not only to map the three-dimensional distribution of molecules within cells and organelles but also to observe the spatial relationships and partnering of molecules. Template matching was first tested with the structurally similar proteasome and thermosome isolated from cells and rapidly frozen in a frozen-hydrated state (6). A later study came closer to an in situ model by using lipid vesicles filled with macromolecules, so-called phantom cells (17). In these vesicles, macromolecules in the size range of 0.5 to 1 MDa were identified accurately.

Once macromolecules are identified in ET reconstructions, classification and averaging can be performed with the repetitive components similar to those done within the well-developed field of single-particle analysis in order to improve the signal-to-noise ratio and the resolution of the extracted components (4). These molecular averages can subsequently replace the original features, resulting in a hybrid reconstruction with certain features having improved resolution. This hybrid procedure has been used in ET studies of herpes simplex virions (30) and of the nuclear pore complexes in intact nuclei (5). In addition to the issue of improved resolution via averaging, there is the attraction of being able to investigate supercomplexes, e.g., the mitochondrial electron transport chain "respirasome" (15) or functional associations, sometimes called the "interactome" (3), in their working environments inside the cell or organelle.

Teletomography and Databases

Because the higher-voltage TEMs most commonly used for ET are expensive, relatively few are available for the biological sciences (see Table 1). To make ET more accessible to those who would like to take advantage of it for their research, teletomography, or more generally telescience, was developed for the remote operation of microscopes and online image processing (32). One example is the "Telescience Portal" (**https://telescience.ucsd.edu/**), a fully integrated, web-based tool for performing end-to-end ET from acquisition to image processing to analysis of tomographic datasets at remote sites. New capabilities are regularly added to this portal to make it easier for the remote user to operate, to connect to parallel computer clusters, thus progressing to near-real-time reconstruction and high throughput, and to add functionality as new software becomes available.

A natural result of the automation of data acquisition, teletomography, and high-throughput processing has been the explosion in the amount of tomographic data that is now generated. The establishment of dedicated databases for ET has not only aided the organization of these data, but also has provided the capability for datamining (28). One such example of a database resource particularly in the area of ET is the cell-centered database (CCDB) (58, 59, 61). Through the CCDB, researchers can access volume reconstructions, "raw" data that went into the creation of the reconstructions, and segmented products from light and electron microscopies applied to cell and organelle structure and macromolecular distribution. Datamining is made available via keyword searches and enhanced by linkages to other archival databases in biomedical research.

EXAMPLES

There is an extensive base of literature covering studies of biological structure by ET, and we limit our discussion in this review to two systems, mitochondria and the relationship between the endoplasmic reticulum (ER) and the Golgi. The applications of ET to many other cellular structures including motile assemblies (9, 66, 78, 82, 104), microtubules and the mitotic spindle (64, 79, 80, 77, 95), and macromolecular complexes are covered in other excellent recent reviews (45, 68).

Mitochondria

Mitochondria were among the first intracellular organelles to be studied extensively by electron microscopy as the development of techniques of specimen preparation of cells and tissues for TEM paralleled recognition of the importance of mitochondria in energy metabolism. Both Palade and Sjostrand (81, 96) recognized the complex double-membrane topology of mitochondria, and their early work eventually led to the "baffle" model of mitochondrial membrane structure commonly described in textbooks. The baffle model depicts the outer membrane defining the mitochondrial compartment with the inner membrane surrounding the matrix, the innermost mitochondrial compartment. The inner membrane has a much larger surface area than the outer membrane, and in the baffle model the inner membrane is accommodated within the mitochondrial compartment through the formation of broad folds forming baffle-like cristae projecting into the matrix. A serial thin-section study by Daems & Wisse (11) offered an alternative that was not popularly accepted, although it contained many of the same features of current models derived from ET. In studying isolated mitochondria in different respiration states, Hackenbrock (31) observed two conformations controlled by the matrix volume. The orthodox conformation is normally observed in situ and is characterized by a relatively large matrix volume, whereas the condensed conformation is often observed in isolated mitochondria and is characterized by a smaller matrix volume.

Mannella's group (50, 51) was the first to study mitochondrial structure by ET, and they observed a more complex inner membrane structure with both tubular and lamellar cristae but lacking the broad folds characteristic of the baffle model. Their work and that of Perkins et al. have led to the currently accepted crista junction model of orthodox mitochondria (23, 23a, 46, 48, 52, 72, 83, 85, 87, 88). Most mitochondria observed in situ are in the orthodox conformation, with a large matrix volume that pushes one component of the inner membrane called the inner boundary membrane (IBM) against the outer membrane with a narrow space between that is approximately 8 nm wide. The remaining inner membrane projects into the matrix at discrete loci called crista junctions of uniform diameter (**Figure 3**). Multiple tubular cristae appear to merge, forming lamellar cristae. In cells with a large amount of inner membrane the lamellar cristae form parallel stacks reminiscent of the baffles in the earlier mitochondrial models (88) (see **Figure 3** and the supplemental movies of mitochondria models). Steroid-producing human Leydig cells display this stacked lamellar cristae morphology, but with tubular extensions perpendicular to the plane of the lamellae connecting adjacent cristae (90). During isolation from cell homogenates, the matrix of mitochondria generally condenses to varying degrees, altering the inner membrane conformation in these condensed mitochondria. Moderate condensation of the matrix increases the volume of the lamellar cristae, leaving much of the IBM closely apposed to the outer membrane, whereas in more extremely condensed mitochondria the IBM is pulled away from the outer membrane (52, 91, 108).

Although the shapes of mitochondria and cristae are remarkably pleiomorphic, the study of mitochondria, both in situ and isolated, from a variety of cell types and species and with a variety of specimen preparation techniques has led to some consistent structural characteristics (23, 87). The inner membrane of orthodox mitochondria is divided into two topologically distinct components: (*a*) the IBM closely apposed to the outer membrane joined at crista junctions to (*b*) the crista membrane that projects into the matrix. The IBM and crista membrane may also represent a functional compartmentation with enzymes involved in electron transport and ATP synthesis concentrated in the crista membrane compartment (23, 26). The crista junction

Figure 3

Computer-generated models based on segmented tomograms of mitochondria. (*a–c*) Models of a mitochondrion observed in a 0.5-μm-thick section of chick cerebellum prepared by conventional chemical fixation and embedding. (*a*) All cristae shown in yellow, with the inner boundary membrane in turquoise and the outer membrane in dark blue. (*b*) As in (*a*) but with only four representative cristae shown in red, yellow, green, and gray. Reproduced, with permission, from Reference 23a. (*c*) The model is rotated approximately 180° showing the inner boundary membrane in translucent light blue and one crista containing tubular and lamellar components in yellow; note the circular crista junctions joining the crista membrane with the inner boundary membrane, including three with short tubular components (*arrows*). Reproduced, with permission, from Reference 89. (*d*) Mitochondrion reconstructed from a section of brown adipose tissue prepared by conventional chemical fixation showing a series of stacked lamellar cristae that all connect to the inner boundary membrane through short circular crista junctions (not shown). Colors as in (*a*) but with outer membrane and inner boundary membrane rendered translucent. (*e*) Mitochondrion reconstructed from brown adipose tissue prepared by cryo-fixation followed by freeze substitution. Colors as in (*a*), but in this specimen the outer membrane and the inner boundary membrane could not be resolved. Also see the supplemental movies of mitochondria models.

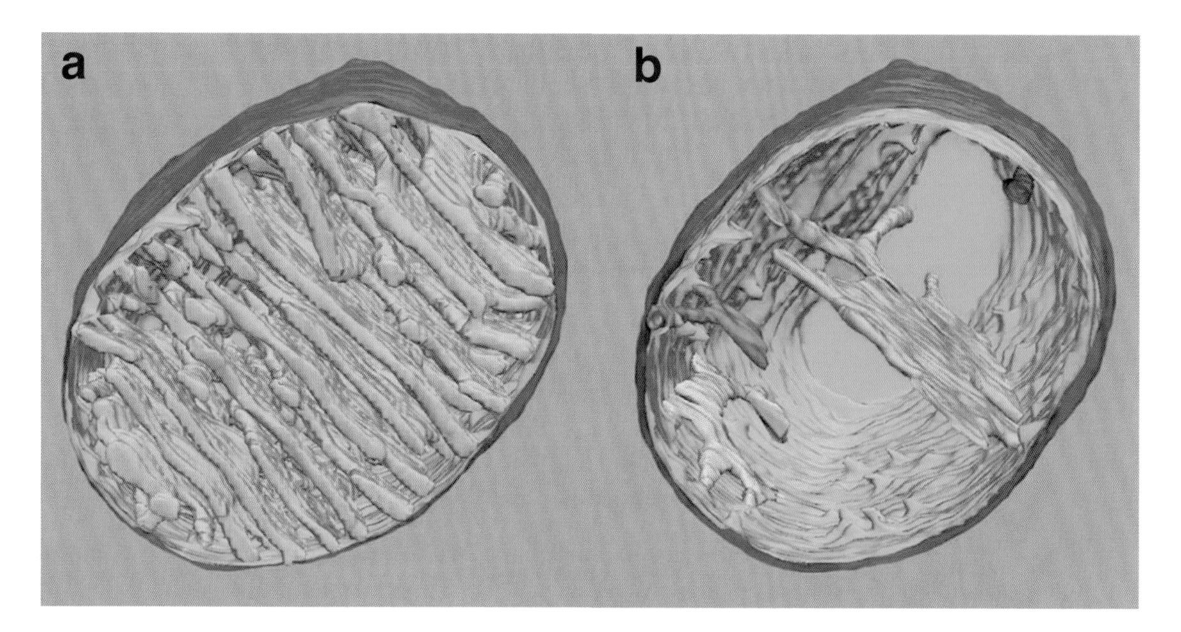

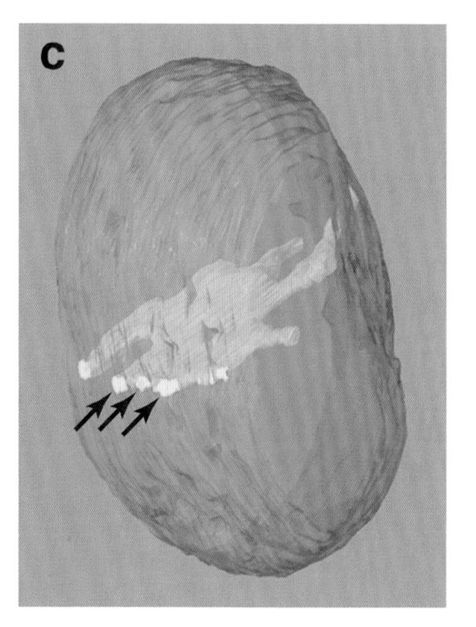

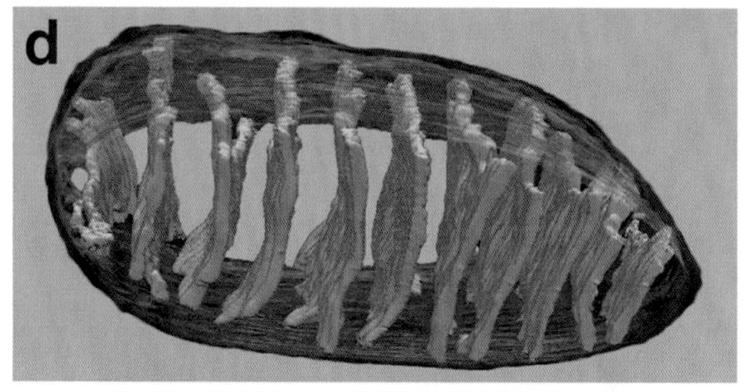

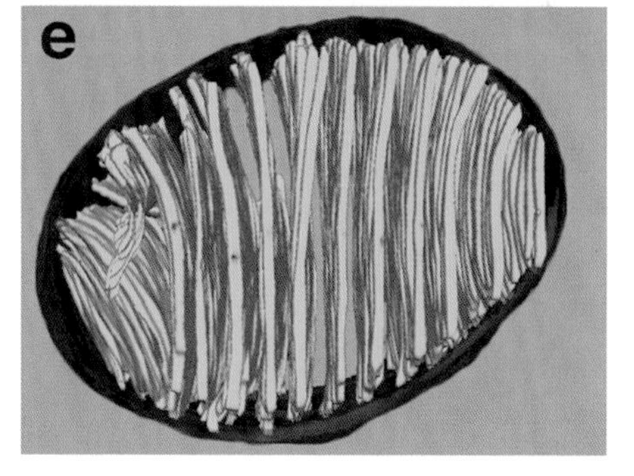

morphology may also support chemical gradients of ADP and other molecules in actively respiring mitochondria (47, 52). With a few exceptions, crista junctions are roughly circular in shape and of fairly uniform maximum diameter, 28 nm ± 6 nm for orthodox mitochondria observed in neuronal tissue, brown adipose tissue, and *Neurospora crassa* (86–88). Notable exceptions are the elongated crista junctions found in *N. crassa* mitochondria (74, 87). The shapes of cristae and the number of crista junctions are variable and dynamic in nature. In studying the changes in mitochondria following the inhibition of protein import in *Neurospora* mitochondria, Perkins et al. (87) observed that the first effect was a loss of crista membrane followed by a decrease in the size of mitochondria. As the area of crista membrane decreased, the number of crista junctions decreased linearly until few crista and crista junctions remained (see supplemental images and movies of *Neurospora* mitochondria). Mannella et al. (52) observed the dynamic nature of the mitochondrial inner membrane conformation in the loss and formation of crista junctions during large-scale swelling followed by condensation of purified yeast mitochondria. The lengths of tubular crista projecting from crista junctions are variable in neural tissue and liver (50, 83) but short in brown adipose tissue and in *N. crassa* (87, 88) (**Figure 3**).

Studies of mitochondria have been conducted mostly on conventionally prepared specimens; however, many of the features observed under these conditions have been confirmed by examination of cryo-fixed specimens prepared by freeze substitution or observed in the frozen-hydrated state. Brown adipose tissue cryo-fixed by freeze slamming followed by freeze substitution was similar to identical samples prepared by conventional chemical fixation (88) (**Figure 3*d*,*e*** and supplemental movies of mitochondria models from brown adipose tissue). Isolated mitochondria imaged in the frozen-hydrated state were in many ways similar to those prepared by conventional chemical fixation, displaying the same distribution of inner membrane into IBM and crista membrane, a well-defined narrow space between the outer membrane and the IBM, and similar contact sites between the outer membrane and the IBM (49, 74). Furthermore, the measurements of membrane dimensions and of the spaces between the outer membrane and inner membrane and the intracristal space are similar. But tomograms of the frozen-hydrated mitochondria also had some notable differences: less wavy membranes, significantly more elliptical crista junctions, and branching tripartite cristae that have not been observed in chemically fixed samples or in samples cryo-fixed and prepared by freeze substitution (74). Although frozen-hydrated specimens are expected to preserve more accurately the true biological structure, in this case the mitochondria were isolated from cells and compressed within a layer of water prior to being frozen, which may cause distortion of structure.

In addition to their central role in eukaryotic energy metabolism, mitochondria also play a critical role in initiating intrinsic apoptosis and in regulating extrinsic apoptosis through the release of cytochrome *c* and other proteins from the intermembrane and intracristal spaces (27). ET is a valuable tool in clarifying possible mechanisms of this release by demonstrating in cell-free and in vivo models of apoptosis that cytochrome *c* release is not effected by mitochondrial swelling and rupture of the outer membrane but apparently through the formation of large pores in the outer membrane through which intact cytochrome *c* can pass (34, 38, 108). Scorrano et al. (94) have observed by ET an inner membrane "remodeling" in purified mitochondria that they believe to be critical in the complete release of cytochrome *c* from the intracristal spaces during apoptosis. More recently, we have observed conversion of typical orthodox mitochondria with crista junctions and lamellar cristae into a form that ET reveals to contain large abundant vesicular cristae during cytochrome *c* release initiated in HeLa cells by treatment with etoposide (M. Sun & T. Frey, unpublished data).

The consistent structural features observed in tomograms of mitochondria have stimulated attempts to create theoretical models predicting the size and shapes of inner membrane structures. Renken et al. (92) created a thermodynamic model predicting the shape of crista junctions as an energetically most probable conformation based on the energy required to bend a membrane as formulated by Helfrich (35), with bending characterized by the two principal curvatures at a point on the membrane surface. The observation that the sizes of crista junctions vary with the matrix volume—condensed mitochondria have larger crista junctions, whereas swollen mitochondria have smaller crista junctions—suggested that the matrix volume controls the size of crista junctions and the diameters of tubular cristae (91, 92). Ponnuswamy et al. modeled the effects of matrix outward pressure on the energetics of crista tubule diameter (89), and on the basis of this model, the differences in crista junction diameters observed by Perkins et al. (84) in the mitochondria of mouse retinal rod and cone cells could be the result of small differences in matrix pressure. Deng et al. (12, 13) observed that the mitochondrial inner membrane in the amoeba *Chaos carolinensis* adopts the lipid cubic phase during fasting, a morphology also observed in *Drosophila* mitochondria following acute oxidative stress (109). The theoretical models ignore the contributions of proteins that are involved in controlling mitochondrial membrane shapes. John et al. (37) found that downregulation of mitofilin, a mitochondrial protein of unknown function, by small interfering RNA dramatically affects inner membrane conformation.

Endomembrane System

The endomembrane system is by far the most intricate organelle system within eukaryotic cells. However, the complex interactions between the ER, Golgi stack, lysosomes, endosomes, and plasma membrane are yielding to careful study by light and electron microscopy, with valuable information coming from ET. Martone et al. produced some of the first tomograms of smooth endoplasmic reticulum (SER), revealing that in Purkinje cell dendrites the SER consists of tubules and cisternae that are highly interconnected and extend throughout the dendritic shaft and the spines. Despite the interconnectivity of the SER, it contains two separate pools of Ca^{2+} released independently by IP_3 and by caffeine (62).

The Golgi apparatus has been the target of numerous tomographic investigations in recent years. Despite more than 100 years of microscopic study of the Golgi apparatus, fundamental questions regarding Golgi biology remain, e.g., what is the structural basis of cargo transport through the Golgi. The Golgi has been a difficult target of study because of its three-dimensional morphological complexity and dynamic nature. The Golgi is described as a ribbon composed of a stack of closely apposed cisternae close to the nucleus surrounded by numerous vesicles and tubular membrane structures. The Golgi stack is polarized with a *cis* face most intimately related to the supply of nascient proteins from the ER and a *trans* face containing more mature glycoproteins; in mammalian cells it commonly consists of seven individual cisternae with multiple *cis* cisternae followed by medial and *trans* cisternae and then the trans Golgi network (TGN). Many competing and partially overlapping models have been published over the years to describe the movement of protein and lipid through the Golgi and the retrieval/retention of Golgi-specific molecules; the principal competing models have been cisternal progression/maturation versus vesicular transport. Careful tomographic analysis, in combination with light microscopy techniques, is modifying the textbook description of traffic within the endomembrane systems via coated vesicles of various types (COPI, COPII, and clathrin) and of the movement through the various cisternae of the Golgi.

Mogelsvang et al. (71) investigated the Golgi structure of the budding yeast *Pichia pastoris*, which possesses multiple Golgi stacks less complex than those of mammalian organisms, each adjacent to the ER and consisting of three to four cisternae identified as *cis* (C1), medial (C2), and *trans*/TGN (C3/C4). In *P. pastoris* it is possible to reconstruct an entire Golgi stack and its associated ER domains within a single tomogram. They used this system to test and confirm three predictions of the cisternal progression model for movement of cargo though the Golgi stack. They found that the precursor of new *cis*-cisternae is a group of COPII and COPI vesicles and that the cisternal morphology and composition changes progressively from *cis*- to medial- to *trans*-cisternae. Finally, the *trans*-cisterna and TGN dissociate from the stack as they mature into secretory carriers. Some of the cisternae displayed the small holes, or fenestrae, found in mammalian Golgi but with varying patterns. C1 contained virtually no fenestrae, while the fenestrae increased from C2 to C4 and were peripherally located in C2 and noticeably larger in C4.

Mironov et al. (70) employed correlated light and electron microscopy and ET to study in fibroblasts the formation and the ultrastructure of carriers transporting a small diffusible cargo, vesicular stomatitis viral protein G (VSVG), and large extended cargo molecules, procollagen, that at >300 nm in length are substantially larger than the largest (85 nm) COPII coated vesicles. In this system they were able to block export and then release the block to monitor formation of carriers and their proximity to ER exit sites (ERES) by labeling of procollagen, VSVG, and COPII. ET revealed, somewhat surprisingly, that despite the large difference in size of the procollagen and VSVG cargoes, they were packaged into four classes of carriers that within each class exhibit the same size and shape characteristics irrespective of the cargo: Type I carriers are distended domains of ER extending from the vicinity of the ERES; Type II carriers are flattened and elongated (>300 nm) saccules protruding from the ER but still connected to it; Type III carriers are distensions larger than 300 nm in length within thin (50–70 nm) tubules; Type IV carriers are larger and more complicated with two to four saccules partially stacked and associated with the ER. The principal differences between procollagen carriers and VSVG carriers are in their associations with COPI and COPII. Type I and Type II procollagen carriers are not associated with either coat protein, although they are found in the vicinity of the ERES that is characterized by the presence of COPII. In contrast, VSVG is also found within the ERES, while Type I VSVG carriers are associated with COPII and Type II and Type IV VSVG carriers are associated with COPI. Irrespective of the type of cargo and the associated coat proteins, the carrier maturation process appears to be Type I → Type II → Type III or Type IV. The formation of carriers does not appear to require budding and fusion of small COPII vesicles, but COPII is required for export, perhaps functioning in concentrating other proteins to create an appropriate environment for export from the ER.

Compared with the Golgi of budding yeast described above, characterization of the structure and dynamic function of the mammalian Golgi ribbon is a much greater challenge given its size and complexity (54). Ladinsky et al. (40, 41) have calculated a three-dimensional model of a portion of the Golgi ribbon found in normal rat kidney (NRK) cells by serial tomography of four 250-nm-thick sections; features of this reconstruction were confirmed in less complete studies of Golgi in other cells. The specimens were cryo-fixed by plunge freezing in cryo-protectant

Figure 4

Surface rendered membranes of the Golgi ribbon produced from electron tomograms calculated from four serial sections of a fast frozen and freeze substituted normal rat kidney cell; reproduced, with permission, from Reference 41. Color coding is as indicated in panel *c*. (*a*) Cross-section of the Golgi ribbon showing the seven cisternae, C1 (*green*) – C7 (*red*); the *cis*-ER in light blue and the ERGIC in yellow are on top. (*b*) View of the ribbon from the *cis*-side with the *cis*-ER removed showing the various elements of the ERGIC apposed to C1. (*c*) Each layer of the ER and Golgi ribbon in succession beginning with the *cis*-ER. Small purple spheres are ribosomes. Scale bar is 250 nm. Also see images and movies of the Golgi ribbon at **http://bio3d.colorado.edu/** and the movie *Golgi* cis *Elements* (follow the Supplemental Material link from the Annual Reviews home page at **http://www.annualreviews.org**).

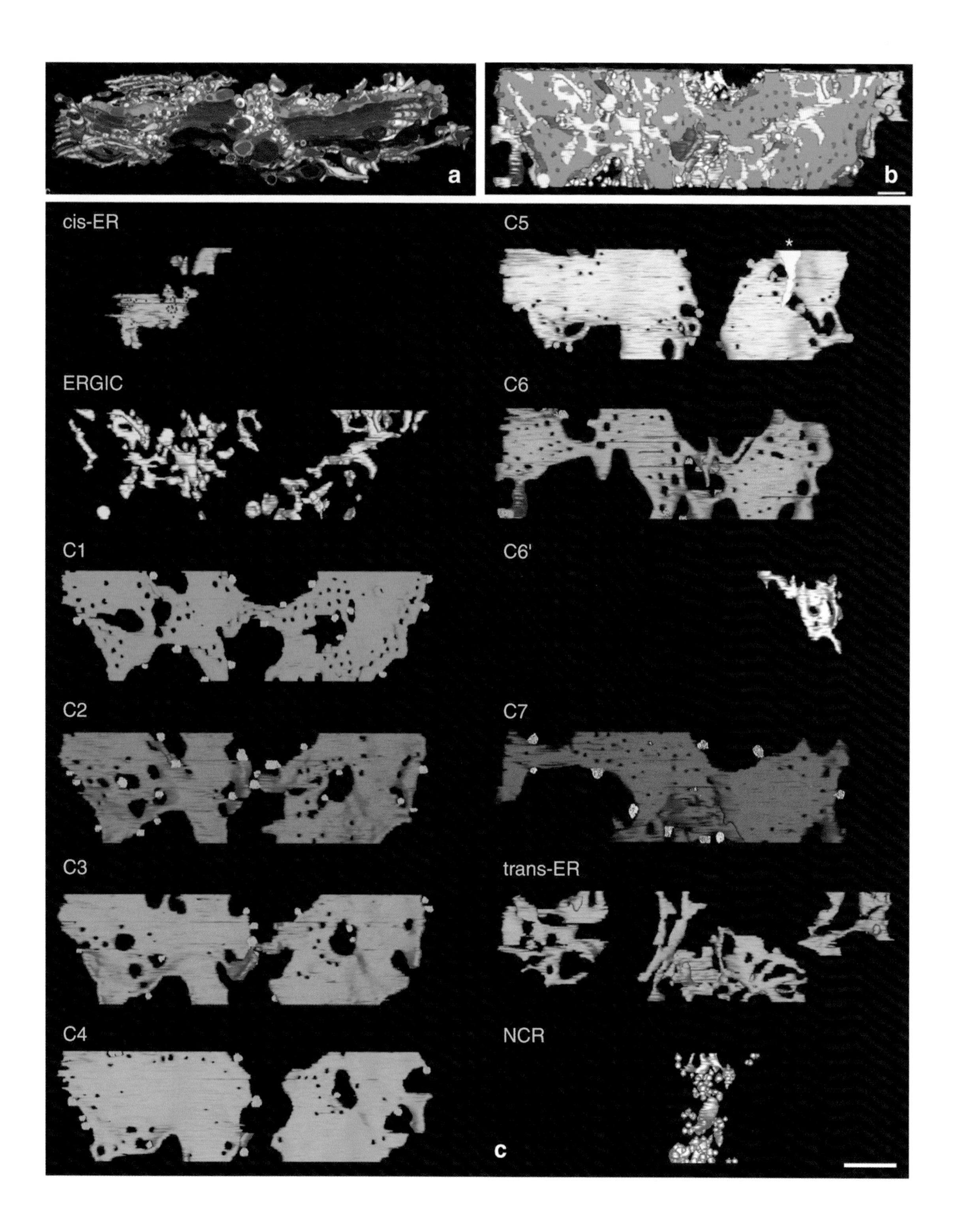
a
b
cis-ER
C5
*
ERGIC
C6
C1
C6'
C2
C7
C3
trans-ER
C4
NCR
c

followed by RF/FS. The piece of the Golgi ribbon reconstructed contained seven cisternae, C1 through C7, in two groups separated by a "noncompact region" (NCR) filled with vesicles and larger polymorphic components (**Figure 4**). C7 was more continuous than the other cisternae, continuing across the NCR without interruption. C1 and C2 were identified as *cis*-cisternae, C3 and C4 as medial-, C5 as both medial- and *trans*-, and C6 and C7 as *trans*-cisternae. The ER-Golgi intermediate compartment (ERGIC) contained branched tubular structures and flatted sacs located between the *cis*-ER and C1; the ERGIC is similar to the vesicular-tubular clusters identified in other studies and may be a precursor to a new *cis*-Golgi cisterna (56, 105). C1 through C6 have large and small holes, with the large holes of adjacent cisternae lined up to form "wells" containing vesicles (**Figure 4*c***). The small holes, or fenestrae, are also found in C7 and do not line up in adjacent cisternae. The function of the fenestrae is not clear, but they could serve to regulate the thickness of the cisternae, to facilitate budding of vesicles by increasing the local membrane curvature, or to provide pathways for diffusion of membrane lipids and proteins from one side of a cisterna to the other.

Many tubular structures connecting both equivalent and nonequivalent cisternae are observed in Golgi tomograms. Bridging tubules connect equivalent cisternae across the NCR, but cisternae are also observed to have both *trans*- and *cis*-projecting tubules that in some cases bypass adjacent cisternae, indicating that cargo may not pass through each of the seven cisternae during its journey from the ER to the plasma membrane or to the lysosomal/endosomal compartment. One significant observation was that clathrin-coated vesicles appear to bud from C7 and that nonclathrin-coated vesicles bud from C5 and C6. Thus, vesicles destined for the lysosomal/endosomal compartment apparently are sorted and transported from the *trans*-most cisterna, C7, while secretory and plasma membrane components are transported from earlier cisternae of the *trans*-Golgi, C5 and C6, both of which extend tubules into the *trans*-ER. The *trans*-ER comes in close contact with C7, although ribosomes are only present on the side of the *trans*-ER away from C7. The purpose of this close interaction is not clear but may involve transfer of lipids between the two organelles similar to that proposed in close interactions between ER and mitochondria (106). When transport was inhibited by reducing the temperature to 20°C the cisternae underwent structural changes, but the total number remained the same. The cisternae in temperature-blocked cells had fewer fenestrae, budding profiles, and tubules, and the three *trans*-cisternae exhibited bulging domains that are probably reservoirs of material blocked by the lowered temperature (42).

Using serial tomography, Marsh et al. (55) studied HIT-T15 cells actively secreting insulin and obtained similar results. The Golgi ribbon in these cells again displayed seven cisternae, but in this case C1 was highly fenestrated with a complex tubular network that appeared to be intermediate between the ERGIC and the C1 cisterna in the NRK cells studied by Ladinsky et al. (41, 42), with C2 appearing to serve as a template for the formation of a new C1. Thus, the ERGIC in NRK cells in the Ladinsky et al. study and C1 in the Marsh et al. study may represent different stages in the formation of a new *cis*-Golgi cisterna, and the C1 observed by Marsh et al. may represent the *cis*-Golgi network (CGN) observed in other studies. Connections between cisternae have been observed in cells experiencing a wave of cargo, either VSVG or procollagen (105), passing through the Golgi ribbon and in cells constitutively secreting insulin (57). In the former study cargo was accumulated at a temperature that blocked transport, and the block was then released to create a wave of cargo passing through the Golgi ribbon. This wave caused the generation of additional Golgi cisternae, the total of which increased from five to seven. The wave also generated tubular intercisternal connections at the rims of the successive cisternae, and in some cases longer tubules skipped the adjacent cisterna joining the one after; these intercisternal connections could also be observed under more physiological (partially perturbed or unperturbed) conditions. The intercisternal connections are permeable to Golgi enzymes and lipids, but not to VSVG or procollagen, which appeared to traverse the Golgi stack by cisternal progression. Thus, the intercisternal connections may function in retrograde and possibly anterograde transport of Golgi enzymes and lipids but not in transport of maturational cargo.

The results of electron tomographic studies of Golgi structure and its interactions with the ER provide support for all models of Golgi transport including cisternal progression/maturation, vesicular transport, and the role of tubular structures (54), and further reveal direct connections between nonequivalent cisternae. Thus, the mechanism for Golgi transport appears to be as complex as the structure itself. The careful application of ET with complementary techniques including correlated light microscopy and specific labeling of key components provides promise that we will eventually understand this complex process in detail.

SUMMARY POINTS

1. Electron tomography provides three-dimensional structural information in a wide range of biological systems in the "mesoscale" between 1 nm^3 and 100 μm^3.
2. Conventional chemical fixation and cryo-fixation followed by freeze substitution can yield tomograms that provide new insights into the structure and function of organelles such as mitochondria and the endomembrane system, endoplasmic reticulum, and Golgi.
3. The high resolution of electron tomography and the push to preserve samples in near-native states (frozen-hydrated) produce three-dimensional reconstructions that reveal the details of organelle structure approaching macromolecular dimensions.
4. Automated transmission electron microscope operation, new software packages for seamless image processing and analysis, and telescience open the field to the nonexpert.
5. Electron tomography is revealing never before characterized structural architectures of organelles that provide insight into their complex functions.

FUTURE ISSUES TO BE RESOLVED

1. Better methods of cryo-sectioning are required in order to calculate tomograms from sections of frozen-hydrated cells and tissues producing three-dimensional structures without fixation and staining.
2. Higher-resolution tomograms are required to facilitate identification of macromolecular complexes within electron tomograms of cells and organelles.
3. Autotracking and/or alignment without fiducial marks to automate precise alignment of images within a tilt series will decrease the time needed to calculate an electron tomogram.

4. Autosegmentation methods are required in order to decrease the time required to obtain an interpreted three-dimensional model from an electron tomogram.
5. Corrections for distortions and limitations inherent in electron optical systems that degrade tomographic reconstructions and revealed by large-area detectors, e.g., 8 K × 8 K, will provide more reliable higher-resolution tomograms.

ACKNOWLEDGMENTS

The authors acknowledge Dr. Maryann Martone at UCSD for valuable suggestions and for proofreading the manuscript and Dr. Mark S. Ladinsky at the University of Colorado for providing the originals used in **Figure 4**. Support was provided to TGF by the San Diego Foundation and to GAP and MHE by the NIH National Center for Research Resources Grant No. P41 RR04050.

LITERATURE CITED

1. Al-Amoudi A, Norlen LP, Dubochet J. 2004. Cryo-electron microscopy of vitreous sections of native biological cells and tissues. *J. Struct. Biol.* 148:131–35
2. Bajaj C, Yu Z, Auer M. 2003. Volumetric feature extraction and visualization of tomographic molecular imaging. *J. Struct. Biol.* 144:132–43
3. Baumeister W. 2002. Electron tomography: towards visualizing the molecular organization of the cytoplasm. *Curr. Opin. Struct. Biol.* 12:679–84
4. Baumeister W. 2005. From proteomic inventory to architecture. *FEBS Lett.* 579:933–37
5. Beck M, Forster F, Ecke M, Plitzko JM, Melchior F, et al. 2004. Nuclear pore complex structure and dynamics revealed by cryoelectron tomography. *Science* 306:1387–90
6. Bohm J, Frangakis AS, Hegerl R, Nickell S, Typke D, Baumeister W. 2000. Toward detecting and identifying macromolecules in a cellular context: template matching applied to electron tomograms. *Proc. Natl. Acad. Sci. USA* 97:14245–50
7. Boudier T, Lechaire JP, Frebourg G, Messaoudi C, Mory C, et al. 2005. A public software for energy filtering transmission electron tomography (EFTET-J): application to the study of granular inclusions in bacteria from Riftia pachyptila. *J. Struct. Biol.* 151:151–59
8. Bouwer JC, Mackey MR, Lawrence A, Deerinck TJ, Jones YZ, et al. 2004. Automated most-probable loss tomography of thick selectively stained biological specimens with quantitative measurement of resolution improvement. *J. Struct. Biol.* 148:297–306
9. Chen LF, Winkler H, Reedy MK, Reedy MC, Taylor KA. 2002. Molecular modeling of averaged rigor crossbridges from tomograms of insect flight muscle. *J. Struct. Biol.* 138:92–104
10. Crowther RA, DeRosier DJ, Klug A. 1970. The reconstruction of a three-dimensional structure from projections and its application to electron microscopy. *Proc. R. Soc. London Ser. A* 317:319–40
11. Daems WT, Wisse E. 1966. Shape and attachment of the cristae mitochondriales in mouse hepatic cell mitochondria. *J. Ultrastruct. Res.* 16:123–40
12. Deng Y, Kohlwein SD, Mannella CA. 2002. Fasting induces cyanide-resistant respiration and oxidative stress in the amoeba Chaos carolinensis: implications for the cubic structural transition in mitochondrial membranes. *Protoplasma* 219:160–67

13. Deng Y, Marko M, Buttle KF, Leith A, Mieczkowski M, Mannella CA. 1999. Cubic membrane structure in amoeba (Chaos carolinensis) mitochondria determined by electron microscopic tomography. *J. Struct. Biol.* 127:231–39
14. Dubochet J, Adrian M, Chang JJ, Homo JC, Lepault J, et al. 1988. Cryo-electron microscopy of vitrified specimens. *Q. Rev. Biophys.* 21:129–228
15. Dudkina NV, Eubel H, Keegstra W, Boekema EJ, Braun HP. 2005. Structure of a mitochondrial supercomplex formed by respiratory-chain complexes I and III. *Proc. Natl. Acad. Sci. USA* 102:3225–29
16. Fernandez JJ, Li S. 2003. An improved algorithm for anisotropic nonlinear diffusion for denoising cryo-tomograms. *J. Struct. Biol.* 144:152–61
17. Frangakis AS, Bohm J, Forster F, Nickell S, Nicastro D, et al. 2002. Identification of macromolecular complexes in cryoelectron tomograms of phantom cells. *Proc. Natl. Acad. Sci. USA* 99:14153–58
18. Frangakis AS, Forster F. 2004. Computational exploration of structural information from cryo-electron tomograms. *Curr. Opin. Struct. Biol.* 14:325–31
19. Frank J. 1992. *Electron Tomography*. New York: Plenum. 399 pp.
20. Frank J. 1996. *Three-Dimensional Electron Microscopy of Macromolecular Assemblies*. San Diego: Academic. 342 pp.
21. Frank J, Radermacher M, Penczek P, Zhu J, Li Y, et al. 1996. SPIDER and WEB: processing and visualization of images in 3D electron microscopy and related fields. *J. Struct. Biol.* 116:190–99
22. Frank J, Wagenknecht T, McEwen BF, Marko M, Hsieh CE, Mannella CA. 2002. Three-dimensional imaging of biological complexity. *J. Struct. Biol.* 138:85–91
23. Frey T, Renken C, Perkins G. 2002. Insight into mitochondrial structure and function from electron tomography. *Biochim. Biophys. Acta* 1555:196–203
23a. Frey TG, Mannella CA. 2000. The internal structure of mitochondria. *Trends Biochem. Sci.* 25:319–24
24. Gaietta G, Deerinck TJ, Adams SR, Bouwer J, Tour O, et al. 2002. Multicolor and electron microscopic imaging of connexin trafficking. *Science* 296:503–7
25. Giepmans BN, Deerinck TJ, Smarr BL, Jones YZ, Ellisman MH. 2005. Correlated light and electron microscopic imaging of multiple endogenous proteins using quantum dots. *Nat. Methods* 2:743–49
26. Gilkerson RW, Selker JM, Capaldi RA. 2003. The cristal membrane of mitochondria is the principal site of oxidative phosphorylation. *FEBS Lett.* 546:355–58
27. Green DR, Kroemer G. 2004. The pathophysiology of mitochondrial cell death. *Science* 305:626–29
28. Grethe JS, Baru C, Gupta A, James M, Ludaescher B, et al. 2005. Biomedical informatics research network: building a national collaboratory to hasten the derivation of new understanding and treatment of disease. *Stud. Health Technol. Inform.* 112:100–9
29. Griffin BA, Adams SR, Tsien RY. 1998. Specific covalent labeling of recombinant protein molecules inside live cells. *Science* 281:269–72
30. Grunewald K, Desai P, Winkler DC, Heymann JB, Belnap DM, et al. 2003. Three-dimensional structure of herpes simplex virus from cryo-electron tomography. *Science* 302:1396–98
31. Hackenbrock CR. 1966. Ultrastructural bases for metabolically linked mechanical activity in mitochondria. I. Reversible ultrastructural changes with change in metabolic steady state in isolated liver mitochondria. *J. Cell Biol.* 30:269–97
32. Hadida-Hassan M, Young SJ, Peltier ST, Wong M, Lamont S, Ellisman MH. 1999. Web-based telemicroscopy. *J. Struct. Biol.* 125:235–45

33. Harlow ML, Ress D, Stoschek A, Marshall RM, McMahan UJ. 2001. The architecture of active zone material at the frog's neuromuscular junction. *Nature* 409:479–84
34. He L, Perkins GA, Poblenz AT, Harris JB, Hung M, et al. 2003. Bcl-xL overexpression blocks bax-mediated mitochondrial contact site formation and apoptosis in rod photoreceptors of lead-exposed mice. *Proc. Natl. Acad. Sci. USA* 100:1022–27
35. Helfrich W. 1973. Elastic properties of lipid bilayers: theory and possible experiments. *Z. Naturforsch. C* 28:693–703
36. Hsieh CE, Marko M, Frank J, Mannella CA. 2002. Electron tomographic analysis of frozen-hydrated tissue sections. *J. Struct. Biol.* 138:63–73
37. John GB, Shang Y, Li L, Renken C, Mannella CA, et al. 2005. The mitochondrial inner membrane protein mitofilin controls cristae morphology. *Mol. Biol. Cell* 16:1543–54
38. Kluck RM, Exposti MD, Perkins G, Renken C, Kuwana T, et al. 1999. The pro-apoptotic proteins, Bid and Bax, cause a limited permeabilization of the mitochondrial outer membrane that is enhanced by cytosol. *J. Cell Biol.* 147:809–22
39. Kremer JR, Mastronarde DN, McIntosh JR. 1996. Computer visualization of three-dimensional image data using IMOD. *J. Struct. Biol.* 116:71–76
40. Ladinsky MS, Kremer JR, Furcinitti PS, McIntosh JR, Howell KE. 1994. HVEM tomography of the trans-Golgi network: structural insights and identification of a lace-like vesicle coat. *J. Cell Biol.* 127:29–38
41. Ladinsky MS, Mastronarde DN, McIntosh JR, Howell KE, Staehelin LA. 1999. Golgi structure in three dimensions: functional insights from the normal rat kidney cell. *J. Cell Biol.* 144:1135–49
42. Ladinsky MS, Wu CC, McIntosh S, McIntosh JR, Howell KE. 2002. Structure of the Golgi and distribution of reporter molecules at 20 degrees C reveals the complexity of the exit compartments. *Mol. Biol. Cell* 13:2810–25
43. Leapman RD, Kocsis E, Zhang G, Talbot TL, Laquerriere P. 2004. Three-dimensional distributions of elements in biological samples by energy-filtered electron tomography. *Ultramicroscopy* 100:115–25
44. Li Y, Leith A, Frank J. 1997. Tinkerbell—a tool for interactive segmentation of 3D data. *J. Struct. Biol.* 120:266–75
45. Lucic V, Forster F, Baumeister W. 2005. Structural studies by electron tomography: from cells to molecules. *Annu. Rev. Biochem.* 74:833–65
46. Mannella CA. 2000. Our changing views of mitochondria. *J. Bioenerg. Biomembr.* 32:1–4
47. Mannella CA. 2006. The relevance of mitochondrial membrane topology to mitochondrial function. *Biochim. Biophys. Acta.* 1762:140–47
48. Mannella CA, Buttle K, Rath BK, Marko M. 1998. Electron microscopic tomography of rat-liver mitochondria and their interaction with the endoplasmic reticulum. *BioFactors* 8:225
49. Mannella CA, Hsieh CE, Marko M. 1999. Electron microscopic tomography of whole, frozen-hydrated rat-liver mitochondria at 400 kV. In *Proceedings of Microscopy and Microanalysis*, ed. GW Bailey, WG Jerome, S McKernan, JF Mansfield, RL Price, pp. 416–17. New York: Springer-Verlag
50. Mannella CA, Marko M, Buttle K. 1997. Reconsidering mitochondrial structure: new views of an old organelle. *Trends Biochem. Sci.* 22:37–38
51. Mannella CA, Marko M, Penczek P, Barnard D, Frank J. 1994. The internal compartmentation of rat-liver mitochondria: tomographic study using the high-voltage transmission electron microscope. *Microsc. Res. Tech.* 27:278–83

52. Mannella CA, Pfeiffer DR, Bradshaw PC, Moraru II, Slepchenko B, et al. 2001. Topology of the mitochondrial inner membrane: dynamics and bioenergetic implications. *IUBMB Life* 52:93–100
53. Marco S, Boudier T, Messaoudi C, Rigaud JL. 2004. Electron tomography of biological samples. *Biochemistry* 69:1219–25
54. Marsh BJ. 2005. Lessons from tomographic studies of the mammalian Golgi. *Biochim. Biophys. Acta* 1744:273–92
55. Marsh BJ, Mastronarde DN, Buttle KF, Howell KE, McIntosh JR. 2001. Organellar relationships in the Golgi region of the pancreatic beta cell line, HIT-T15, visualized by high resolution electron tomography. *Proc. Natl. Acad. Sci. USA* 98:2399–406
56. Marsh BJ, Mastronarde DN, McIntosh JR, Howell KE. 2001. Structural evidence for multiple transport mechanisms through the Golgi in the pancreatic beta-cell line, HIT-T15. *Biochem. Soc. Trans.* 29:461–67
57. Marsh BJ, Volkmann N, McIntosh JR, Howell KE. 2004. Direct continuities between cisternae at different levels of the Golgi complex in glucose-stimulated mouse islet beta cells. *Proc. Natl. Acad. Sci. USA* 101:5565–70
58. Martone ME, Gupta A, Ellisman MH. 2004. E-neuroscience: challenges and triumphs in integrating distributed data from molecules to brains. *Nat. Neurosci.* 7:467–72
59. Martone ME, Gupta A, Wong M, Qian X, Sosinsky G, et al. 2002. A cell-centered database for electron tomographic data. *J. Struct. Biol.* 138:145–55
60. Martone ME, Hu BR, Ellisman MH. 2000. Alterations of hippocampal postsynaptic densities following transient ischemia. *Hippocampus* 10:610–16
61. Martone ME, Zhang S, Gupta A, Qian X, He H, et al. 2003. The cell-centered database: a database for multiscale structural and protein localization data from light and electron microscopy. *Neuroinformatics* 1:379–95
62. Martone ME, Zhang Y, Simpliciano VM, Carragher BO, Ellisman MH. 1993. Three-dimensional visualization of the smooth endoplasmic reticulum in Purkinje cell dendrites. *J. Neurosci.* 13:4636–46
63. Mastronarde DN. 1997. Dual-axis tomography: an approach with alignment methods that preserve resolution. *J. Struct. Biol.* 120:343–52
64. McEwen BF, Marko M. 1999. Three-dimensional transmission electron microscopy and its application to mitosis research. *Methods Cell Biol.* 61:81–111
65. McEwen BF, Marko M. 2001. The emergence of electron tomography as an important tool for investigating cellular ultrastructure. *J. Histochem. Cytochem.* 49:553–64
66. McEwen BF, Marko M, Hsieh CE, Mannella C. 2002. Use of frozen-hydrated axonemes to assess imaging parameters and resolution limits in cryoelectron tomography. *J. Struct. Biol.* 138:47–57
67. McIntosh JR. 2001. Electron microscopy of cells: a new beginning for a new century. *J. Cell Biol.* 153:F25–32
68. McIntosh R, Nicastro D, Mastronarde D. 2005. New views of cells in 3D: an introduction to electron tomography. *Trends Cell Biol.* 15:43–51
69. Medalia O, Weber I, Frangakis AS, Nicastro D, Gerisch G, Baumeister W. 2002. Macromolecular architecture in eukaryotic cells visualized by cryoelectron tomography. *Science* 298:1209–13
70. Mironov AA, Mironov AA Jr, Beznoussenko GV, Trucco A, Lupetti P, et al. 2003. ER-to-Golgi carriers arise through direct en bloc protrusion and multistage maturation of specialized ER exit domains. *Dev. Cell* 5:583–94

71. Mogelsvang S, Gomez-Ospina N, Soderholm J, Glick BS, Staehelin LA. 2003. Tomographic evidence for continuous turnover of Golgi cisternae in Pichia pastoris. *Mol. Biol. Cell* 14:2277–91
72. Morau II, Slepchenko BM, Mannella CA, Loew LM. 2000. Role of cristae morphology in regulating mitochondrial adenine nucleotide and H+ dynamics. *Biophys. J.* 78:194A
73. Murk JL, Posthuma G, Koster AJ, Geuze HJ, Verkleij AJ, et al. 2003. Influence of aldehyde fixation on the morphology of endosomes and lysosomes: quantitative analysis and electron tomography. *J. Microsc.* 212:81–90
74. Nicastro D, Frangakis AS, Typke D, Baumeister W. 2000. Cryo-electron tomography of Neurospora mitochondria. *J. Struct. Biol.* 129:48–56
75. Nickell S, Forster F, Linaroudis A, Net WD, Beck F, et al. 2005. TOM software toolbox: acquisition and analysis for electron tomography. *J. Struct. Biol.* 149:227–34
76. Ofverstedt LG, Zhang K, Isaksson LA, Bricogne G, Skoglund U. 1997. Automated correlation and averaging of three-dimensional reconstructions obtained by electron tomography. *J. Struct. Biol.* 120:329–42
77. Otegui MS, Mastronarde DN, Kang BH, Bednarek SY, Staehelin LA. 2001. Three-dimensional analysis of syncytial-type cell plates during endosperm cellularization visualized by high resolution electron tomography. *Plant Cell* 13:2033–51
78. O'Toole ET, Giddings TH, McIntosh JR, Dutcher SK. 2003. Three-dimensional organization of basal bodies from wild-type and delta-tubulin deletion strains of Chlamydomonas reinhardtii. *Mol. Biol. Cell* 14:2999–3012
79. O'Toole ET, McDonald KL, Mantler J, McIntosh JR, Hyman AA, Muller-Reichert T. 2003. Morphologically distinct microtubule ends in the mitotic centrosome of Caenorhabditis elegans. *J. Cell Biol.* 163:451–56
80. O'Toole ET, Winey M, McIntosh JR. 1999. High-voltage electron tomography of spindle pole bodies and early mitotic spindles in the yeast Saccharomyces cerevisiae. *Mol. Biol. Cell* 10:2017–31
81. Palade G. 1952. The fine structure of mitochondria. *Anat. Rec.* 114:427–51
82. Pascual-Montano A, Taylor KA, Winkler H, Pascual-Marqui RD, Carazo JM. 2002. Quantitative self-organizing maps for clustering electron tomograms. *J. Struct. Biol.* 138:114–22
83. Perkins G, Renken C, Martone ME, Young SJ, Ellisman M, Frey T. 1997. Electron tomography of neuronal mitochondria: three-dimensional structure and organization of cristae and membrane contacts. *J. Struct. Biol.* 119:260–72
84. Perkins GA, Ellisman MH, Fox DA. 2003. Three-dimensional analysis of mouse rod and cone mitochondrial cristae architecture: bioenergetic and functional implications. *Mol. Vis.* 9:60–73
85. Perkins GA, Frey TG. 2000. Recent structural insight into mitochondria gained by microscopy. *Micron* 31:97–111
86. Perkins GA, Renken CW, Song JY, Frey TG, Young SJ, et al. 1997. Electron tomography of large, multicomponent biological structures. *J. Struct. Biol.* 120:219–27
87. Perkins GA, Renken CW, van der Klei IJ, Ellisman MH, Neupert W, Frey TG. 2001. Electron tomography of mitochondria after the arrest of protein import associated with Tom 19 depletion. *Eur. J. Cell Biol.* 80:139–50
88. Perkins GA, Song JY, Tarsa L, Deerinck TJ, Ellisman MH, Frey TG. 1998. Electron tomography of mitochondria from brown adipocytes reveals crista junctions. *J. Bioenerg. Biomembr.* 30:431–42
89. Ponnuswamy A, Nulton J, Mahaffy JM, Salamon P, Frey TG, Baljon ARC. 2005. Modelling tubular shapes of mitochondrial membranes. *Physical Biol.* 2:73–79

90. Prince FP, Buttle KF. 2004. Mitochondrial structure in steroid-producing cells: three-dimensional reconstruction of human Leydig cell mitochondria by electron microscopic tomography. *Anat. Rec.* 278A:454–61
91. Renken C. 2004. *The Structure of Mitochondria*. San Diego, CA: San Diego State Univ. Press. 200 pp.
92. Renken C, Siragusa G, Perkins G, Washington L, Nulton J, et al. 2002. A thermodynamic model describing the nature of the crista junction: a structural motif in the mitochondrion. *J. Struct. Biol.* 138:137–44
93. Ress DB, Harlow ML, Marshall RM, McMahan UJ. 2004. Methods for generating high-resolution structural models from electron microscope tomography data. *Structure* 12:1763–74
94. Scorrano L, Ashiya M, Buttle K, Weiler S, Oakes SA, et al. 2002. A distinct pathway remodels mitochondrial cristae and mobilizes cytochrome c during apoptosis. *Dev. Cell* 2:55–67
95. Segui-Simarro JM, Austin JR 2nd, White EA, Staehelin LA. 2004. Electron tomographic analysis of somatic cell plate formation in meristematic cells of Arabidopsis preserved by high-pressure freezing. *Plant Cell* 16:836–56
96. Sjostrand FS. 1956. The ultrastructure of cells as revealed by the electron microscope. *Int. Rev. Cytol.* 5:455–533
97. Skoglund U, Ofverstedt LG, Burnett RM, Bricogne G. 1996. Maximum-entropy three-dimensional reconstruction with deconvolution of the contrast transfer function: a test application with adenovirus. *J. Struct. Biol.* 117:173–88
98. Sorzano CO, Marabini R, Velazquez-Muriel J, Bilbao-Castro JR, Scheres SH, et al. 2004. XMIPP: a new generation of an open-source image processing package for electron microscopy. *J. Struct. Biol.* 148:194–204
99. Soto GE, Young SJ, Martone ME, Deerinck TJ, Lamont S, et al. 1994. Serial section electron tomography: a method for three-dimensional reconstruction of large structures. *Neuroimage* 1:230–43
100. Staffs TNaE. 2002. Breakthrough of the year. *Science* 298:2297–303
101. Steven AC, Aebi U. 2003. The next ice age: cryo-electron tomography of intact cells. *Trends Cell Biol.* 13:107–10
102. Subramaniam S, Milne JL. 2004. Three-dimensional electron microscopy at molecular resolution. *Annu. Rev. Biophys. Biomol. Struct.* 33:141–55
103. Takaoka A, Yoshida K, Mori H, Hayashi S, Young SJ, Ellisman MH. 2000. International telemicroscopy with a 3 MV ultrahigh voltage electron microscope. *Ultramicroscopy* 83:93–101
104. Taylor KA, Schmitz H, Reedy MC, Goldman YE, Franzini-Armstrong C, et al. 1999. Tomographic 3D reconstruction of quick-frozen, Ca2+-activated contracting insect flight muscle. *Cell* 99:421–31
105. Trucco A, Polishchuk RS, Martella O, Pentima AD, Fusella A, et al. 2004. Secretory traffic triggers the formation of tubular continuities across Golgi sub-compartments. *Nat. Cell Biol.* 6:1071–81
106. Vance JE, Shiao YJ. 1996. Intracellular trafficking of phospholipids: import of phosphatidylserine into mitochondria. *Anticancer Res.* 16:1333–39
107. Volkmann N. 2002. A novel three-dimensional variant of the watershed transform for segmentation of electron density maps. *J. Struct. Biol.* 138:123–29
108. von Ahsen O, Renken C, Perkins G, Kluck RM, Bossy-Wetzel E, Newmeyer DD. 2000. Preservation of mitochondrial structure and function after Bid- or Bax-mediated cytochrome c release. *J. Cell Biol.* 150:1027–36

109. Walker DW, Benzer S. 2004. Mitochondrial "swirls" induced by oxygen stress and in the Drosophila mutant hyperswirl. *Proc. Natl. Acad. Sci. USA* 101:10290–95
110. Winkler H, Taylor KA. 2003. Focus gradient correction applied to tilt series image data used in electron tomography. *J. Struct. Biol.* 143:24–32
111. Zheng QS, Braunfeld MB, Sedat JW, Agard DA. 2004. An improved strategy for automated electron microscopic tomography. *J. Struct. Biol.* 147:91–101
112. Ziese U, Janssen AH, Murk JL, Geerts WJ, Van der Krift T, et al. 2002. Automated high-throughput electron tomography by pre-calibration of image shifts. *J. Microsc.* 205:187–200
113. Al-Amoudi A, Dubochet J, Gnaegi H, Luthi W, Studer D. 2003. An oscillating cryo-knife reduces cutting-induced deformation of vitreous ultrathin sections. *J. Microsc.* 212:26–33

RELATED REVIEWS

Ribosome Dynamics: Insights from Atomic Structure Modeling into Cryo-Electron Microscopy Maps
Frank J, Mitra K.
Annual Review of Biophysics and Biomolecular Structure. 2006. Volume 35, pages 299–317

Electron Cryomicroscopy of Spliceosomal Components
Stark H, Lührmann R
Annual Review of Biophysics and Biomolecular Structure. 2006. Volume 35, pages 435–457

Expanding the Genetic Code

Lei Wang,[1] Jianming Xie,[2] and Peter G. Schultz[2]

[1]The Jack H. Skirball Center for Chemical Biology & Proteomics, The Salk Institute for Biological Studies, La Jolla, California 92037

[2]Department of Chemistry and The Skaggs Institute for Chemical Biology, The Scripps Research Institute, La Jolla, California 92037; email: schultz@scripps.edu

Annu. Rev. Biophys. Biomol. Struct. 2006. 35:225–49

First published online as a Review in Advance on January 13, 2006

The *Annual Review of Biophysics and Biomolecular Structure* is online at biophys.annualreviews.org

doi: 10.1146/annurev.biophys.35.101105.121507

1056-8700/06/0609-0225$20.00

Key Words

amino acid, directed evolution, protein engineering, aminoacyl-tRNA synthetase

Abstract

Recently, a general method was developed that makes it possible to genetically encode unnatural amino acids with diverse physical, chemical, or biological properties in *Escherichia coli*, yeast, and mammalian cells. More than 30 unnatural amino acids have been incorporated into proteins with high fidelity and efficiency by means of a unique codon and corresponding tRNA/aminoacyl-tRNA synthetase pair. These include fluorescent, glycosylated, metal-ion-binding, and redox-active amino acids, as well as amino acids with unique chemical and photochemical reactivity. This methodology provides a powerful tool both for exploring protein structure and function in vitro and in vivo and for generating proteins with new or enhanced properties.

Contents

INTRODUCTION

With the rare exceptions of selenocysteine (9) and pyrrolysine (72), the genetic codes of all known organisms specify the same 20 amino acid building blocks. Although a number of arguments have been put forth to explain the nature and number of amino acids in the code, it is clear that proteins require many additional chemistries, beyond the limited number of functional groups contained in the 20 amino acids, to carry out their natural functions. Thus while a 20 amino acid code is sufficient for life, it may by no means be ideal. The development of a method that allows one to genetically encode additional amino acids might enable the evolution of proteins, or even entire organisms, with new or enhanced properties.

Genetic code: consists of 64 nucleotide triplets (A, U, C, and G) that encode the 20 amino acids used in protein translation, and is universal in all known organisms

Unnatural amino acids: amino acids not specified by the existing genetic code for protein synthesis

Such methodology would also provide powerful new tools to probe protein structure and function both in vitro and in vivo. Although conventional site-directed mutagenesis has dramatically expanded our ability to manipulate protein structure, the number of substitutions that can be made remains limited. Ideally, one would like to be able to make precise changes in the steric and electronic properties of an amino acid (e.g., acidity, size, redox potential, nucleophilicity, and polarity) or introduce spectroscopic probes, posttranslational modifications, metal chelators, photoaffinity labels, and other chemical moieties at unique sites in a protein. Here, we describe an approach that makes it possible for the first time to uniquely encode additional amino acids with novel physical, chemical, or biological properties in both prokaryotic and eukaryotic organisms with high fidelity and efficiency.

BACKGROUND

Both chemical and biosynthetic strategies have been developed to incorporate unnatural amino acids into proteins. The former are simple and straightforward but are often limited by the homogeneity and/or size of the protein that can be synthesized. For example, chemical modification of amino acid side chains (13) can lead to nonselective and nonquantitative derivatization. Moreover, only a limited number of residues can be chemically modified with exogenous agents. Solid-phase peptide synthesis allows a large number of modifications to be made to protein structures but is generally limited to peptides and smaller proteins owing to the decreased yield and purity associated with the synthesis of larger proteins (46). The recent development of chemical (21) and intein-mediated (31, 36) peptide ligation allows the semisynthesis of larger proteins, but substitutions are confined largely to the N or C terminus.

A general in vitro biosynthetic method for incorporating unnatural amino acids into proteins uses nonsense or frameshift suppressor tRNAs that are chemically misacylated with the amino acid (6, 40, 62). This method can be used to synthesize proteins of virtually any size with the novel amino acid located at any site designated by the corresponding codon. Indeed, more than 80 novel amino acids have been incorporated into proteins by this methodology (18). However, the protein in cell-free translation systems yields are low and the generation of aminoacyl-tRNA is relatively complex.

The ability to incorporate unnatural amino acids directly into proteins in vivo offers considerable advantages over both chemical and in vitro biosynthetic strategies, including higher yields, fidelity, and technical ease; it also allows one to study protein structure and function both in vitro and in vivo. One such method involves growing bacteria in media in which a close structural analogue is substituted for the corresponding genetically encoded amino acid. For example, replacement of methionine with selenomethionine (38), or a common amino acid with its ^{15}N-labeled analogue (60), has been used extensively in crystallographic and NMR studies of protein structure, respectively. However, nonquantitative substitution and substitution at multiple sites throughout the protein (and proteome) can limit the utility of this technique. The use of strains that are auxotrophic in a particular amino acid minimizes the competition between the unnatural amino acid and its natural counterpart (42), but the former must nonetheless be a close analogue of the common amino acid. In some cases it is also possible to relax the substrate specificity of the aminoacyl-tRNA synthetases through active-site mutations (44) or to attenuate the proofreading activity of the synthetase (26, 74). Finally, microinjection of a chemically misacylated amber suppressor tRNA and the corresponding mutant mRNA into *Xenopus laevis* oocytes has led to the selective incorporation of unnatural amino acids into proteins (7). Unfortunately, protein yields are again low because the tRNA is chemically acylated in vitro and cannot be reacylated in vivo.

METHODOLOGY

To cotranslationally introduce an unnatural amino acid at a defined site in a protein directly in a living organism, one requires a unique tRNA-codon pair, a corresponding aminoacyl-tRNA synthetase, and significant intracellular levels of the unnatural amino acid. To ensure that the unnatural amino acid is incorporated uniquely at the site specified by its codon, the tRNA must be constructed such that it is not recognized by the endogenous aminoacyl-tRNA synthetases of the host, but functions efficiently in translation (an orthogonal tRNA). Moreover, this tRNA must deliver the novel amino acid in response to a unique codon that does not encode any of the common 20 amino acids. Another requirement for high fidelity is that the cognate aminoacyl-tRNA synthetase (an orthogonal synthetase) aminoacylates the orthogonal tRNA but none of the endogenous tRNAs. In addition, this synthetase must aminoacylate the tRNA with only the desired unnatural amino acid and no endogenous amino acids. Likewise, the unnatural amino acid cannot be a substrate for the endogenous synthetases if it is to be incorporated uniquely in response to its cognate codon. Finally, the amino acid must be transported efficiently into the cytoplasm when added to the growth medium, or biosynthesized by the host, and be stable to endogenous metabolic enzymes.

Aminoacyl-tRNA synthetase: an enzyme that catalyzes the attachment of a specific amino acid to the acceptor stem at the 3′ end of the cognate tRNA

An Orthogonal tRNA-Codon Pair

Efforts to develop such an approach focused first on *Escherichia coli* because of its ease of genetic manipulation and the large body of knowledge around its translational machinery. To uniquely specify an unnatural amino acid, one can use either nonsense (triplet) codons or frameshift (quadruplet) codons, or even construct a bacterium in which redundant codons and their corresponding tRNAs are deleted from the genome. Initially the amber nonsense codon (UAG) was used to specify the unnatural amino acid because it is the least-used stop codon in *E. coli* and *Saccharomyces cerevisiae*. Moreover, some *E. coli* strains contain natural amber suppressor tRNAs that efficiently incorporate common amino acids without significantly affecting growth rates

Orthogonal tRNA/aminoacyl-tRNA synthetase pair: the tRNA is not recognized by the endogenous aminoacyl-tRNA synthetases of the host but functions efficiently in translation; the cognate aminoacyl-tRNA synthetase aminoacylates the orthogonal tRNA but none of the endogenous tRNAs

***Mj*:** *Methanococcus jannaschii*

TyrRS: tyrosyl-tRNA synthetase

(8, 35). Amber suppressors can also be engineered, and natural or engineered amber suppressors have been used routinely for conventional protein mutagenesis in *E. coli* (63) and for the in vitro introduction of unnatural amino acids into proteins using chemically aminoacylated suppressor tRNAs. In addition, nonsense suppressors also exist in mammalian cells (27, 43) and yeast (48, 71).

In theory it should be possible to evolve an orthogonal tRNA/aminoacyl-tRNA synthetase pair from an existing bacterial pair for use in *E. coli*. This was first attempted by mutating an *E. coli* glutaminyl tRNA to a translationally competent suppressor tRNA that is no longer recognized by its cognate synthetase. Although an orthogonal tRNA was successfully generated (56), we were not able to alter the specificity of the corresponding glutaminyl-tRNA synthetase to selectively recognize this new suppressor tRNA and not the wild-type tRNA (which would result in misincorporation of the unnatural amino acid at glutamine sites) (55). An alternative strategy for generating orthogonal tRNA/aminoacyl-tRNA synthetase pairs in bacteria makes use of orthologues from archaeal bacteria and eukaryotic organisms. Some prokaryotic tRNA/aminoacyl-tRNA synthetase pairs do not cross-react to any significant degree with their eukaryotic counterparts, as a result of differences in tRNA identity elements, primarily in the acceptor stem and variable arm (25, 53). Therefore, it should be possible to generate an orthogonal tRNA/synthetase pair for *E. coli* by importing a pair from an eukaryotic organism. Although we found that the yeast amber suppressor $tRNA^{Gln}_{CUA}$ (*Sc* $tRNA^{Gln}_{CUA}$) and cognate glutaminyl-tRNA synthetase are orthogonal in *E. coli* (57), no mutant yeast GlnRS could be evolved that aminoacylates the *Sc* $tRNA^{Gln}_{CUA}$ with an unnatural amino acid in *E. coli* (possibly owing to the low intrinsic expression and/or activity of the synthetase in bacteria). Efforts to use yeast tRNA/synthetase pairs specific for various hydrophobic amino acids, as well as pairs from *Homo sapiens*, were also unsuccessful (L. Wang & P. G. Schultz, unpublished results).

Attention was then focused on archaea as a source of orthogonal tRNA/synthetase pairs for use in *E. coli*. Archaeal aminoacyl-tRNA synthetases are more similar to their eukaryotic than prokaryotic counterparts (45, 53), but unlike synthetases from eukaryotic cells, which often express poorly or have low activities in *E. coli*, synthetases from archaea can be expressed efficiently in *E. coli* in their active forms. Moreover, early work (53) indicated that most tRNAs from the halophile *Halobacterium cutirebrum* cannot be charged by *E. coli* aminoacyl-tRNA synthetases. The first orthogonal *E. coli* tRNA/synthetase pair to be generated from archaeal bacteria was derived from the tyrosyl pair from *Methanococcus jannaschii* (79). Previous experiments showed that the major recognition elements of *M. jannaschii* $tRNA^{Tyr}$ include the discriminator base A73 and the first base pair, C1-G72, in the acceptor stem; the anticodon triplet participates only weakly in identity determination (32). This pattern is the same as that of yeast $tRNA^{Tyr}$ but differs from that of *E. coli* $tRNA^{Tyr}$; the latter uses A73, G1-C72, a long variable arm, and the anticodon as identity elements (**Figure 1*a***). The *M. jannaschii* tyrosyl-tRNA synthetase (*Mj* TyrRS) also has a minimalist anticodon loop binding domain (73), which should make it possible to change the anticodon loop of its cognate tRNA to CUA with little loss in affinity by the synthetase. Finally, this aminoacyl-tRNA synthetase does not have an editing mechanism that could deacylate the attached unnatural amino acid. Indeed, an amber suppressor *M. jannaschii* $tRNA^{Tyr}_{CUA}$ (*Mj* $tRNA^{Tyr}_{CUA}$) and cognate *Mj* TyrRS were shown to function efficiently in *E. coli*, but unfortunately some degree of aminoacylation of this tRNA by endogenous *E. coli* synthetases was observed (81).

A general strategy for the evolution of orthogonal tRNAs in *E. coli* from heterologous precursors was therefore developed (82). This method consists of a combination of negative

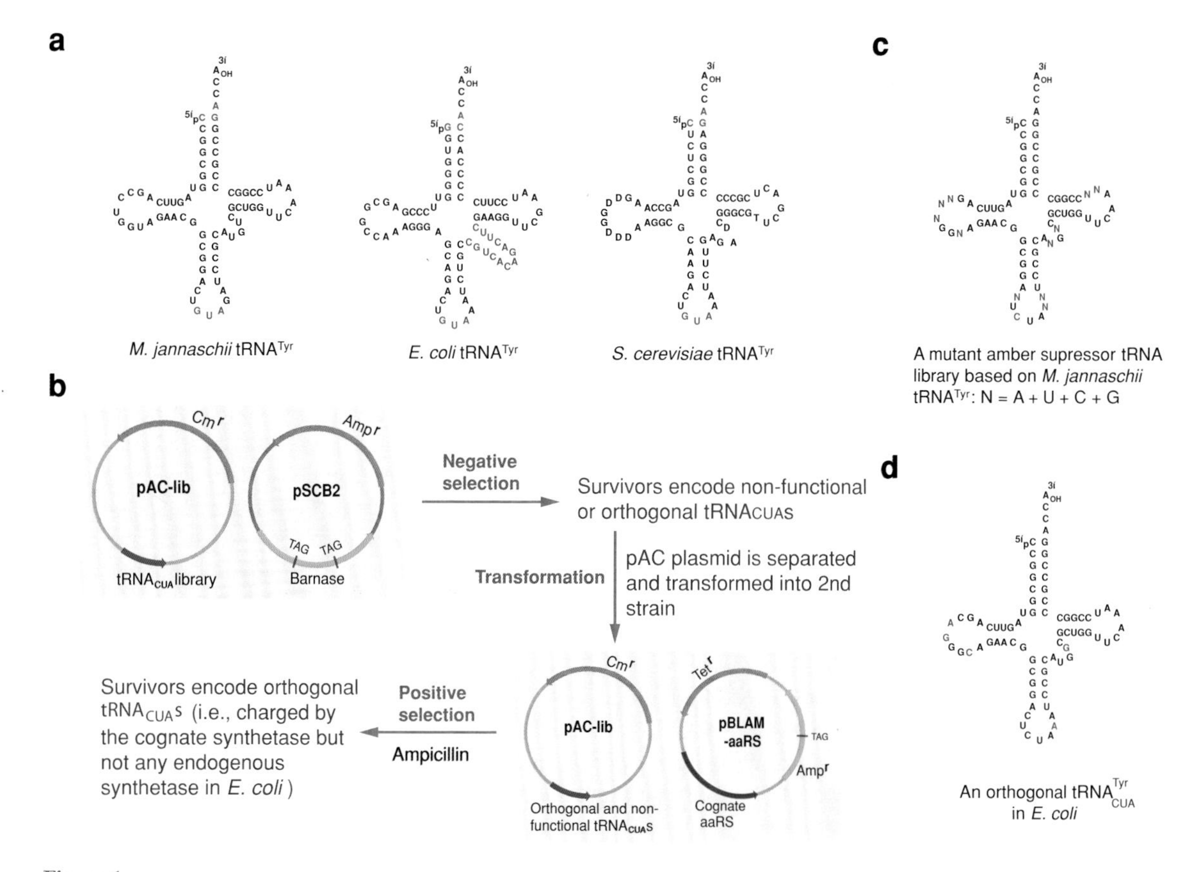

Figure 1

(*a*) Comparison of sequences of $tRNA^{Tyr}$ from three different species. The major identity elements are in red. (*b*) Directed evolution of orthogonal amber suppressor $tRNA^{Tyr}_{CUA}$ in *E. coli* by alternating negative and positive selections. (*c*) A library of amber suppressor $tRNA^{Tyr}$ was generated by randomizing 11 nucleotides (*red*) of *M. jannaschii* $tRNA^{Tyr}_{CUA}$ that do not interact directly with the cognate TyrRS. (*d*) The orthogonal amber suppressor *Mj* $tRNA^{Tyr}_{CUA}$ (changed nucleotides are in *red*).

and positive selections with a library of tRNA mutants (derived from a heterologous suppressor tRNA) in the absence and presence of the cognate synthetase, respectively (**Figure 1*b***). In the negative selection, the tRNA library is introduced into *E. coli* along with a mutant barnase gene in which amber nonsense codons are introduced at sites permissive to substitution by other amino acids. When a member of the suppressor tRNA library is aminoacylated by an endogenous *E. coli* synthetase (i.e., it is not orthogonal to the *E. coli* synthetases), the amber codons are suppressed and the ribonuclease barnase is produced, resulting in cell death. Only cells harboring orthogonal or nonfunctional tRNAs can survive. All tRNAs from surviving clones are then subjected to a positive selection in the presence of the cognate heterologous synthetase and a β-lactamase gene with an amber codon at a permissive site. tRNAs that can function in translation and are good substrates for the cognate heterologous synthetase are selected on the basis of their ability to suppress the amber codon and produce active β-lactamase. Therefore, only tRNAs that

CAT: chloramphenicol acetyl transferase

OMeTyr: *O*-methyl-L-tyrosine

(*a*) are not substrates for endogenous *E. coli* synthetases, (*b*) can be aminoacylated by the synthetase of interest, and (*c*) function in translation will survive both selections.

This approach was applied to the *Mj* $tRNA^{Tyr}_{CUA}$ to further reduce recognition of this tRNA by endogenous *E. coli* synthetases, while preserving activity with both the cognate synthetase and translational machinery (82). Eleven nucleotides of *Mj* $tRNA^{Tyr}_{CUA}$ that do not interact directly with the *Mj* TyrRS were randomly mutated (based on consensus analysis of archaeal $tRNA^{Tyr}$ sequences) to generate a suppressor tRNA library (**Figure 1*c***). This tRNA library was passed through rounds of negative and positive selections to afford a functional, orthogonal tRNA ($mutRNA^{Tyr}_{CUA}$) that functions efficiently with *Mj* TyrRS to translate the amber codon (**Figure 1*d***). Additional orthogonal tRNA/synthetase pairs have since been generated and include a $tRNA^{Asp}_{CUA}$/AspRS pair derived from yeast (65) and an *E. coli* initiator $tRNA^{fMet}_{CUA}$/yeast TyrRS pair (52). Orthogonal suppressor tRNAs can also be derived from consensus sequences of multiple archaeal tRNAs and then improved with the above selections (4). This approach has been used to evolve an orthogonal *Methanococcus thermoautotrophicum* $tRNA^{Leu}_{CUA}$/LeuRS pair (3), an orthogonal *Methanosarcina mazei* $tRNA^{Glu}_{CUA}$/GluRS pair (68), and an orthogonal *Pyrococcus horikoshii* $tRNA^{Lys}_{CUA}$/LysRS pair (4).

An Orthogonal Aminoacyl-tRNA Synthetase

Next, it was necessary to alter the substrate specificity of the orthogonal *Mj* TyrRS to charge its cognate orthogonal tRNA with only the desired unnatural amino acid and none of the common 20 amino acids. Our goal was to develop a general scheme for evolving the specificity of aminoacyl-tRNA synthetases that is independent of the structure of the amino acid of interest. One such approach involves generating a focused library of synthetase active-site mutants and again using a combination of positive and negative selections to evolve synthetases that aminoacylate the cognate tRNA with the unnatural and no endogenous amino acids (79, 83). Libraries of synthetase variants were generated by randomizing five or six residues in the substrate binding pocket of the synthetase, based on an analysis of the X-ray crystal structure of the synthetase (or a homologue) complexed with its cognate amino acid or aminoacyl adenylate (**Figure 2*a***). To identify synthetase variants that specifically recognize the unnatural amino acid and no endogenous host amino acid, the synthetase libraries were subjected to alternating rounds of positive and negative selections (**Figure 2*b***). In *E. coli*, the positive selection is based on resistance to chloramphenicol conferred by suppression of an amber mutation at a permissive site in the chloramphenicol acetyl transferase (CAT) gene; the negative selection uses the barnase gene with amber mutations at permissive sites. When the library of synthetase mutants is passed through the positive selection in the presence of the unnatural amino acid, those cells with mutant synthetases that can acylate the tRNA with either the unnatural amino acid or an endogenous amino acid survive. Plasmids encoding active mutant synthetases are then transformed into the negative selection strain, and selections are carried out in the absence of the unnatural amino acid. Those cells containing mutant synthetases that recognize endogenous amino acids incorporate the latter in response to the amber codons in the barnase gene and die. Repeated rounds of positive and negative selections lead to the isolation of mutant synthetases that can specifically incorporate the unnatural amino acid in response to the amber codon.

This selection scheme was first used to evolve a *Mj* TyrRS mutant capable of selectively inserting *O*-methyl-L-tyrosine (OMeTyr) into proteins in *E. coli* in response to the amber codon (79). A library of synthetase mutants was generated by randomizing the five active-site residues Tyr-32,

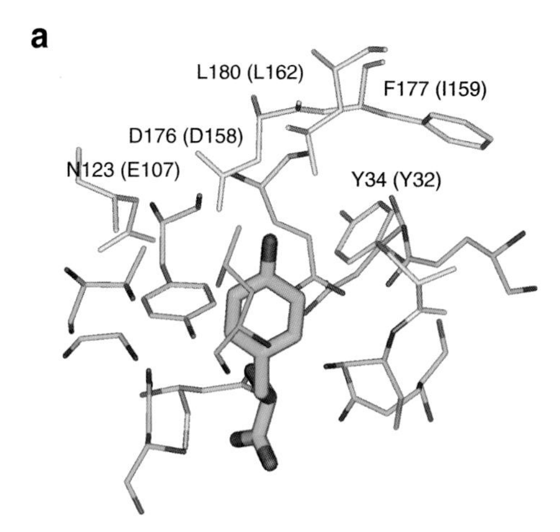

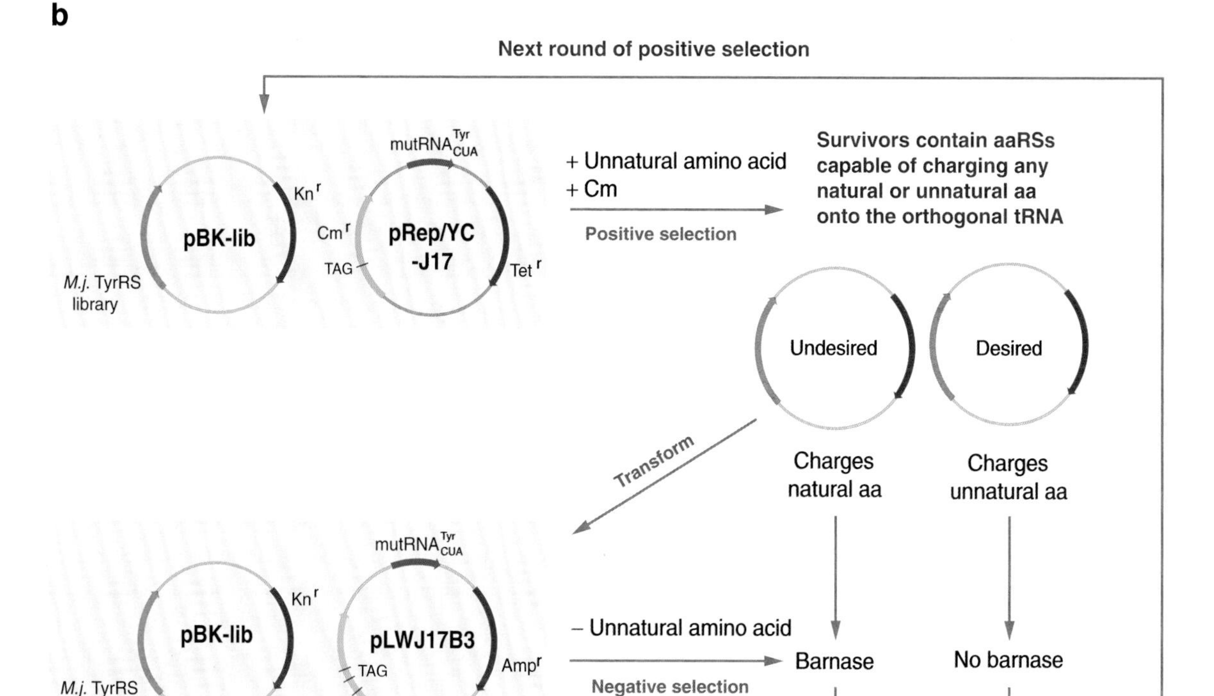

Figure 2

Modification of the amino acid specificity of an orthogonal *M. jannaschii* TyrRS (*Mj* TyrRS) in *E. coli*. (*a*) A library of *Mj* TyrRS mutants was generated by randomizing five residues (in parentheses) in the tyrosine binding site. (*b*) A general positive and negative selection scheme for evolving synthetase variants specific for an unnatural amino acid in *E. coli*. Cm, chloramphenicol.

GFP: green fluorescent protein

*p*AcPhe: *p*-acetylphenylalanine

Glu-107, Asp-158, Ile-159, and Leu-162 based on the crystal structure of the homologous *Bacillus stearothermophilus* TyrRS–tyrosyl adenylate complex (**Figure 2*a***) [more recently, the X-ray structure of *Mj* TyrRS itself was solved (49, 90), allowing the generation of improved libraries]. This library was then subjected to rounds of positive and negative selection to afford clones that survived at high chloramphenicol concentrations in the presence of OMeTyr and at low chloramphenicol concentrations in its absence. Incorporation of OMeTyr by the evolved *Mj*$tRNA_{CUA}$/synthetase pair was assayed directly by suppression of amber codons in *E. coli* dihydrofolate reductase (DHFR). In the presence of OMeTyr and the mutant *Mj* $tRNA_{CUA}$/synthetase pair, 5 to 10 mg/liter of protein was purified from minimal media. In the absence of any one component (synthetase, tRNA, or amino acid) no protein could be detected by Western blot analysis, silver stain, or coomassie stain. High-resolution mass analysis of intact protein and tryptic digests confirmed that only OMeTyr is incorporated in response to the amber codon, and OMeTyr is incorporated at no other site in the protein.

An alternative selection scheme makes use of an amber-T7/GFPuv (a type of green fluorescent protein) instead of an amber-barnase reporter in the negative selection (69). Suppression of amber codons introduced at permissive sites in T7 RNA polymerase produces full-length T7 RNA polymerase, which drives the expression of GFPuv. In this approach both the amber-CAT reporter and amber-T7/GFPuv reporter are encoded in a single plasmid. After positive selection, surviving cells are grown in the absence of both the unnatural amino acid and chloramphenicol. Cells containing mutant synthetases that can acylate the tRNA with any of the 20 common amino acids express GFPuv, whereas cells containing mutant synthetases that can acylate the tRNA only with the unnatural amino acid do not. These nonfluorescent cells are sorted using fluorescence-activated cell-sorting. One advantage of this latter method is that both reporters are contained within a single genetic construct, eliminating the need for plasmid shuttling between positive and negative selections. Other selection schemes have been pursued, including cell surface and phage display systems, but these are less general (i.e., require capture reagents specific for the amino acid of interest) or not as efficient (66).

More than 30 unnatural amino acids have been incorporated into proteins in *E. coli*. In general, for most unnatural amino acids, suppression efficiencies range from 25% to 75% of wild-type protein and translational fidelity is >99%. In an optimized system, approximately 1 g/liter of mutant protein containing the unnatural amino acid *p*-acetylphenylalanine (*p*AcPhe) was produced (H. Cho & T. Daniel, unpublished results).

Encoding Unnatural Amino Acids in Eukaryotic Cells

In addition to having distinct identity elements, eukaryotic tRNAs differ from bacterial and archaeal tRNAs in transcription and modification. Eukaryotic tRNAs have internal A- and B-box sequences required for transcription, and the 3′-CCA is added enzymatically rather than encoded in the gene as in *E. coli* (34, 54). In addition, eukaryotic tRNAs are transcribed in the nucleus and must be exported to the cytoplasm via an exportin-tRNA-dependent process (5). Consequently, a new family of orthogonal tRNA/synthetase pairs was generated to genetically encode unnatural amino acids in eukaryotic organisms. These include an *E. coli* $tRNA^{Tyr}_{CUA}$/TyrRS pair (29, 30, 53) and a human initiator $tRNA^{fMet}_{CUA}$/*E. coli* GlnRS pair (52) that are orthogonal in yeast. In addition, a modified *Bacillus subtilis* $tRNA^{Trp}_{CUA}$/TrpRS (91) and a *B. stearothermophilus* $tRNA^{Tyr}_{CUA}$/*E. coli* TyrRS pair (67) were found to be orthogonal opal and amber suppressors, respectively, in mammalian cells.

To selectively introduce an unnatural amino acid into proteins in eukaryotes, Yokoyama and coworkers (47) screened a collection of designed active-site variants of *E. coli* TyrRS in a wheat germ translation system and discovered a mutant synthetase that uses 3-iodotyrosine more effectively than it uses tyrosine. This mutant synthetase was used with the *B. stearothermophilus* $tRNA^{Tyr}_{CUA}$ to incorporate 3-iodotyrosine into proteins in mammalian cells (67). Similarly, an orthogonal *B. subtilis* $tRNA^{Trp}_{UCA}$/TrpRS pair has been used to selectively introduce 5-hydroxytryptophan (5-HTTP) into proteins in 293T cells (91). On the basis of the crystal structure of the homologous *B. stearothermophilus* TrpRS, a synthetase mutant was generated that selectively charges 5-HTPP. This mutant synthetase and its cognate orthogonal tRNA were used to suppress the opal nonsense codon in the *foldon* gene in the presence of 5-HTPP (91). Indeed, expression of full-length protein was seen only in the presence of 5-HTPP. Electrospray mass spectrometry of the mutant foldon protein verified site-specific incorporation of 5-HTPP with a fidelity of >97%. The yield of the 5-HTPP mutant protein was approximately 100 μg/liter of culture, compared with that of about 1 mg/liter for wild-type protein.

To develop a general selection scheme in yeast to evolve synthetases specific for unnatural amino acids (analogous to those used in *E. coli*), a selection strain of *S. cerevisiae* that contains the transcriptional activator protein GAL4 was created in which codons at two permissive sites were mutated to amber nonsense codons (14) (**Figure 3**). Suppression of these amber codons leads to the production of full-length GAL4, which in turn drives transcription of genomic GAL4-responsive *his3*, *ura3*, and *lacZ* reporter genes. Expression of *HIS3* and *URA3* complements the histidine and uracil auxotrophy in this strain and provides a positive selection for clones expressing active tRNA/synthetase pairs. On the other hand, addition of 5-fluoroorotic acid (5-FOA), which is converted to a toxic product by URA3, results in the death of cells expressing active tRNA/synthetase pairs. In the absence of the unnatural amino acid, this serves as a negative selection to remove synthetases specific to endogenous amino acids. Like GFP, the *lacZ* reporter can serve as an additional chromogenic marker to identify active synthetase from inactive ones.

FOA: 5-fluoroorotic acid

This selection scheme has allowed us to evolve orthogonal *E. coli* $tRNA^{Tyr}_{CUA}$/TyrRS pairs that have been used to incorporate more than 10 unnatural amino acids into proteins in yeast (14, 22). A synthetase library (10^8 in size) was similarly constructed by randomizing five active-site residues in *E. coli* TyrRS. Mutant synthetases were identified after several rounds of positive and negative selection that incorporate a number of unnatural amino acids into proteins, with yields corresponding to 20% to 40% of those of wild-type protein and expression levels up to 75 mg/liter. A similar approach has been used to evolve orthogonal *E. coli* leucyl $tRNA_{CUA}$/LeuRS pairs that selectively incorporate photochromic and fluorescent amino acids into proteins in yeast (86). These same mutant aminoacyl-tRNA synthetases that were evolved in yeast (to accept *p*-azidophenylalanine, *p*-benzoyl-L-phenylalanine, *p*-iodophenylalanine, *p*-acetylphenylalanine, and *p*-methoxyphenylalanine) have been used together with a *B. stearothermophilus* amber suppressor $tRNA^{Tyr}_{CUA}$ to selectively insert unnatural amino acids into proteins in mammalian cells (P. G. Schultz, unpublished results), albeit currently in low yield.

Additional Codons to Specify Unnatural Amino Acids

It should also be possible to use quadruplet codons and cognate suppressor tRNAs with expanded anticodon loops to specify additional amino acids. There are many examples of naturally occurring +1 frameshift suppressors (11, 19, 64). Moreover, genetic selections have been used to identify efficient four- and five-base codon suppressor tRNAs from

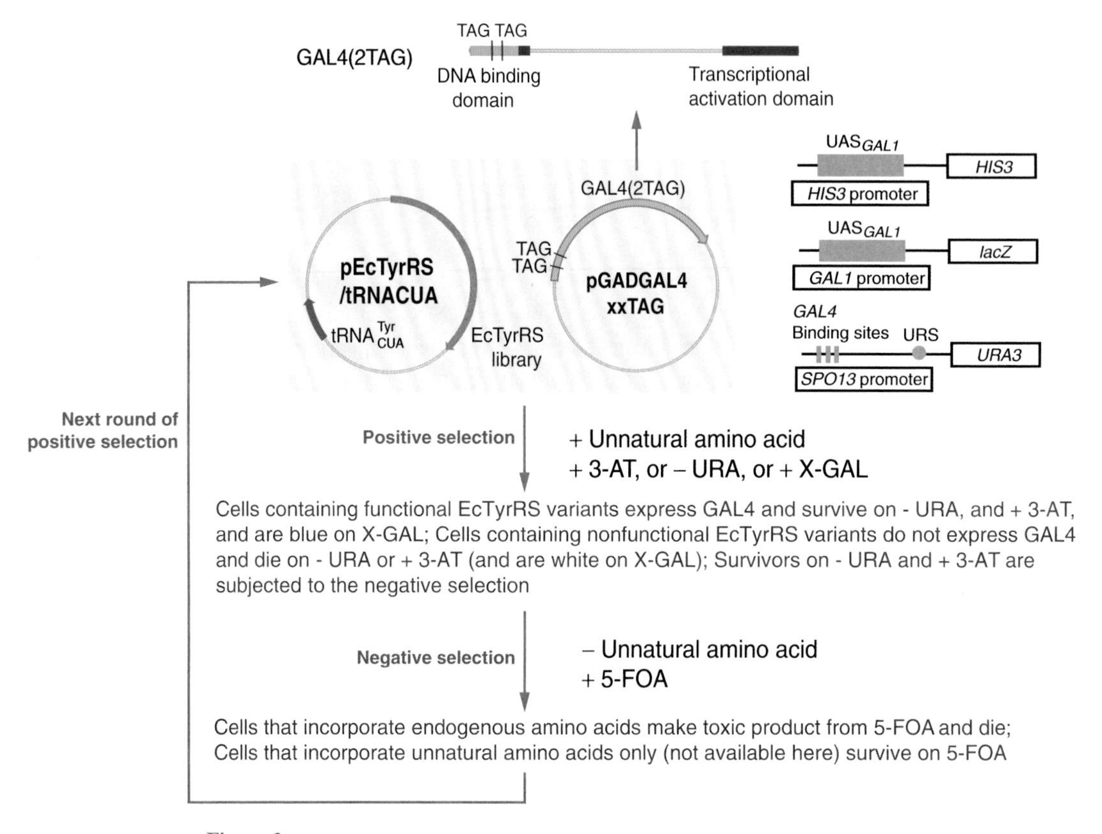

Figure 3

A general positive and negative selection scheme for evolving synthetase variants specific for unnatural amino acids in yeast. 5-FOA, 5-fluoroorotic acid.

mutant tRNA libraries (2, 59). The generation of an orthogonal tRNA/synthetase pair that decodes four-base codons requires either that the tRNA anticodon loop is not involved in tRNA/synthetase recognition or that the corresponding anticodon binding site of the synthetase can be modified to recognize a specific four-nucleotide anticodon sequence. Recently, an orthogonal four-base suppressor tRNA-synthetase pair was generated from $tRNA^{Lys}_{UCCU}$/LysRS of *Pyrococcus horikoshii*. The resultant orthogonal pair efficiently and selectively incorporates the unnatural amino acid homoglutamine into proteins in *E. coli* in response to the quadruplet codon AGGA (4). Frameshift suppression with homoglutamine does not significantly affect protein yields or cell growth rates, and it is mutually orthogonal with amber suppression. This has allowed the simultaneous incorporation of two unnatural amino acids at distinct sites within a single protein, suggesting that neither the number of available triplet codons nor the translational machinery itself represents a significant barrier to further expansion of the code.

An alternative approach that can be used to generate additional codons that uniquely encode unnatural amino acids involves eliminating degenerate codon-tRNA pairs from the *E. coli* genome. To this end we have shown that it is possible to remove four codons from a number of genes, including genes essential for growth, without affecting *E. coli* growth rates.

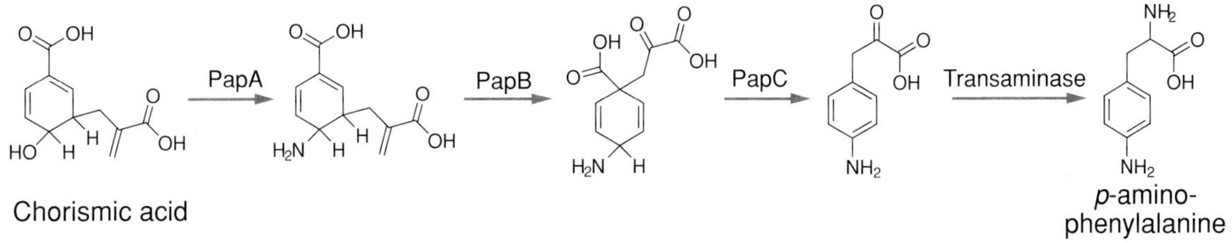

Figure 4

The biosynthesis of an unnatural amino acid, *p*-aminophenylalanine, in *E. coli*. Proteins PapA, PapB, and PapC convert chorismate to *p*-aminophenylpyruvic acid, and *E. coli* aromatic aminotransferase completes the biosynthesis to afford *p*-aminophenylalanine.

We are currently evaluating methods for the efficient construction of a "codon-deleted" *E. coli* genome.

Amino Acid Transport and Biosynthesis

We have found from measurements of cytoplasmic levels of amino acids added to the growth medium that a large number of amino acids are efficiently transported to the *E. coli* cytoplasm in millimolar concentrations. Highly charged or hydrophilic amino acids may require derivatization (e.g., esterification, acylation) with groups that hydrolyze in the cytoplasm. Metabolically labile amino acids or analogues (e.g., α-hydroxy acids, *N*-methyl amino acids) may require strains in which specific metabolic enzymes are deleted.

An alternative to adding exogenous amino acids to the growth media involves engineering a pathway for the biosynthesis of the unnatural amino acid directly in the host organism. For example, a completely autonomous 21-amino-acid bacterium has been generated that contains genes for the biosynthesis of *p*-amino-L-phenylalanine (*p*AF) from simple carbon sources, an aminoacyl-tRNA synthetase that uses *p*AF (and no other endogenous amino acids), and a tRNA that delivers *p*AF into proteins in response to the amber codon (61). *p*AF was biosynthesized from the metabolic intermediate, chorismic acid, using the *papA*, *papB*, and *papC* genes from *Streptomyces venezuelae* in combination with a nonspecific *E. coli* transaminase (**Figure 4**). *E. coli* containing these genes produced *p*AF at levels comparable to those of the other aromatic amino acids and had normal growth rates. In the presence of a *p*AF-specific, orthogonal $\text{mutRNA}^{\text{Tyr}}_{\text{CUA}}$/TyrRS pair, *E. coli* transformed with *papA-C* produced mutant proteins containing *p*AF at sites encoded by the amber codon with excellent yield and fidelity (61). In addition to *p*AF, it should be possible to biosynthesize and genetically encode other amino acids in vivo as well, including methylated, acetylated, and glycosylated amino acids.

UNNATURAL AMINO ACIDS AND APPLICATIONS

The above methodology has been used successfully to add more than 30 unnatural amino acids to the genetic codes of *E. coli*, yeast, or mammalian cells (**Figure 5**). Many of these unnatural amino acids have novel properties that are useful for a variety of biochemical and cellular studies of protein structure and function. For example, unnatural amino acids with chemically reactive groups have been genetically encoded in bacteria and/or yeast (16, 23, 85, 93), including *p*-acetylphenylalanine **1** and *m*-acetylphenylalanine **2**, *p*-(3-oxobutanoyl)-L-phenylalanine **3**, *p*-(2-amino-3-hydroxyethyl)phenylalanine **4**,

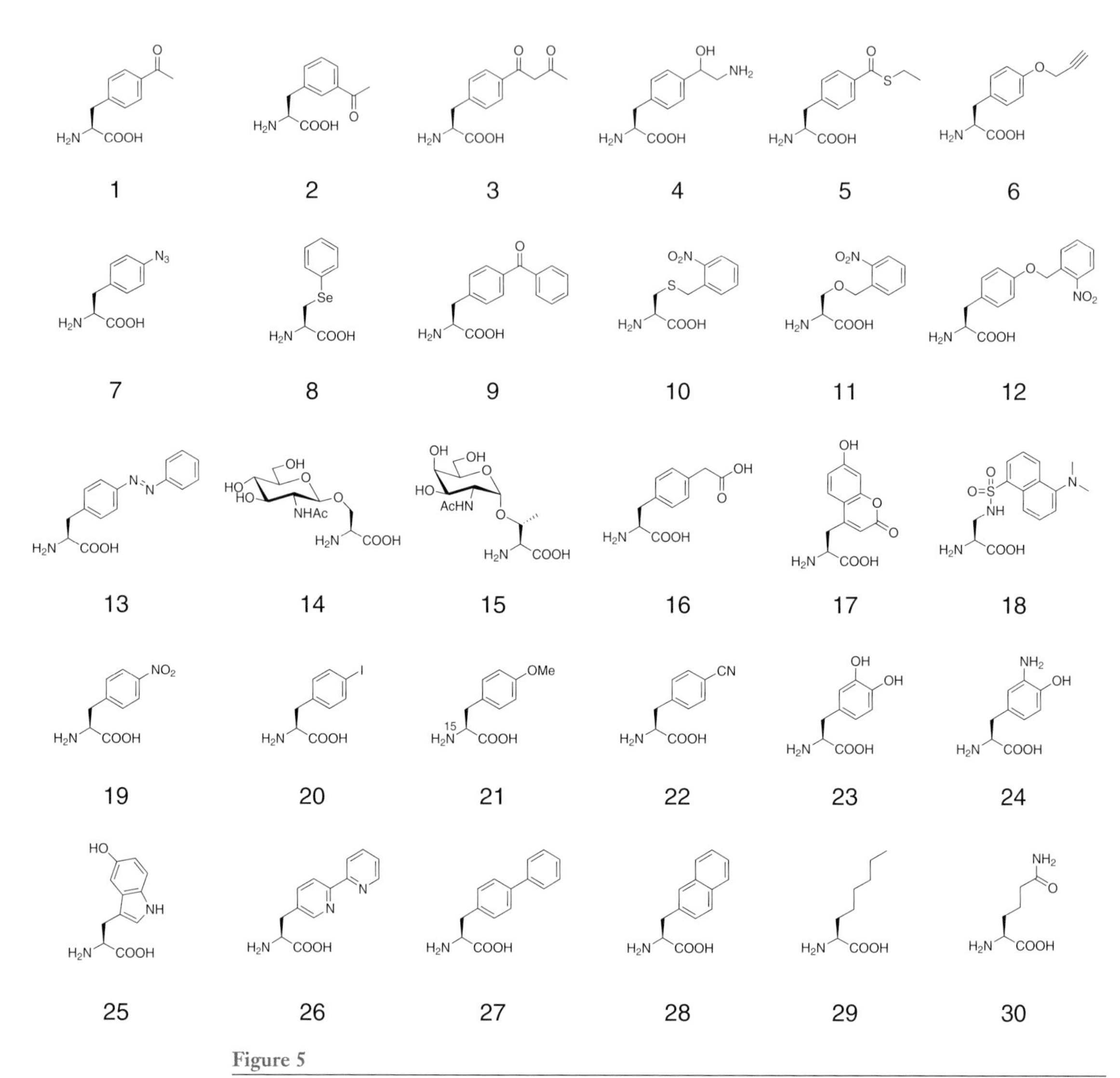

Figure 5

A list of representative unnatural amino acids that have been added to the genetic codes of *E. coli*, yeast, or mammalian cells.

p-ethylthiocarbonyl-phenylalanine **5**, *p*-propargyloxyphenylalanine **6**, and *p*-azidophenylalanine **7**. By virtue of their unique reactivity, these amino acids can be used to selectively modify native proteins under mild conditions with a variety of reagents in the absence of protecting groups. For example, the keto and β-diketo moieties selectively react with both hydrazides and alkoxyamines (85, 93); the azide can be selectively modified by a copper(I)-catalyzed [3+2] cycloaddition reaction with an alkyne derivative (23) or by a Staudinger conjugation reaction with a phosphine derivative (76); and the thioester group can be used in a modified chemical ligation reaction to derivatize protein side chains (and may allow the facile generation of cyclic and branched proteins and peptides). Similarly, a phenylselenide containing amino acid **8** has been introduced selectively into proteins,

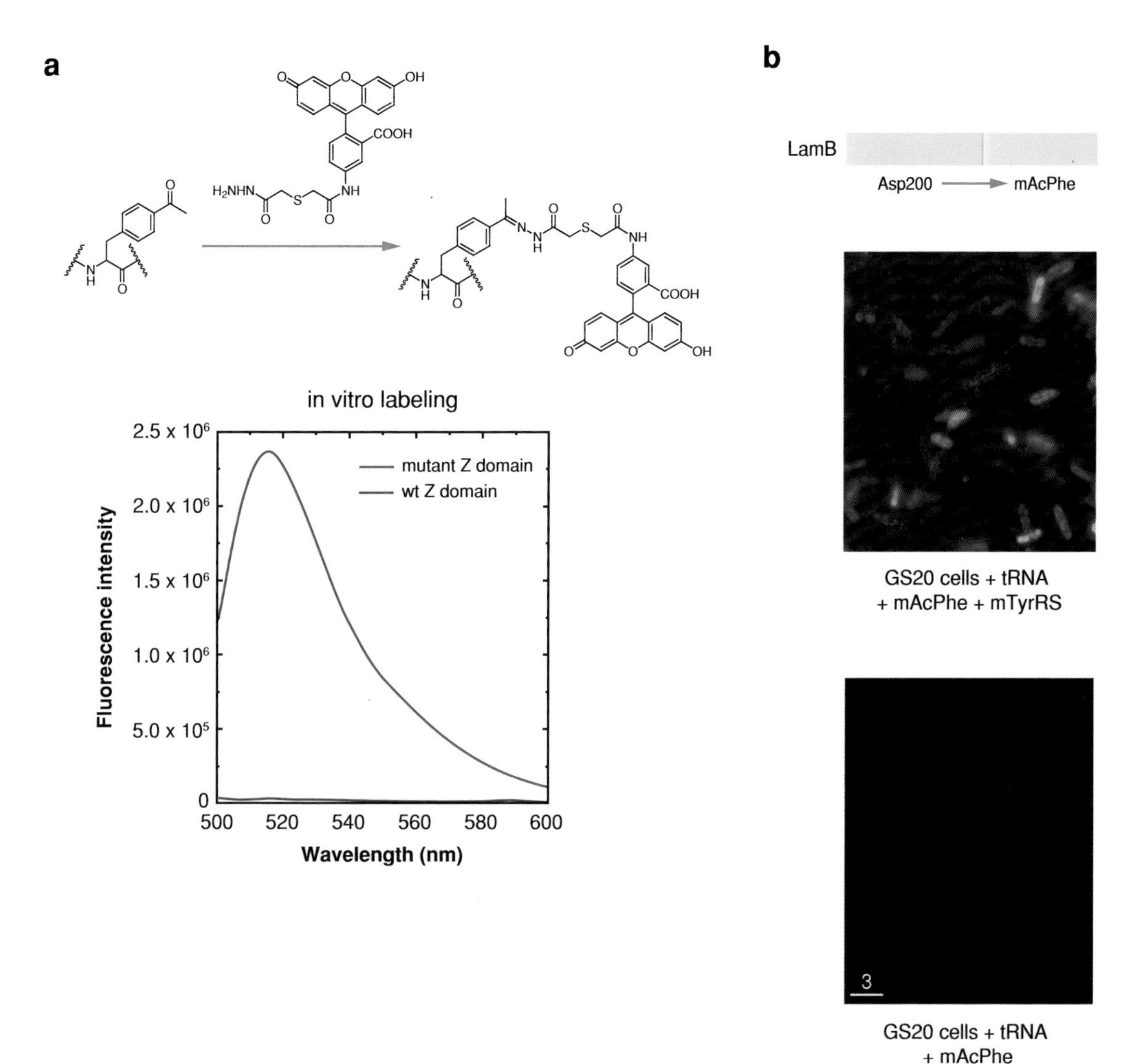

Figure 6

Site-specific modification of proteins containing a keto amino acid. (*a*) In vitro labeling of mutant Z domain that contains *p*-acetyl-L-phenylalanine with fluorescein hydrazide. (*b*) In vivo labeling of *E. coli* surface-expressed LamB mutant that contains *m*-acetyl-L-phenylalanine (*m*AcPhe) with the nonmembrane-permeable fluorescent dyes Cascade blue hydrazide (*blue*), Alexa568 hydrazide (*red*), and Alexa647 hydrazide (*green*).

which, after oxidation and β-elimination, allows intramolecular or intermolecular thiol conjugation reactions (P. G. Schultz, unpublished results).

These orthogonal chemistries have been used to selectively modify proteins with a number of fluorophores (85, 93) (**Figure 6*a***), tags, and other exogenous reagents (23, 85, 93). In one example, a mutant human growth hormone was site-specifically modified in high yield with polyethylene glycol (PEG) to afford a protein that retained wild-type activity but had a considerably improved half-life in serum

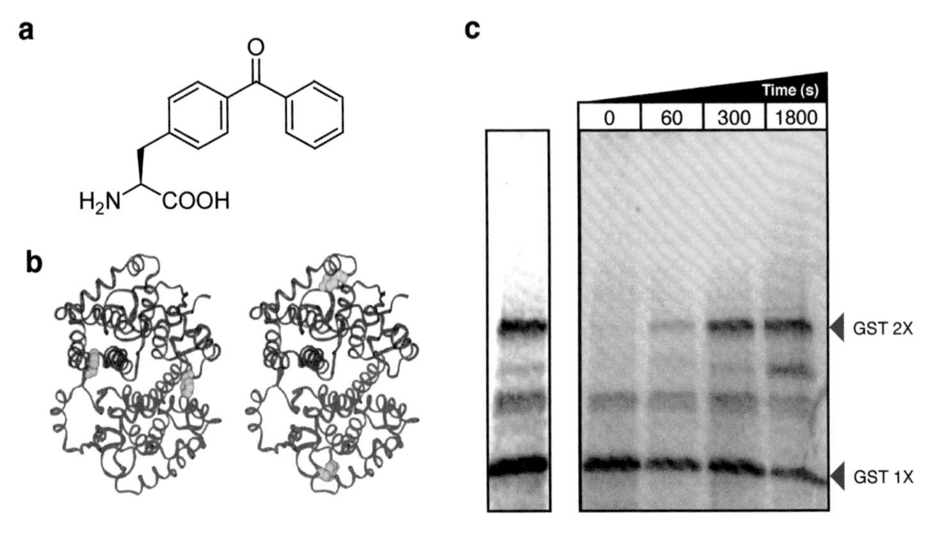

Figure 7

Site-specific incorporation of a photocrosslinker into proteins for mapping protein-protein interactions in vivo. (*a*) The chemical structure of *p*-benzoyl-L-phenylalanine (*p*Bpa). (*b*) Residue Phe-52 or Tyr-198 in *Schistosoma japonica* glutathione S-transferase (SjGST) was substituted by *p*Bpa. Monomers of the dimer are shown in blue and red. The side chain of Phe-52 is shown in orange for each monomer (*left*). The side chain of residue Tyr-198 is shown in orange (*right*). (*c*) The covalent dimerization of SjGST (Phe52*p*Bpa) in vivo upon irradiation at 365 nm.

PEG: polyethylene glycol

Grb2: growth factor receptor-bound protein 2

(H. Cho & T. Daniel, unpublished results). This work is being extended to other therapeutic proteins, as well as the generation of homo- and heterodimeric PEG-linked Fab fragments, and may allow the synthesis of chemically modified proteins and peptides with unprecedented control over selectivity, homogeneity, and chemical structure. In a second example a series of fluorescent dyes were selectively introduced at one site of a *m*-acetylphenylalanine mutant of the membrane protein lamB in live bacteria (**Figure 6*b***). In a third example of generating homogeneous glycoprotein mimetics, amino-oxy-containing sugars were selectively introduced at defined sites in proteins that were subsequently elaborated by adding additional saccharides to the pendant sugar with glycosyltransferases (58).

Two unnatural amino acids, *p*-azidophenylalanine **7** and *p*-benzoylphenylalanine **9**, have side chains that can be photocrosslinked both in vitro and in vivo (15–17). Benzophenone is a particularly useful photocrosslinker because it absorbs at relatively long wavelengths (~360 nm), and the excited state inserts efficiently into carbon-hydrogen bonds (33). Indeed, a mutant homodimeric glutathione S-transferase with *p*-benzoylphenylalanine substituted site-specifically at the dimer interface could be cross-linked in greater than 50% yield in the cytoplasm of *E. coli* (**Figure 7**). More recently, Yokoyama and coworkers showed that *p*-benzoylphenylalanine can be selectively incorporated into human Grb2 protein in mammalian Chinese hamster ovary cells and cross-linked with the epidermal growth factor receptor upon exposure of cells to 365-nm light (39). These amino acids should be useful for in vitro and in vivo probes of protein-protein and protein–nucleic acid interactions, studies of protein structure and dynamics, and the identification of receptors for orphan ligands.

Photocaged cysteine, serine, and tyrosine amino acids **10–12** have also been site-specifically introduced into proteins using this

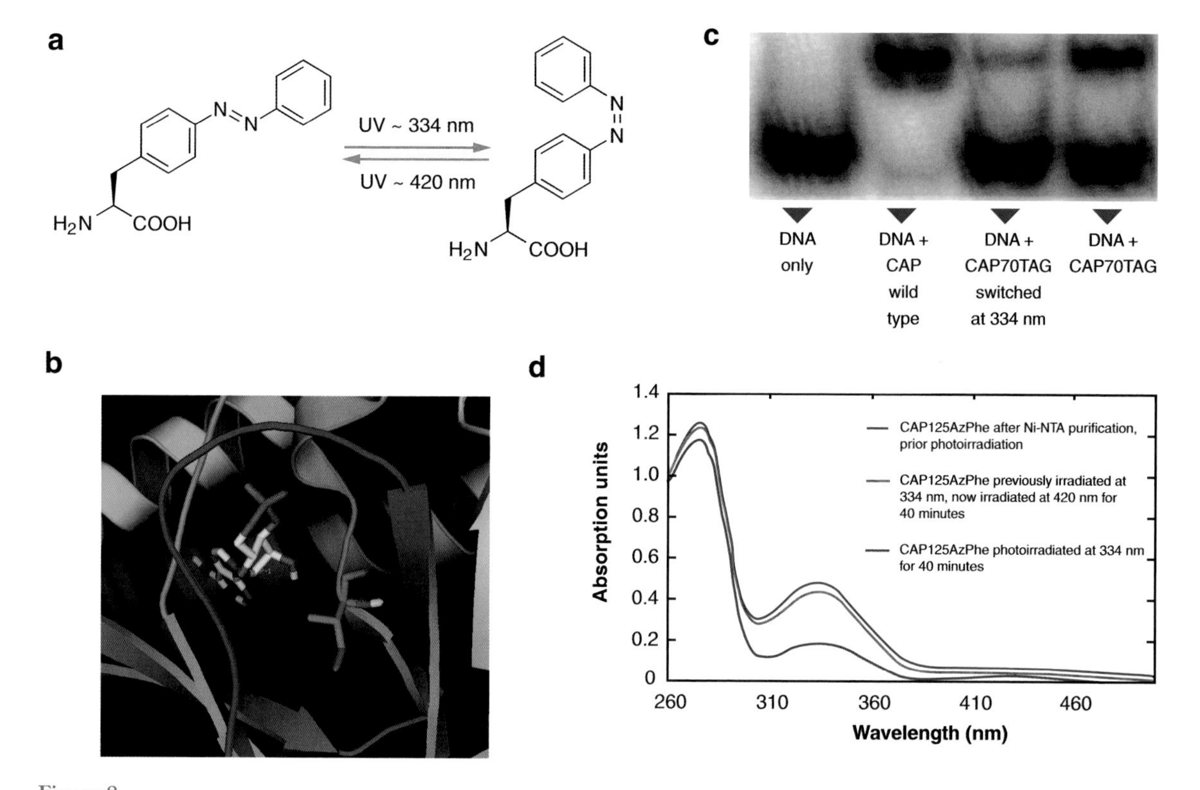

Figure 8

Photomodulation of mutant *E. coli* transcriptional factor catabolite activator protein (CAP)–DNA interactions. (*a*) The photochromic amino acid *p*-azophenyl-phenylalanine. (*b*) Ile-70 in the cAMP binding site of CAP was mutated to *p*-azophenyl-phenylalanine. (*c*) The affinity of mutant CAP for its promoter was regulated by photo-isomerizing *trans p*-azophenyl-phenylalanine to the *cis* isomer. (*d*) Absorption spectra of the mutant CAP (CAP125TAG or CAP125AzPhe; 50 μM) after purification and upon irradiation.

methodology (86). The side chain hydroxy or thiol groups of these amino acids are blocked by nitrobenzyl groups, which can be photochemically removed either in vivo or in vitro upon irradiation with 330 nm light. In one example the active-site cysteine of the proapoptotic cysteine protease caspase 3 was substituted with nitrobenzyl cysteine to afford inactive protein, which could be photoactivated with greater than 70% efficiency upon photolysis. In a related experiment the photochromic amino acid, *p*-azophenyl-phenylalanine **13**, was site-specifically introduced into the cAMP binding site of the *E. coli* transcription factor catabolite activator protein (CAP) (10). Irradiation of this amino acid at 334-nm light converts the *trans* form of the amino acid to predominantly the *cis* form. Because the two isomers differ significantly in structure and dipole, the *cis* and *trans* **13** mutants have different affinities for cAMP, and as a result, the affinity of this mutant CAP for its promoter can be photoregulated (**Figure 8**). Similarly, it should be possible to photomodulate the activity of other enzymes (e.g., kinase, phosphatases) as well as receptors, ion channels, and transcription factors either in vitro or in vivo.

CAP: transcription factor catabolite activator protein

This methodology has also been used to incorporate glycosylated amino acids into proteins in *E. coli* to produce homogeneous glycoproteins (**Figure 9**). Protein

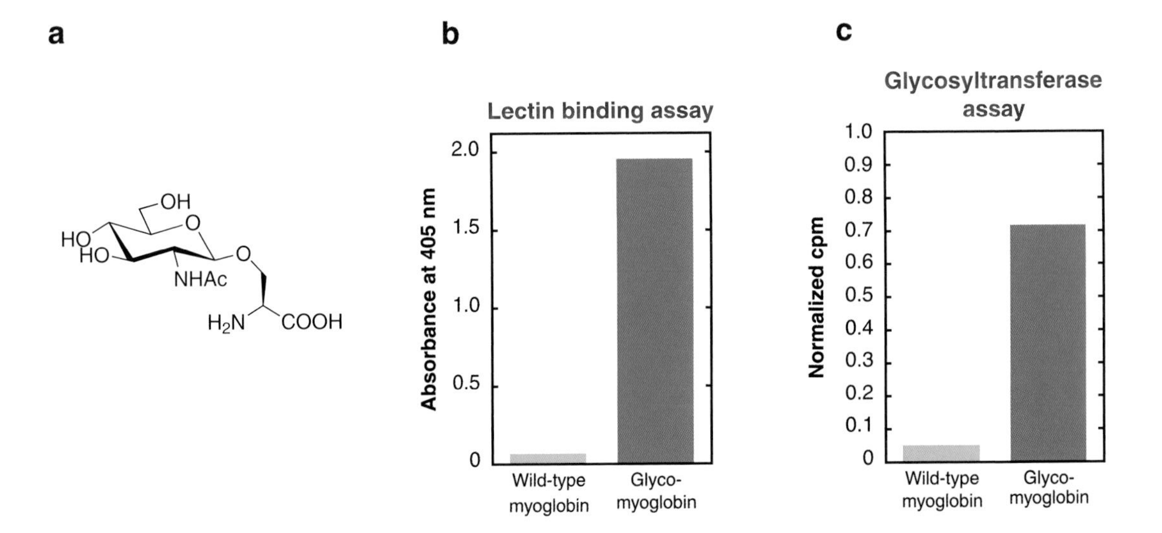

Figure 9

Cotranslational incorporation of a glycosylated amino acid into proteins in *E. coli*. (*a*) The chemical structure of β-*N*-acetylglucosamine-*O*-serine (β-GlcNAc-Ser). (*b*) Lectin binding assay with wild-type myoglobin and mutant myoglobin containing β-GlcNAc-Ser. (*c*) Analysis of the galactosyltransferase reaction with wild-type myoglobin and mutant myoglobin containing β-GlcNAc-Ser. UDP-[H^3]-galactose was used as the glycosyl donor.

glycosylation is an essential posttranslational modification in eukaryotic cells (28), but existing methods for generating glycoproteins often impose restrictions on the size, quantity, and/or purity of the glycoproteins produced (20, 37, 70, 78). To provide a general solution to this problem, mutant synthetases were evolved that site-specifically incorporate β-*N*-acetylglucosamine-*O*-serine (β-GlcNAc-Ser) **14** into proteins in *E. coli* with high translational fidelity (92). An *O*-GlcNAc-Ser mutant myoglobin generated by this method could be elaborated further to form more complex saccharides with galactosyltransferases (**Figure 9*c***). A similar approach was used to selectively introduce α-GalNAc-Thr **15** into proteins (89), and it is currently being extended to a number of other *O*- and *N*-linked sugars. This methodology should make it possible to genetically encode other posttranslationally modified amino acids including methylated and acetylated lysine, and phosphorylated tyrosine, serine, and threonine. Indeed, *p*-carboxylmethyl-L-phenylalanine **16** has recently been successfully introduced into proteins in *E. coli* as a stable mimetic of phosphotyrosine (P. G. Schultz, unpublished results).

Unnatural amino acids that can serve as probes of protein structure and function both in vivo and in vitro have also been genetically encoded in *E. coli* and yeast. For example, fluorescent amino acids with 7-hydroxycoumarin **17** and dansyl side chains **18** have been selectively incorporated into proteins, providing small fluorescent probes for the direct visualization of protein conformational changes, localization, and intermolecular interactions. These amino acids may offer an advantage over GFP and its derivatives by virtue of their smaller size and the fact that they can be introduced throughout a protein at any site. For example, the dansyl containing amino acid **18** was used to spectroscopically monitor the unfolding of superoxide dismutase. A combination of fluorescent amino acids and/or GFPs may also facilitate

FRET measurements in vitro or in vivo. Indeed, nitrophenylalanine **19** can be selectively introduced into proteins and used to quench tryptophan fluorescence. In another example the heavy-atom-containing amino acid *p*-iodo-L-phenylalanine **20** was genetically encoded in both *E. coli* and yeast and used for single-wavelength anomalous dispersion phasing in protein crystallography (with a laboratory CuKα X-ray source). This amino acid caused little structural perturbation when substituted for Phe in the core of T4 lysozyme (88). In yet another example, ^{15}N-labeled *O*-methyltyrosine **21** has been selectively incorporated into proteins as a site-selective NMR probe (24). Finally, *p*-cyanophenylalanine **22** was recently successfully incorporated into proteins (P. G. Schultz, unpublished results). Because the nitrile group has a distinct vibrational frequency, it should be a useful infrared probe of local environment and protein dynamics.

Another novel amino acid that has been incorporated into proteins is dihydroxyphenylalanine (DHP) **23** (1). DHP can undergo 2-electron oxidation to the corresponding quinone and can be used to both probe and manipulate electron transfer processes in proteins. Similarly, a second redox-active amino acid, 3-amino-L-tyrosine **24**, was selectively incorporated into proteins in *E. coli*. This amino acid can act as a radical trap owing to the stability of its oxidized semiquinone form or can serve as a unique handle for chemical modification of proteins (41). The redox-active amino acid 5-HTTP **25** has also been selectively incorporated into proteins in mammalian cells in response to an opal codon (91). This amino acid undergoes electrochemical oxidation to form an efficient protein cross-linking agent. Unnatural amino acids may also be used to perturb the electronic properties of a protein. For example, introduction of several unnatural amino acids in place of Tyr-65 in GFP led to changes in absorbance and fluorescence spectra and altered quantum yields (84).

In another series of experiments, a bipyridyl amino acid **26** has been selectively incorporated into proteins (P. G. Schultz, unpublished results). In this case we could not evolve a synthetase specific for **26** directly, but rather first evolved a synthetase specific for the biphenyl analogue **27**. Generation of a second-generation library from the latter in conjunction with subsequent rounds of positive and negative selections afforded a synthetase specific for **26**. This amino acid can be used to (*a*) introduce redox-active or electrophilic metal ions into proteins, (*b*) form fluorescent metal ion complexes, or (*c*) mediate the metal-ion-dependent dimerization of proteins containing **26**. Other representative unnatural amino acids that have been added to the genetic code of *E. coli* or yeast include L-3-(2-naphthyl)alanine **28** (80), α-aminocaprylic acid **29** (86), and L-homoglutamine **30** (4). The large number of structurally diverse amino acids that have been genetically encoded to date suggest that the translational machinery is rather tolerant of side chain structure. Thus, it should be possible to add additional building blocks to the genetic code including spin labels, electron transfer mediators, infrared and near-infrared probes, α-hydroxy acids, and *N*-alkyl amino acids.

STRUCTURAL STUDIES OF MUTANT SYNTHETASES

We have been surprisingly successful in identifying mutant aminoacyl-tRNA synthetases with altered specificities (79, 83, 87). Crystallographic studies of these mutant synthetases provide an opportunity to examine the origins and evolution of amino acid specificity in this important class of enzymes. For example, in the case of an *Mj* TyrRS mutant specific for *p*AcPhe (**Figure 10*a***), there are significant changes in the active site that lead to altered hydrogen bonding and packing interactions that favor binding of the unnatural amino acid relative to tyrosine (77). Active-site D158G and Y32L mutations remove two hydrogen bonds to the tyrosine side chain hydroxyl group, which would be expected to

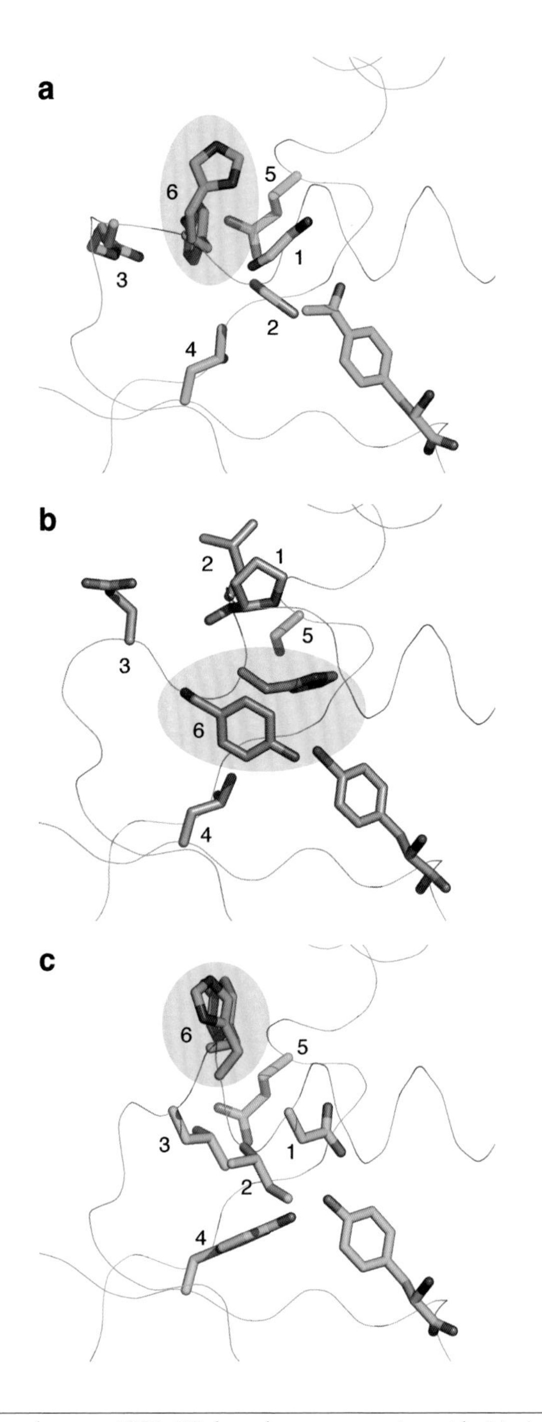

Figure 10

Structure of wild-type and mutant *Mj* TyrRSs bound to cognate amino acids. (*a*) *p*AcPhe synthetase bound to *p*-acetylphenylalanine (*p*AcPhe). (*b*) *p*BrPhe synthetase bound to *p*-bromophenylalanine. (*c*) Wild-type *Mj* TyrRS bound to tyrosine. 1: residue 158; 2: residue 159; 3: residue 162; 4: residue 32; 5: residue 107; 6: residues 160 and 161.

dramatically reduce binding of the natural substrate to the enzyme. The D158G mutation also deepens the binding pocket to accommodate the *para* substituent of *p*AcPhe, while Y32L forms a suitable hydrophobic packing surface for the acetyl methyl group. In addition, the side chain carbonyl oxygen of *p*AcPhe forms a hydrogen bond to Gln-109. In the wild-type *Mj* TyrRS structure (49, 90), the side chain of Asp-158 disrupts this hydrogen bond vector; in the *p*AcPhe-specific synthetase the D158G mutation removes this intervening side chain. The D158G mutation also alters the backbone structure of helix α8. The glycine residue truncates the C terminus of helix α8 by four residues, resulting in a new C-terminal cap. Several intrahelical main chain hydrogen bonds are broken by premature termination of helix α8; however, a short 3_{10}-helix formed in the new structure is stabilized by side chain/main chain and main chain/main chain hydrogen bonds. This altered helix α8 conformation facilitates reconfiguration of active side chains to selectively bind *p*AcPhe.

Similar side chain and backbone conformational changes were observed in the structure of a mutant *Mj* TyrRS specific for *p*BrPhe (which has the mutations Y32L, E107S, D158P, I159L, and L162E) (P. G. Schultz, unpublished results) (**Figure 10*b***). The D158P and Y32L mutations remove two hydrogen bonds to the hydroxyl oxygen of tyrosine. The Y32L mutation also increases the size of the active-site cavity and results in van der Waals interactions between the alkyl side chain and the bromine atom of the bound *p*BrPhe. The D158P mutation also terminates helix α8 and, as a result, induces formation of a subsequent short 3_{10}-helix by residues 157–161, which again is stabilized by side chain and main chain hydrogen bonds. As a result of the D158P mutation and the resulting 3_{10}-helix, significant translational and rotational movements of several active residues expand the active-site cavity [compared with the wild-type structure (49, 90)] to form new van der Waals interactions with the *p*BrPhe side chain. Preliminary results in the case of an *Mj* TyrRS mutant specific for naphthylalanine show that similar backbone conformational changes again reconfigure the active site to bind the unnatural amino acid.

In both of these mutant enzymes, a handful of random mutations alter the substrate specificity of the *Mj* TyrRS by changing the pattern of hydrogen bonding and packing interactions with bound substrate. These mutations lead to changes in both side chain and backbone conformations, indicating a high degree of structural plasticity in the active site of the enzyme. The structural differences between these mutant and wild-type TyrRS are more pronounced than those found in previously described mutant aminoacyl-tRNA synthetases that recognize unnatural or unusual amino acids (12, 50, 51). Indeed, mutations that alter enzyme specificity generally have little effect on backbone configuration. The structural changes observed in these mutant synthetases are likely due to the sites of mutation and the method used to generate the libraries. Structural changes in the solvent-exposed α8 helix minimize the number of required compensatory structural changes on the exposed face, as bulk water can easily reorganize to solvate the exposed residues. In addition, the use of a focused library of mutated residues in close proximity to each other increases the likelihood of selecting compensatory mutations. In contrast, mutations that arise from DNA shuffling are recombined from independent selection experiments, and the resulting set of mutations may not be cooperative. Random-point mutations also have a low probability of being cooperative owing to the limited size of sequence space that can be explored.

CONCLUSIONS AND PERSPECTIVES

The approach described above has proven remarkably effective in allowing us to add a large number of structurally diverse amino acids to the genetic codes of both prokaryotic and eukaryotic organisms. Coincidently, it has recently been shown that nature has evolved a similar strategy using an orthogonal amber suppressor tRNA/synthetase pair derived from lysine to genetically encode the unnatural amino acid pyrrolysine in *Methanosarcina barkeri* (72). Future work in this field will likely focus on expanding the nature and number of amino acids that can be genetically encoded in both prokaryotic and eukaryotic organisms, including multicellular organisms. Additional orthogonal pairs that suppress three- and four-base codons are also being developed. It may even be possible to delete rare redundant codons from the *E. coli* genome and use them instead to encode unnatural amino acids. Considerable improvements in the yields of mutant proteins have been realized in *E. coli* and yeast, but further modifications to the structures of the tRNAs and synthetases, optimization of expression levels, and genomic mutations to the host organism (e.g., ribosome, transporters) are likely to further increase yields, especially in mammalian systems.

The ability to genetically encode unnatural amino acids should provide powerful probes, both in vitro and in vivo, of protein structure and function. It may also allow the design or evolution of proteins with novel properties. Examples might include the rational design of glycosylated or PEGylated therapeutic proteins with improved pharmacological properties, fluorescent proteins that act as sensors of small molecules and protein-protein interactions in the cell, or proteins whose activity can be photoregulated in vivo. One may be able to select for peptides and proteins with enhanced function from libraries of unnatural amino acid mutants. For example, we recently showed that it is possible to incorporate unnatural amino acids into phage-displayed peptides (75); a peptide with enhanced affinity for streptavidin was isolated that contained an unnatural amino acid. It should also be possible to incorporate nonamino acid building blocks into proteins or perhaps even create biopolymers with entirely unnatural backbones. Finally, the ability to add novel amino acids to the genetic codes of organisms should allow us to test experimentally whether there is an evolutionary advantage for organisms with more than the 20 genetically encoded amino acids.

SUMMARY POINTS

1. A general method was developed that makes it possible to genetically encode unnatural amino acids in *E. coli*, yeast, and mammalian cells.
2. More than 30 unnatural amino acids have been cotranslationally incorporated into proteins with high fidelity and efficiency by means of unique nonsense (triplet) or frameshift (quadruplet) codons and the corresponding tRNA/aminoacyl-tRNA synthetase pairs.
3. Unnatural amino acids that have been genetically encoded include those containing spectroscopic probes, posttranslational modifications, metal chelators, photoaffinity labels, functional groups with unique reactivity, and other chemical moieties.
4. Orthogonal tRNA/aminoacyl-tRNA synthetase pairs were derived from heterologous pairs with distinct tRNA identity elements.
5. The amino acid substrate specificity of the orthogonal aminoacyl-tRNA synthetase was altered by alternating positive and negative genetic selections using a large library of active-site mutants of the synthetase.

6. The ability to incorporate amino acids with defined steric and electronic properties at unique sites into proteins will provide powerful new tools for exploring protein structure and function both in vitro and in vivo, and for generating proteins with novel properties.

ACKNOWLEDGMENTS

This work was supported by National Institutes of Health grant GM-62159 and Department of Energy grant DE-FG03-00ER46051 to P.G.S.

LITERATURE CITED

1. Alfonta L, Zhang Z, Uryu S, Loo JA, Schultz PG. 2003. Site-specific incorporation of a redox-active amino acid into proteins. *J. Am. Chem. Soc.* 125:14662–63
2. Anderson JC, Magliery TJ, Schultz PG. 2002. Exploring the limits of codon and anticodon size. *Chem. Biol.* 9:237–44
3. Anderson JC, Schultz PG. 2003. Adaptation of an orthogonal archaeal leucyl-tRNA and synthetase pair for four-base, amber, and opal suppression. *Biochemistry* 42:9598–608
4. **Anderson JC, Wu N, Santoro SW, Lakshman V, King DS, Schultz PG. 2004. An expanded genetic code with a functional quadruplet codon. *Proc. Natl. Acad. Sci. USA* 101:7566–71**
5. Arts GJ, Kuersten S, Romby P, Ehresmann B, Mattaj IW. 1998. The role of exportin-t in selective nuclear export of mature tRNAs. *EMBO J.* 17:7430–41
6. Bain JD, Glabe CG, Dix TA, Chamberlin AR, Diala ES. 1989. Biosynthetic site-specific incorporation of a non-natural amino acid into a polypeptide. *J. Am. Chem. Soc.* 111:8013–14
7. Beene DL, Dougherty DA, Lester HA. 2003. Unnatural amino acid mutagenesis in mapping ion channel function. *Curr. Opin. Neurobiol.* 13:264–70
8. Benzer S, Champe SP. 1962. A change from nonsense to sense in the genetic code. *Proc. Natl. Acad. Sci. USA* 48:1114–21
9. Bock A, Forchhammer K, Heider J, Leinfelder W, Sawers G, et al. 1991. Selenocysteine: the 21st amino acid. *Mol. Microbiol.* 5:515–20
10. Bose M, Groff D, Xie J, Brustad E, Schultz PG. 2006. The incorporation of a photoisomerizable amino acid into proteins in *E. coli*. *J. Am. Chem. Soc.* 128:388–89
11. Bossi L, Roth JR. 1981. Four-base codons ACCA, ACCU and ACCC are recognized by frameshift suppressor sufJ. *Cell* 25:489–96
12. Buddha MR, Crane BR. 2005. Structure and activity of an aminoacyl-tRNA synthetase that charges tRNA with nitro-tryptophan. *Nat. Struct. Mol. Biol.* 12:274–75
13. Carne AF. 1994. Chemical modification of proteins. *Methods Mol. Biol.* 32:311–20
14. **Chin JW, Cropp TA, Anderson JC, Mukherji M, Zhang Z, Schultz PG. 2003. An expanded eukaryotic genetic code. *Science* 301:964–67**
15. **Chin JW, Martin AB, King DS, Wang L, Schultz PG. 2002. Addition of a photocrosslinking amino acid to the genetic code of *Escherichia coli*. *Proc. Natl. Acad. Sci. USA* 99:11020–24**
16. Chin JW, Santoro SW, Martin AB, King DS, Wang L, Schultz PG. 2002. Addition of *p*-azido-L-phenylalanine to the genetic code of *Escherichia coli*. *J. Am. Chem. Soc.* 124:9026–27

4. A combination of amber (TAG) and frameshift (AGGA) suppression was used to incorporate two unnatural amino acids into a protein simultaneously with high fidelity.

14. This report expands this methodology to the genetic code of yeast. Five unnatural amino acids were genetically encoded in yeast in response to the amber nonsense codon.

15. A photocrosslinking amino acid, *p*Bpa, was incorporated into proteins in *E. coli*. It should be useful in probing protein-protein and protein-nucleic acid interactions both in vitro and in vivo.

17. Chin JW, Schultz PG. 2002. In vivo photocrosslinking with unnatural amino acid mutagenesis. *Chembiochem* 3:1135–37
18. Cornish VW, Mendel D, Schultz PG. 1995. Probing protein structure and function with an expanded genetic code. *Angew Chem. Int. Ed. Engl.* 34:621–33
19. Curran JF, Yarus M. 1987. Reading frame selection and transfer RNA anticodon loop stacking. *Science* 238:1545–50
20. Davis BG. 2002. Synthesis of glycoproteins. *Chem. Rev.* 102:579–602
21. Dawson PE, Kent SB. 2000. Synthesis of native proteins by chemical ligation. *Annu. Rev. Biochem.* 69:923–60
22. Deiters A, Cropp TA, Mukherji M, Chin JW, Anderson JC, Schultz PG. 2003. Adding amino acids with novel reactivity to the genetic code of *Saccharomyces cerevisiae*. *J. Am. Chem. Soc.* 125:11782–83
23. Deiters A, Cropp TA, Summerer D, Mukherji M, Schultz PG. 2004. Site-specific PEGylation of proteins containing unnatural amino acids. *Bioorg. Med. Chem. Lett.* 14:5743–45
24. Deiters A, Geierstanger BH, Schultz PG. 2005. Site-specific in vivo labeling of proteins for NMR studies. *Chembiochem* 6:55–58
25. Doctor BP, Mudd JA. 1963. Species specificity of amino acid acceptor ribonucleic acid and aminoacyl soluble ribonucleic acid synthetases. *J. Biol. Chem.* 238:3677–81
26. Doring V, Mootz HD, Nangle LA, Hendrickson TL, de Crecy-Lagard V, et al. 2001. Enlarging the amino acid set of *Escherichia coli* by infiltration of the valine coding pathway. *Science* 292:501–4
27. Drabkin HJ, Park HJ, RajBhandary UL. 1996. Amber suppression in mammalian cells dependent upon expression of an *Escherichia coli* aminoacyl-tRNA synthetase gene. *Mol. Cell. Biol.* 16:907–13
28. Dwek RA. 1996. Glycobiology: toward understanding the function of sugars. *Chem. Rev.* 96:683–720
29. Edwards H, Schimmel P. 1990. A bacterial amber suppressor in *Saccharomyces cerevisiae* is selectively recognized by a bacterial aminoacyl-tRNA synthetase. *Mol. Cell. Biol.* 10:1633–41
30. Edwards H, Trezeguet V, Schimmel P. 1991. An *Escherichia coli* tyrosine transfer RNA is a leucine-specific transfer RNA in the yeast *Saccharomyces cerevisiae*. *Proc. Natl. Acad. Sci. USA* 88:1153–56
31. Evans TC Jr, Xu MQ. 1999. Intein-mediated protein ligation: harnessing nature's escape artists. *Biopolymers* 51:333–42
32. Fechter P, Rudinger-Thirion J, Tukalo M, Giege R. 2001. Major tyrosine identity determinants in *Methanococcus jannaschii* and *Saccharomyces cerevisiae* tRNA(Tyr) are conserved but expressed differently. *Eur. J. Biochem.* 268:761–67
33. Galardy RE, Craig LC, Printz MP. 1973. Benzophenone triplet: a new photochemical probe of biological ligand-receptor interactions. *Nat. New Biol.* 242:127–28
34. Galli G, Hofstetter H, Birnstiel ML. 1981. Two conserved sequence blocks within eukaryotic tRNA genes are major promoter elements. *Nature* 294:626–31
35. Garen A, Siddiqi O. 1962. Suppression of mutations in the alkaline phosphatase structural cistron of *E. coli*. *Proc. Natl. Acad. Sci. USA* 48:1121–27
36. Giriat I, Muir TW. 2003. Protein semi-synthesis in living cells. *J. Am. Chem. Soc.* 125:7180–81
37. Hamilton SR, Bobrowicz P, Bobrowicz B, Davidson RC, Li H, et al. 2003. Production of complex human glycoproteins in yeast. *Science* 301:1244–46

38. Hendrickson WA, Horton JR, LeMaster DM. 1990. Selenomethionyl proteins produced for analysis by multiwavelength anomalous diffraction (MAD): a vehicle for direct determination of three-dimensional structure. *EMBO J.* 9:1665–72

39. Hino N, Okazaki Y, Kobayashi T, Hayashi A, Sakamoto K, Yokoyama S. 2005. Protein photo-cross-linking in mammalian cells by site-specific incorporation of a photoreactive amino acid. *Nat. Methods* 2:201–6

40. Hohsaka T, Kajihara D, Ashizuka Y, Murakami H, Sisido M. 1999. Efficient incorporation of nonnatural amino acids with large aromatic groups into streptavidin in in vitro protein synthesizing systems. *J. Am. Chem. Soc.* 121:34–40

41. Hooker JM, Kovacs EW, Francis MB. 2004. Interior surface modification of bacteriophage MS2. *J. Am. Chem. Soc.* 126:3718–19

42. Hortin G, Boime I. 1983. Applications of amino acid analogs for studying co- and post-translational modifications of proteins. *Methods Enzymol.* 96:777–84

43. Hudziak RM, Laski FA, RajBhandary UL, Sharp PA, Capecchi MR. 1982. Establishment of mammalian cell lines containing multiple nonsense mutations and functional suppressor tRNA genes. *Cell* 31:137–46

44. Ibba M, Hennecke H. 1995. Relaxing the substrate specificity of an aminoacyl-tRNA synthetase allows in vitro and in vivo synthesis of proteins containing unnatural amino acids. *FEBS Lett.* 364:272–75

45. Ibba M, Soll D. 2000. Aminoacyl-tRNA synthesis. *Annu. Rev. Biochem.* 69:617–50

46. Kent SB. 1988. Chemical synthesis of peptides and proteins. *Annu. Rev. Biochem.* 57:957–89

47. Kiga D, Sakamoto K, Kodama K, Kigawa T, Matsuda T, et al. 2002. An engineered *Escherichia coli* tyrosyl-tRNA synthetase for site-specific incorporation of an unnatural amino acid into proteins in eukaryotic translation and its application in a wheat germ cell-free system. *Proc. Natl. Acad. Sci. USA* 99:9715–20

48. Kim D, Johnson J. 1988. Construction, expression, and function of a new yeast amber suppressor, tRNATrpA. *J. Biol. Chem.* 263:7316–21

49. Kobayashi T, Nureki O, Ishitani R, Yaremchuk A, Tukalo M, et al. 2003. Structural basis for orthogonal tRNA specificities of tyrosyl-tRNA synthetases for genetic code expansion. *Nat. Struct. Biol.* 10:425–32

50. Kobayashi T, Sakamoto K, Takimura T, Sekine R, Kelly VP, et al. 2005. Structural basis of nonnatural amino acid recognition by an engineered aminoacyl-tRNA synthetase for genetic code expansion. *Proc. Natl. Acad. Sci. USA* 102:1366–71

51. Kobayashi T, Takimura T, Sekine R, Vincent K, Kamata K, et al. 2005. Structural snapshots of the KMSKS loop rearrangement for amino acid activation by bacterial tyrosyl-tRNA synthetase. *J. Mol. Biol.* 346:105–17

52. Kowal AK, Kohrer C, RajBhandary UL. 2001. Twenty-first aminoacyl-tRNA synthetase-suppressor tRNA pairs for possible use in site-specific incorporation of amino acid analogues into proteins in eukaryotes and in eubacteria. *Proc. Natl. Acad. Sci. USA* 98:2268–73

53. Kwok Y, Wong JT. 1980. Evolutionary relationship between *Halobacterium cutirubrum* and eukaryotes determined by use of aminoacyl-tRNA synthetases as phylogenetic probes. *Can. J. Biochem.* 58:213–18

54. Li F, Xiong Y, Wang J, Cho HD, Tomita K, et al. 2002. Crystal structures of the *Bacillus stearothermophilus* CCA-adding enzyme and its complexes with ATP or CTP. *Cell* 111:815–24

55. Liu DR, Magliery TJ, Pastrnak M, Schultz PG. 1997. Engineering a tRNA and aminoacyl-tRNA synthetase for the site-specific incorporation of unnatural amino acids into proteins in vivo. *Proc. Natl. Acad. Sci. USA* 94:10092–97

39. A photocrosslinking amino acid was incorporated into proteins in mammalian cells.

56. Liu DR, Magliery TJ, Schultz PG. 1997. Characterization of an 'orthogonal' suppressor tRNA derived from *E. coli* tRNA2(Gln). *Chem. Biol.* 4:685–91

57. Liu DR, Schultz PG. 1999. Progress toward the evolution of an organism with an expanded genetic code. *Proc. Natl. Acad. Sci. USA* 96:4780–85

58. Liu H, Wang L, Brock A, Wong CH, Schultz PG. 2003. A method for the generation of glycoprotein mimetics. *J. Am. Chem. Soc.* 125:1702–3

59. Magliery TJ, Anderson JC, Schultz PG. 2001. Expanding the genetic code: selection of efficient suppressors of four-base codons and identification of "shifty" four-base codons with a library approach in *Escherichia coli*. *J. Mol. Biol.* 307:755–69

60. McIntosh LP, Dahlquist FW. 1990. Biosynthetic incorporation of ^{15}N and ^{13}C for assignment and interpretation of nuclear magnetic resonance spectra of proteins. *Q. Rev. Biophys.* 23:1–38

61. Mehl RA, Anderson JC, Santoro SW, Wang L, Martin AB, et al. 2003. Generation of a bacterium with a 21 amino acid genetic code. *J. Am. Chem. Soc.* 125:935–39

62. Noren CJ, Anthony-Cahill SJ, Griffith MC, Schultz PG. 1989. A general method for site-specific incorporation of unnatural amino acids into proteins. *Science* 244:182–88

63. Normanly J, Kleina LG, Masson JM, Abelson J, Miller JH. 1990. Construction of *Escherichia coli* amber suppressor tRNA genes. III. Determination of tRNA specificity. *J. Mol. Biol.* 213:719–26

64. O'Connor M. 2002. Insertions in the anticodon loop of tRNA1Gln(sufG) and tRNA(Lys) promote quadruplet decoding of CAAA. *Nucleic Acids Res.* 30:1985–90

65. Pastrnak M, Magliery TJ, Schultz PG. 2000. A new orthogonal suppressor tRNA/aminoacyl-tRNA synthetase pair for evolving an organism with an expanded genetic code. *Helv. Chim. Acta* 83:2277–86

66. Pastrnak M, Schultz PG. 2001. Phage selection for site-specific incorporation of unnatural amino acids into proteins in vivo. *Bioorg. Med. Chem.* 9:2373–79

67. Sakamoto K, Hayashi A, Sakamoto A, Kiga D, Nakayama H, et al. 2002. Site-specific incorporation of an unnatural amino acid into proteins in mammalian cells. *Nucleic Acids Res.* 30:4692–99

68. Santoro SW, Anderson JC, Lakshman V, Schultz PG. 2003. An archaebacteria-derived glutamyl-tRNA synthetase and tRNA pair for unnatural amino acid mutagenesis of proteins in *Escherichia coli*. *Nucleic Acids Res.* 31:6700–9

69. Santoro SW, Wang L, Herberich B, King DS, Schultz PG. 2002. An efficient system for the evolution of aminoacyl-tRNA synthetase specificity. *Nat. Biotechnol.* 20:1044–50

70. Sears P, Wong CH. 2001. Toward automated synthesis of oligosaccharides and glycoproteins. *Science* 291:2344–50

71. Sherman F. 1982. Suppression in the yeast *Saccharomyces cerevisiae*. In *The Molecular Biology of the Yeast* Saccharomyces: *Metabolism and Gene Expression*, ed. JN Strathern, EW Jones, JR Broach, pp. 463–86. Cold Spring Harbor, NY: Cold Spring Harbor Laboratory

72. Srinivasan G, James CM, Krzycki JA. 2002. Pyrrolysine encoded by UAG in Archaea: charging of a UAG-decoding specialized tRNA. *Science* 296:1459–62

73. Steer BA, Schimmel P. 1999. Major anticodon-binding region missing from an archaebacterial tRNA synthetase. *J. Biol. Chem.* 274:35601–6

74. Tang Y, Tirrell DA. 2002. Attenuation of the editing activity of the *Escherichia coli* leucyl-tRNA synthetase allows incorporation of novel amino acids into proteins in vivo. *Biochemistry* 41:10635–45

75. Tian F, Tsao ML, Schultz PG. 2004. A phage display system with unnatural amino acids. *J. Am. Chem. Soc.* 126:15962–63

61. Describes the generation of a completely autonomous bacterium with a 21-amino-acid genetic code that can biosynthesize a nonstandard amino acid (*p*AF) and selectively incorporate it into proteins.

77. X-ray crystallographic studies of a mutant aminoacyl-tRNA synthetase showed significant structural plasticity in the amino acid side chain and protein backbone conformations in the amino acid binding site.

76. Tsao M, Tian F, Schultz PG. 2005. The Staudinger modification of proteins containing azido amino acids. *Chembiochem* 6:2147–49

77. Turner JM, Graziano J, Spraggon G, Schultz PG. 2005. Structural characterization of a *p*-acetylphenylalanyl aminoacyl-tRNA synthetase. *J. Am. Chem. Soc.* 127:14976–77

78. Wacker M, Linton D, Hitchen PG, Nita-Lazar M, Haslam SM, et al. 2002. N-linked glycosylation in *Campylobacter jejuni* and its functional transfer into *E. coli*. *Science* 298:1790–3

79. Wang L, Brock A, Herberich B, Schultz PG. 2001. Expanding the genetic code of *Escherichia coli*. *Science* 292:498–500

80. Wang L, Brock A, Schultz PG. 2002. Adding L-3-(2-Naphthyl)alanine to the genetic code of *E. coli*. *J. Am. Chem. Soc.* 124:1836–37

81. Wang L, Magliery TJ, Liu DR, Schultz PG. 2000. A new functional suppressor tRNA/aminoacyl-tRNA synthetase pair for the in vivo incorporation of unnatural amino acids into proteins. *J. Am. Chem. Soc.* 122:5010–11

82. Wang L, Schultz PG. 2001. A general approach for the generation of orthogonal tRNAs. *Chem. Biol.* 8:883–90

83. Wang L, Schultz PG. 2004. Expanding the genetic code. *Angew Chem. Int. Ed. Engl.* 44:34–66

84. Wang L, Xie J, Deniz AA, Schultz PG. 2003. Unnatural amino acid mutagenesis of green fluorescent protein. *J. Org. Chem.* 68:174–76

85. Wang L, Zhang Z, Brock A, Schultz PG. 2003. Addition of the keto functional group to the genetic code of *Escherichia coli*. *Proc. Natl. Acad. Sci. USA* 100:56–61

86. Wu N, Deiters A, Cropp TA, King D, Schultz PG. 2004. A genetically encoded photocaged amino acid. *J. Am. Chem. Soc.* 126:14306–7

87. Xie J, Schultz PG. 2005. An expanding genetic code. *Methods* 36:227–38

88. Xie J, Wang L, Wu N, Brock A, Spraggon G, Schultz PG. 2004. The site-specific incorporation of *p*-iodo-L-phenylalanine into proteins for structure determination. *Nat. Biotechnol.* 22:1297–301

89. Xu R, Hanson SR, Zhang Z, Yang YY, Schultz PG, Wong CH. 2004. Site-specific incorporation of the mucin-type *N*-acetylgalactosamine-alpha-*O*-threonine into protein in *Escherichia coli*. *J. Am. Chem. Soc.* 126:15654–55

90. Zhang Y, Wang L, Schultz PG, Wilson IA. 2005. Crystal structures of apo wild-type *M. jannaschii* tyrosyl-tRNA synthetase (TyrRS) and an engineered TyrRS specific for *O*-methyl-L-tyrosine. *Protein Sci.* 14:1340–49

91. Zhang Z, Alfonta L, Tian F, Bursulaya B, Uryu S, et al. 2004. Selective incorporation of 5-hydroxytryptophan into proteins in mammalian cells. *Proc. Natl. Acad. Sci. USA* 101:8882–87

92. Zhang Z, Gildersleeve J, Yang YY, Xu R, Loo JA, et al. 2004. A new strategy for the synthesis of glycoproteins. *Science* 303:371–73

93. Zhang Z, Smith BA, Wang L, Brock A, Cho C, Schultz PG. 2003. A new strategy for the site-specific modification of proteins in vivo. *Biochemistry* 42:6735–46

79. The first report that an unnatural amino acid (OMeTyr) could be successfully added to the genetic code of *E. coli* by means of amber-suppression and an engineered orthogonal *M. jannaschii* TyrRS-tRNA pair.

85. A keto amino acid, *p*AcPhe, was incorporated into proteins. The unique reactivity of the introduced keto group allows the selective modification of proteins.

88. A heavy-atom-containing amino acid, iodophenylalanine, was selectively incorporated into proteins, affording a reliable method to prepare iodinated proteins to facilitate SAD phasing in crystallography.

92. The incorporation of a glycosylated amino acid directly into proteins provides a promising tool for the synthesis of homogenous glycoproteins.

Radiolytic Protein Footprinting with Mass Spectrometry to Probe the Structure of Macromolecular Complexes

Keiji Takamoto and Mark R. Chance

Case Center for Proteomics, Case Western Reserve University, Cleveland, Ohio 44106; email: mark.chance@case.edu

Annu. Rev. Biophys. Biomol. Struct. 2006. 35:251–76

First published online as a Review in Advance on February 28, 2006

The *Annual Review of Biophysics and Biomolecular Structure* is online at biophys.annualreviews.org

doi: 10.1146/annurev.biophys.35.040405.102050

1056-8700/06/0609-0251$20.00

Key Words

hydroxyl radical footprinting, computational modeling, protein-protein interactions, mass spectrometry, radiolysis, Fenton chemistry

Abstract

Structural proteomics approaches using mass spectrometry are increasingly used in biology to examine the composition and structure of macromolecules. Hydroxyl radical–mediated protein footprinting using mass spectrometry has recently been developed to define structure, assembly, and conformational changes of macromolecules in solution based on measurements of reactivity of amino acid side chain groups with covalent modification reagents. Accurate measurements of side chain reactivity are achieved using quantitative liquid-chromatography-coupled mass spectrometry, whereas the side chain modification sites are identified using tandem mass spectrometry. In addition, the use of footprinting data in conjunction with computational modeling approaches is a powerful new method for testing and refining structural models of macromolecules and their complexes. In this review, we discuss the basic chemistry of hydroxyl radical reactions with peptides and proteins, highlight various approaches to map protein structure using radical oxidation methods, and describe state-of-the-art approaches to combine computational and footprinting data.

Contents

INTRODUCTION

Advances in protein structure determination and computational modeling mediated by structural genomics initiatives throughout the world promise to correlate sequence and structure for most protein domains within the next five years (10, 15, 16). Coincident with progress toward this milestone is the realization of the importance of macromolecular interactions and even the fundamental significance of large macromolecular complexes in all major normal and aberrant biological functions (29). Solving the structure and connecting structure to function for these large complexes are two of the most important challenges in structural biology today. Unlike solving the structure of protein domains or short nucleic acids that contain tertiary structure, this effort is far from high throughput and likely involves a combination of computational and experimental approaches tailored specifically to the problem at hand. This hybrid approach involves traditional crystallography or NMR approaches, where this higher-resolution data on complexes or their subcomponents are computationally coupled to lower-resolution data from methods such as cryo-electron microscopy, electron tomography, cross-linking, fluorescence resonance energy transfer, and spin label EPR methods (34, 36, 91).

Another well-established method for probing macromolecular structure in solution is footprinting, a term coined by Galas and Schmitz in their landmark papers on the subject (26, 92). Initially the technique was invented to map the sites of DNA-protein interaction (9, 26, 52); the protein-dependent attenuation of DNA reactivity toward dimethyl sulfate modification and DNase I digestion marked the binding site through high-resolution gel readouts of the data. The tool quickly became a standard method for probing DNA-protein interaction (26, 30, 92, 115) and determining RNA structure and dynamics (75, 119). Hydroxyl radical has been accepted as an ideal reagent for nucleic acid footprinting experiments after Tullius and Dombroski (113) introduced Fe(II)-ethylendiaminetetraacetate (EDTA) Fenton-Haber-Weiss chemistry as the radical generator. The use of gamma rays as a radiolysis source to generate hydroxyl radicals was subsequently established. The introduction of high-flux synchrotron X rays for hydroxyl radical generation enabled millisecond time-scale footprinting, which was used to examine Mg^{2+}-dependent RNA folding (18, 94).

The development of protein footprinting techniques lagged behind that of nucleic acids, as the first report of protein footprinting did not appear until 1988 (100). Paralleling the development of nuclease cleavage methods for DNA, this early approach utilized limited proteolysis of proteins followed by SDS polyacrylamide gel electrophoresis (SDS-PAGE) to separate the cleaved fragments. Compared with nucleic acids gel methods, which have single-nucleotide resolution, protein gel methods are cruder, providing less spatial resolution in the probe of structure. The development of tagged Fe(II)-EDTA (84) and free Fe(II)-EDTA Fenton chemistry (46) was an improvement; however, backbone cleavage with Fenton, although nonspecific, is relatively inefficient. The development of new analytical technologies for examining proteins, particularly the use of mass spectrometry (MS) for the detection of cleaved fragments or irreversibly modified peptides after proteolysis, improved the spatial resolution of the technique considerably (45, 104).

Although the hydroxyl radical has attractive features in that solvent accessibility is conveniently probed, the intrinsic inefficiency of backbone scission is a serious drawback for the conduct of footprinting experiments. However, examination of the well-established literature concerning metal-catalyzed oxidation and radiolysis of peptides and proteins suggests that side chain modification by hydroxyl radical should be efficient and rapid (27). The chemistry of amino acid and peptide oxidation using MS detection was investigated subsequent to radiolysis (32, 70). These experiments showed that in dilute aqueous solution oxidative modification of side chains is observed in considerable excess compared with backbone scission or cross-linking. Tandem mass spectrometry (MS/MS) methods were found to be ideal for examining the specific sites of oxidation (13, 64, 70). Thus the reactivity of side chains to hydroxyl radical attack and the attenuation of this reactivity as a function of ligand binding, unfolding, or macromolecular interactions signal surface accessibility changes at the defined sites in question.

Over the past five years, extensive studies of the radiolysis chemistry of amino acids, peptides, and proteins using MS-based detection and comparisons of these recent data to the extensive literature available (33, 78, 123–127) have resulted in the development of hydroxyl-radical-mediated protein footprinting as an effective method for probing protein structure (13, 31, 34, 37–39, 62–64, 69, 87). In addition to radiolytic approaches, hydroxyl radical protein footprinting methods based on chemical generation of hydroxyl radicals by both Fenton chemistry (96) and UV photolysis of hydrogen peroxide have been described (97, 98). This review describes the current state of hydroxyl radical protein footprinting and outlines likely directions of its future development. "Hydroxyl radical–mediated protein footprinting" as discussed in this review emphasizes modification-based footprinting approaches; we do not include extensive discussions of "cleavage-based footprinting" methods in this review.

Hydroxyl radical: highly reactive species used for footprinting techniques

EDTA: ethylendiaminetetraacetate

Radiolysis: the chemical reaction caused by the absorption of high-energy photons

SDS-PAGE: SDS polyacrylamide gel electrophoresis

Fenton chemistry: well-known chemical reaction used to generate hydroxyl radical based on transition metal (Fe^{2+})-mediated homolysis of peroxides

Tandem mass spectrometry (MS/MS): technique used to identify peptide sequences in which the peptide ion is separated by first-stage MS and then introduced into collision cell or reaction chamber to break down ions

CHEMISTRY OF HYDROXYL RADICAL FOOTPRINTING

Hydroxyl radicals have significant advantages over other footprinting reagents. Although selected chemical modification and cleavage reagents have been used successfully for protein footprinting (45, 60, 104, 111, 118), they

Fenton chemistry

Reductant (ascorbate, thiol)

$$Fe^{2+} + H_2O_2 \longrightarrow Fe^{3+} + {}^{-}OH + {}^{\cdot}OH$$

Radiolysis of water

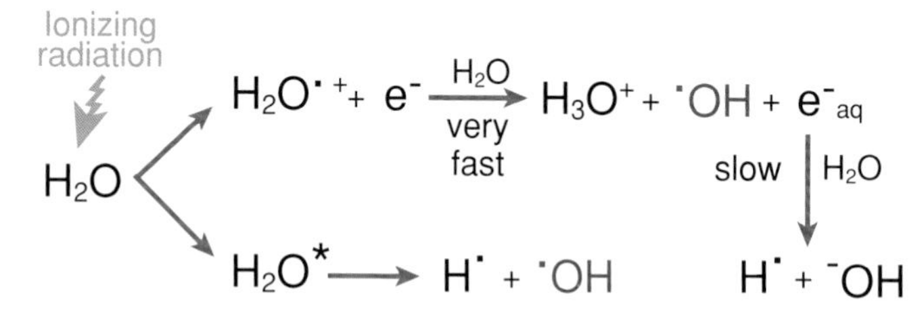

Figure 1

Schematic representation of hydroxyl radical generation by Fenton chemistry and radiolysis of water by ionizing radiations. EDTA chelates Fe^{2+} to provide an electrically neutral species, preventing specific interaction with proteins. It is not directly involved in chemistry.

are often limited by the number of reactive sites available or by the large size of the reagent compared with hydroxyl radicals. The major advantages of hydroxyl radicals are as follows. First, these radicals have van der Waals area and solvent properties similar to those of water molecules. This makes hydroxyl radicals ideal as solvent accessibility probes. Second, they are highly reactive species that have well-understood chemical selectivity. This is of great importance for a protein footprinting strategy that attempts to increase the number of target sites to maximize the available structural information. Third, they can be generated safely and conveniently under a wide range of solution conditions.

Generation of Hydroxyl Radicals

Hydroxyl radical is generated mainly using two methods (other methods can be used as well and are described below) (7). The first method is transition metal-dependent chemical generation from peroxide and the second method is radiolysis of water (**Figure 1**). Each method has advantages over the other. Among the chemical methods, the most frequently used is Fe(II)-EDTA Fenton chemistry (113), which requires commonly available chemicals that are cheap and easy to handle. The EDTA chelate tends to neutralize the transition metal's positive charge; however, specific interactions of the chelate with macromolecules can bias the local reactivity observed. On the other hand, radiolysis of water using gamma or X rays generates radicals isotropically in the solution and does not require the addition of any chemicals, although buffering of any solution for biochemical conditions of interest is desirable. Radiolysis can also be used to carry out footprinting in vivo (81). If high-flux X rays from synchrotron sources are employed, millisecond exposures are sufficient for footprinting such that millisecond time resolution footprinting can be carried out (94).

Reactions of Hydroxyl Radicals with Peptides and Proteins

Initially, hydroxyl radicals for protein footprinting were used to nonspecifically cleave the backbone (2, 3, 24, 46, 48, 83) in conjunction with SDS-PAGE to separate the cleavage products. The chemistry involves the hydrogen abstraction from the Cα carbon and subsequent reaction of the radical species with oxygen leading to backbone cleavage, ultimately generating a new N-terminal end (**Scheme 1**) (27, 28). However, the reaction

Scheme 1

The reaction mechanism of peptide backbone scission by hydroxyl radical attack under aerobic conditions.

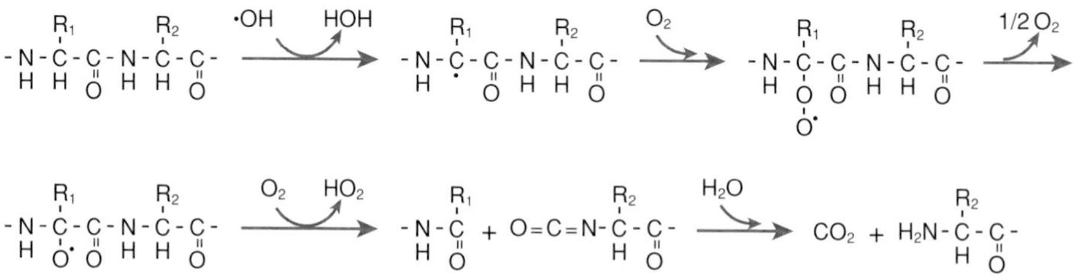

of side chains with hydroxyl radicals occurs at rates 10 to 1000 times faster than the abstraction of hydrogen from the Cα carbon (27, 70, 124). Thus, side chains are preferable as probes for the study of protein structure. Quantitative mass spectrometric analyses of the modified protein fragments provide a footprinting approach that has been used to probe protein structure and protein-ligand and protein-protein interactions (34, 37–39, 62–64, 69, 87, 96–98).

To define the side chain probe set for protein footprinting, the reactions of proteins and amino acids have been investigated extensively using MS under aerobic (27, 70, 123–127) or anaerobic conditions (27, 33, 70, 78). The reactions of hydroxyl radicals with aliphatic side chains are initiated by hydrogen abstraction from side chain carbon atoms. Under aerobic conditions this carbon radical eventually reacts with oxygen and forms hydroxyl or ketone groups (in Ala the side chain is converted to an aldehyde) through a peroxide radical intermediate with mass increases of 16 or 14 Da, respectively (**Figure 2**). Reactions of hydroxyl radical with aromatic rings involve the addition of hydroxyl radical to the ring; subsequent reaction of the radical species with oxygen results in +16-Da or higher integer multiple (e.g., +32, +48) increases in mass. Acidic residues react with hydroxyl radical and lose CO_2, resulting in an aldehyde with an overall reduction in mass of 30 Da. The basic amino acids also react with hydroxyl radical and give a rise to unique products. The reaction of lysine is identical to that of aliphatic residues. Histidine undergoes a complicated set of reactions, including ring opening, which generates multiple products. Arginine can suffer +16 or +14-Da mass changes when the β- or γ-carbon experiences hydrogen abstraction. However, when the δ-carbon is attacked, a guanidino elimination reaction follows, resulting in a product with a 43-Da reduction in mass. This reaction has consequences for the mass spectrometric analysis, as a strong positive charge is lost in the reaction; this may influence digestion with trypsin and make detection of reaction product difficult in positive ion mode MS.

The sulfur-containing residues are highly reactive toward hydroxyl radicals. The study of these residues revealed that Met is converted to sulfoxide (+16 Da) fairly easy (but not to +32 with equal efficiency) and that Cys is converted primarily to the negatively charged cysteinic acid (+48 Da), which is not easily detected in positive ion modes. **Figure 2** summarizes the reactions of hydroxyl radical with many of the amino acid side chains. The uniqueness of the chemical products can in some cases identify the oxidation sites on the basis of characteristic mass signatures.

The usefulness of a side chain in footprinting experiments depends on both its ability to react with hydroxyl radicals and the ease of detection of the reaction products in MS experiments (124). The relative reactivity of the side chains under aerobic conditions using MS detection is as follows: Cys > Met > Trp > Tyr > Phe > Cystine > His > Leu ~ Ile > Arg ~ Lys ~ Val > Ser ~ Thr ~ Pro > Gln ~ Glu > Asp ~ Asn > Ala > Gly. The residues Gly, Ala, Asp, and Asn are not useful as probes because of their low reactivity (Gly is similar in reactivity to backbone carbon atoms). Ser and Thr, though as reactive as Pro, a probe known to be useful, have reaction products that are difficult to detect. Thus, 14 of 20 residues can be routinely used for protein footprinting. These 14 residues cover ~65% of the sequence of the typical protein; thus structural resolution of the technique is reasonably good (127).

The sulfur-containing residues (Met and Cys) are susceptible to secondary oxidation after hydroxyl radical exposure; radiolysis can generate less-reactive and long-lived radical species such as hydrogen peroxide and related peroxide radicals. If these species are not quenched, the observed oxidation extent of peptides can increase during sample processing, including the protease digestion and chromatographic steps. In particular,

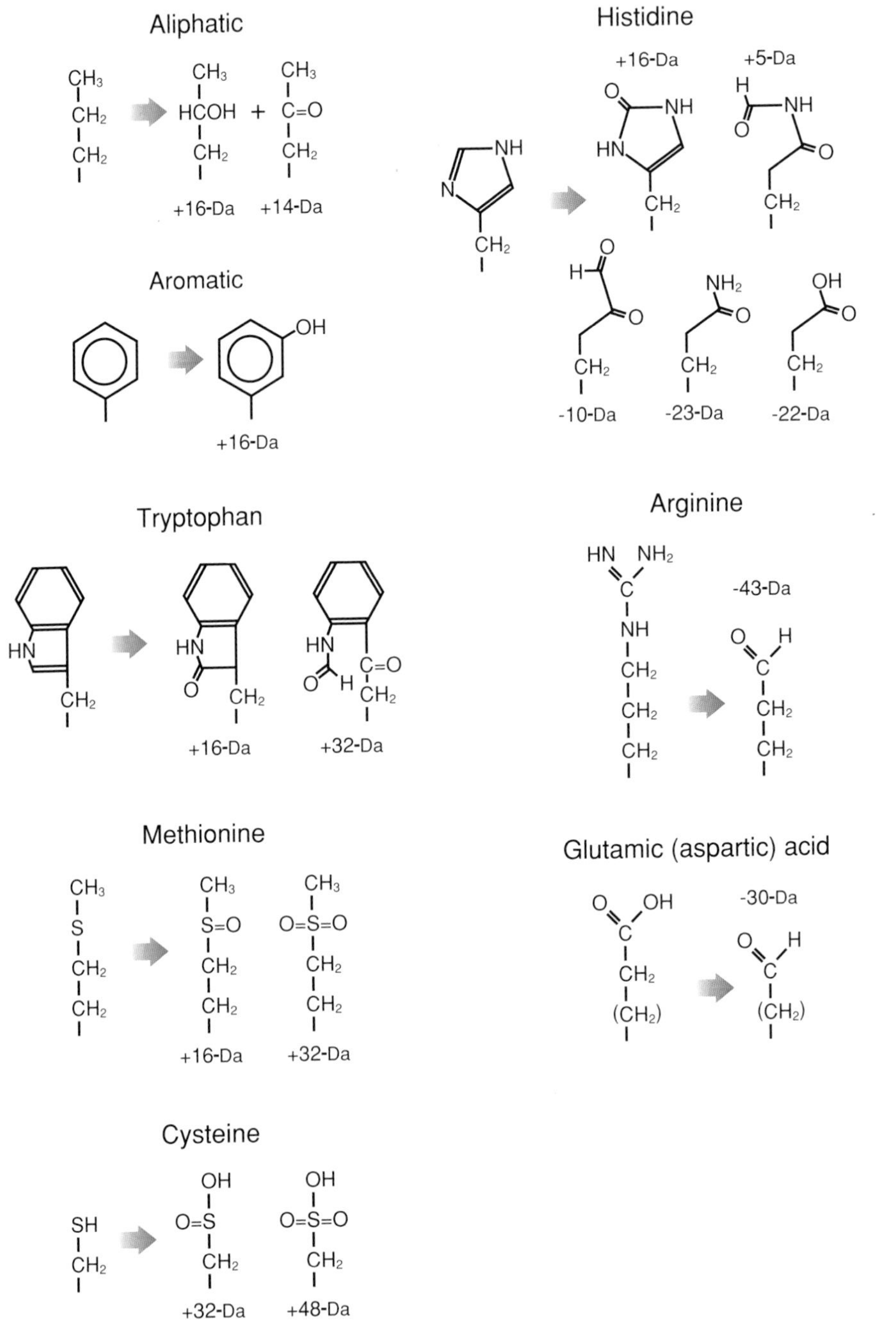

Figure 2
The reaction products by amino acid side chains oxidation subsequent to radiolysis. Only major products and some minor products are shown. Lysine undergoes a reaction identical to that of aliphatic side chains.

methionine is known for its susceptibility to such oxidation. Also, in Fenton chemistry, hydrogen peroxide must be removed or quenched right after reaction completion. Such secondary oxidation can be avoided by the addition of reducing species such as methionine in its amino acid or amide forms (in excess) to compete with unwanted secondary oxidation of methionine and cysteine residues in digested peptides or protein species (126).

Alternatively catalase can be added to scavenge excess peroxide.

MASS SPECTROMETRY METHODS FOR QUANTITATIVE FOOTPRINTING

Mass spectrometric readouts of oxidized proteins provide the analytical basis for determining solvent accessibility according to peptide reactivity and provide structural resolution by MS/MS methods. Accurate determination of the extent of oxidation provides the basis for sensitive measures of structural changes in ligand binding and in protein interactions. We outline the specific protocols required in order to achieve quantitative results from hydroxyl radical footprinting experiments.

Quantification of Peptide Oxidation with LC-MS

The target protein subjected to hydroxyl radical oxidation is analyzed by MS subsequent to protease digestion. These procedures are based on those developed for deuterium exchange experiments (50, 59). As the radiolytic modifications of proteins are covalent and stable (unlike in deuterium exchange), subsequent analyses are straightforward and relatively easy to carry out. One result of this is that samples can be frozen after footprinting for later analysis. Both specific and nonspecific proteases can be selected or used in tandem at a range of temperatures or digestion times to generate peptide fragments of interest. In addition, strong reducing agents [e.g., TCEP; Tris(2-carboxyethyl)phosphine] can be used to digest heavily disulfide-bonded species with long incubation times.

Proteolysis is generally performed individually with a set of specific proteases such as trypsin, Asp-N, or Glu-C in order to maximize sequence coverage. Complete coverage for large proteins using MS is not easy owing to inadequate digestion and poor ionization of certain peptides. However, 80% coverage or more is generally achievable. The resultant mixture of protein fragments (digested peptides) is analyzed by liquid-chromatography (LC)-MS (**Figure 3**). Reverse-phase high-performance liquid chromatography (HPLC) separates the digested peptides and also separates radiolytically modified peptides from their unmodified parent peptides. The modified peptides are generally eluted earlier than the unmodified peptides for simple alcohol or keto additions; however, the identity of all modified and unmodified peptides should be confirmed by MS/MS. The abundance of each peptide is calculated from the selected ion chromatogram (SIC) peak area. As SIC monitors the intensity of specific m/z (mass-to-charge ratio) ions as a function of retention time, integration of the peak area in SIC gives a total current of selected ion. However, a set range of the retention time (determined empirically) within the chromatogram is used for each experiment to ensure consistency of quantitation for the modified and unmodified species. Also, multiple modified species with separate or multiple modifications may be present; these should all be quantified and added into the sum total modified. Then, the fraction of unmodified peptide is calculated from the integrated intensity values of the modified and unmodified peptides. To provide oxidation rate data that emphasize the intact population, we calculate the fraction unmodified peptide and monitor the loss of the unmodified fraction. The dose response curves are generated by the fraction unmodified peptide calculated at multiple time points of oxidation. Multiple independent experiments are globally fit to provide the oxidation rate constants using nonlinear regression. The data obey a pseudo first-order reaction; deviation from these kinetics is observed at increased exposure times and indicates overoxidation of the sample. The longest time points that evidence such overoxidation should be removed from the fits to provide the best data. Although the full dose response experiment is labor intensive, the rate constants calculated by this protocol are reliable measures of peptide reactivity.

Hydroxyl radical footprinting: methods used to probe solvent-accessible surface area of macromolecules

LC-MS: liquid-chromatography/mass spectrometry

HPLC: high-performance liquid chromatography

Selected ion chromatogram (SIC): the time course record of ion intensity with particular m/z value

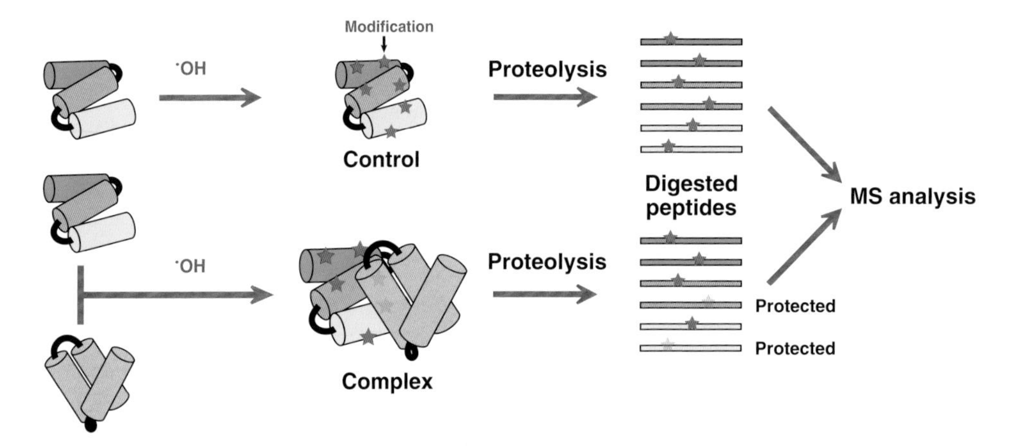
Exposure to hydroxyl radicals and subsequent proteolysis
Modification
·OH
Control
Proteolysis
Digested peptides
MS analysis
·OH
Complex
Proteolysis
Protected
Protected

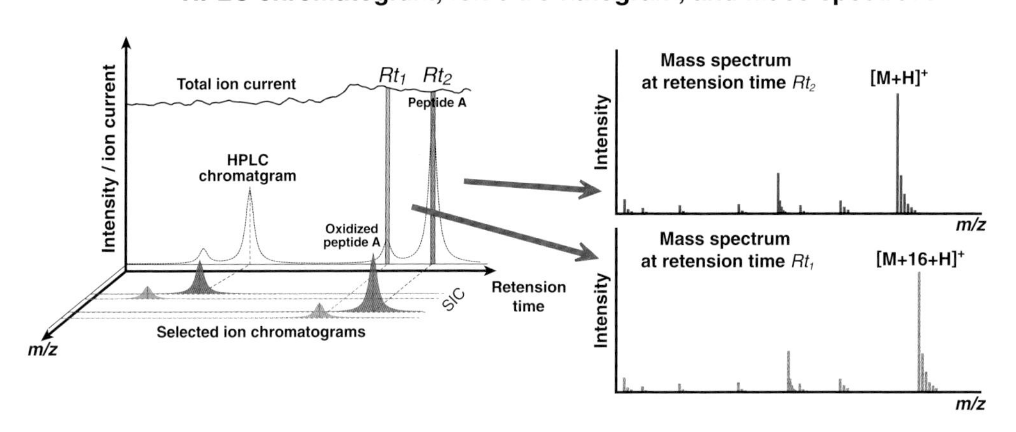
HPLC chromatogram, ion chromatogram, and mass spectrum
Intensity / ion current
Total ion current
Rt1 Rt2
Peptide A
HPLC chromatgram
Oxidized peptide A
SIC
Retension time
Selected ion chromatograms
m/z
Mass spectrum at retension time Rt2
[M+H]+
Intensity
m/z
Mass spectrum at retension time Rt1
[M+16+H]+
Intensity
m/z

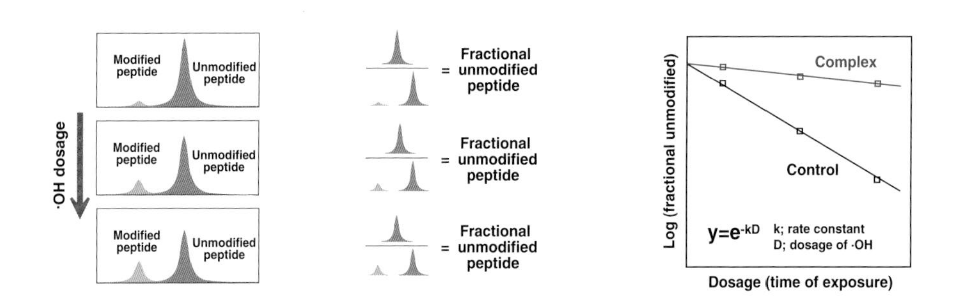
Calculation of rate constants from dose response curve
Modified peptide
Unmodified peptide
·OH dosage
Fractional = unmodified peptide
Log (fractional unmodified)
Complex
Control
y=e-kD
k; rate constant
D; dosage of ·OH
Dosage (time of exposure)

Confirmation of Peptide Identity and Determination of Modification Site by MS/MS

An important aspect of analysis is, of course, confirmation that a peak with a particular m/z value belongs to a peptide expected from the sequence based on the expected digestion pattern. In addition to confirmation of peptide identity, structural resolution requires one to determine the modification site(s) on the peptide. The fragmentation of peptides by collision-induced dissociation (CID) is studied extensively and reliable for this purpose (54, 56, 105, 108, 129). The translational kinetic energy of selected ions is converted to internal energy of the ions (for low-energy collisions this energy is mostly vibrational), inducing cleavage of the peptide backbone. The fragments are termed a-, b-, and c-series (for N-terminal fragments) or x-, y-, and z-series (for C-terminal fragments) depending on the location of the bond cleavage (89). The most frequently observed ion types in low-energy collisions are b- and y-series ions, depending on the position of charged residues in the sequence. By examining the mass differentials of ion peaks observed in the MS/MS spectrum, one can deduce the sequence of peptide (with exceptions for isobaric amino acids or posttranslational modifications). Thus, by performing an MS/MS scan for a particular m/z ion, the identity of peptide can be determined or confirmed. If the peptide has a modification site generated by oxidation, MS/MS scan also identifies the footprinting probe residue. For example, as shown in **Figure 4**, the peptide with sequence ASDFGHK would have a y-series of fragment ions with m/z of y_1 147.1, y_2 284.2, y_3 341.2, y_4 488.3, y_5 603.3, and y_6 690.3 from a singly charged precursor ion of m/z 761.4. These m/z values give mass differences of 137, 57, 147, 115, 87, and 71 that correspond to residues H, G, F, D, S, and A, respectively. If the peptide has one oxidation adduct due to alcohol formation, the selected precursor ion will have m/z 777.4 for the singly charged ion (761.4 + 16). If one observes an MS/MS spectrum with y_1 147.1, y_2 284.2, y_3 341.2, and y_4 + 16 504.3, this indicates that the +16 modification is on the fourth residue of the peptide, as the first three y-type ions are unshifted compared with tandem spectra of the unmodified ion whereas the y_4 ion retains the modification and is shifted. Observation of additional fragment ions of y_5 + 16 603.3 and y_6 690.3 + 16 supports this assignment. That the fourth residue from the C terminus is F, a residue highly likely to be oxidized by +16, also supports the assignment of F as the probe residue. The actual spectrum can be more complex, as modifications may occur at multiple residues. Also, the doubly charged, modified ion at 388.7 can be selected and fragmented to monitor the b-type ions in this case, for which b_1-b_3 should be unshifted relative to the unmodified ion and b_4-b_7 would retain the modification.

A summary of the experiment is shown in **Figure 4**. We can determine the modification site and confirm the identity of peptide; however, the CID process and spectrum are semiquantitative at best, so that it is risky to

Collision-induced dissociation (CID): the technique used for MS/MS to generate fragment ions in which translational and kinetic energy of ions are converted to internal energy by collision with gas within the collision chamber

Figure 3

Hydroxyl radical footprinting: data collection and data analysis. Top panel: Protein is exposed to hydroxyl radical and modified covalently. The resulting protein sample is then digested by protease or chemical cleavage to fragments that are suitable in size for mass spectrometry. The experiment is carried out for each individual protein and for the protein complex. In a tight binding interface, some regions are protected from hydroxyl radical attack. Middle panel: Peptides are separated by liquid chromatography and introduced into a mass analyzer. The selected ion chromatograms (SIC) are constructed for each ion (with particular mass) as a function of retention time. By monitoring the mass and time, we know what species appears at what retention time. By integrating peak areas in SIC, we can calculate the total indicated ion abundance. Bottom panel: The determinations of modification rates are performed by calculating the "loss of intact peptide" in order to maximize the interrogation of intact material.

Nomenclature of fragment ions

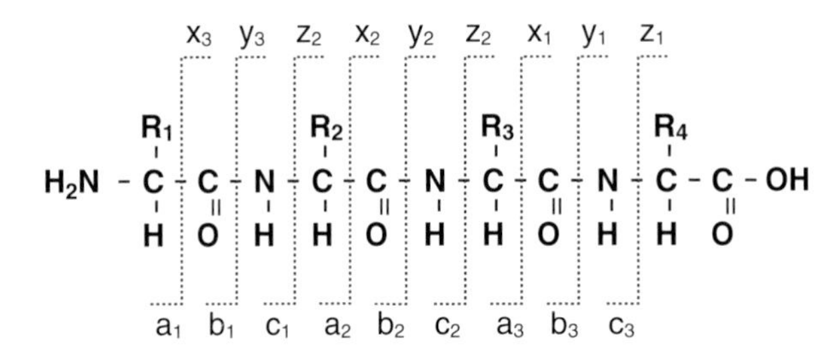

MS/MS spectrum of CID fragment ions
(Identification of peptide and determination of modification sites)

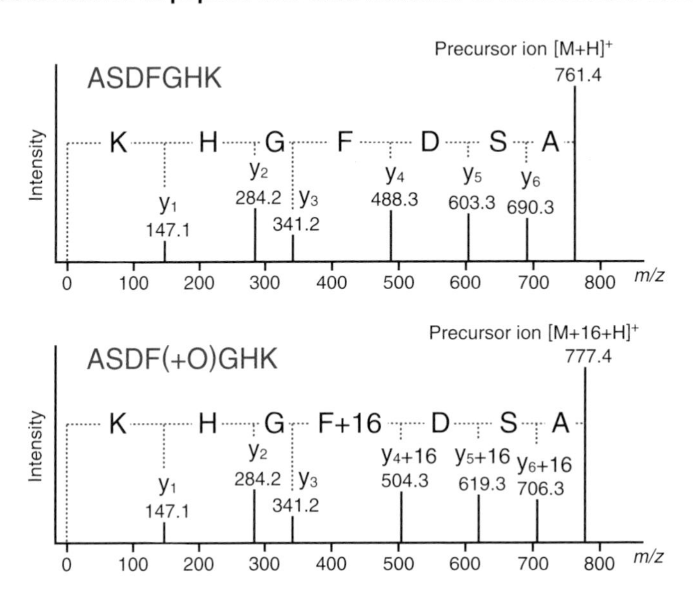

Estimation of contribution of each modification site
in the case of multiple sites in peptides

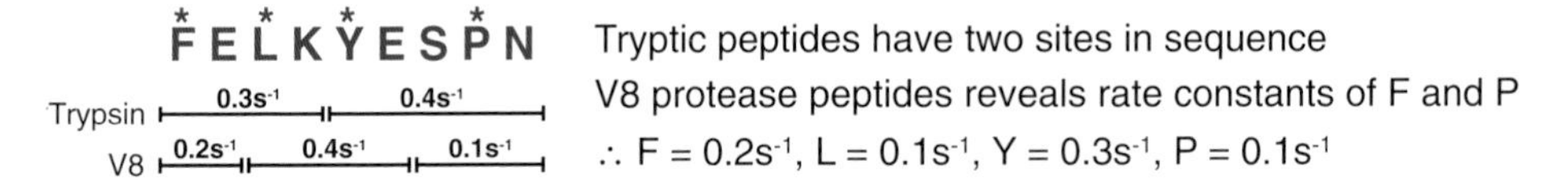

Figure 4

Analyses by MS/MS. Top panel: The nomenclature for fragment ions. The a-, b-, c-series and x-, y-, z-series of ions are N- and C-terminal fragments, respectively. The numbers 1, 2, 3 represent how many residues are in the fragment from N or C terminus of the peptide. Middle panel: Interpretation of MS/MS data. The schemes illustrate the "ideal" CID spectra for hypothetical peptide ASDFGHK. The upper spectrum shows the cartoon of CID data of unmodified peptide, and the lower panel illustrates the CID spectrum of modified peptide. By comparing two spectra with sequence, one can determine the location of the modification site. Bottom panel: The possible determination of contributions of modification sites in the case of multiple modifications is observed.

estimate the relative contribution of multiple residues in the case of multiple modification sites. Multiple enzymatic digestion can give some clues as to the relative contribution of individual residues. If, as a result of peptide mapping, using different enzymes digestion sites can isolate subfragments, it is sometimes possible to estimate the contribution from each site. The bottom panel in **Figure 4** explains one possible approach along these lines. However, it is highly sequence dependent and not always applicable.

Examples of Footprinting to Examine Protein Structure

In this section we show how hydroxyl radical footprinting methods are applied to probe protein structure and protein interactions. The method is still relatively new compared with hydroxyl radical nucleic acids footprinting; however, it is growing in acceptance and use. Different groups have developed variations of the technique to adapt to specific biological problems of interest. Many of these are discussed below.

Radiolytic Footprinting of Cytochrome *c*

Anderson's group has been developing an anaerobic radiolysis method that utilizes nonsolvent-exchangeable H/D exchange mediated by hydroxyl radical oxidation (32, 33, 78). Radiolysis leads to hydrogen abstraction from Cα (or side chain carbon atoms) and forms radicals at these positions. This radical is then repaired by dithiothreitol that has been equilibrated with D_2O under anaerobic conditions. The –SH group has been replaced as –SD; thus when the deuteron repairs the radical a nonexchangeable deuterium addition event has occurred providing a mass tag for the solvent-accessible site (**Scheme 2**). This data was compared to the analysis of radiolytic modification of cytochrome *c* under aerobic conditions (79). The study also includes analyses of oxidation mechanisms of certain residues. Anderson and coworkers utilized 50% ^{18}O-labeled H_2O as a source of radicals generated by gamma-ray irradiation to differentiate the reaction directly by hydroxyl radical (contains ^{18}O) and through oxygen (mainly ^{16}O). As expected from previous literature (70), no +18-Da oxidation of Tyr was observed, indicating that the oxidation source is molecular oxygen. They also found a +14-Da modification product of Tyr that was presumably the carbonyl form. They concluded that the oxidation of Tyr itself is mediated by the addition of radical but that carbonyl oxygen is exchanged with solvent during subsequent analyses.

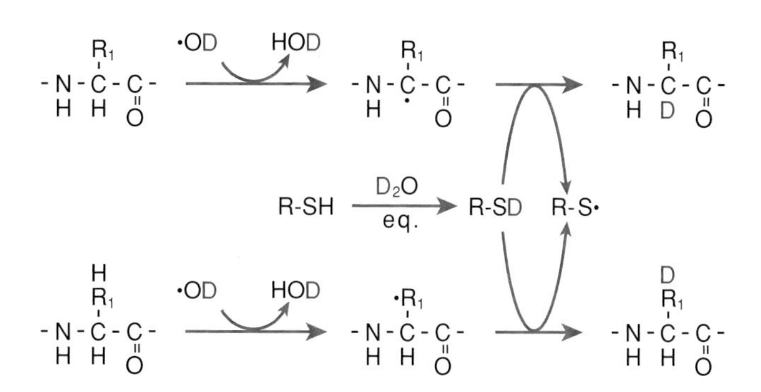

Scheme 2

The mechanism of hydroxyl radical–mediated deuterium exchange of nonsolvent exchangeable backbone/side chain hydrogen atoms.

Radiolysis by Electric Discharge within ESI Ion Source

Maleknia et al. (71) have utilized an electrospray ionization (ESI) ion source as a hydroxyl radical generator. This electrochemical process is a third method that generates hydroxyl radicals. The high voltage given to the electrospray tip (8 kV) while using oxygen as a nebulizer gas in the ESI ion source produces hydroxyl radicals through generation of active oxygen species by electrochemical reactions. The chemistry of the reactions with amino acids is apparently the same as the chemistry of aerobic hydroxyl radical reactions. The advantage of this technique is that no special instruments are needed to perform the modification reaction. Especially in the case of intact protein, it is just a routine ionization reaction without any additional

ESI: electrospray ionization

ES-FT-ICR-MS: electrospray Fourier transform ion cyclotron resonance mass spectrometry

experiments. The group successfully studied some examples that illustrate interactions of peptide and protein (121, 122). In order to retrieve the information of modification sites, the sprayed sample from the ESI ion source needle needs to be collected and condensed. This is a major drawback of this technique. There is also the question of whether the structures of proteins are native under such a severe electrochemical condition within the aerosol. In spite of these limitations, it is a useful technique that can be performed within the regular ESI ion source.

Fenton/Photochemical Hydroxyl Radical Footprinting

Hettich's group is also developing hydroxyl radical–based footprinting techniques. Fenton chemistry (96) and photochemical-based approaches (97, 98) have been reported by the group. Fenton chemistry is a convenient method that generates hydroxyl radicals without special equipment and is suited for the static analyses of structures of macromolecules. This technique has been used for structural analyses of nucleic acids for years (4, 12, 43, 44, 106, 112, 113, 119); however, they are the first group to apply Fenton chemistry–based hydroxyl radical footprinting to proteins. They probed apomyoglobin solution structure using Fenton chemistry and analyzed the products by ES-FT-ICR-MS (electrospray Fourier transform ion cyclotron resonance mass spectrometry) or LC-MS/MS. They reported that apomyoglobin's structure by NMR is consistent with their footprinting experimental results. NMR studies showed an overall structure for apomyoglobin that is globular and rigid except in selected regions. Interestingly, although W7 and L11 are not solvent accessible for X-ray crystal structures of native myoglobin, Hettich and coworkers observed oxidative modifications for these residues. As NMR analyses reported that the region that covers these two residues is disordered in solution, it is clear from footprinting that the residues are much more accessible than in the myoglobin crystal structure, consistent with the NMR data.

In spite of the success of this study, they allow that the Fenton chemistry requirement of Fe^{2+} with EDTA and relatively high concentration of ascorbate may affect the structure of proteins for some experiments. With those concerns as background, they have used a photochemical method for generating hydroxyl radicals (97, 98). UV-induced homolysis of hydrogen peroxide is well known and was adopted as a means of hydroxyl radical generation. Hettich and coworkers took full advantage of state-of-the-art FT-MS (Fourier transform mass spectrometry) instrumentation and investigated the relationship between accessible surface area and the extent of oxidation. The relation between the two was seen to be linear for both the entire protein (until cooperative oxidation is observed at high doses) and peptides digested by protease. They (97) reported that peptides missing in LC-MS/MS experiments can be found in direct infusion ES-FT-MS owing to the higher resolution and wider dynamic range. As mass spectrometers are continuing to evolve, it is feasible to expect that detection limits and sequence coverage will improve. The latest work from this group is also intriguing (98). They compared the footprinting data of guanidine-HCl denatured protein with that of native protein. The footprinting results were used as structural restraints to examine structures that are computationally predicted for Sml1P protein. This subject is discussed further below.

Probing Actin Structure and Interactions by Synchrotron X-Ray Footprinting

Our group has been examining actin structure by utilizing synchrotron X-ray footprinting. Actin is fundamental for eukaryotic cell motility and its interconversion to actin filaments is tightly regulated by binding to a number of actin binding proteins (49). In spite of its importance, the structure of free monomeric

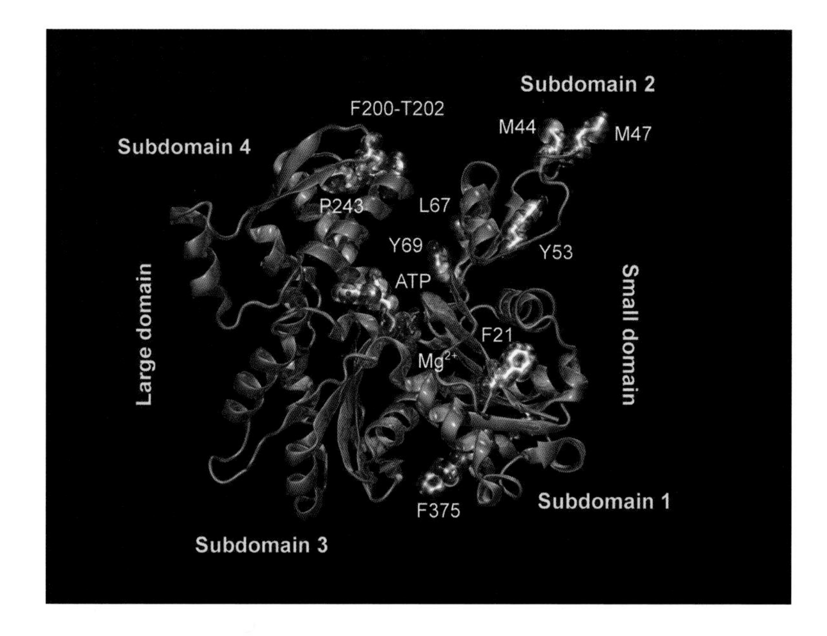

Figure 5

The three-dimensional structure of actin. The ATP, metal ion, and residues that experience the protection upon exchange of metal ion from Ca^{2+} to Mg^{2+} are indicated by sticks and the surrounding spheres.

actin bound to the biologically relevant Mg^{2+} ion has not been determined by X-ray crystallography or NMR. There is a good reason for the failure of these structural determinations. The nature of actin prevents the attempt because it polymerizes at relatively low concentrations (in the few micromolar range for Mg^{2+} actin, depending on solution conditions). Also, actin expressed in *Escherichia coli* is toxic, making the production of labeled material for NMR experiments difficult and expensive. Thus, methods such as footprinting become quite valuable in providing structural information for actin and its complexes with binding proteins (35, 38). **Figure 5** shows the overall structure of the actin monomer. The actin monomer is composed of two domains, and each domain is subdivided into two subdomains. The nucleotide binding pocket and metal binding site are located at the bottom of the large cleft. Although numerous X-ray crystal structures of monomeric actin are bound to Mg^{2+} nucleotide, all of them include other bound ligands or proteins that prevent polymerization. These structures are similar to the structure of monomeric actin bound to Ca^{2+} nucleotide, which requires much higher actin concentrations for polymerization.

The structural differences observed between Ca^{2+}-ATP and Mg^{2+}-ATP actin using hydroxyl radical footprinting are supported by numerous other biophysical and biochemical experimental results such as limited proteolysis (19, 103) and fluorescence studies (25, 74, 116, 130). The characteristic difference in structure of Mg^{2+} compared with the Ca^{2+} form of monomeric actin revealed by footprinting is closure of its large cleft. The sites of protection for Mg^{2+} versus Ca^{2+} actin are located primarily at the edge or within the large cleft (**Figure 5**). Protection sites located on residues Met-44 and Met-47 of subdomain 2 and residues 200–202 and Pro-243 of subdomain 4 are suggestive of contact formation at the cleft edges. The high degree of protection (>80% reduction in reactivity) within the large cleft residues Leu-67 and Tyr-69 strongly indicates the burial of those residues by complete closure of large cleft. There are also observed changes in hydroxyl radical reactivity in N-terminal (Phe-21) and C-terminal (His-371, Cys-374, and Phe-375) regions that are part of subdomain 1 but are linked to other subdomains. Because the other regions of the protein do not show changes in reactivity with

GS1: gelsolin segment 1

Comparative modeling: computational modeling techniques that take advantage of sequence and structural similarities

Ab initio modeling: computational modeling techniques for determining structure in which structural models are computed by physical parameters with a certain degree of simplification (force field)

hydroxyl radicals, it is unlikely that the protein structure undergoes large-scale reorganization. Although the N- and C-terminal regions' changes in reactivity are not fully understood, it is clear that the large cleft experiences Mg^{2+}-dependent closure, thus explaining hydroxyl radical footprinting results. The ability of footprinting to reveal metal-ion-dependent structural changes was also revealed in a study of Ca^{2+}-dependent gelsolin activation (62, 63).

Our study of actin also involves the examination of protein-protein interactions. Actin interacts with a number of proteins that control biological processes such as elongation or severing of F-actin filaments. In the course of our actin monomer structure studies, the structures of Ca^{2+}- and Mg^{2+}-bound actin monomer interacting with gelsolin segment 1 (GS1) were investigated. Protections were observed at positions where contacts are formed in crystal structures, confirming that the structure of actin/GS1 complex in solution is similar to the observed crystal structures. Another actin protein of interest is the actin monomer's interaction with cofilin. Cofilin binds tightly to actin monomers and filaments and facilitates filament severing. It has been predicted to bind to actin at the site of GS1 binding. However, our recent data indicate an entirely different binding mode (A. Kamal & M. R. Chance, unpublished data). The complementary binding surface of actin on cofilin has been previously mapped (38). With this new protection data on actin, a model of the actin/cofilin complex may be constructed.

The Future: Hybrid Approaches that Combine Experimental and Computational Data

We have described examples of hydroxyl radical protein footprinting section and their use in mapping solvent-accessible surface area. The underlying chemistry has been thoroughly investigated, and the method works well for probing structural changes and protein-protein interactions. Computational modeling, both comparative modeling and threading, can predict a wide range of protein structures and can operate to predict structure data on a large scale (15, 16, 22). Although its accuracy of prediction is being improved, experimental data can supplement the predictions, especially when appropriate templates are not available or are distantly related to the target protein (128). To speed structure solution, there is a growing interest in hybrid methods that represent a marriage between experimental and computational approaches. More precisely, we wish to include explicit experimental constraints within the computational structure-prediction and modeling programs.

Hydroxyl radical footprinting data have served as excellent constraints for RNA structure predictions (85, 86, 110). Similar approaches can be applied in the case of protein footprinting data. These approaches are well precedented for proteins; for example, numerous studies combine the experimental and computational approaches using NMR dipolar-coupling data (23, 39, 51, 55, 77, 82, 88, 98, 117, 120). However, there are few published works that take this approach using hydroxyl radical footprinting data.

Structure Modeling by Footprinting and Ab Initio Modeling

As mentioned above, Hettich's group has carried out footprinting experiments on the Sml1p protein. They combined hydroxyl radical footprinting data and an ab initio modeling strategy. Sml1p is a small protein composed of 104 amino acids but does not share enough sequence similarity with any other protein families for accurate comparative modeling (76, 90, 109). Threading methods (57, 61) have also failed to provide a reasonably good template structure (42). Thus, their only choice for computational modeling was an ab initio strategy (6, 61, 102). Using the HMMSTR (11)/Rosetta (101) server, they retrieved five structures and compared the structures with hydroxyl radical footprinting data in order to assess the models. Because they compared footprinting data of native and denatured structures, they could easily assign the residues within the hydrophobic cores, e.g., these residues were reactive only in the denatured state. This is a strong constraint to evaluate predicted structures. They found none of the models to be consistent with experimental data, which is not surprising because it is often difficult to predict hydrophobic contacts by ab initio methods (1). Because partial NMR data that reported the secondary structure was available, they reconstructed the model manually. The final result was satisfactory. This work is novel in terms of computational modeling using an ab initio modeling approach and filtering the predicted structure by experimental data. The difficulty lies in computational modeling, not in filtering predicted structures. As the authors pointed out, it is important to integrate the experimental constraints into the modeling process itself instead of using them as a filter.

Structure Modeling of Nucleic Acid–Protein Interactions

The human adenovirus protease (AVP) plays an important role in the control of infectious virion synthesis. The protease requires two cofactors for complete activation and activity. One cofactor is an 11-residue, highly positively charged peptide called pVIc and the other is viral DNA. Although each cofactor can activate the protease by itself, when combined, the increase of k_{cat}/K_m approaches 30,000-fold. The crystal structure of protease and the protease/peptide complex has been solved (5, 72); however, the structure of the complex with viral DNA and the ternary complex have not. In order to understand the mechanism of activation, it is important to locate the binding site of DNA. Thus, hydroxyl radical footprinting was used to locate the DNA binding site on the protein surface. The AVP was exposed to the synchrotron X ray in the presence or absence of DNA or pVIc and also as a ternary complex with both cofactors. Strong DNA-dependent protection sites were identified at residues Phe-86, Pro-101, Tyr-175, and Pro-183. Those residues experience almost complete burial upon DNA binding. Another fascinating finding was that some residues are more exposed to solvent by DNA binding. Those residues (Cys-40, Trp-55, Met-56, and Trp-60) are located near the active site within the cleft between two domains of AVP. Thus, DNA binding causes the opening of the cleft and possibly increases substrate access to the active site. Also, conserved positively charged residues on the protein surface are likely to interact with DNA phosphates through electrostatics and conserved aromatics on the surface. From these observations, it was concluded that DNA interacts with AVP and pVIc by base stacking with aromatic residues that are protected upon DNA binding and also through electrostatic interactions between basic residues in AVP and pVIc. A model of DNA was manually built on the AVP/pVIc cocrystal structure to satisfy the hydroxyl radical footprinting data and locate the backbone phosphate and conserved basic residues using program O (58) by rotating the DNA backbone angles. The model (**Figure 6**) satisfies the observed protections mediated by DNA binding and also gives an insight into how DNA binding induces changes in the active-site cleft. The DNA wraps around AVP/pVIc and works like a strap, pulling the two domains open. Binding of pVIc alone causes some degree of changes in the structure and viral DNA alone can bind to AVP. Together, pVIc provides more interaction sites for DNA such as a highly positively charged C-terminal side and C-terminal Phe for stacking interaction. Thus, the interaction is stronger and the effect is dramatic in terms of the allosteric changes observed by footprinting that are correlated with significant changes in enzymatic activity. Modeling based on the footprinting data provided a proposed DNA contact surface and explained the allosteric mechanisms of activation.

AVP: human adenovirus protease

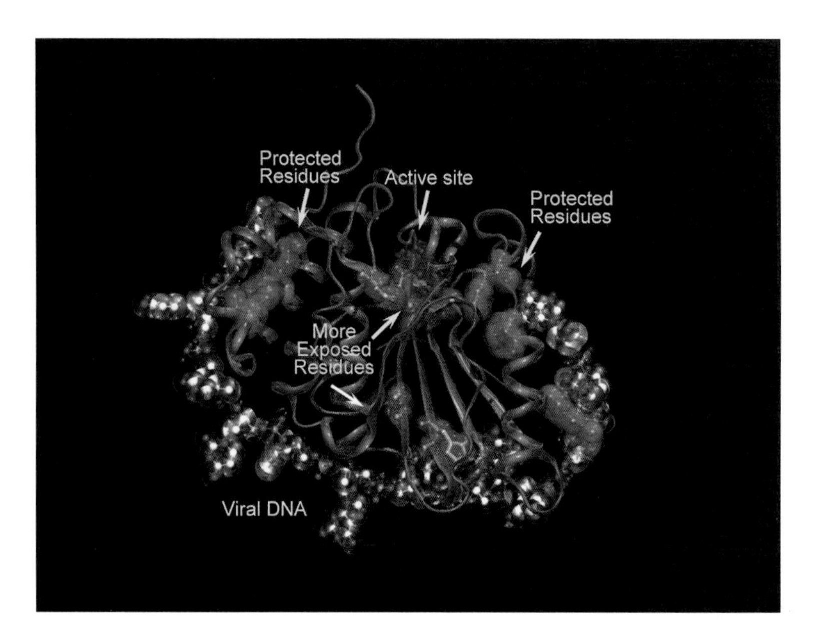

Figure 6

The three-dimensional model of AVP bound to pVIc and DNA cofactors. The residues that are more exposed to solvent upon activation are colored pink. DNA is illustrated as a stick-and-ball model within glass spheres.

Structure Modeling of Mg^{2+}-G-Actin

The structure of Mg^{2+}-G-actin has not been solved. We have probed the solution structure of Mg^{2+}-G-actin utilizing hydroxyl radical footprinting and concluded that the large cleft between two domains must be closed. This conclusion has many biological consequences, as it directly affects processes such as F-actin assembly and ATPase activity. In order to visualize the closed structure of actin, we have conducted a thorough inspection of solved crystal structures, analyzed the deviations within these solved structures using molecular viewer VMD (53) and in-house Tcl scripts for VMD (K. Takamoto & M. R. Chance, unpublished results), and built a model consistent with crystallographic parameters and footprinting data (K. Takamoto & M. R. Chance, unpublished results). The structure of actin is composed of two domains, the large and small domains. The small domain is subdivided into SD1 and SD2. The significant motions of SD2 relative to SD1 and some interesting differences in the geometry of side chains within the nucleotide binding pocket are observed by analyses of these crystal structures. Also, the relative positions of the large and small domains can vary. Those motions are rigid-body hinge movements rather than shear movements or reorganization of the subdomains. This observation is consistent with hydroxyl radical footprinting results that do not indicate any large-scale reorganization within the subdomains.

The protections within the large cleft between two domains (Leu-67 and Tyr-69) can be explained only by rigid-body movement of SD2. Also, the fact that blockage of the small cleft between SD1 and SD3 prohibits polymerization suggests that relative movement of large and small domains also plays an important role for polymerization. This is true for binding of large proteins such as GS1 (73) and vitamin D binding protein (80) and for smaller ligands (66, 107) that should not block the assembly of monomers. Thus, the strategy for the modeling included (*a*) move domain/subdomains as rigid bodies, (*b*) move SD2 relative to SD1, and (*c*) move the large domain relative to SD1 (within small domain). Those movements must satisfy hydroxyl radical footprinting results. The large domain moves toward SD1, and SD2 moves toward the large domain. We have generated

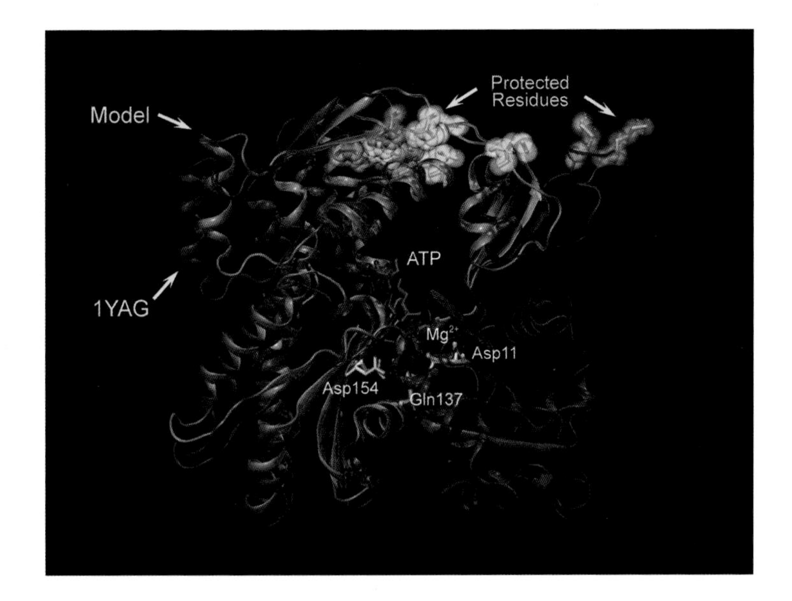

Figure 7

The comparison of the crystal structure (1YAG; *gold*) and model of Mg^{2+}-G-actin (*silver*). Protections within SD4 and D-loop are shown as glowing spheres with sticks inside. The pink spheres indicate the positions in crystal structure 1YAG and the cyan spheres indicate the residues within the model.

a series of movements by combining two possible movements and have filtered them with respect to how the initial models satisfied the footprinting data. The best structure was then optimized by scripts within VMD and using MODELLER 7v7 (90) to correct stereochemical parameters. The resulting structural model is shown in **Figure 7**.

There are some insights revealed by the model. The large domain has moved toward SD1 in order to satisfy the hydroxyl radical data. This domain motion changes the geometry within the nucleotide binding cleft and brings the side chains closer to the metal ion, which is more consistent with the shorter bond lengths likely for Mg^{2+} binding. The coordination states of metal and oxygen atoms within those side chains change from outer-sphere to inner-sphere coordinations, and we suggest this is one driving force for domain movement. The change in geometry also changes the relative positions of Asp-154 and γ-phosphate, making Asp-154 a prime candidate for the catalytic residue as suggested (93). This study shows the potential power of hydroxyl radical footprinting to provide unique structural information relevant to the biological function of actin.

CONCLUSION

Time-Resolved Methods

Similar to the developments that led from static nucleic acid footprinting to millisecond time-resolved approaches (8, 95), protein footprinting needs to be extended such that seconds to milliseconds timescale reactions could be routinely monitored. Protein footprinting, using either cleavage methods (46) or modification (63), can provide reliable biophysical isotherms in equilibrium experiments. The next step would involve delivering the radical dose in a short time by using an appropriate mixing device (8, 95). Fenton approaches are suitable for monitoring reactions that are on the minutes timescale, peroxynitrite as a radical source could reduce the timescales to seconds, and synchrotron experiments can push the timescale to milliseconds (8, 95). Alternatively, photolysis of peroxide, if developed with modern UV laser technology, could attain seconds to millisecond timescales as well.

Hydroxyl Radical Site Sources

One disadvantage of bulk footprinting experiments is that the footprinting data does not have a three-dimensional context. Tethered footprinting approaches developed by Datwyler & Meares (20, 21) were designed to generate radicals from a specific site; the migration of these radicals to adjacent sites provided proximity information and helped map protein-protein interfaces. The mapping was relatively imprecise because these methods used gels. A more recent set of experiments by the Vachet group (67, 68) has used the intrinsic metal atom of proteins along with oxidation and MS to map residues in the metal binding site on the basis of observed side chain reactivity. A union of these two methods, in which sites are engineered in proteins (typically by introducing single Cys residues) and coupled to Fenton-type reagents, may provide sources of radicals that map intra-atomic distances within a protein and interatomic distances within a complex. A challenge of such approaches would be to develop calibration experiments such that the reactivity could be converted into soft distance constraints. If successful, such technologies could be competitive with cross-linking approaches.

Computational Approaches

This review highlights some of the first published attempts to explicitly use solution footprinting data combined with computational methods to refine structure or evaluate competing structural models (39, 41, 98). To make these approaches more valuable, progress in two separate areas is needed. First, although the relationship of solvent accessibility and side chain reactivity is demonstrable, its quantitative basis is far from being well understood. Progress in relating measured changes in reactivity to explicit changes in solvent accessibility would provide specific constraints for protein modeling. In addition, modeling programs could be improved to allow for a more flexible inclusion of surface accessibility and/or distance constraints, particularly with respect to providing seamless user control over the weighting factors applied to specific constraint terms in the modeling process (98). As these improvements accrue, oxidative footprinting approaches using MS will play an increasingly important role in understanding the structure and dynamics of macromolecules and their assemblies.

SUMMARY POINTS

1. The hydroxyl radicals are generated mainly by two methods: transition metal-mediated chemical methods (e.g., Fenton chemistry) or radiolysis of water by X or gamma rays.
2. The hydroxyl radicals react with peptide/protein side chains 10 to 1000 times faster than the backbone and form stable covalent modification products.
3. The reaction products of hydroxyl radicals are digested by proteases, analyzed by LC-MS, and sequenced by MS/MS.
4. The surface accessibility information is retrieved in quantitative fashion by calculation of rate constants from dose-response curves.
5. The footprinting data are a source of constraints for computational modeling of macromolecules.
6. Experimental and computational approaches in combination are an emerging new technique for protein structure prediction.

NOTE ADDED IN PROOF

Recently, two groups have reported generation of hydroxyl radicals by homolysis of hydrogen peroxide using a nanosecond UV-laser pulse (131, 132). These methods are stated to alleviate potential concerns of radical-induced conformational changes to protein structure. Although the rapid generation of radicals has advantages, particularly for development of time-resolved footprinting approaches, the timescale of stable product formation is much longer than initial radical attack. In general all chemical oxidation steps must be carefully considered in the development of covalent labeling approaches.

ACKNOWLEDGMENTS

This research is supported in part by The Biomedical Technology Centers Program of the National Institutes for Biomedical Imaging and Bioengineering (P41-EB-01979).

LITERATURE CITED

1. Aloy P, Stark A, Hadley C, Russell RB. 2003. Predictions without templates: new folds, secondary structure, and contacts in CASP5. *Proteins* 53(Suppl. 6):436–56
2. Baichoo N, Heyduk T. 1997. Mapping conformational changes in a protein: application of a protein footprinting technique to cAMP-induced conformational changes in cAMP receptor protein. *Biochemistry* 36:10830–36
3. Baichoo N, Heyduk T. 1999. Mapping cyclic nucleotide-induced conformational changes in cyclicAMP receptor protein by a protein footprinting technique using different chemical proteases. *Protein Sci.* 8:518–28
4. Balasubramanian B, Pogozelski W, Tullius T. 1998. DNA strand breaking by the hydroxyl radical is governed by the accessible surface areas of the hydrogen atoms of the DNA backbone. *Proc. Natl. Acad. Sci. USA* 95:9738–43
5. Baniecki ML, McGrath WJ, Dauter Z, Mangel WF. 2002. Adenovirus proteinase: crystallization and preliminary X-ray diffraction studies to atomic resolution. *Acta Crystallogr. D* 58:1462–64
6. Bradley P, Chivian D, Meiler J, Misura KM, Rohl CA, et al. 2003. Rosetta predictions in CASP5: successes, failures, and prospects for complete automation. *Proteins* 6(Suppl. 6):457–68
7. Breen AP, Murphy JA. 1995. Reactions of oxyl radicals with DNA. *Free Radic. Biol. Med.* 18:1033–77
8. Brenowitz M, Chance MR, Dhavan G, Takamoto K. 2002. Probing the structural dynamics of nucleic acids by quantitative time-resolved and equilibrium hydroxyl radical 'footprinting'. *Curr. Opin. Struct. Biol.* 12:648–53
9. **Brenowitz M, Senear DF, Shea MA, Ackers GK. 1986. "Footprint" titrations yield valid thermodynamic isotherms. *Proc. Natl. Acad. Sci. USA* 83:8462–66**

9. First article to demonstrate that thermodynamic parameters can be derived correctly from footprinting data.

10. Burley SK, Almo SC, Bonanno JB, Capel M, Chance MR, et al. 1999. Structural genomics: beyond the human genome project. *Nat. Genet.* 23:151–57
11. Bystroff C, Thorsson V, Baker D. 2000. HMMSTR: a hidden Markov model for local sequence-structure correlations in proteins. *J. Mol. Biol.* 301:173–90
12. Celander DW, Cech TR. 1990. Iron(II)-ethylenediaminetetraacetic acid catalyzed cleavage of RNA and DNA oligonucleotides: similar reactivity toward single- and double-stranded forms. *Biochemistry* 29:1355–61

13. Chance MR. 2001. Unfolding of apomyoglobin examined by synchrotron footprinting. *Biochem. Biophys. Res. Commun.* 287:614–21
14. Deleted in proof
15. Chance MR, Bresnick AR, Burley SK, Jiang JS, Lima CD, et al. 2002. Structural genomics: a pipeline for providing structures for the biologist. *Protein Sci.* 11:723–38
16. Chance MR, Fiser A, Sali A, Pieper U, Eswar N, et al. 2004. High-throughput computational and experimental techniques in structural genomics. *Genome Res.* 14:2145–54
17. Deleted in proof
18. Chance MR, Sclavi B, Woodson SA, Brenowitz M. 1997. Examining the conformational dynamics of macromolecules with time-resolved synchrotron X-ray 'footprinting'. *Structure* 5:865–69
19. Chen X, Peng J, Pedram M, Swenson CA, Rubenstein PA. 1995. The effect of the S14A mutation on the conformation and thermostability of *Saccharomyces cerevisiae* G-actin and its interaction with adenine nucleotides. *J. Biol. Chem.* 270:11415–23
20. Datwyler SA, Meares CF. 2000. Protein-protein interactions mapped by artificial proteases: where sigma factors bind to RNA polymerase. *Trends Biochem. Sci.* 25:408–14
21. Datwyler SA, Meares CF. 2001. Artificial iron-dependent proteases. *Met. Ions Biol. Syst.* 38:213–54
22. Fiser A. 2004. Protein structure modeling in the proteomics era. *Expert Rev. Proteomics* 1:97–110
23. Fowler CB, Pogozheva ID, Lomize AL, LeVine H 3rd, Mosberg HI. 2004. Complex of an active mu-opioid receptor with a cyclic peptide agonist modeled from experimental constraints. *Biochemistry* 43:15796–810
24. Frank O, Schwanbeck R, Wisniewski JR. 1998. Protein footprinting reveals specific binding modes of a high mobility group protein I to DNAs of different conformation. *J. Biol. Chem.* 273:20015–20
25. Frieden C, Patane K. 1985. Differences in G-actin containing bound ATP or ADP: the Mg^{2+}-induced conformational change requires ATP. *Biochemistry* 24:4192–96
26. Galas DJ, Schmitz A. 1978. DNAse footprinting: a simple method for the detection of protein-DNA binding specificity. *Nucleic Acids Res.* 5:3157–70
27. Garrison WM. 1987. Reaction mechanisms in the radiolysis of peptides, polypeptides, and proteins. *Chem. Rev.* 87:381–88
28. Garrison WM, Kland-English MJ, Sokol HA, Jayko ME. 1970. Radiolytic degradation of the peptide main chain in dilute aqueous solution containing oxygen. *J. Phys. Chem.* 74:4506–9
29. Gavin AC. 2005. Keystone symposia: proteomics and bioinformatics and systems and biology. *Expert Rev. Proteomics* 2:291–93
30. Germino J, Bastia D. 1983. Interaction of the plasmid R6K-encoded replication initiator protein with its binding sites on DNA. *Cell* 34:125–34
31. Goldsmith SC, Guan JQ, Almo S, Chance MR. 2001. Synchrotron protein footprinting: a technique to investigate protein-protein interactions. *J. Biomol. Struct. Dyn.* 19:405–18
32. Goshe MB, Anderson VE. 1999. Hydroxyl radical-induced hydrogen/deuterium exchange in amino acid carbon-hydrogen bonds. *Radiat. Res.* 151:50–58
33. **Goshe MB, Chen YH, Anderson VE. 2000. Identification of the sites of hydroxyl radical reaction with peptides by hydrogen/deuterium exchange: prevalence of reactions with the side chains. *Biochemistry* 39:1761–70**
34. Guan JQ, Almo SC, Chance MR. 2004. Synchrotron radiolysis and mass spectrometry: a new approach to research on the actin cytoskeleton. *Acc. Chem. Res.* 37:221–29

33. A unique approach in protein footprinting by hydroxyl radical-induced hydrogen/deutrium exchange of side chain and backbone atoms.

35. Guan JQ, Almo SC, Reisler E, Chance MR. 2003. Structural reorganization of proteins revealed by radiolysis and mass spectrometry: G-actin solution structure is divalent cation dependent. *Biochemistry* 42:11992–2000
36. Guan JQ, Chance MR. 2005. Structural proteomics of macromolecular assemblies using oxidative footprinting and mass spectrometry. *Trends Biochem. Sci.* 30:583–92
37. Guan JQ, Takamoto K, Almo SC, Reisler E, Chance MR. 2005. Structure and dynamics of the actin filament. *Biochemistry* 44:3166–75
38. Guan JQ, Vorobiev S, Almo SC, Chance MR. 2002. Mapping the G-actin binding surface of cofilin using synchrotron protein footprinting. *Biochemistry* 41:5765–75
39. Gupta S, Mangel WF, McGrath WJ, Perek JL, Lee DW, et al. 2004. DNA binding provides a molecular strap activating the adenovirus proteinase. *Mol. Cell Proteomics* 3:950–59
40. Deleted in proof
41. Gupta S, Mangel WF, Sullivan M, Takamoto KG, Chance MR. 2005. Mapping a functional viral protein in solution using synchrotron X-ray footprinting technology. *Synch. Radiat. News* 18:25–34
42. Gupta V, Peterson CB, Dice LT, Uchiki T, Racca J, et al. 2004. Sml1p is a dimer in solution: characterization of denaturation and renaturation of recombinant Sml1p. *Biochemistry* 43:8568–78
43. Hampel KJ, Burke JM. 2001. A conformational change in the "loop E-like" motif of the hairpin ribozyme is coincidental with domain docking and is essential for catalysis. *Biochemistry* 40:3723–29
44. Hampel KJ, Burke JM. 2001. Time-resolved hydroxyl radical footprinting of RNA using Fe(II)-EDTA. *Methods* 23:233–39
45. Hanai R, Wang JC. 1994. Protein footprinting by the combined use of reversible and irreversible lysine modifications. *Proc. Natl. Acad. Sci. USA* 91:11904–8
46. Heyduk E, Heyduk T. 1994. Mapping protein domains involved in macromolecular interactions: a novel protein footprinting approach. *Biochemistry* 33:9643–50
47. Deleted in proof
48. Heyduk T, Heyduk E, Severinov K, Tang H, Ebright RH. 1996. Determinants of RNA polymerase alpha subunit for interaction with beta, beta′, and sigma subunits: hydroxyl-radical protein footprinting. *Proc. Natl. Acad. Sci. USA* 93:10162–66
49. Higgs HN, Pollard TD. 2001. Regulation of actin filament network formation through ARP2/3 complex: activation by a diverse array of proteins. *Annu. Rev. Biochem.* 70:649–76
50. Hoofnagle AN, Resing KA, Ahn NG. 2003. Protein analysis by hydrogen exchange mass spectrometry. *Annu. Rev. Biophys. Biomol. Struct.* 32:1–25
51. Huang W, Osman R, Gershengorn MC. 2005. Agonist-induced conformational changes in thyrotropin-releasing hormone receptor type I: disulfide cross-linking and molecular modeling approaches. *Biochemistry* 44:2419–31
52. Humayun Z, Kleid D, Ptashne M. 1977. Sites of contact between lambda operators and lambda repressor. *Nucleic Acids Res.* 4:1595–607
53. Humphrey W, Dalke A, Schulten K. 1996. VMD: visual molecular dynamics. *J. Mol. Graph.* 14:33–38
54. Hunt DF, Yates JR 3rd, Shabanowitz J, Winston S, Hauer CR. 1986. Protein sequencing by tandem mass spectrometry. *Proc. Natl. Acad. Sci. USA* 83:6233–37
55. Jain NU, Wyckoff TJ, Raetz CR, Prestegard JH. 2004. Rapid analysis of large protein-protein complexes using NMR-derived orientational constraints: the 95 kDa complex of LpxA with acyl carrier protein. *J. Mol. Biol.* 343:1379–89

35. Demonstrates the usefulness of hydroxyl radical footprinting technique as powerful tool for proteins that cannot be studied by conventional techniques such as crystallography and NMR.

39. First report to describe the use of hydroxyl radical footprinting data as specific experimental constraints for computational modeling processes.

56. Johnson RS, Martin SA, Biemann K, Stults JT, Watson JT. 1987. Novel fragmentation process of peptides by collision-induced decomposition in a tandem mass spectrometer: differentiation of leucine and isoleucine. *Anal. Chem.* 59:2621–25
57. Jones DT, Taylor WR, Thornton JM. 1992. A new approach to protein fold recognition. *Nature* 358:86–89
58. Jones TA, Zou JY, Cowan SW, Kjeldgaard M. 1991. Improved methods for building protein models in electron density maps and the location of errors in these models. *Acta Crystallogr. A* 47:110–19
59. Katta V, Chait BT. 1991. Conformational changes in proteins probed by hydrogen-exchange electrospray-ionization mass spectrometry. *Rapid Commun. Mass Spectrom.* 5:214–17
60. Kim YJ, Pannell LK, Sackett DL. 2004. Mass spectrometric measurement of differential reactivity of cysteine to localize protein-ligand binding sites. Application to tubulin-binding drugs. *Anal. Biochem.* 332:376–83
61. Kinch LN, Wrabl JO, Krishna SS, Majumdar I, Sadreyev RI, et al. 2003. CASP5 assessment of fold recognition target predictions. *Proteins* 53:395–409
62. Kiselar JG, Janmey PA, Almo SC, Chance MR. 2003. Structural analysis of gelsolin using synchrotron protein footprinting. *Mol. Cell Proteomics* 2:1120–32
63. Kiselar JG, Janmey PA, Almo SC, Chance MR. 2003. Visualizing the Ca2+-dependent activation of gelsolin by using synchrotron footprinting. *Proc. Natl. Acad. Sci. USA* 100:3942–47
64. Kiselar JG, Maleknia SD, Sullivan M, Downard KM, Chance MR. 2002. Hydroxyl radical probe of protein surfaces using synchrotron X-ray radiolysis and mass spectrometry. *Int. J. Radiat. Biol.* 78:101–14
65. Deleted in proof
66. Klenchin VA, Allingham JS, King R, Tanaka J, Marriott G, Rayment I. 2003. Trisoxazole macrolide toxins mimic the binding of actin-capping proteins to actin. *Nat. Struct. Biol.* 10:1058–63
67. Lim J, Vachet RW. 2003. Development of a methodology based on metal-catalyzed oxidation reactions and mass spectrometry to determine the metal binding sites in copper metalloproteins. *Anal. Chem.* 75:1164–72
68. Lim J, Vachet RW. 2004. Using mass spectrometry to study copper-protein binding under native and non-native conditions: beta-2-microglobulin. *Anal. Chem.* 76:3498–504
69. Liu R, Guan JQ, Zak O, Aisen P, Chance MR. 2003. Structural reorganization of the transferrin C-lobe and transferrin receptor upon complex formation: the C-lobe binds to the receptor helical domain. *Biochemistry* 42:12447–54
70. **Maleknia SD, Brenowitz M, Chance MR. 1999. Millisecond radiolytic modification of peptides by synchrotron X-rays identified by mass spectrometry. *Anal. Chem.* 71:8965–73**

70. First report of oxidation mechanisms and MS approaches that formed the basis of hydroxyl radical protein footprinting.

71. Maleknia SD, Chance MR, Downard KM. 1999. Electrospray-assisted modification of proteins: a radical probe of protein structure. *Rapid Commun. Mass Spectrom.* 13:2352–58
72. McGrath WJ, Ding J, Didwania A, Sweet RM, Mangel WF. 2003. Crystallographic structure at 1.6-A resolution of the human adenovirus proteinase in a covalent complex with its 11-amino-acid peptide cofactor: insights on a new fold. *Biochim. Biophys. Acta* 1648:1–11
73. McLaughlin PJ, Gooch JT, Mannherz HG, Weeds AG. 1993. Structure of gelsolin segment 1-actin complex and the mechanism of filament severing. *Nature* 364:685–92

74. Moraczewska J, Wawro B, Seguro K, Strzelecka-Golaszewska H. 1999. Divalent cation-, nucleotide-, and polymerization-dependent changes in the conformation of subdomain 2 of actin. *Biophys. J.* 77:373–85
75. Motoki I, Yosinari S, Watanabe K, Nishikawa K. 1991. Structural analyses on yeast tRNA(Tyr) and its complex with tyrosyl-tRNA synthetase by the use of hydroxyl radical 'footprinting'. *Nucleic Acids Symp. Ser.* (25):173–74
76. Moult J. 1996. The current state of the art in protein structure prediction. *Curr. Opin. Biotechnol.* 7:422–27
77. Nanda V, DeGrado WF. 2005. Automated use of mutagenesis data in structure prediction. *Proteins* 59:454–66
78. Nukuna BN, Goshe MB, Anderson VE. 2001. Sites of hydroxyl radical reaction with amino acids identified by (2)H NMR detection of induced (1)H/(2)H exchange. *J. Am. Chem. Soc.* 123:1208–14
79. Nukuna BN, Sun G, Anderson VE. 2004. Hydroxyl radical oxidation of cytochrome c by aerobic radiolysis. *Free Radic. Biol. Med.* 37:1203–13
80. Otterbein LR, Cosio C, Graceffa P, Dominguez R. 2002. Crystal structures of the vitamin D-binding protein and its complex with actin: structural basis of the actin-scavenger system. *Proc. Natl. Acad. Sci. USA* 99:8003–8
81. Ottinger LM, Tullius TD. 2000. High-resolution in vivo footprinting of a protein-DNA complex using gamma-radiation. *J. Am. Chem. Soc.* 122:5901–2
82. Prestegard JH, Bougault CM, Kishore AI. 2004. Residual dipolar couplings in structure determination of biomolecules. *Chem. Rev.* 104:3519–40
83. Rana TM, Meares CF. 1990. Specific cleavage of a protein by an attached iron chelate. *J. Am. Chem. Soc.* 112:2457–58
84. Rana TM, Meares CF. 1991. Transfer of oxygen from an artificial protease to peptide carbon during proteolysis. *Proc. Natl. Acad. Sci. USA* 88:10578–82
85. Rangan P, Masquida B, Westhof E, Woodson SA. 2003. Assembly of core helices and rapid tertiary folding of a small bacterial group I ribozyme. *Proc. Natl. Acad. Sci. USA* 100:1574–79
86. Rangan P, Masquida B, Westhof E, Woodson SA. 2004. Architecture and folding mechanism of the Azoarcus group I pre-tRNA. *J. Mol. Biol.* 339:41–51
87. Rashidzadeh H, Khrapunov S, Chance MR, Brenowitz M. 2003. Solution structure and interdomain interactions of the *Saccharomyces cerevisiae* "TATA binding protein" (TBP) probed by radiolytic protein footprinting. *Biochemistry* 42:3655–65
88. Riddle DS, Grantcharova VP, Santiago JV, Alm E, Ruczinski I, Baker D. 1999. Experiment and theory highlight role of native state topology in SH3 folding. *Nat. Struct. Biol.* 6:1016–24
89. Roepstorff P, Fohlman J. 1984. Proposal for a common nomenclature for sequence ions in mass spectra of peptides. *Biomed. Mass. Spectrom.* 11:601
90. Sali A, Blundell TL. 1993. Comparative protein modelling by satisfaction of spatial restraints. *J. Mol. Biol.* 234:779–815
91. Sali A, Glaeser R, Earnest T, Baumeister W. 2003. From words to literature in structural proteomics. *Nature* 422:216–25
92. Schmitz A, Galas DJ. 1980. Sequence-specific interactions of the tight-binding I12-X86 lac repressor with non-operator DNA. *Nucleic Acids Res.* 8:487–506
93. Schüler H. 2001. ATPase activity and conformational changes in the regulation of actin. *Biochim. Biophys. Acta* 1549:137–47

94. A breakthrough in hydroxyl radical footprinting technology. By using synchrotron X-ray beam, millisecond time resolution footprinting was achieved.

100. First report of protein footprinting to probe protein-protein interaction by limited proteolysis.

94. **Sclavi B, Sullivan M, Chance MR, Brenowitz M, Woodson SA. 1998. RNA folding at millisecond intervals by synchrotron hydroxyl radical footprinting. *Science* 279:1940–43**
95. Sclavi B, Woodson S, Sullivan M, Chance M, Brenowitz M. 1998. Following the folding of RNA with time-resolved synchrotron X-ray footprinting. *Methods Enzymol.* 295:379–402
96. Sharp JS, Becker JM, Hettich RL. 2003. Protein surface mapping by chemical oxidation: structural analysis by mass spectrometry. *Anal. Biochem.* 313:216–25
97. Sharp JS, Becker JM, Hettich RL. 2004. Analysis of protein solvent accessible surfaces by photochemical oxidation and mass spectrometry. *Anal. Chem.* 76:672–83
98. Sharp JS, Guo JT, Uchiki T, Xu Y, Dealwis C, Hettich RL. 2005. Photochemical surface mapping of C14S-Sml1p for constrained computational modeling of protein structure. *Anal. Biochem.* 340:201–12
99. Deleted in proof
100. **Sheshberadaran H, Payne LG. 1988. Protein antigen-monoclonal antibody contact sites investigated by limited proteolysis of monoclonal antibody-bound antigen: protein "footprinting". *Proc. Natl. Acad. Sci. USA* 85:1–5**
101. Simons KT, Ruczinski I, Kooperberg C, Fox BA, Bystroff C, Baker D. 1999. Improved recognition of native-like protein structures using a combination of sequence-dependent and sequence-independent features of proteins. *Proteins* 34:82–95
102. Skolnick J, Zhang Y, Arakaki AK, Kolinski A, Boniecki M, et al. 2003. TOUCHSTONE: a unified approach to protein structure prediction. *Proteins* 5:469–79
103. Strzelecka-Golaszewska H, Moraczewska J, Khaitlina SY, Mossakowska M. 1993. Localization of the tightly bound divalent-cation-dependent and nucleotide-dependent conformation changes in G-actin using limited proteolytic digestion. *Eur. J. Biochem.* 211:731–42
104. Suckau D, Mak M, Przybylski M. 1992. Protein surface topology-probing by selective chemical modification and mass spectrometric peptide mapping. *Proc. Natl. Acad. Sci. USA* 89:5630–34
105. Tabb DL, Huang Y, Wysocki VH, Yates JR 3rd. 2004. Influence of basic residue content on fragment ion peak intensities in low-energy collision-induced dissociation spectra of peptides. *Anal. Chem.* 76:1243–48
106. Takamoto K, He Q, Morris S, Chance MR, Brenowitz M. 2002. Monovalent cations mediate formation of native tertiary structure of the *Tetrahymena thermophila* ribozyme. *Nat. Struct. Biol.* 9:928–33
107. Tanaka J, Yan Y, Choi J, Bai J, Klenchin VA, et al. 2003. Biomolecular mimicry in the actin cytoskeleton: mechanisms underlying the cytotoxicity of kabiramide C and related macrolides. *Proc. Natl. Acad. Sci. USA* 100:13851–56
108. Tang XJ, Thibault P, Boyd RK. 1993. Fragmentation reactions of multiply-protonated peptides and implications for sequencing by tandem mass spectrometry with low-energy collision-induced dissociation. *Anal. Chem.* 65:2824–34
109. Tramontano A, Morea V. 2003. Assessment of homology-based predictions in CASP5. *Proteins* 53:352–68
110. Tsai HY, Masquida B, Biswas R, Westhof E, Gopalan V. 2003. Molecular modeling of the three-dimensional structure of the bacterial RNase P holoenzyme. *J. Mol. Biol.* 325:661–75
111. Tu BP, Wang JC. 1999. Protein footprinting at cysteines: probing ATP-modulated contacts in cysteine-substitution mutants of yeast DNA topoisomerase II. *Proc. Natl. Acad. Sci. USA* 96:4862–67

112. Tullius TD, Dombroski BA. 1985. Iron(II) EDTA used to measure the helical twist along any DNA molecule. *Science* 230:679–81
113. Tullius TD, Dombroski BA. 1986. Hydroxyl radical "footprinting": high-resolution information about DNA-protein contacts and application to lambda repressor and Cro protein. *Proc. Natl. Acad. Sci. USA* 83:5469–73
114. Deleted in proof
115. Tullius TD, Dombroski BA, Churchill ME, Kam L. 1987. Hydroxyl radical footprinting: a high-resolution method for mapping protein-DNA contacts. *Methods Enzymol.* 155:537–58
116. Valentin-Ranc C, Carlier MF. 1991. Role of ATP-bound divalent metal ion in the conformation and function of actin. Comparison of Mg-ATP, Ca-ATP, and metal ion-free ATP-actin. *J. Biol. Chem.* 266:7668–75
117. van Dijk AD, Boelens R, Bonvin AM. 2005. Data-driven docking for the study of biomolecular complexes. *FEBS J.* 272:293–312
118. Wang X, Kim SH, Ablonczy Z, Crouch RK, Knapp DR. 2004. Probing rhodopsin-transducin interactions by surface modification and mass spectrometry. *Biochemistry* 43:11153–62
119. Wang XD, Padgett R. 1989. Hydroxyl radical "footprinting" of RNA: application to pre-mRNA splicing complexes. *Proc. Natl. Acad. Sci. USA* 86:7795–99
120. Wen X, Yuan Y, Kuntz DA, Rose DR, Pinto BM. 2005. A combined STD-NMR/molecular modeling protocol for predicting the binding modes of the glycosidase inhibitors kifunensine and salacinol to Golgi alpha-mannosidase II. *Biochemistry* 44:6729–37
121. Wong JW, Maleknia SD, Downard KM. 2003. Study of the ribonuclease-S-protein-peptide complex using a radical probe and electrospray ionization mass spectrometry. *Anal. Chem.* 75:1557–63
122. Wong JW, Maleknia SD, Downard KM. 2005. Hydroxyl radical probe of the calmodulin-melittin complex interface by electrospray ionization mass spectrometry. *J. Am. Soc. Mass Spectrom.* 16:225–33
123. Xu G, Chance MR. 2004. Radiolytic modification of acidic amino acid residues in peptides: probes for examining protein-protein interactions. *Anal. Chem.* 76:1213–21
124. Xu G, Chance MR. 2005. Radiolytic modification and reactivity of amino acid residues serving as structural probes for protein footprinting. *Anal. Chem.* 77:4549–55
125. Xu G, Chance MR. 2005. Radiolytic modification of sulfur-containing amino acid residues in model peptides: fundamental studies for protein footprinting. *Anal. Chem.* 77:2437–49
126. Xu G, Kiselar J, He Q, Chance MR. 2005. Secondary reactions and strategies to improve quantitative protein footprinting. *Anal. Chem.* 77:3029–37
127. Xu G, Takamoto K, Chance MR. 2003. Radiolytic modification of basic amino acid residues in peptides: probes for examining protein-protein interactions. *Anal. Chem.* 75:6995–7007
128. Xu Y, Xu D, Crawford OH, Einstein JR. 2000. A computational method for NMR-constrained protein threading. *J. Comput. Biol.* 7:449–67
129. Yuan M, Namikoshi M, Otsuki A, Rinehart KL, Sivonen K, Watanabe MF. 1999. Low-energy collisionally activated decomposition and structural characterization of cyclic heptapeptide microcystins by electrospray ionization mass spectrometry. *J. Mass Spectrom.* 34:33–43
130. Zimmerle CT, Patane K, Frieden C. 1987. Divalent cation binding to the high- and low-affinity sites on G-actin. *Biochemistry* 26:6545–52

113. Established solution-based hydroxyl radical footprinting as standard technique for DNA-protein interaction analyses.

124. Comprehensive guide for the basic chemistry of hydroxyl radical protein footprinting coupled with MS detection.

131. Aye TT, Low TY, Sze SK. 2005. Nanosecond laser-induced photochemical oxidation method for protein surface mapping with mass spectrometry. *Anal. Chem.* 77:5814–22

132. Hambly DM, Gross ML. 2005. Laser flash photolysis of hydrogen peroxide to oxidize protein solvent-accessible residues on the microsecond timescale. *J. Am. Soc. Mass. Spectrom.* 16:2057–63

The ESCRT Complexes: Structure and Mechanism of a Membrane-Trafficking Network*

James H. Hurley[1] and Scott D. Emr[2]

[1]Laboratory of Molecular Biology, National Institute of Diabetes and Digestive and Kidney Diseases, National Institutes of Health, U.S. Department of Health and Human Services, Bethesda, Maryland 20892-0580; email: hurley@helix.nih.gov

[2]Department of Cellular and Molecular Medicine and Howard Hughes Medical Institute, University of California at San Diego, La Jolla, California 92093-0668; email: semr@ucsd.edu

Annu. Rev. Biophys. Biomol. Struct. 2006. 35:277–98

First published online as a Review in Advance on February 7, 2006

The *Annual Review of Biophysics and Biomolecular Structure* is online at biophys.annualreviews.org

doi: 10.1146/annurev.biophys.35.040405.102126

1056-8700/06/0609-0277$20.00

Key Words

ubiquitin, sorting, endocytosis, lysosomes

Abstract

The ESCRT complexes and associated proteins comprise a major pathway for the lysosomal degradation of transmembrane proteins and are critical for receptor downregulation, budding of the HIV virus, and other normal and pathological cell processes. The ESCRT system is conserved from yeast to humans. The ESCRT complexes form a network that recruits monoubiquitinated proteins and drives their internalization into lumenal vesicles within a type of endosome known as a multivesicular body. The structures and interactions of many of the components have been determined over the past three years, revealing mechanisms for membrane and cargo recruitment and for complex assembly.

Contents

Proteasome: large multiprotein complex that proteolyses polyubiquitinated proteins, typically soluble proteins or membrane proteins that are substrates of ER-associated degradation

Ubiquitin (Ub): a highly conserved 76-amino-acid protein that can be covalently attached to itself or other proteins through an isopeptide bond with its terminal carboxylate

INTRODUCTION

Targeted degradation is a fundamental mechanism of protein regulation and quality control (39, 44, 80). Soluble proteins are marked for degradation by the proteasome by their modification with polymeric chains of ubiquitin (Ub), a highly conserved 76-amino-acid protein. These polyubiquitin chains are added by a series of enzymes known as E1, E2, and E3. Polyubiquitin chains target proteins so modified to the regulatory subunit of the proteasome and thence to degradation by the catalytic subunit. This entire process is tightly regulated and highly specific, as errors can lead to inappropriate degradation or failure to degrade appropriate substrates resulting in severe consequences for the cell.

A strikingly different yet equally elaborate process governs the degradation of many, if not most, transmembrane proteins. These proteins are covalently modified by a single Ub moiety, as opposed to a polyubiquitin chain (36, 40, 54, 87, 98). Monoubiquitinated membrane proteins are recognized by a series of receptors that contain specific monoubiquitin-binding domains. These receptors target monoubiquitinated membrane proteins through a series of trafficking steps that ultimately deliver them to their destruction in the lysosome. Unlike the proteasome, the lysosome is a membrane-delimited organelle. Lysosomal proteases and lipases efficiently degrade small internal vesicles loaded with membrane proteins. This delivery system has other functions in addition to protein degradation. Many lysosomal and vacuolar hydrolases arrive via this pathway, and the pathway also is used for antigen presentation. Much attention has focused on the pathway because it is coopted by HIV and other enveloped retroviruses in order to bud from cells (34, 66, 91, 101, 111).

The MVB Sorting Pathway

The Ub-dependent downregulation of activated signaling receptors at the lysosome requires sorting of the receptors at the endosome into a unique class of vesicles that invaginate into the interior of the endosome. The endosomal compartment containing these vesicles is called the multivesicular body (MVB). Pioneering electron microscopy studies showed that ferritin-conjugated epidermal growth factor (EGF) bound to the EGF receptor is sorted into the lumenal vesicles of MVBs en route to the lysosome (35, 37). Fusion of the MVB with the lysosomal membrane results in delivery of the lumenal MVB vesicles and their contents into the lysosome, where the vesicles and the transmembrane receptor are degraded. This unique mechanism enables cells to degrade entire transmembrane proteins as well as lipids in the MVB vesicles. Membrane proteins that are excluded from the inner MVB vesicles remain within the limiting membrane of the

MVB. Following fusion with the lysosome, these proteins are transferred to the limiting membrane of the lysosome. Recent studies have demonstrated that Ub serves as a signal for efficient sorting into the MVB transport pathway (6, 53, 88, 108).

Studies in mammalian cells have revealed critical roles for MVBs in such seemingly distinct processes as growth factor receptor downregulation, antigen presentation, and retroviral budding. However, the simple yeast *Saccharomyces cerevisiae* has served as an important model system for the discovery of the molecular machinery essential for MVB sorting (53). An unexpectedly large number of protein complexes have been identified that directly bind to Ub-modified cargo and also appear to direct the complex process of receptor sorting and MVB vesicle formation (see **Table 1**). The conservation of these components in other organisms including humans highlights the importance of this transport route in all eukaryotic cells.

Class E Vps Proteins

Genetic studies in yeast have identified more than 60 gene products involved in vacuolar protein sorting (Vps). These genes encode transport components that function at distinct stages of protein traffic between the Golgi complex and the vacuole. A subset of the Vps proteins, the class E Vps proteins, functions in the MVB sorting pathway (17, 53, 76). At present, 17 class E *VPS* genes have been identified (**Table 1**) (**Figure 1**). Class E *vps* mutants accumulate endosomal membranes and exhibit defects in the formation of MVB vesicles. The characterization of these proteins has resulted in the identification of three high-molecular-weight cytoplasmic protein complexes that function in the MVB sorting pathway. These complexes are called the ESCRT (endosomal sorting complex required for transport) complexes I, II, and III (41, 54, 73). The ESCRT machinery is required for the formation of MVBs. This process occurs at the late endosomal compartment,

Endosome: an intracellular vesicle involved in transfer of proteins and lipids between different organelles

Multivesicular body (MVB): an endosome containing internal vesicles

Vps: vacuolar protein sorting

Class E VPS: phenotype manifesting alterations in multivesicular bodies

ESCRT: endosomal sorting complex required for transport

Table 1 Class E Vps proteins and complexes

Complex	Yeast protein	Human protein	Motifs	Binds to
Vps27-Hse1 complex	Vps27	HRS	UIM, FYVE, VHS	Ubiquitin, PI(3)P, ESCRT-II (Vps23)
	Hse1	STAM1, STAM2	UIM, VHS, SH3	Ubiquitin
ESCRT-I complex	Vps23	TSG101	UEV	Ubiquitin, Vps27
	Vps28	VPS28		
	Vps37	VPS37A, B, C, D	Coiled-coil	
ESCRT-II complex	Vps22	EAP30	Coiled-coil, WH	
	Vps25	EAP25	PPXY, WH	ESCRT-III (Vps20)
	Vps36	EAP45	GLUE, NZF, WH	Ubiquitin
ESCRT-III complex	Vps2/Did4	CHMP2A, B	Charged, coiled-coil	
	Vps20	CHMP6	Charged, coiled-coil	ESCRT-II (Vps25)
	Vps24	CHMP3	Charged, coiled-coil	
	Snf7/Vps32	CHMP4A, B, C	Charged, coiled-coil	
Vps4 complex	Vps4	VPS4A, B	AAA ATPase, MIT	ESCRT-III
Other MVB proteins	Bro1/Vps31	ALIX/AIP1	Bro1	LBPA, Doa4, ESCRT-III
	Vps60/Mos10	CHMP5	Charged, coiled-coil	ESCRT-III
	Fti1/Did2	CHMP1A, B	Charged, coiled-coil	ESCRT-III
	Vta1	LIP5		Vps4
	Vps44/Nhx1	SLC9A6	Sodium/proton exchanger	

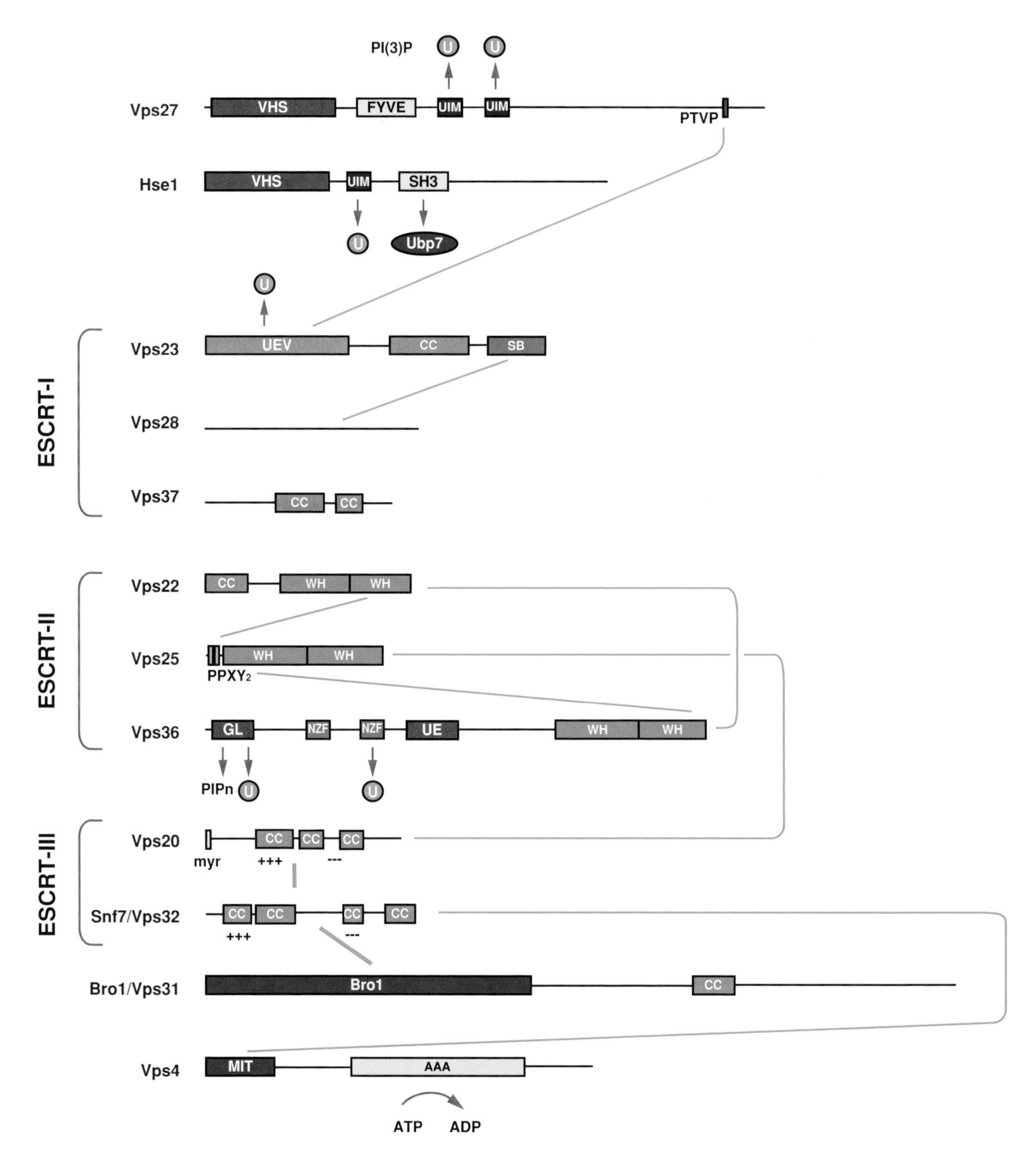

Figure 1

Domain structure and interactions in the ESCRT network. Protein-protein interactions within the network are indicated by solid blue lines. Interactions with lipids, ubiquitin moieties, and other proteins are shown with gray arrows. For simplicity, only two of the ESCRT-III subunits, Vps20 and Snf7, are shown. The GLUE domain of human Vps36 binds to PIP_3; the lipid specificity of the GLUE domain of yeast Vps36 has not been characterized.

where the limiting membrane invaginates and buds small vesicles into its lumen, giving rise to the characteristic morphology of numerous intralumenal vesicles within a larger membrane-enclosed endosome (28, 31, 76).

The yeast class E protein Vps27 and its mammalian homologue (yeast nomenclature is used throughout for simplicity; see **Table 1**) are required for protein sorting in the MVB pathway. Vps27 forms a complex with the Hse1 protein in the cytoplasm (5, 10, 12, 57, 84) and binds directly to monoubiquitinated cargo (10, 12, 84). ESCRT-I (Vps23, Vps28, and Vps37) also interacts with ubiquitinated cargo (14, 32, 53). Genetic studies indicate that ESCRT-II (Vps22, Vps25, and Vps36) acts downstream of ESCRT-I; however, the mechanism of this interaction is not yet known (6). ESCRT-III is composed of two major functional subcomplexes (Vps20/Snf7 and Vps2/Vps24 in yeast) that localize to endosomal membranes. ESCRT-III components fail to localize to endosomes in cells lacking ESCRT-II, suggesting a role for ESCRT-II in the recruitment and assembly of ESCRT-III at the endosome (6). Together, the ESCRT complexes appear to function in both cargo sorting and MVB vesicle formation. The genetic and biochemical data argue for an ordered reaction sequence; Vps27 recruits ESCRT-I, which in turn recruits ESCRT-II, which then recruits ESCRT-III to the endosome.

In addition to the complexes described above, there are several other important players. Vps4 is an AAA-type ATPase (8, 9) that plays a critical role in the MVB sorting pathway by catalyzing the dissociation of all three ESCRT complexes from the endosome (6, 9, 53). Inactivation of Vps4 results in the accumulation of the ESCRT machinery on the surface of the endosome (class E compartment). Purified Vps4 assembles into a homo-oligomer when loaded with ATP and is recruited to the endosome via the ESCRT-III complex (6, 9). The enzyme Doa4 (2, 6) deubiquitinates MVB cargoes prior to sorting into MVB vesicles and also is recruited to the ESCRT-III complex. Doa4 is recruited via an interaction with Bro1 (56, 63, 77), another class E Vps protein (Vps31).

AAA: ATPases associated with diverse cellular activities

UIM: ubiquitin-interacting motif

PI(3)P: phosphatidylinositol 3-phosphate

FYVE: Fab1, YOTB, Vac1, EEA1

THE VPS27/HSE1 COMPLEX

The Vps27/Hse1 complex binds Ub via ubiquitin-interacting motifs (UIMs) in both subunits (12, 14, 52, 72, 81, 84, 97). Vps27 possesses a FYVE domain (for Fab1, YGL023, Vps27, and EEA1) that binds to the endosomal lipid phosphatidylinositol 3-phosphate (PI(3)P) (20, 33, 69, 78, 90, 100). Endosomal PI(3)P recruits the Vps27/Hse1 complex to the endosome. Vps27 is a docking site for the ESCRT-I complex and thereby initiates the MVB sorting reaction at the limiting membrane of the endosome. ESCRT-I physically interacts with membrane-bound Vps27/HRS through a motif in the COOH-terminal portion of Vps27 (4, 10, 13, 23, 55, 83). Therefore, the Vps27 protein appears to direct the compartment-specific activation of MVB sorting, and Vps27 function is regulated by specific interactions with both PI(3)P and ubiquitinated cargo at the late endosome. Hse1 and STAM (signal transducing adaptor molecule) contain an src homology-3 (SH3) domain. Vps27, Hse1, and their mammalian homologues contain predicted helical regions that are essential for formation of the complex (86).

The extreme C terminus of Vps27 contains the sequence LIEL, which binds to a groove in the clathrin N-terminal β-propeller domain (106). The human homologue of Vps27 also binds to clathrin via this motif (85). Planar clathrin lattices have been seen on endosomes containing Hrs (89). Clathrin may serve to cluster Vps27 and cargo at sites that will later invaginate to form the MVB vesicles. The human homologues of Vps27 and Hse1 are phosphorylated on Tyr residues, but the functional significance of this is unknown (57).

Vps27 FYVE Domain and Membrane Localization

The FYVE domain that is responsible for localizing the Vps27/Hse1 complex is a compact 60-amino-acid Zn^{2+}-finger that selectively recognizes PI(3)P in preference to all other phosphoinositides. The crystal structure of the Vps27 FYVE domain (69) showed how such a small domain could selectively bind membrane-embedded PI(3)P (**Figure 2**). A basic RHHCR motif on the first β–strand provides all but one of the basic ligands for the acidic PI(3)P headgroup, as shown by the structure of the complex of the EEA1 FYVE domain with inositol (1,3)-bisphosphate [Ins(1,3)P_2] (27, 70). The His residues with their relatively short side chains are proximal to the 3-phosphate group, and the Arg residues are more distal, allowing room for their longer side chains to reach the headgroup.

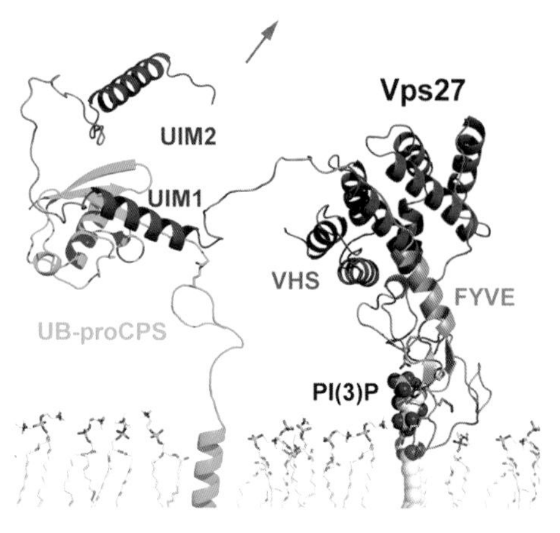

Figure 2

Structure of Vps27. Structures are shown where available for Vps27 proteins [FYVE (69), UIM1-ubiquitin complex (103), UIM1-UIM2 (103)], otherwise modeled on the basis of the closely related structure of the Hrs-VHS domain (64). Linker regions between the domains were generated arbitrarily and are shown only to indicate the length of the segment. The C-terminal putative Hse1-binding domain and the extended region of Vps27 are not shown. Ubiquitin is shown fused via an isopeptide bond between Gly-76 and Lys-8 of the prototypical cargo pro-carboxypeptidase S, and the transmembrane helix of pro-CPS was modeled as an ideal helix. Membrane docking of Vps27 is based on the computationally predicted optimal docking mode for the FYVE domain.

The FYVE domain, like most other membrane-lipid targeting domains (25, 46, 47, 61), is anchored to cell membranes by a combination of specific lipid binding and by less specific electrostatic and hydrophobic forces (26, 27, 58–60, 69). The Vps27 FYVE domain has low affinity (K_d = 90 μM) for the soluble inositol (1,3)-bisphosphate, which corresponds to the headgroup of PI(3)P (15). The affinity rises to K_d = 30 nM when PI(3)P is presented in a realistic model of an endosomal membrane (15). This 3000-fold gain in affinity is explained by a hydrophobic protrusion (turret loop) of the FYVE domain that inserts into the hydrophobic core of the membrane. The depth and angle of membrane penetration dictate how the FYVE domain will be oriented relative to the bilayer and what other surfaces are available for interactions. These parameters have been analyzed by spectroscopy (59), computational simulations (26), and by structure-based modeling (27, 64, 69). While there is no consensus model for a single membrane-binding geometry for all FYVE domains, a consistent general picture has emerged for the best-studied FYVE domains, including that of Vps27. For all PI(3)P-binding FYVE domains, the turret loop is buried in the membrane. The Vps27 FYVE domain is best described as binding PI(3)P in a "side-on" orientation (26, 69), while the EEA1 FYVE domain is tilted with its PI(3)P binding face close to the membrane in a partially "face-on" manner (26, 27). Modeling suggests that differences in the distribution of charged residues on the FYVE domain surface control differences in membrane docking. Given their variability, it seems likely that FYVE domains can wobble in situ and are not likely to be rigidly constrained by membrane forces alone. Such flexibility might make FYVE domain proteins well adapted to function in dynamic protein networks, in which complexes are rapidly formed and broken down and the context of the domain's interactions is subject to constant change.

VHS Domains in the Vps27/Hse1 Complex

Both subunits of the complex have N-terminal Vps27/Hrs/STAM (VHS) domains whose function is intriguing but still unknown. VHS domains are octahelical bundles (64, 68, 71, 95, 118) that are present only at the extreme N termini of the proteins that contain them. Although the VHS domains of the GGA trafficking adaptors bind directly to acidic cluster-dileucine motifs in cargo proteins in a groove between two helices of the VHS domain (68, 71, 95, 118), critical residues in this groove are altered in the VHS domains of the Vps27/Hse1 complex, and the ligands for these VHS domains are unknown. Some nonubiquitinated G-protein-coupled receptors are trafficked by the human homologue of Vps27 (43), and its VHS domain is one candidate for recognizing such cargo (38).

UIMs and Recognition of Ubiquitinated Cargo

UIMs are short helical motifs first identified on the basis of homology to the Ub-recognition sequence in the proteasome subunit S5a (45). Hse1 has one UIM, and Vps27 has two. These UIMs bind monoubiquitin with low affinity, with K_d values in the range of 200 μM to 2 mM (30). The second UIM of Vps27 has a ~10-fold-lower affinity for Ub than the first UIM (30). The structure of the second Vps27 UIM has been determined alone (30), showing that the UIM comprises a single helix. The structure of a Vps27 tandem UIM construct has been determined in complex with Ub by NMR (103). Only the first UIM (UIM-1) was found to participate in complex formation in the structure, consistent with its higher affinity. The second UIM was found to flop about freely in solution. The 28-residue linker between the two UIMs is completely disordered, and the two UIMs appear to sample all accessible conformational space relative to each other.

VHS: Vps27, Hrs, STAM

The Vps27 UIM-1 binds to the Leu-8, Ile-44, Val-70 hydrophobic patch on the Ub surface. This is the same surface that is recognized by all other Ub-binding domains characterized to date. The UIM-1 contains a single six-residue hydrophobic strip that interacts with Ub. The signature C-terminal Ser of the UIM forms a hydrogen bond with the main chain of Ub, and conserved N-terminal Glu residues form salt bridges with Ub Arg-42 and Arg-72. The 400 Å^2 surface area buried in the complex is smaller than that seen in most protein-protein complexes but is typical of what we have come to expect for low-affinity Ub-binding domain complexes.

SH3 Domain of Hse1 and STAM

The SH3 domain of STAM recruits the deubiquitinating enzyme ubiquitin isopeptidase Y (UBPY) (49). The biological rationale for a deubiquitination event at an early stage of entry into the ESCRT pathway is unclear, as the ESCRT proteins interact with ubiquitinated cargo. It seems possible that the STAM/UBPY interaction functions as an off switch to direct cargo out of the pathway or to inactivate components of the ESCRT machinery that are ubiquitinated. The SH3 domain of STAM recognizes a noncanonical SH3-binding motif within UBPY of the form PX(V/I)(D/N)RXXKP (48) with a relatively low affinity of 27 μM, and the structure of this complex has been determined. The SH3 domain of Hse1 has been proposed to recruit Upb7 on the basis of a large-scale proteomic study of yeast SH3 domains (107), although Ubp7 does not contain the PX(V/I)(D/N)RXXKP motif, and the interaction has yet to be confirmed.

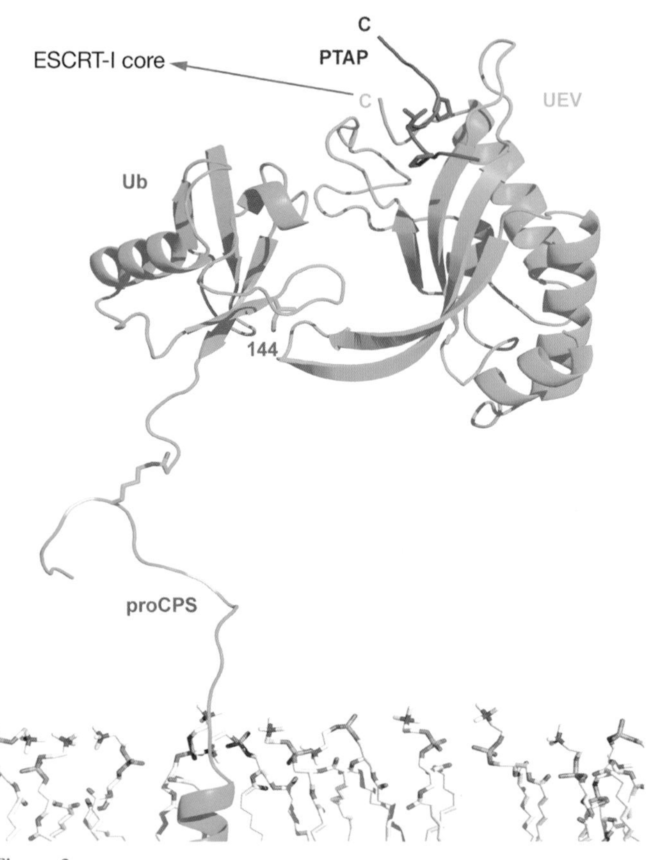

Figure 3

Structure of ESCRT-I. Structures are shown for the UEV domain of human Vps23 in complex with ubiquitin (102) and the HIV-1 p6 PTAP-containing peptide (82). The remainder of the ESCRT-I structure is not yet available.

THE ESCRT-I COMPLEX

The ESCRT-I complex binds directly to monoubiquitinated protein cargo through its UEV domain, a catalytically inactive variant of an Ub-conjugating enzyme (53). ESCRT-I interacts with Vps27 and a number of other cellular proteins, including the mammalian counterpart of Bro1 (101, 111) and the Ub ligase Tal, (3) via their P(S/T)XP sequences, as described above. Intense interest has centered on this motif since the discovery that HIV and certain other enveloped retroviruses contain this motif in their envelope proteins and use it to hijack the MVB sorting machinery to bud from host cells (24, 32, 65, 74, 109).

UEV: ubiquitin E2 variant

These P(S/T)XP sequences also bind to the ubiquitin E2 variant (UEV) domain. Tsg101 and Vps23 contain a C-terminal "steadiness box" (29) that is important for stability in vivo. The functions of the other domains and motifs within the ESCRT-I complex have yet to be worked out.

The UEV Domain of Vps23

The cargo recruitment end of the ESCRT-I complex is the UEV domain, which is responsible for binding both to monoubiquitin moieties and to P(S/T)XP motifs (**Figure 3**). The structures of the UEV domains of Vps23 (105) and Tsg101 (102) have been determined in complex with Ub. In both structures, two different regions of the UEV domains contact Ub. The β1-β2 "tongue" contacts the Ile-44 hydrophobic patch of Ub, the same region involved in contacts with the Vps27 UIM and with other monoubiquitin binding domains. The loop between the α3 and $\alpha 3'$ helices forms a "lip" that contacts a hydrophilic site centered on Gln-62 of Ub (102, 105). In contrast to the Ile-44 site, the Gln-62 site does not participate in most other known Ub/Ub-binding domain interactions. Even though the UEV domain was discovered as a catalytically inactive homolog of Ub-conjugating enzyme, Ub binds to the UEV domain in a completely different manner.

The structure of the Tsg101 UEV domain in complex with the PTAP peptide of HIV-1 p6 shows how P(S/T)XP sequences are recognized by the ESCRT-I complex (82). The second Pro of the P(S/T)XP is a particularly critical element. The XP sequence, together with the first C-terminal flanking residues, forms one turn of a type II polyproline helix. This conformation is also seen in canonical SH3 and WW (tryptophan-tryptophan) domain complexes with Pro-based peptide motifs (117). The UEV domain recognizes the second Pro by using an aromatic pocket reminiscent of Pro-recognition pockets in SH3 and WW domains. The first Pro is in an extended conformation and binds in a shallow pocket. The Thr hydroxyl appears to hydrogen bond to both main chain and side chain residues, although this was not fully defined in the NMR structure. The concept of using peptidomimetics directed at this site to block HIV release has attracted considerable interest.

THE ESCRT-II COMPLEX

ESCRT-II forms a nexus between ubiquitinated cargo, the endosomal membrane, and the ESCRT-I and -III complexes (7). Its structural organization is currently the best understood of the three ESCRT complexes (**Figure 4**). The complex contains two Vps25 subunits, one Vps22 subunit, and one Vps36 subunit (42, 104). The N-terminal two thirds of Vps36 contain a series of domains [GLUE (99) and NZF (1)] that interact with membranes and cargo and perhaps have other functions. Vps22, Vps25, and the C-terminal third of Vps36 form a tightly organized core that contains two binding sites for the ESCRT-III subunit Vps20 (104).

NZF: Npl4 zinc finger

WH: winged helix

The ESCRT-II Core

The ESCRT-II core is Y-shaped, with one Vps25 subunit forming the stalk. One of the branches is formed by the second Vps25 subunit, and the other branch is formed by the subcomplex consisting of Vps22 and the C-terminal third of Vps36 (42, 104). Although the three subunits do not have primary sequence homology to each other, each subunit consists of two repeats of a winged helix (WH) domain. WH domains are typically found in DNA-binding proteins, where the wing (a loop between two of the β-strands) contacts the nucleic acid. Vps22 and Vps36 form a tightly bound subcomplex. The second WH

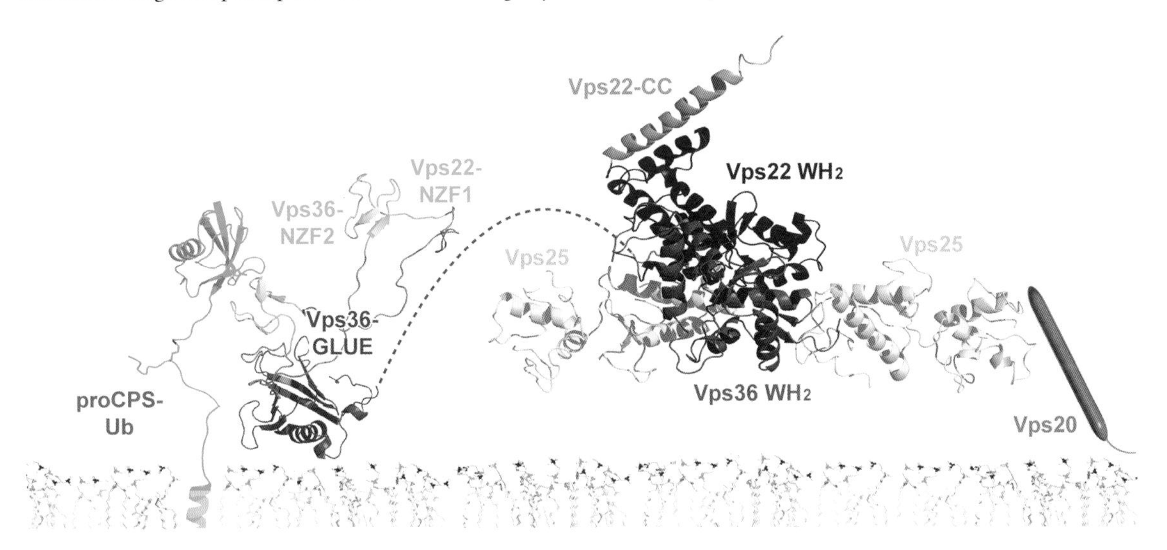

Figure 4

Structure of ESCRT-II. The structure of the ESCRT-II core (42) is shown docked to a membrane on the basis of the interaction between its Vps25 subunit and the membrane-bound ESCRT-III subunit Vps20 (104) (myristoyl group on Vps20 not shown for simplicity). The uncomplexed NZF1 domain and the NZF2-ubiquitin complex are modeled on the basis of the Npl4-NZF structure and ubiquitin complex (1). The GLUE domain (a variant of the GRAM domain, which is in turn a variant of the PH domain) is modeled and docked to the membrane on the basis of the GRAM domain of the lipid phosphatase MTMR2 (11). The GLUE domain of human Vps36 binds to PI(3,4,5)P_3 and ubiquitin at sites that have yet to be determined (99), and are not shown. The binding properties of the yeast Vps36 GLUE domain have yet to be reported. The dashed line between the GLUE domain and Vps36 core winged helix (WH) region indicates residues 290–395 of Vps36, whose structure is unknown.

domains (WH2) of Vps22 and Vps36 bind to the Vps25 subunits through closely equivalent interactions. Vps22 and Vps36 present an aromatic cage to the Vps25 subunit. The N terminus of Vps25 contains two repeats of the sequence PPXY. Sequences of this type are better known for binding to WW domains (a polyproline-rich peptide motif binding domain that is unrelated to the WH domain, despite the similar abbreviation) in a conformation in which the Tyr side chain is exposed and presented to the WW domain. In Vps25, the Tyr side chain is buried, and the diPro sequence interacts with the aromatic cages in the WH2 motifs of Vps22 and Vps36. The Tyr of the PPXY motif of Vps25 is buried even in the isolated subunit unbound to the rest of the complex (113), suggesting it is unlikely to become exposed in a conformation available for WW domain binding. All of the intersubunit contacts are required for the complex to mediate carboxypeptidase S sorting in yeast (42).

Vps36 NZF Domains

Yeast Vps36, but not its human counterpart, contains two Npl4 zinc finger (NZF) motifs in the region N-terminal to the core WH domains. The Npl4 NZF domain was first shown to bind Ub, and this interaction has been characterized structurally (1). The NZF domain binds to the same Ile-44 patch on Ub as do the UIM and UEV domains. Thus, none of these domains can bind to Ub simultaneously. The second NZF domain of Vps36 binds to Ub, and the Ub-binding site on the NZF domain is important for the sorting function of Vps36. The function of the first NZF domain is unknown.

Vps36 GLUE Domain

The absence of NZF or other known Ub-binding domains in human Vps36 and the other human ESCRT-II subunits led to a puzzle: How could the human ESCRT-II complex function in sorting ubiquitinated cargo without a Ub-binding domain? This seeming paradox was resolved with the discovery of the GLUE (Gram-like ubiquitin) domain near the N terminus of the human protein (99). Vps36 also contains a GLUE domain, which is split into two segments by the first NZF domain. GLUE domains bind not only Ub but also the lipid phosphatidylinositol (3,4,5)-triphosphate (99). The structure of the GLUE domain has not been determined, but on the basis of distant sequence similarity it appears to be similar to the phospholipid-binding pleckstrin homology domain.

THE ESCRT-III COMPLEX

The ESCRT-III complex consists of several subunits that are homologous to each other, highly charged, and contain predicted coiled-coil regions (6). Four subunits, Vps2, Vps20, Vps24, and Snf7, are essential for sorting function in yeast (6) (**Figure 4**). Two other proteins, Vps60 and Did2, share homology to the other ESCRT-III subunits but are less critical for function (6). Isolated ESCRT-III subunits are cytoplasmic in yeast (6) and soluble, at least at low concentrations, in vitro (56, 62). One of the subunits, Vps20, is myristoylated, but in isolation is nevertheless cytosolic. Thus the myristoyl group alone seems insufficient to drive it to the membrane. Upon association, the subunits form a large and tightly membrane-bound assembly of indeterminate stoichiometry (**Figure 5**). It is currently thought that this assembly comprises an oligomer of ESCRT-III complexes on the membrane surface.

The major link between the ESCRT-III complex and upstream complexes occurs via ESCRT-II in yeast (16). This interaction is conserved in humans (111). However, human Bro1 bridges Vps23 of ESCRT-I with Snf7 of ESCRT-III and provides a second link. The Vps20 subunit of ESCRT-III (16, 111) binds directly to Vps25 of yeast and human ESCRT-II (104, 116). The Vps2/Vps24 subcomplex of ESCRT-III recruits Vps4 in yeast (6). However, Snf7 and other ESCRT-III

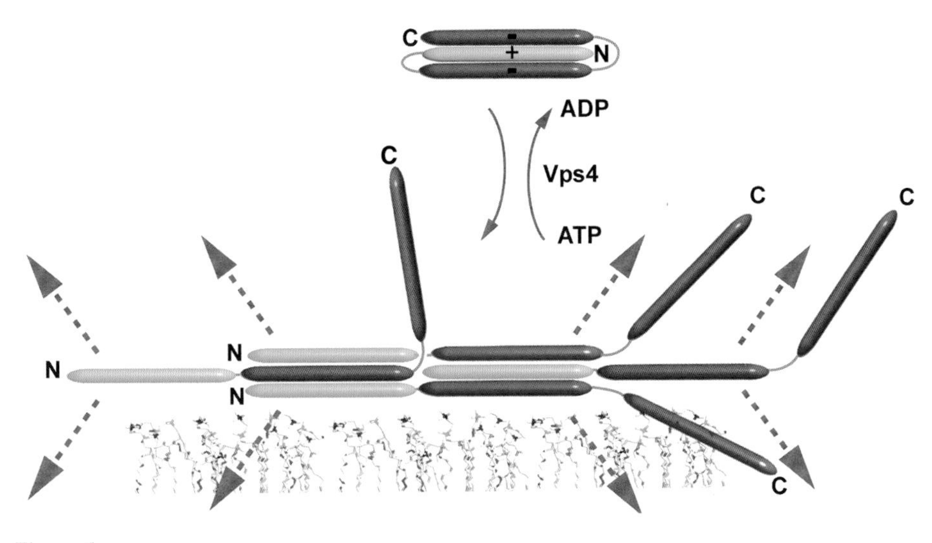

Figure 5

Structure of ESCRT-III. The structure of ESCRT-III is unknown. This cartoon shows a tripartite variant of the bipartite model proposed by Hanson and colleagues (62). The C-terminal anionic region of ESCRT-III subunits is roughly twice as large as the N-terminal basic region and appears to have functions both in oligomerization of ESCRT-III on membranes and in binding to other proteins, such as Vps4. Gray arrows indicate directions in which the oligomeric array can grow.

proteins can also bind to Vps4 (16, 111), suggesting the ESCRT-III/Vps4 interaction is not stringently specific. Vps60 binds to the class E Vps protein Vta1 in yeast (16, 96) and humans (112). Human Vps24 has been proposed to bind to $PI(3,5)P_2$ (114). In summary, all ESCRT-III subunits have a conserved primary structure and interact with each other in a similar manner, whereas the individual subunits retain specific interactions of their own.

The N-terminal one third or so of ESCRT-III subunits are highly basic, whereas the C-terminal two thirds are acidic. Both regions contain predicted coiled-coils. The simple model for the monomeric form of the subunit is that the basic and acidic regions form an antiparallel coiled-coil pair with each other, stabilized by electrostatic interactions between the two halves. Experimental evidence for this model comes from a comparative analysis of human Snf7 and Vps24 and their fragments (62). The basic N-terminal regions bind tightly to membranes even in isolation (62). The C-terminal regions, in contrast, bind to human Vps4 (62, 94). There is also evidence that the N-terminal basic regions can interact not only with membranes but also with other proteins (116). These data suggest that monomeric ESCRT-III subunits are in a closed conformation in which the N- and C-terminal portions are tightly associated with each other and are not available for interactions with membranes or other proteins.

What initiates formation of the insoluble ESCRT-III complex from its soluble subunits? In a working model, interactions with other proteins compete with the intraprotein interaction between the N- and C-terminal regions of ESCRT-III subunits and thereby liberate the N-terminal portion for interactions with membranes. This triggers the membrane localization of the subunit. In a working model, the membrane-bound form of the ESCRT-III subunit is in a more open conformation, more available for interactions with other subunits, and thus disposed to form a polymeric assembly on the membrane (**Figure 5**). The formation of the ESCRT-III complex is independent of Vps4 and ATP (6), and ESCRT-III-like aggregates of isolated recombinant subunits have been observed to form spontaneously (62). In vivo formation of the ESCRT-III complex likely follows the binding of isolated ESCRT-III subunits to ESCRT-II, Bro1, or other factors that allosterically promote the complexation-competent conformation. These considerations point to factors such as ESCRT-II and Bro1 as the likely initiators of ESCRT-III complexation.

The membrane association of ESCRT-III is essentially irreversible in the absence of an energy input: Vps4 must hydrolyze ATP to solubilize the membrane-bound ESCRT-III assembly, once formed. In this respect, ESCRT-III proteins are analogous to the SNAREs of membrane fusion, which must be separated by the action of *N*-ethylmaleimide sensitive factor (NSF), once tight complexes have been formed. In contrast to ESCRT-III subunits, which cycle on and off membranes, SNAREs contain transmembrane regions and are permanently tethered to membranes (18).

BRO1

Bro1 is a monomeric protein that is intimately associated with the ESCRT complexes (**Figure 5**). Bro1 is involved in deubiquitination of the general amino acid permease Gap1 (75) and is required for trafficking of carboxypeptidase S via the MVB pathway (77). Bro1 recruits Doa4, which deubiquitinates MVB cargo proteins (63). Bro1 is recruited to MVBs through its interactions with the ESCRT-III subunit Snf7, also known as Vps32 (16, 77).

The human homolog of Bro1 interacts with the human counterpart of Snf7 (50, 51, 101, 111); hence this key interaction is conserved from yeast to human. It contains a C-terminal Pro-rich region that interacts with the endocytic proteins SETA (22, 92), endophilins (21), and the human Vps23 subunit of ESCRT-I (101, 111). In contrast, Bro1 lacks a P(S/T)XP sequence and does not interact strongly with Vps23 (16). The human homologue of Bro1 is a key player in retroviral budding, because it interacts with HIV-1 and other retroviral proteins containing the sequence motif YPXL (34, 66, 101, 111). Little is known about the function of YPXL motif host proteins that bind to Bro1 homologues, although one protein, PacC, in *Aspergillus* has been described (110).

Bro1 Domain

Bro1 and several other late endosomal proteins share a conserved N-terminal Bro1 domain. The Bro1 domain consists of roughly 370 residues and has a complex structure that is built around a core helical solenoid similar to tetratricopeptide repeat domains (56) (**Figure 6**). The Bro1 domain is necessary and sufficient for binding to the ESCRT-III subunit Snf7 and for the recruitment of Bro1 to late endosomes (56). Snf7 binds to a conserved hydrophobic patch on Bro1 that is required for protein complex formation and for the protein sorting function of Bro1 (56). The Bro1 domain of the Bro1 protein does not contain the binding site for Doa4 (56). A second conserved hydrophobic patch at one tip of the domain is not critical for sorting, and its function has yet to be determined. The structure resembles a boomerang with its concave face filled in, and it is tempting to speculate that the convex face could mediate the putative ability of the human Bro1 counterpart to sense negative curvature in invaginating lumenal vesicles (67); however, this idea has not been tested for the Bro1 domain.

VPS4

Vps4 is responsible for the ATP-dependent disassembly of the ESCRT complexes (8, 9). Vps4 is a homo-oligomer in yeast, and consists of an N-terminal MIT domain and a central AAA ATPase domain (**Figure 7**). Two Vps4 isoforms in human, Vps4A and Vps4B, can hetero-oligomerize with each other. AAA ATPases are ubiquitous disassembly machines

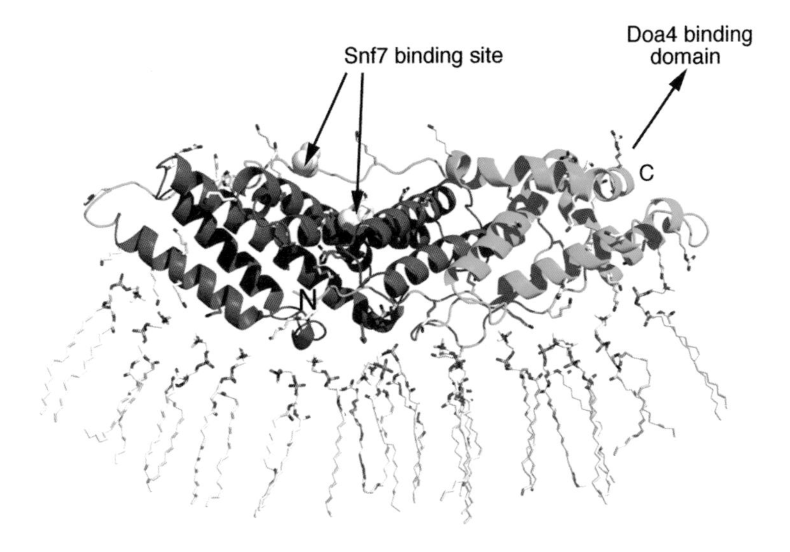

Figure 6

Structure of Bro1. The structure of the Bro1 domain (56), which comprises the N-terminal half of Bro1, is shown docked to a negatively curved membrane. The interaction of this domain with negatively curved membranes is suggested by the shape of the structure and by the properties of the full-length human Bro1 homologue; however, this interaction and docking mode are speculative and have yet to be directly tested.

that are ring hexamers and function in a wide range of cell processes (19). The structure of the AAA domain of Vps4 has been determined (93) and is similar in outline to the structures of other AAA ATPases such as NSF and p97. The N-terminal MIT domain [first called an ESP domain (79)] of Vps4 binds to the human ortholog of the ESCRT-III-like protein Did2 and to other ESCRT-III subunits as well (94, 115). The structure of a MIT domain from human Vps4 has been determined and has been shown to be a three-helix bundle (94) that contains the equivalent of 1.5 tetratricopeptide repeats, reminiscent of the Bro1 domain. The structure of the Vps4 monomer has been used to model a hexamer that assembles in the presence of ATP. This hexamer contains a central pore. In a working mechanism for ESCRT-III disassembly, supported by mutational analysis of the modeled pore, individual membrane-bound ESCRT-III subunits are fed through the pore into solution and so converted to their monomeric, soluble state.

CONCLUSIONS AND FUTURE PERSPECTIVES

MIT: microtubule interacting and trafficking

In gross terms, the ESCRT systems have two major functions: the recruitment of cargo to MVBs, followed by its internalization via the invagination of lumenal vesicles (**Figure 8**). Much has been learned in the past five years about the structures, interactions, and ordered assembly of the ESCRT machinery, and the mechanism of cargo recruitment. The presence of Ub-binding domains in Vps27/Hse1, ESCRT-I, and ESCRT-II has led to the suggestion that there is a serial handoff of ubiquitinated cargo from one complex to the next. Favoring this model, structural analysis shows that the Ub-binding domains interact with just one region of Ub, the Ile-44 patch. This offers an elegant mechanism to prevent more than one complex from interacting with Ub at a given time. However, no direct evidence of hand-off is available. In an alternative model, the presence of multiple Ub-binding sites in the Vps27/ESCRT-I/ESCRT-II "complex of complexes" offers an attractive mechanism for

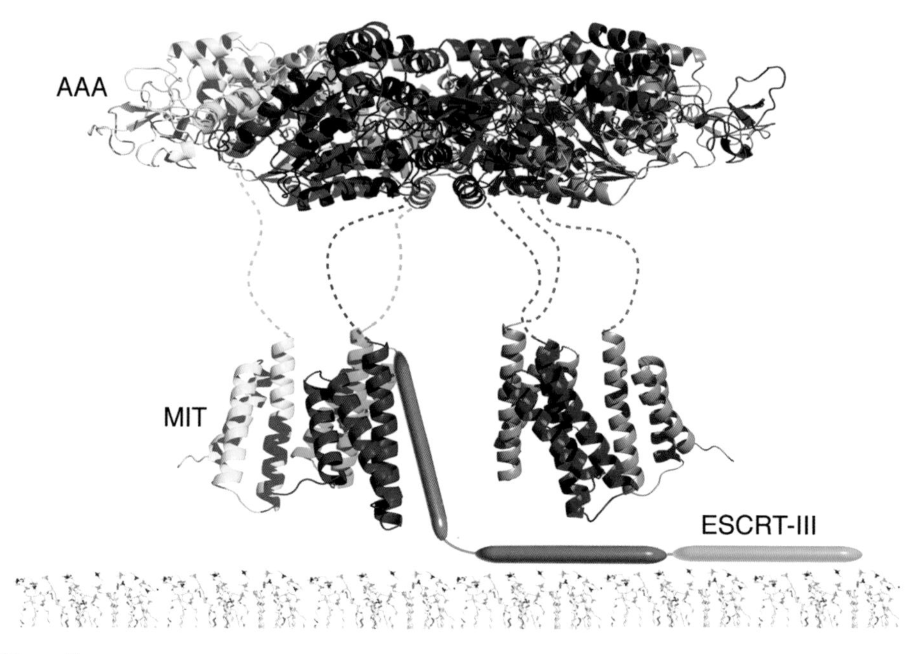

Figure 7

Structure of Vps4. The modeled structure of the Vps4 AAA domain hexamer was modeled (93) and is shown together with the corresponding six copies of the MIT domain (94). For simplicity, a single subunit of the ESCRT-III complex is shown engaged to a single MIT domain. The ESCRT-III subunit is thought to be fed through the central pore in the hexamer (not shown in this view) as ATP is hydrolyzed by the AAA domain.

receptor clustering by simultaneous binding. The presence of two Ub-binding domains in Vps27 and ESCRT-II seems more consistent with a clustering model than with a hand-off model, as there would be little reason to orchestrate hand-off within a single complex. The two models are not mutually exclusive, because hand-off might be operative at one stage in the pathway and clustering at another.

Despite many advances in other aspects of ESCRT function, the fundamental mechanism of vesicle budding remains a matter of conjecture. MVBs from mammalian cells have been reported to be rich in the phospholipid 2,2′ lysobisphosphatidic acid (2,2′ LBPA), which has been suggested to be a key driving force for membrane deformation leading to vesicle invagination (67). On the other hand, no evidence has surfaced for the presence of 2,2′ LBPA in yeast, and no Vps proteins have emerged as candidate regulators of LBPA levels. Given the high conservation of the protein machinery involved, it would be surprising if completely different mechanisms were responsible for the invagination of lumenal vesicles in yeast and mammals. It seems likely that changes in vesicular pH and/or ionic strength are important. The Na^+/H^+ antiporter Vps44 is the only transmembrane protein among the class E Vps proteins and remains the most mysterious in terms of its role in budding. It will be important to determine whether Vps44 drives pH or ionic strength changes in the lumen of the MVB and/or the budding vesicles and, if it does, what proteins act as the downstream effectors of these changes.

One of the major motivations for studying the ESCRT pathway is the discovery that this pathway is central to HIV budding, presenting an unprecedented number of new

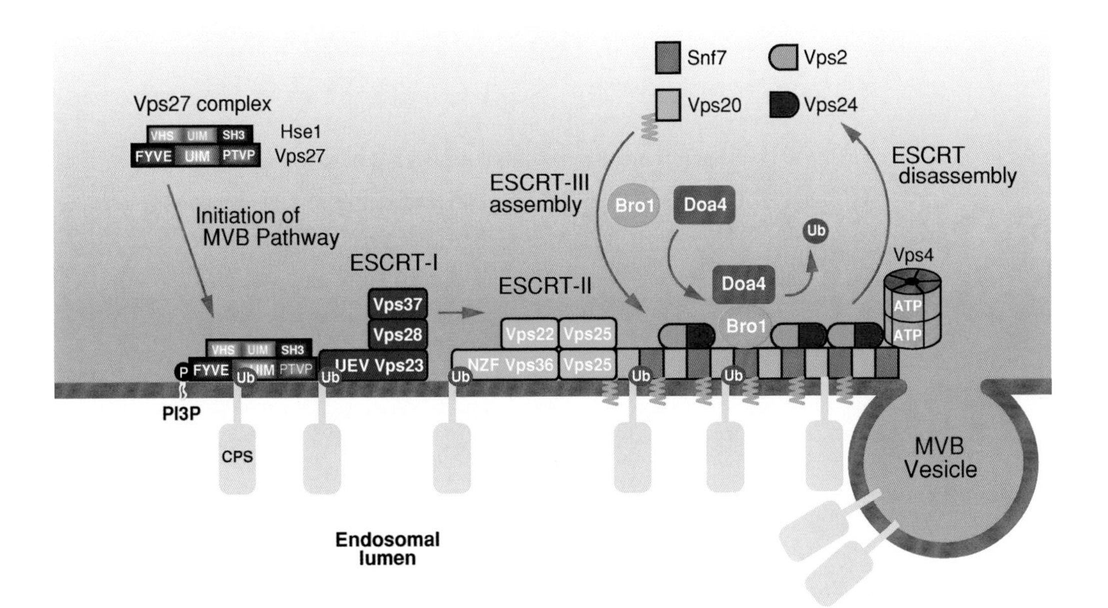

Figure 8

The ESCRT complexes in MVB sorting. The Vps27 protein complex initiates the MVB-sorting process. It is targeted to endosomal membranes via its FYVE domain that binds PI(3)P, and its UIM domains, which bind ubiquitinated MVB cargo such as carboxypeptidase S (CPS). Vps27 subsequently recruits and activates the ESCRT-I complex via the P(S/T)XP motif in the C-terminal domain of Vps27 that interacts with the UEV domain of Vps23 in ESCRT-I. Ubiquitinated cargo is recognized by ESCRT-I (via the UEV domain of Vps23) and by ESCRT-II (via the NZF domain in Vps36). ESCRT-III is required for concentration of cargoes into MVB vesicles and coordinates the association of accessory factors such as Bro1 and the Doa4-deubiquitinating enzyme that removes ubiquitin from cargo. The AAA-type ATPase Vps4 plays a critical role in catalyzing the dissociation of the ESCRT complexes. Together, these proteins appear to direct MVB vesicle formation, cargo sorting into MVB vesicles, and vesicle fission. See text for further details.

potential therapeutic targets. A number of proteins in the pathway are essential for budding and are incorporated directly into HIV virions (101, 111). The usefulness of pathway members as targets depends on the relative sensitivity of virus and host to inhibition of the ESCRT pathway. The locus of HIV budding through this pathway remains controversial. The mechanism by which nascent virions avoid the fate of most other MVB cargo—destruction in the lysosome—remains unknown, and its identification would present an exceptionally interesting target.

The emerging picture of ESCRT assembly on endosomal membranes suggests the formation of arrays of inexact stoichiometry. Dynamic protein networks that assemble on membranes are central to signal transduction and subcellular trafficking. Their complexity, kinetic fragility, membrane localization, and inexact stoichiometry present great challenges to obtaining a precise structural and mechanistic understanding. The payoff will be equally great, perhaps providing a roadmap for analysis of many analogous membrane-bound signaling and trafficking systems.

SUMMARY POINTS

1. The yeast class E *VPS* genes and their human homologues code for a network of proteins that comprise the ESCRT complexes and associated proteins.
2. The Vps27/Hse1, ESCRT-I, ESCRT-II, and ESCRT-III complexes are recruited to endosomal membranes in an ordered manner. Recruitment is initiated by the lipid PI(3)P and the presence of monoubiquitinated transmembrane proteins.
3. The role of the ESCRT network is to (*a*) recruit and cluster monoubiquitinated cargo, and (*b*) drive the formation, invagination, and fission of cargo-containing vesicles into the lumen of the MVB.
4. The structural basis for PI(3)P recruitment via FYVE domains and Ub recruitment via UIM, UEV, and NZF domains has been determined.
5. Vps27/Hse1, ESCRT-I, and ESCRT-II are soluble protein complexes. The structure of the ESCRT-II core complex has been determined.
6. The ESCRT-III complex is thought to be a tightly membrane-associated array. In concert with the ATPase Vps4, ESCRT-III may be involved in mechanical deformation of the membrane to drive the invagination and fission of intralumenal vesicles into MVBs.
7. The ESCRT network is essential for budding of the HIV virus, and structural studies of these complexes present potential targets for development of new anti-HIV therapeutics.
8. The mechanism of invagination is essentially unknown and presents a major challenge for mechanistic studies.

ACKNOWLEDGMENTS

We thank W. Sundquist and C. Hill for sharing unpublished coordinates and data, and Y. Ye, J. Kim, J. Sun, W. Smith, and D. Hurt for comments on the manuscript. Research in the Hurley lab is supported by the Intramural Research Program of the NIH, NIDDK, and IATAP. S.D.E. is supported as an investigator of the Howard Hughes Medical Institute.

LITERATURE CITED

1. Alam SL, Sun J, Payne M, Welch BD, Blake BK, et al. 2004. Ubiquitin interactions of NZF zinc fingers. *EMBO J.* 23:1411–21
2. Amerik AY, Li SJ, Hochstrasser M. 2000. Analysis of the deubiquitinating enzymes of the yeast *Saccharomyces cerevisiae*. *Biol. Chem.* 381:981–92
3. Amit I, Yakir L, Katz M, Zwang Y, Marmor MD, et al. 2004. Tal, a Tsg101-specific E3 ubiquitin ligase, regulates receptor endocytosis and retrovirus budding. *Genes Dev.* 18:1737–52
4. Antonyak MA, McNeill CJ, Wakshlag JJ, Boehm JE, Cerione RA. 2003. Activation of the Ras-ERK pathway inhibits retinoic acid-induced stimulation of tissue transglutaminase expression in NIH3T3 cells. *J. Biol. Chem.* 278:15859–66
5. Asao H, Sasaki Y, Arita T, Tanaka N, Endo K, et al. 1997. Hrs is associated with STAM, a signal-transducing adaptor molecule: its suppressive effect on cytokine-induced cell growth. *J. Biol. Chem.* 272:32785–91

6. Babst M, Katzmann DJ, Estepa-Sabal EJ, Meerloo T, Emr SD. 2002. ESCRT-III: an endosome-associated heterooligomeric protein complex required for MVB sorting. *Dev. Cell* 3:271–82
7. Babst M, Katzmann DJ, Snyder WB, Wendland B, Emr SD. 2002. Endosome-associated complex, ESCRT-II, recruits transport machinery for protein sorting at the multivesicular body. *Dev. Cell* 3:283–89
8. Babst M, Sato TK, Banta LM, Emr SD. 1997. Endosomal transport function in yeast requires a novel AAA-type ATPase, Vps4p. *EMBO J.* 16:1820–31
9. Babst M, Wendland B, Estepa EJ, Emr SD. 1998. The Vps4p AAA ATPase regulates membrane association of a Vps protein complex required for normal endosome function. *EMBO J.* 17:2982–93
10. Bache KG, Brech A, Mehlum A, Stenmark H. 2003. Hrs regulates multivesicular body formation via ESCRT recruitment to endosomes. *J. Cell Biol.* 162:435–42
11. Begley MJ, Taylor GS, Kim SA, Veine DM, Dixon JE, Stuckey JA. 2003. Crystal structure of a phosphoinositide phosphatase, MTMR2: insights into myotubular myopathy and Charcot-Marie-Tooth syndrome. *Mol. Cell* 12:1391–402
12. Bilodeau PS, Urbanowski JL, Winistorfer SC, Piper RC. 2002. The Vps27p-Hse1p complex binds ubiquitin and mediates endosomal protein sorting. *Nat. Cell Biol.* 4:534–39
13. Bilodeau PS, Winistorfer SC, Kearney WR, Robertson AD, Piper RC. 2003. Vps27-Hse1 and ESCRT-I complexes cooperate to increase efficiency of sorting ubiquitinated proteins at the endosome. *J. Cell Biol.* 163:237–43
14. Bishop N, Horman A, Woodman P. 2002. Mammalian class E Vps proteins recognize ubiquitin and act in the removal of endosomal protein-ubiquitin conjugates. *J. Cell Biol.* 157:91–101
15. Blatner N, Stahelin RV, Diraviyam K, Hawkins PT, Hong W, et al. 2004. The molecular basis of the differential subcellular localization of FYVE domains. *J. Biol. Chem.* 279:53818–27
16. Bowers K, Lottridge J, Helliwell SB, Goldthwaite LM, Luzio JP, Stevens TH. 2004. Protein-protein interactions of ESCRT complexes in the yeast *Saccharomyces cerevisiae*. *Traffic* 5:194–210
17. Bowers K, Stevens TH. 2005. Protein transport from the late Golgi to the vacuole in the yeast *Saccharomyces cerevisiae*. *Biochim. Biophys. Acta* 1744:438–54
18. Brunger AT. 2001. Structure of proteins involved in synaptic vesicle fusion in neurons. *Annu. Rev. Biophys. Biomol. Struct.* 30:157–71
19. Brunger AT, DeLaBarre B. 2003. NSF and p97/VCP: similar at first, different at last. *FEBS Lett.* 555:126–33
20. Burd CG, Emr SD. 1998. Phosphatidylinositol(3)-phosphate signaling mediated by specific binding to RING FYVE domains. *Mol. Cell* 2:157–62
21. Chatellard-Causse C, Blot B, Cristina N, Torch S, Missotten M, Sadoul R. 2002. Alix (ALG-2-interacting protein X), a protein involved in apoptosis, binds to endophilins and induces cytoplasmic vacuolization. *J. Biol. Chem.* 277:29108–15
22. Chen B, Borinstein SC, Gillis J, Sykes VW, Bogler O. 2000. The glioma-associated protein SETA interacts with AIP1/Alix and AZIG-2 and modulates apoptosis in astrocytes. *J. Biol. Chem.* 275:19275–81
23. Davies BA, Topp JD, Sfeir AJ, Katzmann DJ, Carney DS, et al. 2003. Vps9p CUE domain ubiquitin binding is required for efficient endocytic protein traffic. *J. Biol. Chem.* 278:19826–33

24. Demirov DG, Ono A, Orenstein JM, Freed EO. 2002. Overexpression of the N-terminal domain of TSG101 inhibits HIV-1 budding by blocking late domain function. *Proc. Natl. Acad. Sci. USA* 99:955–60
25. DiNitto JP, Cronin TC, Lambright DG. 2003. Membrane recognition and targeting by lipid-binding domains. *Sci. STKE* re 16:1–15
26. Diraviyam K, Stahelin RV, Cho W, Murray D. 2003. Computer modeling of the membrane interaction of FYVE domains. *J. Mol. Biol.* 328:721–36
27. Dumas JJ, Merithew E, Sudharshan E, Rajamani D, Hayes S, et al. 2001. Multivalent endosome targeting by homodimeric EEA1. *Mol. Cell* 8:947–58
28. Felder S, Miller K, Moehren G, Ullrich A, Schlessinger J, Hopkins CR. 1990. Kinase activity controls the sorting of the epidermal growth factor receptor within the multivesicular body. *Cell* 61:623–34
29. Feng GH, Lih CJ, Cohen SN. 2000. TSG101 protein steady-state level is regulated posttranslationally by an evolutionarily conserved COOH-terminal sequence. *Cancer Res.* 60:1736–41
30. Fisher RD, Wang B, Alam SL, Higginson DS, Robinson H, et al. 2003. Structure and ubiquitin binding of the ubiquitin-interacting motif. *J. Biol. Chem.* 278:28976–84
31. Futter CE, Pearse A, Hewlett LJ, Hopkins CR. 1996. Multivesicular endosomes containing internalized EGF-EGF receptor complexes mature and then fuse directly with lysosomes. *J. Cell Biol.* 132:1011–23
32. Garrus JE, von Schwedler UK, Pornillos OW, Morham SG, Zavitz KH, et al. 2001. Tsg101 and the vacuolar protein sorting pathway are essential for HIV-1 budding. *Cell* 107:55–65
33. Gaullier JM, Simonsen A, D'Arrigo A, Bremnes B, Stenmark H, Aasland R. 1998. FYVE fingers bind Ptdins(3)P. *Nature* 394:432–33
34. Goila-Gaur R, Demirov DG, Orenstein JM, Ono A, Freed EO. 2003. Defects in human immunodeficiency virus budding and endosomal sorting induced by TSG101 overexpression. *J. Virol.* 77:6507–19
35. Gorden P, Carpentier JL, Cohen S, Orci L. 1978. Epidermal growth factor: morphological demonstration of binding, internalization, and lysosomal association in human fibroblasts. *Proc. Natl. Acad. Sci. USA* 75:5025–29
36. Haglund K, Di Fiore PP, Dikic I. 2003. Distinct monoubiquitin signals in receptor endocytosis. *Trends Biochem. Sci.* 28:598–603
37. Haigler HT, McKanna JA, Cohen S. 1979. Direct visualization of the binding and internalization of a ferritin conjugate of epidermal growth factor in human carcinoma cells A-431. *J. Cell Biol.* 81:382–95
38. Hanyaloglu AC, McCullagh E, von Zastrow M. 2005. Essential role of Hrs in a recycling mechanism mediating functional resensitization of cell signaling. *EMBO J.* 24:2265–83
39. Hershko A, Ciechanover A, Varshavsky A. 2000. The ubiquitin system. *Nat. Med.* 6:1073–81
40. Hicke L. 2001. Protein regulation by monoubiquitin. *Nat. Rev. Mol. Cell Biol.* 2:195–201
41. Hicke L, Dunn R. 2003. Regulation of membrane protein transport by ubiquitin and ubiquitin-binding proteins. *Annu. Rev. Cell. Dev. Biol.* 19:141–72
42. Hierro A, Sun J, Rusnak AS, Kim J, Prag G, et al. 2004. Structure of the ESCRT-II endosomal trafficking complex. *Nature* 431:221–25
43. Hislop JN, Marley A, von Zastrow M. 2004. Role of mammalian vacuolar protein-sorting proteins in endocytic trafficking of a non-ubiquitinated G protein-coupled receptor to lysosomes. *J. Biol. Chem.* 279:22522–31

44. Hochstrasser M. 2000. Evolution and function of ubiquitin-like protein-conjugation systems. *Nat. Cell Biol.* 2:E153–57
45. Hofmann K, Falquet L. 2001. A ubiquitin-interacting motif conserved in components of the proteasomal and lysosomal protein degradation systems. *Trends Biochem. Sci.* 26:347–50
46. Hurley JH, Meyer T. 2001. Subcellular targeting by membrane lipids. *Curr. Opin. Cell Biol.* 13:146–52
47. Hurley JH, Misra S. 2000. Signaling and subcellular targeting by membrane-binding domains. *Annu. Rev. Biophys. Biomol. Struct.* 29:49–79
48. Kaneko T, Kumasaka T, Ganbe T, Sato T, Miyazawa K, et al. 2003. Structural insight into modest binding of a non-PXXP ligand to the signal transducing adaptor molecule-2 Src homology 3 domain. *J. Biol. Chem.* 278:48162–68
49. Kato M, Miyazawa K, Kitamura N. 2000. A deubiquitinating enzyme UBPY interacts with the Src homology 3 domain of Hrs-binding protein via a novel binding motif PX(V/I)(D/N)RXXKP. *J. Biol. Chem.* 275:37481–87
50. Katoh K, Shibata H, Hatta K, Maki M. 2004. CHMP4b is a major binding partner of the ALG-2-interacting protein Alix among the three CHMP4 isoforms. *Arch. Biochem. Biophys.* 421:159–65
51. Katoh K, Shibata H, Suzuki H, Nara A, Ishidoh K, et al. 2003. The ALG-2-interacting protein Alix associates with CHMP4b, a human homologue of yeast Snf7 that is involved in multivesicular body sorting. *J. Biol. Chem.* 278:39104–13
52. Katz M, Shtiegman K, Tal-Or P, Yakir L, Mosesson Y, et al. 2002. Ligand-independent degradation of epidermal growth factor receptor involves receptor ubiquitylation and hgs, an adaptor whose ubiquitin-interacting motif targets ubiquitylation by Nedd4. *Traffic* 3:740–51
53. Katzmann DJ, Babst M, Emr SD. 2001. Ubiquitin-dependent sorting into the multivesicular body pathway requires the function of a conserved endosomal protein sorting complex, ESCRT-I. *Cell* 106:145–55
54. Katzmann DJ, Odorizzi G, Emr SD. 2002. Receptor downregulation and multivesicular-body sorting. *Nat. Rev. Mol. Cell Biol.* 3:893–905
55. Katzmann DJ, Stefan CJ, Babst M, Emr SD. 2003. Vps27 recruits ESCRT machinery to endosomes during MVB sorting. *J. Cell Biol.* 162:413–23
56. Kim J, Sitaraman S, Hierro A, Beach BM, Odorizzi G, Hurley JH. 2005. Structural basis for endosomal targeting by the Bro1 domain. *Dev. Cell* 8:937–47
57. Komada M, Kitamura N. 2005. The Hrs/STAM complex in the downregulation of receptor tyrosine kinases. *J. Biochem.* 137:1–8
58. Kutateladze T, Overduin M. 2001. Structural mechanism of endosome docking by the FYVE domain. *Science* 291:1793–96
59. Kutateladze TG, Capelluto DGS, Ferguson CG, Cheever ML, Kutateladze AG, et al. 2004. Multivalent mechanism of membrane insertion by the FYVE domain. *J. Biol. Chem.* 279:3050–57
60. Kutateladze TG, Ogburn KD, Watson WT, de Beer T, Emr SD, et al. 1999. Phosphatidylinositol 3-phosphate recognition by the FYVE domain. *Mol. Cell* 3:805–11
61. Lemmon MA. 2003. Phosphoinositide recognition domains. *Traffic* 4:201–13
62. Lin Y, Kimpler LA, Naismith TV, Lauer JM, Hanson PI. 2005. Interaction of the mammalian endosomal sorting complex required for transport (ESCRT) III protein hSnf7-1 with itself, membranes, and the AAA+ ATPase SKD1. *J. Biol. Chem.* 280:12799–809
63. Luhtala N, Odorizzi G. 2004. Bro1 coordinates deubiquitination in the multivesicular body pathway by recruiting Doa4 to endosomes. *J. Cell Biol.* 166:717–29

64. Mao YX, Nickitenko A, Duan XQ, Lloyd TE, Wu MN, et al. 2000. Crystal structure of the VHS and FYVE tandem domains of Hrs, a protein involved in membrane trafficking and signal transduction. *Cell* 100:447–56
65. Martin-Serrano J, Zang T, Bieniasz PD. 2001. HIV-I and Ebola virus encode small peptide motifs that recruit Tsg101 to sites of particle assembly to facilitate egress. *Nat. Med.* 7:1313–19
66. Martin-Serrano J, Zang T, Bieniasz PD. 2003. Role of ESCRT-I in retroviral budding. *J. Virol.* 77:4794–804
67. Matsuo H, Chevallier J, Mayran N, Le Blanc I, Ferguson C, et al. 2004. Role of LBPA and Alix in multivesicular liposome formation and endosome organization. *Science* 303:531–34
68. Misra S, Beach BM, Hurley JH. 2000. Structure of the VHS domain of human Tom1 (target of myb 1): insights into interactions with proteins and membranes. *Biochemistry* 39:11282–90
69. Misra S, Hurley JH. 1999. Crystal structure of a phosphatidylinositol 3-phosphate-specific membrane-targeting motif, the FYVE domain of Vps27p. *Cell* 97:657–66
70. Misra S, Miller GJ, Hurley JH. 2001. Recognizing phosphatidylinositol 3-phosphate. *Cell* 107:559–62
71. Misra S, Puertollano R, Kato Y, Bonifacino JS, Hurley JH. 2002. Structural basis for acidic-cluster-dileucine sorting-signal recognition by VHS domains. *Nature* 415:933–37
72. Mizuno E, Kawahata K, Kato M, Kitamura N, Komada M. 2003. STAM proteins bind ubiquitinated proteins on the early endosome via the VHS domain and ubiquitin-interacting motif. *Mol. Biol. Cell* 14:3675–89
73. Morita E, Sundquist WI. 2004. Retrovirus budding. *Annu. Rev. Cell Dev. Biol.* 20:395–425
74. Myers EL, Allen JF. 2002. Tsg101, an inactive homologue of ubiquitin ligase E2, interacts specifically with human immunodeficiency virus type 2 Gag polyprotein and results in increased levels of ubiquitinated Gag. *J. Virol.* 76:11226–35
75. Nikko E, Marini AM, Andre B. 2003. Permease recycling and ubiquitination status reveal a particular role for Bro1 in the multivesicular body pathway. *J. Biol. Chem.* 278:50732–43
76. Odorizzi G, Babst M, Emr SD. 1998. Fab1p PtdIns(3)P 5-kinase function essential for protein sorting in the multivesicular body. *Cell* 95:847–58
77. Odorizzi G, Katzmann DJ, Babst M, Audhya A, Emr SD. 2003. Bro1 is an endosome-associated protein that functions in the MVB pathway in *Saccharomyces cerevisiae*. *J. Cell Sci.* 116:1893–903
78. Patki V, Lawe DC, Corvera S, Virbasius JV, Chawla A. 1998. A functional PtdIns(3)P-binding motif. *Nature* 394:433–34
79. Phillips SA, Barr VA, Haft DH, Taylor SI, Haft CR. 2001. Identification and characterization of SNX15, a novel sorting nexin involved in protein trafficking. *J. Biol. Chem.* 276:5074–84
80. Pickart CM. 2001. Mechanisms underlying ubiquitination. *Annu. Rev. Biochem.* 70:503–33
81. Polo S, Sigismund S, Faretta M, Guidi M, Capua MR, et al. 2002. A single motif responsible for ubiquitin recognition and monoubiquitination in endocytic proteins. *Nature* 416:451–55
82. Pornillos O, Alam SL, Davis DR, Sundquist WI. 2002. Structure of the Tsg101 UEV domain in complex with the PTAP motif of the HIV-1 p6 protein. *Nat. Struct. Biol.* 9:812–17
83. Pornillos O, Higginson DS, Stray KM, Fisher RD, Garrus JE, et al. 2003. HIV Gag mimics the Tsg101-recruiting activity of the human Hrs protein. *J. Cell Biol.* 162:425–34

84. Raiborg C, Bache KG, Gillooly DJ, Madshush IH, Stang E, Stenmark H. 2002. Hrs sorts ubiquitinated proteins into clathrin-coated microdomains of early endosomes. *Nat. Cell Biol.* 4:394–98
85. Raiborg C, Bache KG, Mehlum A, Stang E, Stenmark H. 2001. Hrs recruits clathrin to early endosomes. *EMBO J.* 20:5008–21
86. Raiborg C, Bremnes B, Mehlum A, Gillooly DJ, D'Arrigo A, et al. 2001. FYVE and coiled-coil domains determine the specific localisation of Hrs to early endosomes. *J. Cell Sci.* 114:2255–63
87. Raiborg C, Rusten TE, Stenmark H. 2003. Protein sorting into multivesicular endosomes. *Curr. Opin. Cell Biol.* 15:446–55
88. Reggiori F, Pelham HR. 2001. Sorting of proteins into multivesicular bodies: ubiquitin-dependent and -independent targeting. *EMBO J.* 20:5176–86
89. Sachse M, Strous GJ, Klumperman J. 2004. ATPase-deficient hVPS4 impairs formation of internal endosomal vesicles and stabilizes bilayered clathrin coats on endosomal vacuoles. *J. Cell Sci.* 117:1699–708
90. Sankaran VG, Klein DE, Sachdeva MM, Lemmon MA. 2001. High-affinity binding of a FYVE domain to phosphatidylinositol 3-phosphate requires intact phospholipid but not FYVE domain oligomerization. *Biochemistry* 40:8581–87
91. Scarlata S, Carter C. 2003. Role of HIV-1 Gag domains in viral assembly. *Biochem. Biophys. Acta* 1614:62–72
92. Schmidt MHH, Hoeller D, Yu JH, Furnari FB, Cavenee WK, et al. 2004. Alix/AIP1 antagonizes epidermal growth factor receptor downregulation by the Cbl-SETA/CIN85 complex. *Mol. Cell Biol.* 24:8981–93
93. Scott A, Chung H-Y, Gonciarz-Swiatek M, Hill GC, Whitby FG, et al. 2005. Structural and mechanistic studies of VPS4 proteins. *EMBO J.* 24:3658–69
94. Scott A, Gaspar J, Stuchell-Brereton M, Alam SL, Skalicky J, Sundquist WI. 2005. Structure and ESCRT-III protein interactions of the MIT domain of human Vps4A. *Proc. Natl. Acad. Sci. USA* 102:13813–18
95. Shiba T, Takatsu H, Nogi T, Matsugaki N, Kawasaki M, et al. 2002. Structural basis for recognition of acidic-cluster dileucine sequence by GGA1. *Nature* 415:937–41
96. Shiflett SL, Ward DM, Huynh D, Vaughn MB, Simmons JC, Kaplan J. 2004. Characterization of Vta1p, a class E Vps protein in *Saccharomyces cerevisiae*. *J. Biol. Chem.* 279:10982–90
97. Shih SC, Katzmann DJ, Schnell JD, Sutanto M, Emr SD, Hicke L. 2002. Epsins and Vps27p/Hrs contain ubiquitin-binding domains that function in receptor endocytosis. *Nat. Cell Biol.* 4:389–93
98. Sigismund S, Polo S, Di Fiore PP. 2004. Signaling through monoubiquitination. *Curr. Top. Microbiol. Immunol.* 286:149–85
99. Slagsvold T, Aasland R, Hirano S, Bache KG, Raiborg C, et al. 2005. Eap45 in mammalian ESCRT-II binds ubiquitin via a phosphoinositide-interacting GLUE domain. *J. Biol. Chem.* 280:19600–6
100. Stahelin RV, Long F, Diraviyam K, Bruzik KS, Murray D, Cho W. 2002. Phosphatidylinositol 3-phosphate induces the membrane penetration of the FYVE domains of Vps27p and Hrs. *J. Biol. Chem.* 277:26379–88
101. Strack B, Calistri A, Craig S, Popova E, Gottlinger HG. 2003. AIP1/ALIX is a binding partner for HIV-1 p6 and EIAV p9 functioning in virus budding. *Cell* 114:689–99
102. Sundquist WI, Schubert HL, Kelly BN, Hill GC, Holton JM, Hill CP. 2004. Ubiquitin recognition by the human TSG101 protein. *Mol. Cell* 13:783–89

103. Swanson KA, Kang RS, Stamenova SD, Hicke L, Radhakrishnan I. 2003. Solution structure of Vps27 UIM-ubiquitin complex important for endosomal sorting and receptor downregulation. *EMBO J.* 22:4597–606
104. Teo H, Perisic O, Gonzalez B, Williams RL. 2004. ESCRT-II, an endosome-associated complex required for protein sorting: crystal structure and interactions with ESCRT-III and membranes. *Dev. Cell* 7:559–69
105. Teo H, Veprintsev DB, Williams RL. 2004. Structural insights into endosomal sorting complex required for transport (ESCRT-I) recognition of ubiquitinated proteins. *J. Biol. Chem.* 279:28689–96
106. ter Haar E, Harrison SC, Kirchhausen T. 2000. Peptide-in-groove interactions link target proteins to the beta-propeller of clathrin. *Proc. Natl. Acad. Sci. USA* 97:1096–100
107. Tong AHY, Drees B, Nardelli G, Bader GD, Brannetti B, et al. 2002. A combined experimental and computational strategy to define protein interaction networks for peptide recognition modules. *Science* 295:321–24
108. Urbanowski JL, Piper RC. 2001. Ubiquitin sorts proteins into the intralumenal degradative compartment of the late-endosome/vacuole. *Traffic* 2:622–30
109. VerPlank L, Bouamr F, LaGrassa TJ, Agresta B, Kikonyogo A, et al. 2001. Tsg101, a homologue of ubiquitin-conjugating (E2) enzymes, binds the L domain in HIV type 1 Pr55(Gag). *Proc. Natl. Acad. Sci. USA* 98:7724–29
110. Vincent O, Rainbow L, Tilburn J, Arst HN, Penalva MA. 2003. YPXL/I is a protein interaction motif recognized by *Aspergillus* PalA and its human homologue, AIP1/Alix. *Mol. Cell Biol.* 23:1647–55
111. von Schwedler UK, Stuchell M, Muller B, Ward DM, Chung HY, et al. 2003. The protein network of HIV budding. *Cell* 114:701–13
112. Ward DMV, Vaughn MB, Shiflett SL, White PL, Pollock AL, et al. 2005. The role of LIP5 and CHMP5 in multivesicular body formation and HIV-1 budding in mammalian cells. *J. Biol. Chem.* 280:10548–55
113. Wernimont AK, Weissenhorn W. 2004. Crystal structure of subunit Vps25 of the endosomal trafficking complex ESCRT-II. *BMC Struct. Biol.* 4:10
114. Whitley P, Reaves BJ, Hashimoto M, Riley AM, Potter BV, Holman GD. 2003. Identification of mammalian Vps24p as an effector of phosphatidylinositol 3,5-bisphosphate-dependent endosome compartmentalization. *J. Biol. Chem.* 278:38786–95
115. Yeo SCL, Xu LH, Ren JH, Boulton VJ, Wagle MD, et al. 2003. Vps20p and Vta1p interact with Vps4p and function in multivesicular body sorting and endosomal transport in *Saccharomyces cerevisiae*. *J. Cell Sci.* 116:3957–70
116. Yorikawa C, Shibata H, Waguri S, Hatta K, Horii M, et al. 2005. Human CHMP6, a myristoylated ESCRT-III protein, interacts directly with an ESCRT-II component EAP20 and regulates endosomal cargo sorting. *Biochem. J.* 387:17–26
117. Zarrinpar A, Bhattacharyya RP, Lim WA. 2003. The structure and function of proline recognition domains. *Sci STKE* DOI:10.1126/stke.2003.179.re8
118. Zhu GY, He XY, Zhai P, Terzyan S, Tang J, Zhang XJC. 2003. Crystal structure of GGA2 VHS domain and its implication in plasticity in the ligand binding pocket. *FEBS Lett.* 537:171–76

Ribosome Dynamics: Insights from Atomic Structure Modeling into Cryo-Electron Microscopy Maps

Kakoli Mitra and Joachim Frank

Howard Hughes Medical Institute, Wadsworth Center, Empire State Plaza, Albany, New York 12201-0509; email: kmitra@wadsworth.org; joachim@wadsworth.org

Annu. Rev. Biophys. Biomol. Struct. 2006. 35:299–317

The *Annual Review of Biophysics and Biomolecular Structure* is online at biophys.annualreviews.org

doi: 10.1146/annurev.biophys.35.040405.101950

1056-8700/06/0609-0299$20.00

Key Words

macromolecular machine, signal transduction, X-ray structure, atomic model fitting, conformational dynamics

Abstract

Single-particle cryo-electron microscopy (cryo-EM) is the method of choice for studying the dynamics of macromolecular machines both at a phenomenological and, increasingly, at the molecular level, with the advent of high-resolution component X-ray structures and of progressively improving fitting algorithms. Cryo-EM has shed light on the structure of the ribosome during the four steps of translation: initiation, elongation, termination, and recycling. Interpretation of cryo-EM reconstructions of the ribosome in quasi-atomic detail reveals a picture in which the ribosome uses RNA not only to catalyze chemical reactions, but also as a means for signal transduction over large distances.

Contents

OVERVIEW

Certain essential processes in the cell, such as protein translation, folding, and transport across the nuclear membrane, to name just three such processes, require the finely tuned coordination of multiple subprocesses. The cell's ingenious solution to this problem is to localize in space all the different molecules that carry out these subprocesses, generating a defined macromolecular machine, which can be massive. The ribosome in eubacteria is a 2.5-MDa complex composed of RNA and protein (a ribonucleoprotein), and its role is to make protein from component amino acids using the instructions in the genetic code (24, 56). The chaperonin complex, GroEL-GroES, made entirely of proteins (~1 MDa), provides a protective environment for the refolding of misfolded proteins (63). Finally, the nuclear pore receptor is composed entirely of protein and is another massive machine of ~120 MDa. This machine performs the vital task of selectively transporting molecules into and out of the nucleus (reviewed in Reference 35). The phenomenally complex tasks that these macromolecular machines must accomplish repeatedly and efficiently involve the coordinated movement of many parts, just as in a human-made machine. The inherent dynamic nature of these macromolecular machines and their large size have made challenging the elucidation of their structures using methods such as nuclear magnetic resonance (NMR) and X-ray crystallography. Structure determination of a molecule by NMR still has size limitations of around 30 kDa for a full-atom model including side chains, or between 60 and 100 kDa for determination of the global fold (79). Restrictions on size result in part from requirements of high molecular tumbling rates in solution. X-ray crystallography, on the other hand, requires the ordered crystallization (directional packing) of molecules, a challenging feat for a highly dynamic molecule.

A structural technique that is emerging as increasingly powerful for visualizing dynamic, large macromolecular complexes is single-particle cryo-electron microscopy (cryo-EM). Single-particle cryo-EM has two obvious advantages: There are no size limitations, and the complex does not need to be ordered in an array or to be tumbling at any given rate. Single-particle cryo-EM studies of the eubacterial ribosome, with structural resolutions of 8 to 12 Å, have enabled the phenomenological observation of conformational changes associated with ribosome function (see, for example, Reference 25). Molecular machines, such as the ribosome, perform their tasks in a defined sequence of steps, each triggered by an event of ligand binding or dissociation and usually accompanied by conformational changes of the binding partners. The time course of the entire system of interacting molecules can be described as a succession of transitional states. Among existing biophysical imaging techniques, cryo-EM combined with single-particle reconstruction has a special position in that it allows a molecular machine to be visualized in each of its transitional states, provided those states can be efficiently trapped. Neither NMR nor X-ray crystallography is up to this task. Molecular complexes that fall in the category of molecular machines are usually quite large and as such not amenable to NMR. X-ray crystallography is not necessarily restricted by size, as evidenced by the recent success in solving the structure of the ribosome (6, 36, 64, 88, 91), but the occurrence of many conformations and binding states in the time course of the activity of a molecular machine poses large obstacles to crystallization. With the advent of almost complete X-ray structures of the eubacterial ribosome and the progressive improvement of computational methods, it is now becoming possible to interpret these moderate-resolution cryo-EM maps at the atomic level with increasing reliability.

This review first introduces the technique of single-particle cryo-EM and then outlines the computational approach of atomic model docking and refinement into cryo-EM maps. The ribosome is used as an example to illustrate how the combination of X-ray structures with cryo-EM maps can effect the transition from phenomenological observation to understanding the molecular basis of the workings of a macromolecular machine. Finally, we discuss the potential of this method, as revealed in a recent systematic analysis of fitted atomic models into cryo-EM maps. This analysis has provided new insights into signal transduction mechanisms within the RNA scaffold of the ribosome.

SINGLE-PARTICLE CRYO-ELECTRON MICROSCOPY

Basics of the Technique

The structural resolution obtainable with an imaging technique is usually in the range of the wavelength of the electromagnetic or matter waves used for the imaging. This is why the resolution achieved in the light microscope is only within a few hundred nanometers. The de Broglie wavelength of electrons accelerated through a voltage of 100kV is 0.037 Å. However, not even atomic resolution (<3 Å) has been achieved for biological molecules using cryo-EM. There are three reasons for this: (*a*) Lens aberrations restrict the useful aperture to a range corresponding to 1 Å resolution at best. (*b*) Because the elements comprising biological molecules have low atomic numbers—in a relatively narrow range—the contrast in the image is low, resulting in a low signal-to-noise ratio (SNR). (*c*) The interaction of electrons with organic matter causes damage, owing to the breaking of chemical bonds and the generation of free radicals, which in turn cause further damage. Much of this damage is a result of the heat generated upon the interaction of electrons with the biological specimen and can be reduced by almost an order of magnitude if the sample is cooled from room temperature to liquid nitrogen temperatures. Reducing the dose of the electrons to ~10 electrons per Å^2 (with some researchers advocating 1 electron per Å^2) significantly reduces damage to the specimen, yet reduces the SNR further. The low SNR problem is commonly circumvented by averaging the signal from many identical molecules. In a regular, predictable

NMR: nuclear magnetic resonance spectroscopy

X-ray crystallography: determination of the structure of a molecular complex from X-ray diffraction data of ordered crystals of the molecular complex

Single-particle cryo-electron microscopy (cryo-EM): direct imaging of multiple copies of molecular complex in single-particle (noncrystalline) form embedded in a layer of vitreous ice under cryogenic conditions

Molecular machine: a cellular entity composed of several protein and/or RNA molecules for the purpose of carrying out an essential process in the cell in a sequential, processive fashion

Signal-to-noise ratio (SNR): ratio of the strengths of signal to noise in an image, measured as the ratio of their variances

Three-dimensional reconstruction: reconstruction of the three-dimensional image of a molecular complex from its two-dimensional projections

arrangement of molecules, such as in a two-dimensional crystal, this averaging is straightforward, because all molecules are in the same orientation. In single-particle cryo-EM every particle in the frozen-hydrated sample is potentially in a different orientation. Averaging (or three-dimensional reconstruction) in this case necessitates first the determination of the exact relative orientation of each particle.

When a specimen in a near-physiological environment is flash-frozen to liquid nitrogen temperatures (150° K), the water in the specimen turns into vitreous ice, which has properties similar to those of liquid water, owing to the rapid decrease in temperature. This technique for the preparation of frozen-hydrated specimens (1, 21, 22) has been used for the past 20 years in conjunction with visualization by EM and is commonly called cryo-EM. Prior to the invention of the cryo-fixation technique, samples for visualization by EM were prepared using negative staining (9), a technique in which the specimen is stained with a solution of heavy-metal salts and then air-dried such that the outline of the specimen, coated with the metal salt, presents a high contrast to the surrounding. The disadvantages of negative staining are numerous: The specimen is not in its native environment; no interior density variations of the molecule are visualized, since the metal salt coats mostly the outside; and the specimen may be distorted. In a frozen-hydrated specimen, the specimen remains immersed in water, i.e., in vitreous ice; internal variations in density are visible; and there is minimal distortion of specimen shape or structure. The use of flash-freezing the specimen and cryogenic temperatures, in conjunction with low-dose imaging in the transmission electron microscope, and the concept of averaging many images of the specimen, form the basis of today's high-resolution biological electron microscopy.

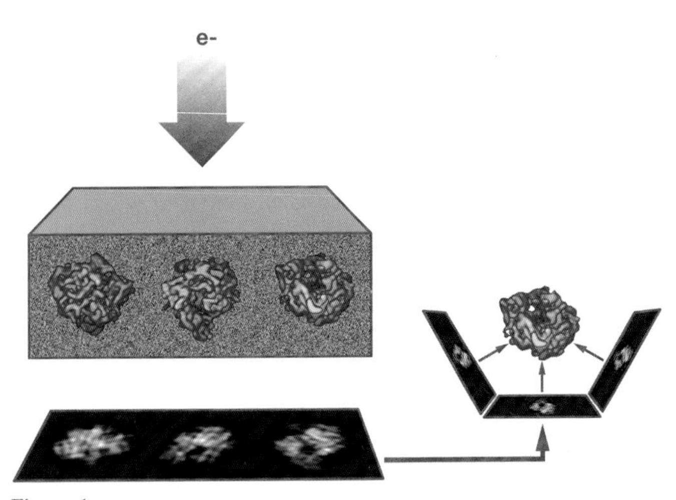

Figure 1

Principle of single-particle reconstruction from images in the transmission electron microscope. In the electron microscope a coherent beam of electrons impinges upon a frozen-hydrated sample (consisting of molecules in random orientations) embedded in a layer of ice, such that two-dimensional projections of the molecules are imaged on a micrograph. A three-dimensional model of the molecule is reconstructed computationally from the two-dimensional projections.

Three-Dimensional Reconstruction from Two-Dimensional Images

In a frozen-hydrated sample of homogeneous, identical particles, each particle is embedded within the vitreous ice layer in a different orientation. The image that is formed of such a layer of particles in a micrograph, recorded on photographic film or a charged-couple device, is a two-dimensional projection of each particle (**Figure 1**). The challenge is thus to reconstruct a three-dimensional image of the particle from different two-dimensional projections. The mathematical theory underlying three-dimensional reconstruction, which goes back to Radon (55), was first put into practice by DeRosier & Klug (20), who reconstructed a bacteriophage tail from electron micrographs. However, unlike the reconstruction of the highly ordered, helical bacteriophage tail, which only requires a single view, the reconstruction of a totally asymmetric particle requires that a large number of views are available and that their relative orientations are known with high accuracy.

For a homogeneous sample, the spatial relationship of the particles in the vitreous ice layer can be described mathematically by a series of rigid-body translations and rotations of a single object. Thus, if the angular distribution of the particles is sufficiently uniform, i.e., if particles are present in orientations that sample a large part of the angular space, then a set of micrographs of the specimen, each showing projections of hundreds of particles, contains all the information necessary to reconstruct the particle in three dimensions.

The problem of determining the relative orientation among all projections is challenging because of the low SNR of the data. If no prior information about the structure of the particle is known, two ab initio methods can be used for orientation determination. (*a*) In the random-conical data collection method (69), which involves the use of an additional tilted view of the specimen, geometric relationships are established among a subset of particles that face the grid in the same orientation. (*b*) In the method of common lines (34, 83), particle images are first classified, and then class averages representing different views of the particle are related to one another following the common lines principle first formulated by Crowther (18). The latter method cannot establish the handedness, however, and requires an extra tilt for this purpose. From a set of projections whose angles have been determined by either of these ab initio methods, a coarse reconstruction can then be computed, which is inaccurate yet contains important shape information allowing higher resolution and accuracy to be achieved through a succession of iterative steps known as angular refinement.

In the refinement, the first, coarse reconstruction serves as an initial reference structure from which two-dimensional projections are computed, which are then compared with the experimental projections, yielding refined angles. With those angles, a new reconstruction is obtained, which is then used as the reference in the next cycle. Refinement continues until no improvement in the reconstructed three-dimensional model is observed or until the orientation angles have stabilized (see References 24a, 26, 54, 84 for detailed treatments of the mathematical and computational procedures for single-particle reconstruction). Currently, resolutions in the range between 10 and 13 Å are routinely obtained, at which detailed features of proteins and RNA are observed at the level of protein domains or RNA backbone (29a). Resolutions in the range of up to 7 Å, at which structural motifs such as alpha-helices start to be discernable (8), still require a large effort. Typically, more than 100,000 particle projections are required to meet this goal. Still larger numbers are needed—an estimated 1,000,000 for the ribosome—to approach 3 Å resolution. However, while the achievement of atomic resolution still awaits improvements in several instrumental and data processing aspects, existing moderate-resolution cryo-EM density maps of a molecular assembly can already be interpreted at the atomic level if X-ray or NMR structures of its components are known. The remainder of this chapter elaborates on the way these hybrid methods of analysis can be put to use.

Angular refinement: the process of iterative refinement of angular assignments to projections, by alternate steps of projection matching and three-dimensional reconstruction

ATOMIC INTERPRETATION OF CRYO-ELECTRON MICROSCOPY MAPS: FITTING HIGH-RESOLUTION X-RAY STRUCTURES

A cryo-EM reconstruction can be interpreted at a level of detail that is greater than the experimental resolution if high-resolution structures of components are available. This is because an atomic model can be placed into a moderate-resolution cryo-EM reconstruction using constrained fitting, with a potential accuracy of placement of four- to fivefold better than the experimental resolution of the cryo-EM map (5, 62). Justification for this accuracy can be made by analogy to X-ray crystal structures for which 0.5 Å model precision is attainable with ~3 Å diffraction data when the modeling is restrained and constrained to agree with standard stereochemistry (23). As a first approximation, individual component structures, such as domains of proteins, can be fitted into cryo-EM reconstructions so that only rigid motions are allowed,

Global search: a rigorous search of all possible rotation and translation angles undertaken to find the best-possible fit of a molecule into a given density map

enabling the placement of collections of atoms to generate quasi-atomic models. In this way, a moderate-resolution map can be interpreted at the atomic level. The first attempts at this approach in the 1990s involved the subjective, manual docking of known structures as rigid bodies into EM reconstructions (5). In the past few years, computational methods have been developed to make this initial fitting more objective (automated) and quantitative, and to go beyond the mere rigid fitting of atomic structures, taking into account conformational changes by flexible fitting as well (reviewed in References 23, 66).

Initial Placement of a Component Atomic Model

Even if the X-ray structure of a component is thought to remain unchanged in the molecule imaged by cryo-EM, the lack of distinct features in a moderate-resolution cryo-EM reconstruction (12–20 Å) makes the unambiguous placement of an atomic model difficult (7, 39). Manual fitting is a viable option, based on prior knowledge and recognition of component shapes. Alternatively, a global computer search must be used. To find the best possible fit, a rigorous search of all possible rotation and translation angles (amounting to six degrees of freedom) needs to be undertaken side by side with a quantification of the fit. Several such global search algorithms for the initial placement of the structure have evolved. In these algorithms the assessment of the fit is performed by comparing the experimental electron density map, i.e., the cryo-EM map, with the map calculated for the model for each grid point in a six-dimensional search space. Moreover, if N independent models (e.g., in the case where the molecule imaged by cryo-EM contains N proteins) are fitted into the cryo-EM map simultaneously, then the configurational space has 6N dimensions. A search through such a vast space must be computationally efficient to be practical. There are two main approaches to accelerating this process: one is through Fourier techniques, and the other by restricting the search to certain subregions defined by a mask.

In a global search the experimental map is compared with a model map by minimization—in the least-squares sense—of the residual, R, in the target function. Comparing electron densities in real space is generally too inefficient for conducting a global search in a large configurational space. Calculations can be performed much faster in reciprocal (Fourier) space than in real space. Such Fourier-space calculations are utilized in existing crystallographic software packages, which can thus be used to optimize the fit of atomic models to EM reconstructions (50).

SITUS is a program that represents the prominent features of a map in skeletonized form, using codebook vectors (15, 90). All possible pairings of the experimental and model or codebook vectors are then searched to find the best mean-squares fit. Distortions can also be introduced into the model to approximate conformational differences between the structure of the component in the model and that in the cryo-EM density. This approach yields the best results if the model can account for the entire cryo-EM density. If only a partial model exists, then the corresponding part in the experimental map has to be masked, increasing the likelihood of a biased outcome. Several errors can occur during a global search: Many local optima for the target function may be found, resulting in multiple solutions for the fitted atomic model; the optimum for the target function lies, in many cases, at the center of low-resolution experimental density; models are rarely fully complete, with missing, unmodeled features being larger components visualized at low resolution. In an effort to mitigate these errors and ambiguities, several approaches have been developed that have in common the use of masking and local normalization. These are implemented in the programs COAN (86, 87), EMFIT (62), DOCKEM (61), and RAMOS (57). Among these, COAN is distinguished by the fact that it considers multiple solutions corresponding to cross-correlation coefficients above a certain significance threshold. This method requires an additional, expert step to select a stereochemically meaningful solution.

The approaches outlined above work best when trying to fit components that are treated as rigid bodies, e.g., an entire protein as one rigid body, into the EM density. The placement of each component is determined independently of surrounding components. These methods are therefore less suitable for large macromolecular complexes with many interconnected, moving components, especially if components need to be treated as flexible entities instead of as rigid bodies. Approaches to fit multiple, interconnected components include the real-space refinement (RSRef) method (16) and normal mode-based flexible fitting (NMFF) (72, 73) (**Figure 2**).

Refinement of Atomic Models of Multiple, Interconnected Components

The RSRef algorithm performs calculations in real space, so that an exhaustive global search is not possible in practice. An initial placement of all components is therefore required, which is then refined to the position of each component within the EM density. In RSRef individual components are treated as rigid bodies; however, flexibility of a multicomponent model can be increased by subdividing the components into smaller rigid bodies, e.g., corresponding to protein domains. RSRef has been developed as a module for utilization in existing crystallographic refinement packages, first in TNT (78), which supported stereochemically restrained least-squares optimization. Recently RSRef has been implemented in the crystallography and NMR system (CNS), which supports simulated-annealing molecular dynamics (11–13) in addition to least-squares optimizations, and torsion angle model parameterization (60) as well as rigid-group optimization. RSRef as a module in TNT has been used to refine atomic models of actomyosin complexes (17) and the ribosome (30) into cryo-EM density maps. A limitation of this application is the resulting often distorted stereochemistry of the fragmentation points between defined rigid bodies (30). Both the stereochemistry at rigid-body junctures and the placement of component atomic models are improved greatly with RSRef implemented in CNS, as illustrated in a recent comparative study of almost complete atomic models of the ribosome fitted into two cryo-EM reconstructions corresponding to different conformations of the ribosome (K. Mitra, C. Schaffitzel, F. Fabiola, M. S. Chapman, N. Ban & J. Frank, manuscript submitted).

High-resolution structures are in most cases obtained for a molecule constrained in its conformation owing to close packing in a crystal, whereas a cryo-EM reconstruction often represents the molecule in its unconstrained, physiological conformation(s). Although the transition from the X-ray crystal to the cryo-EM structure can sometimes be approximated by rigid-body movements of subcomponents of the molecule, which requires a priori knowledge or a good guess about the appropriate fragmentation of the molecule, treating the molecule as a flexible entity may be necessary. Normal-mode analysis (NMA), which uses a simplified elastic network representation of the potential energy function of a biological molecule (77), has been shown to be successful in reproducing large, collective motions of large macromolecules (73, 75), which occur on timescales not accessible via standard molecular dynamics simulations (10, 33). In many cases, a small number of low-frequency normal modes can describe the functional rearrangements of macromolecules (37, 74). Normal mode-based flexible fitting takes advantage of this correspondence and uses the highly collective, low-frequency motions from NMA as search directions in a protocol (41) to flexibly refine an initially placed atomic model into the cryo-EM density (72, 73). The fitting is performed by incrementally deforming the

RSRef: real-space refinement

NMFF: normal mode-based flexible fitting

CNS: crystallography and NMR system

NMA: normal-mode analysis

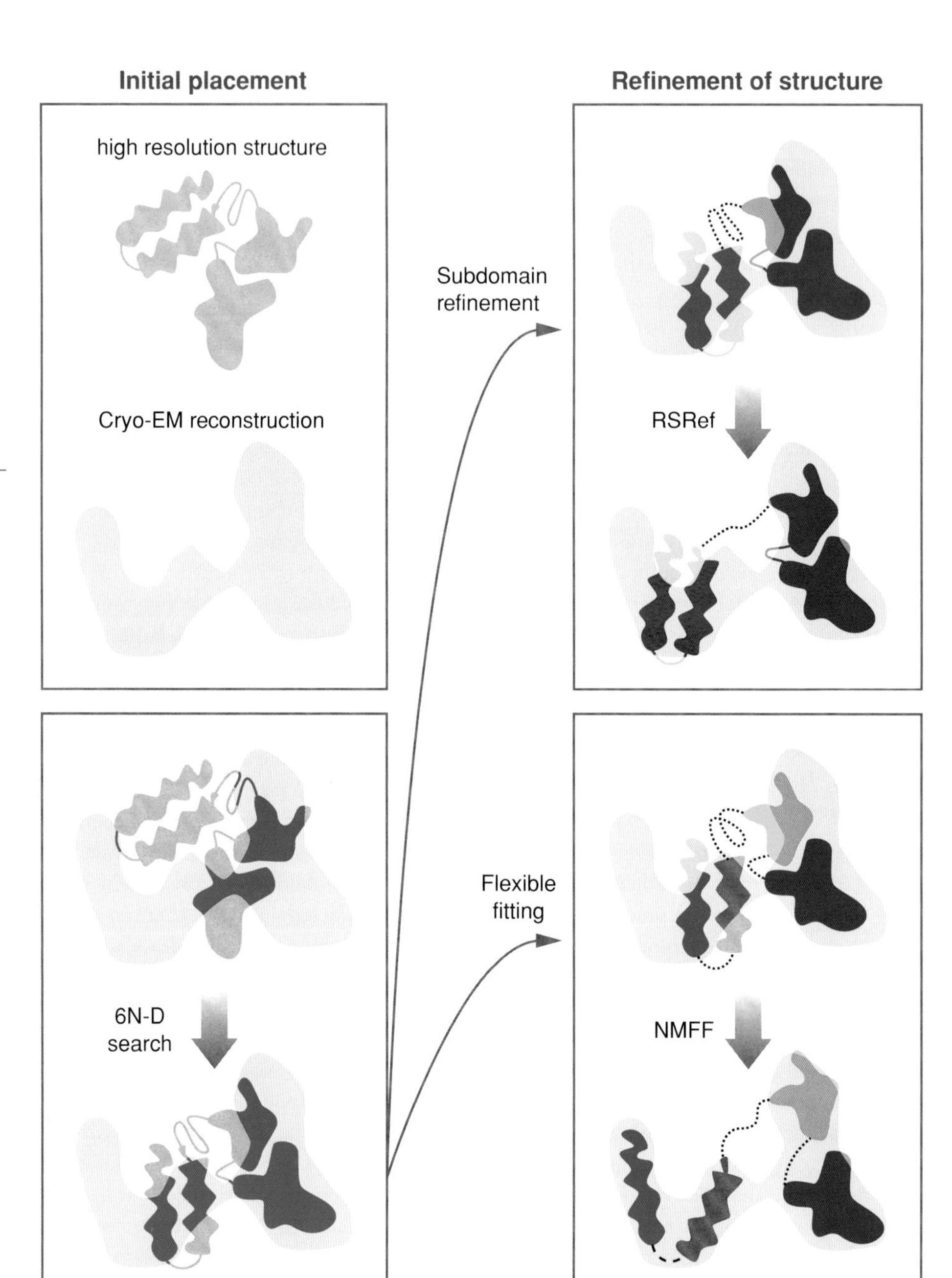

Figure 2

Two-step interpretation of cryo-EM reconstructions at the quasi-atomic level. In the first step, initial placement, the high-resolution atomic model is placed as a single rigid body into the cryo-EM density by manual fitting or computational global search methods. This initially placed model is then refined in the second step, refinement of structure, by flexible-fitting algorithms, in which the atomic model is subdivided into smaller rigid bodies (RSRef) or is flexibly fitted (NMFF).

structure along a set of low-frequency normal modes such that the correlation coefficient between the experimental cryo-EM map and the deformed simulated map increases (72, 73). NMFF has been successfully applied to fitting the X-ray structure of elongation factor G (EF-G) into the cryo-EM density of an EF-G-bound ribosome (81) and also to flexibly fit an *Escherichia coli* RNA polymerase structure (19) into a cryo-EM map (73). Most recently, the structural basis of protein translocation through the protein-conducting channel (PCC) has been elucidated using NMFF of the X-ray structure of a nonfunctional PCC protomer (82) into the cryo-EM density of a functional, dimeric PCC attached to a translating ribosome (45).

THE RIBOSOME: TRANSLATOR OF THE GENETIC CODE

Organization and Function

The ribosome is a macromolecular machine composed of both RNA and protein, i.e., a ribonucleoprotein, that synthesizes proteins from component amino acids that are covalently attached to transfer RNAs [aminoacyl(aa)-tRNA], following the polypeptide sequence encoded on messenger (m)RNA transcribed from DNA. A translating ribosome consists of two ribosomal subunits, the small and large subunits, which preserve between them a space of complex topology, the intersubunit space, that is the site of interaction with tRNA and cofactors of translation. In bacteria the small subunit (or 30S subunit, denoted by its sedimentation coefficient) is composed of a 16S RNA and a number of proteins, and the large subunit (50S subunit) consists of a 23S RNA, a 5S RNA, and several proteins. Together these subunits form the 70S ribosome. The ribosome must fulfill several important functions, in sequential steps, in order to successfully complete the translation of a polypeptide. (*a*) Initiation. The two subunits, along with mRNA, must be assembled to form a 70S ribosome. The ribosome must facilitate recognition of, and binding to, the first codon on the mRNA (usually AUG) by its cognate aa_1-tRNA. (*b*) Elongation. Upon initiation, recognition of the second codon on the mRNA by a cognate aa_2-tRNA (decoding) must occur. Subsequently, the ribosome must catalyze peptide-bond formation between methionine (aa_1) and aa_2. In order for this process to be repeated for downstream codons, the mRNA must move relative to the ribosome, and the aa-tRNAs that are bound to their cognate codons on the mRNA must leave the ribosome for incoming aa-tRNAs (translocation). (*c*) Termination. When the end of a polypeptide sequence is signaled by a stop codon in the ribosome, for which, during routine translation, no cognate aa-tRNAs exist, the ribosome must release the newly synthesized polypeptide chain. (*d*) Recycling. The ribosome must disassemble into its component subunits and release the mRNA so that the translation of a new protein can begin.

Biochemical studies over the past 50 years have elucidated details of these four steps and have also identified the ribosomal components involved. A fundamental insight was that an aa-tRNA binds sequentially to three different sites formed by both the small and large subunits through the progression of elongation. First, the aa_n-tRNA binds to the A (aminoacyl) site, where recognition of the cognate codon in the mRNA occurs. Second, the aa_n-tRNA is moved (translocated) to the adjacent P (peptidyl) site, concomitantly with translocation of the mRNA such that codon-anticodon base-pairing is maintained. This frees the A-site for the incoming aa_{n+1}-tRNA, to which, upon cognate codon recognition, the polypeptide is transferred from the P-site tRNA by formation of a covalent peptide bond between aa_n and aa_{n+1}. Third, the deacylated P-site tRNA (94) is translocated to the E (exit) site to enable its diffusion out of the ribosome. The discovery of a P/E hybrid state, i.e., in which an aa-tRNA was found to be bound to the P-site in the 30S subunit but to the E-site in the 50S subunit, in addition to an A/P hybrid state, led to the hypothesis of a movement of the 30S subunit relative to the 50S subunit during translocation (46) (**Figure 3**), later verified by cryo-EM (25).

Studies on the ribosome also revealed that a number of soluble protein cofactors, by interaction with the ribosome, accelerate processes in all four translation steps. Thus, the initiation factors IF1, IF2, and IF3 aid in the binding of the initiator tRNA (fMet-$tRNA^{fMet}$) to the P site (38, 56). Elongation factor (EF)-Tu associates with the incoming aa-tRNA (forming a ternary complex

Translocation: the movement of tRNA(s) and mRNA relative to the ribosome, a process that is necessary for the translation of the subsequent codon on the mRNA during each elongation cycle

PCC: protein-conducting channel

tRNA: transfer RNA

mRNA: messenger RNA

Intersubunit space: a space of complex topology, preserved between the small and large subunits of the ribosome, that is the site of interaction with tRNA and translational cofactors

EF: elongation factor

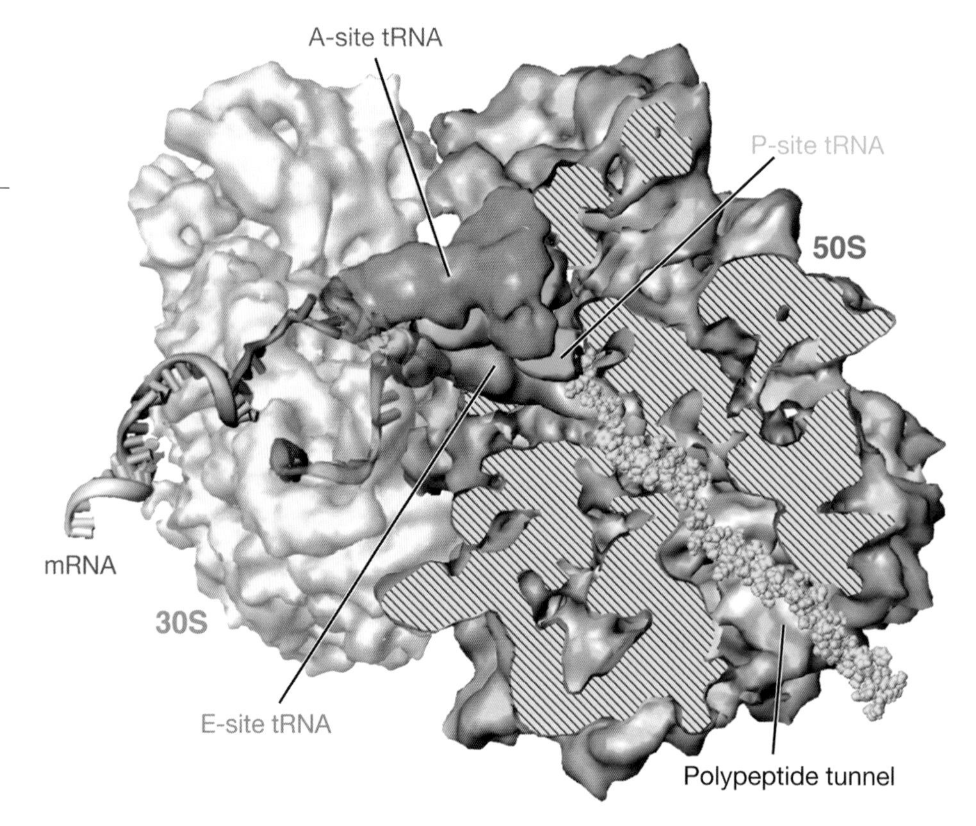

Figure 3

Architecture of the eubacterial 70S ribosome. Shown is a cryo-EM reconstruction of the 70S ribosome, consisting of assembled 30S and 50S subunits, to which A-, P-, and E-site tRNAs are bound. Atomic structures of a stretch of mRNA and the nascent polypeptide chain have been modeled in. A portion of the 50S subunit has been cut away to show the architecture of the polypeptide tunnel.

with GTP) to deliver the aa-tRNA to the A site (46). When cognate codon recognition and accommodation (i.e., the process whereby the aa_n-tRNA is released from EF-Tu and fully incorporated into the A site) has occurred, EF-Tu dissociates from the ribosome. Upon peptide bond formation, which transfers the polypeptide from the P-site polypeptidyl (pp_{n-1})-tRNA to the A-site aa_n-tRNA and is an essentially spontaneous (53), ribosomal RNA-catalyzed (6, 51) reaction, EF-G facilitates translocation of the mRNA and tRNAs. Release factors RF1, RF2, and RF3 are involved in termination and release of the completely translated polypeptide chain (48, 93), and ribosome recycling factor (RRF) and EF-G help dissociate the ribosome into subunits such that a new round of translation can begin (31, 40). Many of these cofactors utilize GTP binding and hydrolysis to regulate their function.

Cryo-Electron Microscopy and the Phenomenological Study of Conformational Dynamics

Crude models of the small- and large-subunit organization of the ribosome were made in the 1970s from EM images of negative-stained specimens (89). Early low-resolution cryo-EM reconstructions of the bacterial 70S ribosome (27, 29, 43, 68) revealed distinct morphological features, such as the central protuberance, and the L1 and L7/L12 stalks in the 50S subunit, the identification of which was facilitated by the earlier placement of ribosomal proteins by antibody labeling (52, 71) and neutron diffraction (47). The first visualization of A-, P-, and E-site tRNAs bound to the ribosome (4, 69), the mRNA tunnel through the 30S subunit (29), as well as binding of elongation factors (2, 3, 69, 70) were all achieved by single-particle cryo-EM. X-ray structures of the 30S and 50S subunits and of the 70S complex helped researchers understand some of the structural details of the peptidyl-transferase center in the 50S subunit, and the decoding process, i.e., how the ribosome recognizes the geometry of codon-anticodon base-pairing and discriminates against noncognate pairings. The path of the mRNA through the 30S subunit in the ribosome has also been visualized using X-ray crystallography (92) and more recently with cryo-EM (45; K. Mitra, C. Schaffitzel, F. Fabiola, M. S. Chapman, N. Ban & J. Frank, manuscript submitted). So far, only two X-ray studies of the 70S ribosome have been conducted, which show the ribosome in a total of three conformations (14, 65, 91, 92). Thus, single-particle cryo-EM has been the main technique that has shed light on the morphological changes accompanying the different steps during translation.

Cryo-EM studies have demonstrated, among other findings, that several factors that are GTPases and act during crucial steps of elongation, release, and recycling all bind in the same funnel-shaped region of the intersubunit space on the side of the L7/L12 stalk base. One of the binding sites of these factors is the GTPase-associated center (GAC) of the large subunit. The GAC, composed of L11 and a 58-nucleotide-long region of RNA, has been observed to move by up to ~10 Å in response to factor binding (28, 80). It is evident from a superposition of the different observed conformations (**Figure 4**) that the main change occurs in the region of kink-turn 42, identified by Klein et al. (42), in the X-ray structure of the 50S subunit [also see the molecular dynamics simulations by Razga et al. (58, 59)]. EF-G is observed to induce a ratchet-like motion of the small subunit against the large subunit (25, 30), which can be rationalized as part of the motion required for translocation. Similar comparisons of cryo-EM maps for successive functional states have shown the high mobility of the L1 stalk and its implication in the ejection of deacylated tRNA (67, 81), the high mobility of the L7/L12 stalk (2), and the change in the position of helix 44 of 16S RNA (85).

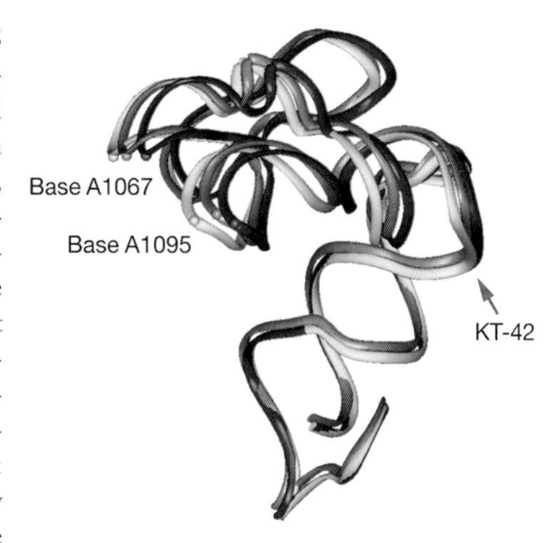

Figure 4

The 58-nucleotide segment of 23S RNA forming the RNA part of the GTPase-associated center, in three positions inferred from cryo-EM maps by real-space refinement. Yellow and gold: ribosome bound with the ternary complex either in the presence of GDP and kiromycin or GDPNP. Red: ribosome without factor bound. The changes in position of the GAC can be traced to the instability of the kink-turn Kt-42 in helix 42, indicated in the figure (adapted from Reference 28).

GAC: GTPase-associated center

TRANSITIONING FROM THE PHENOMENOLOGICAL TO THE ANALYTICAL: UNDERSTANDING THE MOLECULAR BASIS OF RIBOSOME DYNAMICS

Analysis of Conformational Dynamics Through Systematic Comparison of Fitted Atomic Models

Thus far, as we have already stressed, observations about conformational changes in the ribosome have remained at the phenomenological level, i.e., the understanding of the molecular mechanisms underlying ribosome dynamics has been limited. In the case in which the X-ray structures of the soluble cofactors were available, fitting of these atomic models into their corresponding EM densities were performed using a variety of methods ranging from manual docking to NMFF (73). In 2003 an atomic model of the bacterial ribosome was fitted into the cryo-EM reconstructions of a pretranslocational (pre) and posttranslocation (post) 70S ribosome using the RSRef module in TNT (30). This study represented a significant advance, in that it helped identify the conformational changes—in terms of the X-ray structure of the ribosome—associated with one of the key steps of translocation, namely the ratcheting step. Flexible fitting of X-ray structures into a cryo-EM reconstruction surpasses the mere phenomenological description of conformational dynamics to enable a more detailed analysis at the quasi-atomic level: what RNA helices move, what new molecular contacts are made, and what we can say about the role of the proteins in facilitating these movements. In the case of the ratchet motion and its role in translocation, such an analysis has yet to be done.

Recently, a systematic comparison was performed of atomic models fitted into the cryo-EM reconstructions of two ribosomes stalled in elongation: One stalled owing to the absence of EF-G, a prestate ribosome (81), and the other stalled owing to the presence in the polypeptide exit tunnel of the SecM nascent peptide stalling sequence (45). This comparison enabled the elucidation of the molecular mechanism underlying SecM-induced elongation arrest in the ribosome (K. Mitra, C. Schaffitzel, F. Fabiola, M. S. Chapman, N. Ban & J. Frank, manuscript submitted). This analysis paves the way for a more detailed, molecular understanding of ribosome dynamics using cryo-EM reconstructions.

Signal Transduction within the Ribosome RNA Scaffold

Specific sequences in nascent peptides translated by the ribosome can arrest the ribosome during elongation or termination (reviewed in Reference 76). One well-studied example is the stalling motif in the SecM peptide, which through a feedback system regulates the synthesis of SecA, a bacterial motor-protein necessary for posttranslational translocation through the PCC. When translation of the SecM peptide has progressed such that the entire C-terminal stalling sequence is within the polypeptide exit tunnel in the 50S subunit, elongation arrest occurs unless the N-terminal end of the SecM peptide is physically pulled by the SecA-PCC complex. It was shown that arrest occurs owing to interaction of the SecM stalling sequence with the projections of ribosomal proteins L4 and L22, and 23S rRNA bases A^{2058} and $A^{749-753}$ within the polypeptide tunnel (49). How these specific interactions at a localized place inside the polypeptide exit tunnel in the 50S caused elongation arrest remained a complete mystery.

The cryo-EM reconstruction shows that the SecM-stalled ribosome (45; K. Mitra, C. Schaffitzel, F. Fabiola, M. S. Chapman, N. Ban & J. Frank, manuscript submitted) has not undergone ratcheting, which is required for translocation, and that it resembles a pretranslational state ribosome that has not undergone ratcheting because of the absence of EF-G (81). However, notable differences in conformation are observed for the SecM-stalled ribosome, such as the relative inward movements of the L1 stalk and the GTP-associated center of the 50S subunit, and the relative downward movement of the head domain of the 30S subunit toward the tRNAs in the intersubunit space (**Figure 5**). How does a localized SecM interaction within the polypeptide exit tunnel in the 50S effect positional rearrangements in distal regions of the 50S and even in the 30S subunit? The first X-ray structures of the ribosome revealed that the ribosome has an internal core of ribosomal RNA (rRNA) helices packed against each other, with proteins studding the surface and occasionally making forays into the rRNA interior using extended projections (6, 88, 91). When comparing atomic models fitted into the cryo-EM reconstructions of the SecM-stalled and the stalled, pretranslational state ribosome, the level of interconnectedness between these rRNA helices becomes apparent. In the 50S subunit, rRNA regions observed to move are related in two ways to the 23S rRNA bases A^{2058} and $A^{749-753}$, with which SecM interacts: (*a*) proximity in primary sequence, i.e., connected through the phosphoribose backbone, and (*b*) proximity through helix formation via hydrogen bonds (**Figure 5**).

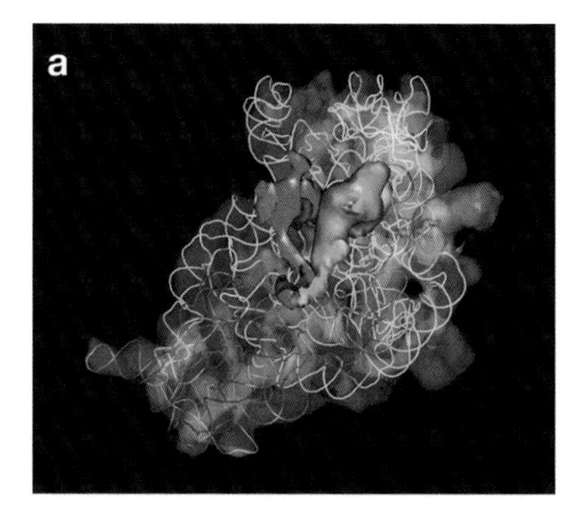

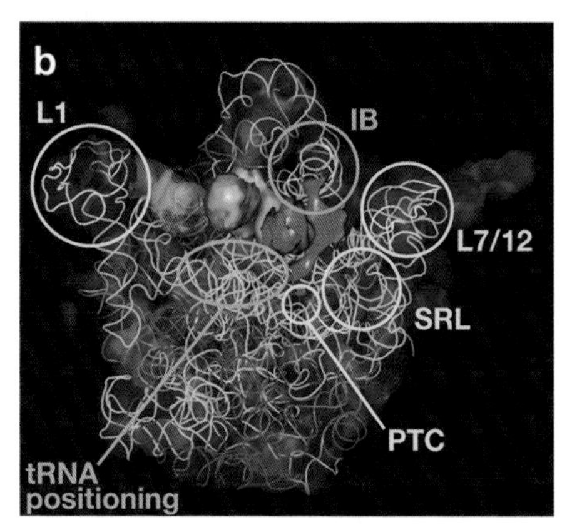

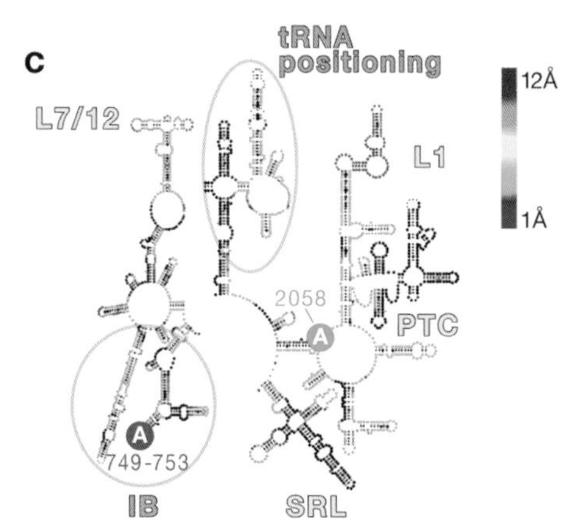

Because of this extensive rRNA interconnectivity, a highly localized SecM interaction with A^{2058} and $A^{749-753}$ is able to propagate through the entire 50S subunit and effect positional rearrangements, in the range of 5 to 14 Å, in distal morphological and functional entities, such as the L1 stalk, the GAC and sarcin-ricin loop (SRL), the peptidyltransferase center (PTC), and the L7/L12 stalk base. Conformational changes in the 50S subunit also affect the geometry of the intersubunit bridges (IB) and the positioning of the tRNAs, which in turn produce a rearrangement of rRNA helices in the 30S subunit. The 30S rRNA elements that are repositioned constitute parts of morphological features such as the head; the platform, which contributes to formation of the mRNA exit site; and the body, which partly forms the mRNA entrance site (**Figure 5**). The net result of this SecM-induced cascade of rRNA rearrangements is that the ribosome locks down on the tRNAs and mRNA, such that translocation cannot occur, even in the presence of EF-G. The elucidation of the structural mechanism underlying SecM-induced elongation

Figure 5

SecM-induced signal propagation within the ribosome RNA scaffold. (*a*, *b*) Atomic *E. coli* ribosome model fitted into the 30S (*a*) and 50S (*b*) subunit of the SecM-stalled complex, viewed from the intersubunit space. (*c*) Interconnectivity of rRNA elements in the 50S subunit in relation to the SecM nascent peptide interaction sites at 23S rRNA bases A^{2058} and $A^{749-753}$. Important functional and morphological regions observed to undergo large positional rearrangements are circled and indicated in the secondary structure representation (*c*) and in the tertiary structure within the cryo-EM reconstruction (*b*). The phosphate backbone (*a*, *b*) and the secondary structure schematic (*c*) of the rRNA are shown colored according to the RMSD between the atomic models fitted into the cryo-EM reconstructions of the two stalled ribosome complexes (see text).

arrest suggests that not only catalytic functions but also internal signal transduction over long distances and even across subunits are mediated by rRNA within the ribosome. If life did indeed evolve out of an RNA world (32), then these findings present the intriguing possibility that the interconnected structure of some ancient RNA molecules enabled them not only to perform catalytic functions, but also internal signaling, e.g., from one end of the polymer to the other. The ribosome, a living fossil, seems to exemplify this dual ability of its RNA.

SUMMARY

Single-particle cryo-EM is the method of choice for studying the dynamics of macromolecular machines both at a phenomenological and, increasingly, at the molecular level. The analysis on the molecular level is facilitated by the advent of high-resolution X-ray structures of ribosomal subunits and of functional ligands, and by the steady improvement and progressive automation of algorithms for fitting and docking. Cryo-EM has shed light on the structure of the ribosome as it changes during the steps of translation: initiation, elongation, termination, and recycling. With the availability of X-ray structures for ribosomal components it is now becoming possible to interpret cryo-EM reconstructions of the ribosome with quasi-atomic detail. A picture ensues in which the ribosome uses RNA not only to catalyze the peptidyl-transferase or decoding process, but also as a medium for signal transduction over large distances within the molecule.

SUMMARY POINTS

1. Single-particle cryo-EM of molecular complexes has two obvious advantages: There are no size limitations, and the complex does not need to be ordered in an array (as in X-ray crystallography) or to be tumbling at any given rate (as in NMR spectroscopy).
2. The damage to a biological sample can be reduced by almost an order of magnitude if the sample is cooled from liquid nitrogen temperatures, forming the basis of the use of cryogenic temperatures in EM of biological samples.
3. Existing moderate-resolution cryo-EM density maps of a molecular assembly can be interpreted at the atomic level if X-ray or NMR structures of its components are known.
4. A cryo-EM reconstruction can be interpreted at the quasi-atomic level using a two-step approach: (*a*) initial, rigid placement of the atomic model into the cryo-EM density, and (*b*) refinement of the model obtained upon initial placement using flexible-fitting algorithms.
5. Single-particle cryo-EM has been the main technique that has shed light on the conformational changes of the ribosome that accompany the different steps during translation.
6. Flexible fitting of X-ray structures into the cryo-EM reconstructions of ribosome complexes surpasses the mere phenomenological description of conformational dynamics and enables a more detailed analysis at the quasi-atomic level.
7. An analysis of cryo-EM reconstructions of ribosome complexes suggests that ribosomal RNA mediates not only catalytic functions, but also internal signal transduction over long distances, and even across subunits, within the ribosome.

ACKNOWLEDGMENTS

We thank J. LeBarron for creating a program for RNA secondary structure display and analysis. We thank M. Watters for assistance with the illustrations. This work was supported by HHMI, NSF DBI 9871347, and NIH grants R37 GM29169 and R01 GM 55440.

LITERATURE CITED

1. Adrian M, Dubochet J, Lepault J, McDowall AW. 1984. Cryoelectron microscopy of viruses. *Nature* 308:32–36
2. Agrawal RK, Heagle AB, Penczek P, Grassucci RA, Frank J. 1999. EF-G-dependent GTP hydrolysis induces translocation accompanied by large conformational changes in the 70S ribosome. *Nat. Struct. Biol.* 6:643–47
3. Agrawal RK, Penczek P, Grassucci RA, Frank J. 1998. Visualization of elongation factor G on the *Escherichia coli* 70S ribosome: the mechanism of translocation. *Proc. Natl. Acad. Sci. USA* 95:6134–38
4. Agrawal RK, Penczek P, Grassucci RA, Li Y, Leith A, et al. 1996. Direct visualization of A-, P-, and E-site transfer RNAs in the *Escherichia coli* ribosome. *Science* 271:1000–2
5. Baker TS, Johnson JE. 1996. Low resolution meets high: towards a resolution continuum from cells to atoms. *Curr. Opin. Struct. Biol.* 6:585–94
6. Ban N, Nissen P, Hansen J, Moore PB, Steitz TA. 2000. The complete atomic structure of the large ribosomal subunit at 2.4 A resolution. *Science* 289:905–20
7. Belnap DM, McDermott BM Jr, Filman DJ, Cheng N, Trus BL, et al. 2000. Three-dimensional structure of poliovirus receptor bound to poliovirus. *Proc. Natl. Acad. Sci. USA* 97:73–78
8. Bottcher B, Wynne SA, Crowther RA. 1997. Determination of the fold of the core protein of hepatitis B virus by electron cryomicroscopy. *Nature* 36:88–91
9. Brenner S, Horne RW. 1959. A negative staining method for high resolution electron microscopy of viruses. *Biochem. Biophys. Acta* 34:103–10
10. Brooks B, Karplus M. 1983. Harmonic dynamics of proteins: normal mode and fluctuations in bovine pancreatic trypsin inhibitor. *Proc. Natl. Acad. Sci. USA* 80:6571–75
11. Brunger AT, Adams PD, Rice LM. 1998. Recent developments for the efficient crystallographic refinement of macromolecular structures. *Curr. Opin. Struct. Biol.* 8:606–11
12. Brunger AT, Adams PD, Rice LM. 1999. Annealing in crystallography: a powerful optimization tool. *Prog. Biophys. Mol. Biol.* 72:135–55
13. Brunger AT, Krukowski A, Erickson J. 1990. Slow-cooling protocols for crystallographic refinement by simulated annealing. *Acta Crystallogr. A* 46:585–93
14. Cate JHD, Yusupov MM, Yusupova GZ, Earnest TN, Noller HN. 1999. X-ray structures of 70S ribosome functional complexes. *Science* 285:2095–104
15. Chacon P, Wriggers W. 2002. Multi-resolution contour-based fitting of macromolecular structures. *J. Mol. Biol.* 317:375–84
16. Chapman MS. 1995. Restrained real-space macromolecular atomic refinement using a new resolution-dependent electron density function. *Acta Crystallogr. A* 51:69–80
17. Chen LF, Blanc E, Chapman MS, Taylor KA. 2001. Real space refinement of acto-myosin structures from sectioned muscle. *J. Struct. Biol.* 133:221–32
18. Crowther RA. 1971. Procedures for three-dimensional reconstruction of spherical viruses by Fourier synthesis from electron micrographs. *Philos. Trans. R. Soc. London B* 261:221–30
19. Darst SA, Opalka N, Chacon P, Polyakov A, Richter C, et al. 2002. Conformational flexibility of bacterial RNA polymerase. *Proc. Natl. Acad. Sci. USA* 99:4296–301

20. DeRosier D, Klug A. 1968. Reconstruction of 3-dimensional structures from electron micrographs. *Nature* 217:130–34

21. Dubochet J, Adrian M, Lepault J, McDowall AW. 1985. Cryo-electron microscopy of vitrified biological specimens. *Trends Biochem. Sci.* 10:143–46

22. Dubochet J, Lepault J, Freeman R, Berriman JA, Homo JC. 1982. Electron microscopy of frozen water and aqueous solutions. *J. Microsc.* 128:219–37

23. Fabiola F, Chapman MS. 2005. Fitting of high-resolution structures into electron microscopy reconstruction images. *Structure* 13:389–400

24. Frank J. 1998. How the ribosome works. *Am. Sci.* 86:428–39

24a. Frank J. 2006. *Three-Dimensional Electron Microscopy of Macromolecular Assemblies*. New York: Oxford Univ. Press

25. Frank J, Agrawal RK. 2000. A ratchet-like inter-subunit reorganization of the ribosome during translocation. *Nature* 406:318–22

26. Frank J, Penczek P, Agrawal RK, Grassucci RA, Heagle AB. 2000. Three-dimensional cryoelectron microscopy of ribosomes. *Methods Enzymol.* 317:276–91

27. Frank J, Penczek P, Grassucci R, Srivastava S. 1991. Three-dimensional reconstruction of the 70S *E. coli* ribosome in ice: the distribution of ribosomal RNA. *J. Cell Biol.* 115:597–605

28. Frank J, Sengupta J, Gao H, Li W, Valle M, et al. 2005. The role of tRNA as a molecular spring in decoding, accommodation, and peptidyl transfer. *FEBS Lett.* 579:959–62

29. Frank J, Zhu J, Penczek P, Li Y, Srivastava S, et al. 1995. A model of protein synthesis based on cryo-electron microscopy of the *E. coli* ribosome. *Nature* 376:441–44

29a. Gabashvili IS, Agrawal RK, Spahn CMT, Grassucci RA, Svergun DI, Frank J. 2000. Solution structure of the *E. coli* ribosome at 11.5 Å resolution. *Cell* 100:537–49

30. Gao H, Sengupta J, Valle M, Korostelev A, Eswar N, et al. 2003. Study of the structural dynamics of the *E. coli* 70S ribosome using real-space refinement. *Cell* 113:789–801

31. Gao N, Zavialov AV, Li W, Sengupta J, Valle M. 2005. Mechanism for the disassembly of the post-termination complex inferred from cryo-EM studies. *Mol. Cell* 18:663–74

32. Gesteland RF, Cech T, Atkins JF, eds. 1999. *The RNA World*. Cold Spring Harbor, NY: Cold Spring Harbor Lab. Press

33. Go N, Noguti T, Nishikawa T. 1983. Dynamics of a small globular protein in terms of low-frequency vibrational modes. *Proc. Natl. Acad. Sci. USA* 80:3696–700

34. Goncharov AB, Vainshtein B, Ryskin AI, Vagin AA. 1987. Three-dimensional reconstruction of arbitrarily oriented particles from their electron photomicrographs. *Sov. Phys. Crystallogr.* 32:504–9

35. Gorlich D, Kutay U. 1999. Transport between the cell nucleus and the cytoplasm. *Annu. Rev. Cell Dev. Biol.* 15:607–60

36. Harms J, Schluenzen F, Zarivach R, Bashan A, Gat S, et al. 2001. High resolution structure of the large ribosomal subunit from mesophilic eubacterium. *Cell* 107:679–88

37. Harrison W. 1984. Variational calculation of the normal modes of a large macromolecule: methods and some initial results. *Biopolymers* 23:2943–49

38. Hartz D, Binkley J, Hollingsworth T, Gold L. 1990. Domains of initiator tRNA and initiation codon crucial for initiator tRNA selection by Escherichia coli IF3. *Genes Dev.* 4:1790–800

39. He Y, Bowman VD, Mueller S, Bator CM, Bella J, et al. 2000. Interaction of the poliovirus receptor with poliovirus. *Proc. Natl. Acad. Sci. USA* 97:79–84

40. Karimi R, Pavlov MY, Buckingham RH, Ehrenberg M. 1999. Novel roles for classical factors at the interface between translation termination and initiation. *Mol. Cell* 3:601–9

21. An informative introduction into the method of specimen preparation and microscopy in cryo-EM by the pioneers of the technique.

23. A clear and concise review on the two-step fitting of high-resolution atomic models into moderate-resolution cryo-EM reconstructions using real-space refinement.

25. First observation, by cryo-EM, of large conformational changes (relative rotation of the subunits) in the ribosome accompanying translocation.

30. Application of real space refinement using rigid bodies to the fitting of an atomic model into cryo-EM reconstructions of the *E. coli* ribosome.

41. Kidera A, Go N. 1990. Refinement of protein dynamic structure: normal mode refinement. *Proc. Natl. Acad. Sci. USA* 87:3718–22
42. Klein DJ, Schmeing TM, Moore PB, Steitz TA. 2001. The kink-turn: a new RNA secondary structure motif. *EMBO J.* 20:4214–21
43. Malhotra A, Penczek P, Agrawal RK, Gabashvili IS, Grassucci RA, et al. 1998. *Escherichia coli* 70S ribosome at 15 Å resolution by cryo-electron microscopy: localization of fMet-$tRNA_f^{Met}$ and fitting of L1 protein. *J. Mol. Biol.* 280:103–16
44. Deleted in proof
45. **Mitra K, Schaffitzel C, Shaikh T, Tama F, Jenni S, et al. 2005. Structure of the *E. coli* protein-conducting channel bound to a translating ribosome. *Nature* 438:318–24**
46. Moazed D, Noller HF. 1989. Intermediate states in the movement of transfer RNA in the ribosome. *Nature* 342:142–48
47. Moore PB. 1980. Scattering studies of the three-dimensional organization of the *E. coli* ribosome. In *Ribosomes: Structure, Function and Genetics*, ed. G Chambliss, GR Craven, J Davies, K Davis, L Kahan, M Nomura, pp. 111–33. Baltimore, MD: Univ. Park Press
48. Nakamura T, Ito K. 2003. Making sense of mimic in translation termination. *Trends Biochem. Sci.* 28:99–105
49. Nakatogawa H, Ito K. 2002. The ribosomal exit tunnel functions as a discriminating gate. *Cell* 108:629–36
50. Navaza J, Lepault J, Rey FA, Alvarez-Rua C, Borge J. 2002. On the fitting of model electron densities into EM reconstructions: a reciprocal-space formulation. *Acta Crystallogr. D* 58:1820–25
51. Noller HF, Hoffarth V, Zimniak L. 1992. Unusual resistance of peptidyl transferase to protein extraction procedures. *Science* 256:1416–19
52. Oakes MI, Scheiman A, Atha T, Shankweiler G, Lake JA. 1990. Ribosome structure: three-dimensional locations of rRNA and proteins. In *The Ribosome, Structure, Function, and Evolution*, ed. A Dahlberg, RA Garrett, PB Moore, D Schlessinger, JR Warner, pp. 180–93. Washington, DC: ASM
53. Pape T, Wintermeyer W, Rodnina M. 1998. Complete kinetic mechanism of elongation factor Tu-dependent binding of aminoacyl-tRNA to the A site of the *E. coli* ribosome. *EMBO J.* 17:7490–97
54. Penczek P, Zhu J, Schröder R, Frank J. 1997. Three-dimensional reconstruction with contrast transfer compensation from defocus series. *Scanning Microsc.* 11:147–54
55. Radon J. 1917. Über die Bestimmung von Funktionen durch ihre Integralwerte längs gewisser Mannigfaltigkeiten. *Ber. Verh. K.-Sächs. Ges. Wiss. Leipz. Math. Phys. Klasse* 69:262–77
56. **Ramakrishnan V. 2002. Ribosome structure and the mechanism of translation. *Cell* 108:557–72**
57. Rath BK, Hegerl A, Leith A, Shaikh TR, Wagenknecht T, Frank J. 2003. Fast 3D motif search of EM density maps using a locally normalized cross-correlation function. *J. Struct. Biol.* 144:95–103
58. Razga F, Koca J, Sponer J, Leontis NB. 2005. Hinge-like motions in RNA kink-turns: the role of the second A-Minor motif and nominally upaired bases. *Biophys. J.* 88:3466–85
59. Razga F, Spackova N, Reblova K, Koca J, Leontis NB, Sponer J. 2004. Ribosomal RNA kink-turn motif: a flexible molecular hinge. *J. Biomol. Struct. Dyn.* 22:183–93
60. Rice LM, Brunger AT. 1994. Torsion angle dynamics: Reduced variable conformational sampling enhances crystallographic structure refinement. *Proteins* 19:277–90
61. Roseman AM. 2000. Docking structures of domains into maps from cryo-electron microscopy using local correlation. *Acta Crystallogr. D* 56:1332–40

45. A successful application of normal mode-based flexible fitting methods to the problem of the conformational dynamics of the protein-conducting channel.

56. A lucid account of the structural basis of translation.

62. Rossmann MG. 2000. Fitting atomic models into electron-microscopy maps. *Acta Crystallogr. D* 56:1341–49
63. Saibil HR. 2000. Molecular chaperones: containers and surfaces for folding, stabilizing or unfolding proteins. *Curr. Opin. Struct. Biol.* 10:251–58
64. Schluenzen F, Tocilj A, Zarivach R, Harms J, Gluehmann M, et al. 2000. Structure of functionally activated small ribosomal subunit at 3.3 A resolution. *Cell* 102:615–23
65. Schuwirth BS, Borovinskyay MA, Hau CW, Zhang W, Vila-Sanjurjo A, Cate JHD. 2005. Structures of the bacterial ribosome at 3.5 Å resolution. *Science* 310:827–34
66. Sengupta J, Frank J. 2006. Cryo-electron microscopy as a tool to study molecular machines. In *Protein Structures: Methods in Protein Structure and Stability Analysis*. Hauppauge, NY: Nova Sci. Publ.
67. Spahn CM, Gomez-Lorenzo MG, Grassucci RA, Jorgensen R, Andersen GR, et al. 2004. Domain movements of elongation factor eEF2 and eukaryotic 80S ribosome facilitate tRNA translocation. *EMBO J.* 23:1008–19
68. Stark H, Mueller F, Orlova EV, Schatz M, Dube P, et al. 1995. The 70S Escherichia coli ribosome at 23 A resolution: fitting the ribosomal RNA. *Structure* 3:815–21
69. Stark H, Orlova EV, Rinke-Appel J, Junke N, Mueller F, et al. 1997. Arrangement of tRNAs in pre- and posttranslocational ribosomes revealed by electron cyromicroscopy. *Cell* 88:19–28
70. Stark H, Rodnina MV, Rinke-Appel J, Brimacombe R, Wintermeyer W, van Heel M. 1997. Visualization of elongation factor Tu on the *Escherichia coli* ribosome. *Nature* 389:403–6
71. Stöffler-Meilicke M, Stoffler G. 1988. Localization of ribosomal proteins on the surface of ribosomal subunits from *Escherichia coli* using immunoelectron microscopy. *Methods Enzymol.* 164:503–20
72. **Tama F, Miyashita O, Brooks CL 3rd. 2004. Flexible multi-scale fitting of atomic structures into low-resolution electron density maps with elastic network normal mode analysis. *J. Mol. Biol.* 337:985–99**

72. Description of a new method of flexible fitting, based on normal mode analysis of a molecule structure, with examples for applications.

73. Tama F, Miyashita O, Brooks CL III. 2004. Normal mode based flexible fitting of high-resolution structure into low-resolution experimental data from cryo-EM. *J. Struct. Biol.* 147:315–26
74. Tama F, Sanejouand YH. 2001. Conformational change of proteins arising from normal mode calculations. *Protein Eng.* 14:1–6
75. Tama F, Valle M, Frank J, Brooks CL III. 2003. Dynamic reorganization of the functionally active ribosome explored by normal mode analysis and cryo-electron microscopy. *Proc. Natl. Acad. Sci. USA* 100:9319–23
76. Tenson T, Ehrenberg M. 2002. Regulatory nascent peptides in the ribosomal tunnel. *Cell* 108:591–94
77. Tirion MM. 1996. Large amplitude elastic motions in proteins from a single-parameter, atomic analysis. *Phys. Rev. Lett.* 77:1905–8
78. Tronrud DE, Ten Eyck LF, Matthews BW. 1987. An efficient general-purpose least-squares refinement program for macromolecular structures. *Acta Crystallogr. A* 43:489–501
79. Tugarinov V, Choy WY, Orekhov VY, Kay LE. 2005. Solution NMR-derived global fold of a monomeric 82-kDa enzyme. *Proc. Natl. Acad. Sci. USA* 102:622–27
80. Valle M, Zavialov A, Li W, Stagg SM, Sengupta J, et al. 2003. Incorporation of aminoacyl-tRNA into the ribosome as seen by cryo-electron microscopy. *Nat. Struct. Biol.* 10:899–906

81. Valle M, Zavialov AV, Sengupta J, Rawat U, Ehrenberg M, Frank J. 2003. Locking and unlocking of ribosomal motions. *Cell* 114:123–34
82. van den Berg B, Clemons WM Jr, Collinson I, Modis Y, Hartmann E, et al. 2004. X-ray structure of a protein-conducting channel. *Nature* 427:36–44
83. van Heel M. 1987. Angular reconstitution: a posteriori assignment of projection directions for 3D reconstruction. *Ultramicroscopy* 21:111–24
84. van Heel M, Gowen B, Matadeen R, Orlova EV, Finn R, et al. 2000. Single-particle electron cryo-microscopy: towards atomic resolution. *Q. Rev. Biophys.* 33:307–69
85. van Loock MS, Alexandrov A, Yu X, Cozzarelli NR, Egelman EH. 2002. SV40 large T antigen hexamer structure: domain organization and DNA-induced conformational changes. *Curr. Biol.* 12:472–76
86. Volkmann N, Hanein D. 1999. Quantitative fitting of atomic models into observed densities derived by electron microscopy. *J. Struct. Biol.* 125:176–84
87. Volkmann N, Hanein D. 2003. Docking of atomic models into reconstructions from electron microscopy. *Methods Enzymol.* 374:204–25
88. Wimberly BT, Brodersen DE, Clemons WM, Morgan-Warren RJ, Carter AP, et al. 2000. Structure of the 30S ribosomal subunit. *Nature* 407:327–39
89. Wittman HG. 1983. Architecture of prokaryotic ribosomes. *Annu. Rev. Biochem.* 52:35–65
90. Wriggers W, Miligan RA, McCammon JA. 1999. Situs: a package for docking crystal structures into low-resolution maps from electron microscopy. *J. Struct. Biol.* 125:185–95
91. Yusupov MM, Yusupova GZ, Baucom A, Lieberman K, Earnest TN, et al. 2001. Crystal structure of the ribosome at 5.5 Å resolution. *Science* 292:883–96
92. Yusupova GZ, Yusupov MM, Cate JHD, Noller HF. 2001. The path of messenger RNA through the ribosome. *Cell* 106:233–41
93. Zavialov AV, Buckingham RH, Ehrenberg M. 2001. A post-termination ribosomal complex is the guanine nucleotide exchange factor for peptide release factor RF3. *Cell* 107:115–24
94. Zavialov AV, Ehrenberg M. 2003. Peptidyl-tRNA regulates the GTPase activity of translation factors. *Cell* 114:113–22

90. The first published atomic model, determined by X-ray crystallography, of an assembled prokaryotic ribosome.

NMR Techniques for Very Large Proteins and RNAs in Solution

Andreas G. Tzakos,[1] Christy R. R. Grace,[2] Peter J. Lukavsky,[1] and Roland Riek[2]

[1]MRC Laboratory of Molecular Biology, Cambridge CB2 2QH, United Kingdom; email: pjl@mrc-lmb.cam.ac.uk

[2]Structural Biology Laboratory, The Salk Institute, La Jolla, California 92037; email: riek@salk.edu

Annu. Rev. Biophys. Biomol. Struct. 2006. 35:319–42

First published online as a Review in Advance on March 6, 2006

The *Annual Review of Biophysics and Biomolecular Structure* is online at biophys.annualreviews.org

doi: 10.1146/annurev.biophys.35.040405.102034

1056-8700/06/0609-0319$20.00

Key Words

TROSY, CRIPT, cross-correlated relaxation, RDCs

Abstract

Three-dimensional structure determination of small proteins and oligonucleotides by solution NMR is established. With the development of novel NMR and labeling techniques, structure determination is now feasible for proteins with a molecular mass of up to ~100 kDa and RNAs of up to 35 kDa. Beyond these molecular masses special techniques and approaches are required for applying NMR as a multiprobe method for structural investigations of proteins and RNAs. It is the aim of this review to summarize the NMR techniques and approaches available to advance the molecular mass limit of NMR both for proteins (up to 1 MDa) and RNAs (up to 100 kDa). Physical pictures of the novel techniques, their experimental applications, as well as labeling and assignment strategies are discussed and accompanied by future perspectives.

Contents

INTRODUCTION

TROSY: transverse relaxation-optimized spectroscopy

RDC: residual dipolar coupling

Nuclear magnetic resonance (NMR) spectroscopy is one of the principal experimental techniques used in structural biology. It allows the determination of atomic-resolution structures of biomolecules and provides insight into their dynamic features and intermolecular interactions. Historically, biomolecular NMR was restrained to study relatively small systems. The introduction of TROSY (transverse relaxation-optimized spectroscopy) (47), RDC (residual dipolar coupling) (3), and labeling strategies (23, 69, 73) opened an avenue for NMR to determine structures of proteins with a molecular mass of up to ~100 kDa and DNA/RNA molecules with a molecular mass of up to ~35 kDa (21, 37, 45, 55, 57). With the development of further methodological advances, structural investigations of even larger systems are now possible (31, 36, 56, 58). These studies include the identification of secondary structures, intermolecular binding sites, conformational changes, relative orientation of domains, or molecules within a macromolecule, and dynamical characterizations of very large biomolecular complexes.

In this review, we discuss the recent technical advances used to investigate these very large protein and RNA systems with molecular masses larger than the above mentioned values. In addition, ideas and stimuli are provided for further developments. Because of the different chemical, physical, and structural nature of nucleic acid molecules and proteins, their studies require different approaches and have different limitations. Therefore, we present the two systems side by side using a similar conceptual framework. This includes a brief introduction, intuitive descriptions of the techniques used, experimental recipes, possible assignment strategies, applications from literature, and future perspectives.

NMR OF VERY LARGE PROTEINS

During an NMR experiment, the signal is lost owing to the relaxation of the magnetization. Transverse relaxation is the main source for this signal loss. For large molecules it increases proportionally with the molecular mass owing to the slow tumbling of the molecule, which is represented by the rotational correlation time τ_c (1). The first prerequisite for measuring a good quality two-dimensional NMR spectrum of very large proteins is therefore the implementation of pulse sequence elements that are transverse relaxation optimized (47) during the entire NMR experiment. A two-dimensional [^{15}N,^{1}H]-correlation experiment consists of two distinct elements, the polarization transfer period during which magnetization is transferred from ^{1}H to ^{15}N and vice versa, and the frequency labeling period, i.e., ^{15}N evolution and ^{1}H acquisition. As we shall see, TROSY can be applied to both these elements, yielding two-dimensional spectra of proteins with a molecular mass of up to 1000 kDa (56).

Physical Picture of TROSY

The principle of TROSY can be described by semiclassical relaxation theory (22, 47). In the following equation a simple intuitive picture of TROSY is given considering a system of two isolated scalar coupled spins, of spin 1/2, ^{1}H and ^{15}N, with a scalar coupling J_{HN} between them (16). Transverse relaxation of this spin system is dominated by the dipole-dipole (DD) relaxation between spins ^{1}H and ^{15}N and by the chemical shift anisotropy (CSA) of each individual spin. The CSA of ^{15}N induces a motion-influenced time-dependent magnetic field $B^{CSA}(t)$ on spin ^{15}N (1):

$$B^{CSA}(t) \propto \gamma_N B_0 \Delta\sigma_N \, [3\cos^2\theta(t) - 1],$$

where $\theta(t)$ is the angle between the external magnetic field B_0 and the principal axis component of the CSA tensor, γ_N is the gyromagnetic ratio of ^{15}N, and $\Delta\sigma_N$ is the CSA part of the chemical shift tensor. $B^{CSA}(t)$ is proportional to the magnetic field of the spectrometer and is modulated with time, owing to Brownian motion of the molecule. This motion-generated fluctuating magnetic field perturbs the local magnetic field of the ^{15}N spin, causing transverse relaxation. In other words, the ^{15}N spin is exposed to many local magnetic fields, which leads to a band of resonance frequencies of the ^{15}N spin, visible as line broadening of the cross peak. Similarly, the dipole-dipole coupling between ^{1}H and ^{15}N spins induces a motion-generated time-dependent magnetic field with the axial principal component parallel to the ^{15}N-^{1}H vector:

$$B^{DD}(t) \propto \gamma_H \gamma_N / r_{HN}^3 [3\cos^2\theta(t) - 1],$$

where $\theta(t)$ is the angle between the external magnetic field B_0 and the ^{15}N-^{1}H vector, γ_H is the gyromagnetic ratio of ^{1}H, and r_{HN} is the internuclear distance. The tumbling of the molecule modulates $B^{DD}(t)$, which leads to transverse relaxation and line broadening, similar to $B^{CSA}(t)$, as discussed above. In contrast to $B^{CSA}(t)$, $B^{DD}(t)$ is independent of B_0, but its sign depends on whether the two spins ^{1}H and ^{15}N are parallel or antiparallel oriented (**Figure 1*a***).

The time-dependent magnetic fields $B^{CSA}(t)$ and $B^{DD}(t)$ both influence the relaxation of spin ^{15}N and show the same angular and time dependence. Thus, at any given time t, depending on whether the ^{1}H spin is parallel or antiparallel to spin ^{15}N, the two fields either add or subtract as shown in **Figure 1*a***. For the multiplet coherence S^{12}, $B^{DD}(t)$ opposes $B^{CSA}(t)$, thereby yielding the favorable narrow linewidth component as shown in **Figure 1*b***. The other multiplet coherence S^{34} relaxes fast because the two time-dependent magnetic fields $B^{DD}(t)$ and $B^{CSA}(t)$ add up together (**Figure 1*b***). An optimal compensation of $B^{CSA}(t)$ with $B^{DD}(t)$, and thus minimal transverse relaxation rates, can be achieved by adjusting the size of $B^{CSA}(t)$ by choosing an optimal magnetic field B_0 of about 1 GHz (47). The compensation is already effective at 500 MHz. The principle of TROSY is in the selection of the favorable multiplet components S^{12} and I^{13} by considering transverse relaxation-optimization along both dimensions, i.e., ^{15}N evolution period and ^{1}H acquisition (47). The described effect between DD and CSA is known as cross-correlation between CSA and DD, because the two relaxation mechanisms interfere (also see below).

In conventional NMR experiments, the multiplet pattern is routinely collapsed by

Transverse relaxation: the process that governs the decay of the transverse magnetization during an NMR experiment

Chemical shift anisotropy (CSA): the anisotropic part of the chemical shift tensor, which is generated by the electron environment

Dipole-dipole coupling (DD): the through-space magnetic interaction between nuclear spins

Transverse relaxation-optimization: suppresses the transverse relaxation by mutual cancellation of two types of relaxation mechanism, most commonly CSA and DD coupling

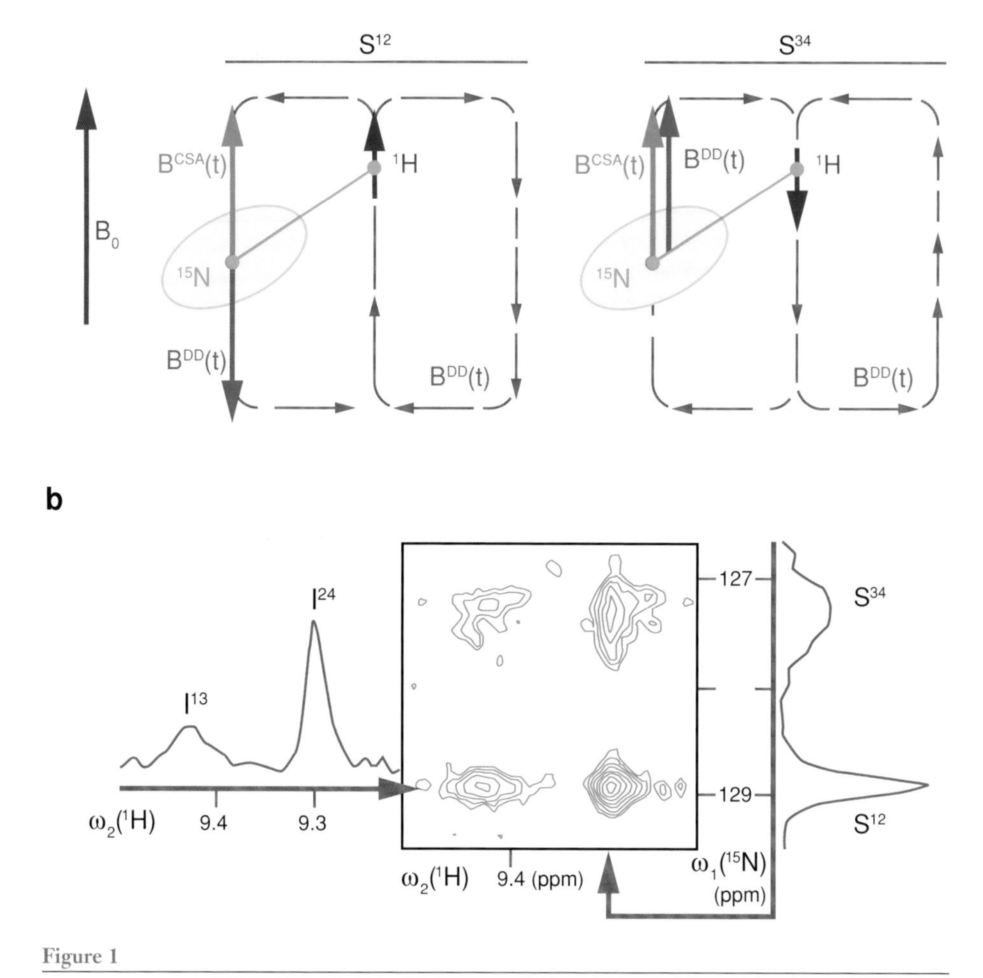

Figure 1

A physical picture of TROSY. (*a*) The interference of the local magnetic fields of $B^{DD}(t)$ with $B^{CSA}(t)$ are shown for both multiplet components of the ^{15}N spin with either the ^{1}H spin up (S^{12}) or down (S^{34}). In the S^{12} coherence the two local magnetic fields are canceling each other, resulting in favorable relaxation properties of the multiplet component as shown in panel *b*. In contrast, in the S^{34} coherence $B^{DD}(t)$ and $B^{CSA}(t)$ add up, causing fast relaxation and concomitantly broad line width and low intensity of the cross peak. B_0 is the static magnetic field. The CSA tensor σ is displayed by an ellipse and the field lines designate the dipole-dipole interaction between the spins ^{1}H and ^{15}N. (*b*) Contour plot and the corresponding cross-sections of a ^{15}N-^{1}H moiety from a conventional two-dimensional [^{15}N,^{1}H]-correlation spectrum without decoupling during evolution are shown.

decoupling the protons. The ^{1}H decoupling flips the spin ^{1}H and the sign of the local magnetic field produced by $B^{DD}(t)$. Thus during ^{15}N evolution, each ^{15}N spin is perturbed by $B^{CSA}(t) + B^{DD}(t)$ during the first half of the evolution time and by $B^{CSA}(t) - B^{DD}(t)$ during the other half of the evolution time, which leads to less favorable relaxation when compared with the relaxation of the component S^{12} selected by TROSY. From a technical perspective, transverse relaxation-optimization is therefore achieved by eliminating the inversion (decoupling) of the scalar coupled spin ^1H during ^{15}N evolution and by eliminating the inversion (decoupling) of ^{15}N during ^1H acquisition.

Physical Picture of Transverse Relaxation-Optimized Polarization Transfer

Transverse relaxation-optimization can also be implemented during the polarization transfer elements: From a technical perspective this is achieved by eliminating the inversion of the scalar coupled ^{15}N spin during the polarization transfer, and from a theoretical perspective this calls for the use of cross-correlated relaxation between DD and CSA. The developed transverse relaxation-optimized polarization transfers are discussed from an intuitive point of view and are compared with INEPT (insensitive nuclei enhancement by polarization transfer) (41), which is the major tool of magnetization for transfer from one type of nucleus to another.

INEPT: insensitive nuclei enhancement by polarization transfer

INEPT is based on transfer via scalar spin-spin couplings creating antiphase magnetization, as depicted in **Figure 2**. With increasing molecular mass (more than ~100 kDa) (58) (**Figure 3*a***), transverse relaxation during the INEPT period becomes a limiting factor

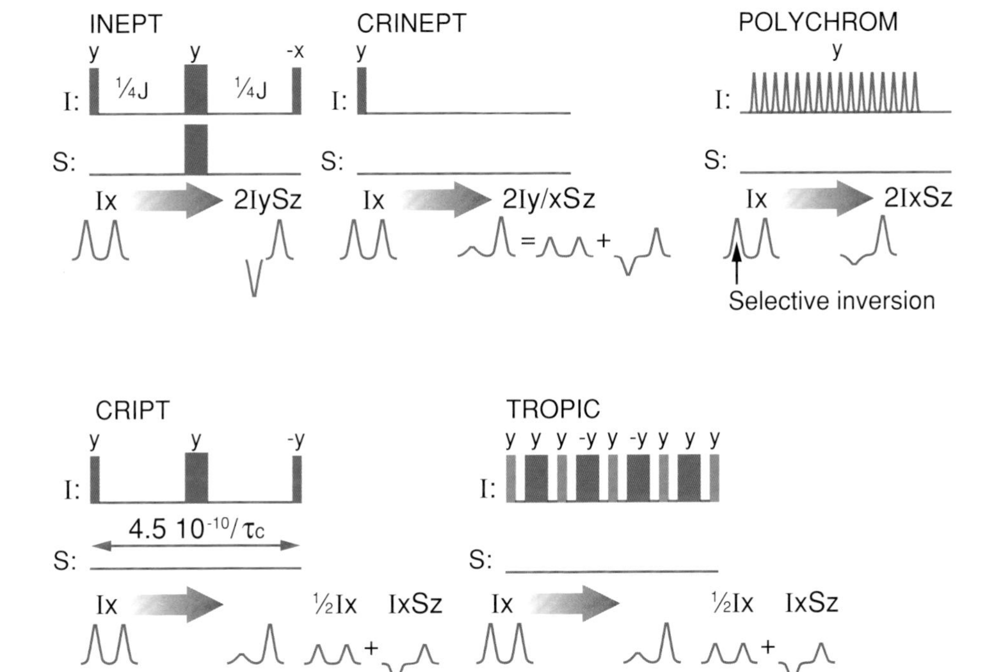

Figure 2

Physical picture of transverse relaxation-optimized polarization transfer periods. Simplified pulse sequence elements of the various polarization transfer schemes are shown. They constitute only part of the full experiment listed in **Table 1**. The narrow and wide blue bars represent nonselective 90° and 180° pulses, the green bars represent 90°/5 pulses. The polychromatic pulse of the POLYCHROM polarization transfer is presented by a multi-gauss shape. The phases of the pulses are labeled accordingly. Below the pulse sequence elements, the corresponding coherence evolutions are given by the product operator formalism and by a graphical representation showing the inphase and antiphase signals. The water-selective pulses as well as pulse field gradients are not shown. For more details we refer to original literature cited in **Table 1**.

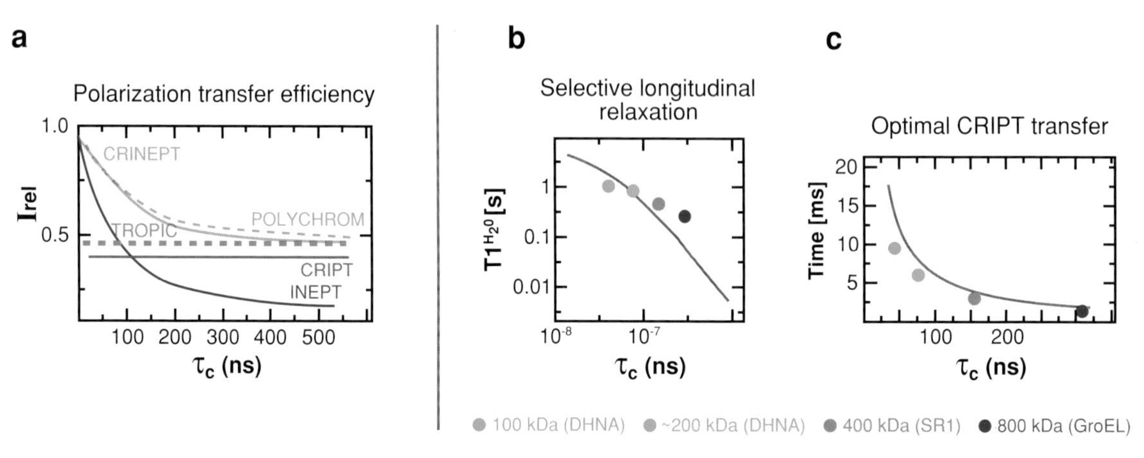

Figure 3

(*a*) Polarization transfer efficiency, (*b*) selective longitudinal relaxation of ^{1}H, and (*c*) optimal CRIPT polarization transfer time in large proteins versus the rotational correlation time τ_c of large molecules. The colored circles represent experimental values measured, and the curves are calculated as described in References 5, 56, 57.

CRIPT: cross-correlated relaxation induced polarization transfer

TROPIC: transverse relaxation optimized polarization transfer induced by cross-correlation effects

that severely compromises sensitivity in two-dimensional experiments. Hence, alternative polarization transfer schemes are required. In CRIPT (cross-correlated relaxation induced polarization transfer) (9, 58, 56), transverse relaxation-optimization during the polarization transfer is applied by omitting the 180° nitrogen pulse during the transfer period, thereby suppressing the scalar coupling-based polarization transfer (**Figure 2**). However, the antiphase magnetization is obtained via cross-correlated relaxation between DD and CSA: The fast-relaxing doublet component has relaxed during the polarization transfer period, whereas the slowly relaxing TROSY component has hardly relaxed (see **Figure 2**). The asymmetric doublet at the end of the CRIPT period can be decomposed into the sum of an inphase and an antiphase doublet with half the intensity of the original resonance line (**Figure 2**). This description visualizes the following properties of CRIPT, which can also be extracted theoretically (6): (*a*) The presence of the antiphase signal indicates that magnetization can be transferred by cross-correlated relaxation. (*b*) However, only half of the magnetization can be transferred using CRIPT (**Figure 3*a***) because the antiphase signal in **Figure 2** has at most half the intensity of the original inphase signal. (*c*) The polarization transfer time depends on how fast the anti-TROSY component of the doublet relaxes. The optimal transfer time is therefore inversely proportional to the rotational correlation time of the molecule τ_c (**Figure 3*c***), and (*d*) the polarization transfer efficiency is τ_c independent (**Figure 3*a***). (*e*) Because the doublet in the CRIPT transfer has the same phase as the initial inphase doublet, the phases of the pulses in CRIPT differ from those in INEPT (**Figure 2**). CRIPT has been implemented in the [^{15}N,^{1}H]-CRIPT-TROSY experiment (**Table 1**) (**Figure 4*a***).

Sensitivity enhancement of CRIPT can be achieved by using optimal control theory (19, 30). The recently proposed TROPIC (transverse relaxation-optimized polarization transfer induced by cross-correlation effects) (19) makes use of optimal control theory by exploiting the fact that the relaxation rates of longitudinal operators H_z and $2H_zN_z$ are much smaller than those of the transverse operators H_x and $2H_xN_z$. In TROPIC, part of the spin coherence is stored in the favorable

Table 1 NMR experiments for very large proteins

NMR experiment	Multiplet selection[a]	Advantages	Disadvantages
1. [^{15}N,^{1}H]-CRINEPT-HMQC-[^{1}H]-TROSY (56)	● ●	Simple to set up Simple water handling Partial multiplet suppression	Low resolution along ^{15}N dimension due to the absence of ^{15}N-TROSY Cross peaks of all molecular weights
2. [^{15}N,^{1}H]-CRIPT-TROSY (56)	○ ● ○ ●	Selects only cross peaks of large molecules	Variable peak intensity due to cross-correlated relaxation Difficult water handling[b]
3. [^{15}N,^{1}H]-TROPIC-TROSY (31)	○ ● ○ ●	Properties as 2 with 20–30% more sensitivity and less variations in cross peak intensity	Properties as 2 Difficult water handling[b]
4. [^{15}N,^{1}H]-CRINEPT-TROSY (56)	○ ●	TROSY component selection Simple water handling Sensitivity equal to 1 and 2	Cross peaks of all molecular weights
5. [^{15}N,^{1}H]-POLYCHROM-TROSY (5)	○ ● ○ ●	20% more sensitive than 1 High resolution along ^{15}N-dimension	Special protocol for spectra processing Cross peaks of all molecular weights
6. [^{13}C^{Methyl},^{1}H] –HMQC (68)	●	Simple set up Side chain information	Special ^{13}C^{methyl} labeling (68)
7. [^{13}C^{Methyl},^{1}H] –TROSY (68)	●	Side chain information High resolution	Special ^{13}C^{methyl} labeling (68)

[a]The four multiplet components of a ^{15}N-^{1}H moiety are shown with closed circles if the peak intensity is positive and with open circles if the peak intensity is negative. Suppressed multiplet components are not shown.
[b]These experiments do not have a WATERGATE-type (52) pulse sequence element that eliminates the residual transverse magnetization of water. These experiments therefore need careful shimming and set up of the water flip back pulses in order to measure them successfully (56, 72).

longitudinal state, whereas the other part of the spin coherence is shuffled to two-spin coherence through cross-correlated relaxation. Experimentally this is achieved by applying hard radio frequency pulses with small flip angles separated by delays and refocusing pulses during which the coherence is converted to the desired two-spin coherence H_xS_z (**Figure 2**). The implementation of TROPIC in the [^{15}N,^{1}H]-TROPIC-TROSY experiment yields 20% to 30% more sensitivity than its CRIPT analog (**Table 1**).

CRINEPT (cross-correlated relaxation-enhanced polarization transfer) presents another signal enhancement strategy over CRIPT (56, 58). CRINEPT combines CRIPT with INEPT by eliminating the 180° ^{1}H chemical shift refocusing pulse during the polarization transfer element, reestablishing scalar coupling-based transfer in the presence of transverse-relaxation optimization (**Figure 2**). The combination of both the scalar coupling–based transfer and cross-correlated relaxation–based transfer results in a significant increase in transfer efficiency, as depicted in **Figure 3*a***. The drawback of CRINEPT is the ^{1}H chemical shift evolution during the polarization transfer period causing severe signal losses when implemented in two-dimensional experiments (approximately a factor of 1.4 to 2 in the CRINEPT-type experiments of **Table 1**) (56, 58).

CRINEPT: cross-correlated relaxation induced INEPT

A conceptually distinct polarization transfer between scalar coupled spins can be achieved by selective inversion or saturation of one of the spin doublet components. This is known as selective population inversion or transfer (28, 44). Selective population inversion is per se transverse relaxation optimized because during the polarization transfer

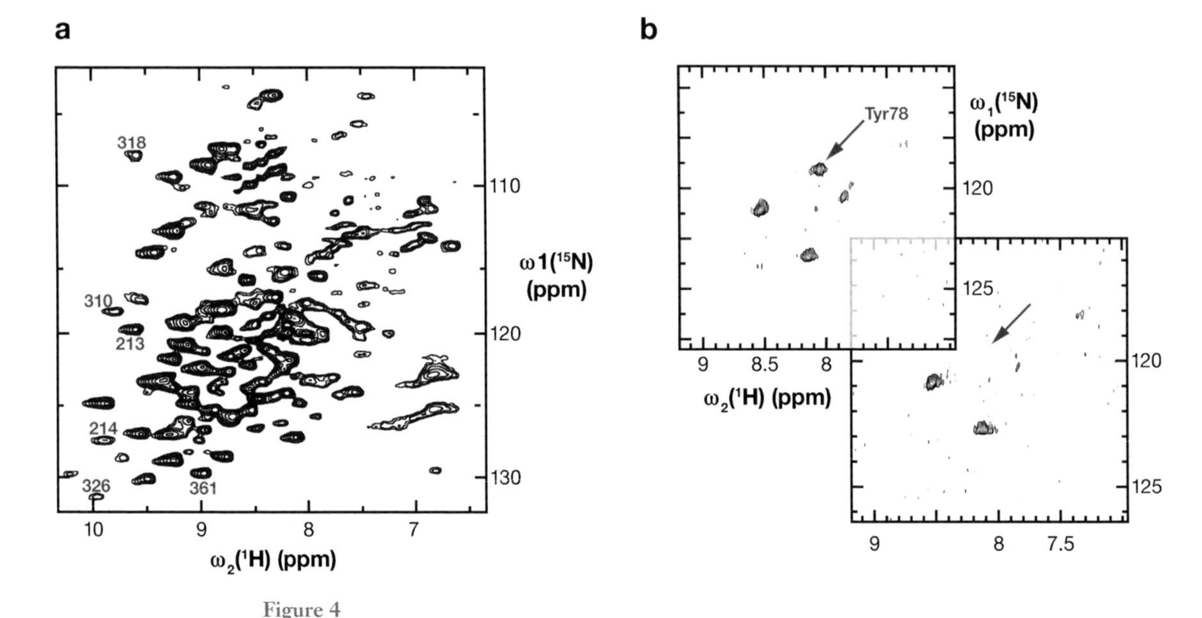

Figure 4

Resonance assignment approaches for (*a*) GroEL and (*b*) KcsA potassium channel. (*a*) The [^{15}N,^{1}H]-CRIPT-TROSY spectrum of GroEL (800-kDa tetradecamer) is shown (56). The resonance assignment was obtained by transfer from the assignment of a single domain of GroEL published by Kobayashi et al. (32). (*b*) The spectrum shown in the top left is a [^{15}N,^{1}H]-TROSY spectrum of ^{15}N-Tyr-labeled KcsA (60-kDa tetramer embedded in 60-kDa detergent micelles). The spectrum shown on the bottom right is from a ^{15}N-Tyr-labeled KcsA (Tyr78Phe) mutant showing the absence of the Tyr-78 cross peak.

element no pulse is applied on ^{15}N (**Figure 2**) (30). Since the coherence transfer efficiency of selective population inversion is almost optimized (30), it is somewhat more sensitive than CRINEPT (**Figure 3*a***). However, selective population inversion is limited to the specific (selective and single) case in which the resonance frequencies of a single spin pair are known. A solution to the broadband application of selective population inversion has been proposed by Bromek et al. (5). Their polarization transfer scheme consists of a polychromatic pulse, which is generated by a group of simultaneous soft pulses of different frequencies (**Figure 2**). The polychromatic pulse sequence is accompanied by a repetitive measurement of the experiment with an identical but frequency-shifted polychromatic pulse, thereby resulting in a signal loss by a factor of ~2. In addition, this approach needs a special protocol for spectrum processing (5).

[^{13}C^{methyl},^{1}H]-TROSY

TROSY is not limited to [^{15}N,^{1}H]-correlation experiments. It has also been applied to aromatic [^{13}C,^{1}H] spin systems (48) and subsequently to methyl groups (either $^{13}CH_3$ or $^{13}CH_2D$) (49, 64, 65, 66, 68) and methylene ($^{13}CH_2$) groups (39). For very large proteins the [^{13}C^{Methyl},^{1}H]-TROSY and [^{13}C^{Methyl},^{1}H]-HMQC experiments should be measured (**6** and **7** in **Table 1**) (66, 68). The successful applications to very large protein complexes with a molecular mass of up to 800 kDa (33, 68) make the [^{13}C^{Methyl},^{1}H]-corrrelation experiments an important alternative to the [^{15}N,^{1}H]-correlation experiments, in particular because the methyl groups are interesting probes for protein structure and dynamics. For more details, the interested reader is referred to reviews about NMR of methyl groups (67) and the cited original literature.

Experimental Recipe: A Practical Guide on How to Measure NMR Spectra of Very Large Proteins

To measure solution NMR of proteins with a molecular mass >100 kDa the following experimental approach is proposed: (*a*) The preparation of a $^{15}N,^{2}H$-labeled protein sample is highly recommended (18). (*b*) The ^{1}H pulse lengths of the protein sample are calibrated by measuring the free induction decay of water using a single hard pulse phase shifted by 45°. The length of the pulse is exactly 180° when the first points of both the real and imaginary parts of the free induction decay are zero (for more details see Reference 56). (*c*) For optimal sensitivity of all the two-dimensional NMR experiments listed in **Table 1**, the water magnetization must be kept along the *z*-axis during the entire duration of the NMR experiments by using water-selective pulses. Each water-selective pulse is optimized individually. Improper water handling results in severe signal loss (56, 72). (*d*) The optimal interscan delay is determined in an experiment with the water magnetization along the *z*-axis [either in a ^{1}H one-dimensional experiment as proposed by Riek et al. (56) or in a two-dimensional experiment of **Table 1**]. The careful water treatment generates a selective $T_1^{H2O}(^{1}H)$, which (to a first approximation and only for large molecules) is inversely proportional to the rotational correlation time of the molecule (**Figure 3*b***).

After these steps, we recommend (*e*) the $[^{15}N,^{1}H]$-CRINEPT-HMQC-$[^{1}H]$-TROSY as the initial two-dimensional experiment to be measured (1 in **Table 1**). This highly sensitive experiment with a partial multiplet selection is easy to set up but has low resolution along the ^{15}N-dimension, since TROSY is not implemented in the indirect dimension. (*f*) Following this, the measurements of several $[^{15}N,^{1}H]$-CRIPT-TROSY experiments (2 in **Table 1**) with different length of polarization transfer times should be measured to determine the optimal polarization transfer time. Because in CRIPT the optimal polarization transfer time is (to a first approximation) inversely proportional to the rotational correlation time of the protein (**Figure 3*c***), this array of experiments will give an estimate of the molecular mass of the species measured and therefore ensure that NMR signals of the species of interest are acquired. Another related and important aspect of the $[^{15}N,^{1}H]$-CRIPT-TROSY is its selection for signals from species with large molecular weights. This is of particular importance because small amounts of impurities with a small molecular weight (i.e., degradation products or single domains of an otherwise multimeric protein) give rise to overproportionally strong cross peaks in the CRINEPT-type experiments but are suppressed largely in the $[^{15}N,^{1}H]$-CRIPT-TROSY. Finally, the $[^{15}N,^{1}H]$-CRIPT-TROSY spectrum comprises high resolution along the ^{15}N-dimension. (*g*) For 20% to 30% sensitivity enhancement the $[^{15}N,^{1}H]$-TROPIC-TROSY (3 in **Table 1**) and $[^{15}N,^{1}H]$-POLYCHROM-TROSY (5 in **Table 1**) experiments are recommended. (*h*) To simplify the spectrum by multiplet selection the $[^{15}N,^{1}H]$-CRINEPT-TROSY (4 in **Table 1**) experiment is recommended. (*i*) The acquisition of $[^{13}C^{Methyl},^{1}H]$-TROSY and $[^{13}C^{Methyl},^{1}H]$-HMQC experiments (6 and 7 in **Table 1**) of a correspondingly labeled sample should also be considered and is a valuable alternative approach (64, 65).

Sequential Assignment Procedures

Obtaining good-quality NMR spectra of large proteins is an essential first step. In addition, assigning resonances to individual nuclei is in most cases indispensable as a basis for detailed structural and dynamical studies.

Unfortunately, the standard sequential assignment procedure with triple resonance experiments (3) is applicable only to systems with a molecular mass up to ~100 kDa (60). There are, however, alternative assignment strategies possible for larger systems:

(*a*) Partial assignment using TROSY-based triple resonance experiments may be feasible up to ~200 kDa (7).

(*b*) While studying a large complex with a molecular mass >100 kDa containing a small protein (with a molecular mass <100 kDa), the sequential assignment of the small free protein is achieved by TROSY-based triple resonance experiments. The binding site of the complex is then elucidated by observing the chemical shift changes of the resonances of the small protein upon complex formation. This strategy has been successfully applied to identify the GroEL (800 kDa) binding site on GroES (70 kDa) (17).

(*c*) While studying a large multidomain protein, the assignment of a free single domain of the protein may be possible using TROSY-based triple resonance experiments. The assignment is then transferred to the full-length protein. This strategy is shown in **Figure 4*a***: Part of the sequential assignment of a domain of GroEL comprising residues 193–335 obtained by Kobayashi et al. (32) is transferred to the $[^{15}N,^{1}H]$-CRIPT-TROSY spectrum of GroEL.

(*d*) Another strategy is based on selective labeling of specific amino acids accompanied by mutagenesis: **Figure 4*b*** shows a $[^{15}N,^{1}H]$ correlation spectrum of a ^{15}N-Tyr-labeled potassium channel KcsA (4*160 amino acids embedded in ~60-kDa detergent micelles comprising ~120 kDa). The comparison with the spectrum of a ^{15}N-Tyr-labeled mutant Tyr78Phe enables the identification of the Tyr-78 cross peak owing to its absence in the mutant spectrum (K. Baker, W. Kwiatkowski, S. Choe & R. Riek, unpublished results).

(*e*) An alternative selective labeling strategy consists of double amino acid–selective labeling with one ^{15}N-labeled amino acid (for example ^{15}N-labeled Leu) and one $^{13}C'$-labeled amino acid (for example $^{13}C'$-labeled Tyr). The measurement of a two-dimensional TROSY-HNCO-type experiment (63) or a two-dimensional difference experiment of **Table 1** with and without $^{13}C'$-decoupling during the ^{15}N evolution reveals a cross peak only if the ^{15}N-labeled amino acid residue has an N-terminal neighbor with the 13C′-labeled residue (in the given example only a Leu cross peak is observed for a Tyr-Leu pair).

(*f*) If the three-dimensional structure of the protein of interest is known, the following assignment strategy of an amino acid–selective ^{15}N-labeled protein sample is possible (51): Several Cys point mutants are constructed and labeled with a paramagnetic spin, which enhances severely the relaxation of spins closer in space (up to a distance of about 1.5 nm). The same sample is selectively ^{15}N-labeled with a particular amino acid. The paramagnetic relaxation-enhanced signal losses of the moieties are determined by comparing the signal intensities of the cross peaks in the two-dimensional correlation spectra measured in the presence and absence of the paramagnetic spin label. Since the paramagnetic relaxation-enhanced signal loss can be transferred to a distance between the Cys and the ^{15}N-labeled residue, the cross peaks can be assigned to individual residues. However, in most of the cases several measurements with several Cys point mutants must be generated.

(*g*) If a ligand binding site at the macromolecule is known, residues close to the binding site can be identified through chemical shift changes of the resonances upon complex formation (53). In combination with amino acid–specific labeling, individual resonances may be assigned.

(*h*) Segmental labeling with inteins combined with amino acid–selective labeling could also be used for assignment (43).

Biological Applications and Future Perspectives

Several NMR applications to very large systems have been published: the identification of the GroEL binding site on GroES (the molecular mass of the complex is 900 kDa) (17), the identification of the binding site of the C3b complement on a portion of the complement receptor type 1 (200 kDa) (5), the identification of the folding state of a protein bound to the chaperone Hsp90 (200 kDa) (59), and a protein bound to GroEL (400 kDa) (27). Furthermore, cross peaks for most of the methyl groups of isoleucine of lysine decarboxylase (810-kDa decameric enzyme) (68) as well as of the heterohepatmeric Arp2/3 actin nucleation complex selectively labeled at one subunit (240 kDa) (33) have been successfully acquired. These applications partially reflect the current potential investigations of solution NMR applied to very large protein systems. The use of NMR as a multiprobe method enables detailed structural studies on conformational equilibria, molecular dynamics, and the relative orientation of protein domains or proteins in a macromolecular complex, as well as the identification of intermolecular interactions at protein-protein, protein-ligand, or protein–nucleic acid interfaces. Structure determinations of such complexes are thereby out of scope.

NMR OF LARGE RNAs

NMR of RNA faces the same challenges as that of proteins. The combination of spectral complexity, relaxation processes due to molecular tumbling (see above), and aggregation problems due to the necessary high concentrations make structure determination difficult. For RNA many of these challenges are even more pronounced compared with proteins. This fact is reflected in the protein data bank (PDB), where only 261 RNA NMR structures have been deposited, whereas the relevant number for proteins is 4421. In addition, protein NMR spectroscopists routinely solve three-dimensional structures of 25-kDa molecules and manage proteins up to 100 kDa, whereas nucleic acid NMR spectroscopists still struggle to handle molecules with masses higher than 15 kDa, the size at which the current NMR methods work efficiently. This differentiation is due to the disparate nature of proteins and nucleic acids. Instead of the 20 different side chains found in proteins, which range from aliphatic to aromatic, acidic to basic, nucleic acids have only 4 chemically similar nucleosides. Therefore, the chemical shift dispersion is more limited in RNA than in proteins, leading to less informative spectra. This is often aggravated by the extended nature of RNA molecules, which results in slower overall tumbling times and therefore shorter transverse relaxation times and increased line widths compared with proteins of similar molecular masses. Thus, it can be challenging to extract enough nuclear Overhauser effect (NOE) and scalar couplings to adequately constrain a RNA conformation locally and consequently to accurately determine its three-dimensional structure.

NOE: nuclear Overhauser effect

Many biologically interesting RNA molecules have therefore not been accessible for high-resolution NMR structure determination because they are far larger than 15 kDa. Instead, large biological RNAs have been reduced to collections of smaller, thermodynamically stable subdomains, such as helices and loops, taken out of their larger structural context. It is evident that the study of such subdomains may lack important tertiary information, since it requires the complex assembly of many weak interactions that are typically observed only in relatively large molecular complexes (>25 kDa). Nonetheless, three RNA structures ranging in size from 75 to 100 nt with molecular masses of 25 to 35 kDa were determined by NMR (10, 14, 36), indicating that it is possible to solve structures of higher-molecular-mass RNAs by NMR. Yet, the largest RNAs solved by NMR to date have predominantly extended helical structures, and condensed packing interactions indicative of extensive tertiary structure are yet to be revealed (15).

NOESY: NOE spectroscopy

HCV: hepatitis C virus

IRES: internal ribosome entry site

RNA Secondary Structure Determination by NMR

Although the secondary structure of RNA can be predicted quite accurately (38), there are examples in which the correct secondary structures of RNAs have been identified only after the utilization of NMR spectroscopy. NMR is a powerful tool to study RNA secondary structure, since the number and type of base pairs as well as their sequential neighbors can be identified in the imino proton "fingerprint" region (between 10 and 15 ppm) of homonuclear two-dimensional NOE spectroscopy (NOESY) spectra recorded in water (26), which is not time demanding (less than 12 h) and does not require isotopic labeling. Because the imino and amino protons often experience exchange processes that make their detection difficult, nonexcitation-type water suppression schemes that do not saturate the solvent, thereby leading to severe loss of signal, are often used. For one-dimensional applications, jump-return or binomial pulses yield the best results, even for exchanging imino and amino protons, but for two-dimensional NOESY spectra the uniform excitation profile of the S-pulse implemented in the two-dimensional S-NOESY is of great advantage and yields the best results (62). On the basis of the imino fingerprint region of a two-dimensional S-NOESY spectrum, we recently established the correct secondary structure of the hepatitis C viral (HCV) internal ribosome entry site (IRES) RNA domain II (36) and unambiguously ruled out two proposed, alternative folds. Two-dimensional NOESY experiments combined with the HNN-COSY (correlation spectroscopy) (12) experiment, which requires isotopically ^{15}N-labeled RNA, can be used for directly observing hydrogen-bonding interactions. For standard Watson–Crick base pairs, this experiment correlates the donor and acceptor nitrogen atoms for hydrogen bonds involving U or G imino protons (12, 46). In the HNN-COSY, magnetization is transferred across the hydrogen bond through the two-bond scalar coupling ($^2J_{NN'}$) (**Figure 5*a***). Numerous derivatives of the original NMR pulse sequence have been described (13, 25, 74), and because many of them employ the TROSY principle (explained in the first part of this review), they can also be applied to larger RNAs, as illustrated by the spectrum of the 73-nt U2-U6 RNA complex shown in **Figure 5*a*** (61). Sashital et al. (61) utilized an HNN-COSY/two-dimensional NOESY combination to refine the secondary structure of the U2-U6 small nuclear RNA from the *Saccharomyces cerevisiae* spliceosome complex. Previous folding models of the U2-U6 complex, based on mutational and cross-linking studies, suggested that the two small nuclear RNAs have the potential to form three intermolecular helices (**Figure 5*b***, left). The 73-nt construct used for the NMR study contained the region in the U2-U6 RNA complex closest to the precursor mRNA binding site (61). The two-dimensional NOESY and HNN-COSY experiments identified 1 G-U wobble, 2 U-U base pairs, and 21 Watson-Crick base pairs, indicating that this RNA forms a four-way helical junction in contrast to the previously suggested models (**Figure 5*b***, right).

Both studies reported above did not rely upon complete resonance assignments. On the contrary, proton-proton correlations obtained from two-dimensional NOESY experiments and hydrogen-bonding interactions from HNN-COSY experiments were utilized to determine the RNA secondary structure. These examples demonstrate the utility of NMR spectroscopy to unambiguously determine the correct secondary structure of larger RNAs, which forms the basis for the design of smaller RNA oligonucleotides that correspond to subdomains within the larger RNA. In the case of the 25-kDa HCV IRES domain II RNA, the correct secondary structure determined by NMR allowed the design

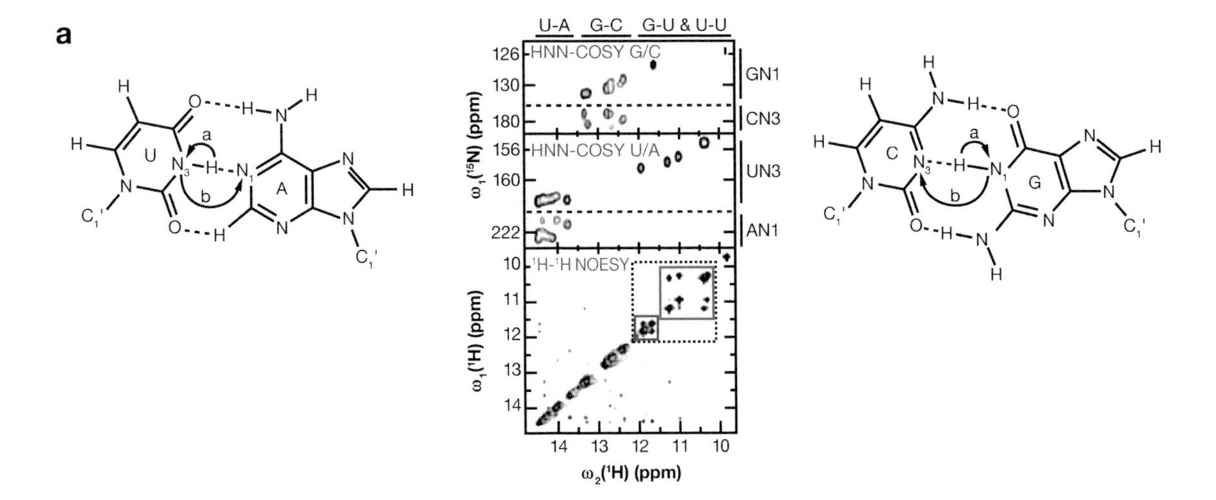

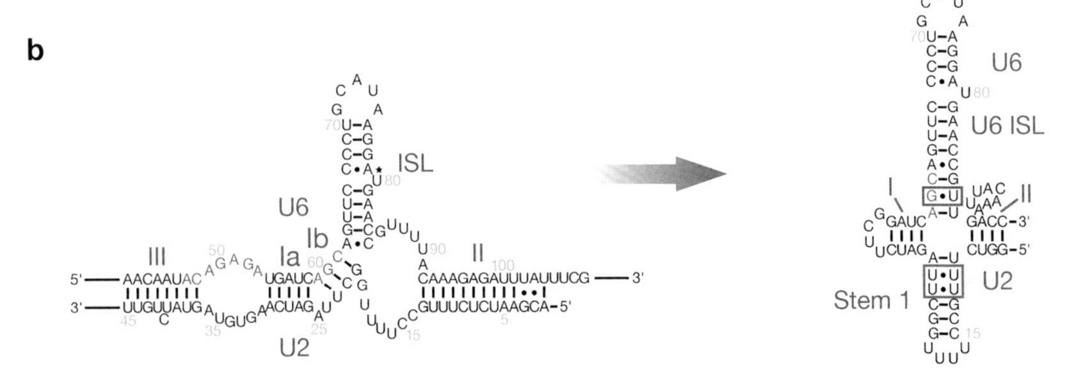

Figure 5

NMR spectroscopy as a tool for secondary structure determination of large RNAs. (*a*) Magnetization transfer steps for the two-dimensional HNN-COSY experiment. Magnetization is transferred from the imino proton to the imino nitrogen (*arrow* "a"). The magnetization is then transferred across the hydrogen bond during a second transfer step (*arrow* "b"). After ^{15}N-frequency labeling, the magnetization is transferred back to the imino proton through the reverse process (34). (*b*) The previously proposed secondary structure of the U2-U6 RNA complex is shown on the left. Simple NMR experiments, the two-dimensional HNN-COSY and the two-dimensional [^{1}H,^{1}H]-NOESY, were used to determine the correct secondary structure shown on the right. Reprinted, with permission, from Reference 61.

of two smaller oligonucleotides comprising domain II, which greatly aided resonance assignments (36). This approach, which we describe in more detail below, has been applied for all the large RNA solution structures determined to date (14, 36), and its size limitation remains to be tested.

Assignment Strategies for Large RNAs

General procedures for the spectral assignment process of RNAs up to 15 kDa have been extensively reviewed (21, 69, 73). For high-resolution structure determination of RNAs, uniformly $^{13}C,^{15}N$-labeled and/or $^{13}C,^{15}N$ base-type-specifically labeled RNA samples are necessary (69, 73) and can be prepared enzymatically using the appropriate $^{13}C,^{15}N$-enriched nucleotide triphosphates (2, 11, 42).

For the assignment of larger RNAs, excision of thermodynamically stable secondary structural elements out of their larger structural context can overcome the problem of increasing spectral overlap, which hampers unambiguous assignments. For the NMR structure determination of the 77-nt HCV IRES domain II (36), the large RNA was dissected into two subdomains, each less than 15 kDa, a molecular mass at which conventional RNA NMR techniques work efficiently.

D'Souza et al. (14) developed a different resonance assignment approach to determine the structure of a 101-nt RNA that contains the "core encapsidation signal" of the Moloney murine leukemia virus (**Figure 6*a***). The approach is based on the principle that nucleotide-type-specific isotopic labeling of the RNA allows the filtering of intranucleotide versus internucleotide NOEs. First, NOESY-correlated signals were grouped according to the nucleotide type with three-dimensional ^{13}C-edited NOESY data obtained for RNA samples containing nucleotide-specific $^{13}C,^{15}N$ isotope labels. Internucleotide and intranucleotide NOEs were then differentiated by comparing the three-dimensional data with four-dimensional $^{13}C,^{13}C$-edited NOESY data obtained for the same samples, in which only intranucleotide NOEs are retained. This way, stretches of sequential nucleotides of a given type were identified, since internucleotide NOEs involving labeled and unlabeled nucleotides were present in the three-dimensional NOESY spectra but absent in the four-dimensional spectra (**Figure 6*c*,*d***, respectively). The same approach failed for the 30-kDa GAAA tetraloop-receptor complex owing to short transverse relaxation times. In this case, more sensitive two-dimensional filtered/edited NOESY experiments were used for the NOE assignments in combination with base-type-specific ^{13}C-labeled RNA (50). These two-dimensional filter/edited experiments select NOEs between labeled and unlabeled nucleotides (50) and, if combined with a selective

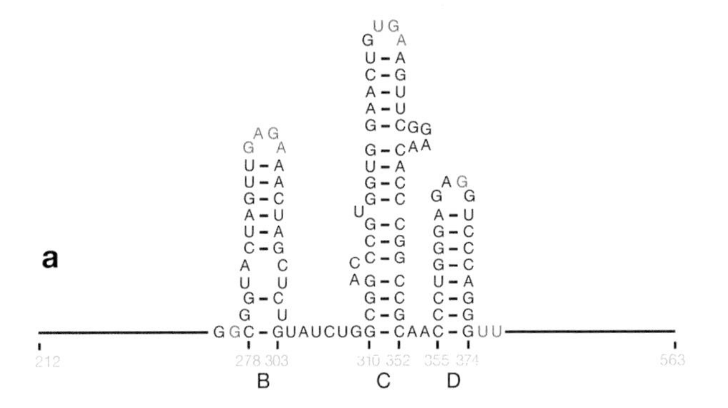

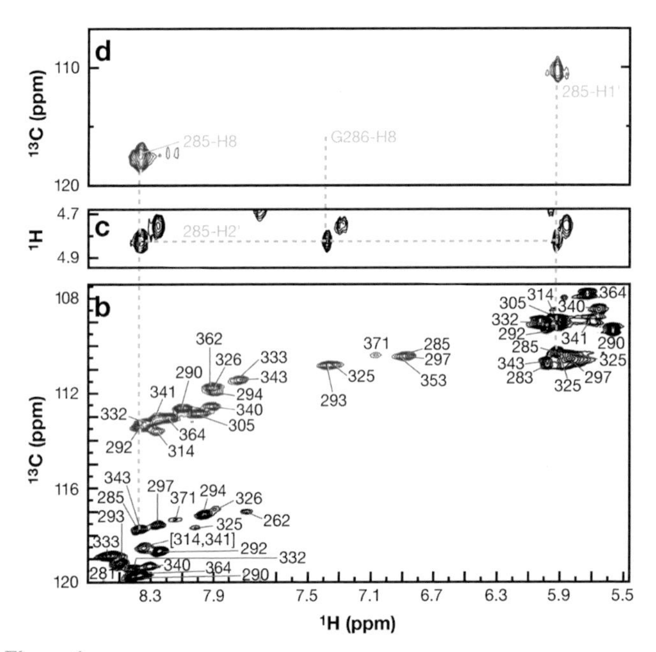

Figure 6

Representation of the assignment strategy followed in the NMR structure determination of the 101-nt core encapsidation signal of the Moloney murine leukemia virus (MLV) (14). (*a*) Secondary structure of the portion of the Ψ-site of MLV, highlighting nucleotides changed from the wild-type sequence. Portions of the two-dimensional (*b*) [^{13}C–^{1}H] HMQC, (*c*) three-dimensional ^{13}C-edited NOESY, and (*d*) four-dimensional $^{13}C,^{13}C$-edited NOESY of a $^{15}N,^{13}C$-adenosine-labeled sample. The novelty of the specific assignment strategy is that NOEs involving unlike (unlabeled) residues are observed in the three-dimensional spectrum but are absent in the four-dimensional data. Reprinted, with permission, from Reference 14.

deuteration strategy for the ribose rings (D3′, D4′, D5′, D5″) and pyrimidine bases (D5), can also overcome the problem of spectral overlap.

Defining the Global Conformation of Large RNAs with RDCs

Structure elucidation of RNA molecules using solely NOE and torsion angle information is inherently difficult because the number of long-range NOEs obtainable from oligonucleotides is smaller than that obtainable from proteins. As a consequence, local RNA structure tends to be well defined but the global structure suffers. Restraints derived from RDCs, which provide unique angular restraints, improve both local and global precision of NMR structures and are therefore especially crucial for RNA structure determination (4, 40). In addition to being used in the refinement of RNA structures, RDC values can also yield valuable, qualitative information about the global shape of the RNA (see below).

For two dipole-coupled nuclei, e.g., carbon and proton, the observable dipolar coupling, D_{CH}, can be expressed as shown in the equation in **Figure 7** (35). The value of D_{CH} depends on the orientation of the bond vector relative to the molecular alignment tensor, which contains the principal components A_{xx}, A_{yy}, and A_{zz}. θ is the angle between the bond vector and the z-axis of the alignment tensor; ϕ is the angle between the projection of the bond vector onto the x-y plane and the x-axis (**Figure 7**). In solution, isotropic tumbling of the molecule averages the value of D_{CH} to zero, but in liquid crystalline media the anisotropic motion of the molecule leads to measurable RDCs. Owing to the extended and more or less regular helical nature, RNA molecules usually align parallel to the z-axis of the alignment tensor, and this has the following consequences for the sign of the measured RDCs: Bond vectors, which are oriented within the x-y plane of the alignment tensor adopt positive values, whereas bond vectors pointing out of the plane adopt negative values.

The structure of HCV IRES domain II aligned relative to its alignment tensor shows that the longest helix of the molecule is oriented parallel to the z-axis and the shorter one is perpendicular to the z-axis (**Figure 7**). In the longest helix, the C1′-H1′ bond vectors are therefore pointing toward the z-axis, consequently displaying negative RDC values (see **Figure 7**: base pair C25-G58). The aromatic C6-H6 and C8-H8 bond vectors of the same base pair are oriented within the x-y plane and therefore adopt positive values. The same situation is encountered for all other aromatic bond vectors (omitted for clarity). The opposite situation is observed for the shorter helical part oriented perpendicular to the z-axis of the alignment tensor (see **Figure 7**: base pair G7-C72). In this part of the RNA molecule, negative RDCs were measured for aromatic residues and positive values obtained for ribose C1′-H1′ bond vectors were consistent with a bend in the global conformation of this RNA. Therefore, simple measurement of a few RDCs from different regions of the RNA molecule can yield information whether or not the RNA adopts an extended conformation. If the local helical structure is known, a few RDCs can be sufficient to determine the angle between different helical regions. Mollova et al. (40) have used this approach to determine the interarm angles of *Escherichia coli* $tRNA^{Val}$. Starting from a modeled $tRNA^{Val}$ derived from the crystal structure of yeast $tRNA^{Phe}$, only 27 RDCs were sufficient to determine an angle of $99 \pm 2^\circ$ between the acceptor arm and anticodon arm. In the future, RDCs could prove to be a valuable tool to define global conformations of large RNAs by assessing the relative orientation of helical regions.

While RDCs allow researchers to define the global shape of a large RNA, this approach still requires that the precise local structure of the subdomains is known. A de novo NMR structure determination of a larger RNA therefore requires the use of

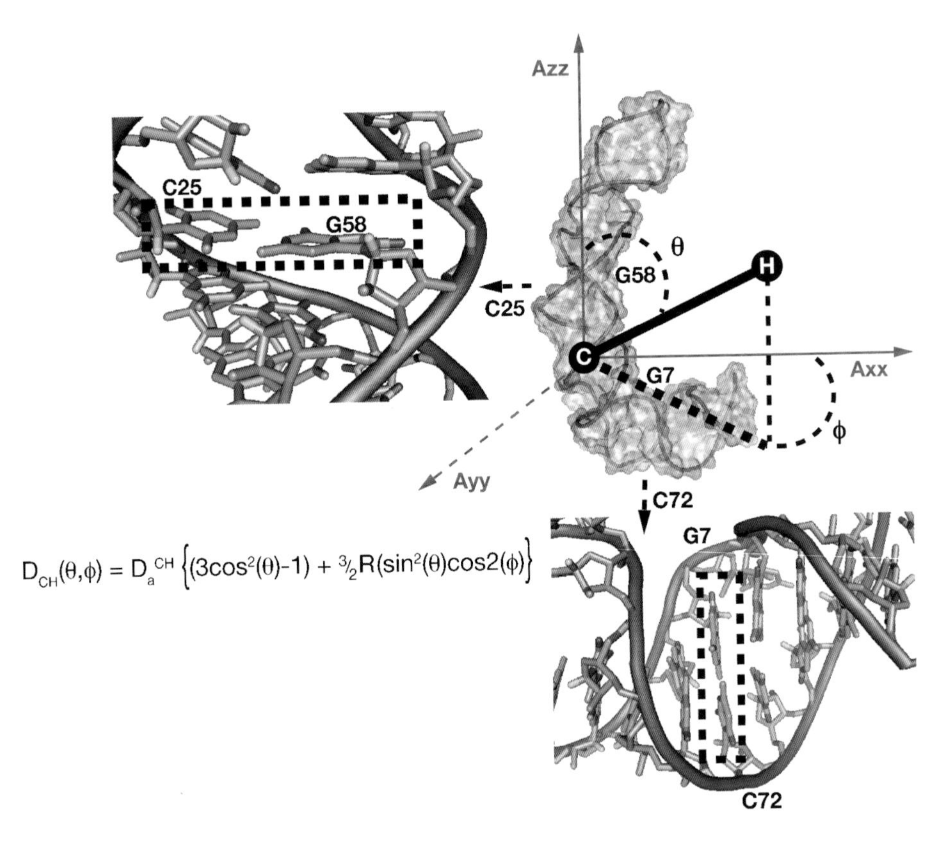

Figure 7

Dependence of RDC values on the orientation of the interdipolar vector (C-H) and the alignment tensor. The structure of HCV IRES domain II (36), represented as a surface plot, is aligned relative to its alignment tensor. A_{xx}, A_{yy}, and A_{zz} are the principal components of the molecular alignment tensor. The longest helix of the molecule is aligned parallel to the *z*-axis and the shorter one is perpendicular. The base pair C25-G58 from the long helix is shown on the left and the base pair G7-C72 from the shorter helix is shown at the bottom. The RNA backbone (*ribbon representation*) is colored purple and the ribose and base rings (*sticks representation*) are colored blue. The ribose protons H1′ are colored green, and the aromatic protons H6 and H8 of the bases are colored red. The molecular alignment tensor was determined by fitting the experiment-derived RDCs to the equation provided in the inset. D_{CH} is the magnitude of the dipolar coupling (Hz) for a certain C-H bond vector, R is the rhombicity, and D_a^{CH} is a scale factor, which subsumes the gyromagnetic ratios of the two nuclei (C and H), their order parameter, and the distance (for a detailed description see Reference 35). Explanation for the rest of the parameters is given in the text.

RDC-derived restraints to refine both the local and global structure of the RNA.

Structure Determination of Large RNAs with RDCs

Researchers routinely determine NMR structures of RNA molecules up to 15 kDa using RDC-derived angular restraints in the refinement protocol to yield better-defined structural ensembles. In practice, measurement of RDCs is straightforward for small RNAs but more difficult for larger RNAs, where resonance overlap can be a problem. RNAs are commonly partially oriented by adding filamentous bacteriophage Pf1 to the NMR sample (10–20 mg ml^{-1}) to measure RDCs from one-bond-separated ^{1}H-^{15}N and ^{1}H-^{13}C spin

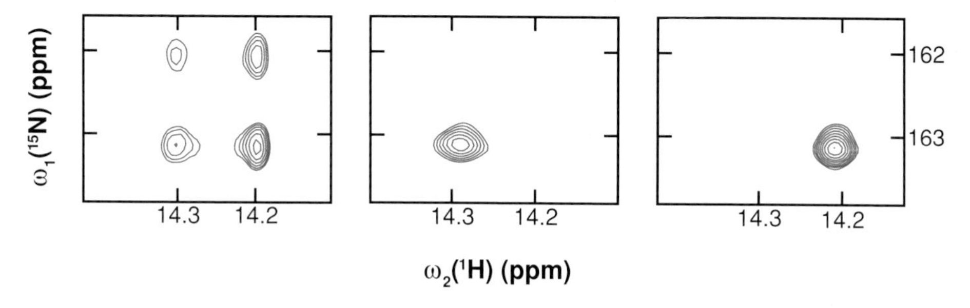

Figure 8

RDC-measurement using [^{15}N,^{1}H]-TROSY. Schematic representation of a uridine N3–H3 multiplet from a two-dimensional nondecoupled [^{15}N,^{1}H]-HSQC experiment of a 74-nt RNA (*left*). The phase cycle of the two-dimensional [^{15}N,^{1}H]-TROSY can be modified to select either the downfield or the upfield component of the downfield ^{15}N–^{1}H doublet (*right*). These experiments have been performed at 800 MHz Bruker Avance^{TM} spectrometer equipped with a cryo probe.

pairs (24). Bacteriophage Pf1 is the ideal liquid crystalline medium for RNA, because negatively charged RNA oligonucleotides are aligned by electrostatic repulsions with the negatively charged phage particles, thereby eliminating the problems of interaction with the alignment media, which would increase the overall tumbling time and, correspondingly, the line widths of RNA resonances (24). One-bond coupling constants in RNA are usually measured in isotropic and aligned media, and RDCs can then be extracted from the difference between the coupling constants (in Hertz) obtained in the two media. Several NMR methods have been developed for the accurate measurement of one-bond couplings (34). Two-dimensional IPAP (inphase and antiphase) HSQC and two-dimensional HSQC coupled in the indirect dimension have been applied successfully to small RNAs but cannot be used for larger RNAs because the upfield component of the ^{15}N–^{1}H or ^{13}C–^{1}H doublet is too broad to allow accurate measurement of the coupling constant. **Figure 8** illustrates a uridine imino N3–H3 cross peak from a [^{15}N, ^{1}H]-HSQC spectrum acquired without decoupling in both dimensions for a 74-nt RNA. Only the downfield components in the ^{15}N dimension yield sufficiently sharp resonances, while the upfield components of the multiplet exhibit severe broadening (**Figure 8**). A similar situation is encountered for the ^{1}H–^{13}C multiplets (37). These problems can be overcome through TROSY-based methods, which allow selection of the sharper, downfield ^{15}N or ^{13}C components of the ^{1}H–^{15}N and ^{1}H–^{13}C multiplets. For the one-bond ^{1}H–^{15}N coupling measurements of the 74-nt RNA, the original two-dimensional WATERGATE-[^{15}N,^{1}H]-TROSY (47) has been used (**Figure 8**).

IPAP: inphase and antiphase HSQC

RDC-derived restraints are crucial for the structure determination of large RNAs, where NMR structures of subdomains can be solved to high resolution, but no structural information is available to define their relative orientation in the context of the larger RNA molecule. An NMR approach that uses RDC-derived angular restraints to improve local structures in RNA and to define the overall shape of the RNA molecule has been applied recently to the high-resolution structure determination of the HCV IRES domain II RNA (36, 37) (**Figure 9**). Dissection of the large RNA into subdomains was required to extract the maximum number of conventional distance and torsion angle restraints, as well as a much larger number of RDC-derived restraints. The combination of these restraints yielded high-resolution subdomain structures, forming the basis for a global refinement of the larger RNA. Local RDCs

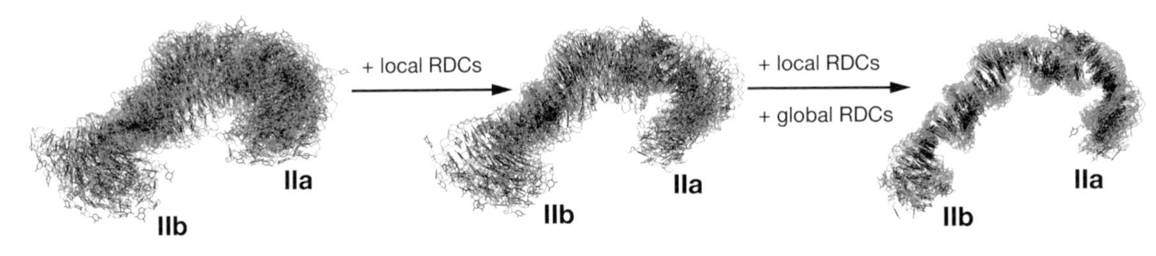

Figure 9

Refinement of the global RNA structure by the use of RDCs. Final ensembles of HCV IRES domain II structures calculated with different sets of RDCs. Bases are black and the ribose-phosphate backbone is gray. The first ensemble of structures calculated with only NOE and torsion angle restraints (rmsd = 7.48 Å). The second ensemble of structures calculated with local RDCs derived from subdomains IIa and IIb (rmsd = 5.79 Å). The final ensemble of structures is calculated with local and global RDCs from domains IIa, IIb, and II (rmsd = 2.18 Å).

from the subdomains IIa and IIb improved the precision of the local domains within the entire domain II RNA but only slightly affected the global conformation (**Figure 9**, middle). A well-defined ensemble of structures was obtained only if global RDCs derived from the entire domain II RNA were used as well (**Figure 9**, right). The same approach has also been used for the structure determination of the 101-nt RNA mΨ and should allow NMR structure determination of even larger RNAs in the future.

Segmental Isotope Labeling: Going from Large to Very Large RNAs

Dissection of large RNAs into subdomains can be helpful for structure determination, but with larger systems resonance overlap does not allow one to extract RDC values sufficient for defining the global conformation. Therefore simplification of NMR spectra using segmental isotope-labeling techniques will play a crucial role in the NMR structure determination of large RNAs. But also beyond structure, segmental labeling is a powerful approach to study domains in the context of an entire, biologically active RNA molecule. Kim et al. (31) have recently used segmental ^{15}N-labeling to address whether the isolated 25-kDa HCV IRES domain II adopts the same fold in the context of the entire 100-kDa HCV IRES. The segmentally labeled RNA was prepared by ligation of ^{15}N-labeled and -unlabeled oligonucleotides using T4 RNA ligase, which joins a 3′-hydroxyl with a 5′-phosphate group. Because in vitro transcription yields RNA oligonucleotides with 5′-phosphate and 3′-hydroxyl groups, various combinations of ligated product could result. The ends of the RNA fragments must therefore be engineered using hammerhead ribozymes, such that the ^{15}N-labeled 5′-fragment contained hydroxyl groups at both ends, while the unlabeled 3′-fragment contained phosphate groups at both ends. This allowed for only one combination of the fragments and yielded a 0.1 mM NMR sample of a segmentally ^{15}N-labeled 100-kDa RNA. Two-dimensional [^{15}N,^{1}H]-HMQC spectra were recorded for both RNAs and chemical shifts of imino ^{1}H and ^{15}N resonances observed for domain II alone and in the context of the 100-kDa HCV IRES RNA were almost identical, confirming that domain II forms an independently folded subdomain in the intact IRES (31) (**Figure 10**).

Future Perspectives for RNA NMR Spectroscopy

The variety of tools currently available in the field of RNA NMR spectroscopy should be combined to shed light on the solution properties of large size RNA structures. The applications range from RNA secondary structure determination using simple

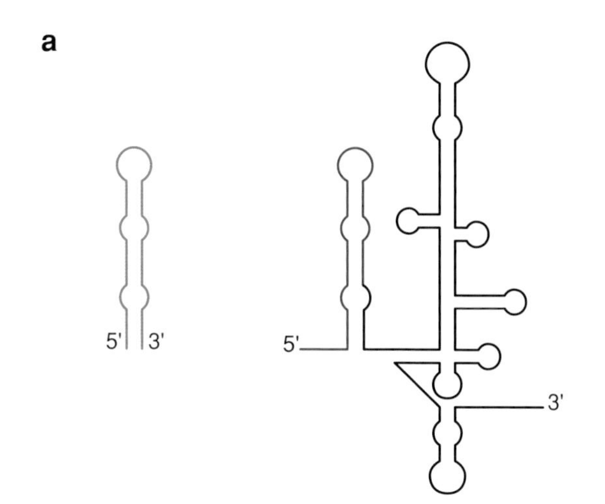

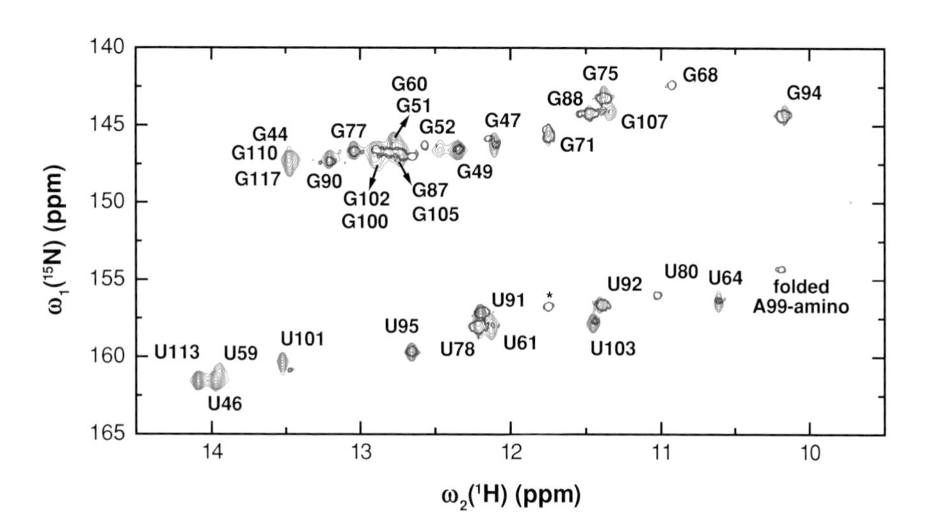

Figure 10

Use of segmental ^{15}N-labeling to study large RNAs. (*a*) IRES domain II (25 kDa) alone is blue, the segmentally (nt 40–104) ^{15}N-labeled IRES (100 kDa) part is red and the unlabeled part is black (31). (*b*) Comparison of the imino regions of $[^{15}N-^{1}H]$ HMQC spectra of 0.54 mM $^{13}C,^{15}N$-labeled domain II (*blue*), and 0.1 mM segmentally ^{15}N-labeled IRES (*red*). An unassigned resonance is marked by an asterisk. Reprinted, with permission, from Reference 36.

NMR techniques, such as two-dimensional S-NOESY and two-dimensional HNN-COSY experiments, to global RNA fold identification using RDCs and a complete structure determination for very large RNAs. The good quality of the two-dimensional HMQC spectrum obtained for the 100-kDa HCV IRES RNA (**Figure 10**) raises hopes that segmental isotope labeling will play a central role in these efforts, since larger systems will require simplified spectra to obtain unambiguous resonance assignments, to extract RDC data, and to probe tertiary interactions within complex RNA folds. Segmental labeling could help determine a sufficient number of RDCs for RNAs up to 100 kDa and thereby help determine their global conformation. Finally, an iterative structure determination of a large RNA starting from its thermodynamic stable subdomains, followed by a subsequent segmental isotope labeling approach in combination with nucleotide-specific ^{13}C- and/or ^{15}N-labeling or even partial deuteration to determine global RDCs, could lead to the structure determination of very large RNAs.

SUMMARY POINTS

1. High-resolution NMR spectroscopy is successfully applied to very large proteins (up to 1000 kDa) and RNAs (up to 100 kDa).
2. Sequential assignment of very large proteins and RNAs is possible.
3. Structural and dynamical information of large biomolecules can be collected by NMR.
4. Structural and dynamical investigations by NMR are based on recently developed NMR and labeling techniques.
5. RDC-derived angular restraints play a crucial role in defining both local and global RNA conformation.
6. Segmental isotope labeling helps reduce the complexity of NMR data often encountered with large RNAs.
7. The variety of RNA NMR tools currently available should be combined in order to determine the structure of large-size and complex RNAs.

ACKNOWLEDGMENTS

A.G.T. is grateful for an *EMBO* long-term fellowship (ALTF: 780-2004). R.R. is a Pew Fellow.

LITERATURE CITED

1. Abragam A. 1961. *The Principles of Nuclear Magnetism*. London: Oxford Univ. Press
2. Batey RT, Battiste JL, Williamson JR. 1995. Preparation of isotopically enriched RNAs for heteronuclear NMR. *Methods Enzymol.* 261:300–22
3. Bax A, Grzesiek S. 1993. Methodological advances in protein NMR. *Acc. Chem. Res.* 26:131–38
4. Bondensgaard K, Mollova ET, Pardi A. 2002. The global conformation of the hammerhead ribozyme determined using residual dipolar couplings. *Biochemistry* 41:11532–42
5. Bromek K, Lee D, Hauhart R, Krych-Goldberg M, Atkinson JP, et al. 2005. Polychromatic selective population inversion for TROSY experiments with large proteins. *J. Am. Chem. Soc.* 127:405–11
6. Brüschweiler R, Ernst RR. 1991. Molecular dynamics modulated by cross-correlated cross relaxation of spins quantized along orthogonal axes. *J. Chem. Phys.* 96:1758–66
7. Chung J, Kroon G. 2003. ^{1}H, ^{15}N, ^{13}C-triple resonance NMR of very large systems at 900 MHz. *J. Magn. Reson.* 163:360–68
8. Deleted in proof
9. Dalvit C. 1992. ^{1}H to ^{15}N polarization transfer via ^{1}H chemical-shift anisotropy: ^{1}H-^{15}N dipole-dipole cross correlation. *J. Magn. Reson.* 97:645–50
10. **Davis JH, Tonelli M, Scott LG, Jaeger L, Williamson JR, Butcher SE. 2005. RNA helical packing in solution: NMR structure of a 30 kDa GAAA tetraloop-receptor complex. *J. Mol. Biol.* 351:371–82**
11. Dieckmann T, Feigon J. 1997. Assignment methodology for larger RNA oligonucleotides: application to an ATP-binding RNA aptamer. *J. Biomol. NMR* 9:259–72
12. Dingley AJ, Grzesiek S. 1998. Direct observation of hydrogen bonds in nucleic acid base pairs by internucleotide $^{2}J_{NN}$ couplings. *J. Am. Chem. Soc.* 120:8293–97

10. Demonstrates the determination of RNA tertiary interactions by NMR.

13. Dingley AJ, Masse JE, Feigon J, Grzesiek S. 2000. Characterization of the hydrogen bond network in guanosine quartets by internucleotide 3hJNC' and 2hJNN scalar couplings. *J. Biomol. NMR* 16:279–89
14. **D'Souza V, Dey A, Habib D, Summers MF. 2004. NMR structure of the 101-nucleotide core encapsidation signal of the Moloney murine leukemia virus. *J. Mol. Biol.* 337:427–42**
15. **D'Souza V, Summers MF. 2004. Structural basis for packaging the dimeric genome of Moloney murine leukemia virus. *Nature* 431:586–89**
16. Farrar TC, Stringfellow TC. 1996. Relaxation of transverse magnetization for coupled spins. In *Encyclopedia of NMR*, ed. DM Grant, RK Harris, 6:4100–7. New York: Wiley
17. Fiaux J, Bertelsen EB, Horwich AL, Wüthrich K. 2002. NMR analysis of a 900 K GroEL/GroES complex. *Nature* 418:207–11
18. Fiaux J, Bertelsen EB, Horwich AL, Wüthrich K. 2004. Uniform and residue-specific 15N-labeling of proteins on a highly deuterated background. *J. Biomol. NMR* 29:289–97
19. Frueh DP, Ito T, Li JS, Wagner G, Glaser SJ, Kaneja N. 2005. Sensitivity enhancement in NMR of macromolecules by application of optimal control theory. *J. Biomol. NMR* 32:23–30
20. Deleted in proof
21. Furtig B, Richter C, Wohnert J, Schwalbe H. 2003. NMR spectroscopy of RNA. *Chembiochemistry* 4:936–62
22. Goldman M. 1984. Interference effects in the relaxation of a pair of unlike spin-1/2 nuclei. *J. Magn. Reson.* 60:437–52
23. Goto NK, Kay LE. 2000. New developments in isotope labeling strategies for protein solution NMR spectroscopy. *Curr. Opin. Struct. Biol.* 10:585–92
24. Hansen MR, Mueller L, Pardi A. 1998. Tunable alignment of macromolecules by filamentous phage yields dipolar coupling interactions. *Nat. Struct. Biol.* 5:1065–74
25. Hennig M, Williamson JR. 2000. Detection of N-H...N hydrogen bonding in RNA via scalar couplings in the absence of observable imino proton resonances. *Nucleic Acids Res.* 28:1585–93
26. Heus H, Pardi A. 1991. Novel proton NMR assignment procedure for RNA duplexes. *J. Am. Chem. Soc.* 113:4360–61
27. Horst R, Bertelsen EB, Fiaux J, Wider G, Horwich AL, Wuthrich K. 2005. From the cover: direct NMR observation of a substrate protein bound to the chaperonin GroEL. *Proc. Natl. Acad. Sci. USA* 102:12748–53
28. Jakobsen HJ, Linde SA, Sorensen S. 1974. Sensitivity enhancement in ^{13}C FT NMR from selective population transfer (SPT) in molecules with degenerate proton transitions. *J. Magn. Reson.* 15:385–88
29. Kay LE, Ikura M, Tschudin R, Bax A. 1990. Three-dimensional triple-resonance NMR spectroscopy of isotopically enriched proteins. *J. Magn. Reson.* 89:496–514
30. **Khaneja N, Li JS, Kehlet C, Luy B, Glaser SJ. 2004. Broadband relaxation-optimized polarization transfer in magnetic resonance. *Proc. Natl. Acad. Sci. USA* 101:14742–47**
31. **Kim I, Lukavsky PJ, Puglisi JD. 2002. NMR study of 100 kDa HCV IRES RNA using segmental isotope labeling. *J. Am. Chem. Soc.* 124:9338–39**
32. Kobayashi N, Freund SM, Chatellier J, Zahn R, Fersht AR. 1999. NMR analysis of the binding of a rhodanese peptide to a minichaperone in solution. *J. Mol. Biol.* 292:181–90
33. Kreishman-Deitrick M, Egile C, Hoyt DW, Ford JJ, Rong L, Rosen MK. 2003. NMR analysis of methyl groups at 100–500 kDa: model systems and Arp2/3 complex. *Biochemistry* 42:8579–86

14. Describes a novel assignment approach for a large RNA.

15. Reports the largest known RNA NMR structure.

30. Applies optimal control theory to optimize polarization transfer.

31. Utilizes segmental isotope labeling for the study of an RNA domain within the context of a large RNA.

34. Latham MP, Brown DJ, McCallum SA, Pardi A. 2005. NMR methods for studying the structure and dynamics of RNA. *Chembiochemisty* 6:1492–505
35. Lipsitz RS, Tjandra N. 2004. Residual dipolar couplings in NMR structure analysis. *Annu. Rev. Biophys. Biomol. Struct.* 33:387–413
36. Lukavsky PJ, Insil K, Otto GA, Puglisi JD. 2003. Structure of HCV IRES domain II determined by NMR. *Nat. Struct. Biol.* 10:1033–38
37. Lukavsky PJ, Puglisi JD. 2005. Structure determination of large biological RNAs. *Methods Enzymol.* 394:399–416
38. Mathews DH, Sabina J, Zuker M, Turner DH. 1999. Expanded sequence dependence of thermodynamic parameters improves prediction of RNA secondary structure. *J. Mol. Biol.* 288:911–40
39. Miclet EE, Williams DC Jr, Clore GM, Bryce DL, Boisbouvier J, Bax A. 2004. Relaxation-optimized NMR spectroscopy of methylene groups in proteins and nucleic acids. *J. Am. Chem. Soc.* 126:10560–70
40. **Mollova E, Hansen MR, Pardi A. 2000. Global structure of RNA determined with residual dipolar couplings. *J. Am. Chem. Soc.* 122:11561–62**

40. Applies RDCs for the determination of the relative orientation of the helical arms in a tRNA.

41. Morris GA, Freeman R. 1979. Enhancement of nuclear magnetic resonance signals by polarization transfer. *J. Am. Chem. Soc.* 101:760–62
42. Nikonowicz EP, Sirr A, Legault P, Jucker FM, Baer LM, Pardi A. 1992. Preparation of ^{13}C and ^{15}N labelled RNAs for heteronuclear multi-dimensional NMR studies. *Nucleic Acids Res.* 20:4507–13
43. Otomo T, Ito N, Kyogoku Y, Yamazaki T. 1999. NMR observation of selected segments in a larger protein: central-segment isotope labeling through intein-mediated ligation. *Biochemistry* 38:16040–44
44. Pachler KGR, Wessels PL. 1973. Selective population inversion (SPI). A pulsed double resonance method in FTNMR spectroscopy equivalent to INDOR. *J. Magn. Reson.* 12:337–39
45. Pervushin K. 2000. Impact of transverse relaxation optimized spectroscopy (TROSY) on NMR as a technique in structural biology. *Q. Rev. Biophys.* 33:161–97
46. Pervushin K, Ono A, Fernandez D, Szyperski T, Kainosho M, Wüthrich K. 1998. NMR scalar couplings across Watson-Crick base pair hydrogen bonds in DNA observed by transverse relaxation-optimized spectroscopy. *Proc. Natl. Acad. Sci. USA* 95:14147–51
47. **Pervushin K, Riek R, Wider G, Wüthrich K. 1997. Attenuated T_2 relaxation by mutual cancellation of dipole-dipole coupling and chemical shift anisotropy indicates an avenue to NMR structures of very large biological macromolecules in solutions. *Proc. Natl. Acad. Sci. USA* 94:12366–71**

47. Introduces TROSY (transverse relaxation-optimized spectroscopy).

48. Pervushin K, Riek R, Wider G, Wüthrich K. 1998. Transverse relaxation-optimized spectroscopy (TROSY) for NMR studies of aromatic spin systems in ^{13}C-labeled proteins. *J. Am. Chem. Soc.* 120:6394–400
49. Pervushin K, Vögeli B. 2003. Observation of individual transitions in magnetically equivalent spin systems. *J. Am. Chem. Soc.* 125:9566–67
50. Peterson RD, Theimer CA, Wu H, Feigon J. 2004. New applications of 2D filtered/edited NOESY for assignment and structure elucidation of RNA and RNA-protein complexes. *J. Biomol. NMR* 28:59–67
51. Pintacuda G, Moshref A, Leonchiks A, Sharipo A, Otting G. 2004. Site-specific labeling with a metal chelator for protein-structure refinement. *J. Biomol. NMR* 29:351–61

52. Piotto M, Saudek V, Sklenar V. 1992. Gradient-tailored excitation for single quantum NMR spectroscopy of aqueous-solution. *J. Biomol. NMR* 2:661–65
53. Reese ML, Dotsch V. 2003. Fast mapping of protein-protein interfaces by NMR spectroscopy. *J. Am. Chem. Soc.* 125:14250–51
54. Deleted in proof
55. Riek R. 2002. TROSY: transverse relaxation optimized spectroscopy in BioNMR in drug research. In *Methods and Principles in Medicinal Chemistry*, ed. O Zerbe, 16:227–41. Weinheim, Ger.: Wiley-VCH
56. **Riek R, Fiaux J, Bertlesen EB, Horwich AL, Wüthrich K. 2002. Solution NMR techniques for large molecular and supramolecular structures. *J. Am. Chem. Soc.* 124:12144–53**
57. Riek R, Pervushin K, Wüthrich K. 2000. TROSY and CRINEPT: NMR with large molecular and supramolecular structures in solution. *Trends Biochem. Sci.* 25:462–68
58. **Riek R, Wider G, Pervushin K, Wüthrich K. 1999. Polarization transfer by cross-correlated relaxation in solution NMR with very large molecules. *Proc. Natl. Acad. Sci. USA* 96:4918–23**
59. Rudiger S, Freund SMV, Veprintsev DB, Fersht AR. 2002. CRINEPT-TROSY NMR reveals p53 core domain bound in an unfolded form to the chaperone Hsp90. *Proc. Natl. Acad. Sci. USA* 99:11085–90
60. Salzmann M, Wider G, Pervushin K, Senn H, Wüthrich K. 1999. TROSY type triple-resonance experiments for sequential assignments of large proteins. *J. Am. Chem. Soc.* 121:844–48
61. Sashital DG, Cornilescu G, McManus CJ, Brow DA, Butcher SE. 2004. U2-U6 RNA folding reveals a group II intron-like domain and a four-helix junction. *Nat. Struct. Mol. Biol.* 11:1237–42
62. Smallcombe SH. 1993. Solvent suppression with symmetrically-shifted pulses. *J. Am. Chem. Soc.* 115:4776–85
63. Trobovic N, Klammt C, Koglin A, Lohr F, Bernhard F, Dotsch V. 2005. Efficient strategy for the rapid backbone assignment of membrane proteins. *J. Am. Chem. Soc.* 127:13504–5
64. Tugarinov V, Kay LE. 2003. Ile, Leu, and Val methyl assignments of the 723-residue malate synthase G using a new labeling strategy and novel NMR methods. *J. Am. Chem. Soc.* 125:13868–78
65. Tugarinov V, Kay LE. 2003. Side chain assignments of Ile $\delta 1$ methyl groups in high molecular weight proteins: an application to a 46 ns tumbling molecule. *J. Am. Chem. Soc.* 125:5701–6
66. Tugarinov V, Kay LE. 2004. An isotope labeling strategy for methyl TROSY. *J. Biomol. NMR* 28:165–72
67. Tugarinov V, Kay LE. 2005. Methyl groups as probes of structure and dynamics in NMR studies of high-molecular weight proteins. *Chembiochemistry* 6:1567–77
68. Tugarinov V, Sprangers R, Kay LE. 2004. Line narrowing in methyl-TROSY using zero-quantum 1H-13C spectroscopy. *J. Am. Chem. Soc.* 126:4921–25
69. Varani G, Aboul-ela F, Allain FHT. 1996. NMR investigations of RNA structure. *Prog. NMR Spectrosc.* 29:51–127
70. Deleted in proof
71. Deleted in proof
72. Wider G. 2005. NMR techniques used with very large biological macromolecules in solution. *Methods Enzymol.* 394:382–98

56. Discusses properties of very large protein systems including the fast protein-selective ^{1}H longitudinal relaxation.

58. Introduces transverse relaxation-optimization during the polarization transfers.

73. Wijmenga SS, van Buuren BNM. 1998. The use of NMR methods in the conformational analysis of nucleic acids. *Prog. NMR Spectrosc.* 32:287–387
74. Wohnert J, Dingley AJ, Stoldt M, Gorlach M, Grzesiek S, Brown LR. 1999. Direct identification of NH. . .N hydrogen bonds in non-canonical base pairs of RNA by NMR spectroscopy. *Nucleic Acids Res.* 27:3104–10

Single-Molecule Analysis of RNA Polymerase Transcription

Lu Bai,[1] Thomas J. Santangelo,[2] and Michelle D. Wang[1]

[1]Department of Physics, Laboratory of Atomic and Solid State Physics, Cornell University, Ithaca, New York 14853; email: lb69@cornell.edu; mdw17@cornell.edu

[2]Department of Microbiology, Ohio State University, Columbus, Ohio 43210; email: santangelo.11@osu.edu

Annu. Rev. Biophys. Biomol. Struct. 2006. 35:343–60

First published online as a Review in Advance on February 17, 2006

The *Annual Review of Biophysics and Biomolecular Structure* is online at biophys.annualreviews.org

doi: 10.1146/annurev.biophys.35.010406.150153

1056-8700/06/0609-0343$20.00

Key Words

initiation, elongation, kinetics, mechanism, pausing

Abstract

The kinetics and mechanisms of transcription are now being investigated by a repertoire of single-molecule techniques, including optical and magnetic tweezers, high-sensitivity fluorescence techniques, and atomic force microscopy. Single-molecule techniques complement traditional biochemical and crystallographic approaches, are capable of detecting the motions and dynamics of individual RNAP molecules and transcription complexes in real time, and make it possible to directly measure RNAP binding to and unwinding of template DNA, as well as RNAP translocation along the DNA during transcript synthesis.

Contents

INTRODUCTION

RNAP: RNA polymerase

Transcription is a crucial step in gene expression and its regulation. Transcription, the synthesis of an RNA transcript complementary to the template DNA, is carried out by DNA-dependent RNA polymerase (RNAP). RNAP is the subject of extensive regulation, dictated not only by interactions between RNAP and regulatory factors, but also by dynamics of the active site and domain movements associated with the transcription process, or cycle. During the past few decades all the phases of transcription have been studied extensively in order to elucidate the mechanism of transcription and its regulation. Before outlining the relatively new mechanical and kinetic details of the transcription cycle elucidated by single-molecule techniques, it is necessary to provide a brief outline of both RNAP structure and function and to summarize the transcription cycle in general.

RNAP Structure

RNAPs occur as both single- and multiple-subunit enzymes. RNAPs from bacteriophages and mitochondria are representative of the single-subunit family; bacterial, archaeal, and eukaryotic nuclear RNAPs constitute the multiple-subunit family. Although the single-subunit and multisubunit RNAPs do not likely share a common ancestor, the available biochemical and structural information from representatives from each family shows that these RNAPs share many characteristics (57). A great body of literature exists, but only those details pertinent to this review are presented here. To date, only two RNAPs, T7 RNAP (single-subunit) and *Escherichia coli* RNAP (multisubunit), have been utilized in single-molecule transcription studies. There is detailed structural information for T7 RNAP (9, 58); however, studies with *E. coli* RNAP typically rely on structural information gathered from other multisubunit RNAPs (10, 11, 38). This review focuses primarily on these two RNAPs.

A gross examination of the currently available structures of RNAP reveals that RNAPs from both families have a main internal channel that can accommodate an 8- to 9-bp RNA/DNA hybrid, a smaller, secondary channel or pore that likely serves as an entry channel for nucleotide triphosphates (NTPs), and an RNA exit channel. The active site is located in the junction of the main channel and secondary channels and contains at least one nucleotide binding site and a tightly bound Mg^{2+}. RNAPs from both families are two-metal ion-dependent enzymes and the second active-site Mg^{2+} is thought to be coordinated with the incoming NTP (56).

The Transcription Cycle

Transcription is traditionally divided into three sequential phases: initiation, elongation, and termination, although termination can be viewed as an alternate pathway branching from elongation (64). During initiation, RNAP recognizes a DNA sequence termed a promoter and melts, or separates, the two strands to form a single-stranded "DNA bubble." The single-subunit RNAPs are capable of recognizing promoters without assistance, whereas multisubunit enzymes rely on separate proteins (typically σ-factors) for promoter recognition. T7 RNAP undergoes a global structural rearrangement following initiation to refold into an elongation complex (58). For multisubunit RNAPs, retention of the initiating σ-factor is not necessary for elongation and the dissociation of σ likely triggers a series of conformational changes during the transition to an elongation complex (35, 36, 38, 40).

The elongation stage of transcription involves movement of RNAP away from the promoter and production of a growing transcript. Elongation proceeds at different rates for the two RNAP families (in vivo: T7 RNAP at $\sim$200 nt s^{-1}; *E. coli* RNAP at $\sim$50 nt s^{-1}) (57). Each nucleotide addition is a competition among elongation, pausing (a transient conformational state incapable of elongation), arrest (a conformational state incapable of elongation without factor-assisted isomerization back to an active complex), and termination (transcript release and enzyme dissociation from the DNA template) (64). The polymerase moves with single-nucleotide steps along the DNA template during elongation, but it is also capable of reverse translocation in the absence of synthesis (termed backtracking) that leads to certain classes of pausing and arrest (4, 26, 27). Termination completes the transcription cycle, recycling RNAP for another promoter recognition event and round of synthesis.

The activity of each family of RNAP is distinct in many phases of transcription, although the transcription cycle is common to both families. Despite some differences during initiation, once an elongation complex is formed, NTP incorporation and the translocation of RNAP along the DNA template are necessary steps for both families of RNAP (13). Both enzymes incorporate individual NTPs into the growing, 3′ end of the transcript, have mechanisms to ensure incorporation of the correct NTP, and, in the unlikely event of misincorporation, mechanisms to remove the incorrectly incorporated nucleotide. Elongation complexes, as well as several intermediates during initiation, are extremely stable, making them good candidates for single-molecule studies.

Advantages and Disadvantages of Single-Molecule Experiments

Although structural studies have provided valuable insights into the structural organization of RNAP "frozen" at different stages of transcription, kinetic aspects of the structural transitions between these snapshots remain obscure. Biochemical studies have filled many gaps, but information regarding many conformational changes associated with each stage of the transcription cycle is still missing as are data on the dynamics of such movements. Many of these questions are better suited to be addressed by single-molecule techniques.

Compared with bulk studies, there are several advantages of single-molecule approaches. First and most importantly, properties measured in bulk studies represent ensemble averages of a population of molecules. Behaviors that are highly unsynchronized among different molecules, such as heterogeneity in population, transient intermediate states, and parallel reaction pathways, are difficult to quantitatively characterize. These problems, in principle, may be overcome by monitoring the motion of individual molecules in real time. Second, single-molecule techniques also provide tools for mechanically manipulating biomolecules, such as stretching and twisting DNA and protein molecules, thus allowing the researcher to manipulate equilibriums between competing reaction pathways in a defined manner.

Single-molecule techniques also have a number of drawbacks. In order to accurately determine kinetics and draw statistically meaningful conclusions, a large dataset of individual single-molecule measurements must be acquired and this can be time

IC: initiation complex

TEC: transcription initiation complex

consuming. In addition, single-molecule assays often introduce perturbations to the system under study (e.g., fluorophore labeling, surface attachment of molecules, and photodamage) that may complicate data interpretation. Finally, all single-molecule approaches are subject to some sort of measurement noise and sometimes require the use of altered reaction conditions (e.g., lowered NTP concentration) in order to achieve sufficient temporal and spatial resolution to probe fast kinetics.

In this article, we provide a rather detailed discussion on how single-molecule techniques have contributed to our better understanding of different phases of transcription. Because of space limitation, we do not focus on the operational principles of these techniques but instead on some of the important findings that have resulted from the use of these techniques.

INITIATION

Background

Initiation is the first phase of transcription, and historically, studies of transcription regulation have focused on initiation. Initiation requires a series of isomerizations, each of which has been biochemically characterized to some extent (33, 41). First, RNAP, either in combination with a σ-factor (e.g., *E. coli* RNAP) or alone (e.g., T7 RNAP), searches for and binds to a promoter sequence to form a so-called closed complex. The initial binding event is with double-stranded or closed DNA. Second, driven by the binding free energy, the complex isomerizes to an open complex by unwinding 10 to 15 base pairs of DNA surrounding the transcription start site. This initiation complex (IC) then undergoes a competition between NTP incorporation and short RNA oligo release, a phenomenon known as abortive initiation. Once the transcript length reaches ~12 nt, the complex enters the processive elongation phase (**Figure 1**).

The reaction pathway described above can be summarized as (41):

$$\mathrm{RNAP} + \mathrm{P} \underset{k_{-1}}{\overset{k_{+1}}{\rightleftharpoons}} \mathrm{RNAP}\cdot\mathrm{P_c} \underset{k_{-2}}{\overset{k_{+2}}{\rightleftharpoons}} \mathrm{RNAP}\cdot\mathrm{P_o} \underset{k_{-3}}{\overset{k_{+3}}{\rightleftharpoons}} \mathrm{IC}_{\leq 12} \overset{k_{+4}}{\rightarrow} \mathrm{TEC}, \qquad 1.$$

where P represents the promoter, $\mathrm{RNAP}\cdot\mathrm{P_c}$ and $\mathrm{RNAP}\cdot\mathrm{P_o}$ represent the closed and open complexes, respectively, $\mathrm{IC}_{\leq 12}$ represents abortive initiation complexes with transcript sizes generally ≤12 nt, and TEC represents the transcription elongation complex. $k_{\pm n}$ are the forward and reverse rate constants between the different states and thus determine the relative population distribution among the states. (This reaction pathway and subsequent pathways in this manuscript are simplified; typically more substates exist than are shown.) Different promoter sequences, DNA supercoiling configurations, and the association of regulatory factors such as other proteins and small molecules affect different $k_{\pm n}$ and thus modulate the overall initiation velocity and efficiency. Therefore characterization of the kinetic processes described in Equation 1 under different conditions is essential in order to elucidate the mechanism of initiation and its regulation.

Thus far single-molecule studies of initiation have allowed direct visualization of static intermediate ICs as well as the kinetic transition processes between the states, and measurement of some of the transition rate constants under various experimental conditions. These results are discussed in the following two sections.

Promoter Search

RNAP must accurately and efficiently locate promoters in spite of the large excess of nonpromoter DNA. It has been suggested that the efficiency of promoter recognition is enhanced by RNAP weakly binding to nonspecific DNA and rapidly transferring between different DNA segments (6). This transfer may include one-dimensional diffusion along the DNA (sliding), microscopic dissociation/reassociation (hopping), and direct transfer between DNA segments. After consideration of this facilitated promoter targeting, the reaction pathway (1) should be expanded as (54):

$$\mathrm{RNAP} + \mathrm{DNA} \underset{k_{\mathrm{off}}}{\overset{k_{\mathrm{on}}}{\rightleftharpoons}} \mathrm{RNAP} \cdot \mathrm{DNA} \underset{k_{-1'}}{\overset{k_{+1'}}{\rightleftharpoons}} \mathrm{RNAP} \cdot \mathrm{P_c} \underset{k_{-2}}{\overset{k_{+2}}{\rightleftharpoons}} \mathrm{RNAP} \cdot \mathrm{P_o} \underset{k_{-3}}{\overset{k_{+3}}{\rightleftharpoons}} \mathrm{IC}_{\leq 12} \overset{k_{+4}}{\rightarrow} \mathrm{TEC}, \quad 2.$$

where RNAP · DNA represents an RNAP and DNA complex. In this notation, the promoter search is considered in the step from RNAP · DNA to RNAP · P_c and is oversimplified as a single-step reaction. The proposed mechanisms for facilitated promoter searches are difficult to verify in bulk because binding, hopping, sliding, and transferring of RNAP may be transient events that are scattered along the DNA. Various single-molecule assays have been developed in order to directly detect these events.

The nonspecific binding of RNAP to DNA was first visualized using fluorescence microscopy by flowing fluorescently labeled *E. coli* RNAP through a bundle of stretched and oriented DNA molecules (23). While maintaining a constant flow rate, the interaction of an RNAP molecule with DNA was identified when its motion deviated from simple Brownian motion with drift. The observed motion of RNAP along the oriented DNA provided evidence for sliding as a

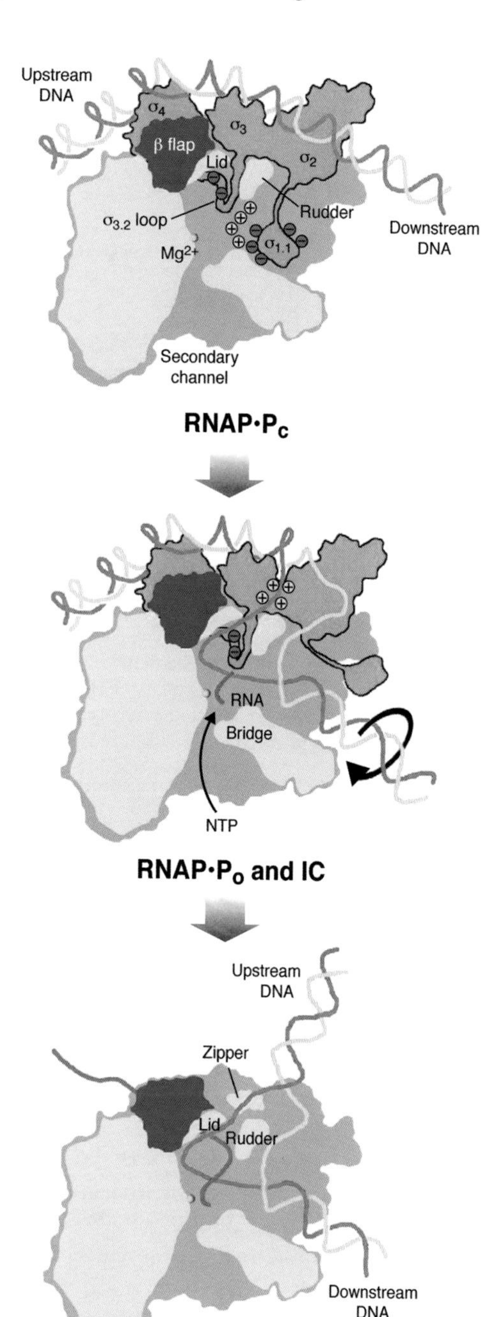

Figure 1

Structural transitions from closed complex (RNAP · P_c; *left panel*), to an open and abortive initiation complex (RNAP · P_o and IC; *center panel*), and to a transcription elongation complex (TEC; *right panel*). Shown are cross-sectional views of *Thermus aquaticus* RNAP holoenzyme (β flap, *blue*; σ, *orange*; rest of RNAP, *gray*; catalytic Mg^{2+}, *yellow sphere*), promoter DNA (template strand, *dark green*; nontemplate strand, *light green*), and the RNA transcript (*red*). Adapted and reprinted from *Current Opinion in Structural Biology*, Volume 13, Issue 1, Murakami KS and Darst SA, *Bacterial RNA polymerase: the wholo story*, Pages 31–39, Figure 3, Copyright © 2003 Elsevier Science Ltd, with permission from Elsevier.

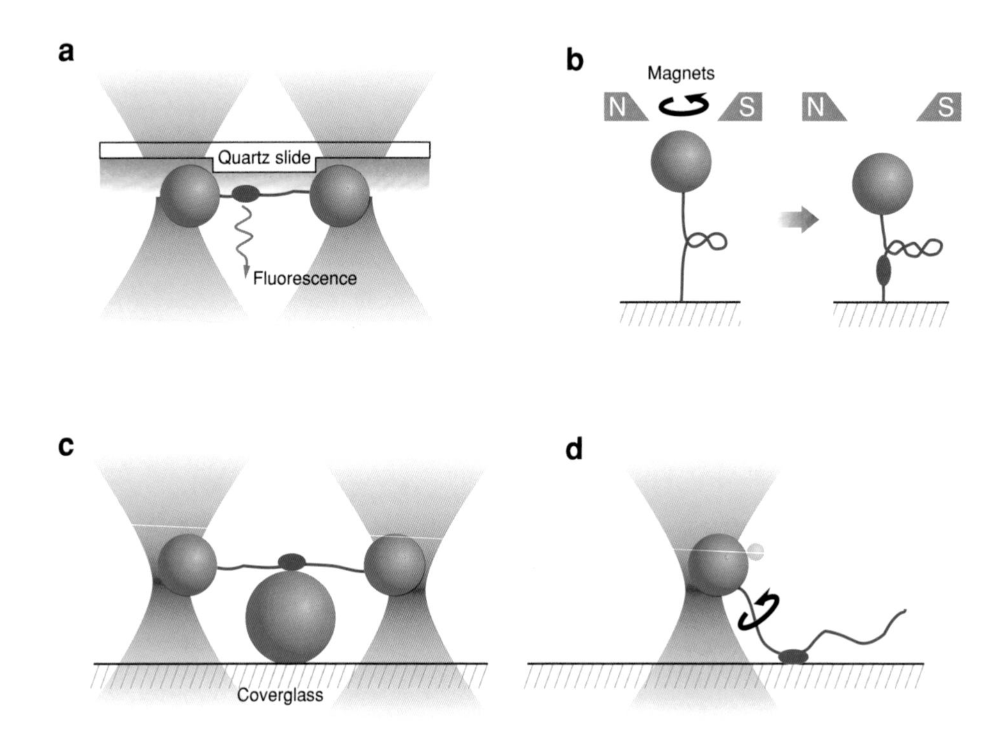

Figure 2

Cartoons of single-molecule experimental configurations used in initiation studies. Panels *a–d* represent the experimental design in Harada et al. (20) (TIRF and dual optical tweezers), Revyakin et al. (43) (magnetic tweezers), Skinner et al. (55) (dual optical tweezers), and Sakata-Sogawa et al. (49) (optical tweezers with rotation detection), respectively (for details see text). Focused laser beam, red; dielectric or magnetic bead, blue; DNA, red; RNAP, purple; small fluorescent bead, orange.

Total internal reflection fluorescence (TIRF): a technique that uses the evanescent field from light that is totally internally reflected at an interface to selectively excite fluorophores that are in very close proximity to the interface

Optical tweezers: an instrument that uses a tightly focused laser beam to trap and exert force on a microscopic dielectric particle

mechanism for promoter search. In a later experiment, the binding of *E. coli* RNAP to a single DNA molecule was observed by the combination of total internal reflection fluorescence (TIRF) with optical trapping (20) (**Figure 2*a***). A DNA molecule was held between two optical traps and individual RNAP molecules were fluorescently labeled. This experimental design decreased background fluorescence and aligned DNA for ease of visualization without the need for flow. The interaction between RNAP and DNA was found to be sequence dependent. RNAP bound more frequently to an AT-rich region than to a GC-rich region and dissociated more slowly at a promoter and promoter-like sequence. The measured association rate corresponds to $k_{on} \sim 10^3$ bp^{-1} M^{-1} s^{-1}. However, since this method only detected RNAP binding and did not differentiate among the three bound species (RNAP · DNA, RNAP · P_c, and RNAP · P_o), the measured dissociation rate of $\sim$1 s^{-1} can only impose some constraints on k_{off}, $k_{\pm 1'}$, and $k_{\pm 2}$. Interestingly, a small fraction of RNAP exhibited random motion along the DNA before dissociation, which provided direct evidence for the linear diffusion of RNAP. The estimated diffusion coefficient of 10^{-10} cm^2 s^{-1} was 1 to 3 orders of magnitude smaller than those predicted by biochemical studies (54). The accuracy of the diffusion measurement was limited likely by the spatial resolution of the technique ($\sim$200 nm or $\sim$600 bp of DNA), which might have exceeded the range of the linear diffusion.

Atomic force microscopy (AFM) is capable of a spatial resolution of ~10 nm. AFM first allowed observation of the binding of individual *E. coli* RNAP molecules to a DNA containing a promoter sequence in aqueous solution (17). Subsequent experiments using a promoterless DNA weakly adsorbed onto a surface demonstrated that *E. coli* RNAP could bind to DNA nonspecifically (18). By comparing sequential images of the same RNAP · DNA complexes and mapping the relative position of RNAP on the DNA as a function of time, RNAP was found to exhibit diffusion, hopping, or intersegment transfer along DNA (8, 18). The time resolution of these AFM studies (of the order of 10 s) made it difficult to measure fast kinetics. Also, the surface adsorption strongly affected RNAP's diffusion properties, and the average lifetime of the RNAP · DNA complex measured in the AFM study (~10 min) was much longer than values reported in other studies (20, 54).

The studies above measured the linear motion of RNAP during promoter search. Because of the double-helical structure of DNA, the translocation of RNAP may involve groove-tracking, which requires the rotation of RNAP around the helical axis of the DNA as it translocates. To determine if RNAP groove-tracks during the promoter search, an experimental setup utilizing a small fluorescent bead attached to a large bead was employed (21, 49) (**Figure 2*d***). DNA was multiply linked at one end to the large bead, which was then optically trapped. The tethered DNA was then dragged across a surface coated with immobilized *E. coli* RNAP molecules. The motion of the small fluorescent bead was analyzed to detect possible RNAP-driven rotation of the DNA. Many factors made these measurements difficult: large rotational Brownian motion of the large bead, low probability for the DNA to be bound by RNAP, bidirectionality of DNA rotation, and short lifetime of the one-dimensional diffusion. Nonetheless, it was argued that on average more coherent rotation of the bead was observed in the presence of RNAP, in support of groove-tracking (49).

Taken together, the experiments discussed above provide strong support for a diffusion-facilitated promoter search model. These studies, however, have some discrepancies in their quantitative values of the binding rate for nonspecific RNAP-DNA interactions, the diffusion coefficient of the RNAP along the DNA, and the lifetime of such diffusion. These values are important in order to understand the efficiency and rate of promoter recognition in vivo and should be further examined in the future. In principle the single-molecule techniques established during these studies could also be configured to study mechanisms of other proteins targeting specific DNA sequences.

Atomic force microscopy (AFM): a scanning probe microscopy that uses the interaction force between a probe tip and a sample to generate high-resolution topographical images of the surface of the sample

σ^{70}-factor: a major σ factor in *E. coli* that directs RNAP to promoters with -10 and -35 elements

Magnetic tweezers: an instrument that uses a magnetic field to generate both force and torque on a microscopic magnetic particle

Open Complex Formation

After RNAP locates a promoter, it isomerizes to form a relatively stable binary complex with a transcription bubble. On strong promoters, open complexes are much more energetically favorable, and the closed-to-open complex transition is essentially irreversible (41). AFM imaging of individual static open complexes of *E. coli* RNAP · σ^{70}-factor showed that the DNA was severely bent by 55° to 88° and its apparent contour length was reduced by ~90 bp, which was interpreted as DNA wrapping around RNAP (42, 47).

Single-molecule techniques have been utilized to follow the formation of open complex. Revyakin et al. (43) specifically probed the transition between closed and open complexes of *E. coli* RNAP by employing a magnetic tweezers setup in which a small change in DNA supercoiling associated with the formation of open complex was amplified as a large end-to-end distance change in the DNA extension (**Figure 2*b***). These studies were capable of monitoring a number of kinetic parameters of promoter formation under a variety of supercoiling states and with different promoter sequences. When NTPs were not present, transcript synthesis was blocked and only a subset of states in Equation 2 were accessible:

$$\text{RNAP} + \text{DNA} \underset{k_{\text{off}}}{\overset{k_{\text{on}}}{\rightleftharpoons}} \text{RNAP} \cdot \text{DNA}$$
$$\underset{k_{-1'}}{\overset{k_{+1'}}{\rightleftharpoons}} \text{RNAP} \cdot \text{P}_{\text{c}} \underset{k_{-2}}{\overset{k_{+2}}{\rightleftharpoons}} \text{RNAP} \cdot \text{P}_{\text{o}}. \quad 3.$$

The lifetime of the unwound (open complex) state directly yielded k_{-2}. For a strong promoter with a negatively supercoiled DNA, k_{-2} was so small that the open complex formation was effectively irreversible. With a strong consensus promoter and positively supercoiled DNA, k_{-2} was $\sim 0.03\ \text{s}^{-1}$. Under the assumptions that RNAP binding and closed promoter formation were in rapid equilibrium and that RNAP $\cdot$ P_c was a much more stable complex than RNAP $\cdot$ DNA, analysis of the time interval between unwinding events yielded $k_{+2} = 0.3\ \text{s}^{-1}$ and the effective equilibrium binding constant $K_\text{B} = k_{\text{on}}k_{+1'}/k_{\text{off}}k_{-1'} = 10^7\ \text{M}^{-1}$. Using this analysis, factors such as ppGpp and the initiating nucleotide were shown to alter the stability of the open complex.

In the study above, upon open complex formation, the DNA extension changes for negative and positive supercoiled DNA were not completely symmetric. This is consistent with the idea that DNA wraps around or is bent by RNAP as proposed by Rivetti et al. (47). This was also supported by results from Harada et al. (20), who found that tension in the DNA decreased the association rate and increased the dissociation rate of RNAP binding.

To our knowledge, there has been only one single-molecule study performed on initiation with T7 RNAP (55). In this experiment, a DNA molecule suspended between two optically trapped beads was held near a surface-immobilized bead sparsely coated with RNAP (**Figure 2*c***). By oscillating one bead with the optical trap, RNAP-DNA binding events were detected when the motion of the two trapped beads became decoupled. Nonspecific binding of RNAP was considered too fast to be detectable so only Equation 1 needed to be considered. The measured RNAP dissociation rate of $2.9\ \text{s}^{-1}$ places some constraints on possible values of k_{-1} and $k_{\pm 2}$. Because the measurements were carried out in the presence of NTPs, there was some probability ($\sim$1% in this study) for RNAP to start elongation, which was detected as a large unidirectional motion of the downstream bead. The transition rate from initiation to elongation ($\sim 0.4\ \text{s}^{-1}$) was significantly slower than the dissociation rate of a bound RNAP complex, consistent with a model in which abortive initiation limits the rate of initiation on strong promoters.

ELONGATION

Background

As the nascent transcript RNA reaches $\sim$13 nt the transcription complex escapes the promoter and enters the elongation phase. During elongation, RNAP, DNA, and RNA form a stable tertiary complex, the TEC, and RNAP moves processively along the DNA template while incorporating complementary NTPs onto the 3′ end of the RNA. An NTP incorporation cycle is composed of multiple reaction steps, including RNAP translocation from the pre- to posttranslocational state, NTP binding, NTP hydrolysis, PPi release, and possible conformational changes of RNAP (13, 28, 45, 63). An AFM-based study showed that DNA contour length was decreased in an elongation complex, consistent with the notion that DNA either is bent by or wraps around the RNAP in elongation complexes. However, the extent of contour length change measured for the elongation complexes was less than that of corresponding open promoter complexes (46).

One of the most fundamental questions in transcription is how RNAP's chemical catalysis is coupled to its mechanical translocation. A number of experimental results from *E. coli* RNAP support a thermal ratchet mechanism in which RNAP can slide back and forth on the DNA template activated by thermal energy and the incorporation of the next nucleotide biases the polymerase forward by one base pair (5, 15, 22, 26, 27, 67). An example of a thermal ratchet model (5, 15) is shown below in which $TEC_{N,\ pre(post)}$ represents the TEC with transcript size N at the pre-(post)translocational mode:

$$TEC_{N,\ pre} \underset{k_{-1}}{\overset{k_1}{\rightleftharpoons}} TEC_{N,\ post} \underset{k_{-2}}{\overset{NTP \searrow k_2}{\rightleftharpoons}} TEC_{N,\ post} \cdot NTP \underset{k_{-3}}{\overset{k_3}{\rightleftharpoons}} TEC_{N+1,\ pre} \cdot PPi \underset{k_{-4}}{\overset{k_4 \nearrow PP_i}{\rightleftharpoons}} TEC_{N+1,\ pre} \quad 4.$$

However, more recent findings from crystallographic data of T7 RNAP are challenging this view (28, 59, 77). Examination of the RNAP structures of elongation intermediates suggests that PP_i release promotes RNAP to undergo a conformational change that induces forward translocation of the polymerase by a single base pair. This conformation-driven translocation was thought to be tightly coupled, with one translocation step for each NTP incorporation cycle. This supported a power-stroke mechanism in which the chemical energy derived from the NTP condensation reaction directly drives the forward translocation of the RNAP along the DNA template. However, it is possible that single-subunit and multisubunit RNAPs utilize distinct mechanisms. More direct kinetic studies using single-molecule and biochemical techniques in conjunction with theoretical studies (5, 15, 60) are beginning to differentiate between these mechanisms of mechanochemical coupling.

Unlike traditional molecular motors (e.g., kinesin and myosin), RNAP moves along a varying substrate because of the varying template DNA sequence. Consequently, transcription does not proceed at a uniform rate and the motion of RNAP is DNA sequence dependent. In particular, RNAP tends to dwell transiently at certain template positions known as pause sites (45, 63, 64). Numerous pause sequences have been shown, or are suspected, to provide regulatory functions such as allowing transcription factors to bind and thereby modify gene expression (31, 48). Other pause sequences that have been detected in vitro have no known biological function but nonetheless reflect the intrinsic sequence dependence of RNAP motion. Biochemical assays have led to the suggestion that transcription pausing results from misalignment of the RNA 3′ end with the RNAP active site (4, 26, 27). Although a large number of pause-inducing sequences are known (30), no consensus sequences have been identified. Recent theoretical work might make it possible to predict some of the pause sites for a given DNA sequence (5).

Many of the pauses studied in traditional bulk biochemical assays are only prominent at low NTP concentrations, which is consistent with a competitive kinetic model in which pausing is an alternative pathway branching from active elongation (64, 65). In this model, the reaction shown in Equation 4 may be rewritten as:

$$\begin{array}{ccc} TEC_{N,\ pre} \underset{k_{-1,-2,-3,-4}}{\overset{k_{1,2,3,4}}{\rightleftharpoons}} \cdots \rightleftharpoons TEC_{N+1,\ pre} & & \text{Main pathway} \\ k_5 \downarrow\uparrow k_{-5} & & \\ \text{Branched pathway} & & 5. \end{array}$$

On the basis of this mechanism, pausing could be caused by a slow rate in the main pathway (e.g., under low NTP concentration) or by a relatively fast rate into the nonproductive branched pathway with a slow rate of returning to the main pathway (e.g., at a pause induced by misalignment of the 3′ end of the RNA). The two pathways are kinetically competitive: a slow rate in the main pathway also increases the probability for RNAP to enter the branched pathway.

Similar to initiation, the rates in the reaction pathway above are sensitive to both internal conditions (e.g., RNAP species, DNA sequence, and RNAP mutation strains) and external conditions (e.g., NTP substrate concentration, temperature, buffer composition, and protein factors) (44). Many single-molecule assays have been carried out to further quantify the reaction pathways under different conditions.

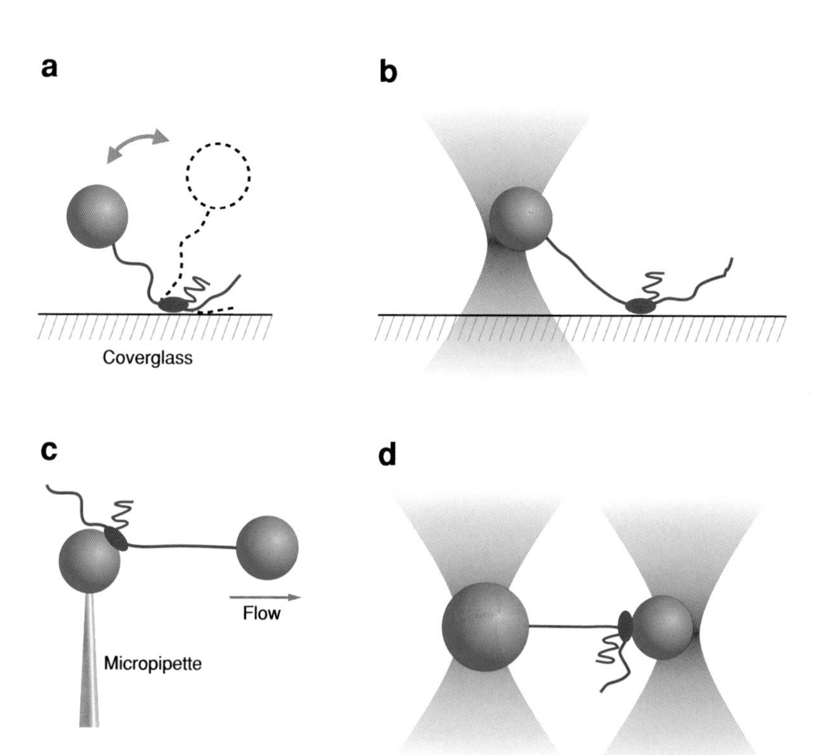

Figure 3

Cartoons of single-molecule experimental configurations used in elongation studies using (*a*) tethered particle motion, (*b*) optical tweezers, (*c*) flow-control video microscopy, and (*d*) dual optical tweezers (for details see text). Focused laser beam, red; dielectric bead, blue; DNA, red; RNAP, purple; RNA, blue.

TPM: tethered particle motion

Below we first discuss general single-molecule approaches for elongation studies, followed by experimental results on active elongation and pausing kinetics.

General Approaches

The initial observations of elongation activity at the single-molecule level came from tethered particle motion (TPM) studies (51, 75) (**Figure 3*a***). In TPM experiments, an RNAP molecule is immobilized onto the surface of a microscope coverglass, and a small particle is tethered to the RNAP by the DNA template that is being transcribed. The tethered particle has a confined Brownian motion with a range indicative of the length of the DNA tether. Therefore the motion of the RNAP along the DNA template can be monitored by the range of the motion of the constrained particle. The TPM method minimally perturbs RNAP activity, and multiple polymerase molecules can be monitored simultaneously via video microscopy to allow high data throughput. However, the large Brownian motion of the tethered bead limits the spatial and temporal resolutions for detecting RNAP motion.

Subsequent studies improved resolution by reducing the Brownian noise of the bead via an external force using optical tweezers (2, 39, 52, 53, 60, 68, 76) (**Figure 3*b***) or a constant flow (12, 14) (**Figure 3*c***). In these studies, an RNAP molecule was attached to a surface (of a microscope coverglass or a bead) while one end of the DNA that was transcribed was attached to another surface in such a way that a bead was tethered by the DNA. Translocation of the polymerase was then monitored via the movement of the bead and the higher resolution of these studies revealed some detailed behaviors of individual RNAP molecules. Movement of an individual RNAP molecule was often interrupted by numerous pauses, even on templates with no known strong regulatory pause sites at saturating NTP concentrations. In these studies, active elongation was differentiated from pausing by setting a threshold to either the instantaneous velocity (12, 14, 39, 68) or the dwell time at each base (2, 53). The threshold needed to be set according to the resolution and data filtering of a given measurement and was thus experiment specific. Transient pausing of short duration (<1 s) often could not be directly resolved, and this inevitably affected the statistics of active elongation velocity and pausing probability. Improved instrumentation and analysis now allow for more precise resolution of RNAP location, including the identification of sequence-dependent pause sites (53), and resolution of backtracking events associated with certain pauses (52).

A single-molecule magnetic tweezers approach also allowed the investigation of the rotational behaviors of RNAP (21). As with the one-dimensional diffusion of RNAP during a promoter search, elongating RNAP is also expected to track the groove of the DNA double helix and rotate relative to the DNA. Using a setup similar to that shown in **Figure 2*d*** except with the bead being pulled up by a single magnet, Harada et al. (21) observed processive unidirectional rotation of a magnetic bead multiply linked to one end of a DNA that was transcribed by a surfaced-immobilized *E. coli* RNAP. These experiments, together with those from initiation studies, are providing increasing evidence that RNAP is a groove-tracking motor molecule.

Active Elongation Kinetics

Compared with bulk methods, single-molecule measurements can more readily separate active elongation from pausing. However, the measured average active elongation velocity of *E. coli* RNAP varied from study to study (10 to 20 bp s^{-1} at room temperature and 1 mM NTPs) even under apparently similar experimental conditions (e.g., temperature, buffer, NTP concentration) (2, 12, 14, 39, 52, 53, 68). The high end of this range is comparable to the rate measured in bulk studies (15 to 20 bp s^{-1}, even including pauses), but the low end is significantly slower. Slight temperature differences may contribute to the inconsistencies because transcription rate was found to have a strong dependence on temperature (1). However, it is unlikely that temperature differences alone can account for the high variability of transcription rates in single-molecule studies. Discrepancies in velocity reported by different labs are due presumably to differences in experimental details and data analysis and need to be carefully resolved.

Single-molecule studies also brought out two other aspects of active elongation. First, Davenport et al. (12) reported a bimodal distribution for active elongation velocity, even for the same RNAP molecule, and suggested that a single RNAP molecule could switch from a more competent to a less competent elongation state. However, other studies did not observe such switching behavior (2, 39, 53, 61). Instead the instantaneous velocity during active elongation was well fit with a single Gaussian distribution (2, 39). Second, it is still controversial whether an RNAP population is homogeneous. While some studies supported uniform kinetics of RNAP among different molecules (2, 53), others provided evidence for a heterogeneous population in which the active elongation velocity varies from molecule to molecule (39, 61). The spread in the velocity, typically 5 to 7 bp s^{-1}, was thought to be larger than the variation arising from the stochastic nature of each NTP incorporation step or experimental conditions, and thus reflected the intrinsic heterogeneity among RNAP molecules. Interestingly, heterogeneous reaction rates within a single molecule or among different molecules of the same population have been reported for other enzymes (19, 32, 70, 71). Single-molecule studies have brought these issues to light and provide a direct way to resolve them.

Optical tweezers (**Figure 3*b***) and the flow technique (**Figure 3*c***) have allowed the application of an assisting or resisting force to

Backtracked pause: transcription pause induced by the reverse translocation of RNAP in the absence of synthesis

the motion of a single RNAP molecule. *E. coli* RNAP is a powerful motor, capable of generating ~25 pN of force (39, 68, 76). The measured force-velocity relation shows that transcription velocity remains nearly constant over a wide range of resisting and assisting forces (+25 pN to −35 pN) under saturating NTP concentration (12, 14, 39, 68, 76). A recent study with T7 RNAP also found a similar force-velocity relation (60). These results indicate that translocation does not limit the rate of transcription at saturating NTP concentration.

Single-molecule experiments are also beginning to elucidate the mechano-chemical coupling mechanism of transcription. Thomen et al. (60) examined the relation of elongation velocity of T7 RNAP and NTP concentration under various forces. Their results were consistent with a thermal ratchet model first proposed by Guajardo & Sousa (15). Extending this simple model, Bai et al. (5) formulated a sequence-dependent thermal ratchet model for transcription by *E. coli* RNAP by constructing a quantitative sequence-dependent transcription energy landscape and performing a full kinetic analysis based on thermodynamic calculations (72). The predictions of this model were consistent with a number of biochemical and single-molecule measurements. Nevertheless, the power-stroke mechanism of transcription can not be ruled out and differentiation between the two models requires future effort.

Pausing Kinetics

Single-molecule studies have provided much information on transcription pausing. In a single-molecule measurement, a transcription pause is characterized by its template sequence, probability of occurrence, and duration. On templates with no known strong pause sites, optical tweezers studies have shown that observed pauses were dominated by those with short durations (<10 s) with 1 mM NTPs (2, 39). The average duration of these pauses was a few seconds, which would have been too short to be detected in conventional bulk transcription studies, for which pauses must be of both significant duration and high enough probability for a significant percentage of the population to be observed at the pause site. The distance between pauses followed a single exponential distribution with a characteristic distance of ~250 bp (2), and this distribution suggested that the pauses either occurred stochastically in a sequence-independent manner, or were sequence dependent but distributed randomly, or occurred at very low probability. An examination of pausing probability versus template position revealed a small but statistically significant variation along the template, suggesting that the pausing should be at least partially sequence dependent (39). A small fraction of pauses detected also had a long duration (>10 s) (2, 39), and long pauses with a duration of 20 to 100 s have also been identified by using a flow force setup (12, 14). These pauses occurred with much lower frequency (~every 5000 bp) and seemed to have a strong sequence dependence. The application of a force had little effect on the short pause kinetics (39) but reduced the long pause efficiency if applied in the assisting direction (14). The exact mechanisms of both short and long pauses and their possible sequence dependence are yet to be fully understood.

The studies above based their statistics on the pauses observed over the whole template. Pausing kinetics at a specific template position, and the effect of assisting or repressive loads, was directly examined by Shundrovsky et al. (53). By using runoff transcription as a well-defined position marker for alignment, the position of RNAP on the DNA template was determined to ~5 bp precision, thus allowing definitive confirmation of sequence-dependent pausing. This resolution represented an ~20-fold enhancement over previous methods of detection (2, 12, 14, 39, 52, 68, 76). At a well-known backtracked pause, ΔtR2, the pause duration decreased significantly with increasing assisting force, providing further evidence for backtracking at this pause site.

Single-molecule studies have also been effective in elucidating the mechanism of a small inhibitor of transcription, the cyclic peptide microcin J25 (3). In the presence of microcin, RNAP paused more frequently but the active elongation velocity was not affected. This finding, in combination with bulk biochemical studies that show that microcin interacts with RNAP in the secondary channel (3, 37), supports a model in which microcin inhibits transcription by binding to the secondary channel of RNAP and thereby blocks the entrance of NTP substrate.

Single-molecule studies have also provided direct evidence for RNAP backtracking during transcription pausing. Backtracked pauses are typically of long duration, and therefore their characterization by single-molecule measurements is often plagued by instrument drift. Shaevitz et al. (52) circumvented this problem by using a dual trap (**Figure 3*d***) so that the long-lived backtracked pauses could be observed with near base pair resolution. The mean backtracking distance was ~5 bp and the pause duration was of the order of 1 min. Consistent with biochemical studies, the addition of GreA and GreB factors, which were known to rescue a backtracked complex by stimulating cleavage of the 3′ end of the RNA (7), decreased the backtracking pause probability.

TERMINATION

During termination, RNAP dissociates from the DNA and releases the transcript. Termination can be triggered by specific DNA sequences (intrinsic termination) or mediated by protein factors (45). An intrinsic terminator sequence encodes an RNA that can form a stem-loop hairpin structure preceding a U-rich segment, and in some way, this sequence (or rather the terminator structure) destabilizes the TEC. The nucleotide sequences in the hairpin and the U-rich region, as well as the spacing between them, strongly affect the termination efficiency. The exact mechanism for intrinsic termination is still under continued investigation. Several models have been proposed, including extensive forward-tracking of RNAP (50, 73), conformational change of RNAP due to the interaction with the RNA hairpin (62, 69), and melting of the RNA/DNA hybrid without physical movement of RNAP (16, 25). Similar to backtracking pause, termination may be an alternative branch pathway competing with active elongation (34, 66).

The termination phase of transcription is the least examined by single-molecule analysis. Only one study has investigated termination, specifically intrinsic termination (74). Using a TPM method (**Figure 3*a***), this study was aimed at differentiating these models by elucidating the reaction pathway leading to termination. This study showed that RNAP molecules that terminated at the intrinsic terminator paused immediately before dissociation, but those molecules that did not terminate lacked an equivalent pause at the terminator sequence. This study suggested that pausing was a necessary intermediate step before dissociation. However, bulk studies have indicated intrinsic termination as a fast process and the TPM data analysis could be complicated by the potential for released DNA to form an RNAP-DNA binary complex indistinguishable from the pausing complex (24). Also, termination on different terminators could have different properties. More studies on various terminators are needed to generalize the conclusion on the termination pathway.

CONCLUSIONS AND PERSPECTIVES

In the past 25 years, single-molecule techniques have been applied to study different phases of transcription carried out by prokaryotic RNAPs. By monitoring individual transcription complexes in real time, researchers study processes that are highly unsynchronized among molecules in a large population

FRET: fluorescence resonance energy transfer

and address questions relating to population heterogeneity. AFM imaging has provided information about the conformation and conformational changes of single-transcription complexes. Other assays using TPM, optical tweezers, magnetic tweezers, and fluorescence techniques have directly detected kinetic processes including RNAP binding, RNAP diffusion during promoter search, transition between closed-to-open IC, and RNAP translocation along DNA. In particular, branched reaction pathways during elongation, which lead to active elongation or pausing of the TEC, could be differentiated and analyzed separately in single-molecule assays. Furthermore, by stretching and twisting DNA, the effect of DNA supercoiling on the transition between closed-to-open IC, and the force dependence of elongation velocity, could be measured. These measurements are difficult, if not impossible, to realize in bulk assays. These studies have begun to shed light on the mechanism of transcription as well as its regulation.

The techniques discussed here typically require extensive signal averaging or filtering over tenths of a second or seconds in order to achieve near nanometer resolution. So far the translocational motion of RNAP during a single-nucleotide addition cycle has not yet been resolved, and it is not clear whether RNAP oscillates between pre- and posttranslocational states or whether translocation occurs only once per NTP incorporation cycle. Also, the dynamics of the conformational changes of a transcription complex upon the transition from initiation to elongation phase and during each elongation step (38, 58) has not been addressed at the single-molecule level. Fluorescence resonance energy transfer (FRET) is sensitive to nanometer-distance changes at a microsecond timescale and has been applied to study structural change during transcription in bulk experiments (35, 36). Recent research has demonstrated that by using alternating laser excitation (ALEX), single-molecule FRET is capable of accurately measuring the distance between the donor and acceptor in a transcription complex (29). An exciting possibility would be to combine the fluorescence and single-molecule mechanical techniques to correlate the short-distance structural changes with the long-distance movement.

SUMMARY POINTS

1. Single-molecule methods, including TPM, optical tweezers, magnetic tweezers, AFM, and fluorescence techniques (TIRF and FRET), have been applied to study transcription.
2. Single-molecule studies of transcription initiation have directly detected important reaction intermediates and measured reaction rates.
3. Single-molecule studies of elongation and termination have revealed kinetics of both active elongation and pausing.
4. By stretching and twisting DNA, the effects of DNA tension and supercoiling on transcription have been studied.

ACKNOWLEDGMENTS

We thank Drs. A. Shundrovsky and R.M. Fulbright for critical reading of the manuscript. This work is supported by grants from the NIH and the Keck Foundation's Distinguished Young Scholar Award to M.D.W.

LITERATURE CITED

1. Abbondanzieri EA, Shaevitz JW, Block SM. 2005. Picocalorimetry of transcription by RNA polymerase. *Biophys. J.* 89:L61–63
2. **Adelman K, La Porta A, Santangelo TJ, Lis JT, Roberts JW, Wang MD. 2002. Single molecule analysis of RNA polymerase elongation reveals uniform kinetic behavior. *Proc. Natl. Acad. Sci. USA* 99:13538–43**
3. Adelman K, Yuzenkova J, La Porta A, Zenkin N, Lee J, et al. 2004. Molecular mechanism of transcription inhibition by peptide antibiotic microcin J25. *Mol. Cell* 14:753–62
4. Artsimovitch I, Landick R. 2000. Pausing by bacterial RNA polymerase is mediated by mechanistically distinct classes of signals. *Proc. Natl. Acad. Sci. USA* 97:7090–95
5. Bai L, Shundrovsky A, Wang MD. 2004. Sequence-dependent kinetic model for transcription elongation by RNA polymerase. *J. Mol. Biol.* 344:335–49
6. Berg OG, Winter RB, von Hippel PH. 1981. Diffusion-driven mechanisms of protein translocation on nucleic acids. 1. Models and theory. *Biochemistry* 20:6929–48
7. Borukhov S, Sagitov V, Goldfarb A. 1993. Transcript cleavage factors from *E. coli*. *Cell* 72:459–66
8. Bustamante C, Guthold M, Zhu X, Yang G. 1999. Facilitated target location on DNA by individual Escherichia coli RNA polymerase molecules observed with the scanning force microscope operating in liquid. *J. Biol. Chem.* 274:16665–68
9. Cheetham GM, Steitz TA. 2000. Insights into transcription: structure and function of single-subunit DNA-dependent RNA polymerases. *Curr. Opin. Struct. Biol.* 10:117–23
10. Cramer P. 2002. Multisubunit RNA polymerases. *Curr. Opin. Struct. Biol.* 12:89–97
11. Darst SA. 2001. Bacterial RNA polymerase. *Curr. Opin. Struct. Biol.* 11:155–62
12. Davenport RJ, Wuite GJ, Landick R, Bustamante C. 2000. Single-molecule study of transcriptional pausing and arrest by *E. coli* RNA polymerase. *Science* 287:2497–500
13. Erie DA, Yager TD, von Hippel PH. 1992. The single-nucleotide addition cycle in transcription: a biophysical and biochemical perspective. *Annu. Rev. Biophys. Biomol. Struct.* 21:379–415
14. Forde NR, Izhaky D, Woodcock GR, Wuite GJ, Bustamante C. 2002. Using mechanical force to probe the mechanism of pausing and arrest during continuous elongation by Escherichia coli RNA polymerase. *Proc. Natl. Acad. Sci. USA* 99:11682–87
15. Guajardo R, Sousa R. 1997. A model for the mechanism of polymerase translocation. *J. Mol. Biol.* 265:8–19
16. Gusarov I, Nudler E. 1999. The mechanism of intrinsic transcription termination. *Mol. Cell* 3:495–504
17. Guthold M, Bezanilla M, Erie DA, Jenkins B, Hansma HG, Bustamante C. 1994. Following the assembly of RNA polymerase-DNA complexes in aqueous solutions with the scanning force microscope. *Proc. Natl. Acad. Sci. USA* 91:12927–31
18. **Guthold M, Zhu X, Rivetti C, Yang G, Thomson NH, et al. 1999. Direct observation of one-dimensional diffusion and transcription by Escherichia coli RNA polymerase. *Biophys. J.* 77:2284–94**
19. Ha T, Zhuang X, Kim HD, Orr JW, Williamson JR, Chu S. 1999. Ligand-induced conformational changes observed in single RNA molecules. *Proc. Natl. Acad. Sci. USA* 96:9077–82
20. **Harada Y, Funatsu T, Murakami K, Nonoyama Y, Ishihama A, Yanagida T. 1999. Single-molecule imaging of RNA polymerase-DNA interactions in real time. *Biophys. J.* 76:709–15**
21. **Harada Y, Ohara O, Takatsuki A, Itoh H, Shimamoto N, Kinosita K Jr. 2001. Direct observation of DNA rotation during transcription by Escherichia coli RNA polymerase. *Nature* 409:113–15**

2. Detailed single-molecule analysis of active elongation and pausing kinetics of *E. coli* RNAP studied using optical tweezers.

18. AFM observation of *E. coli* RNAP's diffusion during a promoter search.

20. Direct visualization of *E. coli* RNAP binding to, dissociation from, and movement along DNA studied using a combination of TIRF and optical tweezers techniques.

21. Detection of *E. coli* RNAP's rotational motion along DNA during elongation, providing evidence that RNAP is a "groove-tracking" motor.

22. Jülicher F, Bruinsma R. 1998. Motion of RNA polymerase along DNA: a stochastic model. *Biophys. J.* 74:1169–85
23. Kabata H, Kurosawa O, Arai I, Washizu M, Margarson SA, et al. 1993. Visualization of single molecules of RNA polymerase sliding along DNA. *Science* 262:1561–63
24. Kashlev M, Komissarova N. 2002. Transcription termination: primary intermediates and secondary adducts. *J. Biol. Chem.* 277:14501–8
25. Komissarova N, Becker J, Solter S, Kireeva M, Kashlev M. 2002. Shortening of RNA:DNA hybrid in the elongation complex of RNA polymerase is a prerequisite for transcription termination. *Mol. Cell* 10:1151–62
26. Komissarova N, Kashlev M. 1997. RNA polymerase switches between inactivated and activated states by translocating back and forth along the DNA and the RNA. *J. Biol. Chem.* 272:15329–38
27. Komissarova N, Kashlev M. 1997. Transcriptional arrest: Escherichia coli RNA polymerase translocates backward, leaving the 3′ end of the RNA intact and extruded. *Proc. Natl. Acad. Sci. USA* 94:1755–60
28. Landick R. 2004. Active-site dynamics in RNA polymerases. *Cell* 116:351–53
29. Lee NK, Kapanidis AN, Wang Y, Michalet X, Mukhopadhyay J, et al. 2005. Accurate FRET measurements within single diffusing biomolecules using alternating-laser excitation. *Biophys. J.* 88:2939–53
30. Levin JR, Chamberlin MJ. 1987. Mapping and characterization of transcriptional pause sites in the early genetic region of bacteriophage T7. *J. Mol. Biol.* 196:61–84
31. Lis J. 1998. Promoter-associated pausing in promoter architecture and postinitiation transcriptional regulation. *Cold Spring Harbor Symp. Quant. Biol.* 63:347–56
32. Lu HP, Xun L, Xie XS. 1998. Single-molecule enzymatic dynamics. *Science* 282:1877–82
33. McClure WR. 1985. Mechanism and control of transcription initiation in prokaryotes. *Annu. Rev. Biochem.* 54:171–204
34. McDowell JC, Roberts JW, Jin DJ, Gross C. 1994. Determination of intrinsic transcription termination efficiency by RNA polymerase elongation rate. *Science* 266:822–25
35. Mekler V, Kortkhonjia E, Mukhopadhyay J, Knight J, Revyakin A, et al. 2002. Structural organization of bacterial RNA polymerase holoenzyme and the RNA polymerase-promoter open complex. *Cell* 108:599–614
36. Mukhopadhyay J, Kapanidis AN, Mekler V, Kortkhonjia E, Ebright YW, Ebright RH. 2001. Translocation of sigma(70) with RNA polymerase during transcription: fluorescence resonance energy transfer assay for movement relative to DNA. *Cell* 106:453–63
37. Mukhopadhyay J, Sineva E, Knight J, Levy RM, Ebright RH. 2004. Antibacterial peptide microcin J25 inhibits transcription by binding within and obstructing the RNA polymerase secondary channel. *Mol. Cell* 14:739–51
38. Murakami KS, Darst SA. 2003. Bacterial RNA polymerases: the wholo story. *Curr. Opin. Struct. Biol.* 13:31–39
39. Neuman KC, Abbondanzieri EA, Landick R, Gelles J, Block SM. 2003. Ubiquitous transcriptional pausing is independent of RNA polymerase backtracking. *Cell* 115:437–47
40. Nickels BE, Garrity SJ, Mekler V, Minakhin L, Severinov K, et al. 2005. The interaction between sigma70 and the beta-flap of Escherichia coli RNA polymerase inhibits extension of nascent RNA during early elongation. *Proc. Natl. Acad. Sci. USA* 102:4488–93
41. Record MT Jr, Reznikoff W, Craig M, McQuade K, Schlax P. 1996. *Escherichia coli* RNA polymerase ($E\sigma^{70}$), promoters, and the kinetics of the steps of transcription initiation. In *Escherichia coli and Salmonella: Cellular and Molecular Biology*, ed. FC Neidhart, 1:792–820. Washington, DC: Am. Soc. Microbiol. Press

42. Rees WA, Keller RW, Vesenka JP, Yang G, Bustamante C. 1993. Evidence of DNA bending in transcription complexes imaged by scanning force microscopy. *Science* 260:1646–49

43. Revyakin A, Ebright RH, Strick TR. 2004. Promoter unwinding and promoter clearance by RNA polymerase: detection by single-molecule DNA nanomanipulation. *Proc. Natl. Acad. Sci. USA* 101:4776–80

44. Rhodes G, Chamberlin MJ. 1974. Ribonucleic acid chain elongation by Escherichia coli ribonucleic acid polymerase. I. Isolation of ternary complexes and the kinetics of elongation. *J. Biol. Chem.* 249:6675–83

45. Richardson JP, Greenblatt J. 1996. Control of RNA chain elongation and termination. In *Escherichia coli and Salmonella: Cellular and Molecular Biology*, ed. FC Neidhart, 1:822–46. Washington, DC: Am. Soc. Microbiol. Press

46. Rivetti C, Codeluppi S, Dieci G, Bustamante C. 2003. Visualizing RNA extrusion and DNA wrapping in transcription elongation complexes of bacterial and eukaryotic RNA polymerases. *J. Mol. Biol.* 326:1413–26

47. Rivetti C, Guthold M, Bustamante C. 1999. Wrapping of DNA around the E. coli RNA polymerase open promoter complex. *EMBO J.* 18:4464–75

48. Roberts JW, Yarnell W, Bartlett E, Guo J, Marr M, et al. 1998. Antitermination by bacteriophage lambda Q protein. *Cold Spring Harbor Symp. Quant. Biol.* 63:319–25

49. Sakata-Sogawa K, Shimamoto N. 2004. RNA polymerase can track a DNA groove during promoter search. *Proc. Natl. Acad. Sci. USA* 101:14731–35

50. Santangelo TJ, Roberts JW. 2004. Forward translocation is the natural pathway of RNA release at an intrinsic terminator. *Mol. Cell* 14:117–26

51. Schafer DA, Gelles J, Sheetz MP, Landick R. 1991. Transcription by single molecules of RNA polymerase observed by light microscopy. *Nature* 352:444–48

52. Shaevitz JW, Abbondanzieri EA, Landick R, Block SM. 2003. Backtracking by single RNA polymerase molecules observed at near-base-pair resolution. *Nature* 426:684–87

53. Shundrovsky A, Santangelo TJ, Roberts JW, Wang MD. 2004. A single-molecule technique to study sequence-dependent transcription pausing. *Biophys. J.* 87:3945–53

54. Singer P, Wu CW. 1987. Promoter search by Escherichia coli RNA polymerase on a circular DNA template. *J. Biol. Chem.* 262:14178–89

55. Skinner GM, Baumann CG, Quinn DM, Molloy JE, Hoggett JG. 2004. Promoter binding, initiation, and elongation by bacteriophage T7 RNA polymerase. A single-molecule view of the transcription cycle. *J. Biol. Chem.* 279:3239–44

56. Sosunov V, Sosunova E, Mustaev A, Bass I, Nikiforov V, Goldfarb A. 2003. Unified two-metal mechanism of RNA synthesis and degradation by RNA polymerase. *EMBO J.* 22:2234–44

57. Sousa R. 1996. Structural and mechanistic relationships between nucleic acid polymerases. *Trends Biochem. Sci.* 21:186–90

58. Steitz TA. 2004. The structural basis of the transition from initiation to elongation phases of transcription, as well as translocation and strand separation, by T7 RNA polymerase. *Curr. Opin. Struct. Biol.* 14:4–9

59. Temiakov D, Patlan V, Anikin M, McAllister WT, Yokoyama S, Vassylyev DG. 2004. Structural basis for substrate selection by t7 RNA polymerase. *Cell* 116:381–91

60. Thomen P, Lopez PJ, Heslot F. 2005. Unravelling the mechanism of RNA-polymerase forward motion by using mechanical force. *Phys. Rev. Lett.* 94:128102–6

43. Detailed kinetic analysis of the transition between closed to open promoter complex formation of *E. coli* RNAP studied using magnetic tweezers.

52. Direct observation of *E. coli* RNAP backtracking during transcriptional pausing studied using dual optical tweezers.

53. Force-dependent kinetic analysis of a sequence-specific pause of *E. coli* RNAP studied using optical tweezers.

60. Analysis of T7 RNAP elongation kinetics in order to probe its mechanochemical coupling mechanism.

61. Tolic-Norrelykke SF, Engh AM, Landick R, Gelles J. 2004. Diversity in the rates of transcript elongation by single RNA polymerase molecules. *J. Biol. Chem.* 279:3292–99
62. Toulokhonov I, Artsimovitch I, Landick R. 2001. Allosteric control of RNA polymerase by a site that contacts nascent RNA hairpins. *Science* 292:730–33
63. Uptain SM, Kane CM, Chamberlin MJ. 1997. Basic mechanisms of transcript elongation and its regulation. *Annu. Rev. Biochem.* 66:117–72
64. von Hippel PH. 1998. An integrated model of the transcription complex in elongation, termination, and editing. *Science* 281:660–65
65. von Hippel PH, Pasman Z. 2002. Reaction pathways in transcript elongation. *Biophys. Chem.* 101–102:401–23
66. von Hippel PH, Yager TD. 1991. Transcript elongation and termination are competitive kinetic processes. *Proc. Natl. Acad. Sci. USA* 88:2307–11
67. Wang HY, Elston T, Mogilner A, Oster G. 1998. Force generation in RNA polymerase. *Biophys. J.* 74:1186–202
68. Wang MD, Schnitzer MJ, Yin H, Landick R, Gelles J, Block SM. 1998. Force and velocity measured for single molecules of RNA polymerase. *Science* 282:902–7
69. Wilson KS, von Hippel PH. 1995. Transcription termination at intrinsic terminators: the role of the RNA hairpin. *Proc. Natl. Acad. Sci. USA* 92:8793–97
70. Wuite GJ, Smith SB, Young M, Keller D, Bustamante C. 2000. Single-molecule studies of the effect of template tension on T7 DNA polymerase activity. *Nature* 404:103–6
71. Xue Q, Yeung ES. 1995. Differences in the chemical reactivity of individual molecules of an enzyme. *Nature* 373:681–83
72. Yager TD, von Hippel PH. 1991. A thermodynamic analysis of RNA transcript elongation and termination in *Escherichia coli*. *Biochemistry* 30:1097–118
73. Yarnell WS, Roberts JW. 1999. Mechanism of intrinsic transcription termination and antitermination. *Science* 284:611–15
74. Yin H, Artsimovitch I, Landick R, Gelles J. 1999. Nonequilibrium mechanism of transcription termination from observations of single RNA polymerase molecules. *Proc. Natl. Acad. Sci. USA* 96:13124–29
75. Yin H, Landick R, Gelles J. 1994. Tethered particle motion method for studying transcript elongation by a single RNA polymerase molecule. *Biophys. J.* 67:2468–78
76. Yin H, Wang MD, Svoboda K, Landick R, Block SM, Gelles J. 1995. Transcription against an applied force. *Science* 270:1653–57
77. Yin YW, Steitz TA. 2004. The structural mechanism of translocation and helicase activity in T7 RNA polymerase. *Cell* 116:393–404

Quantitative Fluorescent Speckle Microscopy of Cytoskeleton Dynamics

Gaudenz Danuser and Clare M. Waterman-Storer

Department of Cell Biology, The Scripps Research Institute, La Jolla, California 92037; email: gdanuser@scripps.edu, waterman@scripps.edu

Annu. Rev. Biophys. Biomol. Struct. 2006. 35:361–87

The *Annual Review of Biophysics and Biomolecular Structure* is online at biophys.annualreviews.org

doi: 10.1146/annurev.biophys.35.040405.102114

1056-8700/06/0609-0361$20.00

Key Words

computer vision, particle tracking, actin, microtubule, focal adhesion

Abstract

Fluorescent speckle microscopy (FSM) is a technology used to analyze the dynamics of macromolecular assemblies in vivo and in vitro. Speckle formation by random association of fluorophores with a macromolecular structure was originally discovered for microtubules. Since then FSM has been expanded to study other cytoskeleton and cytoskeleton-binding proteins. Specialized software has been developed to convert the stochastic speckle image signal into spatiotemporal maps of polymer transport and turnover in living cells. These maps serve as a unique quantitative readout of the dynamic steady state of the cytoskeleton and its responses to molecular and genetic interventions, allowing a systematic study of the mechanisms of cytoskeleton regulation and its effect on cell function. Here, we explain the principles of FSM imaging and signal analysis, outline the biological questions and corresponding methodological advances that have led to the current state of FSM, and give a glimpse of new FSM modalities under development.

Contents

INTRODUCTION

Fluorescent speckle microscopy (FSM): imaging of speckle signals using the different modalities of fluorescence light microscopy

Fluorescent speckle microscopy (FSM) is a method used to analyze the movement and assembly/disassembly dynamics of macromolecular structures in vivo and in vitro (58). As reviewed in References 10 and 57, FSM is derived from the principles of fluorescent analog cytochemistry. There, purified protein is covalently linked to a fluorophore and microinjected or expressed as a GFP fusion in living cells. Fluorescent protein that incorporates into cellular structures is then visualized by epifluorescence light microscopy (40, 54). This classic approach has yielded much information about protein localization and the dynamics of macromolecular assemblies in cells but has been limited in its ability to report protein dynamics because of inherently high background fluorescence from unincorporated and out-of-focus incorporated fluorescent subunits. In addition it is often impossible to detect movement or turnover of subunits within assemblies because of the uniform fluorescent labeling of structures. These problems have been alleviated by the use of laser photobleaching and photoactivation of fluorescence to mark structures in limited cell areas and measure the movement and turnover of subunits in the marked region at steady state (25, 31, 46, 52, 53, 66). Similar to these techniques is the ratiometric method of fluorescence localization after photobleaching (FLAP) (12, 70). FSM provides the same information as these photomarking techniques. In addition, it delivers simultaneous kinetic data in large areas of the cell, offering the capability to detect nonsteady-state molecular dynamics within assemblies at high spatial and temporal resolution. FSM also reduces out-of-focus fluorescence in images and improves the visibility of fluorescently labeled structures and their dynamics in three-dimensional polymer arrays such as the mitotic spindle (27, 28, 59).

In its initial development, FSM utilized wide-field epifluorescence light microscopy and digital imaging with a sensitive, low-noise cooled charge-coupled-device (CCD) camera and was applied to the study of assembly dynamics and movement of microtubules (60). Since then FSM has been transferred to confocal and total internal reflection fluorescence microscopes (1, 2, 16, 27) and has been applied to new biological problems in vivo and in vitro. A critical step in advancing FSM to become a routine method for measuring cytoskeleton flow and turnover was the development of fully automated computer-based tracking and statistical analysis of speckle dynamics. Recently, these models for speckle image analysis have been expanded to infer the transient coupling between two protein assemblies and to probe viscoelastic properties of polymer networks.

In this paper, we review the principles of speckle formation and describe the imaging and analytical requirements for FSM to become a quantitative technique. We heavily rely on FSM data of actin cytoskeleton dynamics in epithelial cell migration, where we believe the integration of FSM imaging and computational image analysis has been brought to the most sophisticated level. The value of FSM as a general technique for the analysis of macromolecular assemblies is documented in a comprehensive table (Table S1; follow the Supplemental Material link from the Annual Reviews home page at **http://www.annualreviews.org**). The paper concludes with an outlook on our newest developments of FSM for the measurement of cytoskeleton-adhesion coupling and of material properties of actin filament networks.

MAP: microtubule-associated protein

Fluorescent speckle: random, diffraction-limited intensity peak in the image signal significantly brighter than the neighboring signal

PRINCIPLES OF SPECKLE IMAGE FORMATION

Stochastic Association of Fluorophores with Microtubules

FSM was discovered by accident when it was noticed in high-resolution images of cells injected with X-rhodamine tubulin that some microtubules exhibited variations in fluorescence intensity along their lattices, i.e., they looked speckled (**Figure 1*a*,*b***) (60). There were several possible interpretations of such images: (*a*) the fluorescent tubulin could preferentially associate with itself, forming bright oligomers or aggregates on the microtubule; (*b*) cellular factors such as organelles or microtubule-associated proteins (MAPs) could be bound to the microtubule and "mask" or quench the fluorescence of some regions; or (*c*) variations in the number of fluorescent tubulin subunits in each resolution-limited image region along the microtubule could occur as the microtubule assembled from a pool of labeled and unlabeled dimers (61).

We discounted the hypothesis that speckles were generated by fluorescent aggregates by showing that labeled tubulin dimers sediment similarly to unlabeled purified dimers in an analytical ultracentrifugation assay. Next, we showed that fluorescent speckle patterns

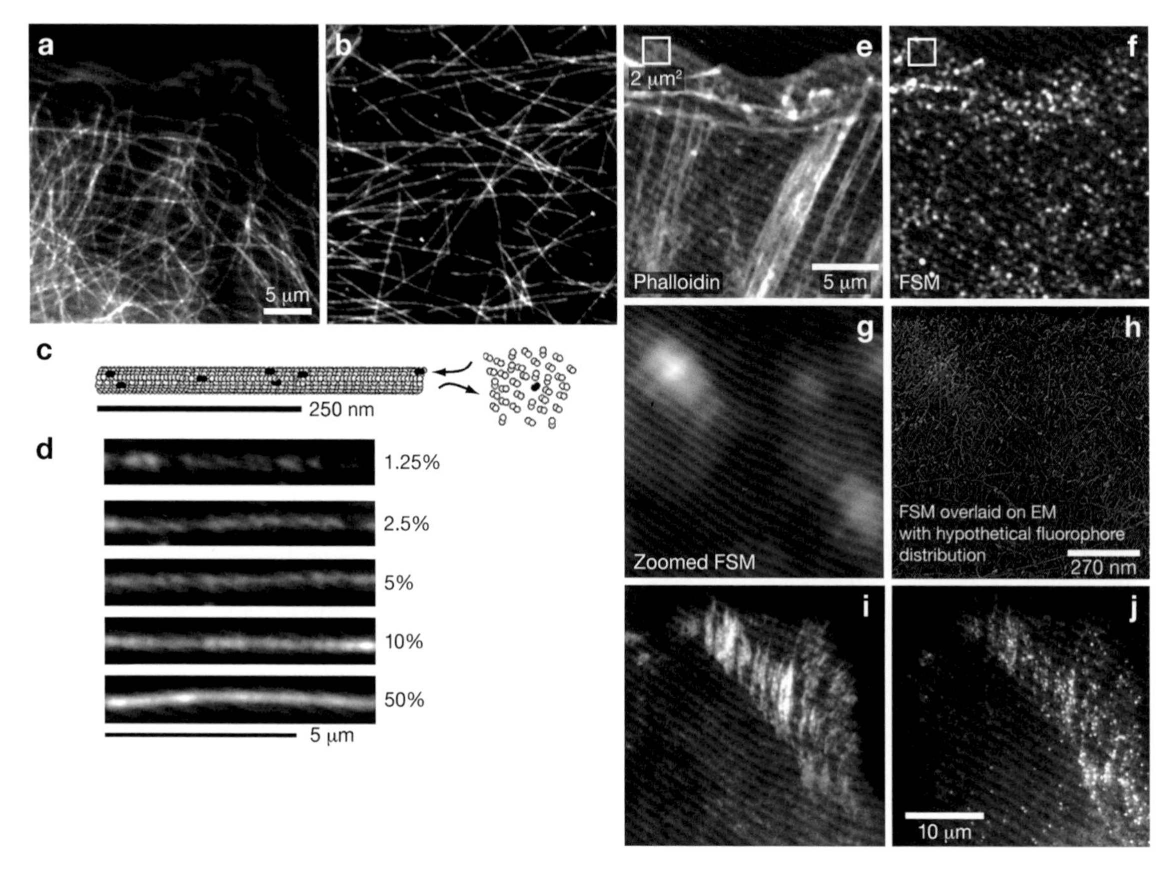

Figure 1

Speckle formation in microtubule and actin cytoskeletal polymers and focal adhesion (FAs). (*a*, *b*) Comparison of random speckle pattern of fluorescence along microtubules for (*a*) a living epithelial cell microinjected with X-rhodamine-labeled tubulin and for (*b*) microtubules assembled in vitro from 5% X-rhodamine-labeled tubulin. Scale bar: 5 μm. (*c*) Model for fluorescent speckle pattern formation in a microtubule grown from a tubulin pool containing a small fraction of labeled dimers. (*d*) Dependence of speckle contrast on the fraction of labeled tubulin dimers. (*e*, *f*) Speckle formation in actin filament networks. An epithelial cell was microinjected with a low level of X-rhodamine-labeled actin, fixed, and stained with Alexa-488 phalloidin. (*e*) Phalloidin image showing the organization of actin filaments in amorphous filament networks and bundles. (*f*) In the single FSM image much of the structural information is lost, but time-lapse FSM series contain dynamic information of filament transport and turnover not accessible with higher-level labeling of the cytoskeleton. (*g*) Close-up of 2 μm × 2 μm window in panels *e* and *f*. (*h*) Colorized speckle signal overlaid onto a quick-freeze deep etch image of the same-sized region of the actin cytoskeleton in the leading edge of a fibroblast (kindly provided by Tatyana Svitkina) with hypothetical fluorophore distribution that could give rise to such speckle pattern. This indicates the scale of FSM compared with ultrastructure of the polymer network and illustrates that a small proportion of the total actin fluoresces and that fluorophores from different filaments contribute to the same speckle. (*i*, *j*) Low-level expression of GFP-tagged FA protein vinculin results in speckled FAs in TIRF images. A cell expressing GFP-vinculin was fixed and immunofluorescently stained with antibodies to (*i*) vinculin to reveal the position of FAs, which in the (*j*) GFP channel appear speckled because of the low level of incorporation of GFP-vinculin.

on microtubules assembled from purified tubulin in vitro were similar to those of microtubules in cells where MAPs and organelles are present (61). Thus, the most plausible explanation for speckle formation in microtubules was that variations existed in the number of fluorescent tubulin subunits in each resolution-limited image region along the microtubule.

To understand how speckles originated, we considered how the images of fluorescent microtubules were formed by the microscope (**Figure 1*c***). Microtubules assemble from α/β tubulin dimers into the 25-nm-diameter cylindrical wall such that there are 1625 dimers per micron (11). The final image results from a convolution of the fluorophore distribution along the microtubule with the point-spread function (PSF) of the microscope. For a two-dimensional treatment of the situation, the in-focus slice of the PSF is given by the Airy disk, for which the radius of the first ring with zero intensity amounts to $r = 0.61\ \lambda\ /\mathrm{NA}$ (20). The parameters λ and NA denote the emission wavelength of the fluorophore and the numerical aperture of the objective lens, respectively. For X-rhodamine-labeled microtubules (620-nm emission), the radius of the Airy disk is 270 nm (NA = 1.4), which corresponds to 440 tubulin dimers. A given fraction of fluorescent dimers, f, produces a mean number of fluorescent dimers $n = 440 \times f$ per PSF. The speckle pattern along the microtubule is produced by variations in the number of fluorescent dimers per PSF relative to this mean. Thus, the contrast of the speckle pattern can be approximated by the ratio of the standard deviation and the mean of a binomial distribution with n elements: $c = \sqrt{n \cdot f \cdot (1 - f)}/(n \cdot f)$. This formula suggests that the contrast c increases with decreasing f and decreases with increasing n, indicating the requirement for optics with the highest NA possible. This behavior is illustrated in **Figure 1*d***. Microtubules assembled from pure tubulin containing 1.25% to 5% labeled dimers exhibit speckles of varying intensity along their length. Microtubules assembled from 10% to 50% labeled tubulin are evenly labeled.

In theory, optimal contrast is obtained where speckles are formed by a single fluorophore per Airy disk (55, 62). For a microtubule, 80% or more speckles arise from a single fluorophore when f is less than 0.1% (10). In practice, however, the lower bound for f is determined by two considerations: (*a*) Technologically, the noise level and sensitivity of the imaging system, instability of the microscope, and the dynamics of the observed process may all deteriorate feasibility of single fluorophore detection. (*b*) Experimentally, too-low fractions result in a very low density of speckles, i.e., in low spatiotemporal sampling of underlying dynamics. We have found by both theory and practice that it is often desirable to image speckles at fractions in the range of 0.5% to 2%. In this range speckles consist of three to eight fluorophores (10).

PSF: point spread function of the microscope

F-actin: filamentous (polymeric) actin

Stochastic Association of Fluorophores in Other Systems: The Platform Model for Speckle Formation

When cells are injected with low levels of fluorescently labeled actin and imaged at high resolution, actin-rich structures appear relatively evenly speckled and do not indicate the architectural organization of the cytoskeleton (**Figure 1*e*,*f***) (41, 51, 55, 58, 63). Similar images can also be obtained by expressing GFP-fused actin at a very low level (23, 55) or by injecting trace amounts of the labeled actin-binding molecule phalloidin (47, 69). In contrast to speckle formation in isolated microtubules, labeled actin subunits associate with a highly cross-linked three-dimensional network of actin filaments (F-actin) (34, 44, 45). The mesh size of an F-actin network in living cells is nearly always below the resolution limit of the light microscope (**Figure 1*g*,*h***). Consequently, unless f is kept extremely low so that only one fluorophore falls into the PSF volume (55), fluorescent speckles arise most likely from subunits distributed across several filaments.

FA: focal adhesion

The same concept of speckle formation has been exploited to visualize molecules making up focal adhesions (FAs) (1). GFP-fusions to FA proteins, including vinculin, talin, paxilin, α-actinin, zyxin, or $\alpha_v$integrin (15), have been expressed in epithelial cells from crippled promoters to achieve very low levels. Labeled proteins assembling with endogenous, unlabeled proteins give FAs a speckled appearance (**Figure 1*i,j***). As with actin networks, speckles represent randomly distributed fluorescent FA proteins that are temporarily clustered in the FA complex within the three-dimensional volume of one PSF.

A speckle is thus defined as a diffraction-limited image region that is significantly higher in fluorophore concentration (reported by fluorescence intensity) than its neighboring diffraction-limited image regions. In addition, for a speckle signal to be detected, the contributing fluorescent molecules must be immobilized within the PSF volume for the 0.1 to 2 s exposure time required by most digital cameras to acquire the dim FSM image. In contrast, unbound, diffusible fluorescent molecules yield an evenly distributed background signal of significantly lower magnitude across the many pixels that the molecules visit during the exposure. In a high-resolution (~250 nm), high-magnification (100 ×) image, actin monomers move ~60 pixels s^{-1} (26). The concept of fluorophore immobilization leading to speckle formation was nicely demonstrated by Watanabe & Mitchison (55). Diffusible GFP expressed in cells at very low levels produced an even distribution of fluorescence, whereas a similar level of GFP-conjugated actin that incorporated into the cytoskeleton produced a speckled image. The same idea was pursued by Bulinski et al. (7) and by Kapoor & Mitchison (24). GFP chimera of the MAP ensconsin (7) and fluorescently tagged tetrameric kinesin motor Eg5 (24) were used to observe the transient binding of these molecules to microtubules in vitro and in vivo.

In summary, speckle signal formation occurs when one or a few fluorescently labeled molecules within a PSF volume associate with a molecular scaffold where no other fluorescent molecules are located. We refer to this scaffold as the speckle platform. Association with the platform occurs when labeled subunits completely incorporate into the platform, as with tubulin or actin, or when they bind to it, as in the case of cytoskeleton-binding proteins or FA molecules. The duration of association must be equal to or longer than the camera exposure of one frame in a time-lapse image sequence.

Naïve Interpretation of Speckle Dynamics

Following the platform model, one would expect the appearance of a speckle to correspond to the local association of subunits with the platform. Conversely, the disappearance of a speckle would mark the local dissociation of subunits. In other words, FSM allows, in principle, the direct kinetic measurement of subunit turnover in space and time via speckle lifetime analysis. In addition, once a speckle is formed, it may undergo motion that indicates the coordinated movement of labeled subunits on the platform and/or the movement of the platform itself.

Computational Models of Speckle Dynamics

The interpretation of speckle dynamics becomes significantly more complicated when individual speckles arise from fluorophores distributed over multiple polymers. To examine how speckle appearance and disappearance relate to the rates of assembly and disassembly of F-actin, we performed Monte Carlo simulations of fluorophore incorporation into growing and shrinking filaments in dense and branched networks and generated synthetic FSM time-lapse sequences (39). The first lesson we learned from this modeling was that the speckle density is independent of whether the network assembles or disassembles; it depends only on how many Airy disks can be resolved per unit area, e.g., 1 μm^2. With NA = 1.4/100 × optics, this amounts to ~4 (approximately 2 × 2), as confirmed by **Figure 2*a***. The graph displays the mean speckle density from five simulations of a network that starts with no fluorophores, assembles for 120 s (inset: mean fluorescence intensity increases), and disassembles for 360 s (inset: mean fluorescence intensity decreases) at equal rates. The density does not change after saturation at 100 s and remains constant despite further addition of fluorophores for another 20 s. This suggests that monomer association can cause an equal number of speckle appearances and disappearances. The same holds in the opposite sense during network disassembly.

Whereas the NA of the optics defines the maximum number of resolvable speckles per unit area, the labeling ratio influences the speckle density indirectly. For multi-fluorophore speckles the ratio f is the main determinant of speckle contrast. When increasing the ratio the speckle density drops because the difference between the peak intensity of a speckle and its surroundings is no longer distinguishable from intensity fluctuations due to noise. Similarly, at labeling ratios where speckles represent the image of single fluorophores ($f < 0.1\%$), a further decrease of f reduces the speckle density proportionally. Across the optimal range of ratios for multi-fluorophore speckles ($0.5\% < f < 3\%$) the density is almost constant. These model predictions were largely confirmed experimentally by Adams et al. (1).

The reason for the constant speckle density in the range of $0.5\% < f < 3\%$ is illustrated in **Figure 2*b***. A speckle is defined as a local image intensity maximum significantly above the surrounding background. The critical intensity difference ΔI_{crit} depends on both the camera noise and the shot noise. The shot noise is by itself a function of the speckle intensity (39). Speckles may appear (speckle birth) for two reasons: the intensity of a local maximum gets brighter because of the association of fluorescent subunits, or the intensity of the surrounding background gets dimmer because of subunit dissociation in the neighborhood. In both cases a speckle birth is detected when the peak-to-background intensity difference exceeds ΔI_{crit}. Analogously, speckles may disappear (speckle death) either because of subunit dissociation in the location of a speckle or because of subunit association in the neighborhood.

Multi-fluorophore speckle: fluorescent speckle with contributions from several fluorophores

Statistical Analysis of Speckle Dynamics

With the classification scheme in **Figure 2*b***, speckles become time-specific, diffraction-limited probes of polymer network turnover. The change in foreground or background intensity that causes the birth or death of a speckle is, on average, proportional to the net number of subunits Δm added to or removed from the PSF volume between two frames. This defined an algorithm for the local measurement of network assembly or disassembly kinetics (39): (*a*) calculation of changes in foreground and background intensities. After detection of a speckle birth/death event, regression lines are fitted to the foreground and background intensities for one time point before, at, and after the event (**Figure 2*c***). Intensity values before birth and after death are extrapolated (39). The line fits provide two estimates a_f and a_b of the slopes of foreground and background intensity variation. They also yield the standard deviations $\sigma(a_f)$ and $\sigma(a_b)$ of the slopes, which are derived from the residuals of the intensity values to the regression line. Noisy data, poorly represented by the regression model, generate large values for σ; intensity values in perfect match with the model result in small values for σ. (*b*) Each of the two slopes is tested for statistical significance. Insignificant intensity changes are discarded. (*c*) If both foreground and background slopes are significant, the one with the higher significance (lower p-value) is selected as the cause of the event. In the example given in

Figure 2c the foreground slope has the higher significance. The magnitude of the more significant slope is recorded as the score of the birth/death event. If neither foreground nor background slope is statistically significant, no score is generated.

Score values represent instantiations of a random variable with an expectation value $\mu = \alpha \Delta m \cdot f$ and variance $\sigma^2 = \alpha^2 \Delta m^2 \cdot f(1 - f)$, where α denotes the unknown intensity of one fluorophore. In addition, the scores are perturbed by noise. However, assuming that the net rate Δm remains constant for a small probing window, the intrinsic score variation and noise are approximately eliminated by averaging all scores falling into the window. The choice of the window size depends on the density of significant scores and the user's preference for spatial or temporal resolution. The larger the number of scores averaged by time integration, the less spatial averaging is required, and vice versa.

Figure 2*d* displays rates of actin assembly (red) and disassembly (green) of the actin

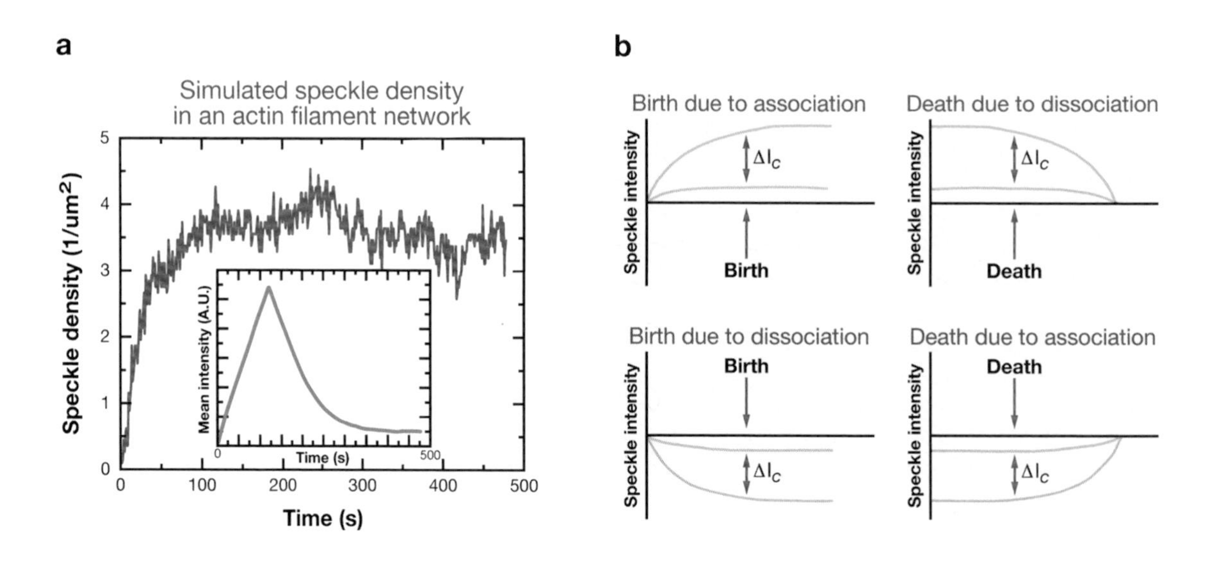

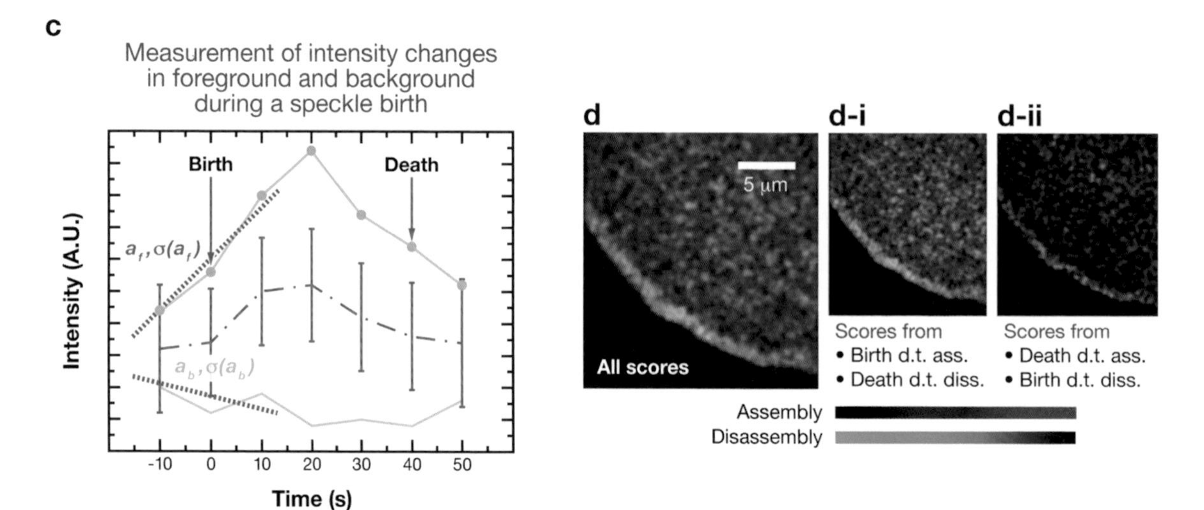

network at the edge of an epithelial cell. Here, score values were averaged over 10 min, reflecting the steady-state turnover. Animated maps resolved at 10-s intervals are provided as online supplementary material (movies M1 and M2; follow the Supplemental Material link from the Annual Reviews home page at **http://www.annualreviews.org**). The two smaller panels indicate the rate distributions calculated from scores extracted from speckle births due to monomer association and from speckle deaths due to monomer dissociation only (**Figure 2*d*-i**), and from births due to monomer dissociation and from deaths due to monomer association only (**Figure 2*d*-ii**). Both panels display the same distribution of loci of strong assembly (for example, the cell edge) and disassembly but at different event densities. **Figure 2*d*-i** corresponds to the naïve interpretation of speckle appearance and disappearance. These events contribute ~70% of all scores. The other ~30% of significant scores is related to the counterintuitive cases of speckle birth and death. Neglecting them would significantly reduce the sample size. How many intuitive versus counterintuitive cases occur depends on the fraction of labeled monomers. The lower the fraction, the fewer counterintuitive cases observed, with a lower boundary defined by the single-fluorophore speckle regime, in which all speckle appearances are due to monomer association and all disappearances are due to monomer dissociation.

Single-fluorophore speckle: fluorescent speckle that consists of an isolated fluorophore generating a detectable signal above background according to the definition of a fluorescent speckle

The processing of only short time intervals around speckle births and deaths focuses the analysis on image events that are more likely to have originated from monomer exchange rather than from intensity fluctuations due to image noise, bleaching, and in- and out-of-focus speckle motion. In addition, the algorithm rejects ~60% of all speckle birth and death events as insignificant (36), i.e., these events are not classifiable as induced by monomer exchange with the certainty the user chooses as the confidence level for the analysis. Bleaching affects all speckle scores and thus can be corrected on the basis of global drifts in the image signal (39). We also showed that, with a NA = 1.4 objective lens, focus drifts smaller than 100 nm over three frames (e.g., 30 nm per 1–5 s) have no effect on the mapping of network turnover. Thus, the statistical model described in this section provides a robust method for calculating spatiotemporal maps of assembly and

Figure 2

Relationship between speckle appearance (birth) and disappearance (death) and the turnover in the underlying macromolecular assembly. (*a*) Simulated speckle density in an actin filament network assembling for 120 s and disassembling for 360 s. Inset: Mean intensity indicating the overall change in bound fluorophore over time. (*b*) Classification of speckle birth and death due to monomer association and dissociation with the network. A speckle appears when the difference between foreground (*pink line*) and background (*light blue line*) is greater than a threshold ΔI_C, which is a function of the camera noise and the shot noise of the signal (39). (*c*) Measurement of intensity changes in foreground (*solid pink line*) and background (*solid light blue line*) during a speckle birth. The entire lifetime of the speckle is shown (40 s). Dash-dotted gray line: Mean between foreground and background; error bars: ΔI_C computed in every time point. Birth and death are defined as the time points at which the intensity difference exceeds ΔI_C for the first and the last time. Red line: Regression line to the foreground intensity values before, at, and after birth. Blue dotted line: Regression line to the background intensity values. The cause of speckle birth is inferred by statistical classification of the two slopes and their standard deviations (see text). The statistically more dominant of the two slopes, if also significant relative to image noise, defines the score of the event (foreground slope in the example given). (*d*) Averaging of scores accumulated over a defined time window of a FSM time-lapse sequence yields maps of net polymerization (*red*) and depolymerization (*green*). Scores from birth due to association and death due to dissociation (*d*-i) or from birth due to dissociation and death due to association (*d*-ii) reveal the same distribution of polymerization and depolymerization. Figure in parts reproduced from (39) and from (37) with permission of *Biophysical Journal* and *Science*.

disassembly of macromolecular structures such as F-actin networks.

TIRFM: total internal reflection fluorescence microscopy

Single-Fluorophore Versus Multi-Fluorophore Speckles

It seems that, in many aspects, FSM would be most powerful if implemented as a single-molecule imaging method, where speckle appearances and disappearances unambiguously signal association and dissociation of fluorescent subunits to the platform (55). However, the much simpler signal analysis is counterweighed by several disadvantages not encountered when using multi-fluorophore speckles. First, establishing that an image contains only single-fluorophore speckles can be challenging, especially when the signal of one fluorophore is close to the noise floor of the imaging system. Our experience with single-fluorophore speckles suggests that, particularly in three-dimensional structures, a large fraction of speckles has residual contributions of at least one other fluorophore. Those mixtures need to be deconvolved or eliminated from the statistics. Second, the imaging of single-fluorophore speckles is much more demanding than multi-fluorophore FSM and requires longer camera exposures to capture the very dim signals, reducing the temporal resolution. Third, in addition to the substantially lower temporal resolution, single-fluorophore FSM offers lower spatial resolution because the density of speckles drops significantly with the extremely low labeling ratio. Fourth, multi-fluorophore speckles distinguish between fast and slow turnover, whereas single-fluorophore speckles deliver on/off information only. To measure rates, single-fluorophore speckle analysis must also rely on spatial and temporal averaging, which further decreases the resolution. Watanabe and Mitchison (55) used single-fluorophore speckle analysis to probe actin network turnover in migrating cells with retrograde flow, where most speckles form at the cell edge, move, and then disappear at a distinct site. It remains an open question whether single-fluorophore speckles can characterize the dynamic equilibrium of a spatially stationary or slowly moving network where polymerization and depolymerization coexist over distances less than 1 μm, as achieved by multi-fluorophore speckle analysis (38, 39). It is also unclear whether single-fluorophore speckle analysis has sufficient temporal resolution to extract the short-term components of the dynamic equilibrium.

REQUIREMENTS FOR SPECKLE IMAGING

Time-lapse FSM requires imaging high-resolution diffraction-limited regions containing 1 to 10 fluorophores and inhibiting fluorescence photobleaching. This requires a sensitive imaging system with little extraneous background fluorescence, efficient light collection, a camera with low noise, high quantum efficiency, high dynamic range, high resolution, and suppression of fluorescence photobleaching with illumination shutters and/or oxygen scavengers (30, 58, 64). In addition, all fluorescently labeled molecules must be functionally competent to bind their platform; otherwise they will contribute to diffusible background and obfuscate the speckle contrast (56). We refer the reader to Gupton & Waterman-Storer (17) for a recent in-depth discussion of the hardware requirements for obtaining FSM images.

Because FSM is achieved by the level of fluorescent protein in the sample, it is adaptable to various modes of high-resolution fluorescence microscopy, such that the specific advantages of each mode can be exploited in combination with the quantitative capabilities of FSM. For example, we have performed FSM on both spinning-disk confocal microscope (2) and total internal reflection fluorescence microscope (TIRFM) systems (1) to gain speckle data in two spectral channels with the specific image advantages of confocal and TIRFM. However, to date FSM has proved incompatible with all commercial laser-scanning confocal microscope

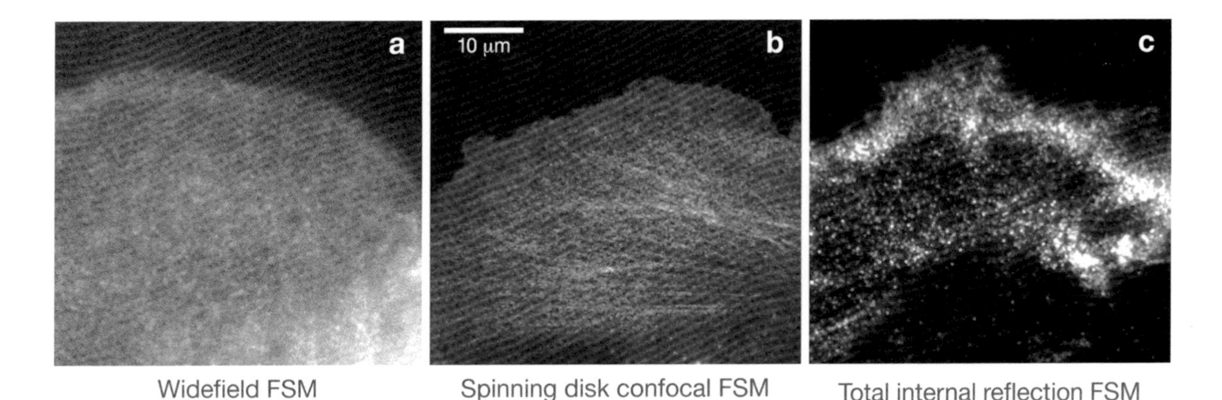

Figure 3

Comparison of X-rhodamine actin FSM images of the edge of migrating Ptk1 epithelial cells using (*a*) wide-field epifluorescence, (*b*) spinning-disk confocal microscopy, and (*c*) total internal reflection fluorescence microscopy. Panels *a* and *b* were acquired using a Nikon 100× 1.4 NA Plan Apo phase contrast objective lens and a 14 bit Hamamatsu Orca II camera with 6.7 micron pixels. Panel *c* was acquired with a Nikon 100× 1.45 NA Plan Apo TIRF objective lens and a 14 bit Hamamatsu Orca II ER with 6.4 micron pixels. Note that speckle contrast and the ability to detect speckles in more central cell regions increases from panels *a* to *c*. Note, however, in the TIRF image that speckles are very bright a few microns back from the edge, likely where the cell is in closer contact with the substrate.

systems available. This is because these instruments use photomultipliers as detectors that are noisy and have a limited dynamic range compared with the low-noise, high dynamic range CCDs used with spinning-disk confocal microscope systems. A comparison of FSM images of the actin cytoskeleton in migrating epithelial cells acquired by wide-field epifluorescence, spinning-disk confocal microscope, and TIRFM is shown in **Figure 3**. Clearly, speckle contrast is improved by reducing out-of-focus fluorescence with either of the last techniques. Contrast in TIRFM images is further improved over the spinning-disk confocal image because the evanescent field excitation depth is reduced to ~50 nm into the specimen. We quantified the effect of the reduced effective imaging volume on modulation and detectability of actin and FA speckles in wide-field epifluorescence and TIRFM. Our analysis showed that TIR-FSM indeed affords major improvements in these parameters over wide-field epifluorescence for imaging macromolecular assemblies at the ventral surface of living cells, both in thin peripheral and thick central cell regions (1).

ANALYSIS OF SPECKLE MOVEMENTS

Tracking Speckle Flow: Early and Recent Developments

In addition to revealing the kinetics of association and dissociation of subunits within a molecular platform, speckles show the movement of the platform. This is evident from raw FSM movies provided as online supplementary material (movies M3 and M4; follow the Supplemental Material link from the Annual Reviews home page at **http://www.annualreviews.org**). The first example shows a speckled actin network, where speckle motion indicates the retrograde flow of the cytoskeleton polymer away from the cell edge. The second example displays speckled microtubules in a meiotic spindle from a *Xenopus laevis* extract. Here, speckles indicate antiparallel, poleward flux of tubulin subunits in the interdigitating microtubule scaffolds of the two half-spindles.

In early applications of FSM, speckle motion was quantified by hand tracking a few speckles, a tedious, error-prone, and

TIR-FSM: total internal reflection fluorescent speckle microscopy

incomplete way of analyzing the wealth of information contained by these images (41, 43, 55). Alternatively, kymographs provided average estimates of speckle velocities (6, 13, 19, 24, 27, 28, 58, 63, 65).

Initial attempts to automate the extraction of more complete speckle flow maps from FSM time-lapse sequences of actin networks relied on correlation-based tracking. The speckled area of a source frame in the movie was divided into small probing windows. Each window was displaced until the normalized cross-correlation of the window with the signal of the target frame, i.e., the next frame in the movie, was maximized. This approach reported the average motion of all the speckles falling into the window. The window size pitted robustness in correlation against spatial resolution. The larger the window, the more unique was the speckle pattern to be recognized in the target frame. On the other hand, larger windows increased the averaging of distinct speckle motions within the window.

Underlying the method of cross-correlation tracking is the assumption that the signal of a probing window, although translocated in space, does not change between source and target frame. In practice, this assumption is always violated because of noise. But, the cross-correlation of two image signals appears to be tolerant toward spatially uncorrelated noise, making it a prime objective function in computer vision tracking (21, 29, 67). The many speckle appearances and disappearances in F-actin networks, however, introduce signal perturbations that cannot be tackled by the cross-correlation function (49). Instead, we developed a particle flow method, in which each speckle was probed separately (49). Speckles were linked between frames by nearest-neighbor assignment in a distance graph, in which conflicts between source speckles competing for the same target speckle were resolved by global optimization (3). Extension of the graph to three frames allowed speckle assignments to be constrained in smooth and unidirectional trajectories, so that speckles moving in antiparallel flow fields could be tracked (49).

Surprisingly, cross-correlation-based tracking was successful in measuring average tubulin flux in mitotic spindles (32). Simulated time-lapse sequences demonstrated that if a significant subpopulation of speckles in the probing window moves jointly, the coherent component of the flow can be estimated even when the rest of the speckles move randomly or, as in the case of the mitotic spindle, a smaller population moves coherently in opposite direction. However, the tracking result will be ambiguous if the window contains multiple, coherently moving speckle subpopulations of equal size. Miyamoto et al. (32) carefully chose windows in the central region of a half-spindle, where the motion of speckles toward the nearer of the two poles dominated speckle motion in the opposite direction and random components. The approach was aided further by several features of the spindle system: Tubulin flux in a spindle is quasi-stationary; speckle appearances and disappearances are concentrated at the spindle midzone and in the pole regions, both of which were excluded from the probing window; and the flow fields were approximately parallel inside the probing window.

Encouraged by these results, we returned to cross-correlation tracking of speckle flow also in F-actin networks (22). The advantage of cross-correlation tracking over particle flow tracking is that there is no requirement to detect the same speckle in at least two consecutive frames. Hence, speckle flows can be tracked in movies with high noise levels and weak speckle contrast (22). To avoid trading correlation stability for spatial resolution, we capitalized on the fact that actin cytoskeleton transport is often stationary on the timescale of minutes. Thus, although the correlation of a single pair of probing windows in source and target frames is ambiguous (**Figure 4*a*-i**), rendering the tracking of speckle flow impossible (**Figure 4*b*-i**), time integration of the correlation function over multiple frame

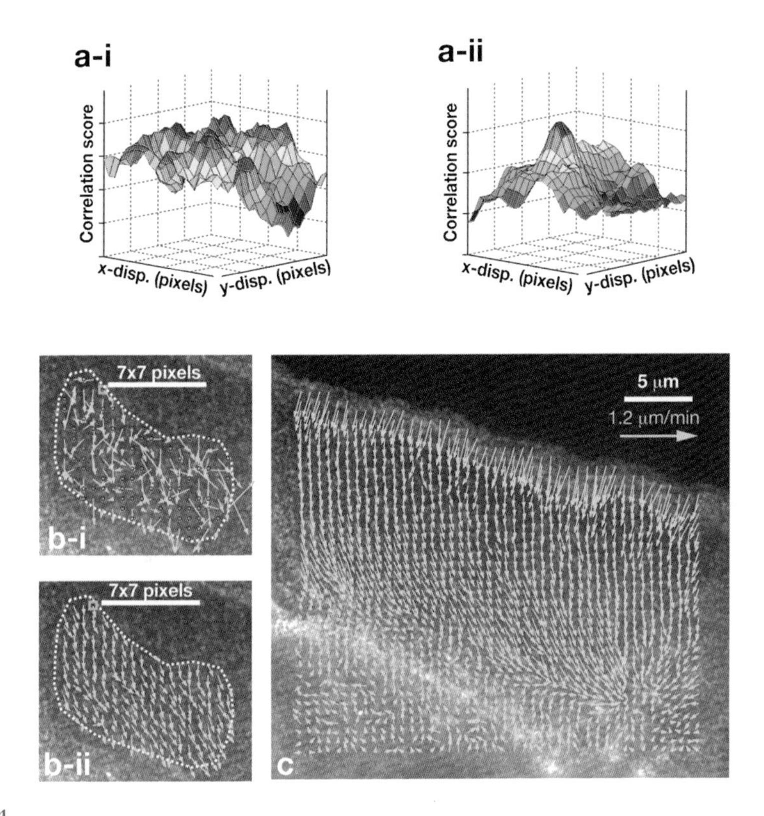

Figure 4

Tracking quasi-stationary speckle flow using multi-frame correlation. (*a*) Cross-correlation for a single frame pair (*a*-i) and for 20 frame pairs integrated (*a*-ii). (*b*) Region of a speckled actin network tracked with a probing window of 7 × 7 pixels (400 × 400 nm) using a single frame pair (*b*-i) and 20 frame pairs (*b*-ii). (*c*) Speckle flow map extracted by integration of the correlation score over 20 frame pairs (movie M3). Speckle flow in this movie is almost stationary, justifying the time integration. Figure reproduced from Reference 22 with permission of *Journal of Microscopy*.

pairs yields robust displacement estimates for probing windows as small as the Airy disk area (**Figure 4*a*-ii, *b*-ii**). **Figure 4*c*** presents a complete high-resolution speckle flow map extracted by integration over 20 frames (~3 min).

Tracking Single-Speckle Trajectories

The extraction of kinetic data according to **Figure 3** requires the accurate localization of speckle birth and death events. For this, we had to devise methods capable of tracking full trajectories at the single-speckle level. The large number (>100,000) of dense speckles poses a significant challenge. Details of the current implementation of single-particle tracking of speckles are described by Ponti et al. (38). Our approach follows the framework of most particle-tracking methods, i.e., detection of speckles as particles on a frame-by-frame basis, and the subsequent assignment of corresponding particles in consecutive frames. Assignment is iterated to close gaps in the trajectories created by short-term instability of the speckle signal. Our implementation of this framework includes two algorithms that address particularities of the speckle signal: (*a*) Speckles are detected in an iterative statistical framework, which accounts for signal overlap between proximal speckles. (*b*) Speckle assignments between consecutive frames are executed in a hybrid approach combining speckle flow and single-speckle tracking. Speckle flow fields are extracted iteratively from previous solutions of single-speckle trajectories and employed to propagate speckle motion between frames prior to establishing the correspondence between thousands of speckle pairs by an efficient numerical implementation of a global nearest-neighbor assignment (4, 8). The very first speckle flow field is obtained by correlation-based tracking (22).

Motion propagation allows us to cope with two problems of FSM data. First, in many cases the magnitude of speckle displacements between two frames significantly exceeds half the distance between speckles. Hence, no solution to the correspondence problem exists without prediction of future speckle locations. Second, speckles undergo sharp spatial gradients in speed and direction of motion. A global propagation scheme discarding regional variations will thus fail, whereas an iterative extraction of the flow field permits a gradually refined trajectory reconstruction in these areas.

Figure 5*a* displays the single-speckle trajectories for speckles initiated in the first 20 frames of the same movie for which speckle flow computation is demonstrated in **Figure 4**. The color-framed close-ups indicate regional differences between trajectories. Window *a*-i contains mostly straight trajectories with an average lifetime of 88 s. The trajectories in window *a*-ii are also straight with an average lifetime of 60 s. In contrast, trajectories in windows *a*-iii and *a*-iv exhibit less directional persistence and have average lifetimes of 65 s and 59 s, respectively. As discussed below, these differences at the level of single-speckle trajectories afforded the segmentation of the cell front into dynamically distinct subregions that correspond to molecularly and functionally distinct actin network modules.

Figure 5*b*,*c* present the steady-state speed of actin network transport and turnover extracted from ~100,000 trajectories. Three different patterns of turnover are recognized that correspond to regions with different average speeds. At the cell edge a ~1-μm-wide band of network assembly (red color; white arrowhead) abuts a ~1-μm-wide band of disassembly (green color; white arrow). The yellow shade in the assembly band indicates that filament polymerization and depolymerization significantly overlap. This 2-μm-wide cell border, which we call the lamellipodium (Lp), exhibits on average the fastest F-actin retrograde flow. Predominant disassembly is found ~10 μm from the cell edge (black arrows), where the speed of F-actin flow is minimal. Here, the retrograde flow of the cell front encounters the anterograde flow of the cell body (B). This region is thus called the convergence zone (C). Between the lamellipodium and the convergence zone is a region called the lamella (L), where assembly and disassembly alternate in a random pattern, accompanied by relatively coherent retrograde flow of moderate speed. The same pattern of network turnover is observed in the cell body.

The high spatial resolution of these turnover maps requires faithful localization of the majority of speckle birth and death events. Trajectory interruptions caused by imprecise tracking introduce pairs of false birth and death events that, unless eliminated by the statistical tests of the score analysis, render inaccurate the measurement of network turnover. Simulated speckle fields with flow characteristics, marker density, and noise similar to real data have demonstrated a 100% success rate of our current single-speckle tracking framework (38). The success of the method on real data cannot be determined because there is no ground truth available. Hand-tracking of speckles is so irreproducible between different operators that the generation of a reliable manual reference data set is impossible. Instead, we demonstrate how incomplete single-speckle trajectories adversely affect the reconstruction of network turnover. **Figure 6*a*** relies on our most current single-speckle tracking package. **Figure 6*b*** presents the

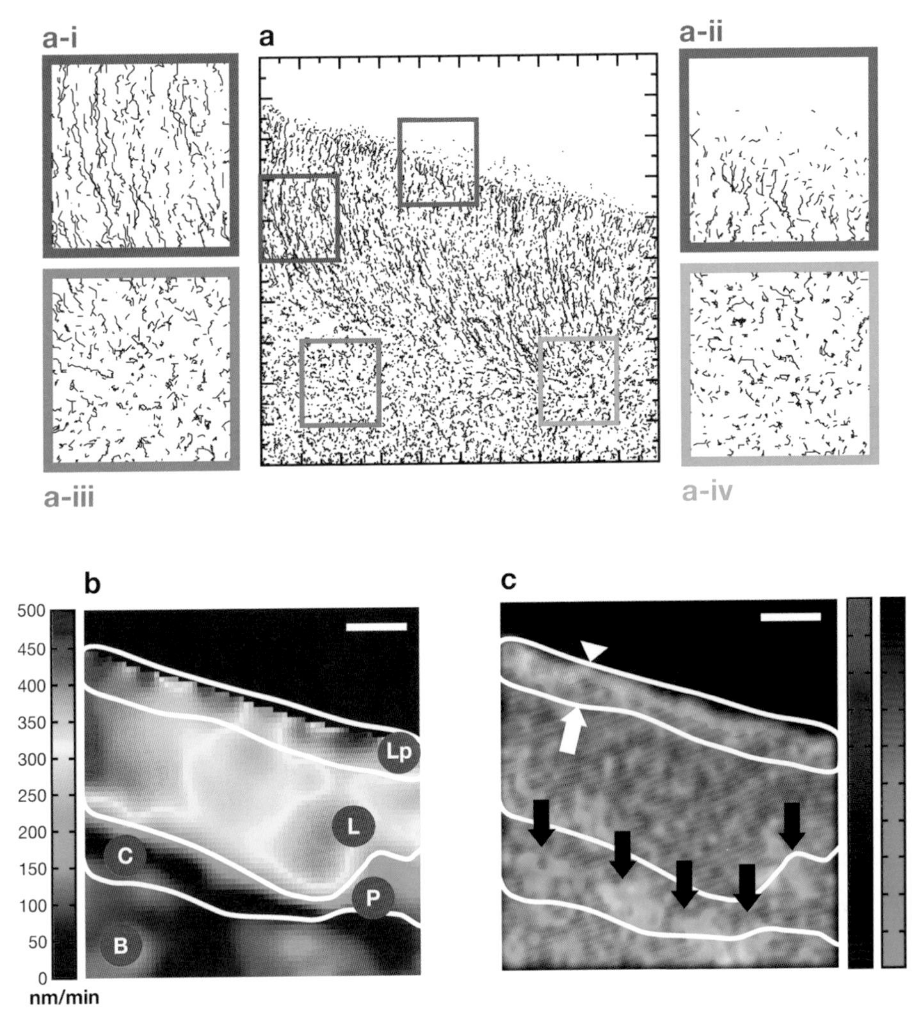

Figure 5

Tracking single-speckle trajectories. (*a*) Trajectories of speckles initiated in the first 20 frames of movie M3. *a*-i to *a*-iv: Close-ups in different areas indicating regional variation in directional persistence, velocity, and lifetime of the trajectories. (*b*) Speed distribution averaged over all 220 frames of the movie. (*c*) Distribution of polymerization (*red channel*) and depolymerization (*green channel*) calculated from scores averaged over 220 frames. Four regions of the actin network with distinct kinematic (motion) and kinetic (turnover) properties can be segmented (see text).

reconstruction of the same turnover map from trajectories tracked without iterative speckle detection and motion propagation. Whereas the maps are similar in large parts of the cell, the less advanced tracking method fails to capture the narrow bands of network assembly and disassembly at the cell edge, where speckles are the densest and a high number of proximate birth and death events confuse the tracking.

Mapping Polymer Turnover Without Speckle Trajectories

It frequently occurs that lower speckle contrast or high image noise do not allow the

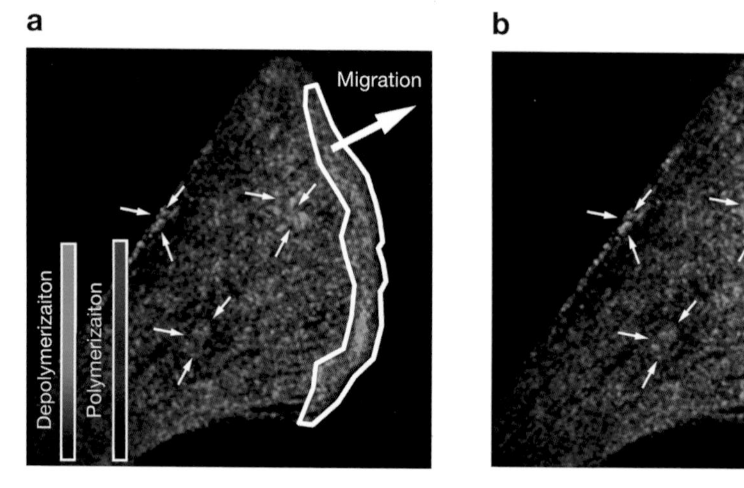

Figure 6

Reconstruction of actin network turnover from trajectories with different tracking quality. (*a*) Application of the model in **Figure 2** to birth and death events determined by the currently most advanced single-speckle tracking method available in our software package. (*b*) Result obtained with a more standard single-particle tracking method that does not include iterative detection of overlapping speckles and motion propagation (see text). Figure reproduced from Reference 22 with permission of *Journal of Microscopy*.

precise identification of single-speckle trajectory endpoints. However, the trackable subsections of the trajectories are usually sufficient to extract the overall structure of speckle flow. In this case an alternative scheme relying on the continuity of the optical density of the speckle field permits the mapping of turnover at lower resolution (48). Shown in **Figure 7**, this method reveals a qualitatively similar organization of actin network assembly and disassembly, but essential features of the turnover patterns that allowed a clear distinction of different regions in **Figure 5*c***, as well as the spatial coexistence of polymerization and depolymerization at the cell edge, are lost with this coarser analysis.

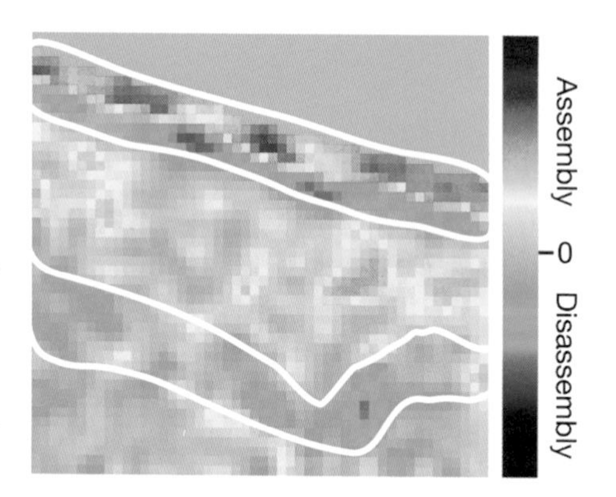

Figure 7

Reconstruction of actin network turnover from speckle flow (**Figure 4*c***) without explicit identification of speckle births and deaths (see text). A spatial organization similar to that in **Figure 5*c*** is mapped; however, essential details in the fine structure of the network turnover are lost. Figure reproduced from Reference 48 with permission of *Proceedings of the National Academy of Sciences, USA*.

APPLICATIONS OF FSM FOR STUDYING PROTEIN DYNAMICS IN VITRO AND IN VIVO

Applications of FSM have thus far focused mostly on the study of actin and microtubule cytoskeleton systems, but other systems have been analyzed with it as well. A summary of the FSM literature can be found in Table S1, which reviews the major biological findings and highlights technical advances in FSM that have been made in these studies. Most of the FSM data analysis has been limited to kymograph measurements of average speckle flow (see above) and to manual tracking of a few hundred speckles to extract lifetime information (55) and selected trajectories of cytoskeleton structures (19, 41, 43). To our knowledge, in addition to those mentioned in the previous section, few efforts have been made outside our labs to systematically exploit the full spatiotemporal information offered by FSM about transport and turnover in molecular assemblies. The most complete quantitative FSM (qFSM) analyses have so far been performed on F-actin cytoskeleton dynamics in migrating epithelial cells. In the following section we summarize some of the most relevant results of these studies to showcase the technical possibilities of qFSM.

SELECTED RESULTS FROM THE STUDY OF ACTIN IN EPITHELIAL CELL MIGRATION

Organization of Actin Cytoskeleton in Four Kinematically and Kinetically Distinct Regions

Figures 4 and **5** indicate the steady-state organization of the F-actin cytoskeleton in four kinematically and kinetically distinct zones: (*a*) the lamellipodium, characterized by fast retrograde flow and two narrow bands of assembly and disassembly resulting from the fast treadmilling of actin between its polymeric and monomeric states (34, 35); (*b*) the lamella, characterized by reduced retrograde flow and assembly and disassembly in random punctate patterns; (*c*) the cell body, characterized by anterograde flow and turnover patterns similar to those of the lamella; and (*d*) the convergence zone, where the flows of the lamella and cell body meet and where strong depolymerization suggests that the lamella and cell body are materially separate structures.

qFSM also delivers nonsteady-state measurements of flow and turnover, revealing distinct variations in the periodicity of turnover between these regions (38). In combination with pharmacological perturbation, qFSM was used to dissect the mechanisms of retrograde flow. We found that lamellipodium flow is independent of myosin motor contraction, whereas lamella flow is blocked by specific inhibition of myosin II activity (37). Also, the lamellipodium and the lamella exhibited different sensitivity to disruption of filament assembly, disassembly, and severing, suggesting that the regional differences could be associated with differential molecular regulation (37). This hypothesis has thus far been confirmed by immunostaining studies (18, 37) and by expression of constitutively active and dominant negative constructs of regulatory proteins (18; M. Machacek, V. Delorme, G. M. Bokoch, C. M. Waterman-Storer & G. Danuser, unpublished data). Here, qFSM provides a critical insight into cytoskeleton dynamic responses to shifted activation of regulatory factors. In summary, these data demonstrate how qFSM can be used to quantitate spatiotemporal modulations of the kinetics and kinematics of molecular assemblies and to identify dynamically distinct structural modules even when they are composed of the same base protein.

Quantitative fluorescent speckle microscopy (qFSM): combination of FSM with statistical analysis of the speckle signal using specialized qFSM software

Correlation of Actin Assembly with a GFP-p34 Signal Indicates Different Function of the Arp2/3 Complex in Lamellipodium and Lamella

Immunolocalization experiments showed that the lamellipodium is enriched in Arp2/3, a protein complex thought to activate network polymerization by nucleating new filaments off preexisting filaments, and in ADF/cofilin, which promotes filament severing and depolymerization (18, 37). Together, these proteins have been described as mediators of actin treadmilling in the lamellipodium (35). However, Arp2/3 stain was also present in punctate patterns in the lamella. We therefore

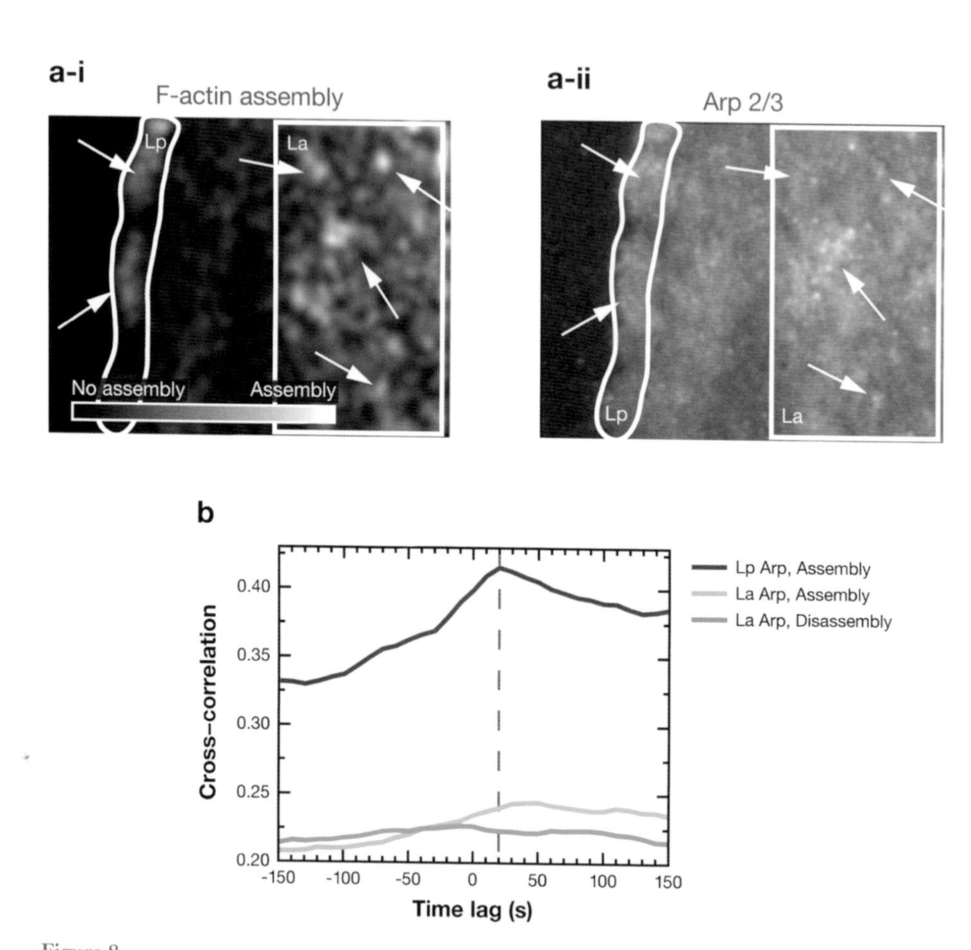

Figure 8

Correlation between actin network assembly, as measured by qFSM, and the signal of the GFP-p34 component of the Arp2/3 complex, a putative regulator of assembly. (*a*) F-actin assembly (*left*) and Arp2/3 distribution (*right*). Arrows point at locations where strong assembly visually correlates with maxima in the Arp2/3 signal. (*b*) Cross-correlations between GFP-p34 signal and assembly in the lamellipodium (*red line*); GFP-p34 signal and assembly in the lamella (*orange line*); and GFP-p34 signal and disassembly in the lamella (*green line*). Figure reproduced from Reference 38 with permission of the *Biophysical Journal*.

speculated that the punctate pattern of lamella assembly could be a direct result of Arp2/3 clustering, as suggested for *Dyctostelium* (5) and yeast (68). To test this hypothesis we correlated time-resolved qFSM F-actin assembly maps with time-lapse image sequences of the GFP-p34 component of the Arp2/3 complex (38). By visual inspection bright signals of net assembly (**Figure 8*a*-i**) appeared to colocalize with bright GFP-p34 signals (**Figure 8*a*-ii**), especially in the lamella (arrows). Cross-correlation of the two maps over time yielded an average correlation of 0.22 with a weak maximum at about +50 s (**Figure 8*b***, orange line). The same correlation was obtained between the GFP-p34 and the disassembly maps (**Figure 8*b***, green line). Simulations of the cross-correlation of random signals confirmed that the correlation value of both comparisons was statistically significant (38). Thus, our data agreed with other studies (5, 68) in that hot spots of lamella network assembly and disassembly tend to colocalize with sites of Arp2/3 accumulation.

Importantly, though, the dynamic analysis revealed the independence of these events in time. This demonstrates the importance of nonsteady-state measurements to probe relationships between molecular processes, as can now be achieved by qFSM.

In the lamellipodium, the cross-correlation between network assembly and GFP-p34 signals was higher and displayed a significant maximum, suggesting a dynamic coupling of F-actin assembly and Arp2/3 aggregation (**Figure 8*b***, red line). The time lag of +20 s implied that the highest rates of assembly precede the maximum of GFP-p34 signal, which is compatible with a model of autocatalytic network assembly by dendritic nucleation (34): A burst of actin polymerization initiates Arp2/3 aggregation by increasing the probability of Arp2/3-mediated filament branching. Increased branching induces exponential network growth until the pool of polymerizable actin monomers is locally depleted. At this point the assembly rate begins to taper off while Arp2/3 continues to associate with preexisting filaments. The peak in Arp2/3 signal is observed when the F-actin network turns from a state of assembly to a state of disassembly. In summary, these data demonstrate how correlating qFSM data with other image cues allows one to examine functional relationships between regulatory factors and the dynamics of effector molecular assemblies.

Coupling of Actin Disassembly and Contraction in the Convergence Zone

A similar spatiotemporal correlation analysis was performed to examine the relationship of actin network depolymerization and contraction in the convergence zone (48). We first established that transient increases in speckle flow convergence are coupled to transient increases in disassembly. This begged the question whether the rate of speckle flow convergence increases because disassembly boosts the efficiency of myosin II motors in contracting a more compliant network or because motor contraction mediates network disassembly. To address this question, we transiently perfused cells with calyculin A, a type II phosphatase inhibitor that increases myosin II activity. Unexpectedly, we reproducibly measured a strong burst of disassembly long before flow convergence was affected. This evidence suggested that myosin II contraction can actively promote depolymerization of F-actin, for example, by breaking filaments. The link between F-actin contractility and turnover has since been confirmed by fluorescence recovery after photobleaching measurements in the contractile ring required for cytokinesis (33). In summary, these data demonstrate the correlation of two qFSM parameters to decipher the relationship between deformation and plasticity of polymer networks inside cells.

Heterogeneity in Speckle Velocity and Lifetime Reveals Spatial Overlap of Lamellipodium and Lamella at the Leading Edge

The transition between the lamellipodium and the lamella is characterized by a narrow band of strong disassembly adjacent to a region of mixed assembly and disassembly and a sharp decrease in retrograde flow velocity (**Figure 5**). Together, these features defined a unique mathematical signature for tracking the boundary between the two regions over time (**Figure 9*a***). In view of the differences of speckle velocities and lifetime between the two regions, we speculated that the same boundary could be tracked by spatial clustering of speckle properties. We predicted that fast, short-living speckles (class 1) would preferentially localize in the lamellipodium, whereas slow, longer-living speckles (class 2) would be dominant in the lamella. To test this hypothesis, we solved a multiobjective optimization problem in which the thresholds of velocity v_{th} and lifetime τ_{th} separating the two classes, as well as the boundary ∂Lp between lamellipodium and lamella,

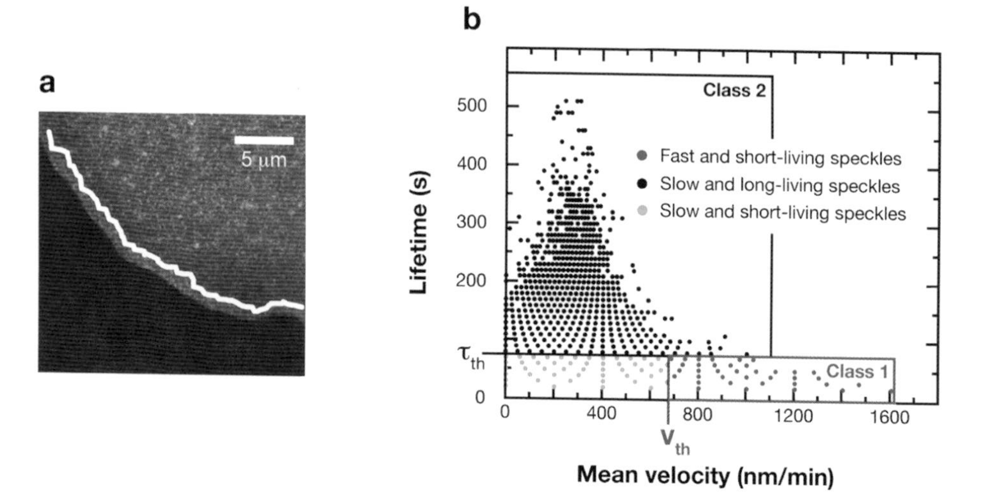

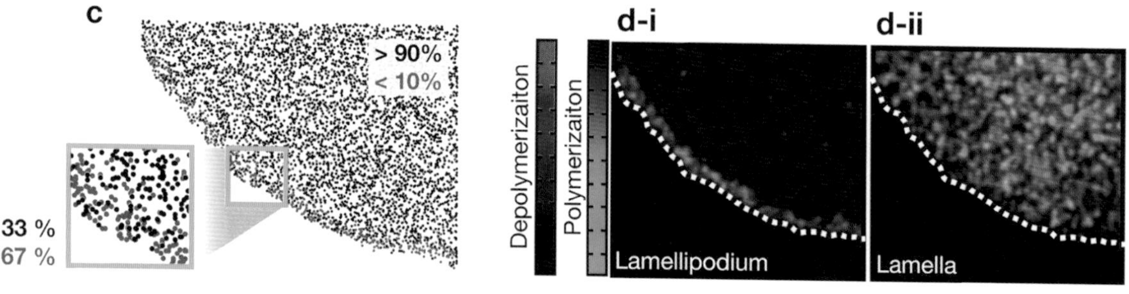

Figure 9

Distinction of two spatially overlapping actin networks based on heterogeneity of single-speckle properties. (*a*) Raw FSM image overlaid with the boundary between lamellipodium and lamella computed from spatial gradients in actin turnover and flow velocity. Animation of this data is provided in movie M5 (follow the Supplemental Material link from the Annual Reviews home page at **http://www.annualreviews.org**). (*b*, *c*) Cluster analysis of speckle lifetime and velocity (see text). (*d*) Class 1 speckles constitute the rapidly treadmilling lamellipodium. Class 2 speckles constitute the lamella with a punctate pattern of random actin turnover. Both networks spatially overlap in the first 2 µm from the cell edge.

were determined subject to the rule $\{\partial \mathrm{Lp}, \tau_{\mathrm{th}}, \nu_{\mathrm{th}}\} = \max(N_1/(N_1 + N_2) \in \mathrm{Lp})\ \&\ \min(N_1/(N_1 + N_2) \in \mathrm{La})$ (**Figure 9*b***). N_1 and N_2 denote the number of speckles in classes 1 and 2, respectively. Our prediction was confirmed in the lamella, where class 1 speckles occupied a statistically insignificant fraction. However, class 2 speckles made up 30% to 40% of the lamellipodium, indicating that in this region speckles with different kinetic and kinematic behavior colocalize. This information was previously lost in the averaged analysis of single-speckle trajectories. When mapping the scores of class 1 and class 2 speckles separately, we discovered that class 1 speckles define the bands of polymerization and depolymerization characteristic of the lamellipodium and that class 2 speckles define the puncta of assembly and disassembly characteristic of the lamella, which reaches all the way to the leading edge. Subsequent experiments specifically disrupting actin treadmilling in the lamellipodium confirmed our finding that the lamellipodium and lamella form two spatially overlapping yet kinetically, kinematically, and molecularly different actin networks (18, 37).

In summary, these data exemplify that qFSM analysis has come to the level at which single-speckle properties can be exploited to probe the heterogeneity of molecular assemblies in space and time.

NEW APPLICATIONS OF FSM

Two-Speckle Microrheology Probes Viscoelastic Properties of Actin Networks

Speckle trajectories probe different dynamic phenomena at different spatial and temporal scales. So far we have exploited the long-range, directed components of trajectories to extract the flow and deformation of F-actin networks induced by molecular forces coordinated over several microns, e.g., the activity of a large number of myosin II motors in the convergence zone. On a shorter spatiotemporal scale, speckle trajectories contain components associated with microscopic deformations of the network that are induced by less coordinated local contractions of small myosin patches (50) and thermal forces. Also, positional fluctuations of speckles relate to the sliding of decoupled filaments, filament bending inside the network, and photometric shifts of the speckle centroids induced by local fluorophore exchange. These fluctuations occur at a length scale shorter than the mesh size of the network and/or are

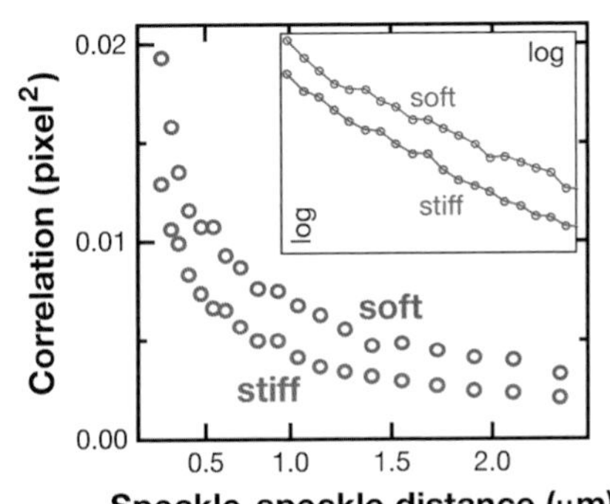

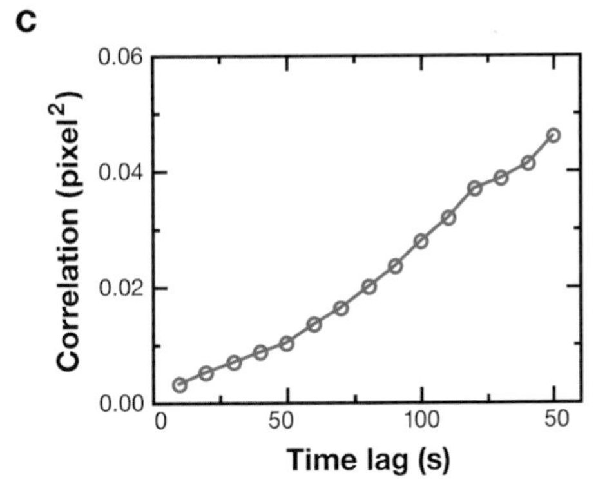

Figure 10

Two-speckle microrheology of F-actin network stiffness inside cells. (*a*) Correlation of random motion of two speckles as a function of their distance r. The curves follow a $1/r$ decay (see inset), as is predicted for a viscoelastic medium. (*b*) Spatial mapping of correlation values. The correlation landscape has a smooth spatial variation at length scales much longer than the size of one probing window. This suggests that regional heterogeneity is not due to noise. (*c*) Correlation as a function of the time lag over which speckle displacements are determined. The linearity of the curve indicates that the correlation is dominated by diffusive motion, which permits direct conversion of the correlation values into parameters describing the viscoelasticity of F-actin networks.

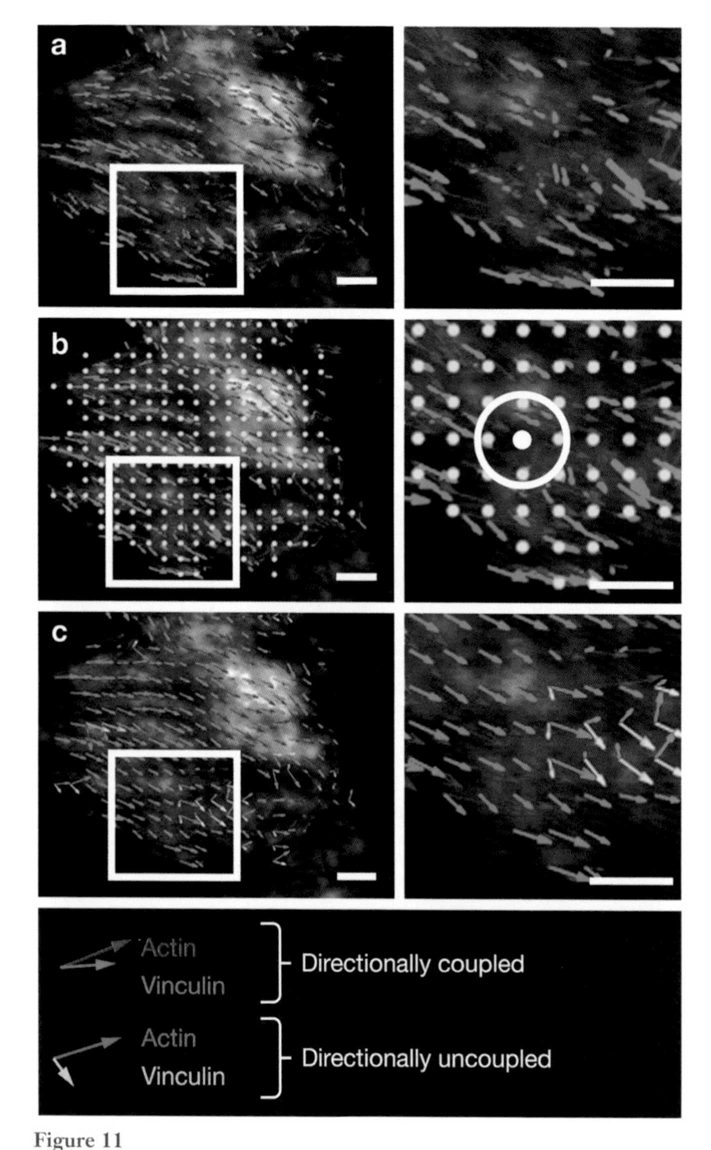

Figure 11

Correlative, multispectral qFSM analysis of F-actin and vinculin. (*a–c*) Processing steps in the measurement of the coupling of F-actin (*red vectors*) and vinculin (*green vectors*) flows, overlaid to the raw speckle image of vinculin (see text for further discussion). Blue (actin) and yellow (vinculin) vectors in panel *c* indicate a region where the two molecules are directionally decoupled. Scale bar: 5 μm.

independent between speckles. When calculating the cross-correlation of trajectories of two speckles separated by a distance greater than the mesh size of the network, these fluctuations cancel out. However, even after directional components are eliminated, the cross-correlation between two-speckle trajectories decays with $1/r$, where r denotes the distance between them (**Figure 10*a***). This behavior is known from two-point microrheology, in which embedded beads instead of speckles are used to track thermal fluctuations in polymer networks (9, 14). Thus, spatially correlated yet undirected components of speckle motion could be used to probe material properties of polymer networks inside a cell with a resolution at the scale of the inter-speckle distance.

This possibility capitalizes on recent enhancements of speckle tracking to an accuracy of approximately one tenth of a pixel even when speckles overlap. Also, we implemented a module that performs correlation analysis in small windows to map out the spatial modulation of material properties (**Figure 10*b***). Whereas in a noncontractile polymer network this map would directly reveal spatial variation of network stiffness, the inhomogeneity could also be the result of spatially variable motor activity. To test for this possibility we computed the correlation as a function of the time over which speckle displacements are tracked. For purely thermal fluctuations the relationship between cross-correlation magnitude and time is linear, whereas for a motor-driven fluctuation one would expect superdiffusive speckle motion (42). **Figure 10*c*** shows that for a time range of 5 to 50 s the relationship is indeed linear. Thus, **Figure 10*b*** may provide a first glimpse of the variation of the stiffness of F-actin networks at the micron scale.

Correlational qFSM of the Dynamic Engagement of Actin Cytoskeleton and Focal Adhesions

We have begun to extend our particle-tracking-based analysis of speckles in one spectral channel to correlative particle tracking in two spectrally distinct channels with the goal of analyzing the kinematic coupling between two macromolecular assemblies. We used this approach to derive initial models of the coupling of the F-actin cytoskeleton to FA proteins in migrating epithelial cells. **Figure 11** illustrates the steps of correlational qFSM on dual-channel TIR-FSM images of F-actin and vinculin speckles. Single-speckle tracking was first applied separately to both channels (**Figure 11*a***). Speckle flows were then mapped onto a common grid to allow comparison of pairs of speckles between the two channels (**Figure 11*b***). Speckle assignment to the closest local grid position is achieved with distance-weighted, locally adaptive interpolation (white circle). Following grid assignment, the correlation between the direction and speed of speckle flow vectors was computed. To quantitate the kinematic coupling of actin and FA molecules, two metrics are currently used. A direction correlation score ranging from –1 to 1 is defined as the cosine of the angle between the two vectors at any one grid point (**Figure 11*c***). A speed correlation score is defined as $(\Sigma\ \nu_{FA} \perp \nu_{act})/|\nu_{act}|)$, where $(\Sigma\ \nu_{FA} \perp \nu_{act})$ denotes the average of the velocity projection of the FA flow field onto the actin vector flow axis, and $|\nu_{act}|$ is the magnitude average of actin vectors. This value thus represents how the speed of FA molecules along the actin axis compares with the speed of F-actin itself. For two identical flow fields, both the direction and speed correlation scores are 1.

Preliminary correlational qFSM studies have already revealed thus far unknown patterns of molecular coupling to FAs. For example, we found a low degree of kinematic coupling between F-actin and the FA proteins integrin, FAK, zyxin, and paxillin, as indicated by low direction and speed correlation scores. In contrast, our analysis revealed a high degree of kinematic coupling between F-actin flow and FA proteins α-actinin, talin, and vinculin. This indicates the power of correlational qFSM for spatiotemporally quantifying the degree of engagement of pairs of colocalized molecules in subcellular macromolecular assemblies.

CONCLUSION

Over the past few years FSM has been advanced to a versatile tool for simultaneously probing the motion, deformation, turnover, and material properties of macromolecular assemblies. In a next step, these parameters will be combined in correlational analyses to establish how assemblies operate as dynamic and plastic structures, enabling a broad variety of cell functions. In parallel, FSM will go multispectral, so that these parameters can be correlated among different macromolecular structures. This requires major modifications to the current qFSM software to cope with the explosion of combinatorial data in two or more simultaneously imaged speckle channels.

Despite the conceptual simplicity, FSM has remained an expert technique used by a relatively small number of research laboratories. Two measures will be taken by our labs in an attempt to spread FSM as a routine mode of light microscopy. First, protocols are under development to generate cell lines that permanently express low levels of fluorescent cytoskeleton proteins optimized for speckle image formation. This will avoid the time-consuming steps of microinjection and the tweaking of transient expression of fluorescent protein and will turn FSM into a robust technique with high quantitative reproducibility. Second, the qFSM software package will be released. The use of cell lines with reproducible levels of fluorescent protein will greatly aid the software distribution. Under standardized conditions many of the processing steps can be furnished with default control parameters so that meaningful results can be obtained without in-depth knowledge of the mathematical algorithms.

Last, the idea of FSM will be promoted by more applications outside the cytoskeleton field. The analyses of FA dynamics make an initial step in this direction. Projects are underway to apply qFSM to studies of the dynamic interaction of clathrin, dynamin, and actin structures during endocytosis, and of DNA repair. Despite the many existing discoveries already made by FSM, as documented by the sizable list of publications in Table S1, it is still a technology in the infancy of its uses.

ACKNOWLEDGMENTS

This research is supported by NIH through the grant R01 GM67230 to C.W.S. and G.D. We are grateful to Ke Hu, Lin Ji, Stephanie Gupton, and Dinah Loerke for sharing unpublished data; and to Kathryn Thompson for a careful reading of the manuscript.

LITERATURE CITED

1. Guidelines for the practical implementation of a total internal reflection light microscope amenable to FSM.

2. Guidelines for the practical implementation of a spinning disk confocal microscope amenable to FSM.

7. Application of FSM to study the transient association of a cytoskeleton-binding protein with the cytoskeleton.

1. **Adams M, Matov A, Yarar D, Gupton S, Danuser G, Waterman-Storer CM. 2004. Signal analysis of total internal reflection fluorescent speckle microscopy (TIR-FSM) and widefield epi-fluorescence FSM of the actin cytoskeleton and focal adhesions in living cells. *J. Microsc.* 216:138–52**
2. **Adams MC, Salmon WC, Gupton SL, Cohan CS, Wittmann T, et al. 2003. A high-speed multispectral spinning-disk confocal microscope system for fluorescent speckle microscopy of living cells. *Methods* 29:29–41**
3. Ahuja RK, Magnanti TM, Orlin JB. 1993. *Network Flows: Theory, Algorithms and Optimization*. Saddle River, NJ: Prentice-Hall. 846 pp.
4. Blackman SS, Popoli R. 1999. *Design and Analysis of Modern Tracking Systems*. Noorwood, MA: Artech House
5. Bretschneider T, Diez S, Anderson K, Heuser JA, Clarke M, et al. 2004. Dynamic actin patterns and Arp2/3 assembly at the substrate-attached surface of motile cells. *Curr. Biol.* 14:1–10
6. Brust-Mascher I, Scholey JM. 2002. Microtubule flux and sliding in mitotic spindles of *Drosophila* embryos. *Mol. Biol. Cell* 13:3967–75
7. **Bulinski JC, Odde DJ, Howell BJ, Salmon TD, Waterman-Storer CM. 2001. Rapid dynamics of the microtubule binding of ensconsin *in vivo*. *J. Cell Sci.* 114:3885–97**
8. Burkard KE, Cela E. 1999. Linear assignment problems and extensions. In *Handbook of Combinatorial Optimization*, ed. DZ Du, PM Pardalos, pp. 75–149. Dordrecht: Kluwer Acad. Publ.
9. Crocker JC, Valentine MT, Weeks ER, Gisler T, Kaplan PD, et al. 2000. Two-point microrheology of inhomogeneous soft materials. *Phys. Rev. Lett.* 85:888–91
10. Danuser G, Waterman-Storer CM. 2003. Fluorescent speckle microscopy: where it came from and where it is going. *J. Microsc.* 211:191–207
11. Desai A, Mitchison TJ. 1997. Microtubule polymerization dynamics. *Annu. Rev. Cell Dev. Biol.* 13:83–117
12. Dunn GA, Dobbie IM, Monypenny J, Holt MR, Zicha D. 2002. Fluorescence localization after photobleaching (FLAP): a new method for studying protein dynamics in living cells. *J. Microsc.* 205:109–12
13. Gaetz J, Kapoor TM. 2004. Dynein/dynactin regulate metaphase spindle length by targeting depolymerizing activities to spindle poles. *J. Cell Biol.* 166:465–71
14. Gardel ML, Shin JH, MacKintosh FC, Mahadevan L, Matsudaira P, Weitz DA. 2004. Elastic behavior of cross-linked and bundled actin networks. *Science* 304:1301–5
15. Geiger B, Bershadsky A, Pankov R, Yamada KM. 2001. Transmembrane extracellular matrix-cytoskeleton crosstalk. *Nat. Rev. Mol. Cell Biol.* 2:793–805

16. Grego S, Cantillana V, Salmon ED. 2001. Microtubule treadmilling *in vitro* investigated by fluorescence speckle and confocal microscopy. *Biophys. J.* 81:66–78
17. **Gupton S, Waterman-Storer CM. 2006. Live-cell fluorescent speckle microscopy (FSM) of actin cytoskeletal dynamics and their perturbation by drug perfusion. In *Cell Biology: A Laboratory Handbook*, ed. J Celis, JV Small, pp. 137–51. Amsterdam: Elsevier**
18. Gupton SL, Anderson KL, Kole TP, Fischer RS, Ponti A, et al. 2005. Cell migration without a lamellipodium: translation of actin dynamics into cell movement mediated by tropomyosin. *J. Cell Biol.* 168:619–31
19. Gupton SL, Salmon WC, Waterman-Storer CM. 2002. Converging populations of F-actin promote breakage of associated microtubules to spatially regulate microtubule turnover in migrating cells. *Curr. Biol.* 12:1891–99
20. Inoue S, Spring KR. 1997. *Video Microscopy: The Fundamentals*. New York/London: Plenum. 764 pp.
21. Jepson AD, Fleet DJ, El-Maraghi TF. 2003. Robust online appearance models for visual tracking. *IEEE Trans. Pattern Anal. Mach. Intell.* 25:1296–311
22. Ji L, Danuser G. 2005. Tracking quasi-stationary flow of weak fluorescent features by multi-frame correlation. *J. Microsc.* 220:150–67
23. Jurado C, Haserick JR, Lee J. 2005. Slipping or gripping? Fluorescent speckle microscopy in fish keratocytes reveals two different mechanisms for generating a retrograde flow of actin. *Mol. Biol. Cell* 16:507–18
24. Kapoor TM, Mitchison TJ. 2001. Eg5 is static in bipolar spindles relative to tubulin: evidence for a static spindle matrix. *J. Cell Biol.* 154:1125–34
25. Lippincott-Schwartz J, Patterson GH. 2003. Development and use of fluorescent protein markers in living cells. *Science* 300:87–91
26. Luby-Phelps K. 2000. Cytoarchitecture and physical properties of cytoplasm: volume, viscosity, diffusion, intracellular surface area. *Int. Rev. Cytol.* 192:189–221
27. Maddox P, Desai A, Oegema K, Mitchison TJ, Salmon ED. 2002. Poleward microtubule flux is a major component of spindle dynamics and anaphase A in mitotic *Drosophila* embryos. *Curr. Biol.* 12:1670–74
28. Maddox P, Straight A, Coughlin P, Mitchison TJ, Salmon ED. 2003. Direct observation of microtubule dynamics at kinetochores in *Xenopus* extract spindles: implications for spindle mechanics. *J. Cell Biol.* 162:377–82
29. Micheli ED, Torre V, Uras S. 1993. The accuracy of the computation of optical flow and of the recovery of motion parameters. *IEEE Trans. Pattern Anal. Mach. Intell.* 15:434–47
30. Mikhailov AV, Gundersen GG. 1995. Centripetal transport of microtubules in motile cells. *Cell Motil. Cytoskelet.* 32:173–86
31. Mitchison TJ. 1989. Polewards microtubule flux in the mitotic spindle: evidence from photoactivation of fluorescence. *J. Cell Biol.* 109:637–52
32. Miyamoto DT, Perlman ZE, Burbank KS, Groen AC, Mitchison TJ. 2004. The kinesin Eg5 drives poleward microtubule flux in *Xenopus laevis* egg extract spindles. *J. Cell Biol.* 167:813–18
33. Murthy K, Wadsworth P. 2005. Myosin-II-dependent localization and dynamics of F-actin during cytokinesis. *Curr. Biol.* 15:724–31
34. Pollard TD, Blanchoin L, Mullins RD. 2000. Molecular mechanisms controlling actin filament dynamics in nonmuscle cells. *Annu. Rev. Biophys. Biomol. Struct.* 29:545–76
35. Pollard TD, Borisy GB. 2003. Cellular motility driven by assembly and disassembly of actin filaments. *Cell* 112:453–65

17. Guidelines for the preparation of live cell samples for FSM and for the realization of perfusion chambers that allow the transient application of small-molecule inhibitors of cytoskeleton functions during FSM imaging.

37. Discovery of two dynamically and molecularly distinct yet spatially overlapping actin cytoskeleton structures in epithelial cells using cluster analysis of single-speckle properties.

39. Statistical model for the extraction of rates of polymer assembly and disassembly by classification of intensity changes during speckle appearance and disappearance events.

41. Application of two-color FSM to examine the interaction of the actin and microtubule cytoskeletons in epithelial cell migration.

43. Application of two-color FSM to examine the interaction of the actin and microtubule cytoskeletons in neuronal growth cones.

55. Imaging and analysis of single-fluorophore speckles to probe actin cytoskeleton dynamics during cell motility.

36. Ponti A. 2004. *High-Resolution Analysis of F-Actin Meshwork Kinetics and Kinematics Using Computational Fluorescent Speckle Microscopy*. Zurich: ETH. 185 pp.
37. **Ponti A, Machacek M, Gupton SL, Waterman-Storer CM, Danuser G. 2004. Two distinct actin networks drive the protrusion of migrating cells. *Science* 305:1782–86**
38. Ponti A, Matov A, Adams M, Gupton S, Waterman-Storer CM, Danuser G. 2005. Periodic patterns of actin turnover in lamellipodia and lamellae of migrating epithelial cells analyzed by quantitative fluorescent speckle microscopy. *Biophys. J.* 89:3456–69
39. **Ponti A, Vallotton P, Salmon WC, Waterman-Storer CM, Danuser G. 2003. Computational analysis of F-actin turnover in cortical actin meshworks using fluorescent speckle microscopy. *Biophys. J.* 84:3336–52**
40. Prasher DC. 1995. Using GFP to see the light. *Trends Genet.* 11:320–23
41. **Salmon WC, Adams MC, Waterman-Storer CM. 2002. Dual-wavelength fluorescent speckle microscopy reveals coupling of microtubule and actin movements in migrating cells. *J. Cell Biol.* 158:31–37**
42. Saxton MJ, Jacobson K. 1997. Single-particle tracking: application to membrane dynamics. *Annu. Rev. Biophys. Biomol. Struct.* 26:373–99
43. **Schaefer AW, Kabir N, Forscher P. 2002. Filopodia and actin arcs guide the assembly and transport of two populations of microtubules with unique dynamic parameters in neuronal growth cones. *J. Cell Biol.* 158:139–52**
44. Small V. 1981. Organization of actin in the leading edge of cultured cells. *J. Cell Biol.* 91:695–705
45. Svitkina TM, Verkhovsky AB, McQuade KM, Borisy GG. 1997. Analysis of the actin-myosin II system in fish epidermal keratocytes: mechanism of cell body translocation. *J. Cell Biol.* 139:397–415
46. Theriot JA, Mitchison TJ. 1991. Actin microfilament dynamics in locomoting cells. *Nature* 352:126–31
47. Vallotton P, Danuser G, Bohnet S, Meister JJ, Verkhovsky A. 2005. Retrograde flow in keratocytes: news from the front. *Mol. Biol. Cell* 16:1223–31
48. Vallotton P, Gupton SL, Waterman-Storer CM, Danuser G. 2004. Simultaneous mapping of filamentous actin flow and turnover in migrating cells by quantitative fluorescent speckle microscopy. *Proc. Natl. Acad. Sci. USA* 101:9660–65
49. Vallotton P, Ponti A, Waterman-Storer CM, Salmon ED, Danuser G. 2003. Recovery, visualization, and analysis of actin and tubulin polymer flow in live cells: a fluorescence speckle microscopy study. *Biophys. J.* 85:1289–306
50. Verkhovsky AB, Svitkina TM, Borisy GG. 1999. Network contraction model for cell translocation and retrograde flow. In *Cell Behaviour: Control and Mechanism of Motility*, ed. JM Lackie, GA Dunn, GE Jones, pp. 207–22. London: Portland
51. Verkhovsky AB, Svitkina TM, Borisy GG. 1999. Self-polarization and directional motility of cytoplasm. *Curr. Biol.* 9:11–20
52. Wadsworth P, Salmon E. 1986. Analysis of the treadmilling model during metaphase of mitosis using fluorescence redistribution after photobleaching. *J. Cell Biol.* 102:1032–38
53. Wang Y. 1985. Exchange of actin subunits at the leading edge of living fibroblasts: possible role of treadmilling. *J. Cell Biol.* 101:597–602
54. Wang YL, Heiple JM, Taylor DL. 1982. Fluorescent analog cytochemistry of contractile proteins. *Methods Cell Biol.* 25:1–11
55. **Watanabe Y, Mitchison TJ. 2002. Single-molecule speckle analysis of actin filament turnover in lamellipodia. *Science* 295:1083–86**

56. Waterman-Storer CM. 2002. Fluorescent speckle microscopy (FSM) of microtubules and actin in living cells. In *Current Protocols in Cell Biology*, ed. JS Bonifacino, M Dasso, JB Harford, J Lippincott-Schwartz, KM Yamada, Unit 4.10. New York: Wiley
57. Waterman-Storer CM, Danuser G. 2002. New direction of fluorescent speckle microscopy. *Curr. Biol.* 12:R633–R40
58. Waterman-Storer CM, Desai A, Bulinski JC, Salmon ED. 1998. Fluorescent speckle microscopy, a method to visualize the dynamics of protein assemblies in living cells. *Curr. Biol.* 8:1227–30
59. Waterman-Storer CM, Desai A, Salmon ED. 1999. Fluorescent speckle microscopy of spindle microtubule assembly and motility in living cells. *Methods Cell Biol.* 61:155–73
60. Waterman-Storer CM, Salmon ED. 1997. Actomyosin-based retrograde flow of microtubules in the lamella of migrating epithelial cells influences microtubule dynamic instability and turnover and is associated with microtubule breakage and treadmilling. *J. Cell Biol.* 139:417–34
61. **Waterman-Storer CM, Salmon ED. 1998. How microtubules get fluorescent speckles. *Biophys. J.* 75:2059–69**

61. Definition and analysis of the stochastic assembly model of how microtubules get speckles.

62. Waterman-Storer CM, Salmon ED. 1999. Fluorescent speckle microscopy of MTs: How low can you go? *FASEB J.* 13:225–30
63. Waterman-Storer CM, Salmon WC, Salmon ED. 2000. Feedback interactions between cell-cell adherens junctions and cytoskeletal dynamics in newt lung epithelial cells. *Mol. Biol. Cell* 11:2471–83
64. Waterman-Storer CM, Sanger JW, Sanger JM. 1993. Dynamics of organelles in the mitotic spindles of living cells: membrane and microtubule interactions. *Cell Motil. Cytoskelet.* 26:19–39
65. Wittmann T, Bokoch GM, Waterman-Storer CM. 2003. Regulation of leading edge microtubule and actin dynamics downstream of Rac1. *J. Cell Biol.* 161:845–51
66. Wolf DE. 1989. Designing, building, and using a fluorescence recovery after photobleaching instrument. *Methods Cell Biol.* 30:271–306
67. Ye M, Haralick RM, Shapiro LG. 2003. Estimating piecewise-smooth optical flow with global matching and graduated optimization. *IEEE Trans. Pattern Anal. Mach. Intell.* 25:1625–30
68. Young ME, Cooper JA, Bridgman PC. 2004. Yeast actin patches are networks of branched actin filaments. *J. Cell Biol.* 166:629–35
69. Zhang X-F, Schaefer AW, Burnette DT, Schoonderwoert VT, Forscher P. 2003. Rho-dependent contractile responses in the neuronal growth cone are independent of classical peripheral retrograde actin flow. *Neuron* 40:931–44
70. Zicha D, Dobbie IM, Holt MR, Monypenny J, Soong DYH, et al. 2003. Rapid actin transport during cell protrusion. *Science* 300:142–45

Water Mediation in Protein Folding and Molecular Recognition

Yaakov Levy and José N. Onuchic

Center for Theoretical Biological Physics and Department of Physics, University of California at San Diego, La Jolla, California 92093; email: jonuchic@ucsd.edu

Annu. Rev. Biophys. Biomol. Struct. 2006. 35:389–415

First published online as a Review in Advance on February 27, 2006

The *Annual Review of Biophysics and Biomolecular Structure* is online at biophys.annualreviews.org

doi: 10.1146/annurev.biophys.35.040405.102134

1056-8700/06/0609-0389$20.00

Key Words

energy landscape, folding funnel, protein association, protein-DNA interactions, solvent effect

Abstract

Water is essential for life in many ways, and without it biomolecules might no longer truly be biomolecules. In particular, water is important to the structure, stability, dynamics, and function of biological macromolecules. In protein folding, water mediates the collapse of the chain and the search for the native topology through a funneled energy landscape. Water actively participates in molecular recognition by mediating the interactions between binding partners and contributes to either enthalpic or entropic stabilization. Accordingly, water must be included in recognition and structure prediction codes to capture specificity. Thus water should not be treated as an inert environment, but rather as an integral and active component of biomolecular systems, where it has both dynamic and structural roles. Focusing on water sheds light on the physics and function of biological machinery and self-assembly and may advance our understanding of the natural design of proteins and nucleic acids.

Contents

Funneled energy landscape: because of minimal frustration, folding can be described as a progressive organization of an ensemble of partially folded structures

TSE: transition state ensemble

PERSPECTIVES AND OVERVIEW

The funneled energy landscape of evolutionarily selected proteins governs their robust ability to efficiently organize the polypeptide chain into a specific structure (16, 86, 105, 106). Theoretical and simulation strategies based on the funnel concept impressively reproduce the experimental characterization of folding mechanisms, intermediate(s), transition state ensemble (TSE), and folding rates (22, 28, 43, 51, 80). The experimental mechanisms and kinetics of protein associations have been recently obtained using a funneled energy landscape (89, 94), illustrating that binding, similar to folding, follows the principle of minimal frustration (17). The funnel landscape idea implies the notion, now well accepted as a general guideline, that because energetic frustration for folding/binding is relatively small in natural sequences, the native topology determines the mechanism of biological self-organization processes. However, it is obvious that the appropriate environment is conditional to all recognition processes in the cell. This review presents a survey of the many ways water properties are exploited in biology with a special emphasis on the dynamic role water has in gating folding and binding, where it actively assists the search through the funneled landscape.

Water is a remarkable chemical compound that has been long appreciated to be absolutely fundamental for life. Indeed, water is one of the four Aristotelian elements and its importance to life is well reflected by its sanctity in many religions and myths. Water is life's true and unique medium, and without it life simply cannot be sustained. It is therefore of high interest to biologists, chemists, physicists, as well as cosmologists (6). Water is the fluid that lubricates the workings of the cell, transporting the materials and molecular machinery and facilitating the chemical reactions. Yet, water also plays an active and complex role in the life of the cell, to the extent that water itself can be considered a biomolecule because without it the cell function would cease to exist. Without water, biomolecules such as proteins and nucleic acids might no longer truly be biomolecules. When dealing with proteins and genes in modern molecular biology, we should not ignore that it is all about the interactions of such molecules in and with water. Truly, water's function in the cell is far beyond that of an inert solvent.

Water has had an active role in the evolution of life, constitutes about 70% of the human body, and covers about 75% of the earth. An astronomer or a nuclear physicist would not find the ubiquity of water in biochemistry a total surprise. Our understanding of the origin of the elements is consistent with their observed abundances in the universe. Clearly, hydrogen (whose name means "water former") and oxygen, which are the most and third-most abundant elements, respectively, are major candidates for chemical combination. The second-most abundant element, helium, is not reactive. Hydrogen peroxide, H_2O_2, is rather unstable. Hydrogen oxide, H_2O, must be present on planets. Yet its abundance alone cannot explain its role in life. This curious molecule has rather extraordinary properties, which are quite different at low and high temperatures. It has often been stated that life depends on these anomalous properties of water (a few examples include its unusually large heat capacity, high melting and boiling points, high thermal conductivity and surface tension, and shrinking on melting). Fluidity seems to be essential for active life, and completely solid-phase life, if it exists, must be very slow in its actions. It is unclear if other liquids (e.g., other hydrides, oxides, or hydrocarbons) can replace water and be compatible with the existence of complex molecules, maintaining their integrity to bear information and self-organize. Clearly, some media seem to present fundamental problems from a physical point of view: For molecular life we need a fluid in which chemical bonds are stable. So far no other milieu has been demonstrated as a viable alternative.

Despite the simple structure of water and its obvious importance, it is still poorly understood and many of its aspects, either as a pure substance or as a solvent, are controversial. An infamous example that highlights both our interest and ignorance of water is the mistaken discovery of "polywater" to explain its perplexing properties. Some have even suggested the notion that water molecules have memory (32, 65). Beyond pure water, the properties of water in the cytoplasm are also a matter of debate. It is usually believed that water in the cell is like bulk water; however, others think that its structure is modified by the presence of many macromolecules and surfaces. Some have postulated that cell water is more like a gel. Others believe that it is strongly inhomogeneous owing to the presence of dissolved ions. Recently, it has been proposed that water very near hydrophobic surfaces is vapor-like. This may suggest that such surfaces are relatively dry and provide an explanation for the long-ranged attraction between hydrophobic surfaces.

The interactions between water and proteins and nucleic acids at the molecular level are also a topic of a major interest with the ultimate goal of understanding cellular function. A variety of experimental methods have been used to study the weak interactions between water and proteins. For example, differential scanning calorimetry, neutron diffraction (125), femtosecond fluorescence (109, 110), NMR spectroscopy (36, 45, 107, 135), and X-ray crystallography measurements are often used to study the binding sites, structure, and dynamics of water. However, some of the methodologies probe water indirectly or have other shortcomings. For example, X-ray crystallography detects only structured and localized water molecules. At present the crystal structures of biological macromolecules are determined after rapid cooling to cryogenic temperatures at which artifacts may be present in the hydration pattern (60). Theoretical and computational approaches can aid and complement the experimental efforts to decipher the interplay between water and biomolecules by providing the microscopic and physical details.

In this review, we discuss the active role that water has in the structures and dynamics of proteins and nucleic acids. Then we discuss the role water has in biological self-assembly processes such as protein folding, protein-protein recognition, and protein-DNA binding. The presented discussion on these biological processes in the context of the interaction of the biomolecules with water is twofold. First, we use these biological processes to illustrate that water is not just an "environment" for biochemical reactions but

Principle of minimal frustration: natural protein sequences are evolutionarily selected to minimize interactions that are in conflict

Frustration: conflicting interactions arising from competition between two or more states that minimize a local part of the free energy

rather it is often an active player, justifying the treatment of water as a "biomolecule." Second, we discuss that by adding the water to the description of these processes, a better understanding of their physical principles can be achieved. Focusing on biomolecules alone while ignoring the environment likely is not sufficient to capture all their properties and binding capabilities.

WATER AND BIOMOLECULE STRUCTURES AND FUNCTIONS

Although not proven, it is widely accepted that water is essential to life. On the basis of this notion water has been called the matrix of life, but it is still questioned whether water has been "fine-tuned" for life and which of its unique properties are essential for life (7, 9). Nevertheless, there is no question that water plays an important role in biomolecular structures, dynamics, and functions, a fact nicely illustrated by treating water as the "twenty-first" amino acid. Indeed, water seems to be key in understanding the interplay between structure and function, which is a central goal in protein science.

Water and Protein Structure and Stability

The hydration forces are responsible for the packing and the three-dimensional structure of proteins, which in many cases is invaluable to the protein bioactivity. The aqueous solution (55 M concentration of water) dictates the hydrophobic force, which is the driving force for protein folding and other biological processes (e.g., aggregation of amphiphilic lipids into bilayers) (15, 37, 68, 116, 130). The conflict between the hydrophobic side chains and the polar nature of the water guides these groups to collapse and be shielded from water by forming a tightly packed core that contains more than 80% of the nonpolar side chains of a typical protein. This conflict is central for protein folding. The hydrophobic interactions, as proposed by Kauzmann (74), are driven by the unfavorable structural entropy decrease that can be caused by forming a large surface area of nonpolar groups with water. Water can additionally drive protein folding by the gain in translational entropy of water molecules bound to the protein in the unfolded state upon their release (62).

Water is essential for protein structure, not only with regard to defining the collapse of the hydrophobic core, but also in maintaining its stability and structure. Affecting the network of hydrogen bonds between water molecules influences the protein stability. Increasing the ordering of water by decreasing the temperature can result in protein denaturation (i.e., cold denaturation). Indeed, the first hydration shell around proteins is ordered and exhibits a density 10% to 20% higher than that exhibited by bulk water (23, 98). These water molecules have longer residence time than water molecules outside the first hydration shell (13, 46, 63, 87, 110, 123). Some of the water molecules are bound at specific locations and can be identified crystallographically and thus are an integral part of the protein structure. Protein crystals, which normally contain substantial amounts of water (up to 70%), show a wide range of nonrandom hydrogen-bonding environments. About 55% of the first hydration shell water is bound to the backbone and the rest to charged side chain. Some of these waters are in fixed positions and are observed every time the structure is determined, whereas others are in nonunique positions and reflect an ensemble of water-protein interactions that hydrate the entire surface and sometimes the protein core (see **Figure 1*a***). The water network around the protein links secondary structure elements and not only determines the fine detail of the structure but also explains how particular molecular vibrations may be preferred. An example of the importance of solvent dynamics and hydrogen bonding to proteins is the capability of some sugars (sucrose and trehalose) to replace water molecules upon dehydration of a variety of microscopic organisms that can

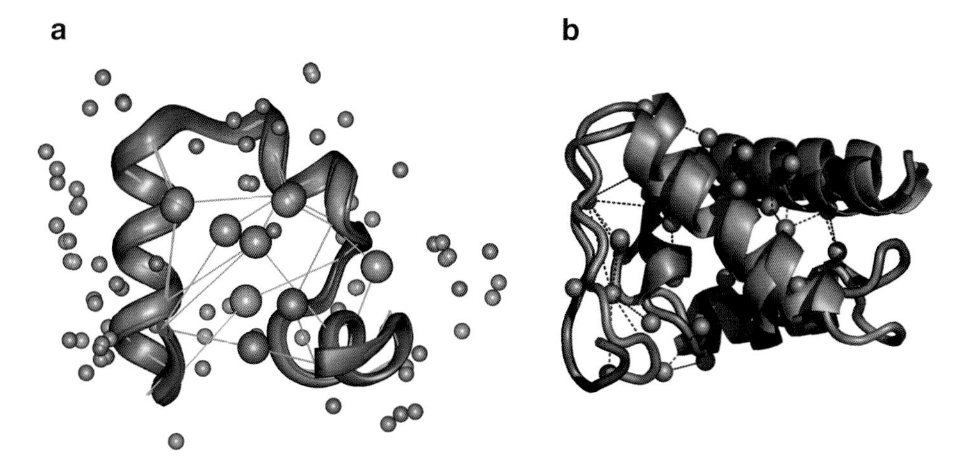

Figure 1

Water maintains protein three-dimensional structures. (*a*) High-resolution X-ray crystal structures of the villin headpiece subdomain at pH 5.1 (*orange*, 1.55 Å resolution, PDB code 1WY4) and pH 7.0 (*gray*, 0.95 Å resolution, PDB code 1WY3) (25). Waters that mediate ternary contacts (*thin lines*) are shown as large spheres, indicating that water is an integral part of the structure and that some waters are conserved in both crystallization conditions. (*b*) The prediction of the structure of CASP5 target T0170 is improved when using an optimized energy function that includes water-mediated interactions. The native and predicted structures are gray and orange, respectively. The virtual waters (defined by distance and the residue solvent accessibility) are shown as spheres.

restore their activity when rehydrated (a phenomenon known as anhydrobiosis) (30). The observation that glucose, for example, is inferior to trehalose in protecting proteins in dry conditions indicates that the preservation ability not only is a consequence of hydrogen bonds but also correlates with their glassy dynamics, which is important for maintaining internal water (124).

Not only do waters interact with the protein surface, but a few water molecules are often found trapped inside internal cavities of the protein (100). These water molecules can interact directly with the protein backbone and side chains in the protein interior (113) or even form clusters of two or more water molecules in hydrophobic cavities (45). The mean residence time is much longer for buried water molecules than for water in the first hydration shell (~500 ps for water in the first hydration shell and 10 ns to 1 ms in internal cavities) (35, 56, 107). Because bound water molecules make important interactions with groups that would otherwise make none, the waters in fixed positions should be considered an integral part of the tertiary structure, and any detailed structural description that does not include them is incomplete. Internal water sites in structurally homologous proteins are highly conserved (140), indicating that introducing buried water may improve the prediction of protein structures. Mutations can affect the number of structural water molecules within the core and disrupt the essential main chain interaction network mediated by ordered water contacts (29), resulting in destabilization. Yet, interior water molecule can escape to the bulk and be replaced by water from the hydration shell (56).

Disrupting the balance between intramolecular interactions within a protein and the hydrogen bond network with the solvent can result in protein destabilization. This principle is routinely used to denature proteins by adding substances such as urea and guanidinium chloride at high-enough concentrations that can unfold proteins (108, 141, 143). The molecular mechanism of protein destabilization by denaturant is a matter of controversy. It is not our goal in this review to cover the different proposals but to argue that the solvent plays an important role in this process. Urea was suggested to promote unfolding by direct hydrogen bonding to the protein's polar groups, which can lead to the screening of intramolecular hydrogen bonds (18, 101, 144). Another scenario is one in which urea interacts with the nonpolar groups to displace a few water molecules from the solvation shell, resulting in a net entropy gain for the water and later unfolding of the protein (82, 150). Several lines of evidence exist for an indirect unfolding mechanism in which urea perturbs the water structure and dynamics by weakening the hydrogen bond network of the water and thus disorders the water structure so that hydrophobic molecules are more easily solvated (10, 55). In contrast to denaturants that destabilize folded proteins, some small organic solutes (osmolytes) are used in nature by a variety of organisms to increase protein stability upon osmotic or water stress, high hydrostatic pressures, and dehydration. The osmolyte trimethylamine-N-oxide, TMAO, was found in molecular dynamics study to stabilize the native state indirectly by ordering and strengthening the water structure, thereby inhibiting unfolding (3, 11, 145).

Water and Protein Dynamics and Function

In addition to being fundamental to protein structure, water is needed for protein function. Increasing hydration was reported to improve the catalytic activity of enzymes (39, 54). Water was suggested to play a role in allosteric regulation at the interface of complex subunits by acting like transmission units. Water as a proton donor and acceptor can also act as a reagent in biochemical processes, illustrating that it can play more than a purely structural role. Water's property as both donor and acceptor of protons is used in nature by the formation of "proton wires" from a chain of water molecules that is used in a variety of proteins (1). However, water's hydrolytic power and high nucleophilicity is disadvantageous in the context of the cell, as it can destroy and oxidize many functional groups. For example, some enzymes shield their substrates from aqueous solvent by taking advantage of conformational changes that close off the active site from contact with bulk solvent. The cell therefore must develop a strategy to prevent water from interfering in certain biochemical reactions such as protein synthesis. Water thus introduces design demands of the cellular machinery that control water activity.

The interplay between the protein environment and its activity likely corresponds to the flexibility of the protein, which is central to the conformational changes required for enzymatic activity. In solution, proteins possess a conformational flexibility that encompasses a wide range of hydration states not seen in the crystal. Water acts as a lubricant, easing the necessary changes of the hydrogen-bonding patterns responsible for fast conformational fluctuations (8, 142). There is high coupling between the protein motions and water dynamics, and it has been suggested that fluctuations of the hydration water can slave the protein dynamics and thus affect its function (2, 48, 99). The interplay between the protein and solvent complexity is an intriguing open question. Simulations and experiments suggest that the glass-like transition of a protein coincides with dynamical changes characteristic of a glass transition in the solvent. The solvent and the protein motions may be intimately coupled, such that as a protein is warmed through its glass transition temperature, the dynamics of the hydration shell awakens motions in the protein. For lysozyme it was suggested, on the basis of molecular dynamics simulations, that water coverage of about 50% of its surface, which corresponds to about 66% coverage of the purely hydrophilic regions, is needed to achieve its dynamics (104). Simulations have shown that adding water to the cavities of bovine pancreatic trypsin inhibitor and barnase makes the proteins more flexible and increases the coupling between the motions of the water and the protein (103). Simulations have shown differences in the hydration between the redox states of cytochrome *c* with larger fluctuations upon oxidation, suggesting that a change in water structure in the hydrophobic pocket is important for undocking after oxidation (4).

Although water is important for function, there are several cases in which enzymes are functional in the absence of bulk water (e.g., halophiles that have adapted to life in high-salt solutions) (77, 78, 85). Addressing whether water is absolutely necessary for any form of life or if water is replaceable is difficult, mainly because life on earth has evolved in the presence of water (7).

Water and Nucleic Acids

As with proteins, the aqueous solution is critical to the conformation and function of nucleic acids (98) (**Figure 2**). Water constrains the conformation of a DNA molecule, as reflected by the transition from B-DNA to A-DNA upon dehydration. DNA undergoes conformational transitions in some polar solvents. The hydration of DNA depends not only on the DNA conformation but also on its sequence (42). The C-G base pairs were found to be more hydrated than T-A base pairs in both A and B conformations in both simulations and experiments (44). As for proteins, the DNA interior is mainly hydrophobic and stabilized by the stacking interactions between the consecutive base pairs, and its surface is rich with hydrophilic groups from the phosphates and sugars (19). While proteins can have hydrophilic residues in the core or hydrophobic residues at the surface, the core of nucleic acids is more uniform, as it is composed of the aromatic bases of each nucleotide. The fundamental forces that cause proteins and nucleic acids to fold to unique structures are the same; however, the energetic contributions from free energies of solvation for DNA are stronger.

Without water to screen the electrostatic repulsions between phosphate groups, the classic double-helical structure of DNA is no longer stable. In addition to hydrating the backbone phosphates, the waters in the grooves are ordered and vital to stability. Because of the regular repeating structure of DNA, hydrating water is held in a cooperative manner along the double helix in both the major and minor grooves. At high humidity at least 25 water molecules per base pair are tightly bound to the DNA. The water molecules are held relatively strongly in the first hydration shell, with residence times of about 1 ns. The water density in the first hydration shell is much larger than that in bulk water and is the outcome of the many strongly solvated sites. Changes in the hydrogen-bonding network between the hydration shell and DNA can assist ligand binding or release of ions. RNA molecules have a greater extent of hydration than DNA because of their extra oxygen atoms (i.e., ribose O2′) and unpaired base sites, suggesting an important role for structured water in RNA-RNA and RNA-protein recognition (40).

Water and Protein Structure Prediction

Physical approaches for predicting protein structure often focus on some heuristic potentials that simply acknowledge the existence of hydrophobic interactions. These models can be called dry models. Wolynes and coworkers (111) have recently incorporated a knowledge-based potential for water-mediated interactions in a Hamiltonian for structure prediction (described below). The many-body water knowledge-based potential revealed that water can stabilize proteins by bridging two hydrophilic or charged residues separated by relatively large distance. There is a substantial improvement in the predicted structures of several α-helical proteins when these water-mediated contacts are included, mainly for those with more than 115 residues.

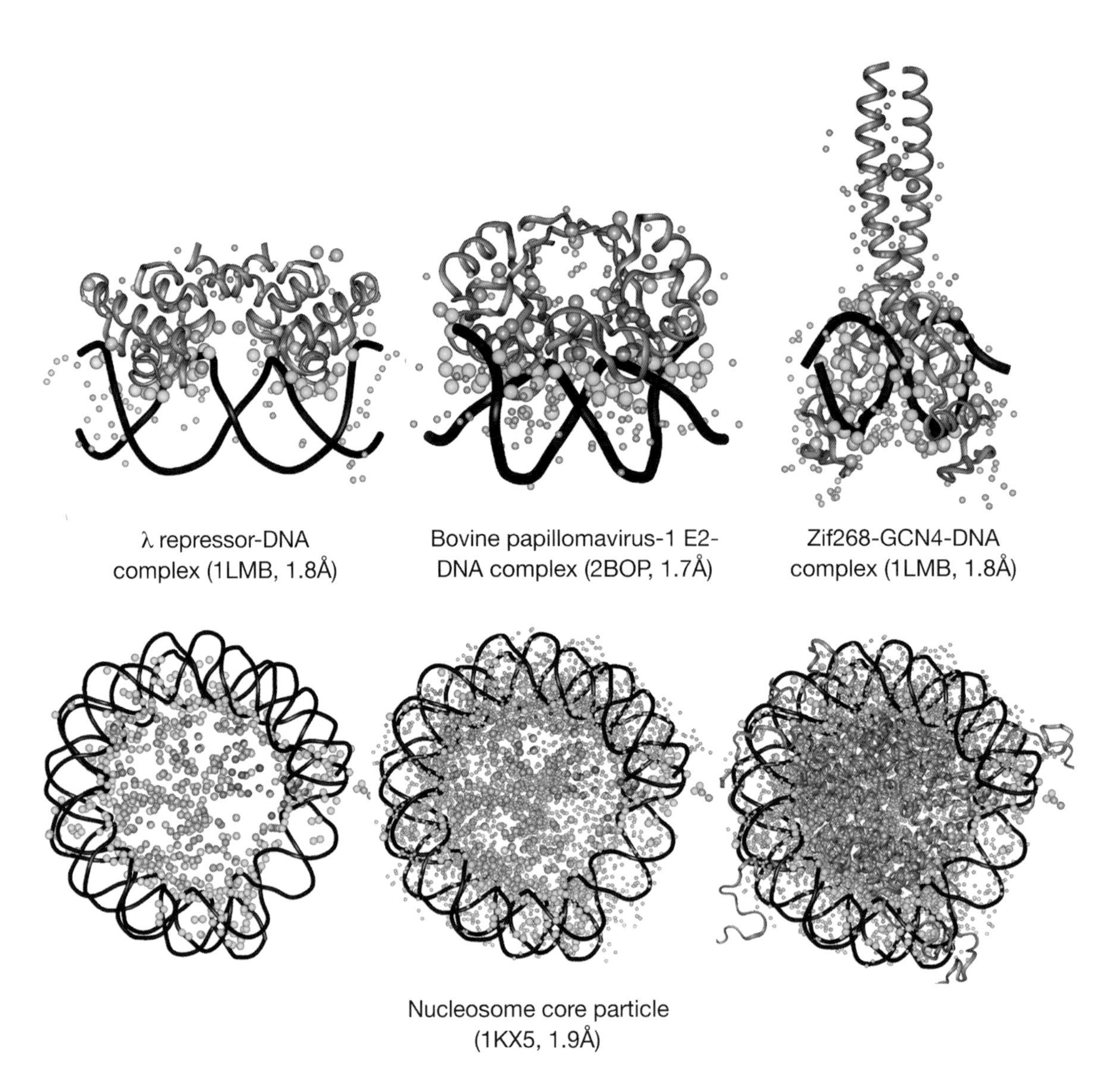

Figure 2

Waters mediate contacts in monomeric proteins, as well as protein-protein and protein-DNA interfaces. The proteins and DNAs are colored orange and purple, respectively. Waters that mediate protein-DNA interactions are shown as cyan spheres and those that mediate protein-protein interactions are shown as blue spheres. Water molecules that bridge intramolecular contacts within a single-protein chain are shown as large gray spheres and all other crystallographic waters are depicted as small spheres. Water is ubiquitous at protein-DNA interfaces (*top panel*) and may contribute to high specificity. The nucleosome is shown three times (*bottom panel*) illustrating the importance of mediation by water (*bottom, left*), its highly hydrated state (*bottom, center*), and the assistance of water in packing the histones (*bottom, right*). These examples nicely illustrate that water molecules are significant in mediating intra- and intermolecular interaction and in particular are extensive in protein-DNA interfaces.

Figure 1*b* shows the predicted structure of target T0170 from the CASP5, indicating the network of water-mediating interactions at the protein surface. Most notably is the improvement in the prediction of the loop structure using the water potential. Water, therefore, may play an important role in stabilizing loop structures and as such may be involved in the recognition of natively disordered proteins, which are often rich with hydrophilic residues. A recent all-atom simulation suggests that fluctuations in loop conformations can strongly affect protein hydration (31). In addition, a statistical study showed that internal water molecules in globular proteins preferentially reside near residues of loops and random coil regions (113). Long-range water-mediated interactions are vital in predicting α/β proteins as well (C. Zong, G. Papoian, J. Ulander, P. G. Wolynes, manuscript in preparation). Upon adding water that bridges solvent-exposed residues, the improved predicted structures suggest that incorporating water-mediated contacts between hydrophilic residues in protein design may result in proteins with enhanced stability.

The wet potential suggests that water does not only entropically drive the interactions of hydrophobic residues but also enthalpically promotes interactions between hydrophilic residues or charged residues, even of like charges. These interactions are important in the early stage of folding to guide the structural search by the formation of long-range contacts. Late events include the formation of short-range contacts and the exclusion of water from the protein interior. Water molecules can guide folding and facilitate packing of supersecondary structural elements by mediating long-range interactions between polar and charged amino acids, highlighting its role for folding and stabilizing large and multi-domain proteins. The water bridges polar groups on the protein surface. The water constrains the conformational freedom of the polypeptide chain and "smoothes" the funneled landscape.

THE DYNAMIC ROLE OF WATER IN PROTEIN FOLDING

Folding mechanisms, TSE, and even intermediates (at least for small- and medium-sized proteins) can be predicted using structure-based (Go) models that include a renormalized effect of solvent free energy (22, 27, 28, 80). This model removes energetic frustration and therefore extracts only the contributions from topological frustration (131). The excellent agreement between theory and experiments suggests that the native fold, or topology, plays the primary role in determining the folding mechanism and kinetics. These models hold true because natural proteins have a sufficiently reduced level of energetic frustration. Nevertheless, without an appropriate solvent description, one cannot explore the microscopic gating of folding by solvent.

Solvent Models in Molecular Dynamics Simulations: Implicit Versus Explicit Models

Introducing solvent effects into molecular dynamics simulations is mandatory to obtain a realistic understanding of biomolecules. The energy landscape of a protein can be drastically affected by changing the environment properties, suggesting different structure and altering the folding thermodynamics and kinetics (90, 138). The direct approach to incorporate water into the simulations of biomolecules is to explicitly include water molecules, which can be modeled in various ways. This approach significantly increases the size of the system by about one order of magnitude compared with the size of the solute alone, making the sampling of the conformation space under physiological conditions on a sensible timescale a nontrivial task. Nevertheless, applying elaborate sampling techniques (e.g., replica exchange sampling) can overcome the sampling limitation in explicit solvent simulations. Using the replica exchange method, Garcia & Onuchic (57) mapped the free-energy landscape for folding of fragment B of protein A of *Staphylococcus aureus* in explicit water, suggesting that folding studies using explicit solvent are difficult but not impossible. The use of explicit water is valuable to gain insight into the discrete role of water in folding; however, the large computational demand imposed by such models does not necessarily equate to higher accuracy, as they include several shortcomings. For example, the water is often represented by rigid three-point charge models such as TIP3P, which has been parameterized to a single temperature (~298 K) and therefore poorly captures the temperature dependence of its properties. Introducing more elaborate models for water (e.g., adding polarization, more charge sites, and bond stretching and bending) (38) will naturally increase the computational demands.

Desolation model: In this model any tertiary native contact can be either direct or mediated by a water molecule with a free-energy barrier separating them

In addition to developing new sampling methodologies of simulation with explicit solvent, another approach is to mimic water effects using simplified models (24, 47, 111, 122, 137). Implicit solvent models yield significant solvent efficiency because the solvent is modeled effectively as a function of the solute configuration alone, and therefore the need to average over the solvent degrees of freedom is overcome. Efforts to develop theories of implicit solvation have been ongoing for some time (47). These models are often based on Poisson-Boltzmann theory (i.e., continuum dielectric models such as generalized Born models) (5, 47), dielectric screening functions (84), or solvent-accessible surface area (41). In recent years, many efforts have been made to improve implicit solvent models, yet large differences still seen between implicit and explicit solvent model calculations question the degree to which implicit solvent models mimic the solvent environment (102, 119, 148).

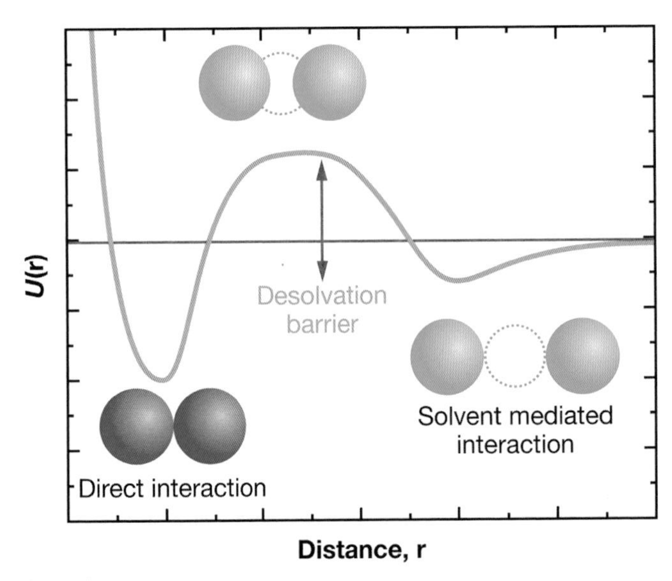

Figure 3

Schematic representation of the potential energy function, U(r), in the desolvation model. In this model, any native interaction between two residues (*spheres*) can either be direct or separated by a water molecule (*light blue dashed circle*). The Cα-Cα distance of two residues that directly interact is defined by the native structure, and when a water molecule separates them the optimal distance increases by the diameter of the water molecule. At the desolvation barrier the water overlaps with the two residues.

It is beyond the scope of this review to cover the different approaches used to implicitly represent solvent effects on biomolecule dynamics and thermodynamics, but we discuss two reduced solvent models designed to incorporate mediating tertiary interactions via water. The desolvation and the water knowledge-based models focus on contact gating by water, but they also show several differences. The desolvation model focuses on the free-energy cost of bringing two nonpolar solutes into contact (24, 64, 114, 117). This free-energy penalty of contact formation is a direct result of the granularity of water molecules in the first hydration shell. Using this idea, Hillson et al. (64) have explained the non-Hammond pressure dependence of folding rates. This model suggests that the solvent gates the formation of individual pairwise contacts in folding. Accordingly, the phenomenological potential for each tertiary contact includes two minima of direct and mediated interactions, which are separated by the desolvation barrier (**Figure 3**). The desolvation model has been applied to protein folding in several studies that are discussed below.

Another approach to model water gating in biomolecular self-organization processes is to construct a knowledge-based potential based on the occurrence of water-mediated interactions in a large nonredundant dataset of biomolecular structures. Because the number of water molecules in X-ray structures is underestimated (even at high resolution), the knowledge-based water potential cannot rely on crystallographic waters, but rather on a more general definition (91). Wolynes and coworkers (111, 112) have recently constructed such wet potentials for folding and binding using a physical bioinformatic approach. In this model, two residues can interact directly or indirectly via a water molecule that serves as a bridge between them. Tertiary interactions mediated by a water molecule exist if the distance between the C_β atoms of the two residues, which are exposed to solvent, is in the range of 6.5 to 9.5 Å (a typical distance between C_β atoms that directly interact is 4.5 to 6.5 Å). As the C_β-C_β distance between residues with large side chains can exceed 6.5 Å, the model also allows each pair of residues to interact directly, with a C_β-C_β distance of 6.5 to 9.5 Å (in a dry environment). The bioinformatics-derived occurrences of residue-residue contacts are optimized by maximizing the ratio of the folding/binding temperatures to the glass-transition temperature (i.e., minimizing frustration effects) (58).

The knowledge-based potential accounts for several central features of water-protein properties: inducing folding by mediating long-range interactions, protein stabilization by the first hydration shell, and the contribution of water to folding cooperativity. **Figure 4** shows the optimized potentials for forming all possible direct and indirect interactions in folding and binding for the 20 amino acids. The folding potential was calculated on the basis of a dataset of monomeric proteins and the water-mediated interactions are between two residues at the protein surface. For binding, however, a dataset of protein complexes has been used and the water mediates interfacial interactions. These potentials, for both folding and binding, illustrate that gating by water is highly favorable for polar and charged residues. This gating by water can increase specificity and stability.

The desolvation model and the wet knowledge-based potential for water-mediated interactions complement each other. Both models are pairwise potentials; however, the knowledge-based model includes cooperativity effect and therefore local frustration. The knowledge-based approach allows both native and nonnative interactions to be mediated by water, and in the current desolvation model only native interactions are treated. Moreover, the desolvation model focuses on the energy penalty in expelling a water molecule that bridges two residues, whereas the knowledge-based potential describes the enthalpic effect of water in stabilizing the native state. Note that both models do not account for buried waters in internal cavities.

Water Expulsion Versus Drying Effects in Folding

Water has a dynamical role in protein folding. A detailed investigation on the dynamical role of water in folding is available for the SH3 domain, for which both minimalist model and atomistic simulations have been used to explore the microscopic properties of water during folding. The desolvation model, which focuses on the energy cost of expelling waters that mediate any native contact, suggests that the folding of the SH3 protein is a two-step process: First, the fully solvated SH3 protein undergoes an initial structural collapse to an overall native topological conformation (funneling landscape), followed by a second transition in which water molecules are cooperatively squeezed out from the hydrophobic core region, resulting in a dry and packed protein (24). Folding, thus, is achieved through a TSE that is highly hydrated but has a native-like structure. The water acts as a lubricant that enables the hydrophobic core to find its optimally packed state, and it can play a role in preventing nonnative contacts from forming.

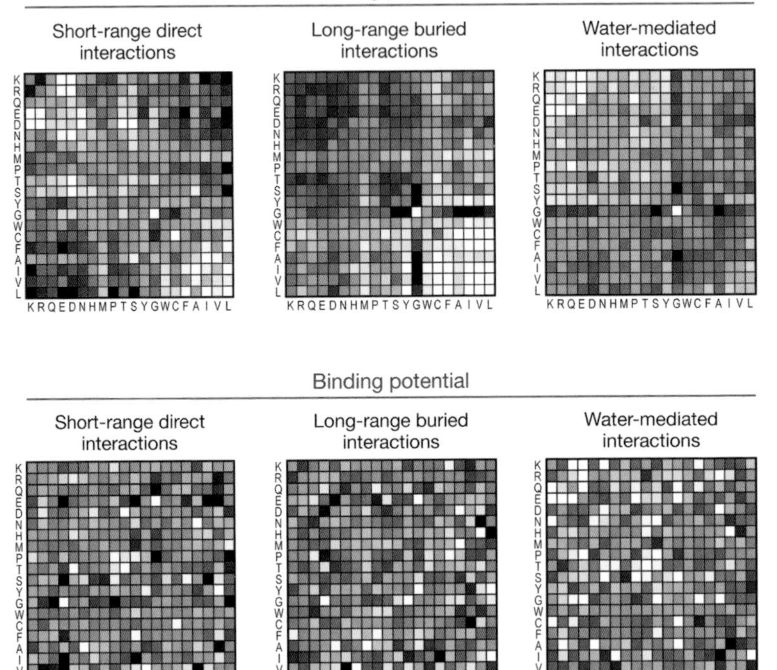

Figure 4

Optimized knowledge-based potentials for folding and binding. Each pair of residues can interact directly or indirectly via a water molecule. Lighter color indicates a more favorable interaction. The potentials illustrate that water-mediated interactions are dominant for both folding and binding yet with a stronger signal for binding processes. Short- and long-range direct contacts occur between residues when the distance between their C_β atoms is 4.5 to 5 Å and 6.5 to 9.5 Å, respectively. Similarly, a water-mediated contact was set to a distance of 6.5 to 9.5 Å with the additional demand of high solvent accessibility.

The desolvation model has correctly reproduced the folding rate of SH3 mutants and provides a microscopic explanation of destabilizing core mutations on the folding rate. Mutating valines 44 and 53, which participate in the folding nucleus, result in slower folding rates because the V44T mutation disrupts the structural search collapse while the V53T mutation hinders the desolvation of the hydrophobic core of the TSE (50).

Adding the pairwise additive desolvation term to the native structure–based model results in increasing the stability of the native conformation (128) and in slowing the folding kinetics for both the SH3 domain (24) and chymotrypsin inhibitor 2 (75). These folding rates are more similar to the experimental folding rates than to those obtained by a model that does not take solvent effects into account. Yet, nonadditivity must likely be incorporated to introduce the many-body nature of solvent-mediated interactions to achieve the level of cooperativity often observed in experiments (20). An initial effort in this direction has been performed by introducing small intrachain pairwise desolvation barriers, which is independent of the interaction stability (96), or increasing the desolvation barrier height (76), resulting in a relatively high enthalpic barrier for folding.

The atomistic simulations of the folding of the SH3 protein domain agree with the folding mechanism of SH3 found using the desolvation model. These studies show that the folding depends on a gradual, few molecules at a time expulsion of water from the collapsed interior and can involve a lubricated hydrophobic core at the late stage of folding (59, 132). Atomistic simulation studies of protein A three-helix bundle (57), protein G (133), and protein L (73) support the role of water as a lubricant for the packing of the hydrophobic core after the formation of the transition state (**Figure 5**). A similar mechanism has been observed in the folding of a 23-residue peptide (119). The latter simulation study has suggested that although water is trapped in the core at the TSE, the TSE is completely defined by the protein and not by the geometry of the water. Accordingly, the folding probability of a given conformation of small peptide was found to be independent in the configuration of the water (119). Moreover, these fully atomistic simulations (57, 119, 132, 133), which are not biased toward direct contacts between the residues, point out that the folded state is not completely dry but that a few core water molecules form hydrogen bonds with the protein backbone.

Thus, water molecules can mediate the search for the protein native topology, in which waters serve a structural role as backbone hydrogen bond bridges between the residues connecting the hydrophobic residues and as water molecules simply residing inside the core. Some of these waters are gradually expelled from the formed core in a later stage after the initial funneling. The active role of water in folding is an outcome of the size of a water molecule, their discrete nature, and the flexibility of proteins as well as the hydrophilicity of the backbone chain. Water might assist the association of two rigid hydrophobic objects by the so-called dewetting (drying) effect, which is characterized by the collective emptying of space around the nucleating sites and the formation of a large vapor bubble (21, 67, 97). Protein hydrophobic collapse via the dewetting mechanism is characterized by a decrease in water density and then by the spontaneously collapse of the core to stabilize the protein by reducing the solvent-accessible area of the core residues. On the other hand, in the expulsion mechanism, core compaction precedes water expulsion. The validity of the dewetting

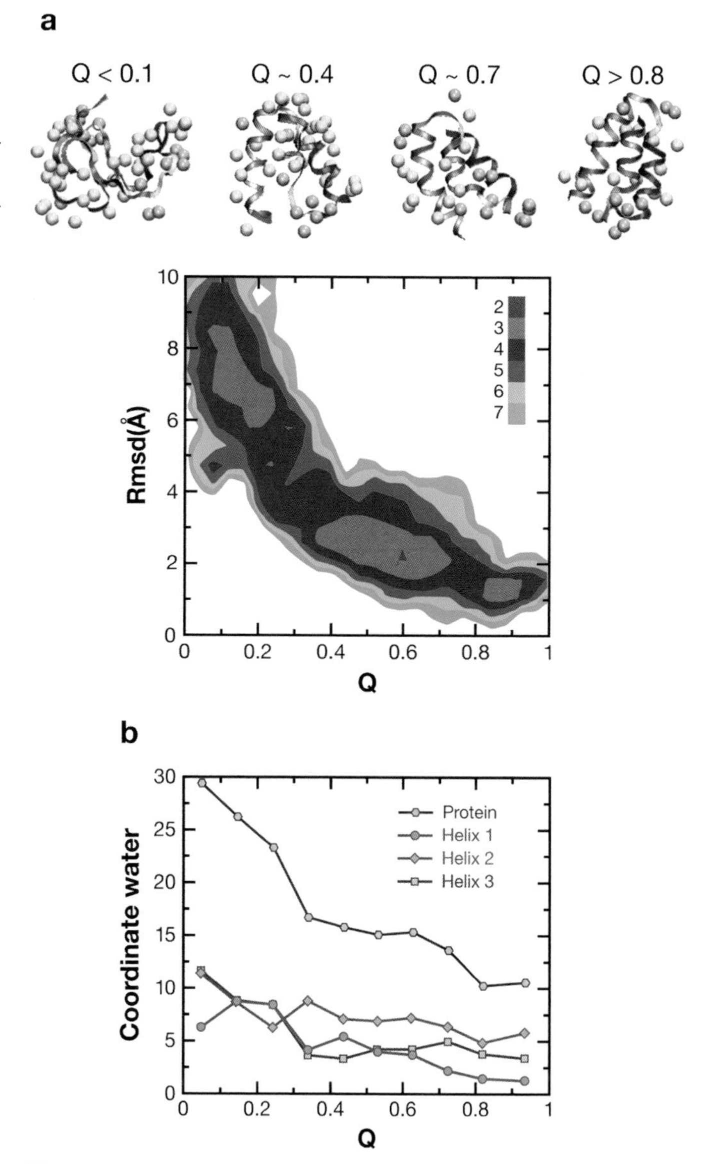

Figure 5

Folding of protein A through a hydrated native-like intermediate. (*a*) Free-energy surface at the transition temperature. (*b*) Average coordination number of water molecules in the helices and the whole protein as a function of Q. The selected conformations at the top illustrate the water expulsion as the folding progresses along the reaction coordinate Q.

Dewetting (drying) effect: water density is decreased and a vapor bubble is formed that drives hydrophobic assembly

Water expulsion: water is gradually expelled from the collapsed interior and assists protein folding by mediating interactions

scenario for proteins has been directly questioned recently by exploring the hydrophobic collapse of several multidomain proteins with hydrophobic simple interfaces using atomistic simulations (66, 95, 149). A signal for the possible drying effect was seen only when the intrinsic properties of the protein chain were turned off. In the association of the two almost-rigid domains of the BphC enzyme, a drying effect was observed only when the electrostatic protein-water forces or attractive van der Waals forces were turned off (149). Similarly, a signal for a drying effect was detected when protein flexibility was drastically suppressed in the association of melittin tetramer and $\alpha_2 D$ homooligomers, although their assembly is experimentally described as coupled folding/binding processes (66, 95).

Water plays an active role in the folding of nucleic acids as well. All-atom molecular dynamics simulations of a RNA hairpin-loop motif showed that, similar to protein folding, RNA folding occurs by hydrophobic collapse via the expulsion mechanism of desolvating central hydrophobic regions after initial nucleation of several base pairs (139). Water-mediated interactions in the folding of small nucleic acids appear to be essential for capturing the correct hydrophobic collapse among other nonspecific collapse events, therefore constituting a structural role. Nucleic acids have more uniform hydrophobic cores than do proteins owing to the aromatic group of the bases compared with hydrophobic residues that are sparsely located along the sequence, suggesting higher cooperative collapse and less trapping of water in the folding of nucleic acids.

WATER IN PROTEIN-PROTEIN BINDING

The forces that drive protein-protein binding are similar to those that drive protein folding, and thus polar and hydrophobic interactions as well as hydrogen bonding dominate both processes. Water, however, as is evident more from their abundance at the interfaces of protein complexes than from the interior of a monomeric protein, is likely to be more dominant for binding and recognition than for folding.

Water in Protein Interfaces

Water is abundant in protein-protein interfaces. Upon assembly, the interfaces of many formed complexes are hydrated and consist of about 10 water molecules per 1000 $Å^2$ of interface area (121). Water molecules at interfaces form hydrogen bonds with the backbone polar groups or charged side chains. Although common, interface hydration is not uniform (87). Protein-protein interfaces exhibit different degrees of solvation and also different spatial distribution patterns. The level of hydration obviously depends on the polarity and geometry of the interface. It was also observed that homooligomers have more hydrated interfaces than do heterooligomers (121). In some complexes the waters are only at the interface rim, whereas in others they cover the entire interface area. The higher hydrophilic nature of interfaces formed when two folded monomeric proteins associate, compared with those formed in coupled folding-binding process (72, 94, 147), may suggest a different role of solvent. Accordingly, the interfaces of complexes formed between subunits that are natively disordered are expected to be dryer than those formed between folded proteins, exhibiting similarities to protein cores. Nonspecific crystal packing interfaces, which are more polar, are often 50% more solvated than the interfaces of protein complexes. Although immobilized solvent is widely observed in X-ray structures, it is likely that the interface solvation is currently underreported, as their detection demands extremely high-resolution structures.

Binding Mechanism Is Governed by the Protein Topology

Protein topology, currently well accepted as a pivotal factor in determining unimolecular folding, also determines many aspects of protein assembly. This notion has been obtained from native topology-based (Go) models, which include only interactions that stabilize the native structure as determined by NMR or X-ray crystallography and thereby capture the protein topology. These models are energetically unfrustrated models (i.e., they do not include nonnative contacts) and correspond to a perfectly funneled energy landscape.

The native topology-based model has been applied recently in several studies to examine the mechanism of protein association (89, 93, 94) and successfully reproduce the experimental classification of homodimers regarding whether monomer folding is prerequisite to monomer association (**Figure 6**). Obligatory homodimers that exhibit two-state thermodynamics are formed by a coupled folding and binding reaction. Transient homodimers, which bind via a thermodynamic intermediate, are formed by the association of already folded monomers. In general, we found that most of the gross and many of the finer features of binding mechanisms can be obtained by Go model simulations. The need to understand the mechanism of binding and its main determinants is well illustrated by our study on the association pathway of dimeric HIV-1 protease (88). These studies have indicated that the monomeric HIV-1 protease is relatively folded in its free form. The binding by association of prefolded monomers suggests a new way to inhibit the protease activity by designing an inhibitor that binds to the monomer and thus prevents dimerization rather than by designing an inhibitor that blocks the active site but eventually becomes ineffective owing to drug resistance.

Native topology-based model: a solvent-averaged energy potential defined by attractive native state interactions and repulsive nonnative interactions; corresponds to a perfectly funneled energy landscape; also called Go model

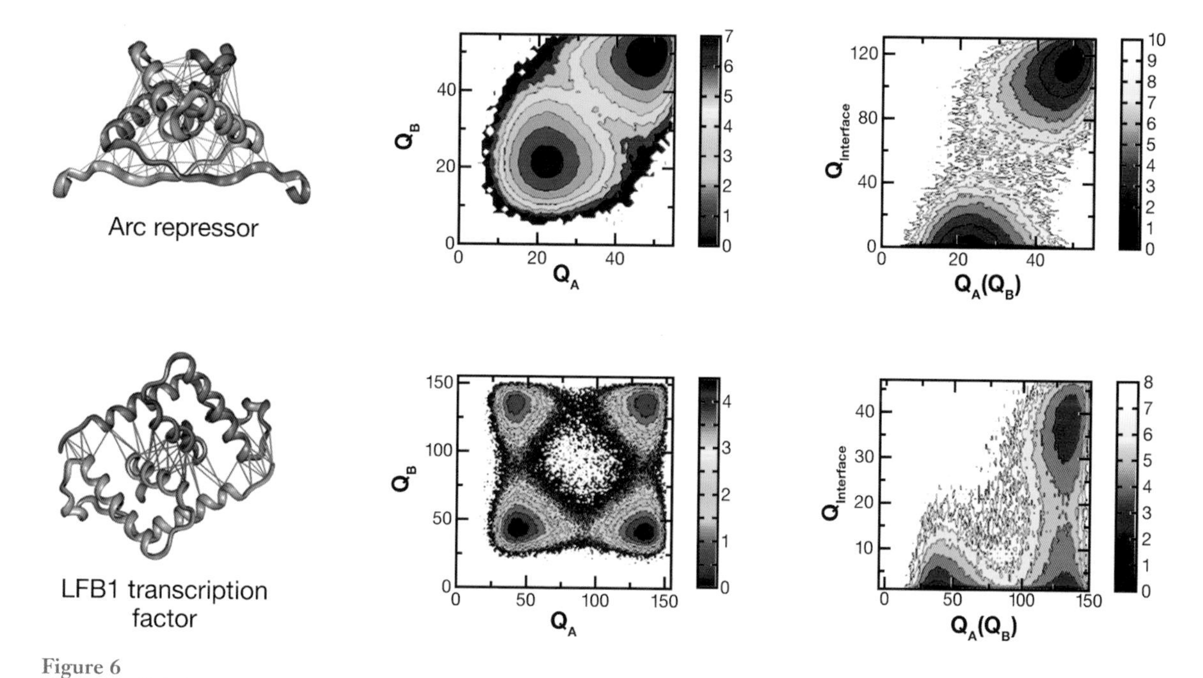

Figure 6

The association mechanism of the dimerization of Arc-repressor (an obligatory homodimer) and LFB1 transcription factor (a transient homodimer). The binding free-energy landscapes are plotted against Q_A and Q_B (the native contacts in monomers A and B, respectively) and $Q_{Interface}$ (the interfacial native contacts). Q_A and Q_B correspond to folding/unfolding events, and $Q_{Interface}$ corresponds to binding/unbinding events. The free-energy surfaces indicate that the native topology-based model reproduced their experimentally determined binding mechanism.

The native topology-based models agree with the experimentally determined binding mechanism regarding the existence of a monomeric intermediate. The validity of the model in studying protein binding is reflected by the good correlation obtained between the computational and experimental Φ values, which measure the degree of structure at the TSE at the residue level. For Arc-repressor and the tetramerization domain of p53 (p53tet), a direct comparison between the simulated and experimental Φ values is available and indicates that the simple Go models capture the nature of the TSE reasonably well. For Arc-repressor there are detailed deviations between the simulated and experimental Φ values of particular residues (reflected by a correlation coefficient of 0.31), but there is an agreement about the overall structure of the TSE. For p53tet, which was experimentally classified as a dimer of dimers, not only did the native-centric model reproduce the association mechanism, but the computational Φ values for the dimerization and tetramerization reactions are in agreement with the experimental ones (**Figure 7**). Note that recently an all-atom molecular dynamics study was done on the dimerization reaction of p53tet (26). The Φ values for the binding TSE from that study, which includes explicit water and nonnative interactions, displays results qualitatively similar to those obtained from the native topology-based model.

The ability of native topology-based models to reproduce the features of binding mechanisms is significant and suggests that the binding TSE and binding mechanism can be obtained by the knowledge of the final complex's structure alone. We have recently found that protein complexes formed by the association of already folded subunits have structural and topological properties different from those with intrinsically unfolded subunits. More specifically, these two classes of complexes differ in the topological properties (i.e., connectivity of residues, average clustering coefficient, and mean shortest path length) of the monomers and the interfaces (89). Nonetheless, one may expect that adding water to binding simulations may reveal the existence of additional binding TSEs that are relatively hydrated as well as high-energy intermediates.

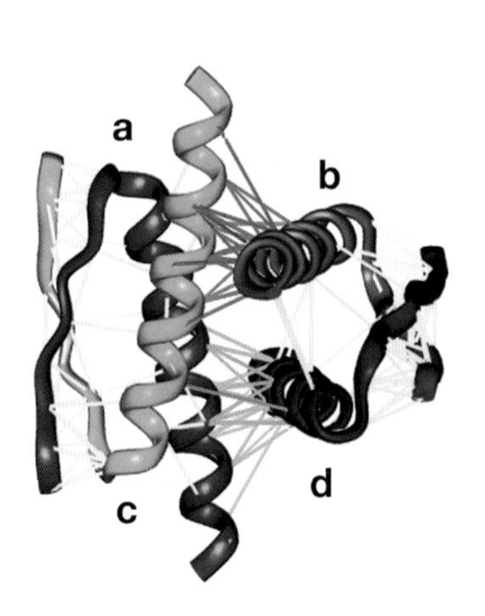

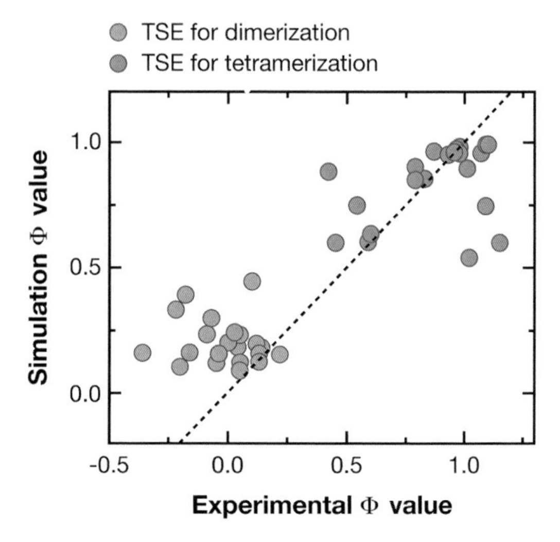

Figure 7

Comparison of the structure of the TSE of the tetramerization of p53tet from simulation and experimental Φ value analysis. The two TSEs in the assembly of the tetramer (dimerization of monomers to form *ac* and *bd* and the dimerization of the dimers) were detected in the native topology-based model.

The Role of Water in Biological Associations and Aggregation

In addition to topology, other factors such as nonnative interactions, electrostatics, and water interaction can affect binding mechanisms and kinetics (92). The abundance of water in complex interfaces, as discussed above, indicates its potential role in binding. Kinetically, water molecules can guide a fully solvated protein to recognize another fully solvated protein (or nucleic acids) by a gradual expulsion of water layers. The native topology-based model effectively takes into account structural water molecules but not dynamic water molecules and thus cannot address the desolvation mechanism of bringing two solvated proteins to form a specific and tight assembly. We had conjectured that our simulations of antibody-antigen complex using the topology-based model poorly reproduced the binding TSE because of lack of water molecules in our model (89). The abundance of water in mediating contacts in other forms of the complex explained the discrepancy between the experimental and computational characterization of the binding transition state. Solvent molecules thus can assist the initial association to form the encounter complex. Alternatively, the main binding transition state, which is squeezed out at a later stage and results in a dry interface, is stabilized by shape complementarity (126). A wet encounter complex and transition state suggests that, similar to folding, proteins bind by a gradual expulsion of the solvent molecules, which is even less complete owing to the hydrophilic nature of many complex interfaces.

Because water is essential for folding and binding, it is certainly important to aggregation as well; however, it is still unclear if its effect is direct or indirect. Dehydration can affect the intricate balance between the protein internal interactions and the interaction with the hydration shell. Destabilization of the weak water-protein interactions can affect protein stability and flexibility and therefore supports conformational changes (81). Molecular dynamics simulation study of the amyloidgenic $A\beta_{16-22}$ peptides has shown that the monomer adopts a β-strand conformation in urea, suggesting that urea at low concentrations may facilitate amyloid formation (61, 79). It was also hypothesized that proteins involved in conformational diseases have a large number of hydrogen bonds not protected against solvent interactions (34, 49). This solvated region of the protein surface was suggested to be structurally more labile with a consequent potential for aggregation.

Water Is Central for Recognition

The common wet nature of protein-protein interfaces may suggest that water is part of the recognition code, as it mediates interactions that are less favorable in its absence. It is plausible that water assists two proteins not only in improving their binding interface but also in discriminating between the potential binding sites. Accordingly, the water smooths the binding funnel. A coarse-grained folding potential showed limited success in describing binding (112). Water can make single or multiple regions of the protein surface more adaptable for binding than other patches. This scenario suggests that water is a kind of molecular glue between protein subunits and can eliminate the number of possible binding modes by contributing to exquisite specificity (12). The water molecules that were part of the hydration shell of the free subunits are much more localized when placed at the interface and can be treated as an integral part of the structure.

The knowledge-based potential for direct and water-mediated interactions for protein binding shows that with the assistance of water molecules some residues are likely to interact (**Figure 4**). These potentials indicate that water-mediated interactions are more central in bimolecular than in unimolecular recognition. These water molecules capture structural information for the formation of the protein complex, or alternatively, they edit empty spaces between the complex subunits and act as extensions of the protein chain. The proposed role of water in binding suggests that water must be included in methods designed to predict protein-binding sites and the complex formed between two or more proteins. Currently, most docking algorithms for predicting protein complexes starting from the free monomers ignore hydration effects upon binding. A docking approach that incorporates discrete water molecules has reported a significant improvement in some cases (118).

Mediating residue-residue interactions by water in protein recognition, which may make protein surfaces more adaptable for binding, can also lead to promiscuous binding (83). In such cases, water acts as a buffer that weakens unfavorable interactions, thereby accommodating various substrates with low specificity (136). It is possible that such weak water-mediated interactions are key for transient protein-protein interactions, which are characterized by smaller and less hydrophobic interfaces. The high adaptability and relatively low energetics of water-mediated interactions are in accordance with the observation that residues that contribute the most to the binding free energy (i.e., "hot spots") are placed in a dry environment (15). Promiscuous binding via water can be the basis of the dynamic protein association needed for signal transduction pathways.

Water-mediated interactions in protein interfaces is suggested to be favorable enthalpically and thus enhance stability in a way that compensates the entropic cost that must be paid for immobilizing interfacial waters. In recent molecular dynamics simulations it was found that water can also enhance binding affinity by a gain in free energy resulting from an increased entropy of the trapped water molecules. Water molecules in bulk have limited freedom due to their participation in a water network, while water molecules inside a slightly nonpolar cavity may have more freedom than in bulk, resulting in higher entropy (115).

WATER IN PROTEIN-DNA RECOGNITION

The tightness and order of the DNA hydration shell and that DNA hydration depends on DNA conformation and sequence indicate that water molecules are an integral part of nucleic acids. This suggests that water can be directly involved in protein-DNA recognition. Indeed, many protein-DNA interfaces are highly solvated (69). These interfaces are much more polar than protein-protein interfaces because of the phosphate groups on the DNA side and the abundance of positively charged groups on the protein side. In addition to direct interactions between proteins and DNA base pairs (i.e., direct hydrogen bonds, van der Waals, electrostatic, and hydrophobic contacts), which are important for sequence-specific recognition, in many cases indirect interactions between residues and the DNA bases exist via water molecules (120) (**Figure 2**). These water molecules are not just "filling spaces"; they mediate recognition and specificity mainly by screening unfavorable electrostatic and hydrogen bonding (70, 127). A large number of water-mediated contacts (mainly between protein and DNA but also within and between the histones) have been found in the structure of the nucleosome core particle, enabling additional interactions between the DNA and the histones and within and between the histones themselves (33) (**Figure 2**).

Mutating the DNA target by perturbing a water site affects the protein binding affinity. It was observed that the DNA hydration pattern is similar in the free and bound states, suggesting that recognition is, in part, due to complementarity of surface hydration (129). It was also suggested that protein atoms involved in binding to DNA occupy positions normally occupied by waters in the free DNA (146). In some cases, water at the interface can exchange with bulk solvent and maintain a partially disordered interface, which can be entropically advantageous. Furthermore, the interfacial disorder can facilitate recognition via the fly-casting mechanism (134), in which the water acts as a molecular glue that increases structural adaptability. In nonspecific protein-DNA complexes more water molecules remain at the interface and lubricate protein sliding on the DNA. Many of these water molecules must be displaced for specific recognition, and the driving force for complex formation would seem primarily entropic (52, 71).

Despite the evidence for the role of water mediation in sequence-specific DNA recognition by proteins, this notion is still underappreciated. Current methods used to decipher protein-DNA recognition codes on a genomic scale rely mainly on the direct contacts between protein residues and the DNA base pairs. An initial approach to construct an optimized knowledge-based potential for protein-DNA binding includes water-mediated interactions in which a virtual water is defined by distance criterion and solvent accessibility of the residue and the base to which it links (**Figure 8**). The wet knowledge-based potential for protein-DNA recognition discriminates between specific and nonspecific DNA sequences (unpublished data). The potential reveals that some water-mediated interactions between the protein residues and the DNA groups are as important as the direct interactions (**Figure 8**). This observation is supported by the finding that perturbing a direct hydrogen bond or a water-mediated interaction at the interface of the papillomavirus E2C–DNA complex results in similar destabilization (53).

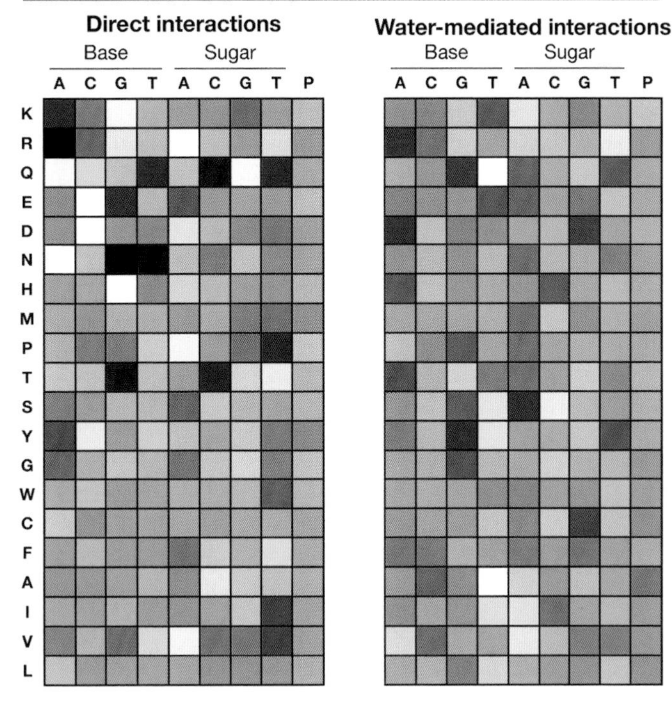

Figure 8

Optimized knowledge-based potentials for protein-DNA binding. Each residue can interact directly or indirectly (mediated by water) with the bases, sugar, and phosphate groups of the four nucleotides. Lighter color indicates a more favorable interaction. A direct contact is defined on the basis of the typical distance found in resolved protein-DNA structures, and the indirect interactions are defined by allowing distances larger by 3 Å than the distances of the direct interactions.

SUMMARY

Whether life can evolve in nonaqueous media is still an open question; however, it is unquestionable that water is central for life on earth. The many roles water plays in biomolecular processes, and particularly the coupling between its motions and the dynamics of proteins and nucleic acids, are currently widely acknowledged, as reflected, for example, by hydrating vacuum simulation, which is done routinely. Nonetheless, water is often treated as an inert environment, yet in many cases it is actually an active player. Therefore, even when we omit water molecules when drawing three-dimensional structures of proteins and nucleic acids for the sake of simplicity, we should keep them in mind.

In this review, we presented a few cases in which water has a dynamic role beyond maintaining the structure of proteins and nucleic acids. For example, water can guide the conformational search in protein folding by gating hydrophobic residues. While our understanding of the role of water solvation in protein folding has improved, the limited successes of implicit solvent models in accurately representing protein stability and dynamics suggest that the physics of the interaction between biomolecules and the solvent is not completely captured. Developing a battery of models to explore the dynamical and structural features of water in the hydration shells, internal cavities, and complex interfaces is vital to understanding the structures of proteins and nucleic acids and their folding and binding processes. Giving attention to water may therefore be beneficial to folding, docking, and structure design efforts.

SUMMARY POINTS

1. Water is highly acknowledged for playing an important role in the structure, stability, dynamics, and function of biological macromolecules. Yet only recently has water been considered an active component rather than an inert environment.
2. Water guides the conformational search in protein folding by gating hydrophobic residues. Several studies reported the existence of wet native-like intermediates. Thus water has a dynamic role in mediating the collapse of the chain and the search for the native topology through a funneled energy landscape.
3. Water can enhance the stability of biological macromolecules. Water-mediated interactions are favorably enthalpic. Alternatively, water residing in hydrophobic pockets can stabilize entropically because it has higher entropy than in bulk water. Both cases indicate that water molecules are not just "filling spaces" but are integral components of the structure.
4. Water can mediate recognition by discriminating between specific and nonspecific binding.
5. Giving attention to water will shed light on the physics of self-assembly and advance our understanding of the natural design of proteins and nucleic acids.

ACKNOWLEDGMENTS

The authors thank Garegin Papoian for providing **Figure 1*b***, Joe Hegler for sharing unpublished data, and Sam Cho and Diego Ferriero for critically reading the manuscript. This work was funded by the NSF-sponsored Center for Theoretical Biological Physics (grants PHY-0216576 and 0225630) with additional support from MCB-0543906. Current address for Dr. Yaakov Levy is Department of Structural Biology, Weizmann Institute of Science, Rehovot 76100, Israel.

LITERATURE CITED

1. Agmon N. 1995. The Grotthuss mechanism. *Chem. Phys. Lett.* 244:456–62
2. Ansari A, Jones CM, Henry ER, Hofrichter J, Eaton WA. 1992. The role of solvent viscosity in the dynamics of protein conformational changes. *Science* 256:1796–98
3. Athawale MV, Dordick JS, Garde S. 2005. Osmolyte trimethylamine-N-oxide does not affect the strength of hydrophobic interactions: origin of osmolyte compatibility. *Biophys. J.* 89:858–66
4. Autenrieth F, Tajkhorshid E, Schulten K, Luthey-Schulten Z. 2004. Role of water in transient cytochrome c(2) docking. *J. Phys. Chem. B* 108:20376–87

5. Baker NA. 2005. Improving implicit solvent simulations: a Poisson-centric view. *Curr. Opin. Struct. Biol.* 15:137–43
6. Ball P. 1999. *H_2O: A Biography of Water*. London: Weidenfeld & Nicolson
7. Ball P. 2005. Water and life: seeking the solution. *Nature* 436:1084–85
8. Barron LD, Hecht L, Wilson G. 1997. The lubricant of life: a proposal that solvent water promotes extremely fast conformational fluctuations in mobile heteropolypeptide structure. *Biochemistry* 36:13143–47
9. Benner SA, Ricardo A, Carrigan MA. 2004. Is there a common chemical model for life in the universe? *Curr. Opin. Chem. Biol.* 8:672–89
10. Bennion BJ, Daggett V. 2003. The molecular basis for the chemical denaturation of proteins by urea. *Proc. Natl. Acad. Sci. USA* 100:5142–47
11. Bennion BJ, Daggett V. 2004. Counteraction of urea-induced protein denaturation by trimethylamine N-oxide: a chemical chaperone at atomic resolution. *Proc. Natl. Acad. Sci. USA* 101:6433–38
12. Bhat TN, Bentley GA, Boulot G, Greene MI, Tello D, et al. 1994. Bound water molecules and conformational stabilization help mediate an antigen-antibody association. *Proc. Natl. Acad. Sci. USA* 91:1089–93
13. Bizzarri AR, Cannistraro S. 2002. Molecular dynamics of water at the protein-solvent interface. *J. Phys. Chem. B* 106:6617–33
14. Blokzijl W, Engberts JBFN. 1993. Hydrophobic effects: opinions and facts. *Angew. Chem. Int. Ed.* 32:1545–79
15. Bogan AA, Thorn KS. 1998. Anatomy of hot spots in protein interfaces. *J. Mol. Biol.* 280:1–9
16. Bryngelson JD, Onuchic JN, Socci ND, Wolynes PG. 1995. Funnels, pathways, and the energy landscape of protein-folding: a synthesis. *Proteins Struct. Funct. Genet.* 21:167–95
17. Bryngelson JD, Wolynes PG. 1987. Spin glasses and the statistical mechanics of protein folding. *Proc. Natl. Acad. Sci. USA* 84:7524–28
18. Caflisch A, Karplus M. 1999. Structural details of urea binding to barnase: a molecular dynamics analysis. *Struct. Fold. Des.* 7:477–88
19. Calladine CR, Drew HR, Luisi BF, Travers AA. 2004. *Understanding DNA: The Molecule and How It Works*. Amsterdam: Elsevier
20. Chan HS, Shimizu S, Kaya H. 2004. Cooperativity principles in protein folding. *Methods Enzymol.* 380:350–79
21. Chandler D. 2005. Interfaces and the driving force of hydrophobic assembly. *Nature* 473:640–47
22. Chavez LL, Onuchic JN, Clementi C. 2004. Quantifying the roughness on the free energy landscape: entropic bottlenecks and protein folding rates. *J. Am. Chem. Soc.* 126:8426–32
23. Cheng YK, Rossky PJ. 1998. Surface topography dependence of biomolecular hydrophobic hydration. *Nature* 392:696–99
24. **Cheung MS, Garcia AE, Onuchic JN. 2002. Protein folding mediated by solvation: Water expulsion and formation of the hydrophobic core occur after the structural collapse. *Proc. Natl. Acad. Sci. USA* 99:685–90**

Using the desolvation model, the authors show that the folding of SH3 domain is assisted by water.

25. Chiu TK, Kubelka J, Herbst-Irmer R, Eaton WA, Hofrichter J, Davies DR. 2005. High-resolution x-ray crystal structures of the villin headpiece subdomain, an ultrafast folding protein. *Proc. Natl. Acad. Sci. USA* 102:7517–22
26. Chong LT, Snow CD, Rhee YM, Pande VS. 2005. Dimerization of the p53 oligomerization domain: identification of a folding nucleus by molecular dynamics simulations. *J. Mol. Biol.* 345:869–78

27. Clementi C, Jennings PA, Onuchic JN. 2000. How native-state topology affects the folding of dihydrofolate reductase and interleukin-1beta. *Proc. Natl. Acad. Sci. USA* 97:5871–76
28. Clementi C, Nymeyer H, Onuchic JN. 2000. Toplogical and energetical factors: What determines the structural details of the transition state ensemble and "en-route" intermediate for protein folding? An investigation of small globular proteins. *J. Mol. Biol.* 298:937–53
29. **Covalt JC, Roy M, Jennings PA. 2001. Core and surface mutations affect folding kinetics, stability and cooperativity in IL-1 beta: Does alteration in buried water play a role? *J. Mol. Biol.* 307:657–69**

Reports that water molecules buried within the protein core are involved in the network of hydrogen-bonding and are important for stabilizing the structure.

30. Crowe JH, Hoekstra FA, Crowe LM. 1992. Anhydrobiosis. *Annu. Rev. Physiol.* 54:579–99
31. Damjanovic A, Garcia-Moreno B, Lattman EE, Garcia AE. 2005. Molecular dynamics study of hydration of the protein interior. *Comput. Phys. Commun.* 169:126–29
32. Davenas E, Beauvais F, Amara J, Oberbaum M, Robinzon B, et al. 1988. Human basophil de-granulation triggered by very dilute antiserum against Ige. *Nature* 333:816–18
33. Davey CA, Sargent DF, Luger K, Maeder AW, Richmond TJ. 2002. Solvent mediated interactions in the structure of the nucleosome core particle at 1.9 Angstrom resolution. *J. Mol. Biol.* 319:1097–113
34. De Simone A, Dodson GG, Verma CS, Zagari A, Fraternali F. 2005. Prion and water: Tight and dynamical hydration sites have a key role in structural stability. *Proc. Natl. Acad. Sci. USA* 102:7535–40
35. Denisov VP, Halle B. 1995. Protein hydration dynamics on aqueous solution: a comparison of bovine pancreatic trypsin inhibitor and ubiquitin by oxygen-17 spin relaxation dispersion. *J. Mol. Biol.* 245:682–97
36. Denisov VP, Jonsson BH, Halle B. 1999. Hydration of denatured and molten globule proteins. *Nat. Struct. Biol.* 6:253–60
37. Dill KA. 1990. Dominant forces in protein folding. *Biochemistry* 29:7133–41
38. **Dill KA, Truskett TM, Vlachy V, Hribar-Lee B. 2005. Modeling water, the hydrophobic effect, and ion solvation. *Annu. Rev. Biophys. Biomol. Struct.* 34:173–99**

Discusses simple physical models of water to describe its thermodynamics and solvation properties.

39. Dunn RV, Daniel RM. 2004. The use of gas-phase substrates to study enzyme catalysis at low hydration. *Philos. Trans. R. Soc. London Ser. B Biol. Sci.* 359:1309–20
40. Egli M, Portmann S, Usman N. 1996. RNA hydration: a detailed look. *Biochemistry* 35:8489–94
41. Eisenberg D, McLachlan AD. 1986. Solvation energy in protein folding and binding. *Nature* 319:199–203
42. Eisenstein M, Shakked Z. 1995. Hydration patterns and intermolecular interactions in A-DNA crystal structures. Implications for DNA recognition. *J. Mol. Biol.* 248:662–78
43. Ejtehadi MR, Avall SP, Plotkin SS. 2004. Three-body interactions improve the prediction of rate and mechanism in protein folding models. *Proc. Natl. Acad. Sci. USA* 101:15088–93
44. Elcock AH, McCammon JA. 1995. Sequence-dependent hydration of DNA: theoretical results. *J. Am. Chem. Soc.* 117:10161–62
45. Ernst JA, Clubb RT, Zhou H-X, Gronenborn AM, Clore GM. 1995. Demonstration of positionally disordered water within a protein hydrophobic cavity by NMR. *Science* 267:1813–17
46. Falconi M, Brunelli M, Pesce A, Ferrario M, Bolognesi M, Desideri A. 2003. Static and dynamic water molecules in Cu,Zn superoxide dismutase. *Proteins Struct. Funct. Genet.* 51:607–15
47. Feig M, Brooks CL. 2004. Recent advances in the development and application of implicit solvent models in biomolecule simulations. *Curr. Opin. Struct. Biol.* 14:217–24

48. **Fenimore PW, Frauenfelder H, McMahon BH, Parak FG. 2002. Slaving: Solvent fluctuations dominate protein dynamics and functions. *Proc. Natl. Acad. Sci. USA* 99:16047–51**
49. Fernandez A, Scheraga HA. 2003. Insufficiently dehydrated hydrogen bonds as determinants of protein interactions. *Proc. Natl. Acad. Sci. USA* 100:113–18
50. Fernandez-Escamilla AM, Cheung MS, Vega MC, Wilmanns M, Onuchic JN, Serrano L. 2004. Solvation in protein folding analysis, combination of theoretical and experimental approches. *Proc. Natl. Acad. Sci. USA* 101:2834–39
51. Ferreiro DU, Cho SS, Komives EA, Wolynes PG. 2005. The energy landscape of modular repeat proteins: Topology determines folding mechanism in the ankyrin family. *J. Mol. Biol.* 354:679–92
52. Ferreiro DU, de Prat-Gay G. 2003. A protein-DNA binding mechanism proceeds through multi-state or two-state parallel pathways. *J. Mol. Biol.* 331:89–99
53. Ferreiro DU, Dellarole M, Nadra AD, de Prat-Gay G. 2005. Free energy contributions to direct readout of a DNA sequence. *J. Biol. Chem.* 280:32480–84
54. Finney JL. 1996. Hydration processes in biological and macromolecular systems. *Faraday Discuss.* (103):1–18
55. Frank H, Franks F. 1968. Structural approach to solvent power of water for hydrocarbons: urea as a structure breaker. *J. Chem. Phys.* 48:4746–57
56. Garcia AE, Hummer G. 2000. Water penetration and escape in proteins. *Proteins Struct. Funct. Genet.* 38:261–72
57. Garcia AE, Onuchic JN. 2003. Folding a protein in a computer: an atomic description of the folding/unfolding of protein A. *Proc. Natl. Acad. Sci. USA* 24:13898–903
58. Goldstein RA, Luthey-Schulten ZA, Wolynes PG. 1992. Protein tertiary structure recognition using optimized Hamiltonians with local interactions. *Proc. Natl. Acad. Sci. USA* 89:9029–33
59. Guo WH, Lampoudi S, Shea JE. 2003. Posttransition state desolvation of the hydrophobic core of the src-SH3 protein domain. *Biophys. J.* 85:61–69
60. Halle B. 2004. Biomolecular cryocrystallography: structural changes during flash-cooling. *Proc. Natl. Acad. Sci. USA* 101:4793–98
61. Hamada D, Dobson CM. 2002. A kinetic study of beta-lactoglobulin amyloid fibril formation promoted by urea. *Protein Sci.* 11:2417–26
62. Harano Y, Kinoshita M. 2004. Large gain in translational entropy of water is a major driving force in protein folding. *Chem. Phys. Lett.* 399:342–48
63. Henchman RH, McCammon JA. 2002. Structural and dynamic properties of water around acetylcholinesterase. *Protein Sci.* 11:2080–90
64. Hillson N, Onuchic JN, Garcia AE. 1999. Pressure-induced protein-folding/unfolding kinetics. *Proc. Natl. Acad. Sci. USA* 96:14848–53
65. Hirst S, Phillips DI, Vines SK, Clark PM, Hales CN. 1993. Reproducibility of the short insulin tolerance test. *Diabet. Med.* 10:839–42
66. Huang Q, Ding SW, Hua CY, Yang HC, Chen CL. 2004. A computer simulation study of water drying at the interface of protein chains. *J. Chem. Phys.* 121:1969–77
67. Huang X, Margulis CJ, Berne BJ. 2003. Dewetting-induced collapse of hydrophobic particles. *Proc. Natl. Acad. Sci. USA* 100:11953–58
68. Hummer G, Garde S, Garcia AE, Pratt LR. 2000. New perspectives on hydrophobic effects. *Chem. Phys.* 258:349–70
69. Janin J. 1999. Wet and dry interfaces: the role of solvent in protein-protein and protein-DNA recognition. *Structure* 7:R277–79

Many motions of myoglobin occur with the temperature dependence of the solvent fluctuations, implying that the solvent fluctuations overwhelm the intrinsic fluctuations of the protein and the hydration shell.

70. Jayaram B, Jain T. 2004. The role of water in protein-DNA recognition. *Annu. Rev. Biophys. Biomol. Struct.* 33:343–61
71. Jen-Jacobson L, Engler LE, Jacobson LA. 2000. Structural and thermodynamic strategies for site-specific DNA binding proteins. *Structure* 8:1015–23
72. Jones S, Thornton JM. 1996. Principles of protein-protein interactions. *Proc. Natl. Acad. Sci. USA* 93:13–20
73. Karanicolas J, Brooks CL. 2003. The structural basis for biphasic kinetics in the folding of the WW domain from a formin-binding protein: lessons for protein design? *Proc. Natl. Acad. Sci. USA* 100:3954–59
74. Kauzmann W. 1959. Some factors in the interpretation of protein denaturation. *Adv. Protein Chem.* 14:1–59
75. Kaya H, Chan HS. 2003. Solvation effects and driving forces for protein thermodynamics and kinetics cooperativity: How adequate is native-centric topological modeling? *J. Mol. Biol.* 326:911–31
76. Kaya H, Liu ZR, Chan HS. 2005. Chevron behavior and isostable enthalpic barriers in protein folding: successes and limitations of simple Go-like modeling. *Biophys. J.* 89:520–35
77. Khmelnitsky YL, Welch SH, Clark DS, Dordick JS. 1994. Salts dramatically enhance activity of enzymes suspended in organic solvents. *J. Am. Chem. Soc.* 116:2647–48
78. Klibanov AM. 2001. Improving enzymes by using them in organic solvents. *Nature* 409:241–46
79. Klimov DK, Straub JE, Thirumalai D. 2004. Aqueous urea solution destabilizes A-beta(16–22) oligomers. *Proc. Natl. Acad. Sci. USA* 101:14760–65
80. Koga N, Takada S. 2001. Roles of native topology and chain-length scaling in protein folding: a simulation study with a Go-like model. *J. Mol. Biol.* 313:171–80
81. Kovacs IA, Szalay MS, Csermely P. 2005. Water and molecular chaperones act as weak links of protein folding networks: Energy landscape and punctuated equilibrium changes point towards a game theory of proteins. *FEBS Lett.* 579:2254–60
82. Kuharski RA, Rossky PJ. 1984. Solvation of hydrophobic species in aqueous urea solution: a molecular dynamics study. *J. Am. Chem. Soc.* 106:5794–800
83. Ladbury JE. 1996. Just add water! The effect of water on the specificity of protein-ligand binding sites and its potential application to drug design. *Chem. Biol.* 3:973–80
84. Lazaridis T, Karplus M. 1999. Effective energy function for protein in solution. *Proteins Struct. Funct. Genet.* 35:133–52
85. Lee MY, Dordick JS. 2002. Enzyme activation for nonaqueous media. *Curr. Opin. Biotechnol.* 13:376–84
86. Leopold PE, Montal M, Onuchic JN. 1992. Protein folding funnels: a kinetic approach to the sequence-structure relationship. *Proc. Natl. Acad. Sci. USA* 89:8721–25
87. Levitt M, Park BH. 1993. Water: Now you see it, now you don't. *Structure* 1:223–26
88. Levy Y, Caflisch A, Onuchic JN, Wolynes PG. 2004. The folding and dimerization of HIV-1 protease: evidence for a stable monomer from simulations. *J. Mol. Biol.* 340:67–79
89. **Levy Y, Cho SS, Onuchic JN, Wolynes PG. 2005. A survey of flexible protein binding mechanisms and their transition states using native topology based energy landscapes. *J. Mol. Biol.* 346:1121–45**
90. **Levy Y, Jortner J, Becker OM. 2001. Solvent effects on the energy landscapes and folding kinetics of polyalanine. *Proc. Natl. Acad. Sci. USA* 98:2188–93**
91. Levy Y, Onuchic JN. 2004. Water and proteins: a love-hate relationship. *Proc. Natl. Acad. Sci. USA* 101:3325–26

The details of protein association mechanisms are predicted by the funnel landscape; calls for water's role in association of large rigid proteins.

Solvation has dramatic effects on the energy landscape of polyalanine. The peptide has different thermodynamic and kinetic behavior at hydrophilic and hydrophobic environments.

92. Levy Y, Onuchic JN. 2006. Mechanism of protein assembly: lessons from minimalist models. *Acc. Chem. Res.* 39:135–42
93. Levy Y, Papoian GA, Onuchic JN, Wolynes PG. 2004. Energy landscape analysis of protein dimers. *Isr. J. Chem.* 44:281–97
94. Levy Y, Wolynes PG, Onuchic JN. 2004. Protein topology determines binding mechanism. *Proc. Natl. Acad. Sci. USA* 101:511–16
95. Liu P, Huang X, Zhou R, Berne BJ. 2005. Observation of a dewetting transition in the collapse of the melittin tetramer. *Nature* 437:159–62
96. Liu ZR, Chan HS. 2005. Desolvation is a likely origin of robust enthalpic barriers to protein folding. *J. Mol. Biol.* 349:872–89
97. Lum K, Chandler D, Weeks JD. 1999. Hydrophobicity at small and large length scales. *J. Chem. Phys. B* 103:4570–77
98. Makarov V, Pettitt BM, Feig M. 2002. Solvation and hydration of proteins and nucleic acids: a theoretical view of simulation and experiment. *Acc. Chem. Res.* 35:376–84
99. Mattos C. 2002. Protein-water interactions in a dynamic world. *Trends Biochem. Sci.* 27:203–8
100. Meyer E. 1992. Internal water molecules and H-bonding in biological macromolecules: a review of structural features with functional implications. *Protein Sci.* 1:1543–62
101. Moglich A, Krieger F, Kiefhaber T. 2005. Molecular basis for the effect of urea and guanidinium chloride on the dynamics of unfolded polypeptide chains. *J. Mol. Biol.* 345:153–62
102. Nymeyer H, Garcia AE. 2003. Simulation of the folding equilibrium of alpha-helical peptides: a comparison of the generalized Born approximation with explicit solvent. *Proc. Natl. Acad. Sci. USA* 100:13934–39
103. Olano LR, Rick SW. 2004. Hydration free energies and entropies for water in protein interiors. *J. Am. Chem. Soc.* 126:7991–8000
104. Oleinikova A, Smolin N, Brovchenko I, Geiger A, Winter R. 2005. Formation of spanning water networks on protein surfaces via 2D percolation transition. *J. Phys. Chem. B* 109:1988–98
105. Onuchic JN, Luthey-Schulten Z, Wolynes PG. 1997. Theory of protein folding: the energy landscape perspective. *Annu. Rev. Phys. Chem.* 48:545–600
106. Onuchic JN, Wolynes PG. 2004. Theory of protein folding. *Curr. Opin. Struct. Biol.* 14:70–75
107. Otting G, Liepinsh E, Wuthrich K. 1991. Protein hydration in aqueous solution. *Science* 254:974–80
108. Pace CN. 1986. Determination and analysis of urea and guanidine hydrochloride denaturation curves. *Methods Enzymol.* 131:266–80
109. Pal SK, Peon J, Zewail AH. 2002. Biological water at the protein surface: dynamical solvation probed directly with femtosecond resolution. *Proc. Natl. Acad. Sci. USA* 99:1763–68
110. Pal SK, Zewail AH. 2004. Dynamics of water in biological recognition. *Chem. Rev.* 104:2099–123
111. Papoian GA, Ulander J, Eastwood ME, Wolynes PG. 2004. Water in protein structure prediction. *Proc. Natl. Acad. Sci. USA* 101:3352–57

Reports a significant improvement in protein structure prediction by adding a water knowledge-based potential to a Hamiltonian for structure prediction.

112. Papoian GA, Ulander J, Wolynes PG. 2003. Role of water mediated interactions in protein-protein recognition landscapes. *J. Am. Chem. Soc.* 125:9170–78
113. Park S, Saven JG. 2005. Statistical and molecular dynamics studies of buried waters in globular proteins. *Proteins Struct. Funct. Bioinform.* 60:450–63

114. Pertsemlidis A, Soper AK, Sorenson JM, Head-Gordon T. 1999. Evidence for microscopic, long-range hydration forces for a hydrophobic amino acid. *Proc. Natl. Acad. Sci. USA* 96:481–86

115. Petrone PM, Garcia AE. 2004. MHC-peptide binding is assisted by bound water molecules. *J. Mol. Biol.* 338:419–35

116. Pratt LR, Pohorille A. 2002. Hydrophobic effects and modeling of biophysical aqueous solution interfaces. *Chem. Rev.* 102:2671–91

117. Rank JA, Baker D. 1997. A desolvation barrier to hydrophobic cluster formation may contribute to the rate-limiting step in protein folding. *Protein Sci.* 6:347–54

118. Rarey M, Kramer B, Lengauer T. 1999. The particle concept: placing discrete water molecules during protein-ligand docking predictions. *Proteins Struct. Funct. Genet.* 34:17–28

119. Rhee YM, Sorin EJ, Jayachandran G, Lindahl E, Pande VS. 2004. Simulations of the role of water in the protein-folding mechanism. *Proc. Natl. Acad. Sci. USA* 101:6456–61

120. Robinson CR, Sligar SG. 1993. Molecular recognition mediated by bound water. A mechanism for star activity of the restriction endonuclease EcoRI. *J. Mol. Biol.* 234:302–6

121. Rodier F, Bahadur RP, Chakrabarti P, Janin J. 2005. Hydration of protein-protein interfaces. *Proteins Struct. Funct. Bioinform.* 60:36–45

122. Roux B, Simonson T. 1999. Implicit solvent models. *Biophys. Chem.* 78:1–20

123. Russo D, Hura G, Head-Gordon T. 2004. Hydration dynamics near a model protein surface. *Biophys. J.* 86:1852–62

124. Sastry GM, Agmon N. 1997. Trehalose prevents myoglobin collapse and preserves its internal mobility. *Biochemistry* 36:7097–108

125. Savage H, Wlodawer E. 1986. Determination of water-structure around bimolecules using X-ray and neutron-diffraction methods. *Methods Enzymol.* 127:162–83

126. Schreiber G. 2002. Kinetic studies of protein-protein interactions. *Curr. Opin. Struct. Biol.* 12:41–47

127. Schwabe JWR. 1997. The role of water in protein DNA interactions. *Curr. Opin. Struct. Biol.* 7:126–34

128. Sessions RB, Thomas GL, Parker MJ. 2004. Water as a conformational editor in protein folding. *J. Mol. Biol.* 343:1125–33

129. Shakked Z, Guzikevichguerstein G, Frolow F, Rabinovich D, Joachimiak A, Sigler PB. 1994. Determinants of repressor-operator recognition from the structure of the Trp operator binding site. *Nature* 368:469–73

130. Sharp KA, Nicholls A, Fine RF, Honig B. 1991. Reconciling the magnitude of the microscopic and macroscopic hydrophobic effects. *Science* 252:106–9

131. Shea JE, Onuchic JN, Brooks CL. 1999. Exploring the origins of topological frustration: design of a minimally frustrated model of fragment B of protein A. *Proc. Natl. Acad. Sci. USA* 96:12512–17

132. Shea J-E, Onuchic JN, Brooks CL. 2002. Probing the folding free energy landscape of the src-SH3 protein domain. *Proc. Natl. Acad. Sci. USA* 25:16064–68

133. Sheinerman FB, Brooks CL 3rd. 1998. Calculations on folding of segment B1 of Streptococcal protein G. *J. Mol. Biol.* 278:439–56

134. Shoemaker BA, Portman JJ, Wolynes PG. 2000. Speeding molecular recognition by using the folding funnel: the fly-casting mechanism. *Proc. Natl. Acad. Sci. USA* 97:8868–73

135. Shortle D. 1999. Protein folding as seen from water's perspective. *Nat. Struct. Biol.* 6:203–5

136. Sleigh SH, Seavers PR, Wilkinson AJ, Ladbury JE, Tame JRH. 1999. Crystallographic and calorimetric analysis of peptide binding to OppA protein. *J. Mol. Biol.* 291:393–415

Reports that water can entropically enhance binding affinity, as bound waters have higher entropy in hydrophobic pockets than in bulk.

Discusses the kinetic pathway of protein-protein association and the kinetic role of water molecules.

Using all-atom simulations, the authors show that the core formation of SH3 domain is coupled to water exclusion, and illustrate the existence of a hydrated native-like intermediate.

137. Sorenson JM, Head-Gordon T. 1998. The importance of hydration for the kinetics and thermodynamics of protein folding: simplified lattice models. *Fold Des.* 3:523–34
138. Sorenson JM, Hura G, Soper AK, Pertsemlidis A, Head-Gordon T. 1999. Determining the role of hydration forces in protein folding. *J. Phys. Chem. B* 103:5413–26
139. Sorin EJ, Nakatani BJ, Rhee YM, Jayachandran G, Vishal V, Pande VS. 2004. Does native state topology determine the RNA folding mechanism? *J. Mol. Biol.* 337:789–97
140. Sreenivasan U, Axelsen PH. 1992. Buried water in homologous serine proteases. *Biochemistry* 31:12785–91
141. Tanford C. 1970. Protein denaturation. Part C. Theoretical models for the mechanism of denaturation. *Adv. Protein Chem.* 24:1–95
142. Tarek M, Tobias DJ. 2002. Role of protein-water hydrogen bond dynamics in the protein dynamical transition. *Phys. Rev. Lett.* 88:138101
143. Timasheff S. 1998. Control of protein stability and reactions by weakly interacting cosolvents: the simplicity of the complicated. *Adv. Protein Chem.* 51:355–432
144. Tobi D, Elber R, Thirumalai D. 2003. The dominant interaction between peptide and urea is electrostatic in nature: a molecular dynamics simulation study. *Biopolymers* 68:359–69
145. Wang AJ, Bolen DW. 1997. A naturally occurring protective system in urea-rich cells: mechanism of osmolyte protection of proteins against urea denaturation. *Biochemistry* 36:9101–8
146. Woda J, Schneider B, Patel K, Mistry K, Berman HM. 1998. An analysis of the relationship between hydration and protein-DNA interactions. *Biophys. J.* 75:2170–77
147. Xu D, Tsai C-J, Nussinov R. 1997. Hydrogen bonds and salt bridges across protein-protein interfaces. *Protein Eng.* 10:999–1012
148. Zhou RH, Berne BJ. 2002. Can a continuum solvent model reproduce the free energy landscape of a beta-hairpin folding in water? *Proc. Natl. Acad. Sci. USA* 99:12777–82
149. Zhou RH, Huang XH, Margulis CJ, Berne BJ. 2004. Hydrophobic collapse in multidomain protein folding. *Science* 305:1605–9
150. Zou Q, Bennion BJ, Daggett V, Murphy KP. 2002. The molecular mechanism of stabilization of proteins by TMAO and its ability to counteract the effects of urea. *J. Am. Chem. Soc.* 124:1192–202

Continuous Membrane-Cytoskeleton Adhesion Requires Continuous Accommodation to Lipid and Cytoskeleton Dynamics

Michael P. Sheetz, Julia E. Sable, and Hans-Günther Döbereiner

Biological Sciences Department, Columbia University, New York, NY, 10027; email: ms2001@columbia.edu

Annu. Rev. Biophys. Biomol. Struct. 2006. 35:417–34

First published online as a Review in Advance on February 28, 2006

The *Annual Review of Biophysics and Biomolecular Structure* is online at biophys.annualreviews.org

doi: 10.1146/annurev.biophys.35.040405.102017

1056-8700/06/0609-0417$20.00

Key Words

membrane tethers, membrane blebs, phosphatidylinositol-4,5-diphosphate (PIP2), spectrin, actin cytoskeleton, membrane tension

Abstract

The plasma membrane of most animal cells conforms to the cytoskeleton and only occasionally separates to form blebs. Previous studies indicated that many weak interactions between cytoskeleton and the lipid bilayer kept the surfaces together to counteract the normal outward pressure of cytoplasm. Either the loss of adhesion strength or the formation of gaps in the cytoskeleton enables the pressure to form blebs. Membrane-associated cytoskeleton proteins, such as spectrin and filamin, can control the movement and aggregation of membrane proteins and lipids, e.g., phosphoinositol phospholipids (PIPs), as well as blebbing. At the same time, lipids (particularly PIPs) and membrane proteins affect cytoskeleton and signaling dynamics. We consider here the roles of the major phosphatidylinositol-4,5-diphosphate (PIP2) binding protein, MARCKS, and PIP2 levels in controlling cytoskeleton dynamics. Further understanding of dynamics will provide important clues about how membrane-cytoskeleton adhesion rapidly adjusts to cytoskeleton and membrane dynamics.

Contents

Phosphatidyl-inositol-4,5-diphosphate (PIP2): critical second messenger in cells that determines the many weak adhesions between the plasma membrane and the cytoskeleton

Actin cytoskeleton: system of interlocked actin filaments that provide cell shape and capacity for directed movement

PLASMA MEMBRANES OF ANIMAL CELLS ARE SHAPED BY CYTOSKELETAL INTERACTIONS

Plasma membranes of most animal cells are much larger in area than needed for the volume of the cells. That excess area allows the cells to adopt a variety of nonspherical shapes that are determined by the cytoskeleton of the cells. In most cases, the membrane conforms to the cytoskeleton, but damage to cells often results in the separation of the membrane from the cytoskeleton in blebs. In newly formed blebs, the membrane proteins and lipids diffuse more rapidly than in the membrane over the cytoskeleton. If the cells are to survive, the blebs will heal with the assembly of actin on the cytoplasmic surface that restores the support of the membrane. In a previous review, the role of membrane-cytoskeleton adhesion as measured by tether force was considered (54). Phosphatidylinositol-4,5-diphosphate (PIP2) is an important regulator of many cytoskeletal reorganization events as well as membrane-cytoskeleton adhesion. In this review, we focus on the roles of PIP2 binding proteins in actin cytoskeleton dynamics and proteins that may have major roles in keeping a continuous bond between the membrane and the cytoskeleton. Both factors are important in preventing bleb formation.

Cytoskeleton adhesion to the membrane defines barriers to lipid and protein diffusion and there is an increasing knowledge of the factors that control adhesion and that may restrict lateral mobility of the membrane components. However, lipid diffusion in biological membranes is only in the range of 3- to 10-fold slower than pure lipid bilayers and those diffusion coefficients are similar to what is observed for supported lipid bilayers (17). Thus, in many situations, there is evidence that an unstressed membrane has a diffusion coefficient that appears to be controlled by cytoskeletal elements through a transmembrane coupling or through a "picket fence" (corral) effect (34, 53). An important aspect that is considered here is that most bonds between the membrane and the cytoskeleton appear to be weak and highly reversible such that diffusion on the second to minute timescale can occur (22, 43).

Lipid rafts have been defined as regions of lipid bilayers that contain many signaling molecules and depend upon sphingolipid-cholesterol complexes. Several reviews of the raft literature have focused on the fact that recent physical studies have found evidence of only very small lipid aggregates in native plasma membranes; the smallest complexes are similar to a boundary lipid layer that was defined by EPR studies some time before (7, 21, 36, 56). The evidence of lipid dependence of cellular activities is most clear in cooperative membrane protein functions, and the boundary lipid layers of aggregated proteins could create a phase in which other proteins would be sequestered either as a result of changing the bilayer thickness or a lipid phase (36a). In either case, protein aggregation may be at the heart of the cooperative phenomenon, and changes in lipid composition could feed back upon aggregate formation. Thus, the fact that cholesterol depletion alters many protein signaling complexes and many functions does not mean that lipid rafts control protein organization. The mechanical factors of membrane lipids constitute lipid-dependent factors that can dramatically affect protein functions or protein-protein interactions (36a). However, in a mixed-lipid bilayer, such as in most biological membranes, the local lipid environment is very heterogeneous and protein aggregates can dramatically influence the nature of the adjacent components. A mutual dependency between membrane-associated proteins and local lipid composition fits with current observations. Similarly, here we are considering how phosphoinositol lipid levels control cytoskeleton assembly and how membrane-associated proteins control phosphoinositol lipid levels in a dynamic interplay.

Recent modeling studies as well as structural analyses have reinforced the view that lipid binding proteins are embedded in the headgroups of the lipids, sometimes reaching into the hydrocarbon interior (41, 66). Hydrophobic amino acids help to hold cationic stretches close to the lipid surface (33). Protein domains such as the C2 domains of protein kinase C (PKC) bind to lipid headgroups and insert at least to the level of the glycerol backbone portion of the bilayer (32). Basic rules govern the interaction of peptides with membrane surfaces, and many protein domains also likely affect diffusion in the plane of the membrane (63, 68).

Cell Morphology, Membrane Area, and Membrane Reservoirs

When considering how cells control their morphology and yet maintain the continuous covering of the plasma membrane, it is clear that the membrane must be flexible and dynamic. Previous measurements indicate that the amount of membrane endocytosed every hour is equivalent to the area of the cell plasma membrane, which is an endocytosis rate of about 1 $\mu m^2\ s^{-1}$ for a normal fibroblast (surface area of about 3600 μm^2) (57). For fibroblasts in the later stages of spreading on a matrix-coated surface, the membrane contact area extends at a rate of about 3 $\mu m^2\ s^{-1}$ (20), which is similar to the steady-state rate of exocytosis and endocytosis. Thus, membrane dynamics can occur on a timescale sufficiently rapid to explain any increases in spread area. In previous articles, it has been suggested that the membrane area is controlled physically by the apparent tension in the plasma membrane (38, 54). For plant cells, the tension is simply a tension in the membrane plane. For animal cells, the apparent tension contains a small in-plane tension component, but the major component is the adhesive energy between the membrane and the cytoskeleton (11). That adhesion energy controls the rates of several membrane functions, including endocytosis, membrane extension, and membrane resealing (54). A major implication of the great adhesion between the membrane and the cytoskeleton is that the membrane conforms to the cytoskeleton and thus the overall shape is defined by the membrane area, the cell volume, and the conformation of the cytoskeleton.

How animal cells control their total plasma membrane area is not well understood at this point (38). The fact that apparent membrane tension as measured by membrane tethers is constant in both animal and plant cells implies that apparent tension is an important

PKC: protein kinase C

Membrane tension: sum of intrinsic in-plane force (T_m) and cytoskeleton adhesion (γ) [$T_m = \sigma + \gamma$]

Membrane tethers: tubular membrane segments produced by pulling forces

Spectrin: a long flexible protein that provides structure and stability to the plasma membrane by linking it to the underlying cytoskeleton

component. In both cell types, there is evidence that the endocytosis rate is controlled by apparent membrane tension (38, 54); in plant cells, the spherical geometry of the cells means that the apparent tension is the actual tension in the membrane. However, in animal cells the control of tension is more complicated, because the pressure of the cytoplasm, the membrane area, and the strength of membrane-cytoskeleton adhesion all contribute to the final apparent tension. If the plasma membrane in animal cells is spread tightly over the cytoskeleton, then the rapid deformations of the cell could cause the membrane to stretch. However, lipid bilayers and plasma membranes are relatively inextensible. Lysis occurs after only 4% stretch and the tension required for lysis is 10 mN m^{-1}, which is 100 to 1000 times the tension typically measured in animal cells (11). Effectively, animal cell membranes are not stretched and membrane needs to be exocytosed or to flow from one region to another as cells spread or change shape. When tethers are formed by laser tweezers, membrane can flow rapidly into the tethers from other regions on the cell surface (9, 10). A surprising finding of many studies of membrane tether forces is that the tether force is constant during tether extension for relatively long lengths of the tethers. This has been described as a membrane reservoir that may be stored in membrane invaginations or deformable extensions (48). After the effective reservoir of membrane is depleted, the tether force rises rapidly and the rate of rise approaches levels expected for membrane stretching. Little is known about what controls the reservoir size; however, there is evidence that multiple tethers draw from the same reservoir whether they are pulled from the same region with an atomic force microscope (AFM) tip (58) or from opposite sides of a cell with a laser tweezers (48). Reservoir size decreases slightly with faster tether pulling rates, which is consistent with a kinetic component such as the trapping of membrane proteins at the base of the tether as membrane flows past the adhesive membrane contacts.

Excess Membrane and Adhesion to the Cytoskeleton

How the cell deals with excess membrane is interesting. This question has been addressed by swelling cells in hypotonic media and then restoring isotonicity (38). Osmotic swelling of cells results in the expansion of the membrane area as measured by changes in membrane capacitance. When the isotonicity is restored, the cell volume shrinks to less than the original volume. Instead of being shed as external vesicles, the excess membrane goes into invaginations in the cytoskeleton that are still connected to the outside of the cell (**Figure 1**). These large vacuole-like dilations (VLDs) have been observed in a variety of cells, from mouse macrophages (J. Heuser, personal communication) to snail neurons (12), and the invaginations are absorbed by the cells within 1 to 2 min. The tether force drops only slightly below control levels when isotonicity is restored, perhaps because the membrane for the tether is being drawn from the cytoskeleton at the site of tether formation. This reinforces the idea that membrane-cytoskeleton adhesion is dynamic and can rapidly adapt to major changes in the cell volume.

Lessons from the Red Blood Cell

The red blood cell plasma membrane is tightly bound to the cytoskeleton, and recent studies have shown that the maturation of a reticulocyte to an erythrocyte involves a fourfold increase in the membrane-cytoskeleton adhesion energy (67). The adhesion energy in the erythrocyte is more than 20-fold higher than that in most fibroblasts. Although the cytoskeleton of the erythrocyte lies in a 5- to 10-nm layer apposed to the plasma membrane, it does contain a cross-linked array of short actin filaments bridged by spectrin tetramers. The lipid binding domains hold spectrin on the surface of the plasma membrane, where it forms a major barrier to membrane protein lateral diffusion (55) by restricting diffusion to small domains of the

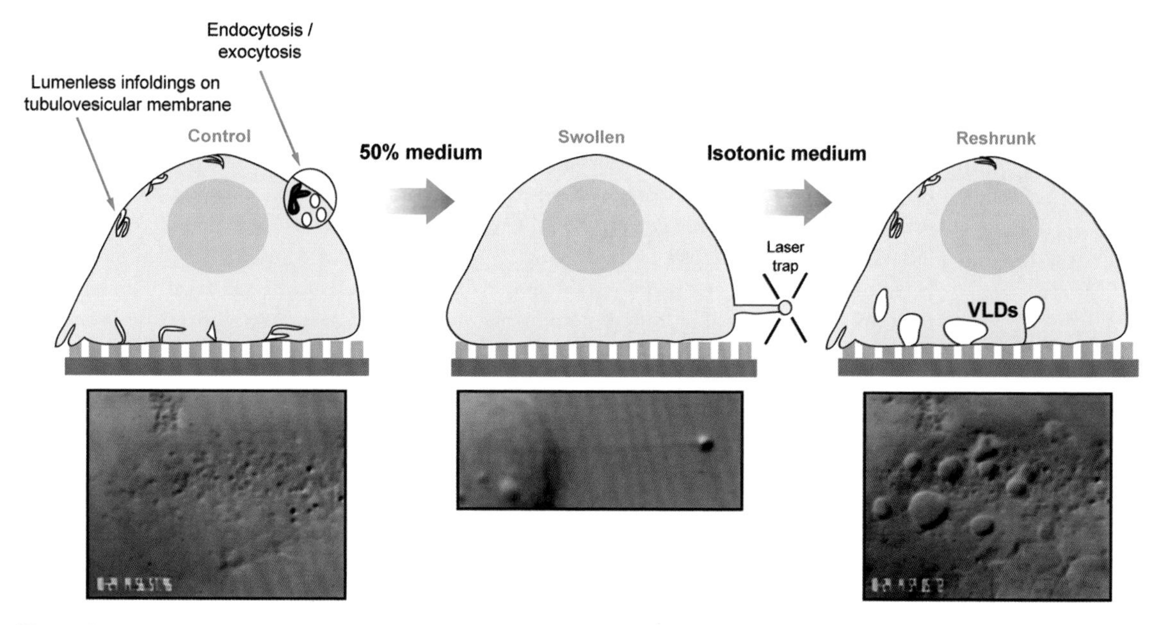

Figure 1

The schematics and accompanying micrographs show how swelling and then reshrinking elicit vacuole-like dilations (VLDs) as invaginations from the substratum. During the swelling process, the plasma membrane area increases 5% to 10%, as measured by capacitance (38). If VLD-bearing neurons are left in isotonic medium, the VLDs disappear in 1 to 2 min as membrane and fluid are reabsorbed into the cytoplasm. If neurons are made to reswell, VLDs disappear by "reversal," as if VLD bilayer material is pulled back to the general plasma membrane surface. Membrane tether force increases dramatically during hypotonic swelling but then drops only slightly below normal levels when isotonic medium is added and VLDs form. Tethers were pulled parallel to the substratum, as shown. Adapted from Reference 12.

plasma membrane (61). The diffusion of fluorescein PE is also restricted in erythrocytes by approximately 10-fold over unsupported lipid bilayers, which may also be due to the spectrin barriers. In support of the idea that both lipid and protein diffusion are inhibited by the same components but to different extents, it was found that deoxygenation causes a threefold increase in the diffusion rate while there is a nearly 10-fold increase in the membrane protein diffusion rate (8). We suggest that spectrin forms a significant barrier to lipid diffusion, even though the inhibition of protein diffusion is significantly greater.

In nonerythroid cells, spectrin is involved in forming boundaries between cellular domains and these boundaries are associated with barriers to membrane protein movement (3). Spectrin appears to have an important role in linking components, since modification of its tetramerization domain is lethal in *Drosophila* (14). Spectrin's tendency to diffuse at the membrane surface, its asymmetric shape, and the network linkage (43) would enable it to rapidly fill in gaps in associated cytoskeletal components and support the membrane.

Lateral Resistance to Membrane Movement

Another aspect of the membrane dynamics is the movement of membrane from one place to another on the cell surface. In growing neurons, there is little resistance to membrane flow across the cytoskeleton, and molecular-level forces can create membrane flow rates

PLC: phospholipase C

Membrane bleb: quasi-spherical membrane segment unsupported by cytoskeleton adhesion

that are 10- to 100-fold greater than axonal extension velocities (10). Thus, mechanical forces of cells are sufficient to move membrane from one site to another if no barriers are present. In neuronal growth cones there is little evidence for lateral barriers to movement, but in more mature neurons the axonal hillock adjacent to the cell body develops a barrier bound to the cytoskeleton that slows lipid movement (40). The differentiation of cell regions then may involve the assembly of cytoskeleton-associated barriers between the domains that slow diffusion and theoretically make it easier for the cell to maintain differences in the compartments over longer times. Even in fibroblasts, polarization of cells for movement and the creation of focal contact regions may involve the assembly of cytoskeleton-associated barriers that could serve to restrict diffusion of proteins and possibly associated lipids. That the membrane reservoir is decreased at higher velocities of membrane movement into the tether suggests that components may be caught at sites of membrane-cytoskeleton adhesion. In the case of bleb formation, the flow of membrane is often rapid and perhaps the bulk lipid flow would concentrate components at adjacent cytoskeletal adhesion sites, thereby slowing further flow of membrane into the bleb. Alternatively, the cytoplasmic pressure could decrease significantly, which would block further expansion of the bleb (5).

CYTOSKELETON ADHESION; TETHER FORCE AND BLEBBING

Adhesion between the membrane and the cytoskeleton is influenced by a number of factors, including the lipid composition, cytoskeleton density and distribution, membrane surface area–to-volume ratio, and internal cellular pressure. There have been two major ways to judge the strength of membrane-cytoskeleton adhesion: bleb formation and tether force. Bleb formation is an indication of the general loss of adhesion strength and/or the formation of large gaps in the cytoskeleton with low adhesion (see Quantitative Analysis of Bleb Formation, above). In the case of tether force, membrane tethers pulled from the plasma membrane by point contacts have a retraction force that is determined primarily by the strength of adhesion between membrane and cytoskeleton in the local domain. Factors that cause an overall decrease in the adhesion between the membrane and the cytoskeleton cause both an increase in blebbing and a decrease in tether force. These factors usually decrease the amount of free PIP2 in the inner plasma membrane by sequestration (50) and can be compensated by overexpression of phosphatidylinositol phosphate kinases (PIPKs) (J. Sable & M. P. Sheetz, unpublished results). Further evidence that supports this theory can be seen when examining the effects of amine anesthetics at the cytoplasmic surface of the plasma membrane, which activate phospholipase C (PLC), decrease PIP2 levels, and subsequently reduce membrane-cytoskeletal adhesion (49).

How Do Apparent Tension and Gaps in the Cytoskeleton Relate to Bleb Formation?

For the cytoskeleton to prevent blebbing, the cytoskeleton must be extensive and cover a significant fraction of the membrane surface (**Figure 2**) (see sidebar, Quantitative Analysis of Bleb Formation). Consequently, it is important to understand which factors are needed to cover the surface. Because the cytoskeleton is dynamic and motile, there is a need to constantly redistribute components involved in adhesion, to limit bleb formation, and to heal blebs by assembling a cytoskeleton on the bleb membrane. Movement of the actin cytoskeleton in lamellipodia is likely to draw elements at the membrane-cytoskeleton interface rearward, leaving regions unsupported. Thus, adhesive elements in lamellipodia must be continually assembled at the leading edge of the lamellipodia to replace adhesive contacts that are moved rearward. When blebs start to form, the contractile force in the cytoskeleton is relieved; however, the underlying cytoskeleton may continue to contract away from the membrane, which would lead to further bleb expansion. In apoptosis, blebs continue to expand and can take up a significant fraction of the cell cytoplasm and membrane.

In order to understand better the major factors that control membrane bleb formation, we consider in a theoretical analysis the conditions that enable the positive cytoplasmic pressure to cause membrane to move into a bleb (see Quantitative Analysis of Bleb Formation). Because the cytoplasm of the cell is normally at a positive pressure relative to the external medium, the plasma membrane always has a tendency to bleb and the pressure creates an expansive tension. Two major factors resist bleb formation: the membrane-bilayer-bending resistance and the apparent membrane tension. A bleb forms when the pressure-dependent expansive tension exceeds the contractile tension from bending stiffness and membrane-cytoskeleton adhesion (for definition of terms, see Quantitative Analysis of Bleb Formation). Because an important part of the bending stiffness is the radius of curvature, the formation of blebs is critically dependent on the size of gaps in the cytoskeleton. On the basis of observed intracellular pressures, the size of the gaps that cause bleb formation are of the order of 0.5 to 1.0 microns and the densities of actin filaments and cytoskeleton-dependent barriers to lateral movement are of the same size scale in some cell regions.

QUANTITATIVE ANALYSIS OF BLEB FORMATION

Cell blebbing can be understood by calculating the stability of a membrane segment spanning across a circular hole with radius a in the cytoskeleton meshwork (see **Figure 2*a***). An internal excess pressure p will tend to bulge the membrane segment outside. This deformation is resisted by the membrane tension σ and the membrane bending elasticity κ, which prefer a flat morphology. Note that $\sigma = \gamma + T_m$, where γ is the cytoskeletal adhesion energy density and T_m is the intrinsic in-plane membrane tension. Lowering membrane tension σ or meshwork density will tend to destabilize the bulge. Assuming a spherical cap with radius R for the outward bulged membrane segment, the total energy E can be written as

$$E(\kappa, \sigma, p, a; R) = \left(\kappa \frac{2}{R^2} + \sigma\right) A(R, a) - pV(R, a),$$

where A is the area of the membrane segment and V is the excess volume of the bulge. Minimization of this energy gives the bulge radius R. However, a finite radius exists only below a critical pressure and above a critical bending modulus. The instability line, which represents the onset of blebbing, is given by

$$\bar{p} = 2 + 8\bar{\kappa},$$

where $\bar{p} = \frac{pa}{\sigma}$ and $\bar{\kappa} = \frac{\kappa}{a^2 \sigma}$ are dimensionless parameters. From this relation, we can calculate the criticial hole radius a_{crit}. For zero bending modulus, we find

$$a_{crit} = \frac{2\sigma}{p}, \quad \kappa = 0.$$

We see that this instability occurs when the hole radius becomes larger than the bulge radius given by the well-known Laplace equation. Using a typical value for the excess pressure $p = 20$ mN/m and membrane tension $\sigma = 0.01$ mN/m, we find $a_{crit} = 1$ µm. In blebbing cells the tension is reduced. With $\sigma = 0.003$ mN/m, we get $a_{crit} = 300$ nm, which is within the range of holes in the cytoskeletal meshwork. For vanishing tension, we get

$$a_{crit} = 2(\kappa/p)^{\frac{1}{3}}, \quad \sigma = 0.$$

Thus a finite bending modulus κ resists blebbing at small tensions. For a bending modulus $\kappa = 2.7\ 10^{-19}$ Nm, we find the

Actin Dynamics and Adhesion; The Roles of PIP2 and MARCKS

Actin cytoskeleton dynamics in terms of the proteins and factors controlling filament assembly and disassembly have been extensively reviewed elsewhere (46). The role of actin dynamics and the uniformity of membrane-cytoskeleton adhesion are indeed complex and

relative large hole radius $a_{crit} = 470$ nm at zero tension. However, depletion of the membrane of cholesterol and long-chain lipids could appreciably lower the bending modulus and foster blebbing. Further, we have restricted bleb morphology to a spherical cap. Allowing for general deformation will lower the critical hole radius. For general values of κ and σ, we have to solve numerically for the cubic root. Using κ and p as given above, one finds $a_{crit} = 1100$ nm for $\sigma = 0.01$ mN/m and $a_{crit} = 600$ nm for $\sigma = 0.003$ mN/m.

In conclusion, by using a simple calculation based on membrane elasticity we have found a blebbing mechanism with critical hole sizes in the range of a few hundred nanometers.

MARCKS: myristoylated alanine-rich C kinase substrate

difficult to understand. If the actin cytoskeleton is continually assembling and disassembling, why aren't there frequent gaps in the cytoskeleton? The related issue is why is the level of PIP2 related directly to the level of actin and inversely to bleb formation (50). Bleb formation is increased when filament assembly is inhibited by blocking barbed filament ends with cytochalasin D. At the leading edge, the assembly of actin has been related to the activity of the WASP-VASP-WAVE families of proteins. Lamellipodial width and actin disassembly are dependent on cofilin activity that is regulated by binding to PIP2 and phosphorylation by Lim kinase (24). Blocking filament disassembly by inhibiting cofilin activity through Lim kinase activation also causes blebbing. Recent studies of apoptosis have shown that caspase 3 cleavage and activation of Lim kinase cause extensive blebbing (62). If cofilin activity is involved in freeing subunits for filament assembly, then the lack of actin filament assembly could be a cause of the loss of support of the membrane with inhibition of cofilin. Others have correlated increased cofilin activity with increased F-actin concentrations and extension of lamellipodia (23). Two possible explanations for cell-increased blebbing are (*a*) a factor associated with actin filaments that inhibits blebbing is not recycled in the absence of cofilin activity, or (*b*) without cofilin activity new filament assembly is inhibited and existing filaments are contracted to the center of the cell, leaving large regions of the lamella unsupported. What are some of the actin-associated proteins that interact with the membrane PIP2 and what role might they have in the process of actin protrusions?

PIP2 is an important regulator of many cytoskeletal reorganization events such as vesicle trafficking, endocytosis, phagocytosis, focal adhesion formation, and cell migration as well as membrane-cytoskeleton adhesion. PIP2 binds to and affects the many actin binding and remodeling proteins such as MARCKS, cofilin, profilin, gelsolin, vinculin, talin, α-actinin, WASP, Arp2/3, and the Rho family of GTPases (15, 37, 45, 69). In general, PIP2 production is thought to occur primarily at the plasma membrane and the total amount of PIP2 in cells is relatively constant. However, local dynamic changes in PIP2 have been observed at sites of phagocytosis and in actin-rich protrusions such as membrane ruffles (4, 6). In all the abovementioned cases of localized changes in PIP2 levels, the initiation step involves hydrolysis of PIP2 to generate the second messengers, inositol phosphate (IP) and diacylglycerol (DAG), which in turn recruit phosphatidylinositol phosphate kinases (PIPKs) to the plasma membrane. DAG production also recruits PKC to the plasma membrane, which subsequently releases MARCKS and thus releases more PIP2 at the plasma membrane. This creates a positive-feedback situation that allows for large localized increases of PIP2 at the plasma membrane. A key question that remains to be elucidated is how local pools of PIP2 accumulate within the cell. Currently, two mechanisms are proposed to explain localized PIP2 accumulation: lateral sequestration and local synthesis. The lateral sequestration model by McLaughlin et al. (37) suggests that PIP2 is sequestered by electrostatic interactions. MARCKS has a high affinity for PIP2 and subsequently has been shown to laterally sequester PIP2 associated with PH domains (65, 66). It is a particularly abundant protein (2 μM in fibroblasts and 10 μM in neurons) that can bind nearly all the plasma membrane PIP2 (each MARCKS can bind three to four PIP2 molecules with a dissociation constant of 10^{-8} M) (37). Because MARCKS-/- fibroblasts do not show increased blebbing or significant changes in tether force, MARCKS is not a major adhesion protein. However, it can compete with the major adhesion proteins, since overexpression of MARCKS causes a decrease in membrane-cytoskeleton adhesion and an increase in bleb formation. The blebs are small, which is consistent with the fact that MARCKS has an actin binding site (J. Sable & M. P. Sheetz, unpublished results).

MARCKS is found in phagosomes and membrane ruffles, sites that contain high levels of PIP2 and actin, implicating MARCKS in PIP2-dependent actin rearrangement processes. MARCKS can function as an actin cross-linking protein when present at low concentrations, at which it forms actin filaments into an actin gel or network; higher concentrations of MARCKS cause actin filament aggregation similar to those found in stress fibers (27). Electron microscopy and biochemical analysis indicate that the majority of MARCKS is associated with the plasma membrane and that once phosphorylated it recycles in a microtubule-dependent mechanism associated with lysosomes (1). MARCKS may be targeted to lysosomes via a conserved sequence. All members of the MARCKS family of proteins contain a conserved MH2 domain of unknown function. A deletion of AA6-140 from MARCKS did not show release from the plasma membrane upon phosphorylation, indicating that this highly acidic MH2 domain could contribute to successful recycling of MARCKS (52).

Myristoylation is also critical for release of myristoylated alanine-rich C kinase substrates (MARCKS) from the plasma membrane. A MARCKS G43 chimera that contains the first seven amino acids from the GAP-23 and is doubly palmitoylated does not leave the plasma membrane after phosphorylation and prevents cell spreading by causing massive blebs (39). This mutant prevented the formation of membrane ruffles and lamellae

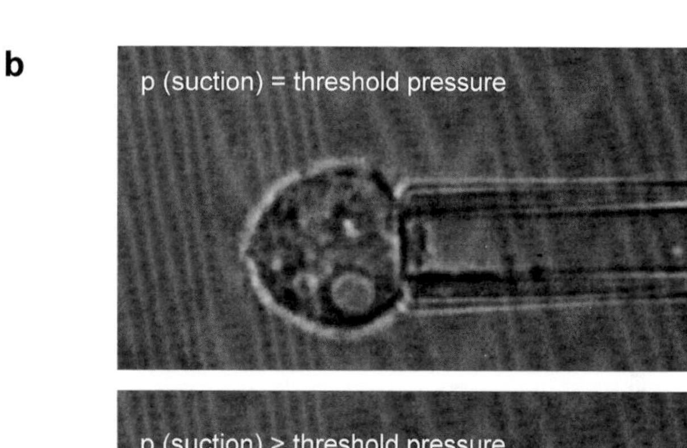

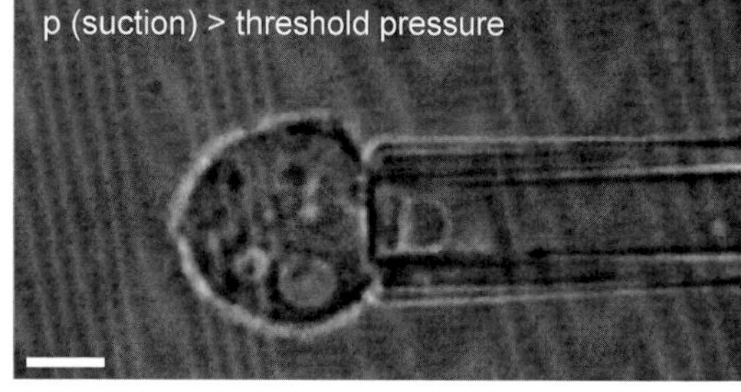

Figure 2

Schematic showing the formation of a pressure-induced bleb from the surface of a cell (*a*) and a micrograph showing a captured cell (*b*). The diagram is the basis for the calculation in Quantitative Analysis of Bleb Formation, above. The advantage of using the pipette to form a bleb is that the cell membrane in the region of bleb formation is not perturbed by a bead or AFM tip, which is usually used to form a tether. In this case, the captured cell has formed a bleb.

for more than 48 hours post spreading and showed decreased cell-substratum adhesion. The control cells for this experiment showed MARCKS and F-actin colocalized around the occasional membrane blebs that formed at early spreading. At intermediate cell spreading both MARCKS and F-actin were localized in membrane ruffles, and in terminal cell spreading both were localized to the leading edge of the lamellae. The G43 mutant also showed MARCKS to be associated with F-actin and talin along the perimeter of the bleb. This study provided evidence to suggest PKC regulation of membrane ruffle formation and cell spreading by activation of the MARCKS myristoyl-electric switch (39). MARCKS can also perhaps regulate the effects of PKC by affecting actin cross-linking and/or sequestering PIP2. Further evidence supporting this hypothesis can be seen with MARCKS overexpression in Chinese hamster ovary cells, which caused actin to localize primarily to the plasma membrane and decrease stress fiber formation (42).

Several recent studies also indicate that MARCKS is involved in sequestration release to maintain localized PIP2 levels at sites of phagocytosis. MARCKS, PKCα, and myosin I colocalize with F-actin and talin in the cortical cytoplasm adjacent to phagocytic cups (1). A different study showed that during pseudopod formation, PIP2 is enriched initially at the nascent phagosomal cup when visualized with the phospholipase C delta pleckstrin homology domain fused to GFP (PLCδ-PH GFP) probe, accompanied by the recruitment of PIPKIα (4). Post particle ingestion, PLCδ-PH GFP was no longer associated with phagosomal membrane. In their discussion, they suggest that the probe could be displaced by a ligand with higher affinity for PIP2. It is then tempting to postulate a revised model for actin rearrangement in phagocytosis due to localized sequestration of PIP2 via accumulation of MARCKS at the forming pseudopod, followed by recruitment of PKC and PIPKIα, which could initiate a positive-feedback mechanism of PIP2 synthesis and degradation (**Figure 3**). In addition, PIP2 can be degraded by phosphatases, which could also account for the decrease in PLC-PH GFP signal post ingestion. The model is simplified but is intended to describe events related to many membrane cytoskeleton rearrangements, such as leading edge actin polymerization, in which initial steps seem to begin with either the hydrolysis of PIP2 and/or localized accumulation of PIP2 for activation or inactivation of critical actin-associated proteins.

A recent comprehensive review of the kinases responsible for synthesis of PIP2 (19) discusses the generation of PIP2 at specific subcellular sites of cytoskeleton rearrangements caused by targeting and activation of specific PIPKs. Note that upon overexpression, all type I PIP kinases are recruited to the plasma membrane and induce actin remodeling, forming actin foci and actin comets.

Figure 3

A hypothetical model for membrane-cytoskeleton adhesion modulated by multiple weak interactions. (*a*) A resting cell's plasma membrane is associated with MARCKS (*purple*), numerous actin-associated proteins (*dark gray*), and actin nucleation proteins such as cofilin (*black*), which is inactive while bound to PIP2. (*b*) Upon stimulation by any number of signaling pathways (such as PKC or GCPR), actin-associated proteins and phosphorylated MARCKS leave the plasma membrane to release a pool of free PIP2. Hydrolysis of free PIP2 via PLC causes activation of cofilin and severing of F-actin to create new barbed end formation and activating new polymerizing activity. Recruitment of type 1 PIPKs to the plasma membrane may occur via vesicle-mediated transport with small GTPases such as Rho, Rac, and Arf. (*c*) A combination of local synthesis of PIP2 by PIP kinases and sequestration by MARCKS and/or other PIP2 binding proteins (*pink and gray complex*) could account for the local accumulation of PIP2 required for defining the sites of activation for the Arp2/3 complex (*orange*) and subsequent lamellar protrusions. Both MARCKS and actin-associated proteins are thought to recycle back to the plasma membrane to stabilize membrane-cytoskeleton adhesion.

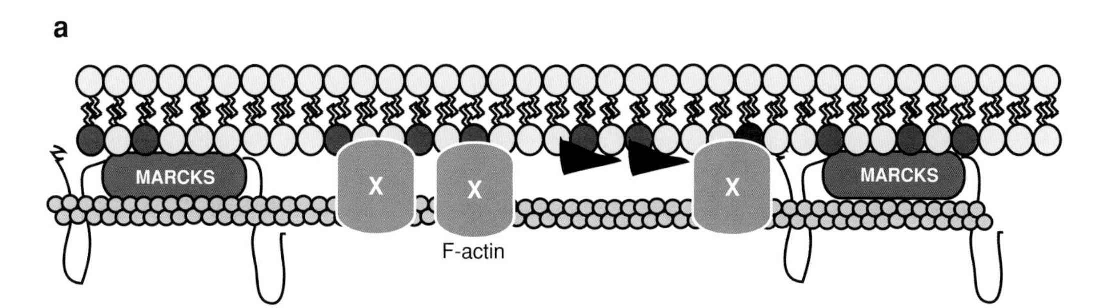
a
MARCKS
X
X
X
MARCKS
F-actin

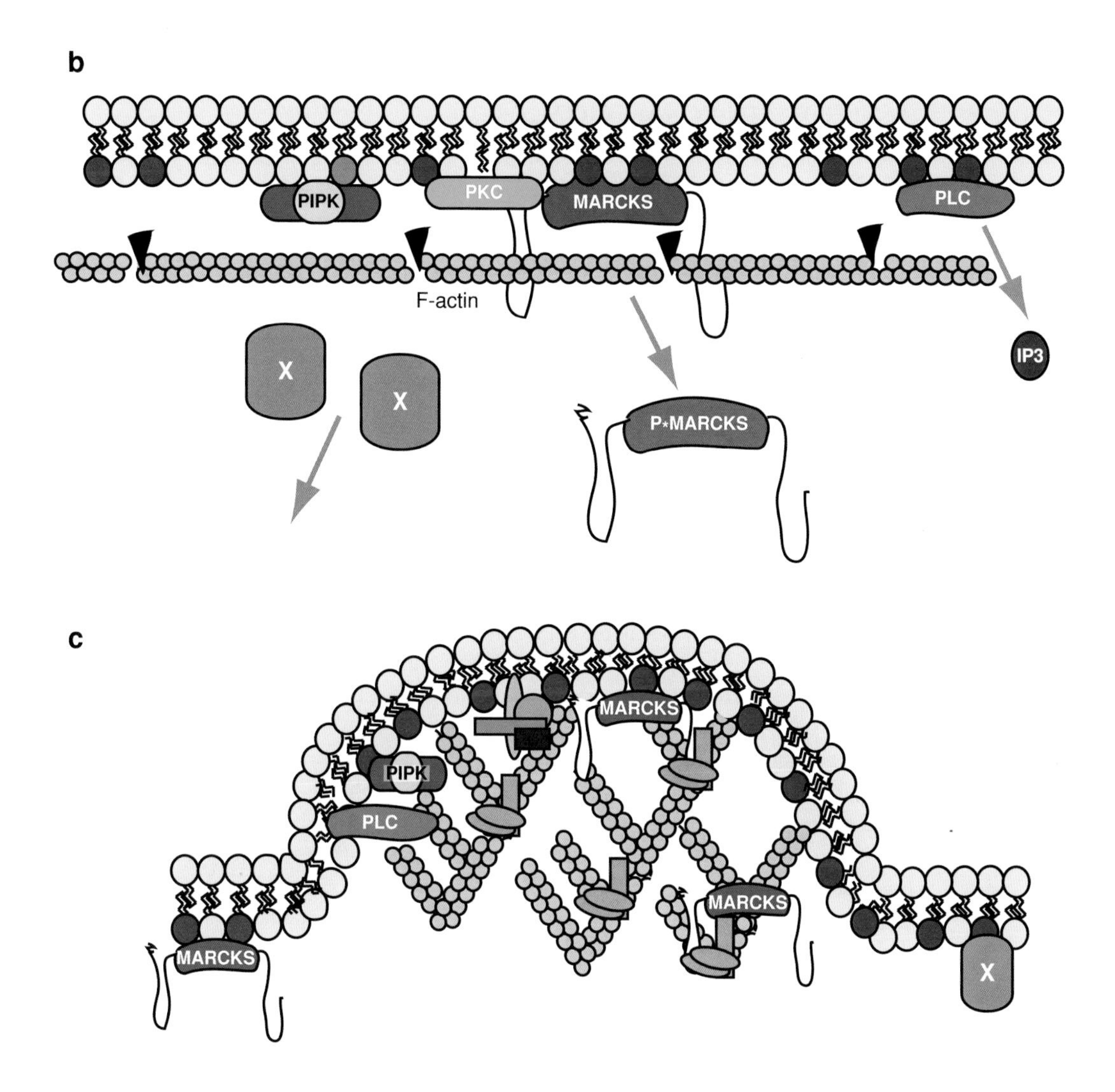
b
PIPK
PKC
MARCKS
PLC
F-actin
X
X
P*MARCKS
IP3
c
MARCKS
PIPK
PLC
MARCKS
MARCKS
X

When compared with endogenous localization, the overexpressed condition may be artifact and should be interpreted with caution. How PIP2 levels are spatially and temporally regulated can then best be understood by a mechanism that accounts not only for localized synthesis of PIP2 via phosphatidylinositol 5′ kinases (type I) (PIP5Ks) but also for sequestration and release of PIP2 via MARCKS protein and PIP2 binding proteins.

PIP5Ks: phosphatidylinositol 5′ kinases (type I)

MEMBRANE-ASSOCIATED PROTEINS INVOLVED IN CYTOSKELTON ADHESION AND DOMAIN FORMATION

The depletion of several membrane-associated proteins, filamin, spectrin, ankyrin, and affixin, causes membrane blebbing. Filamin and spectrin bind directly to lipids and to actin filaments, thus providing adhesive bonds (16, 59), and the structurally similar dystrophins have similar lipid binding properties (35). One of the surprising aspects of spectrin's interaction with lipids is that spectrin prefers binding to phosphatidyl ethanolamine (PE) and to phosphatidyl serine (PS). Further, the PE and PS binding domains are near the ankyrin binding site and ankyrin competes with lipid binding (2, 29). Interaction of nonerythroid spectrin is further affected by the level of PIP2, and the PH domain of beta spectrin brings the C terminus to the surface of the membrane (64).

Spectrin regulates the lateral diffusion coefficient of membrane proteins (55, 61). Because it is associated with the cytoplasmic surface of the plasma membrane, spectrin affects the lateral mobility of membrane lipids as well as proteins. The question is how much will the diffusion coefficients be affected. Spectrin is anchored to transmembrane proteins by ankyrin in erythroid and nonerythroid systems. Because the ankyrin binding site is near the lipid binding domains and competes for PE binding (2, 29), the effect of spectrin on lipid diffusion may be inhibited by ankyrin. However, in the axonal hillock, where a barrier to lipid diffusion has been observed (31), the inhibition of the lipid diffusion correlates with an increase in the concentration of ankyrin and presumably spectrin in the hillock region (18, 40).

Filamin has been linked to a number of signaling processes, including those regulated by the Src family kinase, p56(lck), and PKCα (26). Many protein interactions with filamin have been described, ranging from integral membrane proteins, integrins, and seven-transmembrane receptors, to a variety of soluble kinases. Filamin is believed to be involved in regulating the pathway from insulin receptor activation to the MAP kinase signaling cascade (28). The phosphorylation of filamin is thought to regulate a number of interactions including those with lipids and actin filaments (25, 26, 44, 60). Guanine exchange factors (GEFs) of the Dbl family interact with filamin and also regulate Pak activity. A neuronal GEF, Kal-GEF1 or kalirin, binds to filamin as well as Pak1 and they can be immunoprecipitated as a complex (51). This could explain how the GEF could act on Pak1. SHIP2 binds to filamin and is responsible for lamellipodial spreading of HeLa cells (47). Furthermore, in neurons filamin has a role in organizing dopamine D3R receptors in complexes with beta arrestin, and those complexes are altered upon agonist binding (30).

For the filamin-depleted cell lines, e.g., M2, blebs form at the leading edge and then heal after about 30 s. Many groups have studied M2 blebbing and find that it is a regional process. In the case of local perturbations, the relaxation of the cytoskeleton on one side of the cell with actin-depolymerizing drugs does not inhibit significantly the blebbing frequency on the other side of the cell (5). Charras et al. (5) suggest that the cytoplasm is like a sponge and that there is a barrier to pressure equilibrium from one side to another. In our earlier membrane studies of tether forces on blebs of M2 cells, we observed that bleb formation correlated with a rapid rise in tension in neighboring tethers over the cytoskeleton, which was interpreted at the time as an in-membrane tension increase due to a large bleb forming (13). We also found that the membrane in the blebs could be pulled into the tethers formed on the bleb but that tethers pulled over neighboring regions with cytoskeletal support would not allow membrane to be pulled from the blebs, even though the tether forces were significantly greater over the cytoskeleton. There seemed to be a barrier to lipid movement surrounding the bleb region, or alternatively, the cytoskeleton can be considered as a rigid gel that controls membrane behaviors regionally.

Cytoskeleton Gels and the Organization of Membrane Domains

Typically, regions of the cytoskeleton are coherently organized by a cross-linked array of actin and other filamentous proteins. Such cytoskeletal gels are rigid and resist expansion. Even if they had a high affinity for membrane lipids, they would not be able to draw in additional lipids. However, the extraction of lipids would be difficult. An analogy has been made between the membrane-cytoskeleton adhesion-dependent force on tethers and membrane osmotic pressure. Because the lipids have a lower free energy bound to the cytoskeleton, they try to move back to the cytoskeleton just as water creates an osmotic pressure by trying to move to compartments with higher osmolyte concentrations. Because different cytoskeleton domains can sometimes exist in a single cell, different tether forces can be found in one region versus another (13). Whether or not lipid flow from one cytoskeletally organized domain to another can be inhibited has only been studied in a few cases.

Barriers to Lipid Diffusion

Historically, the first discovery of barriers to lateral diffusion of membrane lipids was in the tight junctions of epithelial cells. Apical lipids were not diffusing to the basolateral surface through the tight junctions and needed to be endocytosed and then recycled to the basolateral surface. This barrier was surprising because the lipid bilayer was continuous through the tight junctions and there was even a gap between the bilayers of adjacent cells. We still don't understand how lipid diffusion is blocked and which proteins constitute the barrier; however, many specialized components are part of the tight junction complex.

In the axonal hillock, there is a barrier to lateral diffusion of outer surface lipids without a neighboring cell that presumably depends upon the cytoskeleton. It is useful to examine this case to determine how the barrier forms. A correlation was made with the concentration of ankyrin in the hillock region. In most cases, ankyrin recruits spectrin and spectrin inhibits membrane protein diffusion through the region. An inhibition of lipid diffusion much greater than can be explained, even by a spectrin concentration as high as that in the erythrocyte, was observed (40). An additional component appears to provide even greater resistance for lipid movement and it will be important to define how the inhibition is created.

SUMMARY

Cell-wide changes in membrane-cytoskeleton adhesion correlate with PIP2 levels but cytoskeletal domains can produce local changes in adhesion strength and blebbing. Analyses of the factors that control blebbing in cells show that, in addition to a general decrease in membrane-cytoskeleton adhesion, the formation of gaps in the adhesive actin cytoskeleton larger than 1 μm can cause blebbing. The dynamic nature of the cytoskeleton in most systems means that domains are transitory, but upon differentiation cells sometimes create long-term domain boundaries. Because of the heterogeneity of lipids and proteins in biological membranes, the lateral aggregation of components can cause major changes in associated components through the creation of aggregate domains. Further analyses of the dynamics of lipids (particularly PIPs) and proteins that link the membrane to the cytoskeleton will be important in defining how adhesion is maintained for long periods with motility and other dynamic processes.

FUTURE ISSUES TO BE RESOLVED

1. How is membrane-cytoskeleton adhesion maintained as local cytoskeletal domains assemble, move, and disassemble?
2. Do cytoskeletal domain boundaries or specific types of domains inhibit the lateral movement of lipids as well as proteins?
3. How are the dynamics of lipids (particularly PIPs) and proteins that link PIP2 to the cytoskeleton involved in maintaining adhesion for long periods with membrane traffic and other dynamic processes?

ACKNOWLEDGMENTS

We thank Dr. Viola Vogel and members of the Sheetz lab for their comments and advice in the writing of this manuscript. This was partially supported by an NIH grant to MPS.

LITERATURE CITED

1. Allen LH, Aderem A. 1995. A role for MARCKS, the alpha isozyme of protein kinase C and myosin I in zymosan phagocytosis by macrophages. *J. Exp. Med.* 182:829–40
2. An X, Guo X, Sum H, Morrow J, Gratzer W, Mohandas N. 2004. Phosphatidylserine binding sites in erythroid spectrin: location and implications for membrane stability. *Biochemistry* 43:310–15
3. Bennett V. 1990. Spectrin: a structural mediator between diverse plasma membrane proteins and the cytoplasm. *Curr. Opin. Cell Biol.* 2:51–56
4. Botelho RJ, Teruel M, Dierckman R, Anderson R, Wells A, et al. 2000. Localized biphasic changes in phosphatidylinositol-4,5-bisphosphate at sites of phagocytosis. *J. Cell Biol.* 151:1353–68
5. Charras GT, Yarrow JC, Horton MA, Mahadevan L, Mitchison TJ. 2005. Non-equilibration of hydrostatic pressure in blebbing cells. *Nature* 435:365–69
6. Chatah NE, Abrams CS. 2001. G-protein-coupled receptor activation induces the membrane translocation and activation of phosphatidylinositol-4-phosphate 5-kinase I alpha by a Rac- and Rho-dependent pathway. *J. Biol. Chem.* 276:34059–65
7. Chen G, Raman P, Bhonagiri P, Strawbridge AB, Pattar GR, Elmendorf JS. 2004. Protective effect of phosphatidylinositol 4,5-bisphosphate against cortical filamentous actin loss and insulin resistance induced by sustained exposure of 3T3-L1 adipocytes to insulin. *J. Biol.Chem.* 279:39705–9
8. Corbett JD, Cho MR, Golan DE. 1994. Deoxygenation affects fluorescence photobleaching recovery measurements of red cell membrane protein lateral mobility. *Biophys. J.* 66:25–30
9. Dai J, Sheetz MP. 1995. Axon membrane flows from the growth cone to the cell body. *Cell* 83:693–701

10. Dai J, Sheetz MP. 1995. Mechanical properties of neuronal growth cone membranes studied by tether formation with laser optical tweezers. *Biophys. J.* 68:988–96
11. Dai J, Sheetz MP. 1999. Membrane tether formation from blebbing cells. *Biophys. J.* 77:3363–70
12. Dai J, Sheetz MP, Wan X, Morris CE. 1998. Membrane tension in swelling and shrinking molluscan neurons. *J. Neurosci.* 18:6681–92
13. Dai J, Ting-Beall HP, Hochmuth RM, Sheetz MP, Titus MA. 1999. Myosin I contributes to the generation of resting cortical tension. *Biophys. J.* 77:1168–76
14. Deng H, Lee JK, Goldstein LS, Branton D. 1995. Drosophila development requires spectrin network formation. *J. Cell Biol.* 128:71–79
15. DesMarais V, Ghosh M, Eddy R, Condeelis J. 2005. Cofilin takes the lead. *J. Cell Sci.* 118:19–26
16. Diakowski W, Prychidny A, Swistak M, Nietubyc M, Bialkowska K, et al. 1999. Brain spectrin (fodrin) interacts with phospholipids as revealed by intrinsic fluorescence quenching and monolayer experiments. *Biochem. J.* 338(Pt. 1):83–90
17. Dietrich C, Bagatolli LA, Volovyk ZN, Thompson NL, Levi M, et al. 2001. Lipid rafts reconstituted in model membranes. *Biophys. J.* 80:1417–28
18. Dotti CG, Poo MM. 2003. Neuronal polarization: building fences for molecular segregation. *Nat. Cell Biol.* 5:591–94
19. Doughman RL, Firestone AJ, Anderson RA. 2003. Phosphatidylinositol phosphate kinases put PI4,5P(2) in its place. *J. Membr. Biol.* 194:77–89
20. Dubin-Thaler BJ, Giannone G, Doebereiner HG, Sheetz MP. 2004. Nanometer analysis of cell spreading on matrix-coated surfaces reveals two distinct cell states and STEPs. *Biophys. J.* 86:1794–806
21. Edidin M. 2003. The state of lipid rafts: from model membranes to cells. *Annu. Rev. Biophys. Biomol. Struct.* 32:257–83
22. Edidin M, Zuniga MC, Sheetz MP. 1994. Truncation mutants define and locate cytoplasmic barriers to lateral mobility of membrane glycoproteins. *Proc. Natl. Acad. Sci. USA* 91:3378–82
23. Ghosh M, Song X, Mouneimne G, Sidani M, Lawrence DS, Condeelis JS. 2004. Cofilin promotes actin polymerization and defines the direction of cell motility. *Science* 304:743–46
24. Giannone G, Dubin-Thaler BJ, Doebereiner HG, Kieffer N, Bresnick AR, Sheetz MP. 2004. Periodic lamellipodial contractions correlate with rearward actin waves. *Cell* 116:431–43
25. Goldmann WH. 2001. Phosphorylation of filamin (ABP-280) regulates the binding to the lipid membrane, integrin, and actin. *Cell Biol. Int.* 25:805–8
26. Goldmann WH. 2002. p56(lck) controls phosphorylation of filamin (ABP-280) and regulates focal adhesion kinase (pp125(FAK)). *Cell Biol. Int.* 26:567–71
27. Hartwig JH, Thelen M, Rosen A, Janmey PA, Nairn AC, Aderem A. 1992. MARCKS is an actin filament crosslinking protein regulated by protein kinase C and calcium-calmodulin. *Nature* 356:618–22
28. He HJ, Kole S, Kwon YK, Crow MT, Bernier M. 2003. Interaction of filamin A with the insulin receptor alters insulin-dependent activation of the mitogen-activated protein kinase pathway. *J. Biol. Chem.* 278:27096–104
29. Hryniewicz-Jankowska A, Bok E, Dubielecka P, Chorzalska A, Diakowski W, et al. 2004. Mapping of an ankyrin-sensitive, phosphatidylethanolamine/phosphatidylcholine mono- and bi-layer binding site in erythroid beta-spectrin. *Biochem. J.* 382:677–85

30. Kim KM, Gainetdinov RR, Laporte SA, Caron MG, Barak LS. 2005. G protein-coupled receptor kinase regulates dopamine D3 receptor signaling by modulating the stability of a receptor-filamin-beta-arrestin complex. A case of autoreceptor regulation. *J. Biol. Chem.* 280:12774–80
31. Kobayashi T, Storrie B, Simons K, Dotti CG. 1992. A functional barrier to movement of lipids in polarized neurons. *Nature* 359:647–50
32. Kohout SC, Corbalan-Garcia S, Gomez-Fernandez JC, Falke JJ. 2003. C2 domain of protein kinase C alpha: elucidation of the membrane docking surface by site-directed fluorescence and spin labeling. *Biochemistry* 42:1254–65
33. Kulkarni S, Das S, Funk CD, Murray D, Cho W. 2002. Molecular basis of the specific subcellular localization of the C2-like domain of 5-lipoxygenase. *J. Biol. Chem.* 277:13167–74
34. Kusumi A, Nakada C, Ritchie K, Murase K, Suzuki K, et al. 2005. Paradigm shift of the plasma membrane concept from the two-dimensional continuum fluid to the partitioned fluid: high-speed single-molecule tracking of membrane molecules. *Annu. Rev. Biophys. Biomol. Struct.* 34:351–78
35. Le Rumeur E, Fichou Y, Pottier S, Gaboriau F, Rondeau-Mouro C, et al. 2003. Interaction of dystrophin rod domain with membrane phospholipids. Evidence of a close proximity between tryptophan residues and lipids. *J. Biol. Chem.* 278:5993–6001
36. McConnell HM, Vrljic M. 2003. Liquid-liquid immiscibility in membranes. *Annu. Rev. Biophys. Biomol. Struct.* 32:469–92
36a. McIntosh TJ, Simon SA. 2006. Roles of bilayer material properties in function and distribution of membrane proteins. *Annu. Rev. Biophys. Biomol. Struct.* 36:177–98
37. McLaughlin S, Wang J, Gambhir A, Murray D. 2002. PIP(2) and proteins: interactions, organization, and information flow. *Annu. Rev. Biophys. Biomol. Struct.* 31:151–75
38. Morris CE, Homann U. 2001. Cell surface area regulation and membrane tension. *J. Membr. Biol.* 179:79–102
39. Myat MM, Anderson S, Allen LA, Aderem A. 1997. MARCKS regulates membrane ruffling and cell spreading. *Curr. Biol.* 7:611–14
40. Nakada C, Ritchie K, Oba Y, Nakamura M, Hotta Y, et al. 2003. Accumulation of anchored proteins forms membrane diffusion barriers during neuronal polarization. *Nat. Cell Biol.* 5:626–32
41. Niggli V. 2001. Structural properties of lipid-binding sites in cytoskeletal proteins. *Trends Biochem. Sci.* 26:604–11
42. Ohmori S, Sakai N, Shirai Y, Yamamoto H, Miyamoto E, et al. 2000. Importance of protein kinase C targeting for the phosphorylation of its substrate, myristoylated alanine-rich C-kinase substrate. *J. Biol. Chem.* 275:26449–57
43. O'Toole PJ, Wolfe C, Ladha S, Cherry RJ. 1999. Rapid diffusion of spectrin bound to a lipid surface. *Biochim. Biophys. Acta* 1419:64–70
44. Pal Sharma C, Goldmann WH. 2004. Phosphorylation of actin-binding protein (ABP-280; filamin) by tyrosine kinase p56lck modulates actin filament cross-linking. *Cell Biol. Int.* 28:935–41
45. Pollard TD, Blanchoin L, Mullins RD. 2000. Molecular mechanisms controlling actin filament dynamics in nonmuscle cells. *Annu. Rev. Biophys. Biomol. Struct.* 29:545–76
46. Pollard TD, Borisy GG. 2003. Cellular motility driven by assembly and disassembly of actin filaments. *Cell* 112:453–65
47. Prasad NK, Decker SJ. 2005. SH2-containing 5′-inositol phosphatase, SHIP2, regulates cytoskeleton organization and ligand-dependent down-regulation of the epidermal growth factor receptor. *J. Biol. Chem.* 280:13129–36

48. Raucher D, Sheetz MP. 1999. Characteristics of a membrane reservoir buffering membrane tension. *Biophys. J.* 77:1992–2002
49. Raucher D, Sheetz MP. 2001. Phospholipase C activation by anesthetics decreases membrane-cytoskeleton adhesion. *J. Cell Sci.* 114:3759–66
50. Raucher D, Stauffer T, Chen W, Shen K, Guo S, et al. 2000. Phosphatidylinositol 4,5-bisphosphate functions as a second messenger that regulates cytoskeleton-plasma membrane adhesion. *Cell* 100:221–28
51. Schiller MR, Blangy A, Huang J, Mains RE, Eipper BA. 2005. Induction of lamellipodia by Kalirin does not require its guanine nucleotide exchange factor activity. *Exp. Cell Res.* 307:402–17
52. Seykora JT, Myat MM, Allen LA, Ravetch JV, Aderem A. 1996. Molecular determinants of the myristoyl-electrostatic switch of MARCKS. *J. Biol. Chem.* 271:18797–802
53. Sheetz MP. 1983. Membrane skeletal dynamics: role in modulation of red cell deformability, mobility of transmembrane proteins, and shape. *Semin. Hematol.* 20:175–88
54. Sheetz MP. 2001. Cell control by membrane-cytoskeleton adhesion. *Nat. Rev. Mol. Cell Biol.* 2:392–96
55. Sheetz MP, Casaly J. 1980. 2,3-Diphosphoglycerate and ATP dissociate erythrocyte membrane skeletons. *J. Biol. Chem.* 255:9955–60
56. Simons K, Vaz WL. 2004. Model systems, lipid rafts, and cell membranes. *Annu. Rev. Biophys. Biomol. Struct.* 33:269–95
57. Steinman RM, Mellman IS, Muller WA, Cohn ZA. 1983. Endocytosis and the recycling of plasma membrane. *J. Cell Biol.* 96:1–27
58. Sun M, Graham JS, Hegedus B, Francoise M, Zhang Y, et al. 2005. Multiple membrane tethers probed by atomic force microscopy. *Biophys. J.* 89:4320–29
59. Tempel M, Goldmann WH, Dietrich C, Niggli V, Weber T, et al. 1994. Insertion of filamin into lipid membranes examined by calorimetry, the film balance technique, and lipid photolabeling. *Biochemistry* 33:12565–72
60. Tigges U, Koch B, Wissing J, Jockusch BM, Ziegler WH. 2003. The F-actin cross-linking and focal adhesion protein filamin A is a ligand and in vivo substrate for protein kinase C alpha. *J. Biol. Chem.* 278:23561–69
61. Tomishige M, Sako Y, Kusumi A. 1998. Regulation mechanism of the lateral diffusion of band 3 in erythrocyte membranes by the membrane skeleton. *J. Cell Biol.* 142:989–1000
62. Tomiyoshi G, Horita Y, Nishita M, Ohashi K, Mizuno K. 2004. Caspase-mediated cleavage and activation of LIM-kinase 1 and its role in apoptotic membrane blebbing. *Genes Cells* 9:591–600
63. Wagner ML, Tamm LK. 2001. Reconstituted syntaxin1a/SNAP25 interacts with negatively charged lipids as measured by lateral diffusion in planar supported bilayers. *Biophys. J.* 81:266–75
64. Wang DS, Shaw G. 1995. The association of the C-terminal region of beta I sigma II spectrin to brain membranes is mediated by a PH domain, does not require membrane proteins, and coincides with a inositol-1,4,5 triphosphate binding site. *Biochem. Biophys. Res. Commun.* 217:608–15
65. Wang J, Arbuzova A, Hangyas-Mihalyne G, McLaughlin S. 2001. The effector domain of myristoylated alanine-rich C kinase substrate binds strongly to phosphatidylinositol 4,5-bisphosphate. *J. Biol. Chem.* 276:5012–19
66. Wang J, Gambhir A, Hangyas-Mihalyne G, Murray D, Golebiewska U, McLaughlin S. 2002. Lateral sequestration of phosphatidylinositol 4,5-bisphosphate by the basic effector domain of myristoylated alanine-rich C kinase substrate is due to nonspecific electrostatic interactions. *J. Biol. Chem.* 277:34401–12

67. Waugh MG, Minogue S, Anderson JS, Dos Santos M, Hsuan JJ. 2001. Signalling and non-caveolar rafts. *Biochem. Soc. Trans.* 29:509–11
68. White SH, Wimley WC. 1999. Membrane protein folding and stability: physical principles. *Annu. Rev. Biophys. Biomol. Struct.* 28:319–65
69. Yin HL, Janmey PA. 2003. Phosphoinositide regulation of the actin cytoskeleton. *Annu. Rev. Physiol.* 65:761–89

Cryo-Electron Microscopy of Spliceosomal Components

Holger Stark and Reinhard Lührmann

Max Planck Institute for Biophysical Chemistry, 37077 Göttingen, Germany;
email: holger.stark@mpibpc.mpg.de; reinhard.luehrmann@mpi-bpc.mpg.de

Annu. Rev. Biophys. Biomol. Struct.
2006. 35:435–57

The *Annual Review of Biophysics and Biomolecular Structure* is online at
biophys.annualreviews.org

doi: 10.1146/
annurev.biophys.35.040405.101953

1056-8700/06/0609-
0435$20.00

Key Words

snRNP, RNA, single-particle analysis, image processing, splicing machinery, pre-mRNA splicing

Abstract

Splicing is an essential step of gene expression in which introns are removed from pre-mRNA to generate mature mRNA that can be translated by the ribosome. This reaction is catalyzed by a large and dynamic macromolecular RNP complex called the spliceosome. The spliceosome is formed by the stepwise integration of five snRNPs composed of U1, U2, U4, U5, and U6 snRNAs and more than 150 proteins binding sequentially to pre-mRNA. To study the structure of this particularly dynamic RNP machine that undergoes many changes in composition and conformation, single-particle cryo-electron microscopy (cryo-EM) is currently the method of choice. In this review, we present the results of these cryo-EM studies along with some new perspectives on structural and functional aspects of splicing, and we outline the perspectives and limitations of the cryo-EM technique in obtaining structural information about macromolecular complexes, such as the spliceosome, involved in splicing.

Contents

INTRODUCTION

Splicing: the removal of nonexpressed regions in pre-mRNA

Intron sequence: sequence of nucleotides in pre-mRNA to be accurately excised by the spliceosome

Exon sequence: sequence of nucleotides in pre-mRNA that are interrupted by introns

Spliceosome: macromolecular complex consisting of U snRNPs and further protein factors that is required for splicing

Most eukaryotic genes are expressed as precursor mRNAs (pre-mRNAs) that are converted to mRNA by splicing, an essential step of gene expression in which intron (noncoding) sequences are removed and exon (coding) sequences are ligated together. The significance of pre-mRNA splicing has recently been bolstered by the results of genome-sequencing projects which indicate that the proteomic complexity of many higher eukaryotes is achieved in part by alternative splicing events that greatly expand the number of unique mRNAs that are generated by an organism (22).

The splicing reaction proceeds in two steps. In the first step, 5′ splice site cleavage and ligation of the intron's 5′ end to the so-called branch site occur concomitantly, and 3′ splice site cleavage (with the resulting excision of the intron) and ligation of the 5′ and 3′ exons take place in the second step. The spliceosome, the complex macromolecular machinery that catalyzes pre-mRNA splicing, is formed from several RNP subunits, termed uridine-rich small nuclear ribonucleoproteins (UsnRNPs), and numerous non-small ribonucleoprotein (snRNP) splicing factors (43). Each UsnRNP particle consists of a UsnRNA molecule complexed with a set of seven Sm or Sm-like proteins and several particle-specific proteins (93). The major spliceosomal UsnRNPs, U1, U2, U4, U5, and U6, are responsible for splicing the vast majority of pre-mRNA introns (so-called U2-type introns). A group of less abundant snRNPs, U11, U12, U4atac, and U6atac, together with U5, are subunits of the so-called minor spliceosome that splices a rare class of pre-mRNA introns, termed U12-type (8).

Spliceosome formation initially involves the interaction of U1 and U2 snRNPs with the 5′splice site and the branch site, respectively, yielding the pre-spliceosome or complex A (**Figure 1**). The U4/U6.U5 tri-snRNP, in which U4 and U6 snRNAs are base-paired, is then stably bound to give complex B, which, however, still has no catalytic center. For catalytic activation of the spliceosome, complex B undergoes a dramatic structural change that involves the dissociation of the intermolecular U4-U6 RNA helices and the formation of an intricate network of interactions between the U6, U2, and pre-messenger RNA molecules, which together constitute part of the catalytic core of the spliceosome. Prior to this RNA rearrangement, U1 snRNP has to dissociate from the 5′ splice site. The activated complex B undergoes the first catalytic step of splicing, which generates complex C. Complex C undergoes the second catalytic step after which the post-spliceosomal intron-containing complex is dismantled and mRNA product is released (49, 74). Thus, the spliceosome is a particularly dynamic RNP machine that undergoes many changes in composition and conformation; this is one factor that makes the determination of the three-dimensional structures of the spliceosome and its modules such a challenge.

Chemically speaking, the splicing reaction is a two-step transesterification reaction

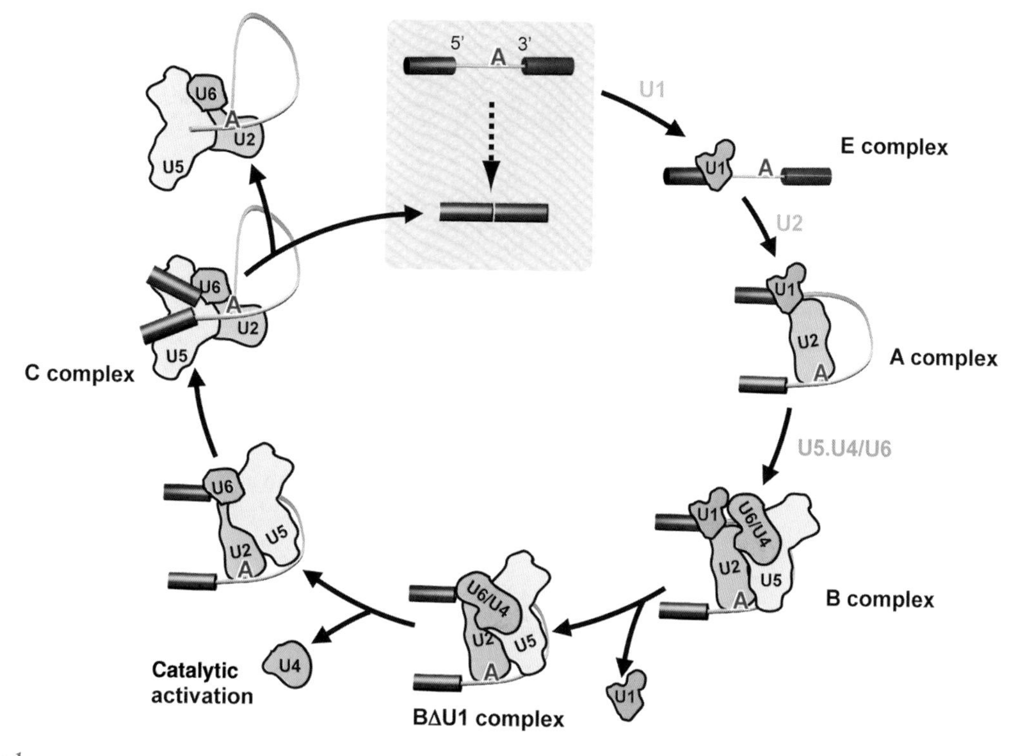

Figure 1

Schematic representation of pre-mRNA splicing and the spliceosome assembly pathway. Introns are excised from pre-mRNA by the spliceosome that is assembled by stepwise integration of U1, U2, and U4/U6.U5 snRNPs.

carried out by the complex C. First, the branchpoint adenosine in the intron performs a nucleophilic attack at the 5′ splice site. The nucleotide at the 5′ splice site, thus freed, attacks the 3′ splice site. This leads to the formation of mature mRNA plus the excised intron in a "lariat" structure. Although it is known from self-splicing introns (41) that these chemical transesterification reactions can be catalyzed by RNA alone, it still remains an open question whether catalysis by the spliceosome involves proteins. Compared with the ribosome, the protein content of snRNPs and spliceosomes is considerably higher (~90% protein). Altogether, splicing requires ~70 different snRNP proteins and more than 100 non-snRNP proteins (30, 60, 94, 99). Among those, proteins with interesting enzymatic functions, such as RNA helicases, GTPases, and peptidyl-prolyl isomerases, can be found, as well as proteins that are important for protein-protein or protein-RNA interactions. The quantitative protein composition of spliceosomal complexes is still not known in detail. In recent years, mass spectrometry has been instrumental in obtaining a better understanding of how the spliceosomal protein pattern changes during its functional cycle, demonstrating the pronounced structural dynamics of this macromolecular machine (26, 29, 38).

Small nuclear ribonucleoprotein (snRNP): complexes that consist of RNA and proteins

WHY USE ELECTRON MICROSCOPY?

The task of gathering structural information about snRNPs and spliceosomes has proved to be difficult. At present, no high-resolution

Single-particle cryo-electron microscopy (cryo-EM): determines the three-dimensional structure of a macromolecular complex by exploiting a large number of individual molecular electron microscopic images obtained at low temperature

structure is available for any of the snRNPs or spliceosomal complexes. X-ray structures of some individual proteins and a few RNA-protein complexes have been obtained (32). Yet why is it so difficult to crystallize entire snRNPs and spliceosomes?

Several factors make the determination of snRNP and spliceosome structures by X-ray crystallography technically demanding. It is highly challenging to prepare individual snRNPs in amounts and concentrations sufficient for crystallization trials. In view of the large number of proteins associated with the individual snRNPs (except U1), reconstitution of snRNPs from purified recombinant proteins and RNAs is similarly time consuming and technically demanding. In addition, these purified complexes are often structurally unstable over the relatively long period needed for crystal growth making the crystallization of purified spliceosomal complexes technically even more challenging.

Under these circumstances, the method of choice for three-dimensional structure determination of snRNPs and spliceosomes is currently single-particle cryo-electron microscopy (cryo-EM). Sample preparation for cryo-EM is quick and usually requires amounts of material that are smaller than those required for X-ray crystallography by a factor of 100 to 1000. This allows structural studies by cryo-EM of snRNPs and spliceosomes prepared and purified by established, small-scale methods. The molecular masses of snRNPs and spliceosomes vary between 240 kDa and ~5 MDa, which is also a good size range for complexes to be studied by single-particle cryo-EM.

A number of splicing-related complexes have been studied by this technique in recent years. In this review, we present the results of these studies along with some new perspectives on structural and functional aspects of splicing, and we outline the perspectives and limitations of the cryo-EM technique in obtaining structural information about macromolecular complexes involved in splicing.

SAMPLE PURIFICATION AND PREPARATION FOR cryo-EM

One of the most crucial steps in structural work on macromolecular complexes is sample purification. The quality and resolution of the final three-dimensional reconstruction is critically dependent upon the sample-purification procedure. In past years, a number of affinity selection techniques have been employed to purify snRNPs and/or spliceosomal complexes from nuclear extract. These include immunoaffinity chromatography (38, 98), affinity selection using biotinylated antisense oligonucleotides (65), selection of pre-mRNAs with randomly incorporated biotinylated nucleotides (48, 60), and binding of aptamer-tagged pre-mRNA by either viral MS2 protein fused to the maltose binding protein (29, 99) or tobramycin (26, 96).

After purification, the material is immediately prepared for single-particle cryo-EM. Owing to stability problems that affect most of the macromolecular complexes involved in splicing, the time that elapses between biochemical purification and sample preparation for EM studies must be minimized. According to practical experience, most of the purified complexes such as spliceosomes cannot be frozen without damaging the sample and thus dramatically reducing the quality of the purified material. It is therefore essential to have the facilities for sample purification and for cryo-EM sample preparation close to one another.

Basically, two different sample preparation methods are generally used. The more elegant method is to vitrify the individual macromolecular complexes by flash-freezing them in liquid ethane or propane (the "native" technique). This keeps the molecules fully hydrated in a thin film of vitrified water (2). This method has been applied successfully to study the structure of other large asymmetrical macromolecular complexes (17, 23, 76) and does not require any additional contrast enhancement, e.g., with heavy-metal salt.

The second method is termed the cryo-negative-staining technique (1, 6, 19, 50), in which the sample is embedded in a mixture of stain, buffer, and sugar and then flash-frozen in liquid ethane or nitrogen. The advantage of the latter method is the greatly enhanced contrast obtained in the recorded images. Its disadvantages are a varying amount of sample flattening and that it still remains unclear to what extent heavy-metal staining affects the structure of the macromolecular complexes.

For technical reasons, most protocols for the purification of snRNPs and spliceosomes require the presence of glycerol to keep the macromolecular complexes intact and to minimize aggregation. In the absence of glycerol, many macromolecular complexes, including snRNPs and spliceosomes, may either dissociate easily or reveal a strong tendency to aggregate. Although cryo-EM of unstained molecules in the presence of high concentrations of salt and sugar has been described (68), cryo preparations of snRNPs in the presence of sugar or glycerol result in intolerable loss of contrast. In contrast to the native technique, cryo-negative staining tolerates the presence of high concentrations of sugar or glycerol and thus can always be used to image spliceosomal complexes. Consequently, most of the cryo-EM work so far has been done with the cryo-negative staining technique, because the highest priority in structural work is to preserve the structures of the molecules under investigation. So far, only few splicing-related complexes have been studied using the native technique (4, 75; B. Sander, M.M. Golas, E.M. Makarov, B. Dube, B. Kastner, R. Lührmann & H. Stark, manuscript in preparation).

DATA ACQUISITION AND PROCESSING IN cryo-EM

No matter which sample-preparation procedure is used, images are always taken while the sample is kept at low temperature by cooling it with liquid nitrogen (or liquid helium) in the electron microscope. Electron-microscopic images can be recorded either on photographic film or on a charged-coupled device (CCD) camera. In a systematic study comparing CCD cameras with photographic film, it was found that CCD images are superior for the initial image-processing analysis and for the reconstruction of three-dimensional macromolecular complex structures (67). The reasons for the significant improvement of image analysis using images taken on a CCD camera are mainly the increased signal-to-noise ratio and the higher reliability of CCD technology in transmitting the image-phase information of spatial frequencies in the region 10 to 25 $Å^{-1}$. Thus, to use a CCD camera in the initial phase of structure determination by single-particle cryo-EM can be highly recommended. As expected, the same study confirmed the results of earlier studies of CCD detectors (69) that had shown the superiority of photographic film in the high-resolution regime. Because photographic film is still more powerful than current CCD technology beyond 7 Å resolution, film remains the preferred image-recording device, pending the development of better detectors.

CCD: charged-coupled device

MSA: multivariate statistical analysis

It is possible to record enough electron-microscopic images in a single working day to obtain 20,000 to 100,000 individual molecular images that can be extracted by applying semi-automated software (36, 55, 64, 90). However, before images of sufficient quality for image processing can be recorded, the conditions for sample purification and specimen preparation have to be optimized; this is very often the most time-consuming step of the entire procedure because of the difficulties inherent in handling snRNP and spliceosome preparations. Once available in the computer, EM images are aligned translationally and rotationally (53, 86) in order to average a set of similar "views" (87) of the molecules that have been identified previously using multivariate statistical analysis (MSA) (16, 84) and classification routines (82). The averaging of 10 to 30 noisy raw images leads to a significant improvement in the signal-to-noise ratio, and this is required for the next step in image processing, which is the assignment of projection angles (58, 83). Because the macromolecular complexes are imaged as noncrystalline material in solution, the molecules may adopt all possible orientations on the EM grid (defined by the three Euler angles).

Single-particle cryo-EM requires a highly isotropic distribution of angular orientations. Lack of isotropy in the angular distribution of molecular images may cause severe image-processing problems and artifacts. In single-particle cryo-EM, the advantage of not having to crystallize the macromolecules is offset by the computational need for correct Euler angle assignment to identify the various orientations represented in the data. The image alignment and Euler angle assignment can thus be considered the computational equivalent to the formation of a well-ordered three-dimensional crystal in X-ray crystallography. Again in analogy to X-ray crystallography, where crystallization is often the most difficult step, it is likewise difficult to determine the correct angular relationship of the various views of a molecule in single-particle cryo-EM. For technical reasons (see below), Euler angle assignment is considerably more problematic in the case of asymmetric macromolecules compared with highly symmetric molecules such as icosahedral viruses.

Once the raw images have been aligned and the Euler angles have been determined correctly, a three-dimensional density distribution of the macromolecular complex can be computed by making use of algorithms based upon weighted backprojection (25), algebraic reconstruction techniques (52), or Fourier inversion techniques (54). Initial three-dimensional structures can subsequently be used in refining the accuracy of the image alignment and Euler angle assignment. This is most commonly done applying projection-matching techniques (53) in which reprojections of a three-dimensional reconstruction are used to realign the entire data set using ever-finer angular sampling of the computed reprojections.

In addition to alignment and Euler angle assignment, other accurate image-processing tools are required to determine the underlying parameters of the contrast transfer function (CTF), such as defocus, astigmatism, amplitude contrast ratio and a noise function (42, 66, 100). Such tools are now available in the various image-processing packages (15, 36, 86) and are applied routinely for structure determination at high resolution.

cryo-EM STRUCTURES OF SPLICEOSOMAL COMPONENTS

U1 snRNP

U1 snRNP consists of the U1 snRNA, the seven common Sm proteins, and three proteins specific to this snRNP (U1-A, U1-70K, U1-C). With its molecular mass of ~240 kDa, the U1 snRNP is the smallest snRNP. It is also by far the most abundant snRNP in the cell and can be isolated in large quantities as a stable complex. For these reasons, it is probably the biochemically best-studied snRNP. Whereas a high-resolution structure of the intact U1 snRNP is not yet available, individual components, including the structure of an RNA-protein complex, have been studied successfully by X-ray crystallography and NMR. The structures of the heteromeric Sm protein complexes, D1, · D2, and B · D3, have been determined by X-ray crystallography, which led to the model of a ring-like arrangement of the seven Sm proteins (33) (**Figure 2**). Furthermore, the crystal structure of the U1-A protein in complex with a fragment of the U1-snRNA (stem-loop II) (46) and the solution NMR structures of U1-C, as well as a fragment of the U1-A protein (3), have been determined (45). Other structural and biochemical information available for U1 snRNP includes a structural model for U1 snRNA (35, 91) and chemical protein-protein (47) and protein-RNA cross-links (81).

The three-dimensional structure of the entire U1 snRNP has been determined by single-particle cryo-EM at 10 to 14 Å resolution (75) (**Figure 2**). Including the available data for U1 snRNP as described above,

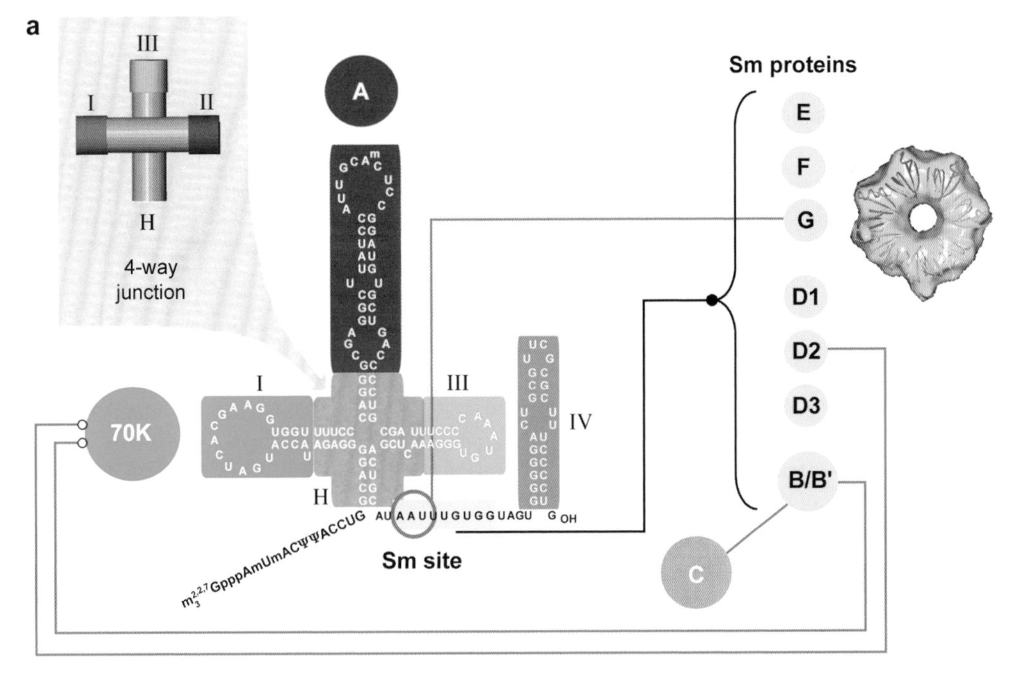

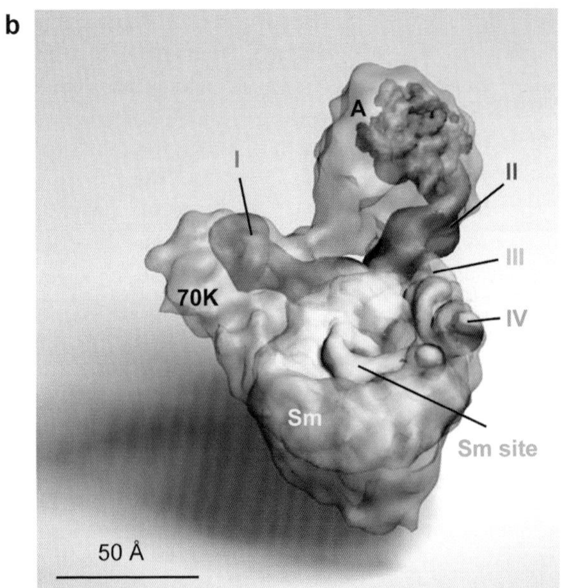

Figure 2

Three-dimensional reconstruction of U1 snRNP. (*a*) Network of biochemical and structural data currently available for U1 snRNP. This body of information was the basis for the interpretation of the three-dimensional arrangement of RNA and proteins in U1 snRNP. The secondary structure of the U1 snRNA is depicted in the same color code that is used for the modeling in panel *b*. The U1-specific proteins 70K and A bind to the RNA stem loops I and II, respectively. The Sm proteins form a seven-membered ring structure. Known cross-links are indicated by orange lines. (*b*) Model of the arrangement of RNA and protein in a 10 to 14 Å resolution structural model of U1 snRNP. The model agrees well with all current biochemical and structural data for U1 snRNP.

SF3b: splicing factor 3b

RRM: RNA recognition motif

a structural model of the three-dimensional arrangement of proteins and RNA has been built (**Figure 2**) that is consistent with the biochemical data available. It shows the U1 snRNP with the two U1-specific proteins A and 70K bound to the upper part of the Sm protein ring. The four-way junction of the RNA is positioned laterally at the bottom of the Sm ring, while the 3′-terminal stem of the U1 snRNA is positioned above of the Sm ring, opposite the U1-70K protein. Although the single-stranded Sm site of the U1 snRNA is not seen directly owing to limited resolution, the arrangement of double-stranded RNA and proteins leads to a model in which the single-stranded Sm site RNA is threaded through the hole in the ring formed by the Sm proteins. This model is supported by the crystal structure of archaea Sm-like proteins in complex with RNA that reveal RNA binding in the central cavity of the Sm protein ring (79, 80).

U2 snRNP and SF3b

17S U2 snRNP has a molecular mass of ~1.2 MDa and consists of the U2 snRNA, the 7 Sm proteins, and approximately 15 U2 snRNP-specific proteins. Most of the U2 snRNP-specific proteins are part of the protein subcomplexes splicing factor 3a and 3b (SF3a and SF3b) that form U2 snRNP by sequential interactions (34). The three-dimensional structure of the U2 snRNP is unknown. However, several structural components of the U2 snRNP have been studied by X-ray crystallography and cryo-EM. The crystal structure of the protein complex U2B″-U2A′ bound to the hairpin loop IV of U2 snRNA has been determined at 2.4 Å resolution (56). The multiprotein subcomplex SF3b has been studied by cryo-EM (19). SF3b plays a key role in the recruitment and tethering of the U2 snRNP to the intronic branch site (10, 57, 97). This 450-kDa complex consists of seven proteins. Most of the SF3b proteins cross-link to pre-mRNA to a region spanning ~25 nucleotides upstream and ~5 nucleotides downstream of the branchpoint (10, 21). The SF3b protein p14 cross-links directly to the branchpoint adenosine of the intron after integration of the U2 snRNP into the pre-spliceosome, and it can be cross-linked to the same point in subsequently formed spliceosomal complexes, including the catalytically active C complex (37, 57, 98). The SF3b complex can be isolated as a stable entity from HeLa nuclear extract by a series of immunopurification steps. The purified SF3b complex (97) can be seen under the electron microscope as a monodisperse particle, well suited for structure determination by single-particle analysis.

The structure of SF3b has been determined from ~30,000 individual macromolecular images of SF3b by the cryo-negative stain technique (19) (**Figure 3**). A resolution level of ~10 Å was obtained, which allowed many fine-structural details to be discerned and led to a detailed assignment of structural elements to their constituent macromolecules and the localization of those proteins that have a known structural fold.

Of course, this strategy works only if the structural domains are sufficiently large, and their shape is sufficiently recognizable, to allow unambiguous placing, even at the intermediate-resolution level obtained. In the case of SF3b, the special architecture of the complex is extremely helpful in that respect. The EM map of SF3b reveals a cage-like structure with some densities protruding outside the complex and a single density protruding inside the complex.

Two of the SF3b proteins (SF3b49 and p14) have been reported to contain RNA recognition motifs (RRMs) (10, 97). Because of the thin wall of the SF3b shell, potential RRM locations can be expected only in the larger elements of density that protrude from this wall. Therefore, all these protruding density elements were tested by investigating how well homology-based RRM models fit into them. Only three regions of density agree well with RRMs with respect to shape, size, mass, volume, and correlation coefficient. One RRM was found pointing into the central

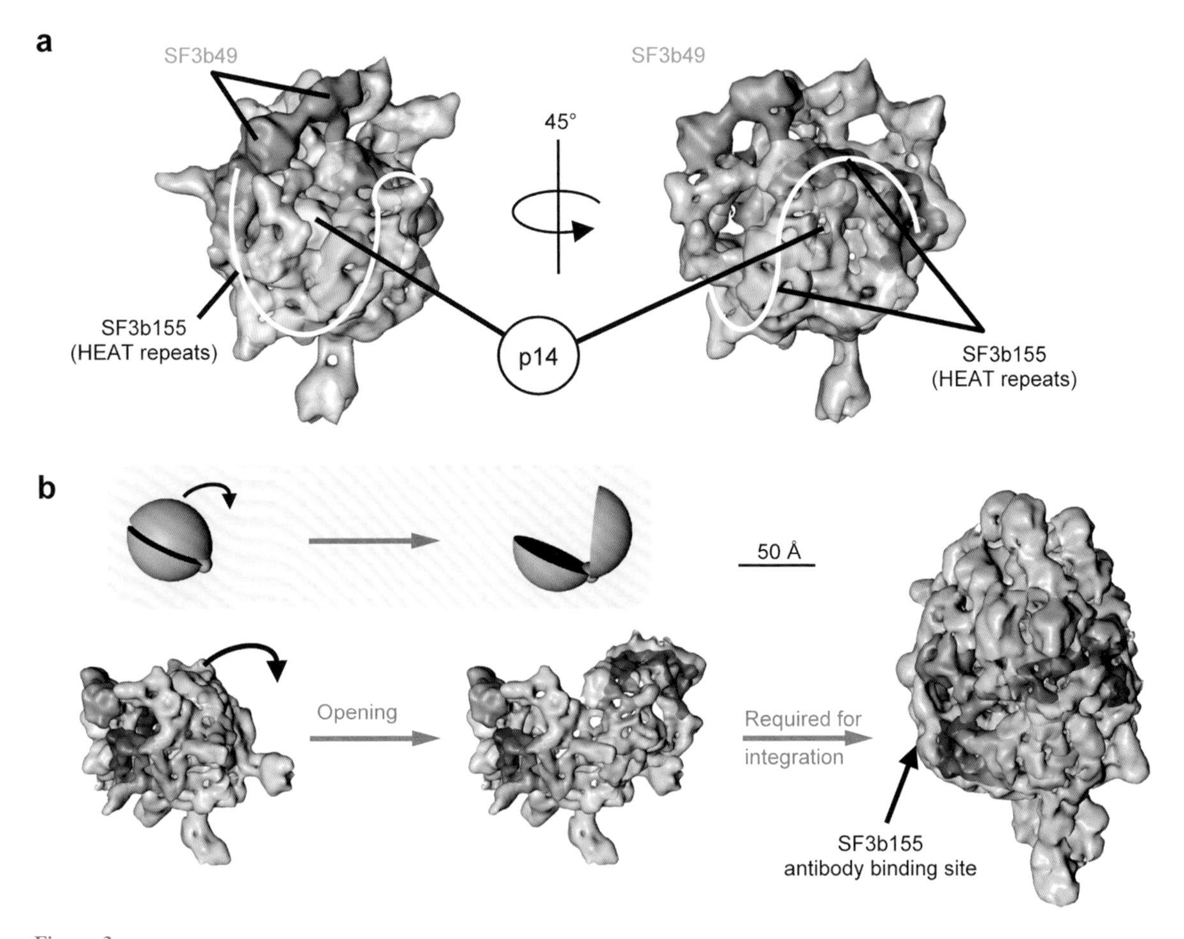

Figure 3

Three-dimensional structure of SF3b and its structural rearrangement upon integration into U11/U12 dimeric snRNP (di-snRNP). (*a*) Three-dimensional reconstruction of the multimeric protein complex SF3b at ~10 Å resolution. Three out of seven proteins were located on the basis of their structural domains and their excellent fit into the density map. Three RRMs belonging to the proteins SF3b49 (2x) and p14 (1x) could be identified clearly. The pattern of 22 tandem helices as present in the HEAT repeats were found wrapped around the SF3b complex in an S-like manner. (*b*) For optimal fit of SF3b into the U11/U12 di-snRNP density, an opening of SF3b is required. The SF3b opening requires the movement of one shell-like half of the complex to match the outer wall of U11/U12 di-snRNP.

cavity of the complex and the SF3b49 RRMs were found next to each other at the outside of the complex (19). Because SF3b49 contains two RRMs, the two density elements at the periphery of SF3b are interpreted as SF3b49 RRMs and the density in the central cavity is interpreted as protein p14. The hidden location of p14 is intriguing in connection with the interaction between SF3b and the branch site. If the interaction between p14 and the branch site occurs via the RRM of p14, how then does the pre-mRNA bind to p14 if the latter is structurally hidden in the shell of the SF3b complex? That the p14 protein can be cross-linked directly to the branchpoint adenosine (57) of a pre-mRNA intron argues for some kind of conformational rearrangement or opening in order to allow integration

HEAT: structural repeat motif consisting of tandem helical arrays

into the larger U2 snRNP, U11/U12 dimeric (di)-snRNP, and/or early spliceosomal complexes.

The SF3b155 protein can be divided into two domains. The N-terminal ~450-amino-acid domain contains several TPR repeats that are phosphorylated during splicing and the C-terminal domain consists of 22 HEAT repeats (92). HEAT repeats have been shown by X-ray crystallography to fold into antiparallel-oriented tandem helical repeats with an overall curved or spiraling structure (11, 12, 24, 40, 88). In the EM density map of SF3b at ~10 Å resolution, it is not possible to unambiguously assign all individual alpha helices with confidence. However, the tandem helical HEAT repeats do form a ladder-like pattern that can be seen clearly in the SF3b density map, even at intermediate resolution. Following this pattern, the 22 HEAT repeats in SF3b have been located. Interestingly, the SF3b HEAT repeats meander around two thirds of the entire complex in an S-like curve (**Figure 3**).

U11/U12 di-snRNP

The U11/U12 di-snRNP is a component of the minor spliceosome (51, 78, 95, 97). U11 and U12 snRNP are functionally equivalent to U1 and U2 snRNP, respectively, in the major spliceosome. The U11 and U12 snRNPs exist in nuclear extract as a stable pre-formed U11/U12 di-snRNP complex that can be isolated as such and bind to the pre-mRNA as a single entity (18). It can thus be regarded as the core of the minor spliceosomal A complex.

The U11/U12 di-snRNP shares with U2 snRNP all seven proteins of the SF3b subunit, while most of the other proteins are unique to U11/U12 (98). The known structure of SF3b can be exploited by fitting it into a three-dimensional map of the U11/U12 di-snRNP.

The three-dimensional structure of U11/U12 di-snRNP has recently been determined by cryo-negative stain EM at a resolution of ~12 Å (20) (**Figure 3*b***). At this level of resolution one cannot expect to see any individual alpha helices, but the pattern of tandem helical arrays of SF3b155 should still be discernible. The first part of the SF3b155 HEAT repeat pattern could indeed be seen directly in the EM density map. By using this and some other structural landmarks, it was possible to locate SF3b in the U11/U12 di-snRNP structure.

A closer look at the U11/U12 di-snRNP density, however, revealed that only one part of SF3b fit well. The other half of the complex showed a severe density mismatch in the region that was not used for the initial fitting into the U11/U12 di-snRNP density, because the diameters of the outer wall of SF3b and U11/U12 di-snRNP are different. The wall of the smaller sphere of SF3b does not fit perfectly into the somewhat larger sphere of U11/U12 di-snRNP. The latter observation led to the idea that SF3b has to undergo a conformational change in its second shell half in order to fit completely into the density of the U11/U12 di-snRNP. This fitting was performed both manually and by automated fitting (9), with largely identical results. A very good fit was found for the second shell half adjacent to the first one in the outer wall of the U11/U12 di-snRNP. This rearrangement of SF3b resembles the opening movement of a mussel, in which one shell half has to open by ~90° (**Figure 3*b***).

The obvious questions arise: What drives this large rearrangement, and what is the functional reason (if any) for such a rearrangement? A possible answer comes from X-ray crystallographic studies of proteins that contain the HEAT repeat. Crystal structures of the nuclear transport factor importin-β have shown that the HEAT domain can adopt different degrees of curvature, depending on the size of its binding partner (11, 12, 88). This possibility for changing the curvature of SF3b155 is exactly what is needed for the structural rearrangement of SF3b to fit it completely into the U11/U12 di-snRNP. The potential opening of the two shell halves may thus be assumed to be driven by, or to drive, a change in the curvature of the SF3b155 HEAT repeat.

In view of the atomic distance of p14 to the branch site adenosine (57, 98) the hidden location of p14 in the central cavity of the SF3b complex might be problematic for pre-mRNA binding. After the structural rearrangement of SF3b upon integration into the U11/U12 di-snRNP complex, the situation is different. The rearrangement also re-positions the p14 RRM, making it accessible from the outside. Most interestingly, not only p14 but also all the other SF3b proteins that have been described to cross-link the pre-mRNA (73) line up in a single groove along the longitudinal axis of the U11/U12 di-snRNP complex. The currently available biochemical cross-linking data, together with the positions of the proteins in the U11/U12 three-dimensional map, suggest a possible manner in which the pre-mRNA may bind along the surface of the complex.

U5 snRNP, U4/U6 snRNP, and U4/U6.U5 snRNP

The largest snRNP that participates in spliceosome assembly is a trimeric snRNP complex (tri-snRNP) consisting of U5 snRNP and the U4/U6 di-snRNP. As is the case for the other snRNP complexes, only little high-resolution structural information is available. The crystal structure of the U4/U6.U5 snRNP protein 15.5K has been determined in complex with the 5′ stem loop of U4 snRNA (89). Another U5 snRNP-specific protein structure, that of the U5-15K protein, has been solved by X-ray crystallography and is essential for splicing in vivo (63). Apart from those, one more crystal structure is known, that of the human U4/U6 snRNP-specific cyclophilin Snu-Cyp-20, both alone and in complex with a fragment of the U4/U6-snRNP protein hPrp4 (61, 62).

EM studies on the U4/U6.U5 snRNP, the U5 snRNP, and the U4/U6 snRNP complexes have been proven difficult, which can be attributed to problems in purifying snRNPs to homogeneity while preventing the particles from aggregating or dissociating. Optimized sample preparation protocols (B. Kastner, M.M. Golas, N. Fischer, B. Sander, D. Boehringer, P. Dube & H. Stark, manuscript in preparation) have recently led to the first insights into the overall architecture of U4/U6.U5 snRNP (B. Sander, M.M. Golas, E.M. Makarov, B. Dube, B. Kastner, R. Lührmann & H. Stark, manuscript in preparation) (**Figure 4**). Note that the structure of U4/U6.U5 snRNP has been determined by unstained cryo-EM at a resolution of ~20 Å as well as by cryo-negative staining, revealing the same overall three-dimensional structure. Its main appearance is an elongated triangle-shaped complex with a length of ~300 Å similar to the first EM images obtained for the yeast U4/U6.U5 snRNP (14). Two large structural domains are separated by a groove in the broader part of the complex. The three-dimensional structure of the U5 snRNP has been determined by cryo-negative staining at ~25 Å resolution and also reveals an elongated (280 Å), approximately pyramidal-shaped structure. In contrast to U4/U6.U5 snRNP, the broader head region of the U5 snRNP is slightly bent, giving the complex a more curved appearance. The structure of U4/U6 snRNP could be determined only by the random conical tilt technique at ~35 Å resolution and reveals two main structural domains linked to each other by a thin connection. Higher-resolution studies of U4/U6 snRNP were not possible because of variations in distance between the two main structural domains.

Having the U4/U6.U5 snRNP structure and its two main structural components (U5 snRNP and U4/U6 snRNP) available, it was demonstrated that the two components can indeed be combined to fit well into the U4/U6.U5 snRNP structure (**Figure 4*a***). In such a fit, the U4/U6 snRNP binds to the broader top region of U5 snRNP. Only one structural domain of U5 snRNP does not fit completely into the U4/U6.U5 density. Some indications from three-dimensional structure

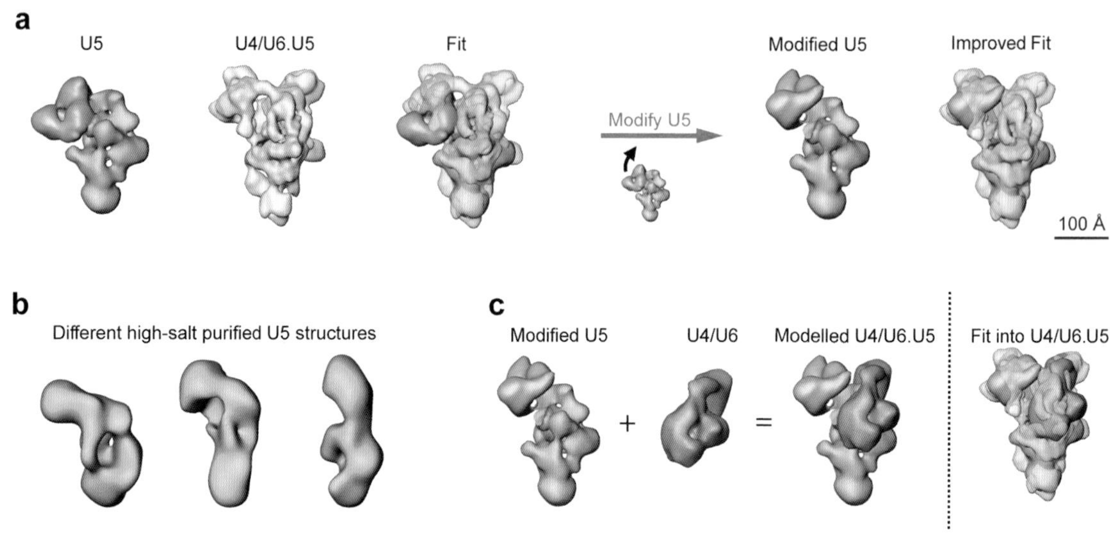

Figure 4

U5, U4/U6, and U4/U6.U5 snRNPs. (*a*) The fit of (low-salt purified) U5 snRNP into the three-dimensional map of U4/U6.U5 snRNP is shown. This fit can be improved by a minor readjustment of the U5 snRNP domain (*blue*). (*b*) Structures of high-salt purified U5 snRNP reveal different curvatures by changing the position of the head domain supporting the required rearrangement of the blue domain in panel *a*. (*c*) Three-dimensional jigsaw puzzle of U5 snRNP and U4/U6 snRNP into a modeled U4/U6.U5 snRNP structure. The modeled structure agrees well with the U4/U6.U5 snRNP structure determined directly by cryo-EM (fit shown at *bottom right*).

RCT: random conical tilt

B△U1: spliceosomal complex comprising U2, U5 and U4/U6 snRNP but lacking U1 snRNP

analysis of a U5 snRNP complex isolated under high-salt conditions provides some ideas how this minor local misfit could be improved (**Figure 4*b***).

In contrast to the standard (low-salt purified) U5 snRNP preparation, high-salt U5 snRNP has been found in many significantly different conformations. A combination of random conical tilt (RCT) (58) experiments followed by a maximum-likelihood alignment of RCT three-dimensional reconstructions and three-dimensional MSA (B. Sander, M. Golas & H. Stark, manuscript in preparation; for technical details see below) revealed differences in the overall curvature of the various U5 snRNP three-dimensional reconstructions computed from the same preparation, indicating an inherent structural flexibility of U5 snRNP under certain experimental conditions. Interestingly, a U5 snRNP structure slightly more elongated than the one observed in low-salt purified U5 snRNP preparations would indeed result in an improved fit of U5 snRNP into the U4/U6.U5 snRNP structure (**Figure 4*a*,*c***).

Spliceosomal Complexes

Several cryo-EM reconstructions of spliceosomal complexes have recently been described. Two of these complexes were assembled in vitro in defined functional states and purified in the presence of heparin to prevent unspecific aggregation.

The three-dimensional structure of a spliceosomal complex B containing U2 and the U4/U6.U5 snRNPs but lacking the U1 snRNP (B△U1) was determined at ~40 Å resolution (5) (**Figure 5**). The maximum dimension of this complex is ~370 Å. The B△U1 complex functionally represents a precatalytic spliceosome with a protein composition that differs significantly from the activated spliceosome (39). The B△U1 three-dimensional reconstruction appeared to be limited in resolution because of a large, flexible domain

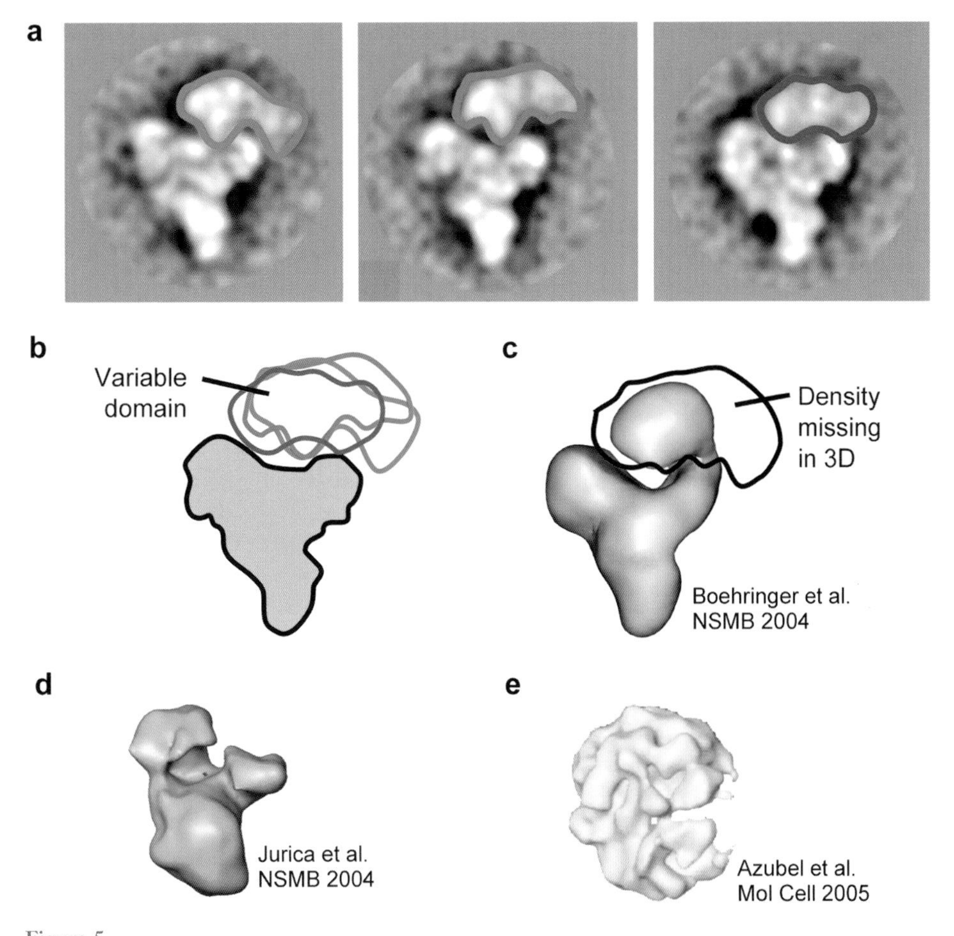

Figure 5

Flexible domain in spliceosome preparation. (*a*) Image analysis of BΔU1 EM images reveals a stable triangular structural domain (bottom part, ~300 Å in diameter) that is flexibly linked to the upper domain. Because of its flexibility, the head domain was found in various different orientations relative to the rest of the particle. (*b*) The different locations of the upper domains are indicated by areas enclosed with differently colored lines. (*c*) Owing to the inherent flexibility of the BΔU1 structural head domain, the overall area of density in the BΔU1 head region is only partially occupied by the reconstructed three-dimensional structure. The maximum diameter of the BΔU1 complex is 370 Å. (*d*) Three-dimensional structure of the spliceosomal complex C determined at 25 Å resolution revealing three distinct structural domains. (*e*) Three-dimensional structure of the individual subunit of the supraspliceosomal complex. The diameter of this structure is ~280 Å.

that is attached to a more rigid triangular domain. The triangular domain itself is similar in size and shape to the U4/U6.U5 snRNP. The flexible "head" domain can be found in many considerably different conformations (**Figure 5*a–c***). The observed heterogeneity of such data sets is a major resolution-limiting factor.

The first three-dimensional structure reported on the spliceosomal complex C shows an elongated structure, ~270 Å in length, with some density protruding from the center (29, 31) (**Figure 5*d***). This complex was arrested before the second chemical step of splicing, in which U2, U5, and U6 snRNA are stably associated. The structure has been described as a core complex C that reveals a relatively open arrangement of three primary domains. The size and shape of this complex is similar to the triangular part of the BΔU1 complex described above.

The third complex has originally been described in the literature as an endogenous large nuclear ribonucleoprotein (lnRNP) complex (72), isolated as a tetramer under mild purification conditions. The U snRNAs as well as several snRNP proteins and splicing factors have been identified in these complexes by blotting experiments and were hence named supraspliceosomal complexes (44). In addition to splicing factors, other RNA-processing components such as RNA-editing factors (59) were also found. The exact biochemical composition and the functional state of the supraspliceosomal complex remain to be determined. In a recent study the tetrameric supraspliceosomal complexes were separated by targeted RNAse H treatment in order to determine the three-dimensional reconstruction of the individual subunits by single-particle cryo-EM. The three-dimensional reconstruction of this complex reveals a sphere-shaped structure ~280 Å in diameter (4) that can be divided into a large and small subunit (**Figure 5*e***). In size and shape this complex displays only a few similarities to the in vitro–assembled spliceosomal complexes described above.

The pronounced dynamics of the spliceosome assembly pathway makes it difficult to predict the level of changes that are to be expected in three-dimensional structure. So far, all spliceosomal complexes studied represent different assembly states. Their structures are thus not directly comparable. Three-dimensional structural work on the spliceosome is still in its infancy, because of the obvious difficulties in purification of defined, biochemically homogeneous complexes that are structurally stable and suitable for three-dimensional structure determination. Future improvements in purification and sample preparation techniques will certainly help researchers obtain higher-resolution structures of the spliceosome. We have recently succeeded in purifying in vitro–assembled precatalytic complex B under mild conditions (in the absence of heparin) and viewed them under the electron microscope. Initial image-processing analysis reveals a significantly more stable complex. Whereas the head domain of the heparin-treated BΔU1 complex is highly flexible, a more rigid conformation is obtained in the non-heparin-treated complex. In class averages this leads to an image of the spliceosome that is slightly larger (~410 Å) owing to the well-defined head domain. Three-dimensional structural analysis of this complex is currently in progress.

In the ribosome field the availability of antibiotics that interfere with the protein synthesis machinery at defined functional steps has been helpful also for three-dimensional structure determination (71, 77). In the case of the spliceosome, such small-molecule inhibitors of defined spliceosomal assembly states will have great promise for future structural investigation. Furthermore, the possibility cannot be excluded that sample preparation procedures for cryo-EM are a major source of sample heterogeneity, because the binding of the macromolecular complex to a support film may induce structural changes or facilitate structural rearrangements. For the spliceosome, and likewise for many other large and fragile macromolecular assemblies, it is thus important to find new methods of specimen preparation that avoid these potential pitfalls.

TECHNICAL ASPECTS

Structure determination of large macromolecular assemblies by single-particle cryo-EM without internal point group symmetry is not a routine matter. The most serious problem is the sample heterogeneity that is almost inevitably present in preparations of macromolecular assemblies with a "machine" function, as these can have several, clearly distinct conformations. Thus, preparations of macromolecular complexes such as the spliceosome may be heterogeneous because they possess moveable domains or simply because of biochemical heterogeneity. In image-processing analysis and three-dimensional reconstruction, systematically different conformers of a macromolecular complex must be separated to avoid errors in structure determination.

Let us examine this in more detail. A complex under investigation may contain a minor domain that either (*a*) is flexible over a wide and continuous range or (*b*) has two or more defined and clearly different conformations. (For simplicity we do not consider the situation in which both factors operate.) In the first case, as long as the major part of the complex is stable, the image alignment will be mainly driven by the stable part, and the assignment of the Euler angles will still be largely correct. The stable parts of the structure will be visualized in the computed three-dimensional map, but the flexible ones will not; these will be "smeared out," and the resolution will be limited accordingly. An example for this case is the ribosome, for which the density of the L7/L12 arm is only partially visible in the reconstructions and the L1 arm also disappears at higher resolution.

In the second case (e.g., the BΔU1 complex as discussed above) molecular images that reveal large structural domains adopting different orientations will no longer align to the same reference even when they are imaged in the same projection direction. The flexible part changes the appearance of the molecule and will consequently influence the assignment of projection angles. As a result the computed three-dimensional reconstructions will be erroneous.

Unfortunately, once such a reconstruction has been arrived at, the commonly used refinement based upon projection-matching is no longer helpful in getting closer to the "true" structure. By applying exclusively projection-matching techniques using an initially wrong three-dimensional model, there is the danger of refining the wrong structure to "high resolution" by overfitting large noisy data sets during the image alignment. One of the most obvious and urgent technical questions is how to ensure that a computed three-dimensional reconstruction is correct. A tool similar to the X-ray crystallographic "free R factor" (7), which is used to validate the results, is still missing in EM.

Because there is no real validation tool available at present, one is compelled to adopt another possibility to ensure a correct three-dimensional structure, i.e., to make sure that the de novo three-dimensional start-up model is correct. In other words, a user-independent method that reliably computes the correct initial three-dimensional structure(s) of a given complex is needed. Such a method would be helpful, because once the correctness of the initial three-dimensional model is ensured, the final structure is likely to be correct as well because in standard projection-matching-based refinements the individual raw images no longer have the freedom to align to anything else but a reprojection calculated from the initially correct three-dimensional model.

In the ideal case of a homogeneous preparation, or in the first case described above, a single start-up model suffices. However, where there is extensive conformational heterogeneity (the second case described above), it is necessary to determine a set of three-dimensional start-up structures that correspond to the adopted conformations of the complex in a given specimen preparation. A method is required that can generate this set of three-dimensional start-up structures without being influenced by any model bias.

One possible solution to this problem is the well-known RCT reconstruction technique (58). The advantage of this method is that the tilt angle is predetermined by the experiment and does not need to be computed. Reference-free alignment algorithms (two-dimensional level) are available (13, 53) that, in combination with MSA, allow one to compute unbiased class averages of any given data set. Individual RCT three-dimensional reconstructions can be determined reliably by using the images that contribute to such class averages. RCT three-dimensional reconstructions additionally reveal the correct handedness of the structure but suffer from the "missing cone" phenomenon and are thus generally limited in resolution. The resolution perpendicular to the tilt axis is considerably worse, which also makes it difficult to compare individual RCT three-dimensional reconstructions. However, by applying a new three-dimensional alignment approach based on maximum likelihood (70; B. Sander, M.M. Golas & H. Stark, manuscript in preparation), the individual RCT reconstructions can be reliably aligned with respect to each other. The advantage of this approach is that the alignment requires only Gaussian noise as a start-up reference and, like unbiased reference-free two-dimensional alignment methods, the three-dimensional alignment of RCT three-dimensional reconstructions is not biased by the choice of a three-dimensional model used as a reference. Subsequent to this alignment, a three-dimensional MSA that can then separate different RCT three-dimensional structures into different classes can be computed (**Figure 4*b***).

Large movements within any macromolecular complex can be detected by this approach without imposing reference bias at any step of the procedure. A set of three-dimensional structures of a heterogeneous sample can be determined and the information can be used for supervised classification schemes to divide the mixed populations of a macromolecular complex into subpopulations of images that belong to the same original three-dimensional object. This technique provides reliable start-up models as well as information about molecular motion in the macromolecular complexes studied. It thus avoids errors introduced into the image-processing analysis and opens the way to study macromolecular dynamics of important macromolecular machines at increasingly higher resolution.

PERSPECTIVES

For structural studies on macromolecular complexes, single-particle cryo-EM is expected to become increasingly important. The combination of the continuing technical improvement in electron-microscopic instrumentation, the availability of fast and affordable computers, and the development of new tools and algorithms for improved image-processing analysis is highly promising for the future of the single-particle approach. Parallel developments in all these fields are required in order to make single-particle cryo-EM a reliable high-resolution structure determination method that can contribute enormously to the understanding of structural dynamics in macromolecular complexes. With the tools and instrumentation already available, it is a matter of time until three-dimensional structures at $\leq$7 Å resolution become routine in single-particle cryo-EM. Even near-atomic resolution structure determination is theoretically possible and will probably be reached in the near future (27, 28, 85).

To reach a very high level of resolution requires gigantic numbers of image statistics to be processed in a reasonable time. The image statistics must be increased even more dramatically (probably beyond 10^6 individual macromolecular images) if this is demanded by heterogeneity of the sample and thus by the corresponding amount of in silico purification required. This leads to a major disadvantage of the technique: The total time needed to record and to process this huge amount of data is tremendous and will by far exceed the time required to grow a crystal and to solve the structure by X-ray crystallography. Wherever there is enough material available in the amounts and quality needed to grow crystals, high-resolution structures are more likely to be obtained by X-ray crystallography. Another advantage of X-ray crystallography is that once crystallization of a given macromolecular complex has been established it can be rapidly repeated, looking at the binding of ligands or other interacting partners. In case of cryo-EM one has to go through the lengthy image-recording/image-processing cycle each time a high-resolution project is started. Collecting the data may take two to four months, even for a single high-resolution project. The automation of EM image recording may help considerably in this respect, but it will not allow cryo-EM to compete with X-ray crystallography in terms of speed in high-resolution structure determination.

At present, the most promising strategy to study the structure of snRNPs and spliceosomes is the hybrid approach of using intermediate-resolution structures determined by cryo-EM and combining them with high-resolution crystal structures of individual components or smaller subcomplexes. Because of their complexity and the limited structural information that is available for the larger snRNPs and spliceosomes, it is part of our strategy to also study smaller snRNP subcomplexes (like SF3b) using cryo-EM in order to interpret the larger complexes, once three-dimensional information for snRNPs and spliceosomes and high-resolution X-ray structures become available. About 70 snRNP proteins and more than 100 non-snRNP proteins are located in the spliceosome and, in the absence of any smaller substructure or label that can be identified at intermediate resolution, interpretation of cryo-EM three-dimensional maps in terms of fitting high-resolution X-ray structures is difficult. The lack of three-dimensional structural information in the splicing field calls for a study of all available snRNP complexes and subcomplexes using many different techniques so that researchers may understand better the dynamics of the spliceosome assembly pathway. Hence, all the cryo-EM snRNP structures and their subcomplexes play an essential part in the structure interpretation of the spliceosome at an intermediate-resolution level and will be used as a scaffold to integrate the available structural and biochemical information to obtain a detailed understanding of the spliceosome at a molecular level.

SUMMARY POINTS

1. Pre-mRNA splicing is an essential step in gene expression.
2. The spliceosome is a highly dynamic macromolecular complex composed of U1, U2, U4, U5, and U6 snRNAs and more than 150 proteins.
3. Single-particle cryo-EM is the method of choice for three-dimensional structure determination owing to problems in purifying snRNPs to homogeneity and in the amounts needed for crystallization.
4. Intermediate resolution three-dimensional structures can be determined by cryo-EM to reveal the overall arrangement of proteins and RNA and to characterize the structural dynamics of spliceosomal components.

NOTE ADDED IN PROOF

Recently, the 2.5 Å crystal structure of a complex containing the U2 snRNP protein SF3bp14 and a peptide derived from the p14-associated protein SF3b155 was reported (101). Fitting of the X-ray structure of the p14.SF3b155 peptide into the EM density assigned for p14 on the basis of its predicted RRM within isolated SF3b (19) was in excellent agreement with respect to the overall shape.

ACKNOWLEDGMENTS

This work was supported by grants from the Federal Ministry of Education and Research, Germany (031U215B and 0311899), to HS and (031U21B) to RL, and by the Fonds der chemischen Industrie and the Ernst Jung Stiftung to RL.

LITERATURE CITED

1. Adrian M, Dubochet J, Fuller SD, Harris JR. 1998. Cryo-negative staining. *Micron* 29:145–60
2. Adrian M, Dubochet J, Lepault J, McDowall AW. 1984. Cryo-electron microscopy of viruses. *Nature* 308:32–36
3. Avis JM, Allain FH, Howe PW, Varani G, Nagai K, Neuhaus D. 1996. Solution structure of the N-terminal RNP domain of U1A protein: the role of C-terminal residues in structure stability and RNA binding. *J. Mol. Biol.* 257:398–411
4. Azubel M, Wolf SG, Sperling J, Sperling R. 2004. Three-dimensional structure of the native spliceosome by cryo-electron microscopy. *Mol. Cell.* 15:833–39
5. **Boehringer D, Makarov EM, Sander B, Makarova OV, Kastner B, et al. 2004. Three-dimensional structure of a pre-catalytic human spliceosomal complex B. *Nat. Struct. Mol. Biol.* 11:463–68**

5. The 3D structure of an in vitro–assembled spliceosomal complex including U2, U4/U6.U5 snRNP but lacking U1 snRNP has been determined by single-particle EM at 40 Å resolution, revealing a 370 Å large complex with one structural domain flexibly bound to a stable triangular body.

6. Bottcher C, Ludwig K, Herrmann A, van Heel M, Stark H. 1999. Structure of influenza haemagglutinin at neutral and at fusogenic pH by electron cryo-microscopy. *FEBS Lett.* 463:255–59
7. Brünger AT. 1997. Free *R* value: cross-validation in crystallography. *Methods Enzymol.* 277:366–96
8. Burge C, Tuschl T, Sharp PA. 1999. *Splicing of Precursors to mRNA by the Spliceosomes.* Cold Spring Harbor, NY: Cold Spring Harbor Press. 525 pp.
9. Chacon P, Wriggers W. 2002. Multi-resolution contour-based fitting of macromolecular structures. *J. Mol. Biol.* 317:375–84
10. Champion-Arnaud P, Reed R. 1994. The prespliceosome components SAP 49 and SAP 145 interact in a complex implicated in tethering U2 snRNP to the branch site. *Genes. Dev.* 8:1974–83
11. Cingolani G, Petosa C, Weis K, Muller CW. 1999. Structure of importin-beta bound to the IBB domain of importin-alpha. *Nature* 399:221–29
12. Cook A, Fernandez E, Lindner D, Ebert J, Schlenstedt G, Conti E. 2005. The structure of the nuclear export receptor Cse1 in its cytosolic state reveals a closed conformation incompatible with cargo binding. *Mol. Cell.* 18:355–67
13. Dube P, Tavares P, Lurz R, van Heel M. 1993. The portal protein of bacteriophage SPP1: a DNA pump with 13-fold symmetry. *EMBO J.* 12:1303–9
14. Fabrizio P, Esser S, Kastner B, Luhrmann R. 1994. Isolation of *S. cerevisiae* snRNPs: comparison of U1 and U4/U6.U5 to their human counterparts. *Science* 264:261–65
15. Frank J, Radermacher M, Penczek P, Zhu J, Li Y, et al. 1996. SPIDER and WEB: processing and visualization of images in 3D electron microscopy and related fields. *J. Struct. Biol.* 116:190–99
16. Frank J, Van Heel M. 1982. Correspondence analysis of aligned images of biological particles. *J. Mol. Biol.* 161:134–37
17. Frank J, Zhu J, Penczek P, Li Y, Srivastava S, et al. 1995. A model of protein synthesis based on cryo-electron microscopy of the *E. coli* ribosome. *Nature* 376:441–44

18. Frilander MJ, Steitz JA. 1999. Initial recognition of U12-dependent introns requires both U11/5′ splice-site and U12/branchpoint interactions. *Genes. Dev.* 13:851–63

19. Golas MM, Sander B, Will CL, Lührmann R, Stark H. 2003. Molecular architecture of the multiprotein splicing factor SF3b. *Science* 300:980–84

20. Golas MM, Sander B, Will CL, Lührmann R, Stark H. 2005. Major conformational change in the complex SF3b upon integration into the spliceosomal U11/U12 di-snRNP as revealed by electron cryomicroscopy. *Mol. Cell.* 17:869–83

21. Gozani O, Feld R, Reed R. 1996. Evidence that sequence-independent binding of highly conserved U2 snRNP proteins upstream of the branch site is required for assembly of spliceosomal complex A. *Genes Dev.* 10:233–43

22. Graveley BR. 2001. Alternative splicing: increasing diversity in the proteomic world. *Trends Genet.* 17:100–7

23. Grigorieff N. 1998. Three-dimensional structure of bovine NADH:ubiquinone oxidoreductase (complex I) at 22 A in ice. *J. Mol. Biol.* 277:1033–46

24. Groves MR, Hanlon N, Turowski P, Hemmings BA, Barford D. 1999. The structure of the protein phosphatase 2A PR65/A subunit reveals the conformation of its 15 tandemly repeated HEAT motifs. *Cell* 96:99–110

25. Harauz G, van Heel M. 1986. Exact filters for general geometry three-dimensional reconstruction. *Optik* 73:146–56

26. Hartmuth K, Urlaub H, Vornlocher HP, Will CL, Gentzel M, et al. 2002. Protein composition of human prespliceosomes isolated by a tobramycin affinity-selection method. *Proc. Natl. Acad. Sci. USA* 99:16719–24

27. Henderson R. 1995. The potential and limitations of neutrons, electrons and X-rays for atomic resolution microscopy of unstained biological molecules. *Q. Rev. Biophys.* 28:171–93

28. Jiang W, Ludtke SJ. 2005. Electron cryomicroscopy of single particles at subnanometer resolution. *Curr. Opin. Struct. Biol.* 15:571–77

29. Jurica MS, Licklider LJ, Gygi SR, Grigorieff N, Moore MJ. 2002. Purification and characterization of native spliceosomes suitable for three-dimensional structural analysis. *RNA* 8:426–39

30. Jurica MS, Moore MJ. 2003. Pre-mRNA splicing: awash in a sea of proteins. *Mol. Cell.* 12:5–14

31. Jurica MS, Sousa D, Moore MJ, Grigorieff N. 2004. Three-dimensional structure of C complex spliceosomes by electron microscopy. *Nat. Struct. Mol. Biol.* 11:265–69

32. Kambach C, Walke S, Nagai K. 1999. Structure and assembly of the spliceosomal small nuclear ribonucleoprotein particles. *Curr. Opin. Struct. Biol.* 9:222–30

33. Kambach C, Walke S, Young R, Avis JM, de la Fortelle E, et al. 1999. Crystal structures of two Sm protein complexes and their implications for the assembly of the spliceosomal snRNPs. *Cell* 96:375–87

34. Kramer A, Gruter P, Groning K, Kastner B. 1999. Combined biochemical and electron microscopic analyses reveal the architecture of the mammalian U2 snRNP. *J. Cell. Biol.* 145:1355–68

35. Krol A, Westhof E, Bach M, Luhrmann R, Ebel JP, Carbon P. 1990. Solution structure of human U1 snRNA. Derivation of a possible three-dimensional model. *Nucleic. Acids Res.* 18:3803–11

36. Ludtke SJ, Baldwin PR, Chiu W. 1999. EMAN: semiautomated software for high-resolution single-particle reconstructions. *J. Struct. Biol.* 128:82–97

19. The 3D structure of SF3b has been determined at 10 Å resolution by single-particle cryo-EM, allowing the identification of protein domains with known structural folds.

20. The 3D structure of the U11/U12 di-snRNP complex has been determined at 12 Å resolution. The multiprotein complex SF3b has been fitted into the U11/U12 di-snRNP density map, revealing a significant conformational rearrangement of SF3b.

21. Site-specific cross-linking identifies interactions between U2 snRNP proteins, including interactions between components of the essential splicing factor SF3b and a region directly upstream of the branch site.

31. The 3D structure of an in vitro–assembled spliceosomal C complex has been determined by single-particle cryo-EM at 35 Å

resolution, revealing three major structural domains.

37. MacMillan AM, Query CC, Allerson CR, Chen S, Verdine GL, Sharp PA. 1994. Dynamic association of proteins with the pre-mRNA branch region. *Genes Dev.* 8:3008–20
38. Makarov EM, Makarova OV, Urlaub H, Gentzel M, Will CL, et al. 2002. Small nuclear ribonucleoprotein remodeling during catalytic activation of the spliceosome. *Science* 298:2205–8
39. Makarova OV, Makarov EM, Urlaub H, Will CL, Gentzel M, et al. 2004. A subset of human 35S U5 proteins, including Prp19, function prior to catalytic step 1 of splicing. *EMBO J.* 23:2381–91
40. Marcotrigiano J, Lomakin IB, Sonenberg N, Pestova TV, Hellen CU, Burley SK. 2001. A conserved HEAT domain within eIF4G directs assembly of the translation initiation machinery. *Mol. Cell.* 7:193–203
41. Michel F, Ferat JL. 1995. Structure and activities of group II introns. *Annu. Rev. Biochem.* 64:435–61
42. Mindell JA, Grigorieff N. 2003. Accurate determination of local defocus and specimen tilt in electron microscopy. *J. Struct. Biol.* 142:334–47
43. Moore MJ, Query CC, Sharp PA. 1993. Splicing of precursors to mRNA by the spliceosome. In *The RNA World*, pp. 303–57. Cold Spring Harbor, NY: Cold Spring Harbor Lab. Press
44. Muller S, Wolpensinger B, Angenitzki M, Engel A, Sperling J, Sperling R. 1998. A suprraspliceosome model for large nuclear ribonucleoprotein particles based on mass determinations by scanning transmission electron microscopy. *J. Mol. Biol.* 283:383–94
45. Muto Y, Pomeranz Krummel D, Oubridge C, Hernandez H, Robinson CV, et al. 2004. The structure and biochemical properties of the human spliceosomal protein U1C. *J. Mol. Biol.* 341:185–98
46. Nagai K, Oubridge C, Jessen TH, Li J, Evans PR. 1990. Crystal structure of the RNA-binding domain of the U1 small nuclear ribonucleoprotein A. *Nature* 348:515–20
47. Nelissen RL, Will CL, van Venrooij WJ, Luhrmann R. 1994. The association of the U1-specific 70K and C proteins with U1 snRNPs is mediated in part by common U snRNP proteins. *EMBO J.* 13:4113–25
48. Neubauer G, King A, Rappsilber J, Calvio C, Watson M, et al. 1998. Mass spectrometry and EST-database searching allows characterization of the multi-protein spliceosome complex. *Nat. Genet.* 20:46–50
49. Nilsen TW. 1994. RNA-RNA interactions in the spliceosome: unraveling the ties that bind. *Cell* 78:1–4
50. Orlova EV, Dube P, Harris JR, Beckman E, Zemlin F, et al. 1997. Structure of keyhole limpet hemocyanin type 1 (KLH1) at 15 A resolution by electron cryomicroscopy and angular reconstitution. *J. Mol. Biol.* 271:417–37
51. Patel AA, Steitz JA. 2003. Splicing double: insights from the second spliceosome. *Nat. Rev. Mol. Cell Biol.* 4:960–70
52. Penczek P, Radermacher M, Frank J. 1992. Three-dimensional reconstruction of single particles embedded in ice. *Ultramicroscopy* 40:33–53
53. Penczek PA, Grassucci RA, Frank J. 1994. The ribosome at improved resolution: new techniques for merging and orientation refinement in 3D cryo-electron microscopy of biological particles. *Ultramicroscopy* 53:251–70
54. Penczek PA, Renka R, Schomberg H. 2004. Gridding-based direct Fourier inversion of the three-dimensional ray transform. *J. Opt. Soc. Am. A Opt. Image Sci. Vis.* 21:499–509
55. Plaisier JR, Koning RI, Koerten HK, van Heel M, Abrahams JP. 2004. TYSON: robust searching, sorting, and selecting of single particles in electron micrographs. *J. Struct. Biol.* 145:76–83

56. Price SR, Evans PR, Nagai K. 1998. Crystal structure of the spliceosomal U2B″-U2A′ protein complex bound to a fragment of U2 small nuclear RNA. *Nature* 394:645–50
57. Query CC, Strobel SA, Sharp PA. 1996. Three recognition events at the branch-site adenine. *EMBO J.* 15:1392–402
58. Radermacher M. 1988. Three-dimensional reconstruction of single particles from random and nonrandom tilt series. *J. Electron. Microsc. Tech.* 9:359–94
59. Raitskin O, Angenitzki M, Sperling J, Sperling R. 2002. Large nuclear RNP particles—the nuclear pre-mRNA processing machine. *J. Struct. Biol.* 140:123–30
60. Rappsilber J, Ryder U, Lamond AI, Mann M. 2002. Large-scale proteomic analysis of the human spliceosome. *Genome. Res.* 12:1231–45
61. Reidt U, Reuter K, Achsel T, Ingelfinger D, Luhrmann R, Ficner R. 2000. Crystal structure of the human U4/U6 small nuclear ribonucleoprotein particle-specific SnuCyp-20, a nuclear cyclophilin. *J. Biol. Chem.* 275:7439–42
62. Reidt U, Wahl MC, Fasshauer D, Horowitz DS, Luhrmann R, Ficner R. 2003. Crystal structure of a complex between human spliceosomal cyclophilin H and a U4/U6 snRNP-60K peptide. *J. Mol. Biol.* 331:45–56
63. Reuter K, Nottrott S, Fabrizio P, Luhrmann R, Ficner R. 1999. Identification, characterization and crystal structure analysis of the human spliceosomal U5 snRNP-specific 15 kD protein. *J. Mol. Biol.* 294:515–25
64. Roseman AM. 2004. FindEM—a fast, efficient program for automatic selection of particles from electron micrographs. *J. Struct. Biol.* 145:91–99
65. Ryder U, Sproat BS, Lamond AI. 1990. Sequence-specific affinity selection of mammalian splicing complexes. *Nucleic. Acids Res.* 18:7373–79
66. Sander B, Golas MM, Stark H. 2003. Automatic CTF correction for single particles based upon multivariate statistical analysis of individual power spectra. *J. Struct. Biol.* 142:392–401
67. Sander B, Golas MM, Stark H. 2005. Advantages of CCD detectors for de novo three-dimensional structure determination in single-particle electron microscopy. *J. Struct. Biol.* 151:92–105
68. Sato C, Ueno Y, Asai K, Takahashi K, Sato M, et al. 2001. The voltage-sensitive sodium channel is a bell-shaped molecule with several cavities. *Nature* 409:1047–51
69. Sherman MB, Brink J, Chiu W. 1996. Performance of a slow-scan CCD camera for macromolecular imaging in a 400 kV electron cryomicroscope. *Micron* 27:129–39
70. Sigworth FJ. 1998. A maximum-likelihood approach to single-particle image refinement. *J. Struct. Biol.* 122:328–39
71. Spahn CM, Gomez-Lorenzo MG, Grassucci RA, Jorgensen R, Andersen GR, et al. 2004. Domain movements of elongation factor eEF2 and the eukaryotic 80S ribosome facilitate tRNA translocation. *EMBO J.* 23:1008–19
72. Sperling R, Spann P, Offen D, Sperling J. 1986. U1, U2, and U6 small nuclear ribonucleoproteins (snRNPs) are associated with large nuclear RNP particles containing transcripts of an amplified gene in vivo. *Proc. Natl. Acad. Sci. USA* 83:6721–25
73. Staknis D, Reed R. 1994. Direct interactions between pre-mRNA and six U2 small nuclear ribonucleoproteins during spliceosome assembly. *Mol. Cell Biol.* 14:2994–3005
74. Staley JP, Guthrie C. 1998. Mechanical devices of the spliceosome: motors, clocks, springs, and things. *Cell* 92:315–26
75. **Stark H, Dube P, Luhrmann R, Kastner B. 2001. Arrangement of RNA and proteins in the spliceosomal U1 small nuclear ribonucleoprotein particle. *Nature* 409:539–42**

75. Single-particle cryo-EM reveals the 3D structure of the spliceosomal U1 snRNP.

76. Stark H, Mueller F, Orlova EV, Schatz M, Dube P, et al. 1995. The 70S *Escherichia coli* ribosome at 23 A resolution: fitting the ribosomal RNA. *Structure* 3:815–21
77. Stark H, Rodnina MV, Wieden HJ, Zemlin F, Wintermeyer W, van Heel M. 2002. Ribosome interactions of aminoacyl-tRNA and elongation factor Tu in the codon-recognition complex. *Nat. Struct. Biol.* 9:849–54
78. Tarn WY, Steitz JA. 1996. A novel spliceosome containing U11, U12, and U5 snRNPs excises a minor class (AT-AC) intron in vitro. *Cell* 84:801–11
79. Thore S, Mayer C, Sauter C, Weeks S, Suck D. 2003. Crystal structures of the *Pyrococcus abyssi* Sm core and its complex with RNA. Common features of RNA binding in archaea and eukarya. *J. Biol. Chem.* 278:1239–47
80. Toro I, Thore S, Mayer C, Basquin J, Seraphin B, Suck D. 2001. RNA binding in an Sm core domain: X-ray structure and functional analysis of an archaeal Sm protein complex. *EMBO J.* 20:2293–303
81. Urlaub H, Raker VA, Kostka S, Luhrmann R. 2001. Sm protein-Sm site RNA interactions within the inner ring of the spliceosomal snRNP core structure. *EMBO J.* 20:187–96
82. van Heel M. 1984. Multivariate statistical classification of noisy images (randomly oriented biological macromolecules). *Ultramicroscopy* 13:165–83
83. van Heel M. 1987. Angular reconstitution: a posteriori assignment of projection directions for 3D reconstruction. *Ultramicroscopy* 21:95–100
84. van Heel M, Frank J. 1981. Use of multivariate statistics in analysing the images of biological macromolecules. *Ultramicroscopy* 6:187–94
85. van Heel M, Gowen B, Matadeen R, Orlova EV, Finn R, et al. 2000. Single-particle electron cryo-microscopy: towards atomic resolution. *Q. Rev. Biophys.* 33:307–69
86. van Heel M, Harauz G, Orlova EV. 1996. A new generation of the IMAGIC image processing system. *J. Struct. Biol.* 116:17–24
87. van Heel M, Stoffler-Meilicke M. 1985. Characteristic views of *E. coli* and *B. stearothermophilus* 30S ribosomal subunits in the electron microscope. *EMBO J.* 4:2389–95
88. Vetter IR, Arndt A, Kutay U, Gorlich D, Wittinghofer A. 1999. Structural view of the Ran-importin beta interaction at 2.3 A resolution. *Cell* 97:635–46
89. Vidovic I, Nottrott S, Hartmuth K, Luhrmann R, Ficner R. 2000. Crystal structure of the spliceosomal 15.5kD protein bound to a U4 snRNA fragment. *Mol. Cell* 6:1331–42
90. Volkmann N. 2004. An approach to automated particle picking from electron micrographs based on reduced representation templates. *J. Struct. Biol.* 145:152–56
91. Walter F, Murchie AI, Duckett DR, Lilley DM. 1998. Global structure of four-way RNA junctions studied using fluorescence resonance energy transfer. *Rna* 4:719–28
92. Wang C, Chua K, Seghezzi W, Lees E, Gozani O, Reed R. 1998. Phosphorylation of spliceosomal protein SAP 155 coupled with splicing catalysis. *Genes Dev.* 12:1409–14
93. Will CL, Luhrmann R. 1997. Protein functions in pre-mRNA splicing. *Curr. Opin. Cell Biol.* 9:320
94. Will CL, Luhrmann R. 2005. Richness of RNA roles in modern RNA world. In *RNA World*, ed. RF Gesteland, TR Cech, JF Atkins, pp. 369–400. Cold Spring Harbor, NY: Cold Spring Harbor Press. 3rd ed.
95. Will CL, Luhrmann R. 2005. Splicing of a rare class of introns by the U12-dependent spliceosome. *Biol. Chem.* 386:713
96. Will CL, Makarov EM, Makarova OV, Luhrmann R. 2005. *Immunoaffinity Purification of Spliceosomal and Small Nuclear Ribonucleoprotein Complexes*. Weinheim, Ger.: Wiley-VCH

97. Will CL, Schneider C, MacMillan AM, Katopodis NF, Neubauer G, et al. 2001. A novel U2 and U11/U12 snRNP protein that associates with the pre-mRNA branch site. *EMBO J.* 20:4536–46
98. Will CL, Urlaub H, Achsel T, Gentzel M, Wilm M, Luhrmann R. 2002. Characterization of novel SF3b and 17S U2 snRNP proteins, including a human Prp5p homolog and an SF3b DEAD-box protein. *EMBO J.* 21:4978–88
99. Zhou Z, Licklider LJ, Gygi SP, Reed R. 2002. Comprehensive proteomic analysis of the human spliceosome. *Nature* 419:182–85
100. Zhou ZH, Hardt S, Wang B, Sherman MB, Jakana J, Chiu W. 1996. CTF determination of images of ice-embedded single particles using a graphics interface. *J. Struct. Biol.* 116:216–22
101. Schellenberg MJ, Edwards RA, Dustin BR, Kent OA, Golas MM, et al. 2006. Crystal structure of a core spliceosomal protein interface. *Proc. Natl. Acad. Sci. USA* 103:1266–71

Mechanotransduction Involving Multimodular Proteins: Converting Force into Biochemical Signals

Viola Vogel

Laboratory for Biologically Oriented Materials, Department of Materials, Swiss Federal Institute of Technology, ETH Zurich, CH-8093 Switzerland;
email: viola.vogel@mat.ethz.ch

Annu. Rev. Biophys. Biomol. Struct. 2006. 35:459–88

First published online as a Review in Advance on February 28, 2006

The *Annual Review of Biophysics and Biomolecular Structure* is online at biophys.annualreviews.org

doi: 10.1146/annurev.biophys.35.040405.102013

1056-8700/06/0609-0459$20.00

Key Words

random coil, α-helical bundles, β-sheets, fibronectin, cell adhesion molecules, titin, spectrin family members

Abstract

Cells can sense and transduce a broad range of mechanical forces into distinct sets of biochemical signals that ultimately regulate cellular processes, including adhesion, proliferation, differentiation, and apoptosis. Deciphering at the nanoscale the design principles by which sensory elements are integrated into structural protein motifs whose conformations can be switched mechanically is crucial to understand the process of transduction of force into biochemical signals that are then integrated to regulate mechanoresponsive pathways. While the major focus in the search for mechanosensory units has been on membrane proteins such as ion channels, integrins, and associated cytoplasmic complexes, a multimodular design of tandem repeats of various structural motifs is ubiquitously found among extracellular matrix proteins, as well as cell adhesion molecules, and among many intracellular players that physically link transmembrane proteins to the contractile cytoskeleton. Single-molecule studies have revealed an unexpected richness of mechanosensory motifs, including force-regulated conformational changes of loop-exposed molecular recognition sites, intermediate states in the unraveling pathway that might either expose cryptic binding or phosphorylation sites, or regions that display enzymatic activity only when unmasked by force. Insights into mechanochemical signal conversion principles will also affect various technological fields, from biotechnology to tissue engineering and drug development.

Contents

MECHANICAL FORCES REGULATE MOLECULAR AND CELLULAR FUNCTIONS

Major innovations in biomedicine and in our understanding of how cells work were made possible over the past few decades by a series of new technologies, from biochemistry in the 1970s, to biotechnology, and finally through the deciphering of the human genome at the end of the 1990s. However, we are only beginning to learn how cells use mechanical forces, thereby complementing biochemical signals in the repertoire of tools used to sense and respond to their environments. Advances in nanotechnology and modern cell biology are beginning to provide the means to investigate many of the physical aspects of complex cellular processes, including cell migration, proliferation, differentiation, and ultimately protein expression. Understanding how cells sense and respond to mechanical forces at the molecular level requires measurements of the mechanical properties of biomolecules, protein unraveling pathways, and a characterization of how the functional states of proteins are changing if under mechanical loads (16, 37, 97, 126, 129, 138). In another review, the mechanosensing of force is discussed in terms of cellular-level mechanoresponses

(139), whereas the focus of this article is primarily on the molecular-level principles of mechanosensing and transduction.

Forces generated by the interplay of cells with their surrounding matrices are transmitted through protein-protein interactions that rely upon the dynamic assembly of physically coupled protein networks that link the extracellular matrix (ECM) via transmembrane proteins to cytoskeletal components (as further reviewed in References 6, 21, 44, 45, 49, 51, 54, 98, 115, 138). At the cellular level, new tools are emerging that probe with high spatial and temporal resolution the forces that cells apply to their environments (5, 11, 30, 73, 99, 103, 130), enabling researchers to learn how the linkage between the ECM and the cytoskeleton is stabilized by mechanical force (6, 23, 42, 55, 57, 84, 109, 132, 136, 140), and to identify major intracellular players that are involved in force sensing and force generation (23, 44, 47, 65, 116, 141) and how they are physically connected to each other.

The Search for Mechanochemical Signal Converters

The search for mechanosensors identified molecules of the focal adhesion complex that are phosphorylated in a force-dependent manner (58, 74, 125, 129, 141), and revealed that the size of focal contacts, their composition, and signaling activity is force dependent (10, 36, 44, 46, 146). Molecular mechanisms are gradually emerging by which cells sense mechanical signals and turn conformational changes into biochemical events that differentially regulate cell signaling pathways (12, 43, 57, 87, 98, 116, 127, 129, 131, 145). For example, myotubes differentiate optimally on substrates with tissue-like stiffness (31) and tissue stiffness could promote malignant behavior by upregulating integrin clustering (98). Cell shape, cytoskeletal tension, and RhoA regulate whether human mesenchymal stem cells differentiate into adipocytes or osteoblasts (87). Thus cells are susceptible not only to biochemical stimuli, but also to functional changes imposed by mechanical forces. Identifying the molecules involved as well as the detailed molecular mechanisms by which mechanical cues are converted into biochemical signals is thus of fundamental interest.

ECM: extracellular matrix

AFM: atomic force microscopy

Why is so little known about molecular mechanosensors and how they regulate the mechanochemical signal conversion? While ample information is available about the equilibrium structures of proteins and their biochemical functions, insight into the relationship between mechanical strain and the resultant conformational changes or exposure of molecular recognition sites is still in its infancy. Gaining insights into the design principles of proteins from a mechanical perspective was made possible only through the advent of new nanotools capable of stretching single molecules, namely atomic force microscopy (AFM) and optical tweezers. The first force measurements on single multimodular proteins were performed in 1997 on the muscle protein titin, which is made of multiple tandem repeats and part of the elastic I-band part of muscle fibers, showing that multimodular proteins do not deform continuously if stretched but that the modules rupture sequentially. In the case of titin, the force versus length plots showed a series of sawtooth peaks that correspond to the unraveling of single titin repeats (62, 106, 134). Computational simulations of the unraveling trajectories soon revealed that the rupture of modules typically coincides with the breakage of force-bearing backbone hydrogen bonds that stabilize the tertiary structure of the protein (80), a prediction later confirmed experimentally (85).

For cells to take advantage of mechanical force to regulate protein function requires that the proteins are embedded in force-bearing molecular networks that are physically coupled to the contractile cytoskeleton. The linkages could be formed by covalent or noncovalent bonds to other elements of the force-bearing structure. The important aspect to be addressed in this review is that many proteins of the ECM as well as those that play key roles in linking the transmembrane

Fn: fibronectin

integrins to the contractile cytoskeleton consist of multiple and often repeating modules strung together to form multimodular proteins. Many proteins of these physically coupled extra- and intracellular force-bearing networks have one or more high-affinity binding sites and/or catalytic sites. The cell adhesion protein fibronectin (Fn), for example, consists of more than 50 modular repeats and is abundant in serum as well as in many matrices (**Figure 1**). Although many molecular recognition sites for other matrix proteins and integrins are present in the loop regions of these β-sandwich motifs, many cryptic binding sites buried within FnIII modules in the folded state have been identified. But what is the physiological significance of these cryptic sites on fibronectin and on other multidomain proteins, as cells do not operate under denaturing conditions? Is it possible that mechanical forces regulate their exposure? And what happens to molecular recognition sites that are already exposed under equilibrium conditions if they are strained by force?

Cells Partially Unravel Proteins, a Process Regulated by Cell Contractility

Before exploring the detailed mechanisms of how force can potentially switch protein functions, it is crucial to know whether cells can generate sufficient forces to actively unravel proteins, or is the major physiological function of protein unraveling to absorb shock waves? Only very few studies indicating that cells can indeed unravel proteins are available owing to the challenge of probing conformational changes of proteins in cell culture. Matrix fibrils of fibroblasts cultured on glass exhibited a wide range of fibronectin conformations, as probed by fluorescence resonance energy transfer (FRET), and suggested partial unraveling of FnIII modules (7). By blocking cell contractility, Baneyx et al. (8) showed that fibronectin unraveling within cell matrix fibrils is controlled by cytoskeletal tension and that unraveling is reversible. Furthermore, fibronectin unraveling in matrix fibrils increases with substrate rigidity (M. Antia, G. Baneyx & V. Vogel, unpublished data), and Rho-mediated cell contractility can expose cryptic sites in fibronectin and induce fibronectin matrix assembly (148). Finally, cells lacking the 32-kDa matricellular glycoprotein SPARC exhibit diminished fibronectin-induced integrin-linked kinase (ILK) activation and ILK-dependent cell contractile signaling, and induced expression of SPARC in SPARC-null fibroblasts restores fibronectin-induced ILK activation, downstream signaling, and fibronectin unraveling (9). It is thus essential to learn how unraveling affects the functional display of the ECM and to identify other molecules that can be partially unraveled by cell-generated tension. This requires the design of new assays that can test how forced unraveling of proteins correlates with their functional changes.

PROTEIN UNRAVELING, SIGNIFICANCE OF INTERMEDIATE STATES, AND SEQUENTIAL MECHANOCHEMICAL SIGNAL CONVERSION

The lack of tools to explore the world of molecular mechanics resulted previously in an underappreciation of the design principles of the proteins involved in forming the force-bearing linkages, as well as of the significance that forces play in regulating molecular and cellular functions. Because the effect of force on the thermodynamics and kinetics of molecular interactions has been reviewed (16, 32), the focus here is to assemble a conceptual framework of the design of multimodular proteins, their mechanical properties, and ultimately the importance of force-regulated functional changes. As function is always tightly coupled to protein structure, high-resolution structural models are essential to decipher the mechanisms by which proteins can be functionally regulated by force. Currently, such models can only be obtained

a

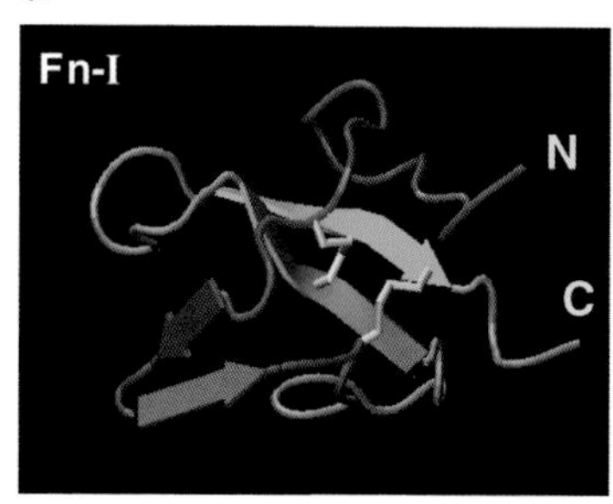

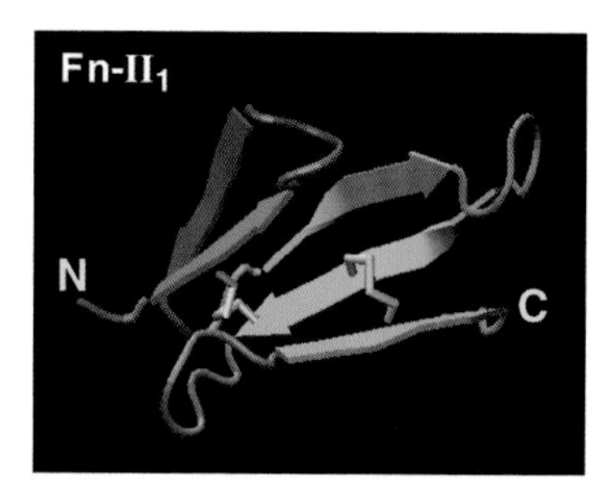

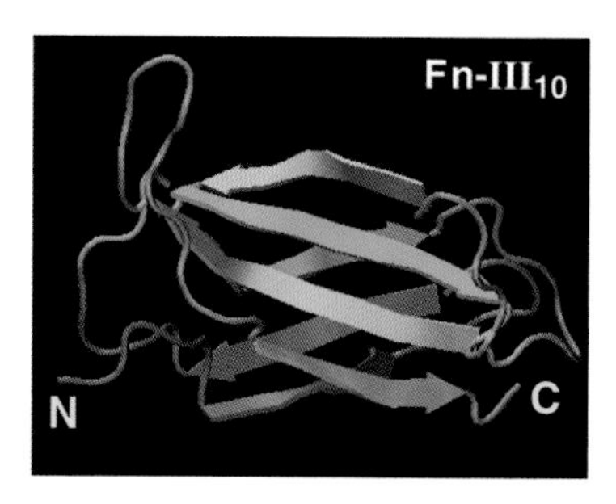

b

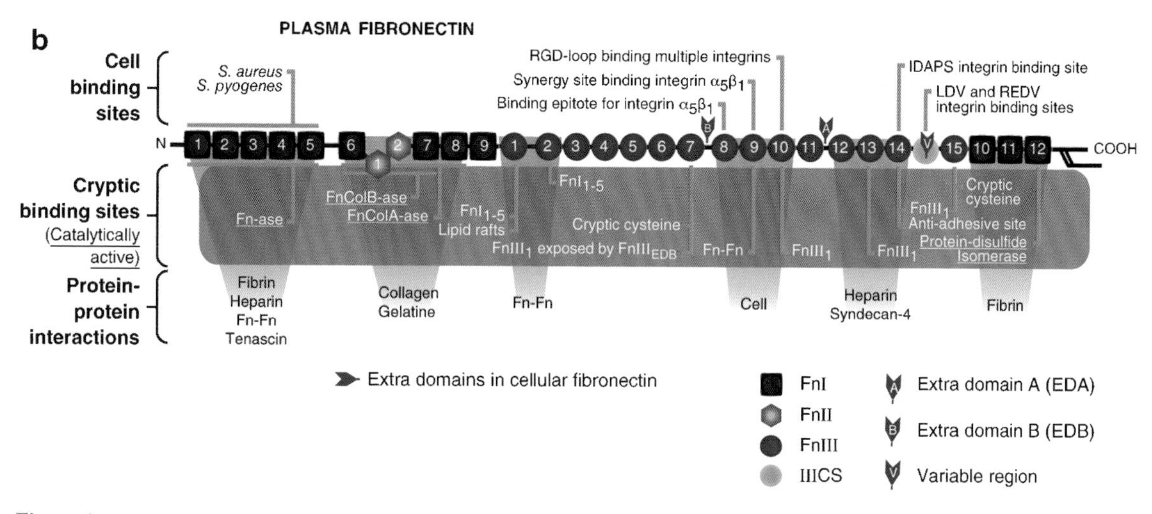

Figure 1

Fibronectin and its molecular recognition and cryptic sites. Fibronectins are dimeric molecules composed of more than 50 repeats of type FnI, FnII, and FnIII, whereby cellular fibronectin may contain additional alternatively spliced modules as indicated. All the repeats are composed of β-sheet motifs, and representative crystal structures are given in the upper panel for FnI (143), $FnII_1$ (100), and $FnIII_{10}$ (28). Fibronectins contain a large number of molecular recognition and cryptic sites (*lower panel*), including the cell binding site RGD (101); the synergy site PHSRN, which is recognized by α5β1 integrins (104, 128); and the sequence IDAPS in the HepII region of fibronectin that supports α4β1-dependent cell adhesion (91). The IDAPS motif implicated in integrin α4β1 binding is at the $FnIII_{13-14}$ junction (124). The cryptic sites include various Fn self-assembly sites whose exposure is needed to induce fibronectin fibrillogenesis (123), a cryptic fragment from $FnIII_1$ that localizes to lipid rafts and stimulates cell growth and contractility (50), and a binding site for tenascin (52). Other cryptic sites with enzymatic activity include FnCol-ase, a metalloprotease in the collagen binding domain of plasma fibronectin capable of digesting gelatin, helical type II and type IV collagen, α- and β-casein, and insulin β-chain (119); as well as a proteinase (Fn-ase) specific to fibronectin, actin, and myosin (120); and a disulfide isomerase (70).

computationally. Various steered molecular dynamics (SMD)-based predictions regarding the unraveling pathways and the location of critical force-bearing hydrogen bonds have been experimentally verified (26, 38, 77, 80, 85, 111). Important to the predictive power of these simulations is that the proteins are solvated in explicit water, as single water molecules play critical roles in protecting or breaking force-bearing bonds (25, 26, 39).

SMD: steered molecular dynamics

Ig: immunoglobulin-like module

In the sections below, the knowledge gained regarding the energy barriers that have to be overcome to unravel modules is discussed, and how the height of these energy barriers can be tuned, for example, by disulfide-crosslinking, amino acid sequence variations, as well as by environmental factors, including pH, ionic strength, and ligand binding. After a few remarks regarding differences in chemical versus mechanical unraveling pathways and protein refolding—even under residual tensile forces—the engineering principles by which the relative mechanical stability of tandem repeats can be regulated are systematically reviewed. For a protein that has several energy barriers that have to be passed before it is completely unraveled, the height of each one of them might have a different force-dependency (26, 32). Caution is thus needed when estimating the mechanical stability at physiological conditions by extrapolation of AFM data dominated by one energy barrier to zero force, unless certainty exists regarding the locations of *all* intermediate states along the unraveling pathway and their barrier shapes (26, 88, 144). The same cautionary note applies when comparing relative mechanical stabilities of different proteins, since the energy barriers have different loading rate-dependencies unless the same structural changes are involved (26). Finally, the given examples focus on describing the structural characteristics of those intermediate states that have been associated with regulatory functions.

Mechanical Versus Chemical Stability

In contrast to protein unraveling in solution, the direction along which stress is applied to a protein may restrict the number of possible unraveling trajectories. In most single-molecule studies of multimodular proteins, the force was applied between the N and C termini, which pointed in opposite directions. They were conducted on proteins integral to force-bearing networks, and the direction along which they were unraveled thus agreed well with the vectoral direction along which they would be strained in their native environments. For the protein titin, which spans the A-bands and I-bands of striated muscles with its 40 immunoglobulin-like (Ig) modules, it was found that the mechanical stability of its Ig modules is dominated by a few critical hydrogen bondings (80, 85) and that mechanical stability is correlated with its kinetic rather than its thermodynamic stability (19). Furthermore, by finding that the unraveling rate estimated from AFM data extrapolated to zero force was about the same as the unraveling rates at 0 M denaturant estimated from kinetic stopped-flow experiments, it was initially suggested that the mechanical and chemical unraveling pathways are similar for Ig modules (20). More detailed SMD and AFM measurements combined with mutational studies, however, revealed that the transition states observed in forced unraveling of titin's Ig_{27} as well as of tenascin's FnIII domains are different from those observed in the absence of force (112), since both Ig and FnIII modules unravel through intermediate states (27, 35, 80, 85).

Refolding Against a Tensile Force

Can biochemical signals be switched on and off reversibly by altering the tensile force? Once unraveled or partially unraveled, how fast can multimodular proteins refold and how do the refolding kinetics relate to the force applied? While it was discovered early that lowering the unraveling force allowed repeats to refold one by one (39, 60–62, 94), rarely leading to misfolded modules, these questions have recently been systematically addressed for ubiquitin by force-clamp AFM (33). Ubiquitin mediates protein degradation, thus allowing cells to clear out unwanted proteins. In those AFM studies, a ubiquitin polyprotein was first unraveled and extended at a high force, and refolding was probed after quenching the force. The duration of a folding trajectory is force dependent and in the case of ubiquitin lasted subseconds at a quenched force of 10 pN, and about 10 s at 50 pN (**Figure 2**). Thus refolding can occur at physiologically significant times and forces (33).

Refolding is key to convert signaling peptide sequences back to their original conformation. No data are available, however, that show how ligand binding, for example, to force-exposed cryptic sites might slow the refolding rate, or whether certain cases exist in which the association of partially unraveled modules either with free ligands or during oligomerization or fibrillogenesis might even prevent refolding.

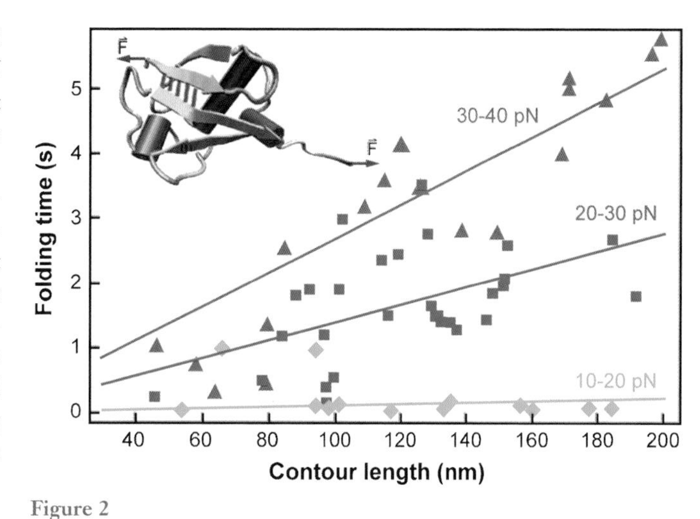

Figure 2

Refolding under an applied force. By using force-clamp AFM, ubiquitin as polyprotein was first unraveled at high force. After partially quenching the force, the duration of the folding collapse is given as a function of the contour length and the applied force. Figure adapted, with permission, from Reference 33.

The Dominant Structural Motifs of Force-Bearing Proteins Found Outside and Inside of the Cell are Different

Major differences in the structural design schemes seem to exist when comparing force-bearing proteins outside and inside the cell. While ECM proteins are composed mostly of tandem β-sandwich motifs, intracellular proteins are often made of repeats of α-helical bundles. The former include fibronectin and cadherins, whereas the latter include spectrin and vinculin. The exception is titin, which is found preferentially in muscle cells, yet its repeats are highly homologous to those of ECM proteins. In optimizing the mechanical response of a protein to best perform its function, does the mechanical response give further insights into the design principles explaining why some proteins are rich in α-helices whereas others have a high β-strand content, or how these secondary motifs are positioned with respect to each other? Drawing attention to this question might give us further insight into the underlying engineering principles by which the level of stress acting on a molecule can be sensed and read quantitatively by cells.

The mechanical stability of a protein is tightly regulated by the positional arrangements of its secondary motifs. Stability depends on how these secondary motifs are oriented with respect to the external force vector and ultimately on the sequence in which force-bearing hydrogen bonds are ruptured. Finally, targeted mutations of either critical amino acids or by disulfide cross-linking can tune the mechanical stability by orders of magnitude.

First, let us consider the idealized case of isolated secondary structural motifs found within globular proteins. Modeling the relative mechanical stability of model peptides forming secondary structural elements revealed that the longitudinal shearing of β-ribbons requires considerably more force than does the lateral unzipping of a β-ribbon (110). Longitudinal shearing of β-ribbons is dominated by the combined strength of the hydrogen bonds that lie perpendicular to the ribbon axis, whereas the unzipping of a β-ribbon requires only one hydrogen bond to be broken at a time. β-sandwich domains with N-terminal and C-terminal strands parallel to each other, but pointing in opposite directions, thus have in general a higher mechanical stability than

CAM: cell adhesion molecule

those with the two terminal strands parallel to each other and pointing in the same direction (81), unless stabilized otherwise. Stretching an α-helix takes less force than the longitudinal shearing of a β-ribbon, but more force than the lateral unzipping of an α-helix dimer (110).

Second, the softest structures are random coil conformations. Titin's cardiac PEVK region, for example, serves as a highly flexible entropic spring (78, 79). PEVK regions are rich in proline (P), glutamate (E), valine (V), and lysine (K) residues and consist of polyproline II helices interspersed with random coil structures (82). Also the EH-segment of about 100 amino acids spliced into the center of myomesin, an M-band component of the sarcomere, acts as a molecular spring with random coil conformations (121). While alternative splicing seems to modulate the elasticity of the EH-domain (121), PEVK exons encode polypeptides of different length, yet with similar elastic properties (114). In this context modules composed of β-sandwich motifs are found preferentially in the extracellular domain [ECM proteins and extracellular domains of cell adhesion molecules (CAMs)], and the more labile α-helical bundles are frequently found among the intracellular proteins involved in mechanotransduction. Finally, the most elastic and softest components, i.e., unstructured sequences, are found inserted into ECM proteins and are also present intracellularly.

While the above considerations are informative, the secondary structural elements experience additional interactions within a globular protein that further regulate their behavior. The β-sandwich motif of titin's Ig modules is more stable than that of FnIII modules (94, 105, 107). The globular RNase barnase composed of α-helices and β-strands has considerably lower resistance to force than titin's Ig27 (13), and the force needed to unravel spectrin's repeats of three antiparallel α-helices is 5 to 10 times smaller than that measured for titin's Ig module (2, 71, 108). Next, it will be discussed how the formation of covalent bonds and hydrogen bonds in strategic positions interlinking secondary structural motifs can alter the unraveling characteristics of proteins.

Tuning the Relative Mechanical Stabilities of Tandem Repeats by Amino Acid Variations

Many proteins that are part of the force-bearing network are composed of tandem repeats. If these repeats possess similar mechanical stabilities, they would rupture stochastically, whereas tuning their relative mechanical stabilities could define the sequence in which they unravel. Thus, differences in the mechanical stability between repeats can signal the magnitude of stress acting on the molecule. Exploiting this scheme for mechanochemical signal conversion is particularly powerful if these repeats have different molecular recognition sites that can be functionally switched in response to partial unraveling. Most multidomain proteins show considerable variations of their amino acid sequences among their tandem repeats (**Figure 3**). This can lead to significant variations in their relative mechanical stabilities as shown for titin's Ig domains (35, 39, 59, 62, 75, 105, 106, 133, 134, 142, 144) and fibronectin's type III modules (26, 27, 80, 92, 93, 96). FnIII modules are ideal to study in great structural detail how sequence variations alter their mechanical stability. FnIII modules are found in 2% of all the mammalian proteins and many protein functions are linked to specific FnIII modules. Thus, nature not only provides us with a large number of FnIII variants whose sequence homology is typically below 20%, but an increasing number of their structures have been resolved in recent years. The biggest surprise was that the FnIII modules share remarkable structural homology despite their low sequence homology, and SMD gave the first systematic insight into the relationship between sequence variations and mechanical stability (26, 27, 37, 92).

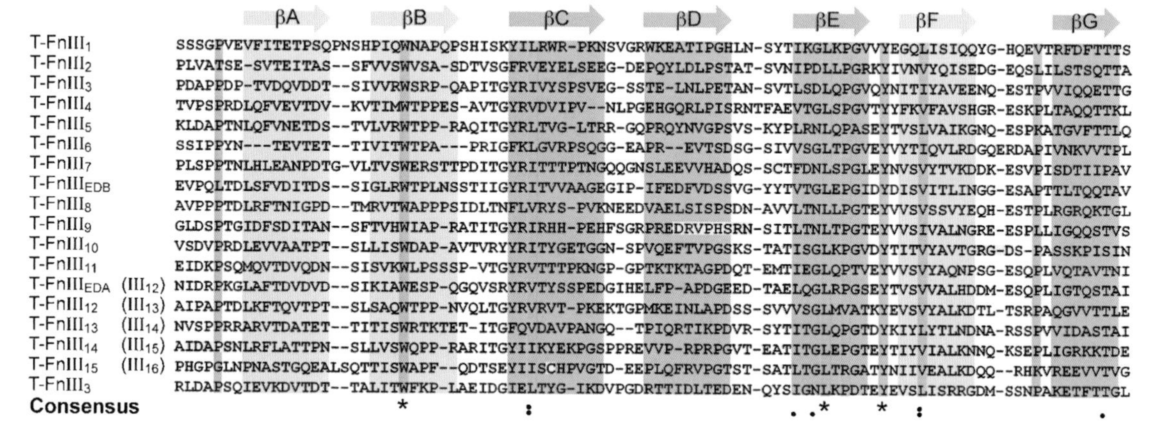

Figure 3

Sequence alignment of human FnIII modules. Despite the high structural homology, only a small fraction of amino acids is highly preserved among different FnIII modules. The numbering of modules is according to a common literature convention that diverges from that used in the database (Swiss-Prot P02751), which is shown in brackets. Tenascin's FnIII module is included in the alignment (T-FnIII$_3$, Swiss-Prot P24521) as is fibronectin extra domain B sequence (EDB, Swiss-Prot P02751). The β-strands are highlighted by green ("lower" strands) and by red ("upper" strands). Conserved residues and similar residues belonging to strongly conserved residues (:) and three sites showing high abundance of proline are shaded in blue. The RGD tripeptide in FnIII$_{10}$, the synergy site in FnIII$_9$, the IDAPS peptide located in FnIII$_{14}$, as well as the buried cysteine residues in FnIII$_7$ and FnIII$_{15}$, which we employ for site-specific FRET labeling, are shaded in yellow. The alignment was done by ClustalW by using the following changes in the default parameters: Window 3, Gap opening 5, Gap extension 0.05, Gap distance 3, Pair gap 1.

Unraveling trajectories for FnIII modules show a rather complex unraveling behavior. Whereas the β-strands of the titin's Ig modules are aligned parallel to each other, the β-strands of the opposing β-sheets of the FnIII modules are twisted with respect to each other (twisted state) as shown in **Figure 4**. Overcoming the first major energy barrier requires that the two opposing β-sheets of Fn-III are rotated against each other such that all β-strands align parallel to the force vector (aligned state) (26, 27). Mutations in the core can alter the unraveling force as shown experimentally for tenascin-FnIII (112). SMD reveals that overcoming this first energy barrier requires that one or two conserved backbone hydrogen bonds connecting the A and the B strands have to be broken (**Figure 5**), which allows one or two water molecules to enter the periphery of the hydrophobic core. Slipping of water between β-sheets was needed to allow all β-strands to align with the external force vector. This transition from the twisted state to the aligned state is the major transition that regulates the relative mechanical stability of the FnIII modules, and all further events in the unraveling pathway proceed from the aligned state (26, 27). Variability in the energy barrier that must be overcome to align the β-strands was correlated to very specific amino acid substitutions. For example, FnIII$_7$ and FnIII$_{EDB}$ have a unique hydrogen bond (backbone to side chain) that mechanically connects the two β-sheets and is formed within the hydrophobic core, whereas most other FnIII modules are mechanically weakened by possessing a proline in that particular location, thus lacking the essential amine hydrogen necessary for hydrogen bond formation (27). FnIII$_{EDB}$ is mechanically weaker than FnIII$_7$ owing to a proline substitution that facilitates water access to this stabilizing hydrogen bond.

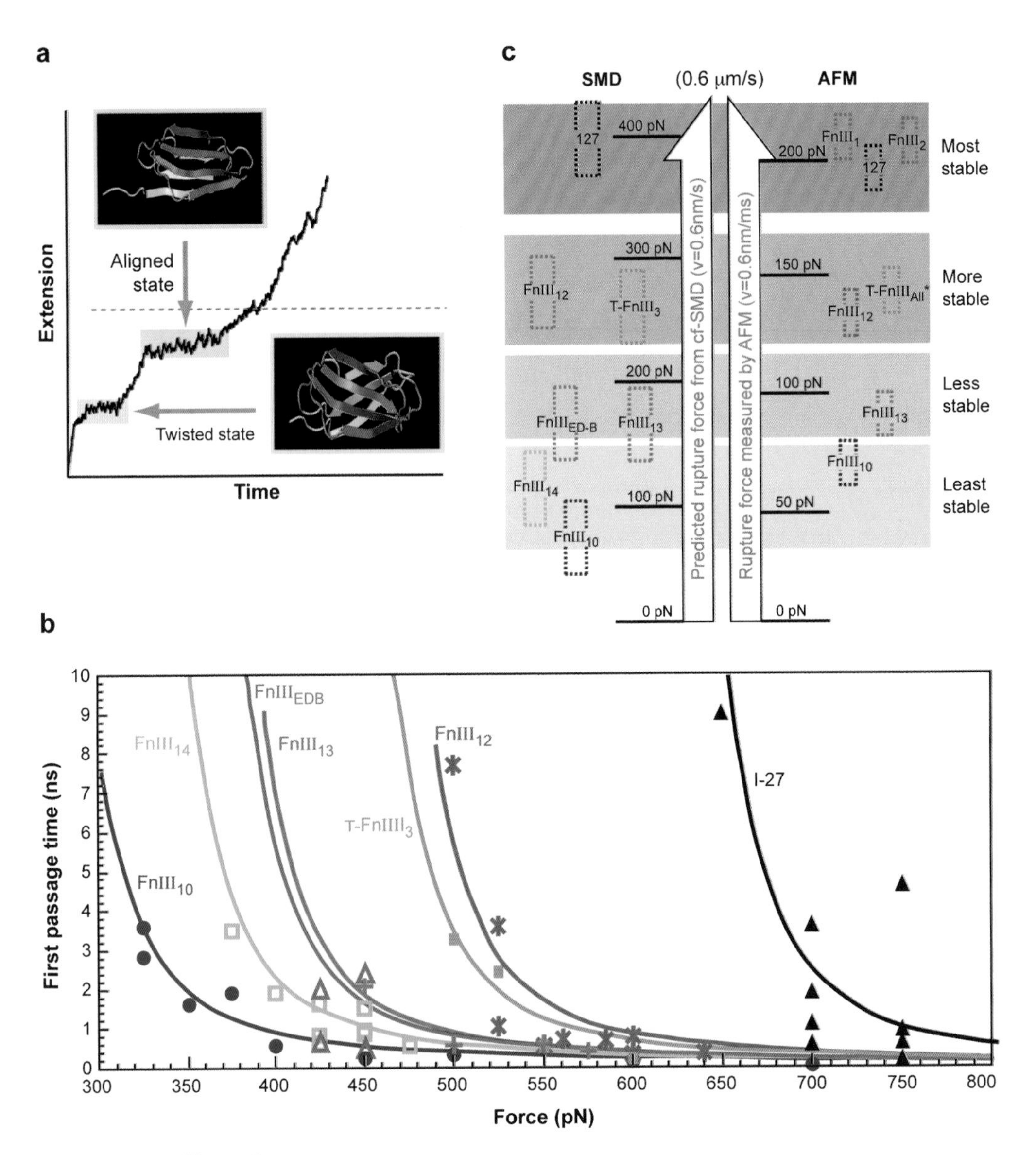

Figure 4

Relative mechanical stabilities of FnIII modules. (*a*) Upon crossing the first major energy barrier, the modules transition from the "twisted" to the "aligned" state, where all β-strands are aligned with the force vector. A second energy barrier has to be passed to break apart the first β-strand. The length of the plateau in the extension time plots, i.e., the first passage time, is indicative of the energy barrier height. The first energy barrier is highly variable among FnIII modules and determines their mechanical stability (26, 27); the second energy barrier is about equal for all FnIII modules studied so far (27). (*b*) The first passage time needed to cross the first major energy barrier (plateau length in panel *a*) was derived from each computational unraveling trajectory for different constant pulling forces and FnIII modules. The fitted curves assume a 3 Å energy barrier width. The mechanical stability decreases as follows: Ig_{27} > $FnIII_{12}$ > T-$FnIII_3$ > $FnIII_{13}$ ∼ $FnIII_{EDB}$ > ∼ $FnIII_{14}$ > $FnIII_{10}$ (26), and a previous study established that $FnIII_7$ > $FnIII_9$ ∼ $FnIII_8$ > $FnIII_{10}$ (27). (*c*) A comparison between forces needed to pass the first major energy barrier at a pulling velocity of 0.6 μm s^{-1}, either calculated from the SMD-derived energy barrier heights (26) or from AFM experiments (93).

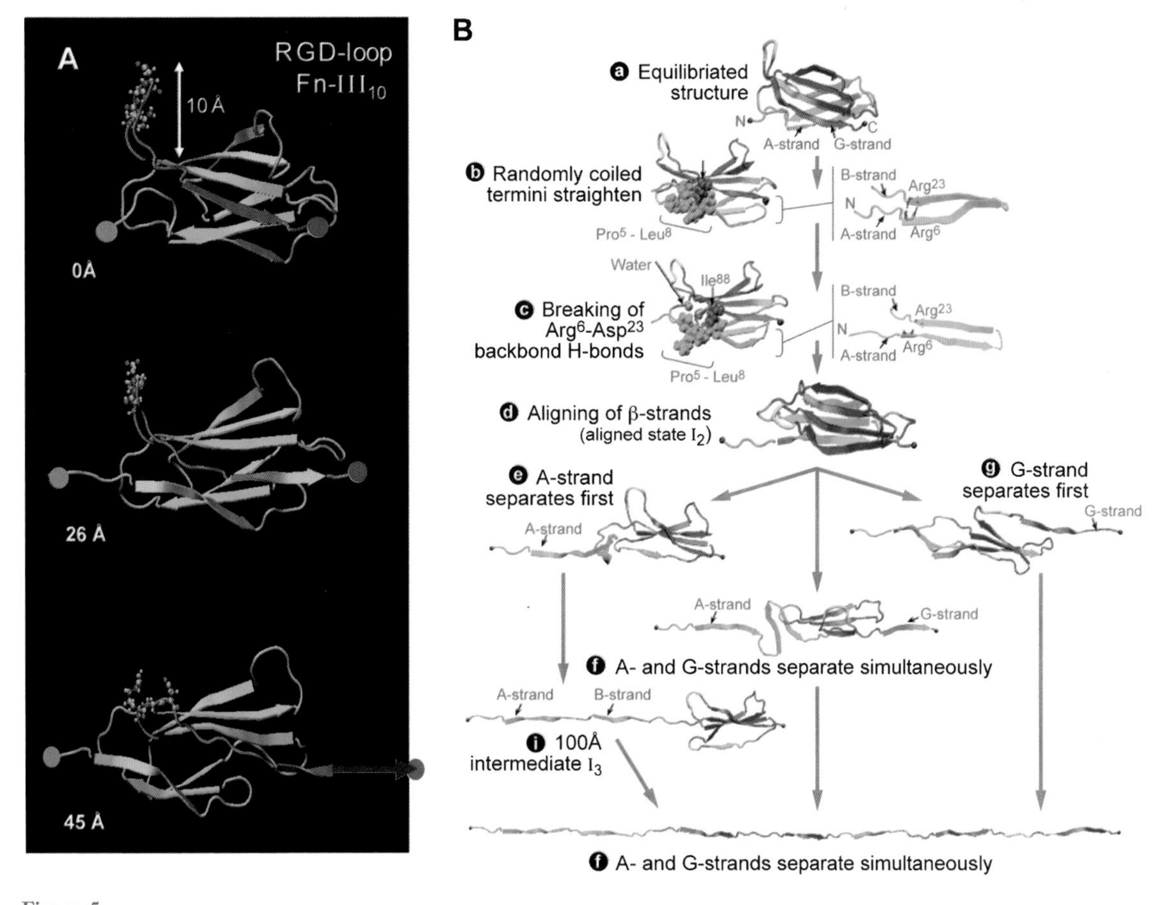

Figure 5

Force-induced conformational changes of the (*A*) RGD cell adhesion site and alternate unraveling pathways for (*B*) Fn-III_{10}. (*A*) The RGD-loop connects the G- and F-strands and is first shortened and then straightened once the G-strand (*red*) breaks away. (*B*) Unraveling pathways for Fn-III_{10} derived from SMD simulations (the upper and lower β-sheets are *red* and *green*, respectively). Upon applying a constant force to Fn-III_{10}, equilibrated in a periodic box (a), the randomly coiled termini are straightened slightly (b), while the overall tertiary structure remained intact. Rupture of the two backbone hydrogen bonds Arg6 and Asp23 connecting the A- and B-strands is the first significant event in unraveling (c) and permits solvation of the hydrophobic core periphery. Upon penetration of water molecules (*blue*), the β-strands between the two β-sheets rotate against each other and align with the force vector (d) of the β-strands and proceed along alternative pathways by separation of first the A-strand (e), or simultaneously the A- and G-strands (f), or first the G-strand (g). Finally, the structure is fully unraveled (h). Panel *b* adapted, with permission, from Reference 38.

This and other examples demonstrate the extent to which the mechanical stability of FnIII is regulated by key hydrogen bonds, initially well shielded from attack by water molecules, that then have to be broken to pass major energy barriers. Strategic point mutations can thus tune the relative mechanical stability of the modules and therefore define the sequence of unraveling events (26, 27, 93) as summarized in **Figure 4**. The mechanical stability decreases as follows: Ig_{27} > $FnIII_{12}$ > T-$FnIII_3$ > $FnIII_{13}$ ~ $FnIII_{EDB}$ > ~ $FnIII_{14}$ > $FnIII_{10}$ (26), and a previous study established that $FnIII_7$ > $FnIII_9$ ~ $FnIII_8$ > $FnIII_{10}$ (27). This could explain why the tandem repeats of fibronectin show high variability in their amino acid sequences, while the sequences are highly preserved among particular FnIII modules of different species. Once this first major energy barrier to align the β-strands is passed, multiple intermediate states exist later in the unraveling pathway, and further unraveling can proceed along various pathways (**Figure 5**) as revealed by SMD and AFM studies (38, 77).

Finally, while all FnIII modules studied so far are mechanically less stable than titin's Ig modules (94, 105, 107), the polycystic kidney disease (PKD) modules, which constitute nearly half of the extracellular region of polycystin-1, resemble FnIII modules structurally yet show an enhanced mechanical strength similar to that of titin's I_{27}. It is proposed that PKDd1 forms a force-stabilized intermediate where additional hydrogen bonds form between the A-B loop and the G-strand; however, this force-stabilized intermediate was observed only at low forces (34). Polycystin-1 is a large membrane-associated protein that interacts with the mechanosensitive ion channel polycystin-2, a member of the mechanosensitive transient receptor potential (TRP) cation channels. Polycystin-1 has been localized to focal adhesion complexes and desmosomes.

Tuning the Mechanical Stability by Environmental Factors: pH, Ionic Strength

Is the relative mechanical stability defined solely by the amino acid sequence, or can it be tuned also by environmental factors? Will the sequence in which they unravel be affected by changes in pH or ionic strength? While most force-bearing bonds identified so far in the literature are backbone hydrogen bonds, Craig et al. (26) found that the side chains, whether polar or nonpolar, whether charged or uncharged, can regulate water access to force-bearing bonds. Because facilitating water access to force-bearing hydrogen bonds weakens mechanical stability, we should expect that pH and ionic strength may further tune the mechanical stability of proteins. To test this hypothesis, the authors protonated and thus neutralized $Asp7^-$, $Asp23^-$, and $Glu9^-$ of $FnIII_{10}$, which surround the first hydrogen bonds that break upon unraveling, and conducted additional SMD simulations. At neutral pH, repulsive interactions between these three negative side chains ease water access to the force-bearing backbone hydrogen bonds of $FnIII_{10}$. Once these side chains are neutralized, a new side chain hydrogen bond formed between the protonated carboxyl groups of Asp23 and Asp7 upon equilibration, which resulted in an enhanced mechanical stability (26). This observation is intriguing because $FnIII_{10}$ also becomes thermodynamically more stable at ~pH 4.7 as a result of protonation of these same three negative amino acids (66). Lower extracellular pH is associated with cells in stressed environments such as wound sites, inflammation, and cancer, or in some cellular organelles, suggesting that the pH dependence of mechanical stability might be a physiologically relevant phenomenon.

Intradomain Disulfide Cross-Linking to Stabilize Repeats or Intermediate States

The formation of disulfide bonds is essential for maintaining the structure and function of many proteins. The vascular cell adhesion molecule-1 (VCAM-1), for example, has seven Ig modules. Each of these Ig modules has a disulfide bond (-S-S-) buried in its core, where the disulfide bond covalently stabilizes about half of the module against unraveling. One half of the module can thus be extended easily by force, whereas the other half is protected by this disulfide bond, which blocks further molecular extension (14). In the case of the disulfide cross-linked Ig domain from the proximal region of the titin I-band, Ig_1, SMD simulations reveal that the Ig_1 domain is protected against external stress mainly through six interstrand hydrogen bonds located between its A- and B-strands, in contrast to the six A′-G hydrogen bonds that break first for Ig_{27}. By bridging the C- and E-strands, the disulfide bond between Cys36 and Cys61 of Ig_1 restricts the extension of Ig_1 to only 220 Å (40). Furthermore, several Ig modules, including ICAM-1, ICAM-2, and human CD2, have a second disulfide bridge at or near the N- or C-terminal loops, further limiting the extent to which these modules can be unraveled by force. However, AFM data suggest that the formation of a disulfide bridge in Ig_1 is a relatively rare event in solution, even under oxidative conditions (75).

Can mechanical strain regulate the redox state of disulfide bonds? For the melanoma cell adhesion molecules, Mel-CAM (also known as MCAM, MUC18, and CD146), the S-S bond is buried in the core of the folded module, and it becomes solvent exposed in the early stages of protein elongation. Accordingly, protonation of these disulfides to SH can indeed be catalyzed by external force (14). Protonation breaks the force-bearing S-S bond that otherwise protects 50% of the domain from unraveling. The finding that force can regulate the redox state of buried disulfide bonds is significant because the disulfide bond of the Ig domains of CAMs is almost universally found to span their cores.

Importance of Intramolecular Domain-Domain Interactions in Regulating Mechanical Stability

Are the mechanical properties of repeats independent of one another or regulated by adjacent domains? The linker chains between tandem β-sandwich repeats typically do not have secondary structure. The mechanical stability depends on linker length, which must be sufficiently long to allow complete and stable folding of the adjacent modules. In the case of titin, the force to unravel Ig_{28} was higher when Ig_{28} was in tandem array (linked naturally) with Ig_{27} than in an engineered Ig_{28} polyprotein. The first interpretation was that the presence of Ig_{27} modified the mechanical properties of Ig_{28} (76). It was later suggested that the linker peptides linking the engineered Ig_{28} polyprotein were too short, and thus Ig_{28} has greater mechanical stability in the presence of unraveled Ig_{27} (122).

In the case of fibronectin, it is suggested that the insertion of the extra domain B (EDB) module can expose an otherwise cryptic site on the neighboring module (17). Does this happen as result of a too short linker chain? The epitope for the antifibronectin monoclonal antibody (mAb) (BC-1), which is specific for EDB-containing fibronectin molecules, is localized on the module $FnIII_7$, which directly precedes the EDB sequence. In contrast, this epitope is masked in fibronectin lacking the EDB sequence (17). If the linker chains are sufficiently long, an increase in stability of approximately 1 kcal mol^{-1} is seen, compared with the monomer if eight tenascin FnIII modules are linked together. This effect is salt and pH dependent, suggesting that the stabilization results from electrostatic interactions, possibly involving charged residues at the interfaces of the domains (112).

In contrast to the peptide chains that connect the β-strand repeats of extracellular proteins, the linkers connecting the α-helical bundles of the spectrin repeats are made of α-helices themselves (**Figure 6**). Spectrin domains are stabilized by each other in equilibrium experiments (83) and can be observed to unravel cooperatively in AFM experiments (71). Upon linker unraveling, two proximal loops from each of the repeats that normally sequester and "protect" the linker from solvent lift away under the applied force, as shown in **Figure 6** (95). The linker connecting the spectrin repeats 2–3 shows five- to sixfold-greater exposure to water than the linker connecting repeats 1–2. Accordingly, for all computationally derived trajectories at the slower pulling rate, the linker between repeats 1–2 never unraveled, whereas the linker between repeats 2–3 always unraveled. The contact between the linker and the sequestering loops is primarily between hydrophobic residues.

EDB: extra domain B

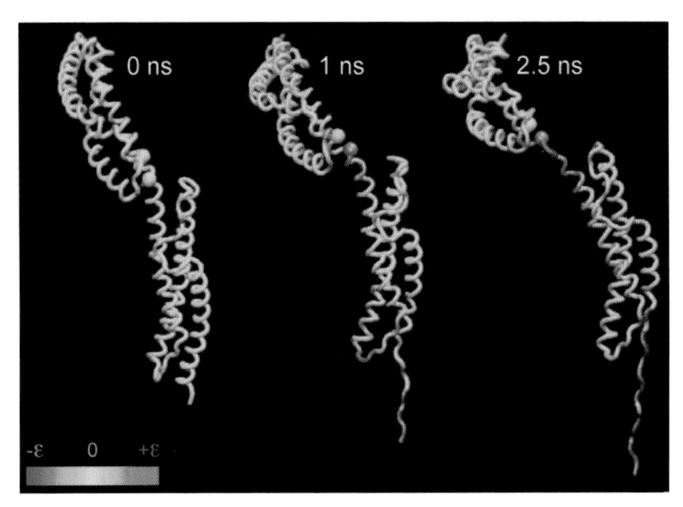

Figure 6

Unraveling a linker between α-helical repeats: SMD snapshots showing the strain in the tandem repeats 2–3 of α-actinin when pulled at 0.5 nm ns^{-1}. With its three-helix bundles, α-actinin belongs to the large family of spectrin-like proteins. A high strain value in the linker is a result of the separation of the protecting loops, which move away from each other in a perpendicular direction. Green represents zero strain, blue negative strain (compression), and red positive strain (extension). Reproduced, with permission, from Reference 95.

Regulating Mechanical Stability by Homophilic Oligomerization

Many of the multimodular proteins oligomerize into larger bundles or fibers, but to what extent do interdomain interactions change the mechanical stability of the modules involved? Experiments suggest that titin molecules may globally associate into oligomers that mechanically behave as independent worm-like chains (WLCs). Although oligomers may form globally via head-to-head association of titin, the constituent molecules otherwise appear independent from each other along their contour. In the case of titin, first results indicate that constituent modules appear to unravel independently from each other (59).

Balancing external forces acting on cells, the intracellular assembly of proteins into force-bearing networks is essential for cell function. About 500 spectrin repeats that have homologous triple-helical repeat motifs in common can be found in the human genome, most commonly in cytoskeletal actin filament-associated proteins, such as spectrins, dystrophin, utrophin, and α-actinins (2). Does lateral association facilitate or oppose forced unraveling? In the case of antiparallel spectrin heterodimers, the associated chains in a dimer can stay together and unravel simultaneously in addition to unraveling independently (**Figure 7**). Strong lateral interactions lead to coupled unraveling of laterally adjacent repeats, whereas weak lateral interactions are surprisingly neutral in their effects on the force to unravel a repeat (72). Interestingly though, the frequency of tandem repeat unraveling events is lower relative to single repeat events, suggesting that tandem unraveling does not propagate as well in two laterally associated heterodimer chains compared with a single chain (72).

Mechanical Strengthening of Repeats by Ligand Binding: Physiological Implications

Are repeats that expose a molecular recognition site in the folded state mechanically stabilized or perturbed upon ligand binding? While very little information is available concerning the affect of ligand binding on the mechanical stability of proteins, in contrast to chemical stability, which is affected by ligation, SMD suggests that the mechanical stability of repeats can be enhanced if the ligand shields critical force-bearing hydrogen bonds from water attacks (26). Heparin binding to $FnIII_{13}$, for example, involves interactions with the same amino acids that play key roles in the early stages of unraveling. The key force-bearing hydrogen bonds between Ser3 and Thr73, and Arg6+ and Arg23+ also show a significant NMR chemical shift upon heparin binding, indicating a decrease in their solvent exposure (113). On the basis of this and other evidence, it was proposed that heparin binding thus stabilizes $FnIII_{13}$ from force-induced unraveling (26).

Ligand-regulated mechanical stability may significantly influence other physiological processes such as protein degradation.

Mechanical unraveling of the target protein, for example, is the rate-limiting step in protein degradation by ATP-dependent proteases. In the cell, proteolytic machines powered by ATP hydrolysis bind proteins with specific peptide tags, denature these substrates, and translocate them into a sequestered compartment for degradation. Destabilization of substrate structure near the degradation tag accelerated degradation and dramatically reduced ATP consumption, revealing an important role for local protein stability in resisting denaturation (63). Furthermore, the mechanical stability of the enzyme dihydrofolate reductase (DHFR) is enhanced severalfold by ligand binding (MTX, NADPH, or DHF), and ligand binding leads to a large reduction in the degradation rate of DHFR (1).

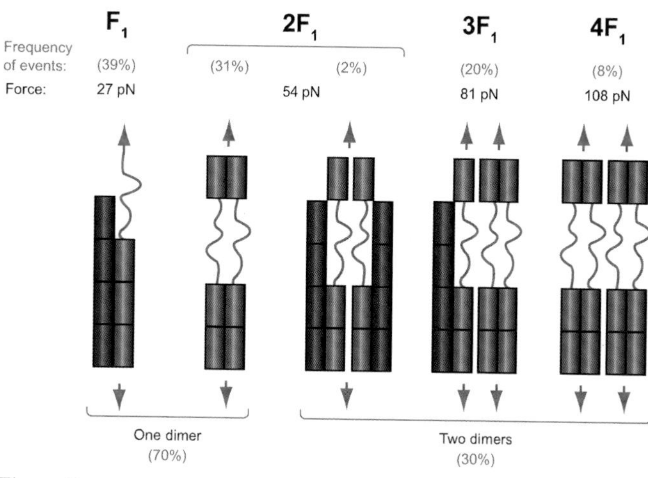

Figure 7

Influence of the lateral association on forced unraveling of antiparallel spectrin heterodimers. Possible dimer unraveling scenarios, frequency of events and forces as obtained by atomic force spectroscopy from α_{18-21}-spectrin and heterodimers of α_{18-21},β_{1-4}-spectrin. Adapted from Reference 72.

FORCE-REGULATED SWITCHING OF PROTEIN FUNCTIONS

Key to transducing a mechanical force into a biochemical signal is that force can switch molecular recognition sites, on or off, or alter their conformations. It is now discussed how the conformations of molecular recognition sites and enzymatic processes might be regulated by force, for example, via the opening of cryptic sites or the opening of autoinhibited binding sites.

Changing the Conformation of Loop-Exposed Molecular Binding Sites by Force

Cell adhesion to fibronectin is mediated by the $FnIII_{10}$ module and its integrin binding motif, Arg78, Gly79, and Asp80 (RGD), which is placed at the apex of the loop connecting β-strands F and G (**Figure 5**). Once the G-strand breaks away from the module upon forced unraveling, the RGD-loop is straightened out (69). A force-induced straightening of the RGD-loop from a tight β-turn into a linear conformation might have far-reaching physiological implications, since a linear RGD-peptide has been shown to have reduced binding strength and selectivity with respect to different members of the integrin family (18, 67, 102, 118, 137).

In addition to the RGD-loop on $FnIII_{10}$, fibronectin is unique in presenting the synergy site on $FnIII_9$ (PHSRN), which is located approximately 32 Å away from the RGD-loop. Only the integrins α5β1 and αIIbβ3 recognize the synergy site, and cell binding to fibronectin is greatly enhanced if those two sides are presented at the right distance (41). The synergy site greatly enhances k_{on} but has little effect on the stability or k_{off} of the complex (128). SMD predicted the existence of an intermediate state where the synergy-RGD distance is increased from 32 Å to approximately 55 Å, as shown in **Figure 8** (68). A 55 Å distance is too large for both sites to cobind the same integrin molecule. Thus at forces too low to unravel the mechanically weaker $FnIII_{10}$, fibronectin if slightly stretched loses its specificity to recognize α5β1 and αIIbβ3 integrins, in which case α5β1 would be outcompeted by αvβ3. Vice versa, cells can no longer distinguish

RGD: tripeptide (Arg-Gly-Asp)

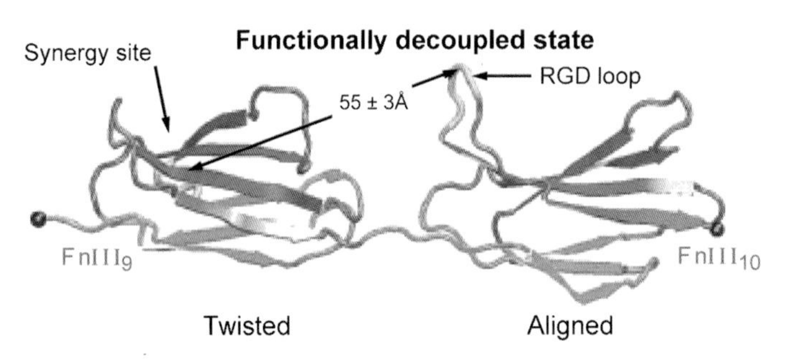

Figure 8

Switching integrin specificity by force. SMD predicts that the distance of the synergy site to the RGD-loop can be switched if $FnIII_{10}$ has transitioned into the aligned state (see **Figure 4**) while $FnIII_9$ remains in the twisted state. Cartoon representations are given for the extensions 20 and 34 Å, respectively. Figure reproduced, with permission, from Reference 68.

fibronectin from other RGD-containing proteins. The α5β1 and αvβ3 integrins play important differential roles in regulating cell proliferation and differentiation. Different conformations of fibronectin when adsorbed to different biomaterial surface chemistries (as further reviewed in References 3, 64) regulate integrin binding specificity with subsequent effects on cell differentiation (64). Taken together, conformational changes imposed by mechanical forces could thus play a potentially significant role in regulating integrin-specific signaling and cell differentiation, a hypothesis that requires experimental validation.

Force-Regulated Exposure of Cryptic Sites: Fn, Cas, Spectrin

Cryptic sites are peptide sequences buried in the native fold. Such sites were typically identified by protein denaturation or discovered as their peptide fragments showed biological activity. Although a large number of cryptic sites have been identified in numerous adhesion molecules (**Figure 1**), an unanswered question regarding their physiological relevance was raised by biologists, since denaturing conditions are rare in the biological context. Much literature now implies that the exposure of cryptic sites might be regulated by mechanical force. The force-regulated exposure of cryptic sites is an effective means to convert defined stress acting on a molecule into biochemical signals. What are the underlying engineering principles that would allow force-regulated exposure while allowing the protein to rapidly refold when the force is reduced? Fulfillment of these two criteria is best accomplished if intermediate states exist along the unraveling pathways. While titin's Ig modules do not show well-pronounced intermediate states once the major energy barrier has been passed (62, 80, 106, 134), the FnIII modules show a rather complex unraveling behavior (38, 68, 77, 96). It is thus interesting that cryptic sites have not been identified within titins Ig modules but are common to and buried within FnIIII modules (53). While addressing the potential relevance of intermediate states of various proteins, it is important to note that proteins might unravel along multiple pathways if stretched and that intrinsic and external factors might regulate the pathways selected.

With respect to the potential role of intermediate states in mechanosensing, the Fn modules are of special interest because many cryptic binding sites have been identified buried within FnIII modules in the folded state, in addition to the many molecular recognition sites presented in the loop regions of these β-sandwich motifs (**Figure 1**). After passing the first major energy barrier, which is common to all FnIII modules studied so far, FnIII modules can unravel along several pathways, each of which has distinctive intermediates (38), as illustrated for $FnIII_{10}$ in **Figure 5**. While initially predicted by SMD, AFM studies have confirmed the divergence of forced-unraveling pathways for $FnIII_{10}$ (38, 77). Along one pathway, for example, the A- and then the B-strand break away first, followed by the unraveling of the remaining β-sandwich structure, whereas the G-strand is the first to separate along the second pathway.

For diverging unraveling pathways, point mutations or other physical or biochemical interactions are particularly effective in regulating which of these pathways is energetically most favored. For example, most of the FnIII modules have a highly conserved proline in the G-strand, thus reducing the number of hydrogen bonds formed between the F- and G-strands (**Figure 9**). By weakening the interactions of the F- and G-strands, this proline facilitates that the G-strand often breaks away first (38). For a few FnIII modules, including $FnIII_1$ and $FnIII_2$, which play major regulatory roles in fibronectin fibrillogenesis (53, 123, 148), the G-strand does not contain this otherwise highly conserved proline residue, thus enhancing the interaction of the F- and G-strands. Consequently, the A- and B-strands break away first, leading to a functionally intact intermediate comprising the C- to G-strand, which is mechanically more stable than the completely folded $FnIII_1$ (37, 93). The structure of this intermediate is analogous to the 76-residue protein anastellin, which, if added to cell culture, induces fibronectin fibrillogenesis (90). Equally interesting is that anastellin displays antitumor, antimetastatic, and antiangiogenic properties in vivo and resembles amyloid fibril precursors (15). While the $FnIII_1$ module is biologically inactive in the fully folded state, breaking away of the A- and B-strands by force might unmask biologically active sites involved in Fn fibrillogenesis. Finally, a binding site for tenascin has been localized to fibronectin's N-terminal 29-kDa heparin/fibrin binding domain. This binding site appears to be cryptic in the whole molecule in solution, but it is exposed on proteolytic fragments and may be exposed when fibronectin is in the extended conformation (52). However, no information is available to determine whether exposure might be force regulated.

While the regulatory role of force on intracellular signaling is well established as reviewed above, little information is available concerning which of the intracellular molecules act as mechanochemical signal converters. First evidence for a force-regulated exposure of cryptic sites was given by removing the membrane of cells anchored to a stretchable substrate using Triton-X as detergent (116, 129). Letting the remaining triton-cytoskeleton react with cytoplasmic cell extracts in an unstretched and stretched state revealed that the focal contact proteins paxillin, focal adhesion kinase (FAK), p130cas, and the protein kinase PKB/Akt (116), as well as C3G and the adapter protein CrkII (129), bind preferentially to stretched triton-cytoskeletons. Triton-cytoskeletons can thus transduce stretching forces into Rap1 activation via tyrosine phosphorylation of Cas by Src family kinases (129). Using an in vitro protein stretch assay, Sawada et al. (117) showed that mechanical unraveling of the Cas substrate domain remarkably enhances its

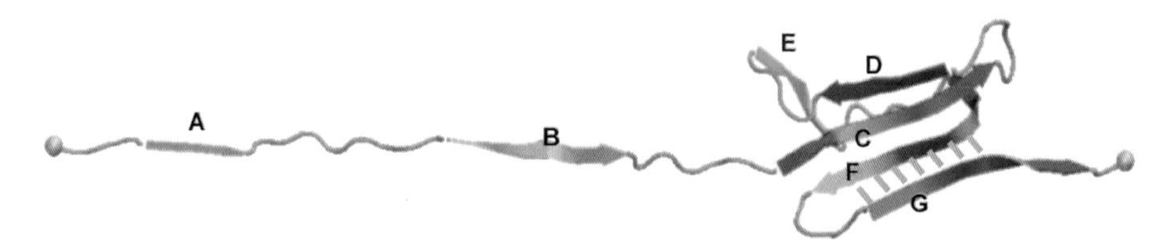

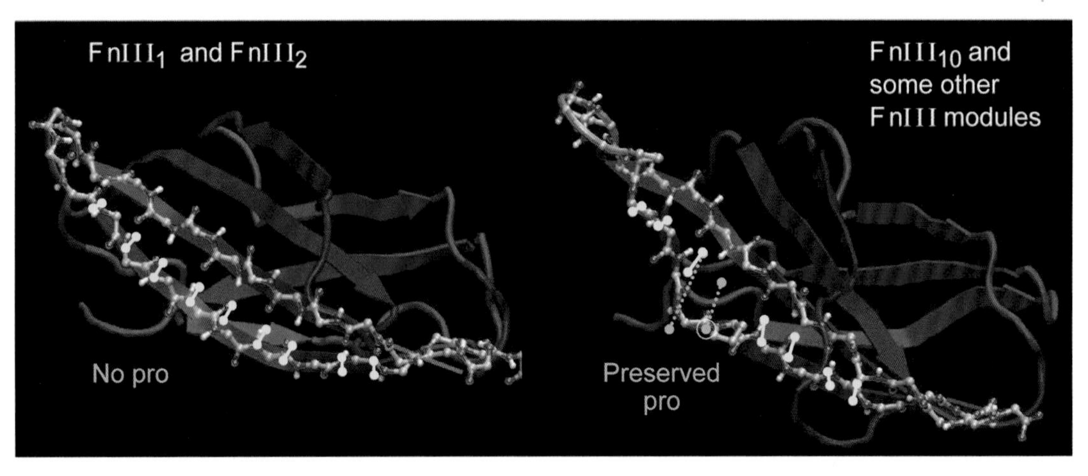

Figure 9

Unmasking anastellin by force. The G-strand of $FnIII_1$ does not contain an otherwise highly conserved proline residue, which weakens the G- and F-strand interactions in other FnIII modules. Consequently, the A- and B-strands break away first, leading to a functionally intact intermediate comprising the C- to G-strand, which is mechanically more stable than the completely folded $FnIII_1$ (37) (93). The structure of this intermediate is analogous to the 76-residue protein anastellin, which, if added to cell culture, induces fibronectin fibrillogenesis (90). Adapted, with permission, from Reference 37.

tyrosine phosphorylation with constant kinase activity and that unraveling of the Cas substrate domain in vivo correlates with regions of higher tensile force.

Which other proteins might have stable intermediates in the unraveling pathway that could expose cryptic sites? The triple helical bundles of spectrin can form stable unraveling intermediates when subjected to external forces, yet mainly mechanical functions have been proposed (2). It is not known whether any of these intermediates of spectrin-like repeats have regulatory functions or whether the variety of potential unraveling pathways (2) are of regulatory significance. Involvement of a cryptic site was implicated in the binding of the actin filament-crosslinking protein α-actinin to a titin-like protein, smitin. The smitin binding sites are located in the smooth muscle α-actinin R2-R3 spectrin-like repeat rod domain and a C-terminal domain formed by cryptic EF-hand structures (22). No information is available whether this cryptic site is force regulated.

Upregulating the Activity of Enzymes Embedded in Force-Bearing Protein Networks

Do cryptic sites exist that display enzymatic activity once exposed by force? A few have been recently characterized for fibronectin (**Figure 1**). The twelfth type I module of fibronectin (FnI_{12}) displays protein-disulfide isomerase activity, and this activity is localized to the C terminus of FnI_{12}. Similar to several other proteins involved in disulfide exchange reactions, including thioredoxin and protein-disulfide isomerase, FnI_{12} contains the Cys-X-X-Cys sequence in its active sites.

However, the protein-disulfide isomerase activity of fibronectin appears to be partially cryptic (70). Additionally, fibronectin contains a cryptic metalloprotease in the collagen binding domain. This domain is included in N-terminal 70-kDa fragment of fibronectin and is capable of digesting gelatin, helical type II and type IV collagen, α- and β-casein, insulin β-chain, and a synthetic Mca-peptide in the presence of Ca^{2+} (119). In addition, the N-terminal heparin/fibrin binding domain of fibronectin contains a cryptic proteinase, i.e., fibronectinase, that specifically acts on actin and myosin. Because this proteinase can be activated if pFN is digested by cathepsin D, and because cathepsin D is released from granulocytes under inflammatory conditions, a possible role of fibronectinase in muscular dystrophy has been suggested (120).

Is it by chance or design that all catalytic sites identified on fibronectin are cryptic? Are they activated upon proteolytic digestion, or might their exposure be regulated by force? All of fibronectin's catalytically active sites identified to date are located on FnI and -II modules. As discussed above, disulfide bonds restrict further unraveling and thus stabilize intermediate states only for the FnI and -II modules, but disulfide bonds are not present within FnIII modules. No data are available whether force can regulate the activities of these cryptic sites.

While most of titin's more than 300 repeats seem to have plain mechanical functions, titin contains a single and structurally distinct module near its C terminus that can display kinase activity. Titin kinase, with its β-sheet-rich smaller lobe and an α-helix-rich bigger lobe, is autoinhibited by its C-terminal regulatory tail (86). Titin kinase activation therefore requires removal of the autoinhibitory tail from the active site. SMD now provides evidence for a mechanically induced opening of the catalytic site without complete domain unraveling, suggesting that it might act as a force sensor (48). Titin kinase is more stable than other proteins containing α-helical secondary structure (13), but it is less stable than the β-sandwich repeats of titin and Fn. In vivo, titin's kinase must be catalytically active because a targeted deletion of the C-terminal end of titin, including the titin kinase domain, impairs myofibrillogenesis (89). The prediction that titin's kinase activity is upregulated by force is also of great interest, since it is structurally homologous to the extracellular regulated kinase (ERK-2) (147). ERK and Rho constitute part of an integrated mechanoregulatory circuit that links matrix stiffness to cytoskeletal tension through integrins to regulate tissue phenotype (24, 29, 98). Integrin clustering is suggested to upregulate ERK activation and increase Rho-associated, coiled-coil containing protein kinase (ROCK)-generated contractility and focal adhesions (98). Furthermore, contractile, epidermal growth factor (EGF)-transformed epithelia with elevated ERK and Rho activity could be phenotypically reverted to tissues lacking focal adhesions if Rho-generated contractility or ERK activity was decreased.

PHYSIOLOGICAL AND TECHNOLOGICAL PERSPECTIVES

A comprehensive survey of the literature illustrates an unexpected richness of structural motifs among multimodular proteins by which mechanical forces can possibly be converted into biochemical signal changes. As in biochemical signaling, we can expect that cells rely not on one mechanotransduction pathway, but on a myriad of molecules translating a range of forces into a multitude of biochemical signals (**Figure 10**). Modules capable of mechanochemical signal conversion are likely found among many ECM molecules. Others are likely found coupled to transmembrane proteins, as well as among the force-bearing intracellular players for which far less is known how force might regulate their functions. Most information derived so far was obtained from single-molecule studies,

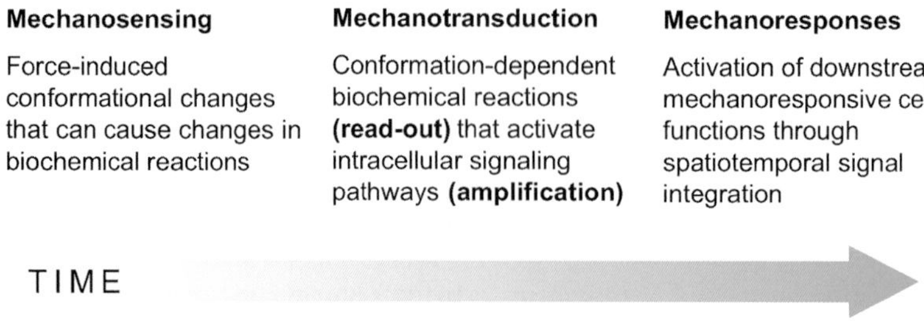

Mechanosensing

Force-induced conformational changes that can cause changes in biochemical reactions

Mechanotransduction

Conformation-dependent biochemical reactions **(read-out)** that activate intracellular signaling pathways **(amplification)**

Mechanoresponses

Activation of downstream mechanoresponsive cell functions through spatiotemporal signal integration

TIME

Figure 10

Translating local mechanical events into global changes in cell function (further reviewed in Reference 139).

including force spectroscopy and computational simulations of the unraveling trajectories. The signature of motifs by which mechanical forces might be converted into biochemical signals combined with high-resolution structural models provide the foundation to start exploring systematically the extent to which these principles are employed by cells to regulate cell functions. After discussing the emerging protein design principles from a mechanical perspective, some concluding and speculative remarks are given, because of the lack of experimental data, on how force might regulate ECM functions and how the new insights might be employed in tissue engineering and drug development.

From an engineering perspective, translating a range of forces into a multitude of biochemical signals is not a trivial task. First, peptide signaling sequences need to be integrated into mechanical switches that respond under the physiologically relevant conditions. If multiple modules are linked together within a protein where each module has one or more specialized functions, a single molecule can translate a range of forces into distinct sets of biochemical functions simply by having the relative energy barriers of its modules tuned with respect to each other. Second, the forces that cells apply to their environment are distributed over an extended network of proteins, complexes, and fibers, and it is the force per protein, i.e., the stress, acting over time that determines whether a module can pass a major energy barrier and unravel. Consequently, which module unravels for a given external force is not regulated solely by its mechanical stability but depends furthermore on the total number of interaction partners and the network density. Balancing mechanical stability and the network density and geometry will determine the biochemical response of the system.

Mechanotransducers might thus be integral parts of many multimodular proteins: The reviewed data suggest that force can regulate a diversity of biochemical signals, from the exposure of cryptic binding and phosphorylation sites, to sites that are enzymatically active only if strained. Central to regulating many of these functionalities by force is that physiologically significant intermediate states exist in the unraveling pathways. Stabilization of intermediate states is challenging from a protein engineering perspective, because the energy barrier to transition the protein into the intermediate state needs to be low enough to unmask relevant sites under physiological conditions, yet further unraveling needs to be prevented, which is done in part via disulfide bridges. For example, all the known catalytically active cryptic sites in fibronectin have been found on the disulfide-stabilized FnI and FnII modules (**Figure 1**). Controlling how the sequence in which biochemical signals are switched with increasing stress levels is not given solely by the structural motifs or their amino acid sequence that encodes them: Relative mechanical stabilities are adaptive and can be regulated by environmental conditions as well, for example, by pH, ligand binding, or other external factors. The ability to regulate the relative mechanical stabilities of the constituents of protein networks by external factors may play a crucial physiological role. While a cell might require a certain set of functionalities to be active under one condition, different responses might be solicited, for example, due to acidification during injury or via ligation of various factors during tissue development and disease. Much future work is also needed to fully comprehend the multifaceted regulatory roles of the ECM and its constituents and how forces might coregulate its functions.

Understanding the principles of mechanotransduction is crucial not only to gain far deeper insights into how cells work, but also to derive new technologies. New technologies that exploit proteins as mechanically regulated switches could be developed. Knowledge about how forces are sensed and transduced into mechanical signals will also innovate how we might treat diseases. New drugs could target intermediate states, or they might be designed to block unraveling or regulate alternate unraveling pathways. Because some point mutations, while having little affect on the equilibrium structure, can have a significant role in regulating mechanical stability and unraveling pathway as discussed above, we might understand better the physiological role of some genetic defects. Finally, expanding on the mechanisms outlined above by which ligands can regulate mechanical stability and protein degradation, one can envision the development of new drugs that regulate mechanosensory or mechanotransducing entities of the cell. Such approaches have received little attention in the past. Drug screening has thus to be conducted not only on proteins under equilibrium, but also on proteins in various force-induced unraveled states. Finally, an enhanced appreciation of the engineering principles of adhesion molecules and of the complexity by which the ECM alters its biochemical display in response to cell contractility might lead to new approaches on how to engineer better the interface between cells and synthetic materials. Consequently, one reason why scaffolds derived from natural tissues or matrix proteins have significantly better clinical performance than their synthetic counterparts (4, 135) might be that cells are mostly deprived from using force to regulate the biochemical display of synthetic matrices. The way in which cells dynamically regulate the biochemical display of their native matrix by force is likely different than if they are in contact with synthetic surfaces and have to respond to the statically exposed surface chemistries.

SUMMARY POINTS

1. In the light that extracellular force-bearing proteins are made primarily of β-sandwich motifs, while many of the proteins that link integrins to the contractile cytoskeleton contain α-helix bundles, this review attempts to piece together a conceptual framework of how extra- and intracellular proteins that contain tandem repeats of various structural motifs can potentially translate a broad range of forces into distinct sets of biochemical signals.
2. An unexpected richness of design principles is emerging by which sensory elements are integrated into structural protein motifs whose conformations can be switched mechanically. This information is crucial to understand the mechanosensation and transduction processes and the subsequent regulation of mechanoresponsive signaling pathways.
3. The importance of intermediate states in mechanochemical signal conversion is emerging, some of which are stabilized against further mechanical unraveling by disulfide bonds. All cryptic sites of fibronectin with catalytic activity, for example, have been found so far on disulfide-stabilized modules, and tensile mechanical forces can regulate the redox state of buried disulfide bonds.

4. Linking modules of different mechanical stability into multimodular macromolecular assemblies enables force-regulated switching of their functionalities, one by one, and has striking physiological advantages, as a broad range of forces can be sensed and translated.

FUTURE DIRECTIONS/UNRESOLVED ISSUES

1. While single-molecule studies have revealed a richness of motifs by which force can potentially be translated into functional changes, major future work is needed to establish the extent to which these motifs are employed by cells under in vivo conditions to regulate cellular processes.
2. It is essential to learn how unraveling affects the functional display of the ECM and to identify other extra- and intracellular molecules that can be partially unraveled by cell-generated tension. This requires the design of new assays that can test how forced unraveling of proteins correlates with functional changes. Much future work is also needed to fully comprehend the multifaceted regulatory roles of the ECM and its constituents and how forces might coregulate its functions.
3. Learning how to switch protein functions by force also has far-reaching potential in biotechnology, tissue engineering, and regenerative medicine, as well as for the development of new drugs that might target proteins stretched into nonequilibrium states.
4. The ultimate goal is to understand the general design principles by which cells sense force in a dynamic fashion, and translate these forces into biochemical signal changes.

ACKNOWLEDGMENTS

Major contributions to the research and the art work presented here by my group members at the University of Washington and at the ETH Zurich, particularly from Andre Krammer, David Craig, Gretchen Baneyx, Meher Antia, Loren Baugh, William Little, Mike Smith, Kris Kubow, Sheila Luna, and Vesa Hytoenen, are deeply appreciated. My gratitude also extends to my long-term collaborators, Klaus Schulten and Mu Gao, and members of the NIH Nanomedicine Center on Biological Mechanics (NIH 2PN2EY016586-02), especially Michael P. Sheetz. This research was made possible by generous support from NIH (NIH 5 R01 GM 49063, 8 R01 EB00249, 3 P50 HG 02360 03S1) and from the ETH Zurich.

LITERATURE CITED

1. Ainavarapu SR, Li L, Badilla CL, Fernandez JM. 2005. Ligand binding modulates the mechanical stability of dihydrofolate reductase (DHFR). *Biophys. J.* 89:3337–44
2. Altmann SM, Grunberg RG, Lenne PF, Ylanne J, Raae A, et al. 2002. Pathways and intermediates in forced unfolding of spectrin repeats. *Structure* 10:1085–96
3. Antia M, Islas LD, Boness DA, Baneyx G, Vogel V. 2006. Single molecule fluorescence studies of surface-adsorbed fibronectin. *Biomaterials* 27: 679–90
4. Badylak SF. 2002. The extracellular matrix as a scaffold for tissue reconstruction. *Semin. Cell Dev. Biol.* 13:377–83

5. Balaban NQ, Schwarz US, Riveline D, Goichberg P, Tzur G, et al. 2001. Force and focal adhesion assembly: a close relationship studied using elastic micropatterned substrates. *Nat. Cell Biol.* 3:466–72
6. Ballestrem C, Geiger B. 2005. Application of microscope-based FRET to study molecular interactions in focal adhesions of live cells. *Methods Mol. Biol.* 294:321–34
7. Baneyx G, Baugh L, Vogel V. 2001. Coexisting conformations of fibronectin in cell culture imaged using fluorescence resonance energy transfer. *Proc. Natl. Acad. Sci. USA* 98:14464–68
8. **Baneyx G, Baugh L, Vogel V. 2002. Fibronectin extension and unfolding within cell matrix fibrils controlled by cytoskeletal tension. *Proc. Natl. Acad. Sci. USA* 99:5139–43**
9. Barker TH, Baneyx G, Cardo-Vila M, Workman GA, Weaver M, et al. 2005. SPARC regulates extracellular matrix organization through its modulation of integrin-linked kinase activity. *J. Biol. Chem.* 280:36483–93
10. Beningo KA, Dembo M, Kaverina I, Small JV, Wang YL. 2001. Nascent focal adhesions are responsible for the generation of strong propulsive forces in migrating fibroblasts. *J. Cell Biol.* 153:881–88
11. Beningo KA, Wang YL. 2002. Flexible substrata for the detection of cellular traction forces. *Trends Cell Biol.* 12:79–84
12. Bershadsky AD, Balaban NQ, Geiger B. 2003. Adhesion-dependent cell mechanosensitivity. *Annu. Rev. Cell Dev. Biol.* 19:677–95
13. Best RB, Li B, Steward A, Daggett V, Clarke J. 2001. Can non-mechanical proteins withstand force? Stretching barnase by atomic force microscopy and molecular dynamics simulation. *Biophys. J.* 81:2344–56
14. Bhasin N, Carl P, Harper S, Feng G, Lu H, et al. 2004. Chemistry on a single protein, vascular cell adhesion molecule-1, during forced unfolding. *J. Biol. Chem.* 279:45865–74
15. Briknarova K, Akerman ME, Hoyt DW, Ruoslahti E, Ely KR. 2003. Anastellin, an Fn3 fragment with fibronectin polymerization activity, resembles amyloid fibril precursors. *J. Mol. Biol.* 332:205–15
16. **Bustamante C, Chemla YR, Forde NR, Izhaky D. 2004. Mechanical processes in biochemistry. *Annu. Rev. Biochem.* 73:705–48**
17. Carnemolla B, Leprini A, Allemanni G, Saginati M, Zardi L. 1992. The inclusion of the type III repeat ED-B in the fibronectin molecule generates conformational modifications that unmask a cryptic sequence. *J. Biol. Chem.* 267:24689–92
18. Carr PA, Erickson HP, Palmer AG 3rd. 1997. Backbone dynamics of homologous fibronectin type III cell adhesion domains from fibronectin and tenascin. *Structure* 5:949–59
19. Carrion-Vazquez M, Oberhauser AF, Fisher TE, Marszalek PE, Li H, Fernandez JM. 2000. Mechanical design of proteins studied by single-molecule force spectroscopy and protein engineering. *Prog. Biophys. Mol. Biol.* 74:63–91
20. Carrion-Vazquez M, Oberhauser AF, Fowler SB, Marszalek PE, Broedel SE, et al. 1999. Mechanical and chemical unfolding of a single protein: a comparison. *Proc. Natl. Acad. Sci. USA* 96:3694–99
21. Charras GT, Horton MA. 2002. Single cell mechanotransduction and its modulation analyzed by atomic force microscope indentation. *Biophys. J.* 82:2970–81
22. Chi RJ, Olenych SG, Kim K, Keller TC 3rd. 2005. Smooth muscle alpha-actinin interaction with smitin. *Int. J. Biochem. Cell Biol.* 37:1470–82
23. Choquet D, Felsenfeld DP, Sheetz MP. 1997. Extracellular matrix rigidity causes strengthening of integrin-cytoskeleton linkages. *Cell* 88:39–48

8. FRET was used to show that fibroblasts stretch and partially unfold fibronectin in their matrix fibrils and that fibronectin refolds if cell contractility is inhibited.

16. Comprehensive review covering the thermodynamic and kinetic aspects of selected force-regulated biochemical processes.

24. Chou FL, Hill JM, Hsieh JC, Pouyssegur J, Brunet A, et al. 2003. PEA-15 binding to ERK1/2 MAPKs is required for its modulation of integrin activation. *J. Biol. Chem.* 278:52587–97
25. Craig D, Gao M, Schulten K, Vogel V. 2004. Structural insights into how the MIDAS ion stabilizes integrin binding to an RGD peptide under force. *Structure* 12:2049–58
26. Craig D, Gao M, Schulten K, Vogel V. 2004. Tuning the mechanical stability of fibronectin type III modules through sequence variations. *Structure* 12:21–30
27. Craig D, Krammer A, Schulten K, Vogel V. 2001. Comparison of the early stages of forced unfolding for fibronectin type III modules. *Proc. Natl. Acad. Sci. USA* 98:5590–95
28. Dickinson CD, Veerapandian B, Dai XP, Hamlin RC, Xuong NH, et al. 1994. Crystal structure of the tenth type III cell adhesion module of human fibronectin. *J. Mol. Biol.* 236:1079–92
29. Dixon RD, Chen Y, Ding F, Khare SD, Prutzman KC, et al. 2004. New insights into FAK signaling and localization based on detection of a FAT domain folding intermediate. *Structure* 12:2161–71
30. du Roure O, Saez A, Buguin A, Austin RH, Chavrier P, et al. 2005. Force mapping in epithelial cell migration. *Proc. Natl. Acad. Sci. USA* 102:2390–95
31. Engler AJ, Griffin MA, Sen S, Bonnemann CG, Sweeney HL, Discher DE. 2004. Myotubes differentiate optimally on substrates with tissue-like stiffness: pathological implications for soft or stiff microenvironments. *J. Cell Biol.* 166:877–87
32. Evans E. 2001. Probing the relation between force—lifetime—and chemistry in single molecular bonds. *Annu. Rev. Biophys. Biomol. Struct.* 30:105–28
33. Fernandez JM, Li H. 2004. Force-clamp spectroscopy monitors the folding trajectory of a single protein. *Science* 303:1674–78
34. Forman JR, Qamar S, Paci E, Sandford RN, Clarke J. 2005. The remarkable mechanical strength of polycystin-1 supports a direct role in mechanotransduction. *J. Mol. Biol.* 349:861–71
35. Fowler SB, Best RB, Toca Herrera JL, Rutherford TJ, Steward A, et al. 2002. Mechanical unfolding of a titin Ig domain: structure of unfolding intermediate revealed by combining AFM, molecular dynamics simulations, NMR and protein engineering. *J. Mol. Biol.* 322:841–49
36. Galbraith CG, Yamada KM, Sheetz MP. 2002. The relationship between force and focal complex development. *J. Cell Biol.* 159:695–705
37. Gao M, Craig D, Lequin O, Campbell ID, Vogel V, Schulten K. 2003. Structure and functional significance of mechanically unfolded fibronectin type III1 intermediates. *Proc. Natl. Acad. Sci. USA* 100:14784–89
38. Gao M, Craig D, Vogel V, Schulten K. 2002. Identifying unfolding intermediates of FN-III(10) by steered molecular dynamics. *J. Mol. Biol.* 323:939–50
39. Gao M, Lu H, Schulten K. 2002. Unfolding of titin domains studied by molecular dynamics simulations. *J. Muscle Res. Cell Motil.* 23:513–21
40. Gao M, Wilmanns M, Schulten K. 2002. Steered molecular dynamics studies of titin I1 domain unfolding. *Biophys. J.* 83:3435–45
41. Garcia AJ, Gallant ND. 2003. Stick and grip: measurement systems and quantitative analyses of integrin-mediated cell adhesion strength. *Cell Biochem. Biophys.* 39:61–73
42. Geiger B, Bershadsky A. 2001. Assembly and mechanosensory function of focal contacts. *Curr. Opin. Cell Biol.* 13:584–92
43. Geiger B, Bershadsky A. 2002. Exploring the neighborhood: adhesion-coupled cell mechanosensors. *Cell* 110:139–42

26. Provides structural insights into how single amino acids can tune the mechanical stability by protecting or exposing force-bearing backbone hydrogen bonds.

32. Reviews how energy barriers that are difficult or impossible to detect in assays of near equilibrium dissociation can determine bond lifetime and strength under rapid detachment.

33. Probes how refolding kinetics of a polyprotein (ubiquitin) depends on the force applied to its terminal ends.

44. Geiger B, Bershadsky A, Pankov R, Yamada KM. 2001. Transmembrane crosstalk between the extracellular matrix and the cytoskeleton. *Nat. Rev. Mol. Cell Biol.* 2:793–805
45. Georges PC, Janmey PA. 2005. Cell type-specific response to growth on soft materials. *J. Appl. Physiol.* 98:1547–53
46. Giannone G, Dubin-Thaler BJ, Dobereiner HG, Kieffer N, Bresnick AR, Sheetz MP. 2004. Periodic lamellipodial contractions correlate with rearward actin waves. *Cell* 116:431–43
47. Giannone G, Jiang G, Sutton DH, Critchley DR, Sheetz MP. 2003. Talin1 is critical for force-dependent reinforcement of initial integrin-cytoskeleton bonds but not tyrosine kinase activation. *J. Cell Biol.* 163:409–19
48. Grater F, Shen J, Jiang H, Gautel M, Grubmuller H. 2005. Mechanically induced titin kinase activation studied by force-probe molecular dynamics simulations. *Biophys. J.* 88:790–804
49. Heidemann SR, Wirtz D. 2004. Towards a regional approach to cell mechanics. *Trends Cell Biol.* 14:160–66
50. Hocking DC, Kowalski K. 2002. A cryptic fragment from fibronectin's III1 module localizes to lipid rafts and stimulates cell growth and contractility. *J. Cell Biol.* 158:175–84
51. Ingber DE, Folkman J. 1989. Mechanochemical switching between growth and differentiation during fibroblast growth factor-stimulated angiogenesis in vitro: role of extracellular matrix. *J. Cell Biol.* 109:317–30
52. Ingham KC, Brew SA, Erickson HP. 2004. Localization of a cryptic binding site for tenascin on fibronectin. *J. Biol. Chem.* 279:28132–35
53. Ingham KC, Brew SA, Huff S, Litvinovich SV. 1997. Cryptic self-association sites in type III modules of fibronectin. *J. Biol. Chem.* 272:1718–24
54. Janmey PA, Weitz DA. 2004. Dealing with mechanics: mechanisms of force transduction in cells. *Trends Biochem. Sci.* 29:364–70
55. Jiang G, Giannone G, Critchley DR, Fukumoto E, Sheetz MP. 2003. Two-picoNewton slip bond between fibronectin and the cytoskeleton depends on talin. *Nature* 424:334–37
56. Deleted in proof
57. Katsumi A, Naoe T, Matsushita T, Kaibuchi K, Schwartz MA. 2005. Integrin activation and matrix binding mediate cellular responses to mechanical stretch. *J. Biol. Chem.* 280:16546–49
58. **Katz BZ, Zamir E, Bershadsky A, Kam Z, Yamada KM, Geiger B. 2000. Physical state of the extracellular matrix regulates the structure and molecular composition of cell-matrix adhesions. *Mol. Biol. Cell* 11:1047–60**
59. Kellermayer MS, Bustamante C, Granzier HL. 2003. Mechanics and structure of titin oligomers explored with atomic force microscopy. *Biochim. Biophys. Acta* 1604:105–14
60. Kellermayer MS, Smith S, Bustamante C, Granzier HL. 2000. Mechanical manipulation of single titin molecules with laser tweezers. *Adv. Exp. Med. Biol.* 481:111–28
61. Kellermayer MS, Smith SB, Bustamante C, Granzier HL. 2001. Mechanical fatigue in repetitively stretched single molecules of titin. *Biophys. J.* 80:852–63
62. Kellermayer MS, Smith SB, Granzier HL, Bustamante C. 1997. Folding-unfolding transitions in single titin molecules characterized with laser tweezers. *Science* 276:1112–16
63. Kenniston JA, Baker TA, Fernandez JM, Sauer RT. 2003. Linkage between ATP consumption and mechanical unfolding during the protein processing reactions of an AAA+ degradation machine. *Cell* 114:511–20
64. **Keselowsky BG, Collard DM, Garcia AJ. 2005. Integrin binding specificity regulates biomaterial surface chemistry effects on cell differentiation. *Proc. Natl. Acad. Sci. USA* 102:5953–57**

58. Demonstrated that the physical state of the ECM can regulate integrin-mediated cytoskeletal assembly and tyrosine phosphorylation to generate two distinct types of cell-matrix adhesions.

64. Authors show that surface chemistry can modulate FN structure, thereby altering integrin adhesion receptor binding and ultimately osteoblastic differentiation.

65. Khan S, Sheetz MP. 1997. Force effects on biochemical kinetics. *Annu. Rev. Biochem.* 66:785–805
66. Koide A, Jordan MR, Horner SR, Batori V, Koide S. 2001. Stabilization of a fibronectin type III domain by the removal of unfavorable electrostatic interactions on the protein surface. *Biochemistry* 40:10326–33
67. Koivunen E, Gay DA, Ruoslahti E. 1993. Selection of peptides binding to the alpha 5 beta 1 integrin from phage display library. *J. Biol. Chem.* 268:20205–10
68. Krammer A, Craig D, Thomas WE, Schulten K, Vogel V. 2002. A structural model for force regulated integrin binding to fibronectin's RGD-synergy site. *Matrix Biol.* 21:139–47
69. Krammer A, Lu H, Isralewitz B, Schulten K, Vogel V. 1999. Forced unfolding of the fibronectin type III module reveals a tensile molecular recognition switch. *Proc. Natl. Acad. Sci. USA* 96:1351–56
70. Langenbach KJ, Sottile J. 1999. Identification of protein-disulfide isomerase activity in fibronectin. *J. Biol. Chem.* 274:7032–38
71. Law R, Carl P, Harper S, Dalhaimer P, Speicher DW, Discher DE. 2003. Cooperativity in forced unfolding of tandem spectrin repeats. *Biophys. J.* 84:533–44
72. Law R, Harper S, Speicher DW, Discher DE. 2004. Influence of lateral association on forced unfolding of antiparallel spectrin heterodimers. *J. Biol. Chem.* 279:16410–16
73. LeDuc P, Ostuni E, Whitesides G, Ingber D. 2002. Use of micropatterned adhesive surfaces for control of cell behavior. *Methods Cell Biol.* 69:385–401
74. Leong L, Hughes PE, Schwartz MA, Ginsberg MH, Shattil SJ. 1995. Integrin signaling: roles for the cytoplasmic tails of alpha IIb beta 3 in the tyrosine phosphorylation of pp125FAK. *J. Cell Sci.* 108(Pt. 12):3817–25
75. Li H, Fernandez JM. 2003. Mechanical design of the first proximal Ig domain of human cardiac titin revealed by single molecule force spectroscopy. *J. Mol. Biol.* 334:75–86
76. Li H, Oberhauser AF, Fowler SB, Clarke J, Fernandez JM. 2000. Atomic force microscopy reveals the mechanical design of a modular protein. *Proc. Natl. Acad. Sci. USA* 97:6527–31
77. Li L, Huang HH, Badilla CL, Fernandez JM. 2005. Mechanical unfolding intermediates observed by single-molecule force spectroscopy in a fibronectin type III module. *J. Mol. Biol.* 345:817–26
78. Linke WA, Kulke M, Li H, Fujita-Becker S, Neagoe C, et al. 2002. PEVK domain of titin: an entropic spring with actin-binding properties. *J. Struct. Biol.* 137:194–205
79. Linke WA, Rudy DE, Centner T, Gautel M, Witt C, et al. 1999. I-band titin in cardiac muscle is a three-element molecular spring and is critical for maintaining thin filament structure. *J. Cell Biol.* 146:631–44
80. Lu H, Isralewitz B, Krammer A, Vogel V, Schulten K. 1998. Unfolding of titin immunoglobulin domains by steered molecular dynamics simulation. *Biophys. J.* 75:662–71
81. Lu H, Schulten K. 1999. Steered molecular dynamics simulations of force-induced protein domain unfolding. *Proteins* 35:453–63
82. Ma K, Kan L, Wang K. 2001. Polyproline II helix is a key structural motif of the elastic PEVK segment of titin. *Biochemistry* 40:3427–38
83. MacDonald RI, Pozharski EV. 2001. Free energies of urea and of thermal unfolding show that two tandem repeats of spectrin are thermodynamically more stable than a single repeat. *Biochemistry* 40:3974–84
84. Mack PJ, Kaazempur-Mofrad MR, Karcher H, Lee RT, Kamm RD. 2004. Force-induced focal adhesion translocation: effects of force amplitude and frequency. *Am. J. Physiol. Cell Physiol.* 287:C954–62

85. Marszalek PE, Lu H, Li H, Carrion-Vazquez M, Oberhauser AF, et al. 1999. Mechanical unfolding intermediates in titin modules. *Nature* 402:100–3
86. Mayans O, van der Ven PF, Wilm M, Mues A, Young P, et al. 1998. Structural basis for activation of the titin kinase domain during myofibrillogenesis. *Nature* 395:863–69
87. McBeath R, Pirone DM, Nelson CM, Bhadriraju K, Chen CS. 2004. Cell shape, cytoskeletal tension, and RhoA regulate stem cell lineage commitment. *Dev. Cell* 6:483–95
88. Merkel R, Nassoy P, Leung A, Ritchie K, Evans E. 1999. Energy landscapes of receptor-ligand bonds explored with dynamic force spectroscopy. *Nature* 397:50–53
89. Miller G, Musa H, Gautel M, Peckham M. 2003. A targeted deletion of the C-terminal end of titin, including the titin kinase domain, impairs myofibrillogenesis. *J. Cell Sci.* 116:4811–19
90. Morla A, Zhang Z, Ruoslahti E. 1994. Superfibronectin is a functionally distinct form of fibronectin. *Nature* 367:193–96
91. Mould AP, Humphries MJ. 1991. Identification of a novel recognition sequence for the integrin alpha 4 beta 1 in the COOH-terminal heparin-binding domain of fibronectin. *EMBO J.* 10:4089–95
92. Ng SP, Rounsevell RW, Steward A, Geierhaas CD, Williams PM, et al. 2005. Mechanical unfolding of TNfn3: the unfolding pathway of a fnIII domain probed by protein engineering, AFM and MD simulation. *J. Mol. Biol.* 350:776–89
93. Oberhauser AF, Badilla-Fernandez C, Carrion-Vazquez M, Fernandez JM. 2002. The mechanical hierarchies of fibronectin observed with single-molecule AFM. *J. Mol. Biol.* 319:433–47
94. Oberhauser AF, Marszalek PE, Erickson HP, Fernandez JM. 1998. The molecular elasticity of the extracellular matrix protein tenascin. *Nature* 393:181–85
95. Ortiz V, Nielsen SO, Klein ML, Discher DE. 2005. Unfolding a linker between helical repeats. *J. Mol. Biol.* 349:638–47
96. Paci E, Karplus M. 1999. Forced unfolding of fibronectin type 3 modules: an analysis by biased molecular dynamics simulations. *J. Mol. Biol.* 288:441–59
97. Pankov R, Yamada KM. 2002. Fibronectin at a glance. *J. Cell Sci.* 115:3861–63
98. Paszek MJ, Zahir N, Johnson KR, Lakins JN, Rozenberg GI, et al. 2005. Tensional homeostasis and the malignant phenotype. *Cancer Cell* 8:241–54
99. Pelham RJ Jr, Wang Y. 1997. Cell locomotion and focal adhesions are regulated by substrate flexibility. *Proc. Natl. Acad. Sci. USA* 94:13661–65
100. Pickford AR, Potts JR, Bright JR, Phan I, Campbell ID. 1997. Solution structure of a type 2 module from fibronectin: implications for the structure and function of the gelatin-binding domain. *Structure* 5:359–70
101. Pierschbacher MD, Ruoslahti E. 1984. Cell attachment activity of fibronectin can be duplicated by small synthetic fragments of the molecule. *Nature* 309:30–33
102. Pierschbacher MD, Ruoslahti E. 1987. Influence of stereochemistry of the sequence Arg-Gly-Asp-Xaa on binding specificity in cell adhesion. *J. Biol. Chem.* 262:17294–98
103. Prechtel K, Bausch AR, Marchi-Artzner V, Kantlehner M, Kessler H, Merkel R. 2002. Dynamic force spectroscopy to probe adhesion strength of living cells. *Phys. Rev. Lett.* 89:028101
104. Redick SD, Settles DL, Briscoe G, Erickson HP. 2000. Defining fibronectin's cell adhesion synergy site by site-directed mutagenesis. *J. Cell Biol.* 149:521–27
105. Rief M, Gautel M, Gaub HE. 2000. Unfolding forces of titin and fibronectin domains directly measured by AFM. *Adv. Exp. Med. Biol.* 481:129–41
106. Rief M, Gautel M, Oesterhelt F, Fernandez JM, Gaub HE. 1997. Reversible unfolding of individual titin immunoglobulin domains by AFM. *Science* 276:1109–12

107. Rief M, Gautel M, Schemmel A, Gaub HE. 1998. The mechanical stability of immunoglobulin and fibronectin III domains in the muscle protein titin measured by atomic force microscopy. *Biophys. J.* 75:3008–14
108. Rief M, Pascual J, Saraste M, Gaub HE. 1999. Single molecule force spectroscopy of spectrin repeats: low unfolding forces in helix bundles. *J. Mol. Biol.* 286:553–61
109. Riveline D, Zamir E, Balaban NQ, Schwarz US, Ishizaki T, et al. 2001. Focal contacts as mechanosensors: externally applied local mechanical force induces growth of focal contacts by an mDia1-dependent and ROCK-independent mechanism. *J. Cell Biol.* 153:1175–86
110. Rohs R, Etchebest C, Lavery R. 1999. Unraveling proteins: a molecular mechanics study. *Biophys. J.* 76:2760–68
111. Rounsevell RW, Clarke J. 2004. FnIII domains: predicting mechanical stability. *Structure* 12:4–5
112. Rounsevell RW, Steward A, Clarke J. 2005. Biophysical investigations of engineered polyproteins: implications for force data. *Biophys. J.* 88:2022–29
113. Sachchidanand, Lequin O, Staunton D, Mulloy B, Forster MJ, et al. 2002. Mapping the heparin-binding site on the 13-14F3 fragment of fibronectin. *J. Biol. Chem.* 277:50629–35
114. Sarkar A, Caamano S, Fernandez JM. 2005. The elasticity of individual titin PEVK exons measured by single molecule atomic force microscopy. *J. Biol. Chem.* 280:6261–64
115. Sawada Y, Nakamura K, Doi K, Takeda K, Tobiume K, et al. 2001. Rap1 is involved in cell stretching modulation of p38 but not ERK or JNK MAP kinase. *J. Cell Sci.* 114:1221–27
116. Sawada Y, Sheetz MP. 2002. Force transduction by Triton cytoskeletons. *J. Cell Biol.* 156:609–15
117. Sawada Y, Tamada M, Dubin-Thaler B, Cherniavskaya O, Sakai R, et al. 2005. *Cas serves as a force receptor through mechanical unfolding-dependent substrate priming*. Presented at Am. Soc. Cell Biol.
118. Scarborough RM, Naughton MA, Teng W, Rose JW, Phillips DR, et al. 1993. Design of potent and specific integrin antagonists. Peptide antagonists with high specificity for glycoprotein IIb-IIIa. *J. Biol. Chem.* 268:1066–73
119. Schnepel J, Tschesche H. 2000. The proteolytic activity of the recombinant cryptic human fibronectin type IV collagenase from *E. coli* expression. *J. Protein Chem.* 19:685–92
120. Schnepel J, Unger J, Tschesche H. 2001. Recombinant cryptic human fibronectinase cleaves actin and myosin: substrate specificity and possible role in muscular dystrophy. *Biol. Chem.* 382:1707–14
121. Schoenauer R, Bertoncini P, Machaidze G, Aebi U, Perriard JC, et al. 2005. Myomesin is a molecular spring with adaptable elasticity. *J. Mol. Biol.* 349:367–79
122. Scott KA, Steward A, Fowler SB, Clarke J. 2002. Titin: a multidomain protein that behaves as the sum of its parts. *J. Mol. Biol.* 315:819–29
123. Sechler JL, Rao H, Cumiskey AM, Vega-Colon I, Smith MS, et al. 2001. A novel fibronectin binding site required for fibronectin fibril growth during matrix assembly. *J. Cell Biol.* 154:1081–88
124. Sharma A, Askari JA, Humphries MJ, Jones EY, Stuart DI. 1999. Crystal structure of a heparin- and integrin-binding segment of human fibronectin. *EMBO J.* 18:1468–79
125. Shi Q, Boettiger D. 2003. A novel mode for integrin-mediated signaling: tethering is required for phosphorylation of FAK Y397. *Mol. Biol. Cell* 14:4306–15
126. Sukharev S, Corey DP. 2004. Mechanosensitive channels: multiplicity of families and gating paradigms. *Sci. STKE* 2004:re4

127. Tafolla E, Wang S, Wong B, Leong J, Kapila YL. 2005. JNK1 and JNK2 oppositely regulate p53 in signaling linked to apoptosis triggered by an altered fibronectin matrix: JNK links FAK and p53. *J. Biol. Chem.* 280:19992–99
128. Takagi J, Strokovich K, Springer TA, Walz T. 2003. Structure of integrin alpha5beta1 in complex with fibronectin. *EMBO J.* 22:4607–15
129. Tamada M, Sheetz MP, Sawada Y. 2004. Activation of a signaling cascade by cytoskeleton stretch. *Dev. Cell* 7:709–18
130. Tan JL, Tien J, Pirone DM, Gray DS, Bhadriraju K, Chen CS. 2003. Cells lying on a bed of microneedles: an approach to isolate mechanical force. *Proc. Natl. Acad. Sci. USA* 100:1484–89
131. Tschumperlin DJ, Dai G, Maly IV, Kikuchi T, Laiho LH, et al. 2004. Mechanotransduction through growth-factor shedding into the extracellular space. *Nature* 429:83–86
132. Tseng Y, Kole TP, Wirtz D. 2002. Micromechanical mapping of live cells by multiple-particle-tracking microrheology. *Biophys. J.* 83:3162–76
133. Tskhovrebova L, Trinick J. 2003. Titin: properties and family relationships. *Nat. Rev. Mol. Cell Biol.* 4:679–89
134. Tskhovrebova L, Trinick J, Sleep JA, Simmons RM. 1997. Elasticity and unfolding of single molecules of the giant muscle protein titin. *Nature* 387:308–12
135. Urech L, Bittermann AG, Hubbell JA, Hall H. 2005. Mechanical properties, proteolytic degradability and biological modifications affect angiogenic process extension into native and modified fibrin matrices in vitro. *Biomaterials* 26:1369–79
136. Vallotton P, Danuser G, Bohnet S, Meister JJ, Verkhovsky AB. 2005. Tracking retrograde flow in keratocytes: news from the front. *Mol. Biol. Cell* 16:1223–31
137. Verrier S, Pallu S, Bareille R, Jonczyk A, Meyer J, et al. 2002. Function of linear and cyclic RGD-containing peptides in osteoprogenitor cells adhesion process. *Biomaterials* 23:585–96
138. Vogel V, Baneyx G. 2003. The tissue engineering puzzle: a molecular perspective. *Annu. Rev. Biomed. Eng.* 5:441–63
139. Vogel V, Sheetz MP. 2006. Local force and geometry sensing regulate cell functions. *Nat. Rev. Mol. Cell Biol.* 7:1–11
140. von Wichert G, Haimovich B, Feng GS, Sheetz MP. 2003. Force-dependent integrin-cytoskeleton linkage formation requires downregulation of focal complex dynamics by Shp2. *EMBO J.* 22:5023–35
141. von Wichert G, Jiang G, Kostic A, De Vos K, Sap J, Sheetz MP. 2003. RPTP-alpha acts as a transducer of mechanical force on alphav/beta3-integrin-cytoskeleton linkages. *J. Cell Biol.* 161:143–53
142. Watanabe K, Muhle-Goll C, Kellermayer MS, Labeit S, Granzier H. 2002. Different molecular mechanics displayed by titin's constitutively and differentially expressed tandem Ig segments. *J. Struct. Biol.* 137:248–58
143. Williams MJ, Phan I, Baron M, Driscoll PC, Campbell ID. 1993. Secondary structure of a pair of fibronectin type 1 modules by two-dimensional nuclear magnetic resonance. *Biochemistry* 32:7388–95
144. Williams PM, Fowler SB, Best RB, Toca-Herrera JL, Scott KA, et al. 2003. Hidden complexity in the mechanical properties of titin. *Nature* 422:446–49
145. Yeung T, Georges PC, Flanagan LA, Marg B, Ortiz M, et al. 2005. Effects of substrate stiffness on cell morphology, cytoskeletal structure, and adhesion. *Cell Motil. Cytoskelet.* 60:24–34
146. Zaidel-Bar R, Cohen M, Addadi L, Geiger B. 2004. Hierarchical assembly of cell-matrix adhesion complexes. *Biochem. Soc. Trans.* 32:416–20

129. After removing the cell membrane, force-dependent tyrosine phosphorylation of the remaining intracellular complexes (triton cytoskeletons) was observed upon addition of cell extracts.

139. Focuses on how local force-induced conformational changes are transduced into biochemical signals which then regulate cellular mechanoresponses.

146. Demonstrates that the formation of matrix adhesions is a hierarchical process consisting of several sequential molecular events.

147. Zhang F, Strand A, Robbins D, Cobb MH, Goldsmith EJ. 1994. Atomic structure of the MAP kinase ERK2 at 2.3 A resolution. *Nature* 367:704–11
148. Zhong C, Chrzanowska-Wodnicka M, Brown J, Shaub A, Belkin AM, Burridge K. 1998. Rho-mediated contractility exposes a cryptic site in fibronectin and induces fibronectin matrix assembly. *J. Cell Biol.* 141:539–51

Subject Index

B

C

D

E

G

H

I

M

N

O

P

S

T

U

V

W

X

Y

Cumulative Indexes

Contributing Authors, Volumes 31–35

I

J

K

L

M

N

O

P

R

S

T

Chapter Titles, Volumes 31–35

Synchrotron Radiation

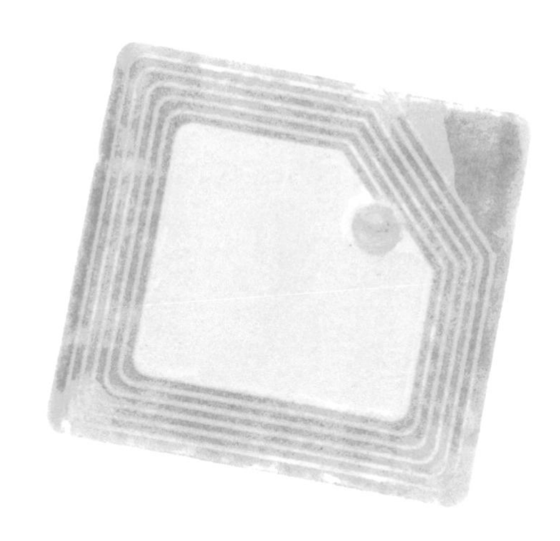

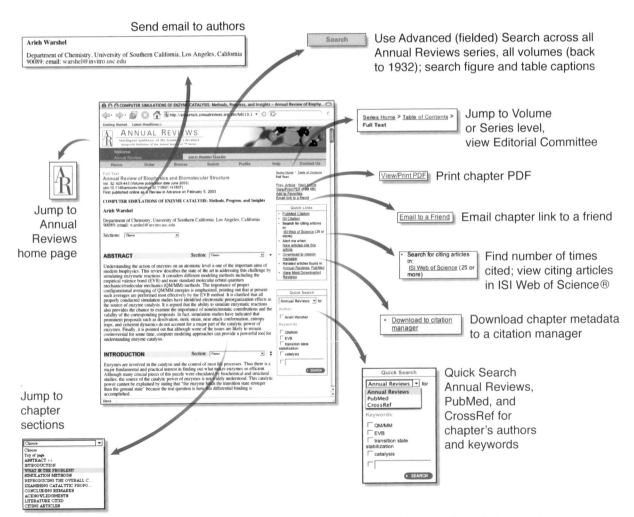